DOE/EIA-0384(2007)

Annual Energy Review 2007

June 2008

Energy Information Administration
Office of Energy Markets and End Use
U.S. Department of Energy
Washington, DC 20585

This report was prepared by the Energy Information Administration, the independent statistical and analytical agency within the U.S. Department of Energy. The information contained herein should be attributed to the Energy Information Administration and should not be construed as advocating or reflecting any policy of the Department of Energy or any other organization.

Contacts

The *Annual Energy Review* (*AER*) is prepared by the Energy Information Administration, Office of Energy Markets and End Use, Integrated Energy Statistics Division, Domestic Energy Statistics Team, under the direction of Katherine E. Seiferlein, 202-586-5695 (kitty.seiferlein@eia.doe.gov). Questions and comments about the *AER* may be referred to Ray Boyer, 202-586-1680 (ray.boyer@eia.doe.gov), the National Energy Information Center, 202-586-8800 (infoctr@eia.doe.gov), or to the following subject specialists:

Subject	Contact	Email	Phone
1. Energy Overview	Ray Boyer	ray.boyer@eia.doe.gov	202-586-1680
2. Energy Consumption by Sector			
Manufacturing Energy Consumption Survey	Robert Adler	robert.adler@eia.doe.gov	202-586-1134
Residential Energy Consumption Survey	Eileen O'Brien	eileen.obrien@eia.doe.gov	202-586-1122
Commercial Buildings Energy Consumption Survey	Joelle Michaels	joelle.michaels@eia.doe.gov	202-586-8952
3. Financial Indicators			
Financial Reporting System	Greg Filas	greg.filas@eia.doe.gov	202-586-1347
4. Energy Resources			
Petroleum and Natural Gas	Robert F. King	robert.king@eia.doe.gov	202-586-4787
Coal	George Warholic	george.warholic@eia.doe.gov	202-586-2307
Uranium	Douglas C. Bonnar	douglas.bonnar@eia.doe.gov	202-586-1085
5. Petroleum	Michael Conner	michael.conner@eia.doe.gov	202-586-1795
Prices	Elizabeth K. Scott	elizabeth.scott@eia.doe.gov	202-586-1258
6. Natural Gas	Joseph Conklin	joseph.conklin@eia.doe.gov	202-586-6664
7. Coal	Paulette Young	paulette.young@eia.doe.gov	202-586-1719
8. Electricity	James R. Diefenderfer	jim.diefenderfer@eia.doe.gov	202-586-2432
9. Nuclear Energy	John R. Moens	john.moens@eia.doe.gov	202-586-1509
10. Renewable Energy	Louise Guey-Lee	louise.guey-lee@eia.doe.gov	202-586-1293
11. International Energy	Michael J. Grillot	michael.grillot@eia.doe.gov	202-586-6577
12. Environmental Indicators			
Greenhouse Gases	Stephen E. Calopedis	stephen.calopedis@eia.doe.gov	202-586-1156
Environmental Equipment	Natalie Ko	natalie.ko@eia.doe.gov	202-586-3139

Preface

Energy analysts often talk in terms of the "supply" and "disposition" of energy. Although these specific terms do not appear in the *Annual Energy Review 2007* (*AER*), their concepts are the foundation of much of the data in this report and their components are central to any energy data analysis.

The main components of energy supply include the amount extracted from the earth and the environment (production), the amount brought into the country from other countries (imports), and the amount taken out of storage to be used by consumers (stock withdrawals). The main components of energy disposition include the amount used by domestic consumers (consumption), the amount sent out of the country (exports), and the amount put into storage (stock additions).

If we had complete knowledge of all energy markets, total supply would exactly equal total disposition in any given time period. In reality, that is not possible in our data because there are far too many transactions occurring in these activities and too many variables related to each activity to be able to make perfect measurements. We come as close as we can, though, by using an extensive set of data surveys and estimation techniques to develop the data series that you see in the *AER*.

Near the end of the data development process for an individual energy resource, a calculation is run to determine the difference between the resource's total supply and total disposition; the difference is called the "balancing item." The balancing item, then, is made up of survey discrepancies, unknown variables, and error. Our goal, of course, is to get this quantity down to the smallest level possible.

AER data organized around supply and disposition concepts are found in the set of tables called "Overviews." Overview tables are provided for total energy, petroleum, natural gas, coal, and electricity. In addition, the *AER* now provides a summary of renewable energy production and consumption, as well as overviews for fuel ethanol and biodiesel.

Many other data series throughout the *AER* support the overviews and show extensive details of the supply and disposition components. For example, Table 7.2 provides the details of coal production—by rank (bituminous coal, subbituminous coal, lignite, or anthracite); by mining method (underground or surface); and by location (East of the Mississippi or West of the Mississippi). Total coal production from Table 7.2 flows into Table 7.1's coal overview. After conversion of data units from short tons into the common measurement of British thermal units (Btu), coal production funnels into the calculation of fossil fuels production and then into total energy production by primary energy source in Table 1.2.

Organizing data around the concepts of supply and disposition aids analysts in the examination of issues related to the Nation's energy resources. From the petroleum overview (Table 5.1), for example, one can see the shift that has occurred over the decades from domestically produced supplies to imported supplies. A similar, but less dramatic and more recent, shift is seen in the natural gas overview (Table 6.1).

The integrated tables and figures for production, consumption, and imports in this report do not include uranium as a primary energy source. For both production and consumption, nuclear-generated electricity is included based on the nuclear plants approximate heat rate.

The *AER*, including complete time series supporting the tables and graphs, is available on the Energy Information Administration's (EIA) Web site at: http://www.eia.doe.gov/aer. EIA continually updates the time series; more recent data for many of the *AER* series can be found in the *Monthly Energy Review* at http://www.eia.doe.gov/mer.

Contents

Tables

Page

Tables

Tables

Page

Tables

Page

Tables

Page

Tables

Figures

Page

Figures

Figures

Page

Figures

Figures

Page

12. Environmental Indicators—*Continued*

Appendix C. U.S. Census Regions and Divisions

Overview

Figure 1. Primary Energy Overview

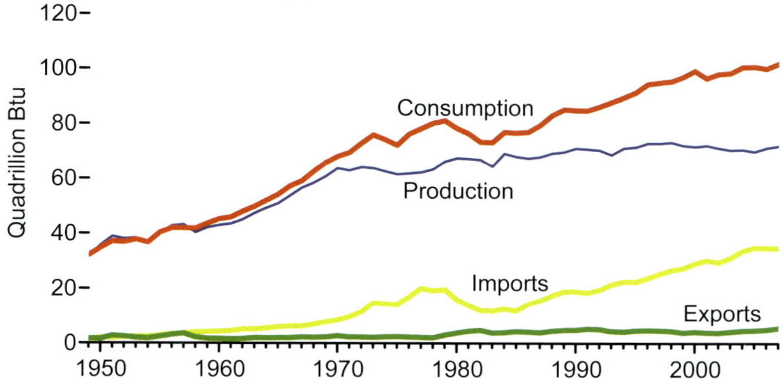

The United States was self-sufficient in energy until the late 1950s when energy consumption began to outpace domestic production. At that point, the Nation began to import more energy to fill the gap. In 2007, net imported energy accounted for 29 percent of all energy consumed.

Figure 2. Energy Consumption per Person

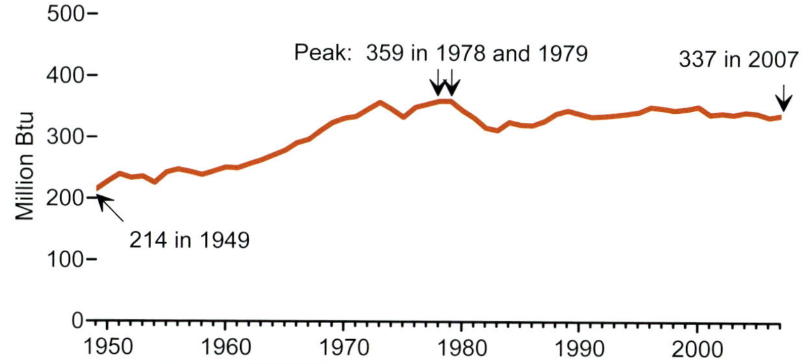

Energy use per person stood at 214 million British thermal units (Btu) in 1949. The rate generally increased until the oil price shocks of the mid-1970s and early 1980s when the trend reversed for a few years. From 1988 on, the rate held fairly steady. In 2007, 337 million Btu of energy were consumed per person, 57 percent above the 1949 rate.

Figure 3. Energy Consumption per Real Dollar of Gross Domestic Product

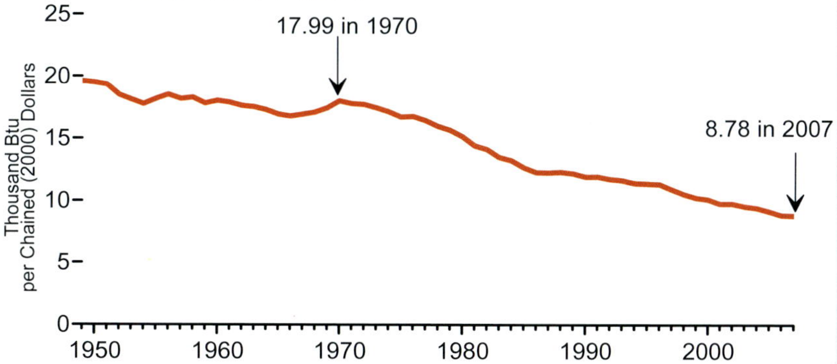

See "Chained Dollars" in Glossary.

After 1970, the amount of energy consumed to produce a dollar's worth of the Nation's output of goods and services trended down. The decline resulted from efficiency improvements and structural changes in the economy. The level in 2007 was 51 percent below that of 1970.

Figure 4. Primary Energy Consumption by Source

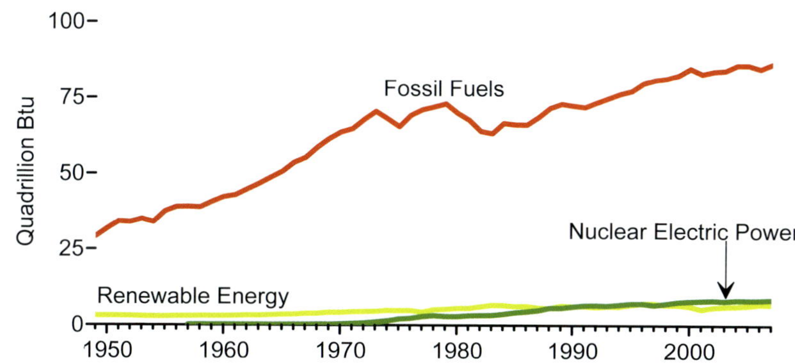

Most energy consumed in the United States comes from fossil fuels. Renewable energy resources supplies a relatively small but steady portion. In the late 1950s, nuclear fuel began to be used to generate electricity, and in most years since 1988, nuclear electric power surpassed renewable energy.

Consumption by Source

Figure 5. Primary Energy Consumption by Source, 1635-2007

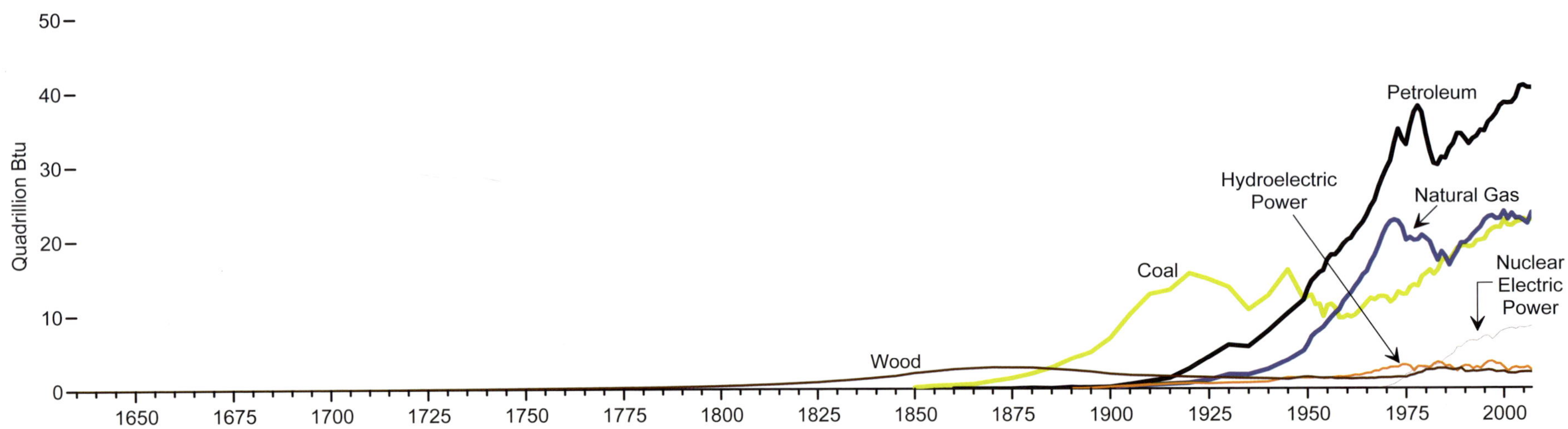

Figure 6. Energy Consumption History and Outlook, 1949-2030

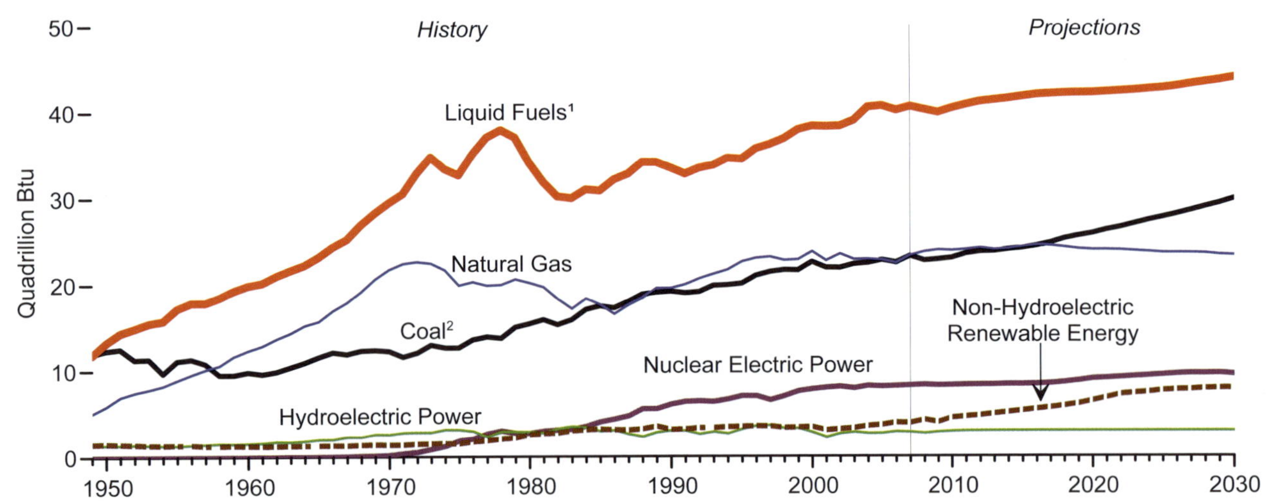

[1] History: petroleum-derived fuels. Projections: petroleum-derived fuels and non-petroleum-derived fuels, such as fuel ethanol, biodiesel, and coal-based synthetic liquids.
[2] Includes net imports of coal coke.

In the long view of American history, wood served as the preeminent form of energy for about half of the Nation's history. Around 1885, coal surpassed wood's usage. Despite its tremendous and rapid expansion, coal was, in turn, overtaken by petroleum in the middle of the 20th century. Natural gas, too, experienced rapid development into the second half of the 20th century, and coal began to expand again. Late in the 20th century still another form of energy, nuclear electric power, was developed and made significant contributions.

While the Nation's energy history is one of large-scale change as new forms of energy were developed, the outlook for the next couple of decades (assuming current laws, regulations, and policies) is for continued growth and reliance on the three major fossil fuels—petroleum, natural gas, and coal—modest expansion in renewable resources, and relatively flat generation from nuclear electric power.

Consumption by Sector

Figure 7. Total Energy Consumption by End-Use Sector

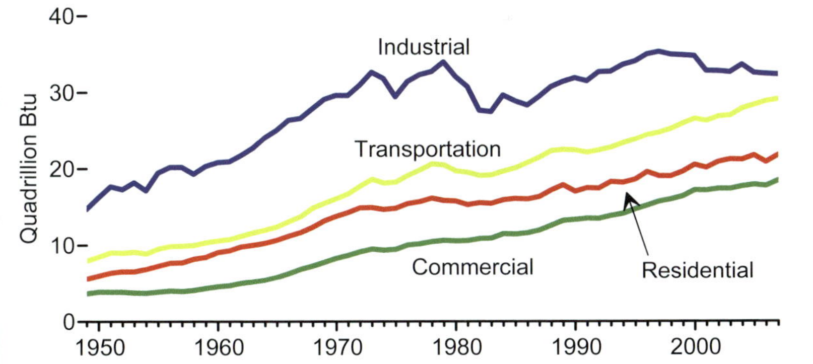

All four major economic sectors of the economy recorded tremendous growth in their use of energy. The industrial sector used the biggest share of total energy and showed the greatest volatility; in particular, steep drops occurred in the sector in 1975 and 1980-1983 largely in response to high oil prices and economic slowdown.

Figure 8. Residential and Commercial Total Energy Consumption, Major Sources

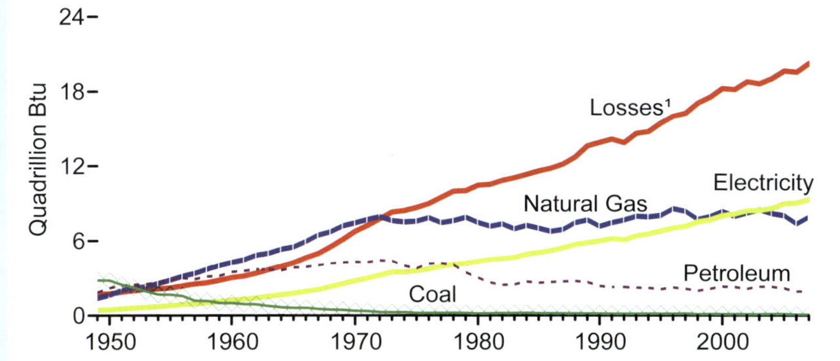

[1] Energy lost during generation, transmission, and distribution of electricity.

In the 1950s and 1960s, coal, which had been important to residential and commercial consumers, was gradually replaced by other forms of energy. Petroleum consumption peaked in the early 1970s. Natural gas consumption grew fast until the early 1970s and then, with mild fluctuations, held fairly steady in the following years. Meanwhile, electricity use (and related losses) expanded dramatically.

Figure 9. Industrial Total Energy Consumption, Major Sources

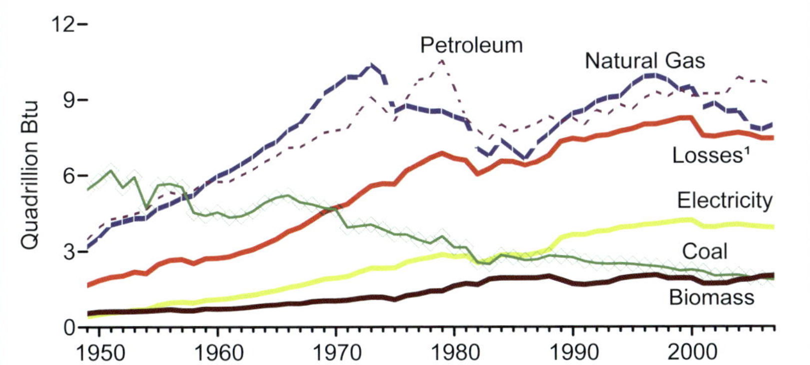

[1] Energy lost during generation, transmission, and distribution of electricity.

Coal, once the predominant form of energy in the industrial sector, gave way to natural gas and petroleum in the late 1950s. Both natural gas and petroleum use expanded rapidly until the early 1970s and then fluctuated widely over the following decades. Use of electricity and biomass trended upward.

Figure 10. Transportation Total Energy Consumption

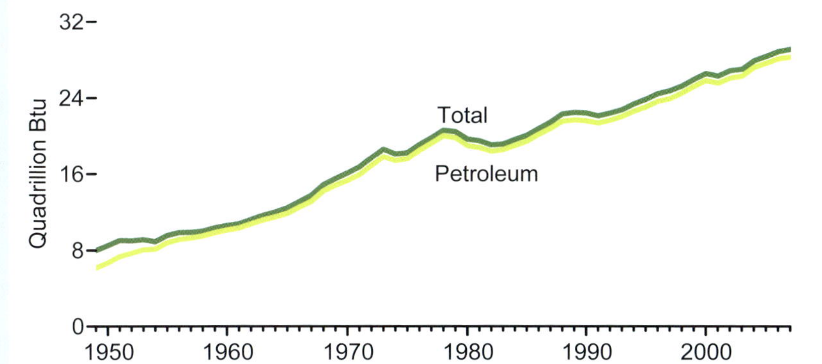

Transportation sector use of energy experienced tremendous growth overall but registered noticeable pauses in 1974, 1979-1982, 1990 and 1991, and 2001. In 2007, petroleum accounted for 95 percent of the sector's energy. In Btu, motor gasoline accounted for 61 percent of all petroleum used in the sector; in barrels, motor gasoline accounted for 64 percent .

Production and Trade

Figure 11. Primary Energy Production by Major Source

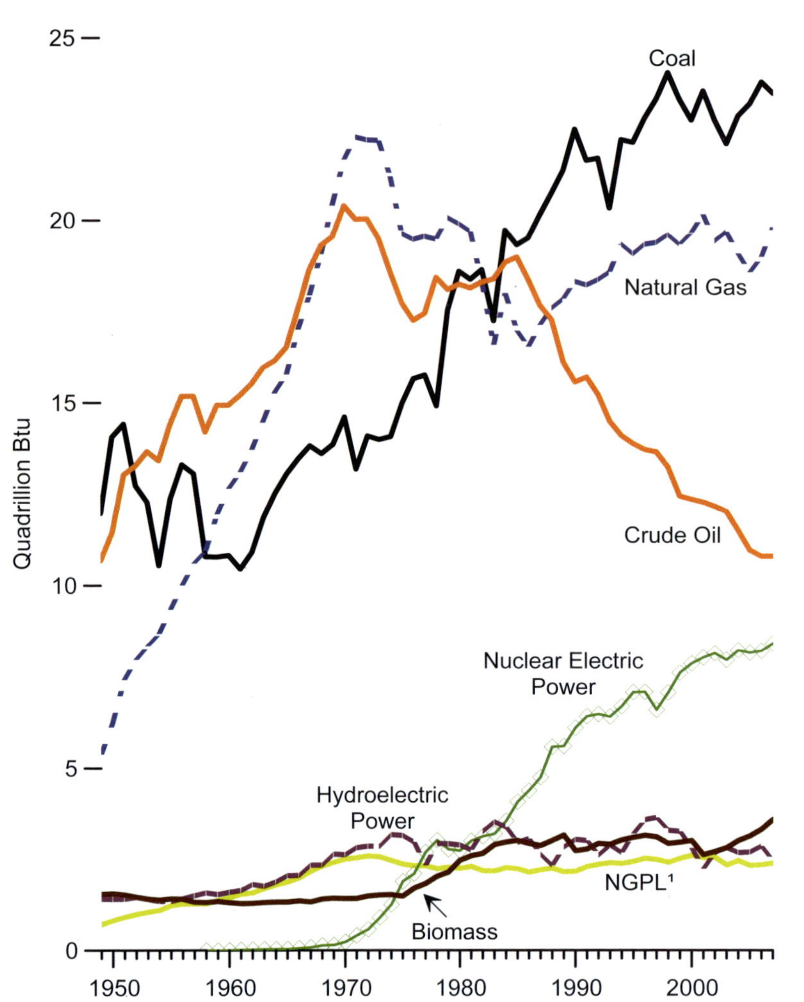

¹ Natural gas plant liquids.

Most energy produced in the United States came from fossil fuels—coal, natural gas, and crude oil. Coal, the leading source at the middle of the 20th century, was surpassed by crude oil and then by natural gas. By the mid-1980s, coal again became the leading energy source produced in the United States, and crude oil declined sharply. In the 1970s, electricity produced from nuclear fuel began to make a significant contribution and expanded rapidly in the following decades.

Figure 12. Production as Share of Consumption for Coal, Natural Gas, and Petroleum

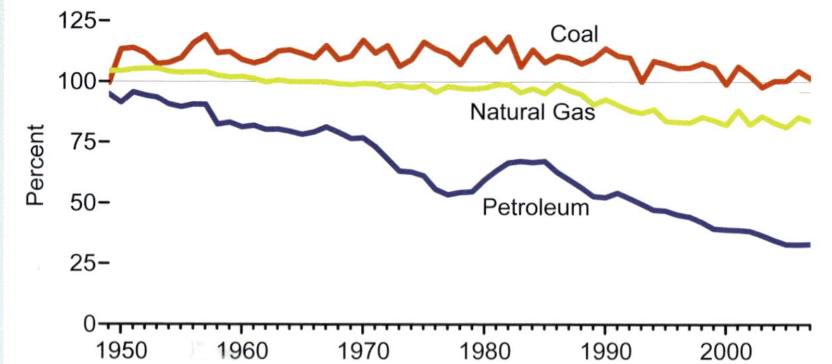

The United States almost always produced more than enough coal for its own requirements. For many years, the United States was also self-sufficient in natural gas, but after 1967, it produced less than it consumed each year. Petroleum production fell far short of domestic demands, requiring the Nation to rely on imported supplies.

Figure 13. Primary Energy Imports and Exports

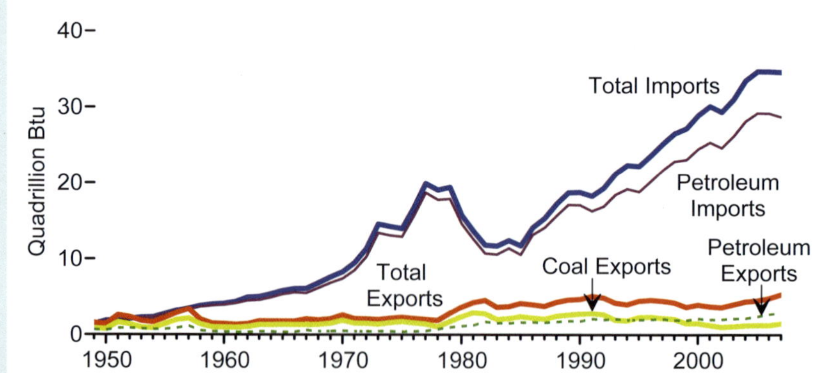

Since the mid-1950s, the Nation imported more energy than it exported. In 2007, the United States imported 35 quadrillion Btu of energy and exported 5 quadrillion Btu. Most imported energy was in the form of petroleum; since 1986, natural gas imports expanded rapidly as well. Through 1992, most exported energy was in the form of coal; after that, petroleum exports often exceeded coal exports.

Petroleum Overview and Crude Oil Production

Figure 14. Petroleum Overview

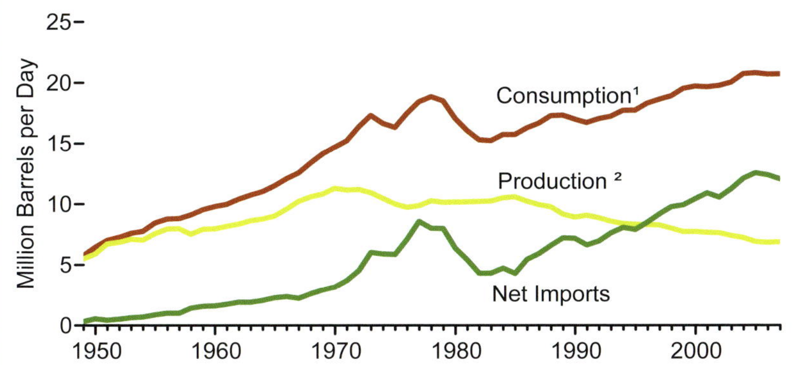

[1] Petroleum products supplied is used as an approximation for consumption.
[2] Crude oil and natural gas plant liquids production.

When U.S. petroleum production peaked at 11.3 million barrels per day in 1970, net imports stood at 3.2 million barrels per day. By 1996, net imports exceeded production. In 2007, production was 6.9 million barrels per day, and net imports were 12.0 million barrels per day.

Figure 15. 48 States and Alaskan Crude Oil Production

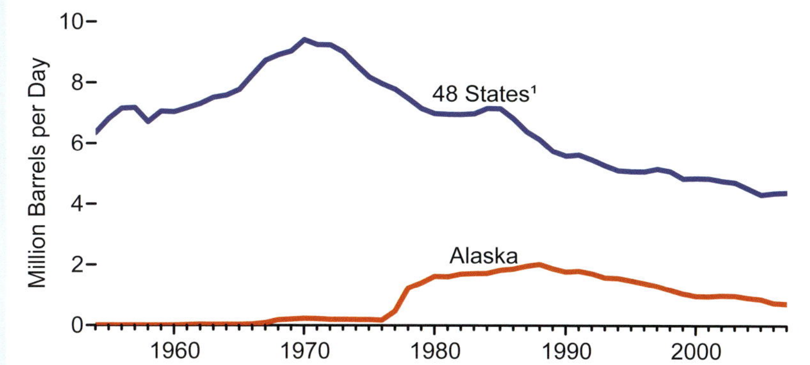

[1] United States excluding Alaska and Hawaii.

Crude oil production peaked in the 48 States at 9.4 million barrels per day in 1970. As production fell in the 48 States, Alaska's production came on line and helped supply U.S. needs. Alaskan production peaked at 2.0 million barrels per day in 1988; in 2007, Alaska's production stood at 36 percent of its peak level.

Figure 16. Crude Oil Well Productivity

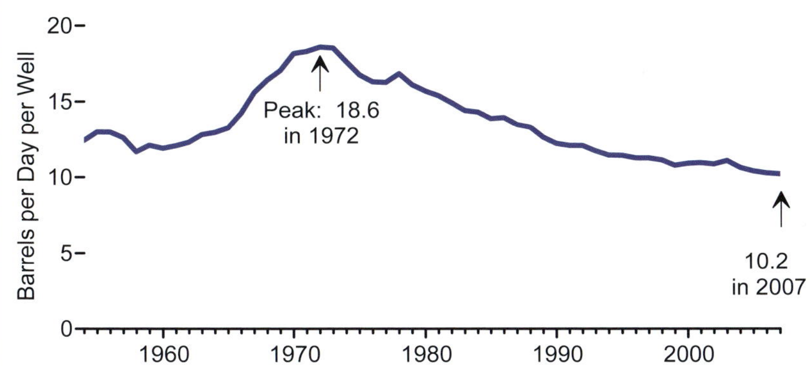

The amount of crude oil produced per day per well rose sharply in the 1960s and reached a peak of 18.6 barrels per day per well in 1972. After 1972, productivity generally declined. The 2007 rate of 10.2 barrels per day per well was 45 percent below the peak and the lowest level since the Energy Information Administration began reporting oil well productivity.

Figure 17. Crude Oil and Natural Gas Rotary Rigs in Operation

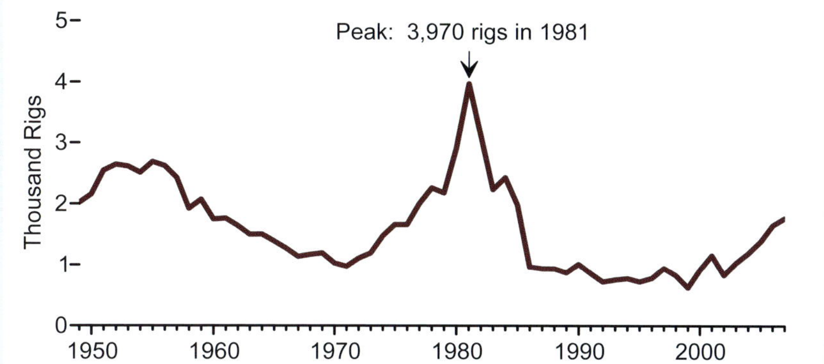

Rotary rig activity declined sharply from 1955 to 1971. After 1971, the number of rigs in operation began to climb again, and a peak of nearly 4 thousand rigs in operation was registered in 1981. In 2007, 1,768 rigs were in operation, more than double the level in 2002, but only 45 percent of the peak level in 1981.

Petroleum Consumption and Prices

Figure 18. Petroleum Consumption[1] by Sector

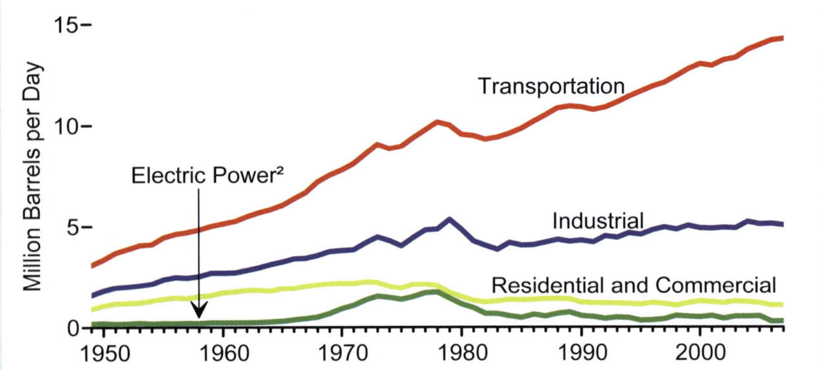

[1] Petroleum products supplied is used as an approximation for consumption.
[2] Through 1988, electric utilities only; after 1988, includes independent power producers.

Transportation was the largest consuming sector of petroleum and the one showing the greatest expansion. In 2007, 14.3 million barrels per day of petroleum products were consumed for transportation purposes, accounting for 69 percent of all petroleum used.

Figure 19. Petroleum Consumption[1] by Selected Product

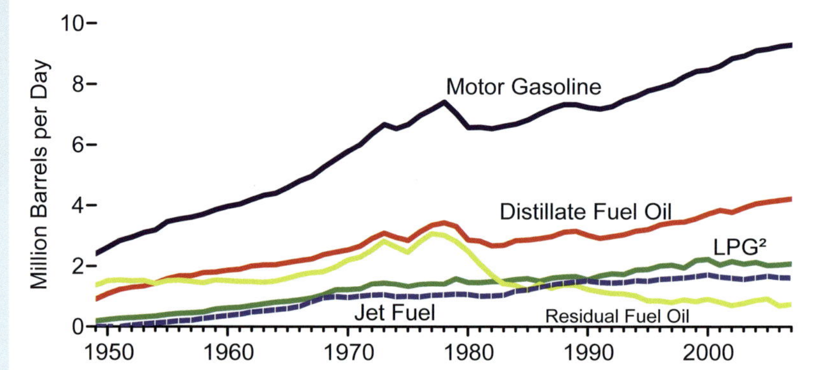

[1] Petroleum products supplied is used as an approximation for consumption.
[2] Liquefied petroleum gases.

Motor gasoline was the single largest petroleum product consumed in the United States. Its consumption stood at 9.3 million barrels per day in 2007, 45 percent of all petroleum consumption. Distillate fuel oil, liquefied petroleum gases (LPG), and jet fuel were other important products. The use of residual fuel oil fell off sharply after 1977.

Figure 20. Crude Oil Refiner Acquisition Cost[1]

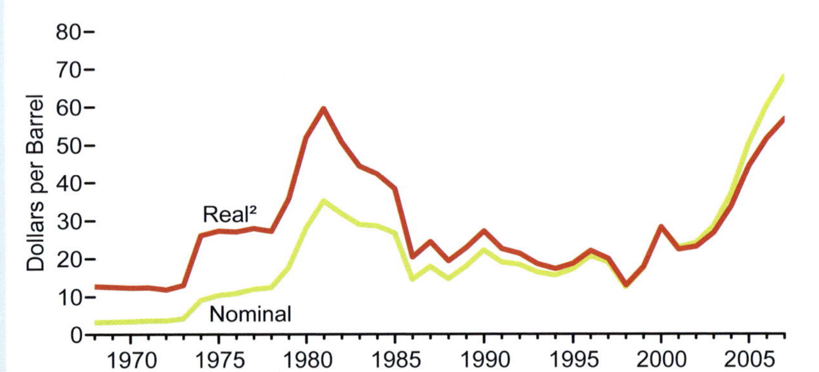

[1] Composite of domestic and imported crude oil. [2] In chained (2000) dollars, calculated by using gross domestic product implicit price deflator. See "Chained Dollars" in Glossary.

Unadjusted for inflation (nominal dollars), the refiner acquisition composite (domestic and foreign) cost of crude oil reached $35.24 per barrel in 1981. Over the years that followed, the price fell dramatically to a low of $12.52 per barrel in 1998 before rising again. The preliminary nominal price reported for 2007 was $67.93 per barrel, a new peak level and up 13 percent over the 2006 price.

Figure 21. Retail Price of Motor Gasoline, All Grades

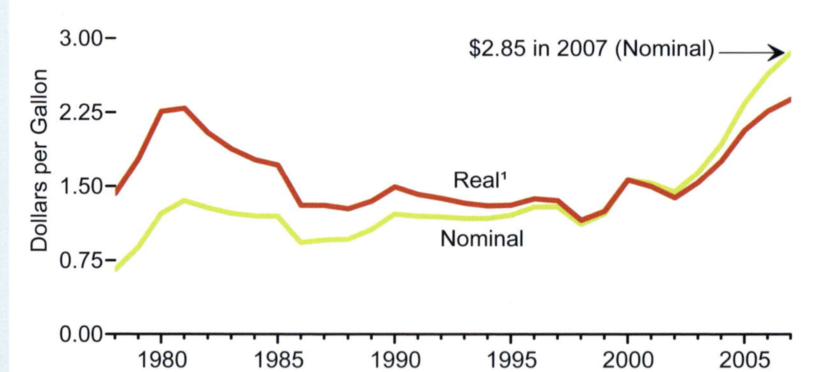

[1] In chained (2000) dollars, calculated by using gross domestic product implicit price deflator. See "Chained Dollars" in Glossary.

In nominal (unadjusted for inflation) dollars, Americans paid an average of 65¢ per gallon for motor gasoline in 1978. The 2007 average price of $2.85 was more than four times the 1978 rate; adjusted for inflation, it was 66 percent higher.

Petroleum Trade

Figure 22. Petroleum Trade

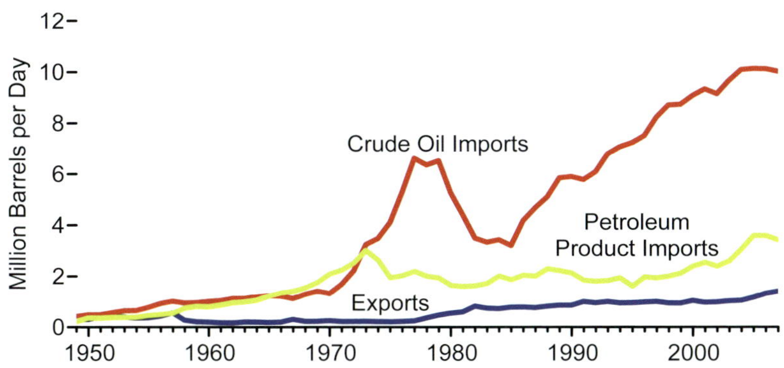

U.S. crude oil imports grew rapidly from mid-20th century until the late 1970s but fell sharply from 1979 to 1985 due to conservation efforts and improved efficiency. After 1985, the upward trend resumed and stood at 10 million barrels per day in 2007. Petroleum product imports were 3.4 million barrels per day in 2007. Exports totaled 1.4 million barrels per day in 2007, mainly in the form of petroleum coke and residual fuel oil.

Figure 23. Imports From OPEC and Non-OPEC Countries

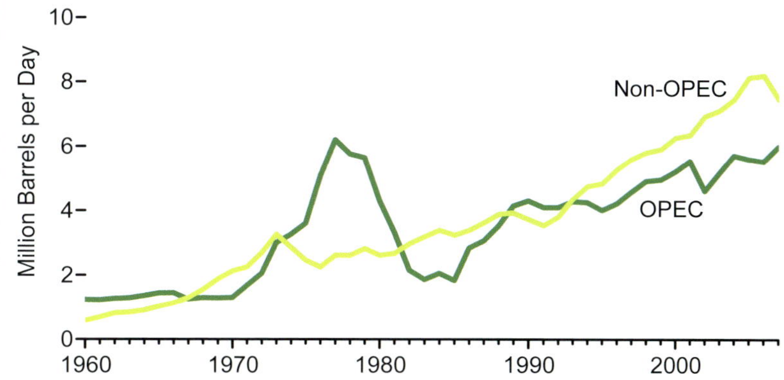

OPEC = Organization of the Petroleum Exporting Countries.

U.S. petroleum imports rose sharply in the 1970s, and reliance on petroleum from the Organization of the Petroleum Exporting Countries (OPEC) grew. In 2007, 45 percent of U.S. petroleum imports came from OPEC countries, down from 70 percent in 1977. After 1992, more petroleum came into the United States from non-OPEC countries than from OPEC countries.

Figure 24. Imports From Selected OPEC Countries

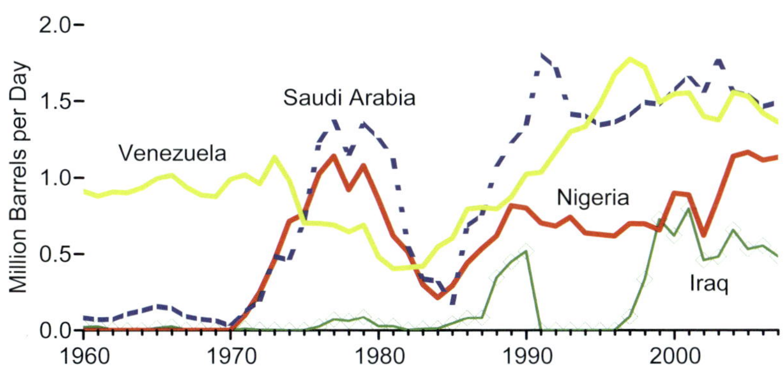

OPEC = Organization of the Petroleum Exporting Countries.

Among OPEC countries, Saudi Arabia, Venezuela, and Nigeria—nations from three different continents—were key suppliers of petroleum to the American market. Each experienced wide fluctuation in the amount of petroleum it sold to the United States over the decades. In 2007, 0.5 million barrels per day of petroleum came into the United States from Iraq.

Figure 25. Imports From Canada and Mexico

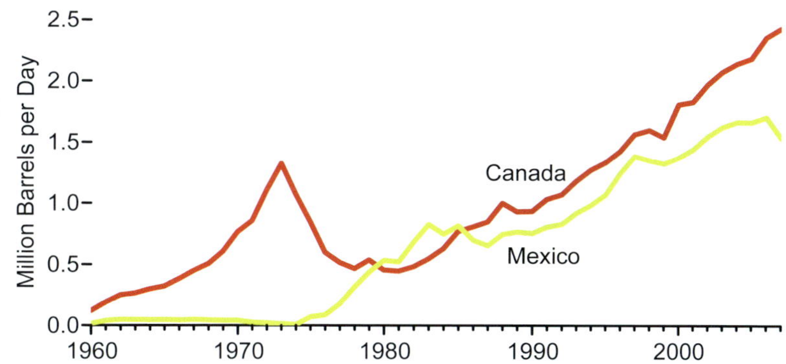

Canada and Mexico were the largest non-OPEC suppliers of petroleum to the United States. In 2007, imports from Canada reached a new high of 2.4 million barrels per day. Imports from Mexico were insignificant until the mid-1970s when they began to play a key role in U.S. supplies. Canadian and Mexican petroleum together accounted for 29 percent of all U.S. imports in 2007.

Petroleum Stocks

Figure 26. Stocks of Crude Oil and Products

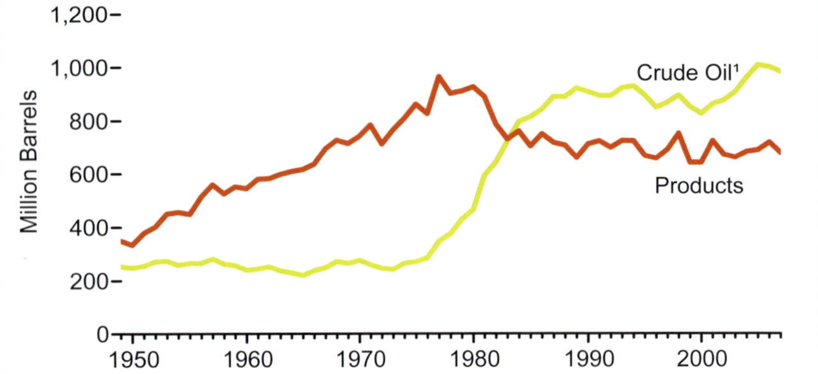

[1] Includes crude oil and lease condensate stored in the Strategic Petroleum Reserve.

Through 1983, the Nation held most of its petroleum storage in the form of products, which were ready for the market. After 1983, most petroleum in storage was in the form of crude oil (including that held by the government in the Strategic Petroleum Reserve) that still needed to be refined into usable end products. At the end of 2007, petroleum stocks totaled 1.7 billion barrels, 59 percent crude oil and 41 percent products.

Figure 27. Total Stocks and the Strategic Petroleum Reserve

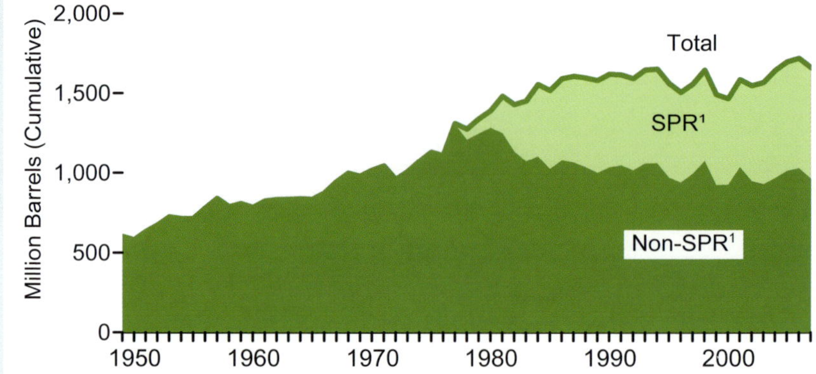

[1] Strategic Petroleum Reserve.

In 1977, the United States began filling the Strategic Petroleum Reserve (SPR), a national reserve of petroleum stocks in case of emergency. At the end of 2007, the SPR held 697 million barrels of crude oil, 42 percent of all U.S. petroleum stocks.

Figure 28. Crude Oil Imports for the SPR[1]

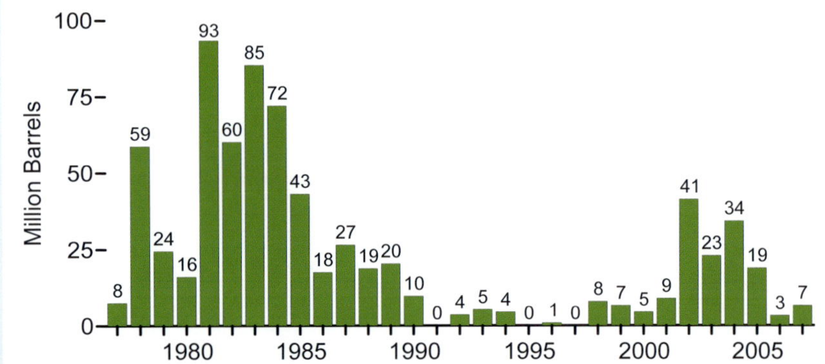

[1] Imported by the SPR and imported by others for the SPR.

Most crude oil in the SPR was imported and came in during the early 1980s. In fact, from 1991 through 1997, only 14 million barrels were imported for the reserve, and in 3 of those years, no oil at all was imported for the reserve. SPR imports picked up again after 1997 and stored another 156 million barrels from 1998 through 2007.

Figure 29. SPR Stocks as Days of Petroleum Net Imports

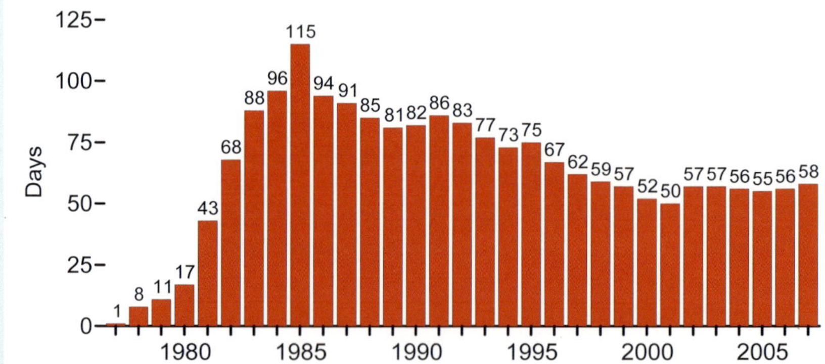

Stocks are often measured by the number of days of total net imports of petroleum that could be met by the reserve in an emergency. The peak level occurred in 1985 when the reserve could have supplied 115 days of petroleum net imports, at the 1985 level. The rate trended down for many years, falling to 50 days in 2001. In 2007, SPR held 58 days of net imports.

Motor Vehicles

Figure 30. Motor Vehicle Indicators

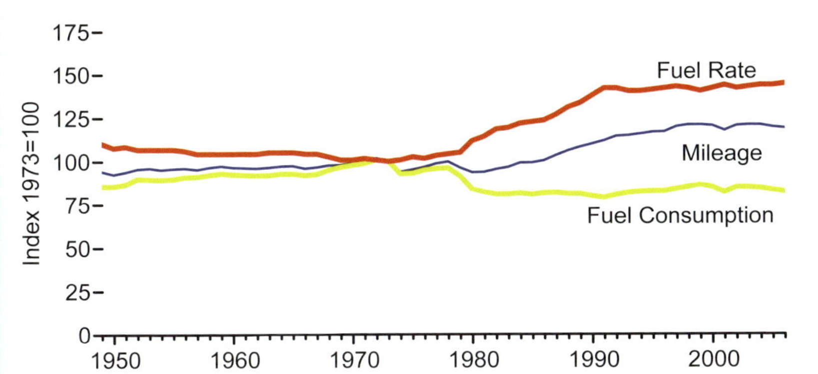

The composite motor vehicle fuel rate (miles per gallon) rose 42 percent from 1973 to 1991 and then varied little over the next 15 years. Mileage (miles per vehicle) grew steadily from 1980 to 1998 and then hovered around 12 thousand miles per vehicle per year through 2006. Fuel consumption (gallons per vehicle) fell 21 percent from 1973 to 1991, regained 9 percent from 1991 to 1999, and then trended down through 2006.

Figure 31. Motor Vehicle Fuel Consumption

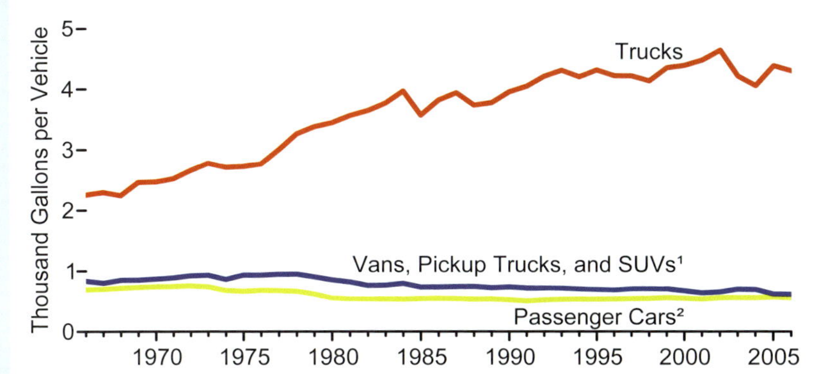

[1] Sport utility vehicle. [2] Motorcycles are included through 1989.

Truck average fuel consumption rates greatly exceeded other vehicle categories' fuel consumption rates and trended upward over time—actually doubling from 2.3 thousand gallons per truck in 1966 to 4.6 thousand gallons per truck in 2002. Passenger car and van, pickup truck, and sport utility vehicle average fuel consumption rates were much lower and generally trended downward.

Figure 32. Motor Vehicle Mileage

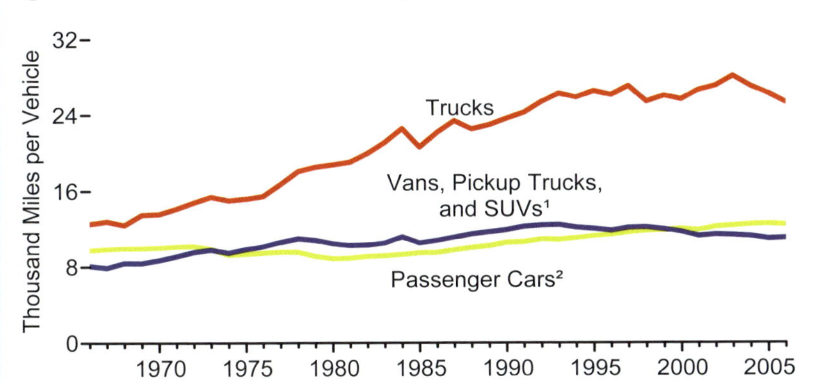

[1] Sport utility vehicle. [2] Motorcycles are included through 1989.

Truck miles traveled per year, which greatly exceeded other vehicle categories, grew by 124 percent from 1966 to 2003, decreased 10 percent from 2003 to 2006, and averaged 25.3 thousand miles per vehicle in 2006. Passenger cars averaged 12.4 thousand miles per vehicle in 2006. Vans, pickup trucks, and sport utility vehicles averaged 11.0 thousand miles per vehicle in 2006.

Figure 33. Motor Vehicle Fuel Rates

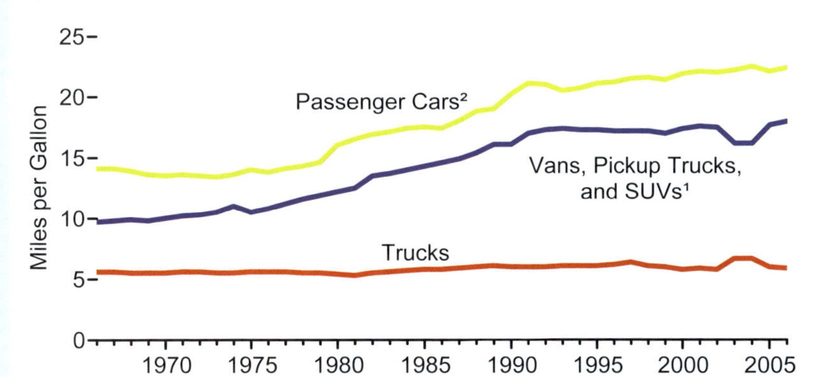

[1] Sport utility vehicle. [2] Motorcycles are included through 1989.

Fuel rates (miles per gallon) for passenger cars and vans, pickup trucks, and SUVs improved noticeably from the late 1970s through the early 1990s. Passenger cars improved further in subsequent years; rates for vans, pickup trucks, and SUVs fell in 2003 and 2004 but recovered in 2005 and 2006. Truck fuel rates, which were much lower than rates for the other vehicle categories, showed far less change over time.

Note: Motor vehicles include passenger cars, motorcycles, vans, pickup trucks, sport utility vehicles, trucks, and buses.

Natural Gas

Figure 34. Natural Gas Overview

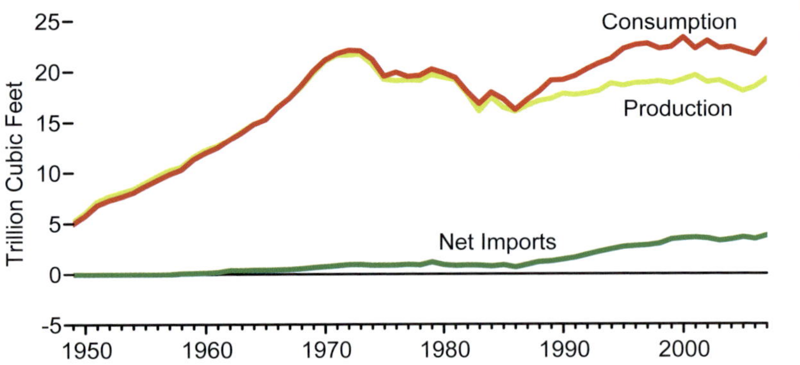

U.S. natural gas production and consumption were nearly in balance through 1986. After that, consumption began to outpace production, and imports of natural gas rose to meet U.S. requirements for the fuel. In 2007, production stood at 19.3 trillion cubic feet (Tcf), net imports at 3.8 Tcf, and consumption at 23.1Tcf.

Figure 35. Natural Gas Well Average Productivity

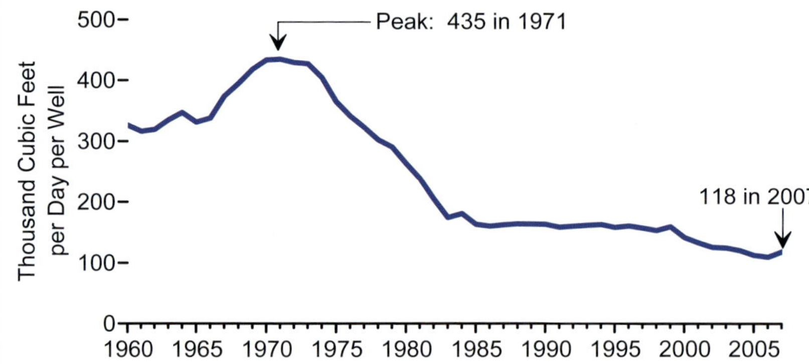

Natural gas well productivity, measured as gross withdrawals per day per well, grew rapidly in the late 1960s, peaked in 1971, and then fell sharply until the mid-1980s. Productivity remained fairly steady from 1985 through 1999, fell annually through 2006, and turned up slightly in 2007.

Figure 36. Net Imports as Share of Consumption

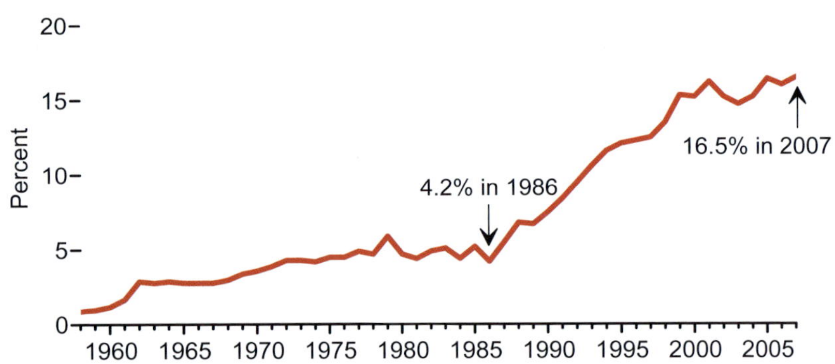

Net imports of natural gas as a share of consumption was in the 4-to-6 percent range from 1970 through 1987. Then, during a period when consumption outpaced production, the share rose from 4.2 percent in 1986 to 16.2 percent in 2001. In 2007, the share was 16.5 percent.

Figure 37. Natural Gas Consumption by Sector

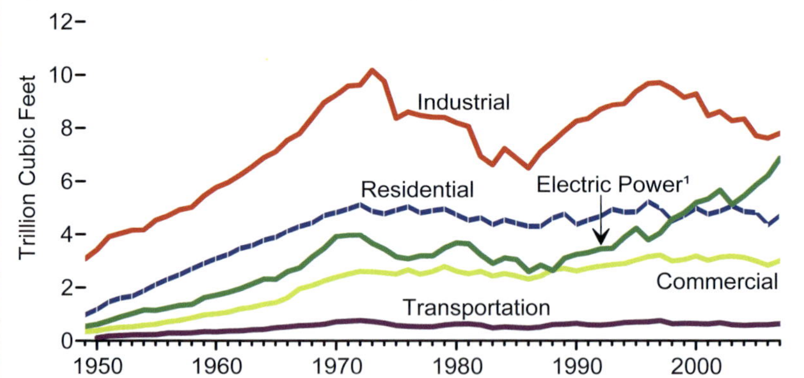

[1] Through 1988, electric utilities only; after 1988, includes independent power producers.

The industrial sector was both the largest consuming sector of natural gas and the sector with the greatest volatility due to variability in industrial output. In 2007, the industrial sector accounted for 34 percent of all natural gas consumption, and the electric power sector accounted for 30 percent.

Coal

Figure 38. Coal Overview

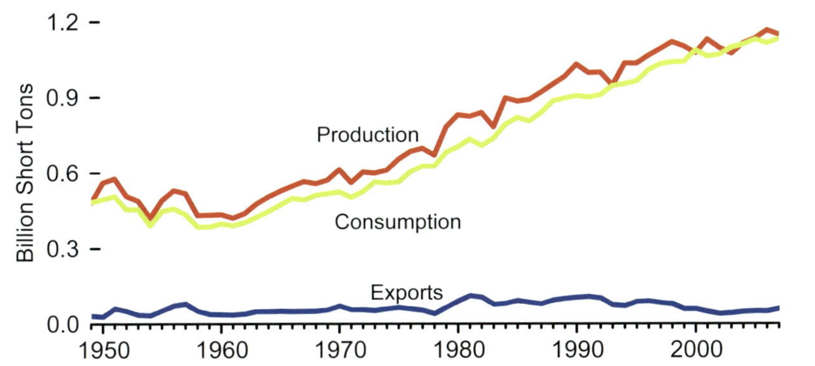

Historically, U.S. coal production usually surpassed U.S. coal consumption. In 2004 and 2005, however, production and consumption were in balance at 1.11 billion short tons in 2004 and 1.13 billion short tons in 2005. In 2006 and 2007, production again slightly exceeded consumption. Exports, which peaked at 113 million short tons in 1981, stood at 59 million short tons in 2007.

Figure 39. Coal Consumption by Sector

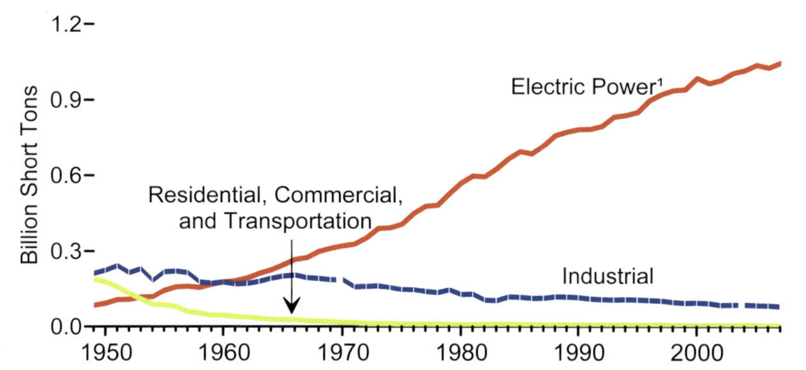

[1] Through 1988, electric utilities only; after 1988, includes independent power producers.

In the 1950s, most coal was consumed in the industrial sector, many homes were still heated by coal, and the transportation sector consumed coal in steam-driven trains and ships. By the 1960s, most coal was used for generating electricity. In 2007, the electric power sector accounted for 93 percent of all coal consumption.

Figure 40. Coal Mining Productivity

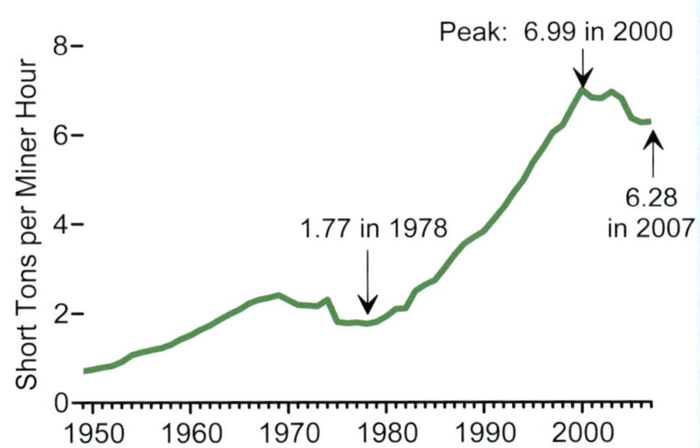

Improved mining technology and the shift toward more surface-mined coal promoted dramatic improvement in productivity from the Nation's mines from 1978 through 2000, but productivity declined in 3 of the last 4 years.

Figure 41. Production by Mining Method

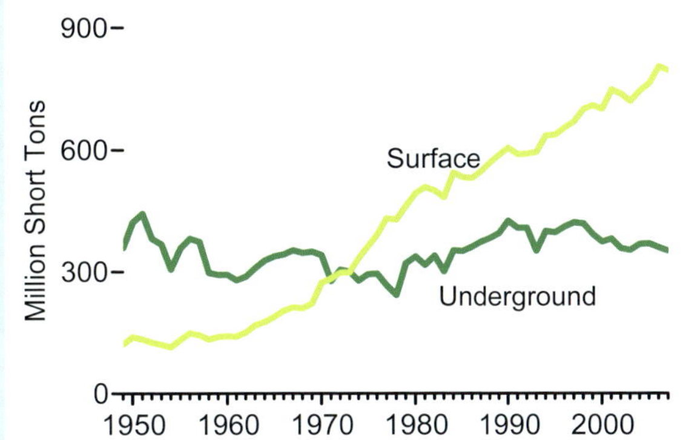

In 1949, one-fourth of U.S. coal came from surface mines; by 1971, more than one-half was surface-mined; and in 2007, 69 percent came from above-ground mines.

Figure 42. Production by Location

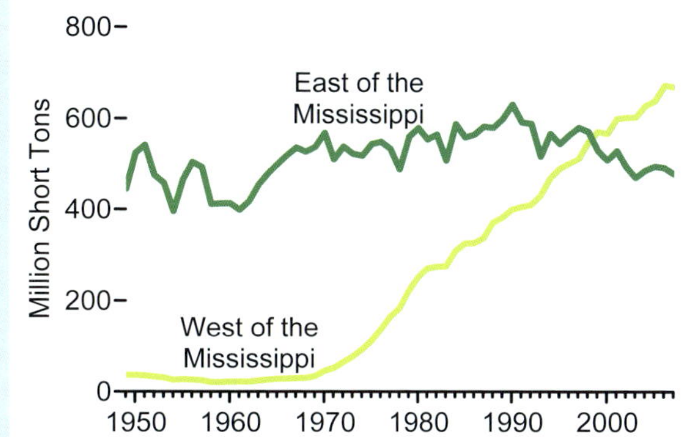

Western coal production expanded tremendously after 1969 and surpassed Eastern production beginning in 1999. In 2007, an estimated 58 percent of U.S. coal came from West of the Mississippi.

Electricity Net Generation and Useful Thermal Output

Figure 43. Electricity Net Generation by Sector

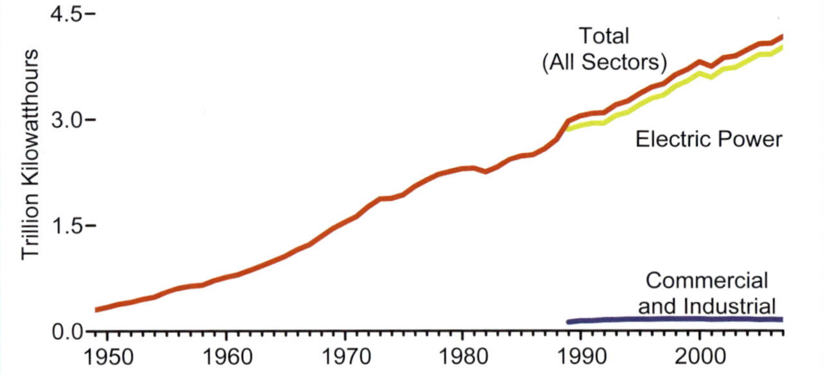

Total electricity net generation in all sectors grew from 0.3 trillion kilowatthours in 1949 to 4.2 trillion kilowatthours in 2007, failing to increase in only 2 years (1982 and 2001) over the entire span. Most generation was in the electric power sector, but some occurred directly in the commercial and industrial sectors.

Figure 44. Major Sources of Total Electricity Net Generation

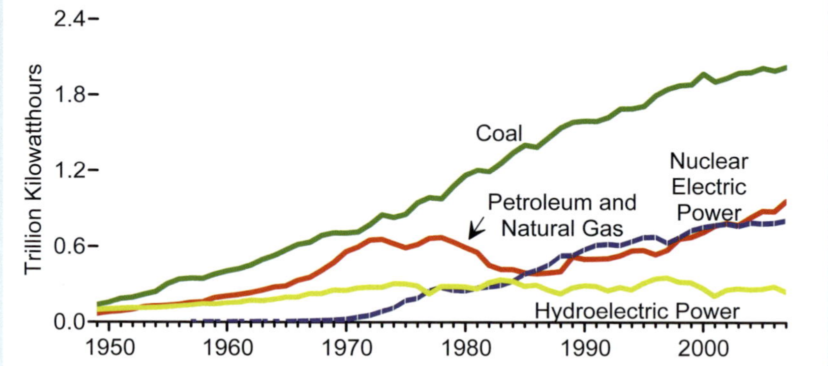

Most electricity net generation came from coal. In 2007, fossil fuels (coal, petroleum, and natural gas) accounted for 72 percent of all net generation, while nuclear electric power contributed 19 percent, and renewable energy resources 8 percent. In 2007, 71 percent of the net generation from renewable energy resources was derived from conventional hydroelectric power.

Figure 45. Electric Power Sector by Plant Type

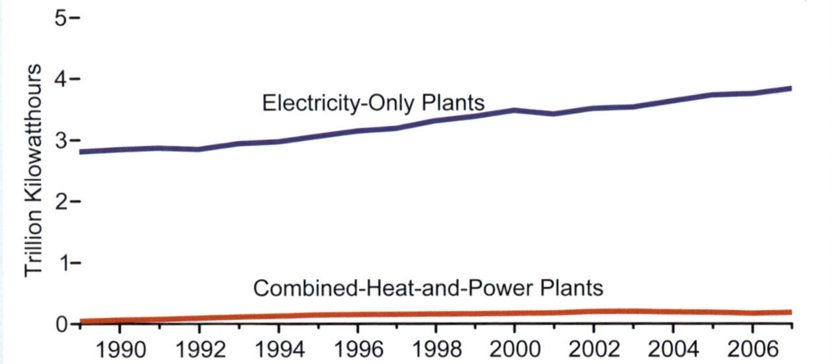

Most generating facilities exist to produce only electricity, but some function as combined-heat-and-power (CHP) plants that produce both electricity and heat from a single heat source. Rather than being wasted, the heat from a CHP plant is used for processes and applications in addition to electricity generation.

Figure 46. Useful Thermal Output at Combined-Heat-and-Power Plants by Sector

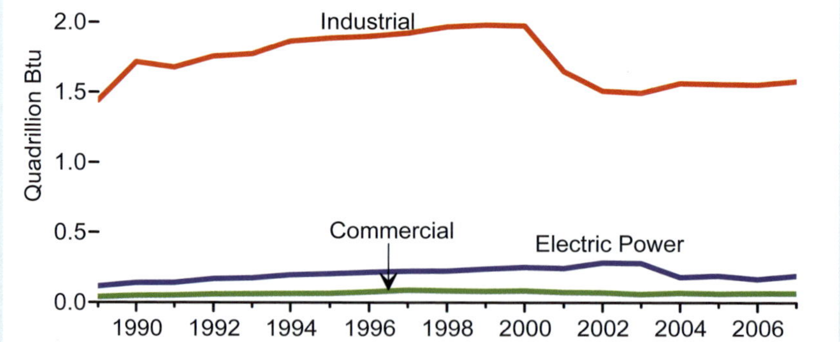

The non-electrical output at a CHP plant is called useful thermal output. Useful thermal output is thermal energy that is available from the plant for use in industrial or commercial processes or heating or cooling applications. In 2007, the industrial sector generated 1.6 quadrillion Btu of useful thermal output; the electric power and commercial sectors generated much smaller quantities.

Electricity Prices, Sales, and Trade

Figure 47. Average Real[1] Retail Prices of Electricity by Sector

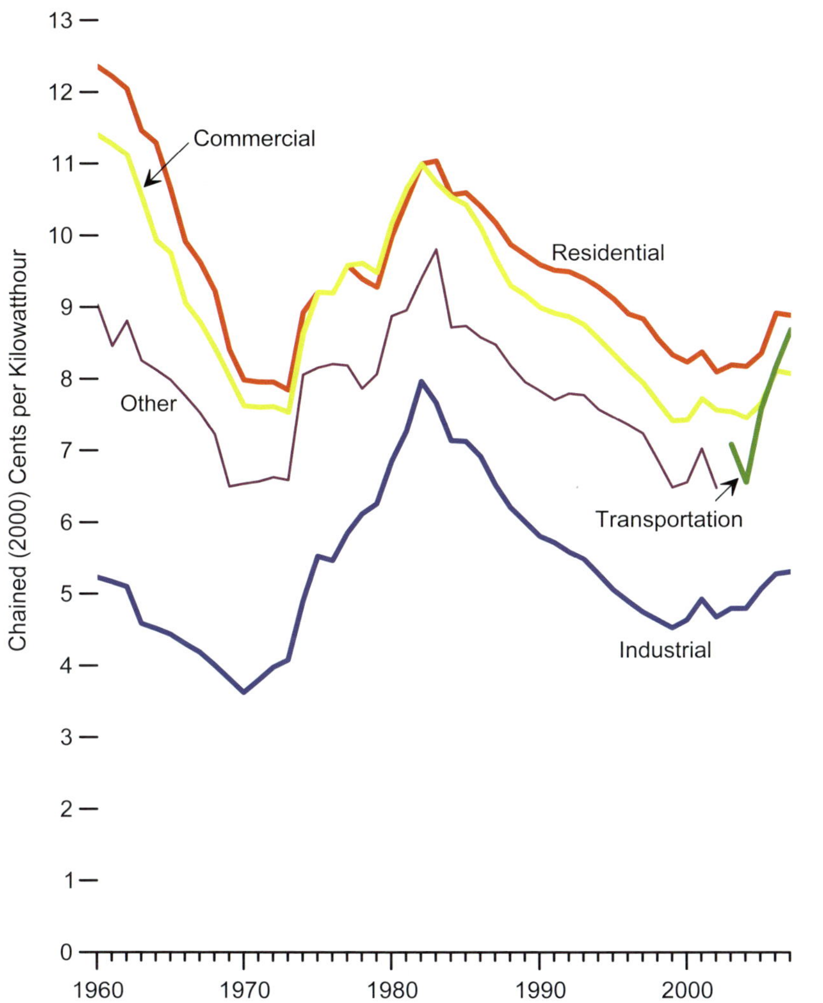

[1] In chained (2000) dollars, calculated by using gross domestic product implicit price deflators. See "Chained Dollars" in Glossary.

Over the decades, industrial consumers paid the lowest rates for electricity; residential customers usually paid the highest prices. Inflation-adjusted prices rose in all sectors in both 2005 and 2006 but remained well below the peak price levels of the mid-1980s. Price trends were mixed in 2007.

Figure 48. Retail Sales by Sector

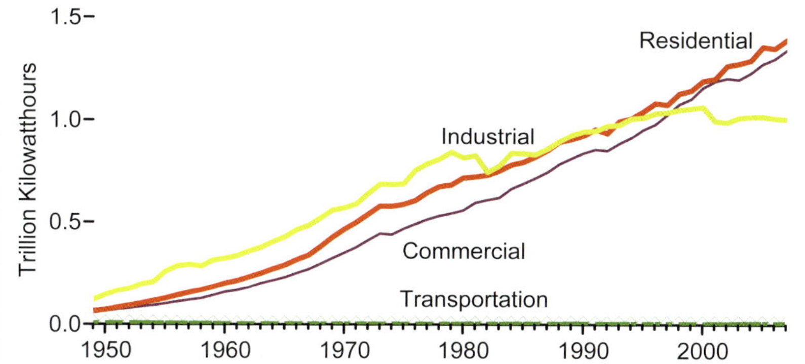

Enormous growth occurred in the amount of electricity sold to the three major sectors—residential, commercial, and industrial. Industrial sector sales showed the greatest volatility. Sales to residences exceeded sales to industrial sites beginning in the early 1990s, and sales to commercial sites surpassed industrial sales beginning in the late 1990s.

Figure 49. Electricity Trade

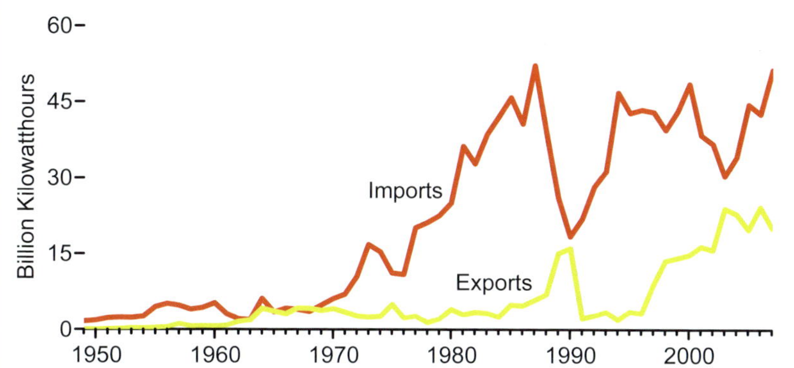

Except for a few years in the 1960s when imported and exported electricity were nearly equal, the United States imported more electricity than it exported. Most electricity trade occurred with Canada; very small exchanges occurred between the United States and Mexico. In 2007, net imported electricity was less than 0.8 percent of all electricity used in the United States.

Nuclear Electric Power

Figure 50. Full-Power Operating Licenses Issued

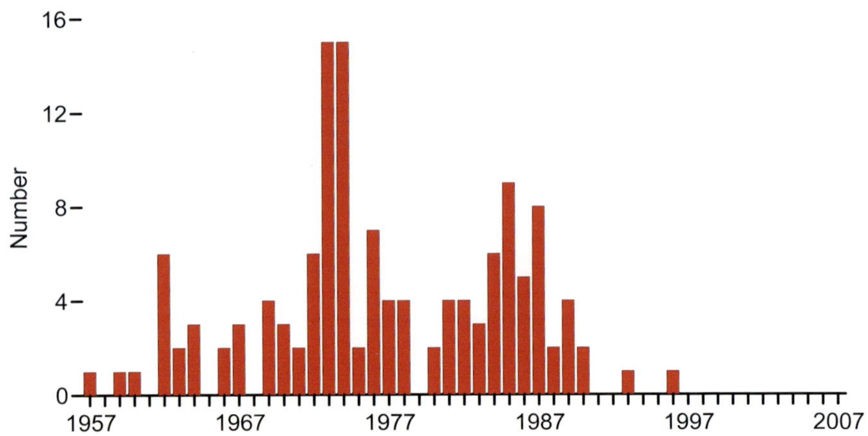

A total of 132 full-power operating licenses, or equivalent permission, were issued in the United States since the industry got its start in the 1950s. Most of the licenses were granted between 1962 and 1990. After 1990, one license was issued in 1993 and one in 1996.

Figure 51. Operable Units

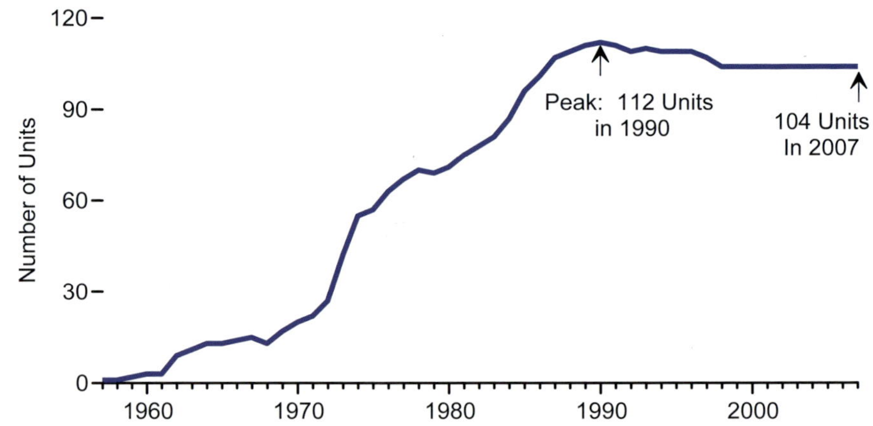

Out of the 132 units that were granted full-power operating licenses, or equivalent permission, over time, 28 were permanently shut down. The largest number of units ever operable in the United States was 112 in 1990. From 1998 through 2007, 104 units were operable.

Figure 52. Nuclear Net Summer Capacity

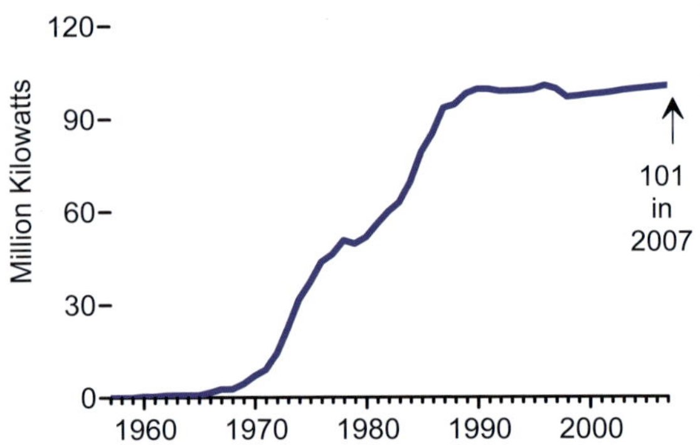

The U.S. nuclear industry's first commercial plant opened in Shippingport, Pennsylvania, in 1957. Nuclear capacity expanded sharply in the 1970s and 1980s. Total net capacity stood at 101 million kilowatts in 2007.

Figure 53. Capacity Factor

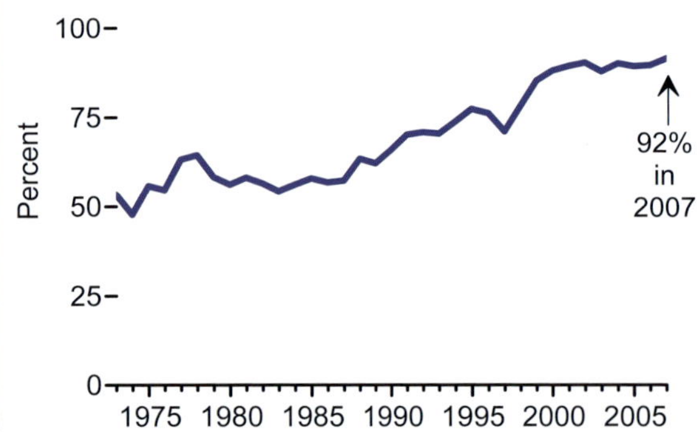

Capacity factors measure actual power generation as a share of maximum possible output. Factors for the industry, which were in the 50-to-60 percent range through the 1980s, generally improved in later years and stood at 92 percent in 2007.

Figure 54. Nuclear Share of Electricity

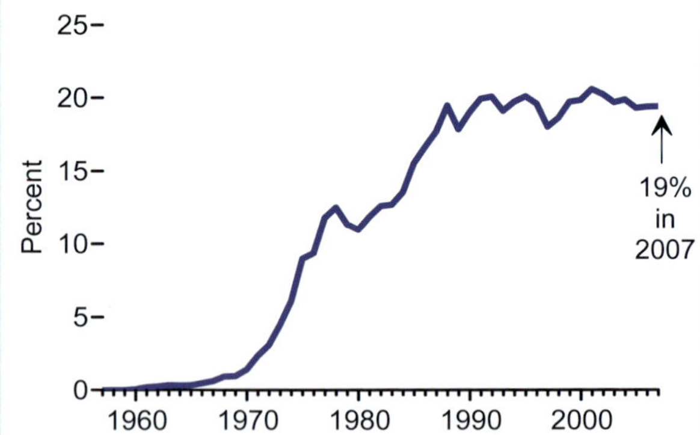

Over the latter part of the last century, nuclear electric power began to play a key role in meeting the Nation's rapidly growing electricity requirements. In 2007, 19 percent of U.S. total electricity net generation came from nuclear electric power.

Renewable Energy

Figure 55. Renewable Energy Total Consumption and Major Sources

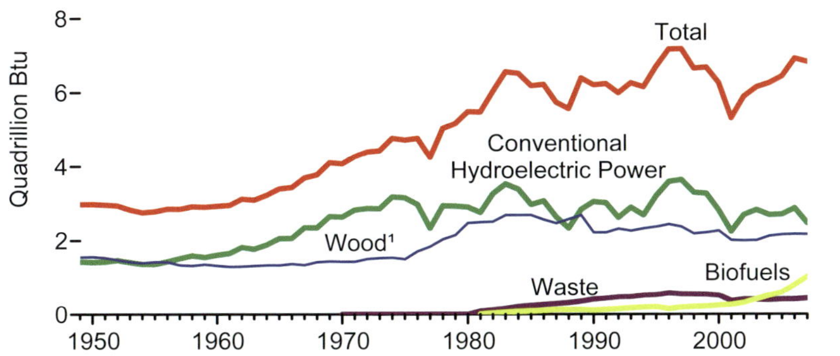

[1] Wood and wood-derived fuels.

Total renewable energy consumption generally followed the pattern of hydroelectric power output, which was the largest component of the total for most of the years shown. In 2007, hydroelectric power accounted for 36 percent of the total. Wood was the next largest source of renewable energy, followed by biofuels, waste, geothermal, wind, and solar/photovoltaic.

Figure 56. Renewable Energy Consumption by Sector

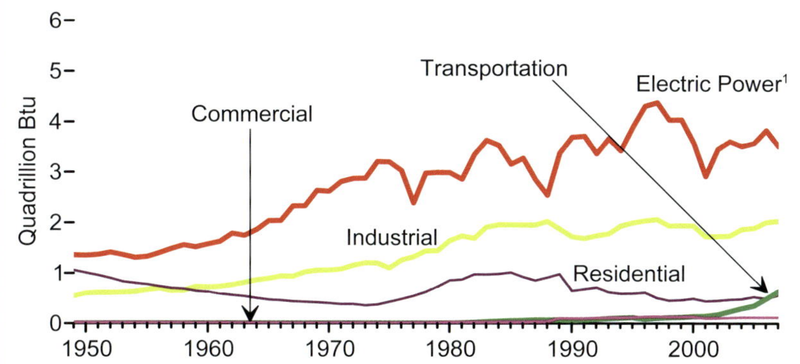

[1] Through 1988, electric utilities only; after 1988, includes independent power producers.

Most renewable energy was consumed by the electric power sector to generate electricity. After 1958, the industrial sector was the second largest consuming sector of renewable energy; the residential sector was the third largest consuming sector of renewable energy until it was exceeded by the transportation sector in 2007.

Figure 57. Biomass Consumption by Sector

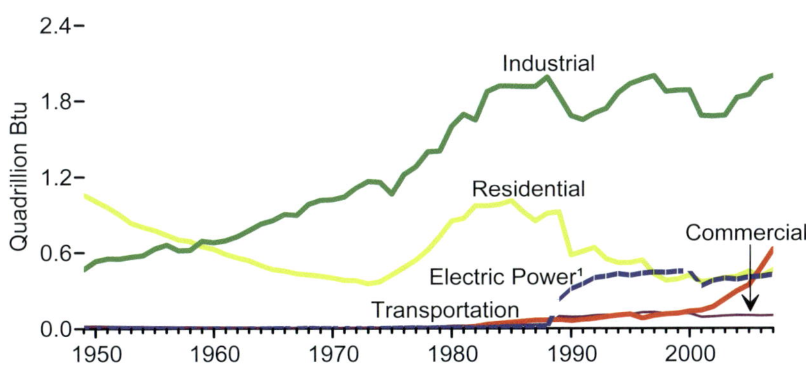

[1] Through 1988, electric utilities only; after 1988, includes independent power producers.

After 1959, the industrial sector consumed the most biomass (wood, waste, fuel ethanol, and biodiesel). Residential use of biomass (wood) fell through 1973, expanded from 1974 through 1985, and then trended downward again. Transportation consumption of biomass (fuel ethanol and biodiesel) expanded after 1996 and, by 2006, exceeded the electric power sector's consumption of biomass (wood and waste).

Figure 58. Solar Collector Shipments and Trade

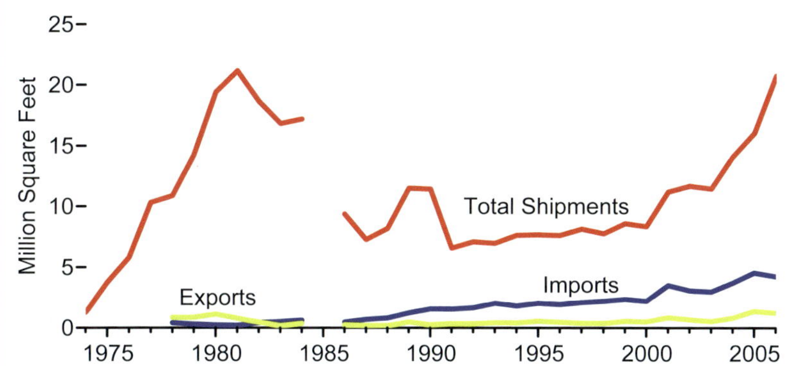

Notes: • Data were not collected for 1985. • Shipments include all domestically manufactured collectors plus imports.

Shipments of solar collectors grew strongly in the 1970s and reached a peak of 21 million square feet in 1981. Uneven performance was recorded over the next decade, followed by a mild upward trend during the 1990s and a bump up in 2001 and again in 2004 through 2006. Imports reached a record level of 4.5 million square feet in 2005.

International Energy

Figure 59. World Primary Energy Production by Source

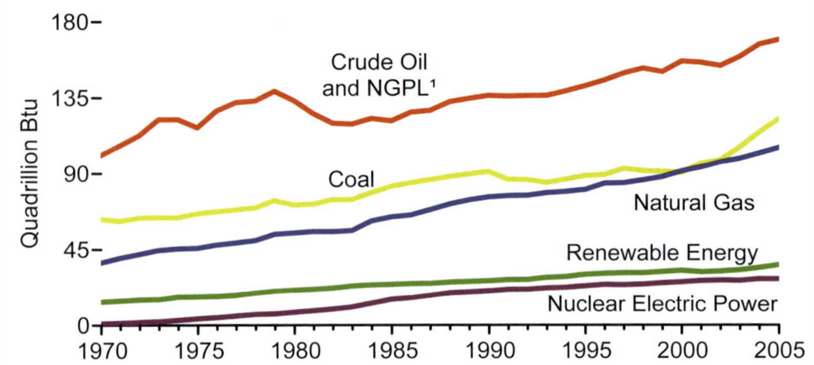

[1] Natural gas plant liquids.

From 1970 to 2005, world primary energy production grew by 114 percent, reaching 460 quadrillion Btu in 2005. Growth occurred in all types of energy. In 2005, fossil fuels accounted for 86 percent of all energy produced worldwide, renewable energy 8 percent, and nuclear electric power 6 percent.

Figure 60. World Primary Energy Production by Region

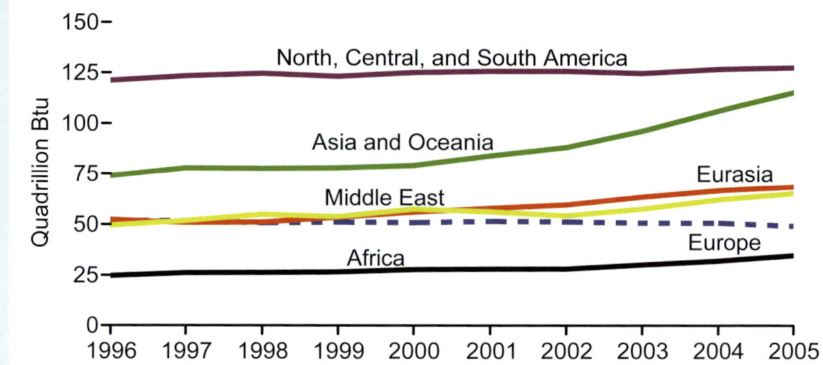

Twenty-eight percent of the 460 quadrillion Btu of energy produced worldwide in 2005 came from North, Central, and South America. The second largest regional energy producer was Asia and Oceania with 25 percent of the world total in 2005.

Figure 61. World Crude Oil Production

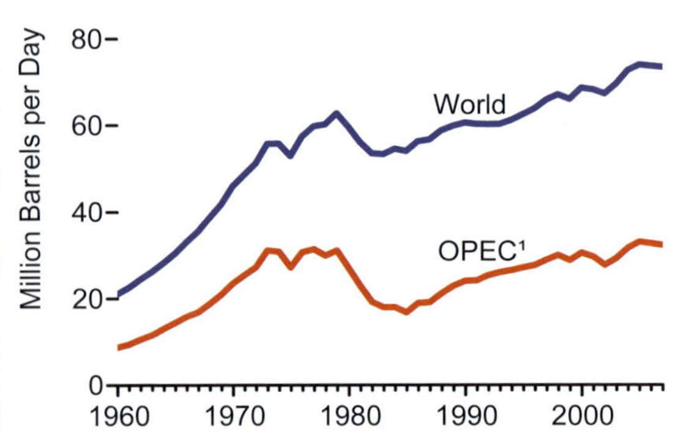

[1] Organization of the Petroleum Exporting Countries.

World crude oil production totaled 73 million barrels per day in 2007, down slightly from the level in 2006. OPEC's share of the world total in 2007 was 44 percent, compared to the peak level of 56 percent in 1973.

Figure 62. Leading Crude Oil Producers

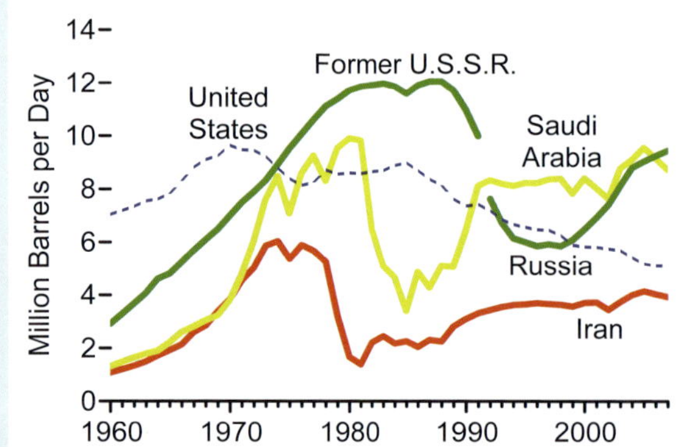

From 1974 through 1991, the former U.S.S.R. was the world's leading crude oil producer. After 1991, Saudi Arabia was the top producer until 2006 when Russia's production exceeded Saudi Arabia's for the first time. U.S. production peaked in 1970 but still ranked third in 2007.

Figure 63. Leading Petroleum Consumers

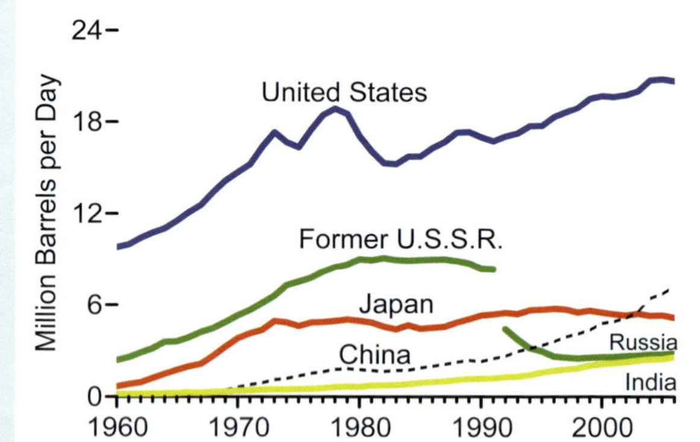

The United States accounted for 24 percent of world petroleum consumption in 2006. China and Japan, the next two leading consumers, together accounted for 15 percent. Russia, Germany, and India were the next largest consumers of petroleum in 2006.

Emissions

Figure 64. Greenhouse Gas Emissions, Based on Global Warming Potential

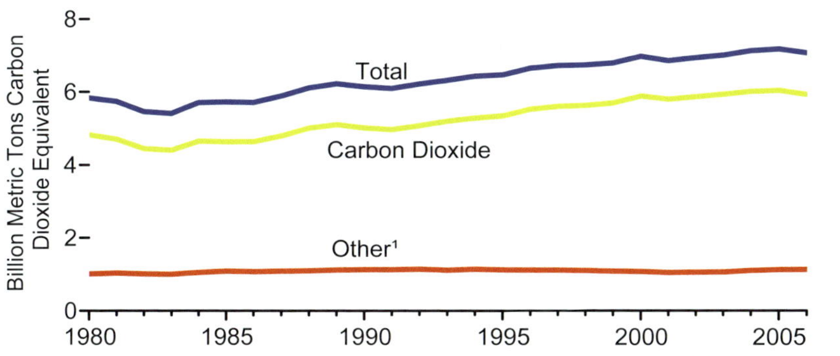

[1] Methane, nitrous oxide, hydrofluorocarbons (HFCs), perfluorocarbons (PFCs), and sulfur hexafluoride (SF_6).

The combustion of fossil fuels—coal, petroleum, and natural gas—to release their energy creates carbon dioxide emissions, the most significant greenhouse gas. Total carbon dioxide emissions stood at 6 billion metric tons of gas in 2006, 18 percent higher than the 1990 level.

Figure 65. Carbon Dioxide Emissions From Energy Use

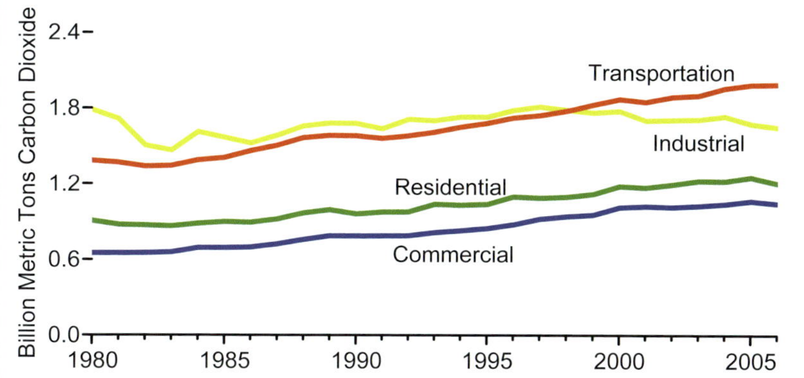

Note: Electric power sector emissions are distributed to the end-use sectors.

Carbon dioxide emitted by the industrial sector fell by 8 percent from 1980 to 2006. By 1999, transportation sector carbon dioxide emissions exceeded industrial sector emissions. Of the major sectors, the commercial sector generated the least carbon dioxide but recorded the largest growth (60 percent) since 1980.

Figure 66. GDP Growth and Carbon Dioxide Emissions

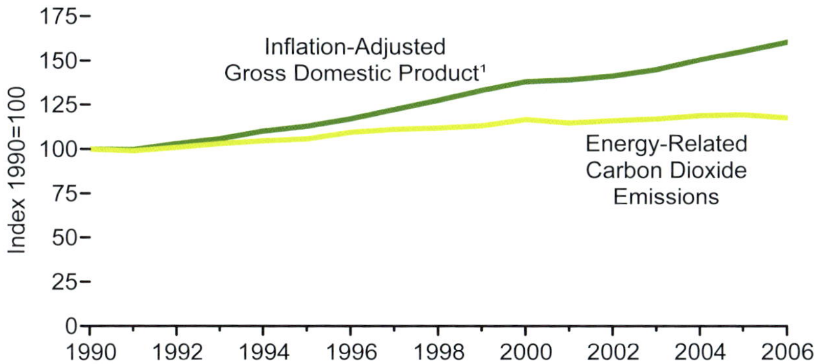

[1] Based on chained (2000) dollars. See "Chained Dollars" in Glossary.

While real gross domestic product (GDP) grew by 59 percent from 1990 to 2006, energy-related carbon dioxide emissions grew by 18 percent. From 2000 to 2006, GDP rose 15 percent, and energy-related carbon dioxide emissions rose by less than 1 percent.

Figure 67. Methane Emissions by Source

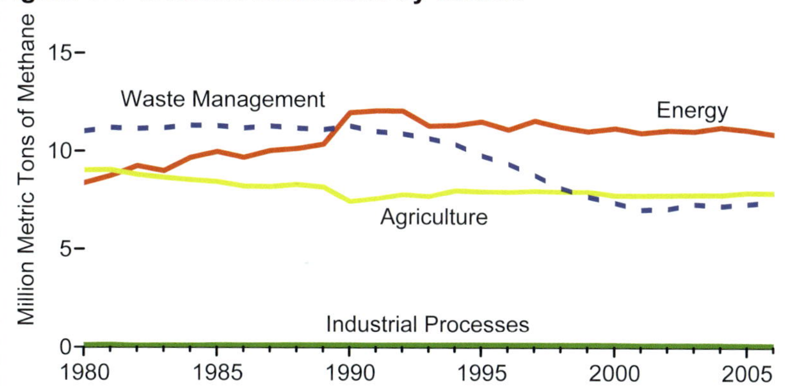

In 2006, methane emissions accounted for 9 percent of total U.S. greenhouse gas emissions, weighted by global warming potential. Most methane emissions came from energy, waste management, and agricultural sources. The production, processing, and distribution of natural gas accounted for 60 percent of the energy-related methane emissions in 2006.

Figure Sources

Data for "Energy Perspectives" figures and text are derived from the following *Annual Energy Review 2007* tables and additional sources:

1. Table 1.1.
2. Table 1.5.
3. Table 1.5.
4. Table 1.3.
5. Tables 1.3, 10.1, and E1.
6. Historical data: Table 1.3. Projections: Energy Information Administration, *Annual Energy Outlook 2008* (March 2008), Reference Case Tables, Table 1, "Total Energy Supply and Disposition Summary."
7. Table 2.1a.
8. Tables 2.1b and 2.1c.
9. Table 2.1d.
10. Tables 2.1e, 5.13c, and 5.14c.
11. Table 1.2.
12. Tables 5.1, 6.1, and 7.1.
13. Table 1.4.
14. Table 5.1.
15. Table 5.2.
16. Table 5.2.
17. Table 4.4.
18. Tables 5.13a, 5.13b, 5.13c, and 5.13d.
19. Table 5.11.
20. Table 5.21.
21. Table 5.24.
22. Tables 5.3 and 5.5.
23. Table 5.4.
24. Table 5.4.
25. Table 5.4.
26. Table 5.16.
27. Table 5.16.
28. Table 5.17.
29. Table 5.17.
30. Table 2.8.
31. Table 2.8.
32. Table 2.8.
33. Table 2.8.
34. Table 6.1.
35. Table 6.4.
36. Table 6.3.
37. Table 6.5.
38. Table 7.1.
39. Table 7.3.
40. Table 7.6.
41. Table 7.2.
42. Table 7.2.
43. Tables 8.2a, 8.2b, and 8.2d.
44. Table 8.2a.
45. Table 8.2c.
46. Tables 8.3b and 8.3c.
47. Table 8.10.
48. Table 8.9.
49. Table 8.1.
50. Table 9.1.
51. Table 9.1.
52. Table 9.2.
53. Table 9.2.
54. Table 9.2.
55. Table 10.1.
56. Tables 10.2a–10.2c.
57. Tables 10.2a–10.2c.
58. Table 10.5.
59. Table 11.1.
60. Table 11.2.
61. Table 11.5.
62. Table 11.5.
63. Table 11.10.
64. Table 12.1.
65. Table 12.2.
66. Tables 1.5 and 12.2.
67. Tables 12.1 and 12.5.

1 Energy Overview

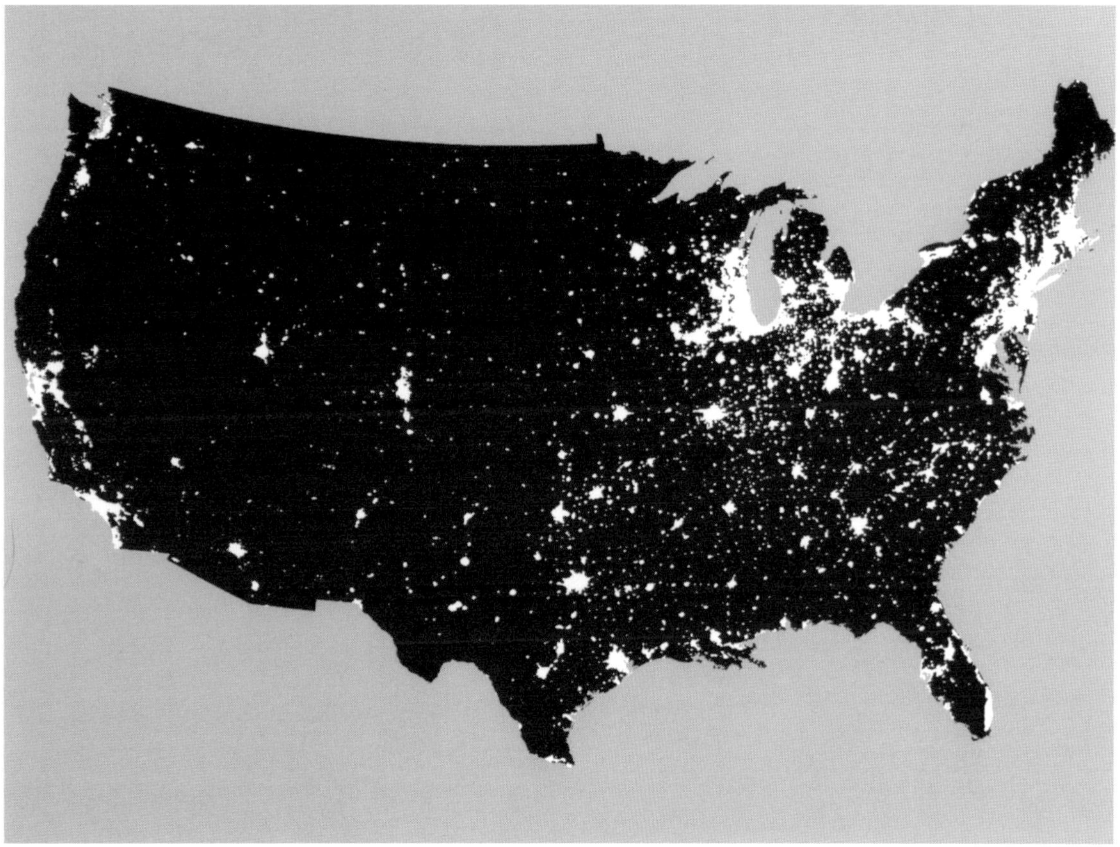

The continental United States at night from orbit. Source: National Oceanic and Atmospheric Administration satellite imagery; mosaic provided by U.S. Geological Survey.

Diagram 1. Energy Flow, 2007
(Quadrillion Btu)

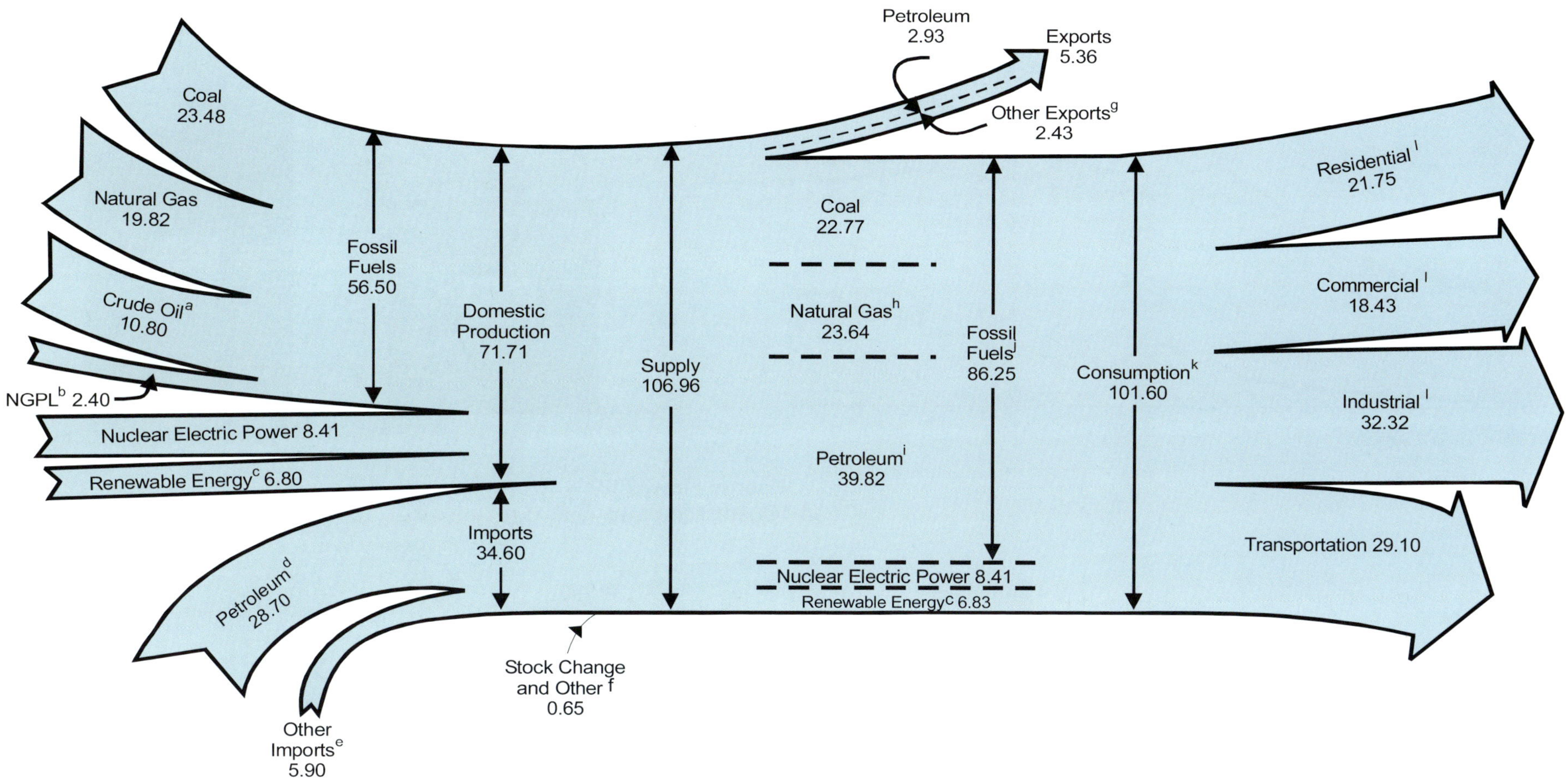

[a] Includes lease condensate.
[b] Natural gas plant liquids.
[c] Conventional hydroelectric power, biomass, geothermal, solar/photovoltaic, and wind.
[d] Crude oil and petroleum products. Includes imports into the Strategic Petroleum Reserve.
[e] Natural gas, coal, coal coke, fuel ethanol, and electricity.
[f] Adjustments, losses, and unaccounted for.
[g] Coal, natural gas, coal coke, and electricity.
[h] Natural gas only; excludes supplemental gaseous fuels.
[i] Petroleum products, including natural gas plant liquids, and crude oil burned as fuel.
[j] Includes 0.03 quadrillion Btu of coal coke net imports.
[k] Includes 0.11 quadrillion Btu of electricity net imports.
[l] Primary consumption, electricity retail sales, and electrical system energy losses, which are allocated to the end-use sectors in proportion to each sector's share of total electricity retail sales. See Note, "Electrical Systems Energy Losses," at end of Section 2.

Notes: • Data are preliminary. • Values are derived from source data prior to rounding for publication. • Totals may not equal sum of components due to independent rounding.

Sources: Tables 1.1, 1.2, 1.3, 1.4, and 2.1a.

Figure 1.1 Primary Energy Overview

Overview, 1949-2007

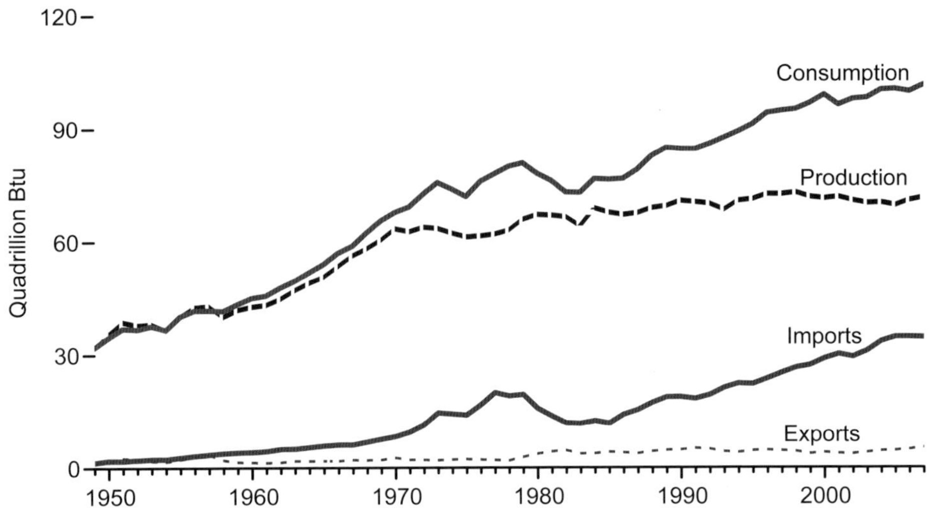

Production and Consumption, 2007

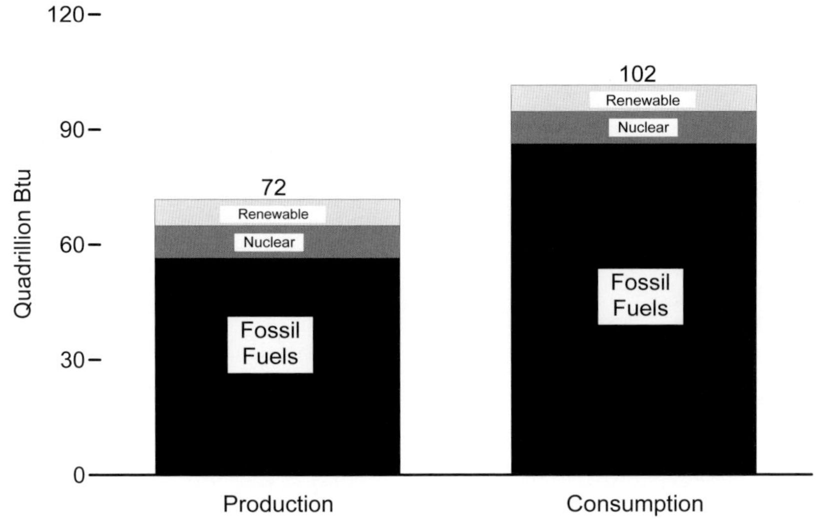

Overview, 2007

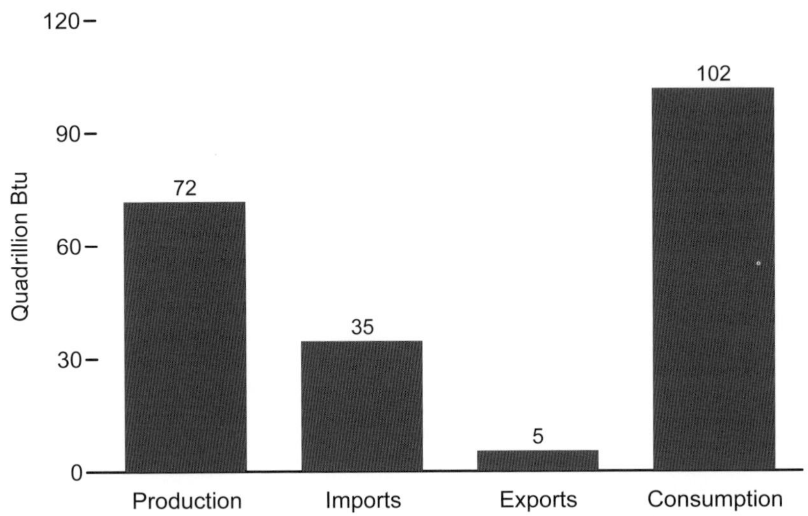

Energy Flow, 2007
(Quadrillion Btu)

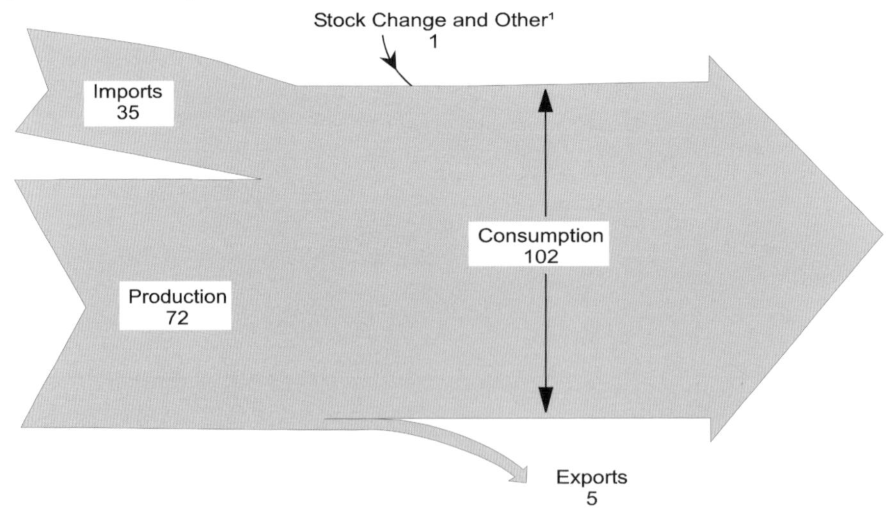

[1] Adjustments, losses, and unaccounted for.

Source: Table 1.1.

Table 1.1 Primary Energy Overview, Selected Years, 1949-2007

(Quadrillion Btu)

Year	Production [1]				Trade					Stock Change and Other [9]	Consumption [2]			
					Imports		Exports		Net Imports [3]					
	Fossil Fuels [4]	Nuclear Electric Power	Renewable Energy [5]	Total	Petroleum [6]	Total [7]	Coal	Total [8]	Total		Fossil Fuels [10]	Nuclear Electric Power	Renewable Energy [5]	Total [11]
1949	28.748	0.000	2.974	31.722	1.427	1.448	0.877	1.592	-0.144	0.403	29.002	0.000	2.974	31.982
1950	32.563	.000	2.978	35.540	1.886	1.913	.786	1.465	.448	-1.372	31.632	.000	2.978	34.616
1955	37.364	.000	2.784	40.148	2.752	2.790	1.465	2.286	.504	-.444	37.410	.000	2.784	40.208
1960	39.869	.006	2.929	42.804	3.999	4.188	1.023	1.477	2.710	-.427	42.137	.006	2.929	45.087
1965	47.235	.043	3.398	50.676	5.402	5.892	1.376	1.829	4.063	-.722	50.577	.043	3.398	54.017
1970	59.186	.239	4.076	63.501	7.470	8.342	1.936	2.632	5.709	-1.367	63.522	.239	4.076	67.844
1971	58.042	.413	4.268	62.723	8.540	9.535	1.546	2.151	7.384	-.818	64.596	.413	4.268	69.289
1972	58.938	.584	4.398	63.920	10.299	11.387	1.531	2.118	9.269	-.485	67.696	.584	4.398	72.704
1973	58.241	.910	4.433	63.585	13.466	14.613	1.425	2.033	12.580	-.456	70.316	.910	4.433	75.708
1974	56.331	1.272	4.769	62.372	13.127	14.304	1.620	2.203	12.101	-.482	67.906	1.272	4.769	73.991
1975	54.733	1.900	4.723	61.357	12.948	14.032	1.761	2.323	11.709	-1.067	65.355	1.900	4.723	71.999
1976	54.723	2.111	4.768	61.602	15.672	16.760	1.597	2.172	14.588	-.178	69.104	2.111	4.768	76.012
1977	55.101	2.702	4.249	62.052	18.756	19.948	1.442	2.052	17.896	-1.948	70.989	2.702	4.249	78.000
1978	55.074	3.024	5.039	63.137	17.824	19.106	1.078	1.920	17.186	-.337	71.856	3.024	5.039	79.986
1979	58.006	2.776	5.166	65.948	17.933	19.460	1.753	2.855	16.605	-1.649	72.892	2.776	5.166	80.903
1980	59.008	2.739	5.485	67.232	14.658	15.796	2.421	3.695	12.101	-1.212	69.826	2.739	5.485	78.122
1981	58.529	3.008	5.477	67.014	12.639	13.719	2.944	4.307	9.412	-.258	67.570	3.008	5.477	76.168
1982	57.458	3.131	6.034	66.623	10.777	11.861	2.787	4.608	7.253	-.724	63.888	3.131	6.034	73.153
1983	54.416	3.203	6.561	64.180	10.647	11.752	2.045	3.693	8.059	.799	63.154	3.203	6.561	73.038
1984	58.849	3.553	6.522	68.924	11.433	12.471	2.151	3.786	8.685	-.894	66.504	3.553	6.522	76.714
1985	57.539	4.076	6.185	67.799	10.609	11.781	2.438	4.196	7.584	1.107	66.091	4.076	6.185	76.491
1986	56.575	4.380	6.223	67.178	13.201	14.151	2.248	4.021	10.130	-.552	66.031	4.380	6.223	76.756
1987	57.167	4.754	5.739	67.659	14.162	15.398	2.093	3.812	11.586	-.073	68.522	4.754	5.739	79.173
1988	57.875	5.587	5.568	69.030	15.747	17.296	2.499	4.366	12.929	.860	71.556	5.587	5.568	82.819
1989	57.483	5.602	6.391	69.476	17.162	18.766	2.637	4.661	14.105	1.362	72.913	5.602	6.391	84.944
1990	58.560	6.104	6.206	70.870	17.117	18.817	2.772	4.752	14.065	-.283	72.333	6.104	6.206	84.652
1991	57.872	6.422	6.238	70.532	16.348	18.335	2.854	5.141	13.194	.881	71.880	6.422	6.238	84.607
1992	57.655	6.479	5.993	70.127	16.968	19.372	2.682	4.937	14.435	1.394	73.397	6.479	5.993	85.956
1993	55.822	6.410	6.263	68.495	18.510	21.273	1.962	4.258	R17.014	R2.094	74.836	6.410	6.262	87.603
1994	58.044	6.694	6.155	70.893	R19.243	R22.390	1.879	4.061	R18.329	R.038	76.258	6.694	6.155	89.260
1995	57.540	7.075	6.703	71.319	R18.881	R22.260	2.318	4.511	R17.750	R2.104	77.258	7.075	6.705	91.173
1996	58.387	7.087	7.167	72.641	R20.284	R23.702	2.368	4.633	R19.069	R2.466	79.783	7.087	7.168	94.175
1997	58.857	6.597	7.180	72.634	21.740	R25.215	2.193	4.514	R20.701	1.430	80.874	6.597	7.178	94.765
1998	59.314	7.068	6.659	73.041	22.908	26.581	2.092	4.299	R22.281	-.139	81.370	7.068	6.657	95.183
1999	57.614	7.610	6.683	71.907	23.133	R27.252	1.525	3.715	R23.537	1.373	82.428	7.610	6.681	96.817
2000	57.366	7.862	6.262	71.490	R24.531	28.973	1.528	4.006	24.967	2.518	84.733	7.862	6.264	98.975
2001	58.541	8.033	5.318	71.892	R25.398	R30.157	1.265	3.770	R26.386	R-1.952	82.903	8.033	R5.316	96.326
2002	56.894	8.143	5.899	R70.936	R24.674	R29.407	1.032	3.668	R25.739	R1.184	83.750	8.143	5.893	97.858
2003	R56.157	7.959	6.149	R70.264	R26.219	R31.061	1.117	4.054	R27.007	R.938	84.078	7.959	R6.150	98.209
2004	R55.914	8.222	6.248	R70.384	R28.196	R33.543	1.253	4.433	R29.110	R.857	85.830	8.222	6.261	100.351
2005	R55.056	8.160	R6.431	R69.647	R29.248	R34.710	1.273	4.561	R30.149	R.710	R85.817	8.160	R6.444	R100.506
2006	R55.940	R8.214	R6.872	R71.025	R29.168	R34.673	1.264	R4.868	R29.805	R-.974	R84.658	R8.214	R6.922	R99.856
2007P	56.499	8.415	6.800	71.713	28.701	34.599	1.507	5.361	29.238	.649	86.248	8.415	6.830	101.600

[1] See Note 1, "Primary Energy Production," at end of section.
[2] See Note 2, "Primary Energy Consumption," at end of section.
[3] Net imports equal imports minus exports. Minus sign indicates exports are greater than imports.
[4] Coal, natural gas (dry), crude oil, and natural gas plant liquids.
[5] See Tables 10.1-10.2c for notes on series components and estimation.
[6] Crude oil and petroleum products. Includes imports into the Strategic Petroleum Reserve.
[7] Also includes natural gas, coal, coal coke, fuel ethanol, and electricity.
[8] Also includes natural gas, petroleum, coal coke, and electricity.
[9] Calculated as consumption and exports minus production and imports. Includes petroleum stock change and adjustments; natural gas net storage withdrawals and balancing item; coal stock change, losses, and unaccounted for; and fuel ethanol stock change.
[10] Coal, coal coke net imports, natural gas, and petroleum.
[11] Also includes electricity net imports.

R=Revised. P=Preliminary.

Note: Totals may not equal sum of components due to independent rounding.

Web Page: For all data beginning in 1949, see http://www.eia.doe.gov/emeu/aer/overview.html.

Sources: Tables 1.2, 1.3, and 1.4.

Figure 1.2 Primary Energy Production by Source

By Source Category, 1949-2007

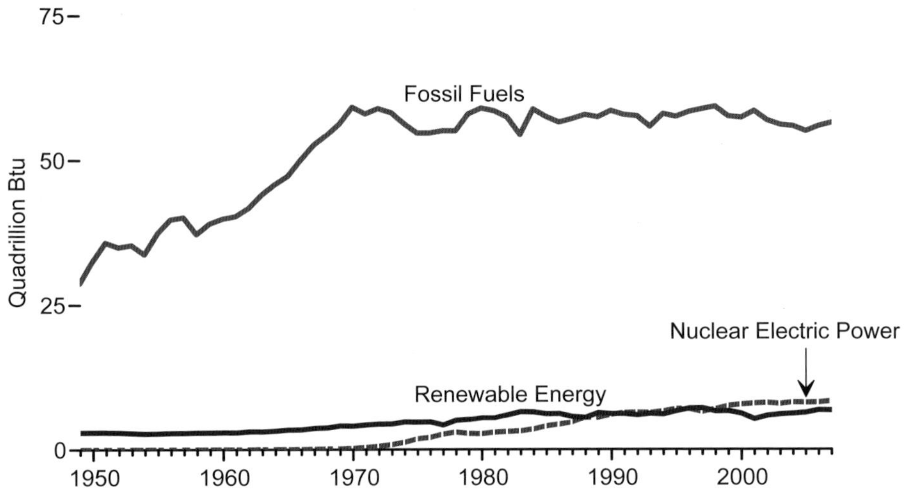

By Source, 2007

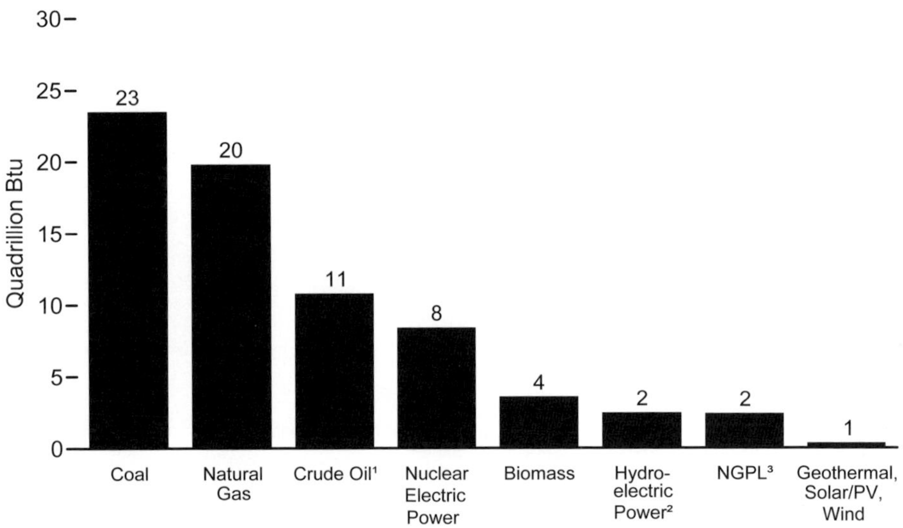

By Major Source, 1949-2007

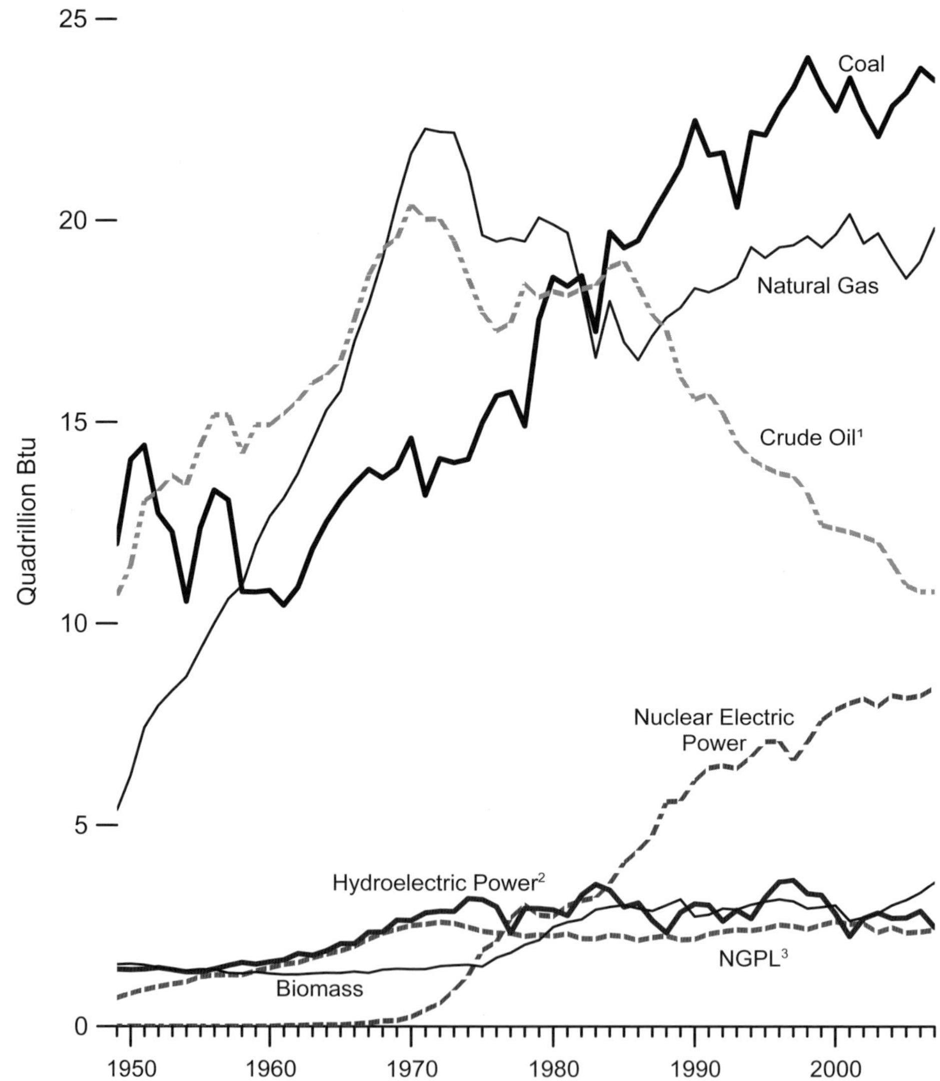

[1] Includes lease condensate.
[2] Conventional hydroelectric power.
[3] Natural gas plant liquids.

Note: Because vertical scales differ, graphs should not be compared.
Source: Table 1.2.

Table 1.2 Primary Energy Production by Source, Selected Years, 1949-2007

(Quadrillion Btu)

Year	Fossil Fuels					Nuclear Electric Power	Renewable Energy [1]						Total
	Coal [2]	Natural Gas (Dry)	Crude Oil [3]	NGPL [4]	Total		Hydro-electric Power [5]	Geothermal	Solar/PV	Wind	Biomass	Total	
1949	11.974	5.377	10.683	0.714	28.748	0.000	1.425	NA	NA	NA	1.549	2.974	31.722
1950	14.060	6.233	11.447	.823	32.563	.000	1.415	NA	NA	NA	1.562	2.978	35.540
1955	12.370	9.345	14.410	1.240	37.364	.000	1.360	NA	NA	NA	1.424	2.784	40.148
1960	10.817	12.656	14.935	1.461	39.869	.006	1.608	.001	NA	NA	1.320	2.929	42.804
1965	13.055	15.775	16.521	1.883	47.235	.043	2.059	.004	NA	NA	1.335	3.398	50.676
1970	14.607	21.666	20.401	2.512	59.186	.239	2.634	.011	NA	NA	1.431	4.076	63.501
1971	13.186	22.280	20.033	2.544	58.042	.413	2.824	.012	NA	NA	1.432	4.268	62.723
1972	14.092	22.208	20.041	2.598	58.938	.584	2.864	.031	NA	NA	1.503	4.398	63.920
1973	13.992	22.187	19.493	2.569	58.241	.910	2.861	.043	NA	NA	1.529	4.433	63.585
1974	14.074	21.210	18.575	2.471	56.331	1.272	3.177	.053	NA	NA	1.540	4.769	62.372
1975	14.989	19.640	17.729	2.374	54.733	1.900	3.155	.070	NA	NA	1.499	4.723	61.357
1976	15.654	19.480	17.262	2.327	54.723	2.111	2.976	.078	NA	NA	1.713	4.768	61.602
1977	15.755	19.565	17.454	2.327	55.101	2.702	2.333	.077	NA	NA	1.838	4.249	62.052
1978	14.910	19.485	18.434	2.245	55.074	3.024	2.937	.064	NA	NA	2.038	5.039	63.137
1979	17.540	20.076	18.104	2.286	58.006	2.776	2.931	.084	NA	NA	2.152	5.166	65.948
1980	18.598	19.908	18.249	2.254	59.008	2.739	2.900	.110	NA	NA	2.476	5.485	67.232
1981	18.377	19.699	18.146	2.307	58.529	3.008	2.758	.123	NA	NA	2.596	5.477	67.014
1982	18.639	18.319	18.309	2.191	57.458	3.131	3.266	.105	NA	NA	2.664	6.034	66.623
1983	17.247	16.593	18.392	2.184	54.416	3.203	3.527	.129	NA	(s)	2.904	6.561	64.180
1984	19.719	18.008	18.848	2.274	58.849	3.553	3.386	.165	(s)	(s)	2.971	6.522	68.924
1985	19.325	16.980	18.992	2.241	57.539	4.076	2.970	.198	(s)	(s)	3.016	6.185	67.799
1986	19.509	16.541	18.376	2.149	56.575	4.380	3.071	.219	(s)	(s)	2.932	6.223	67.178
1987	20.141	17.136	17.675	2.215	57.167	4.754	2.635	.229	(s)	(s)	2.875	5.739	67.659
1988	20.738	17.599	17.279	2.260	57.875	5.587	2.334	.217	(s)	(s)	3.016	5.568	69.030
1989	[2]21.360	17.847	16.117	2.158	57.483	5.602	2.837	.317	.055	.022	3.160	6.391	69.476
1990	22.488	18.326	15.571	2.175	58.560	6.104	3.046	.336	.060	.029	2.735	6.206	70.870
1991	21.636	18.229	15.701	2.306	57.872	6.422	3.016	.346	.063	.031	2.782	6.238	70.532
1992	21.694	18.375	15.223	2.363	57.655	6.479	2.617	.349	.064	.030	2.933	5.993	70.127
1993	20.336	18.584	14.494	2.408	55.822	6.410	2.892	.364	.066	.031	2.910	6.263	68.495
1994	22.202	19.348	14.103	2.391	58.044	6.694	2.683	.338	.069	.036	3.030	6.155	70.893
1995	22.130	19.082	13.887	2.442	57.540	7.075	3.205	.294	.070	.033	3.102	6.703	71.319
1996	22.790	19.344	13.723	2.530	58.387	7.087	3.590	.316	.071	.033	3.157	7.167	72.641
1997	23.310	19.394	13.658	2.495	58.857	6.597	3.640	.325	.070	.034	3.111	7.180	72.634
1998	24.045	19.613	13.235	2.420	59.314	7.068	3.297	.328	.070	.031	2.933	6.659	73.041
1999	23.295	19.341	12.451	2.528	57.614	7.610	3.268	.331	.069	.046	2.969	6.683	71.907
2000	22.735	19.662	12.358	2.611	57.366	7.862	2.811	.317	.066	.057	3.010	6.262	71.490
2001	[2]23.547	20.166	12.282	2.547	58.541	8.033	2.242	.311	.065	.070	2.629	5.318	71.892
2002	22.732	19.439	12.163	2.559	56.894	8.143	2.689	.328	.064	.105	2.712	5.899	[R]70.936
2003	[R]22.094	19.691	12.026	2.346	[R]56.157	7.959	2.825	.331	.064	.115	[R]2.815	6.149	[R]70.264
2004	[R]22.852	19.093	11.503	2.466	[R]55.914	8.222	2.690	.341	.064	.142	3.011	6.248	[R]70.384
2005	[R]23.185	[R]18.574	10.963	2.334	[R]55.056	8.160	2.703	.343	.066	.178	[R]3.141	[R]6.431	[R]69.647
2006	[R]23.790	[R]18.993	[R]10.801	[R]2.356	[R]55.940	[R]8.214	[R]2.869	[R].343	[R].072	[R].264	[R]3.324	[R]6.872	[R]71.025
2007[P]	23.480	19.817	10.802	2.400	56.499	8.415	2.463	.353	.080	.319	3.584	6.800	71.713

[1] Most data are estimates. See Tables 10.1-10.2c for notes on series components and estimation.
[2] Beginning in 1989, includes waste coal supplied. Beginning in 2001, also includes a small amount of refuse recovery. See Table 7.1.
[3] Includes lease condensate.
[4] Natural gas plant liquids.
[5] Conventional hydroelectric power.

R=Revised. P=Preliminary. NA=Not available. (s)=Less than 0.0005 quadrillion Btu.

Notes: • See Note 1, "Primary Energy Production," at end of section. • Totals may not equal sum of components due to independent rounding.

Web Page: For all data beginning in 1949, see http://www.eia.doe.gov/emeu/aer/overview.html.

Sources: Tables 5.1, 6.1, 7.1, 8.2a, 10.1, A2, A4, A5, and A6.

Figure 1.3 Primary Energy Consumption by Source

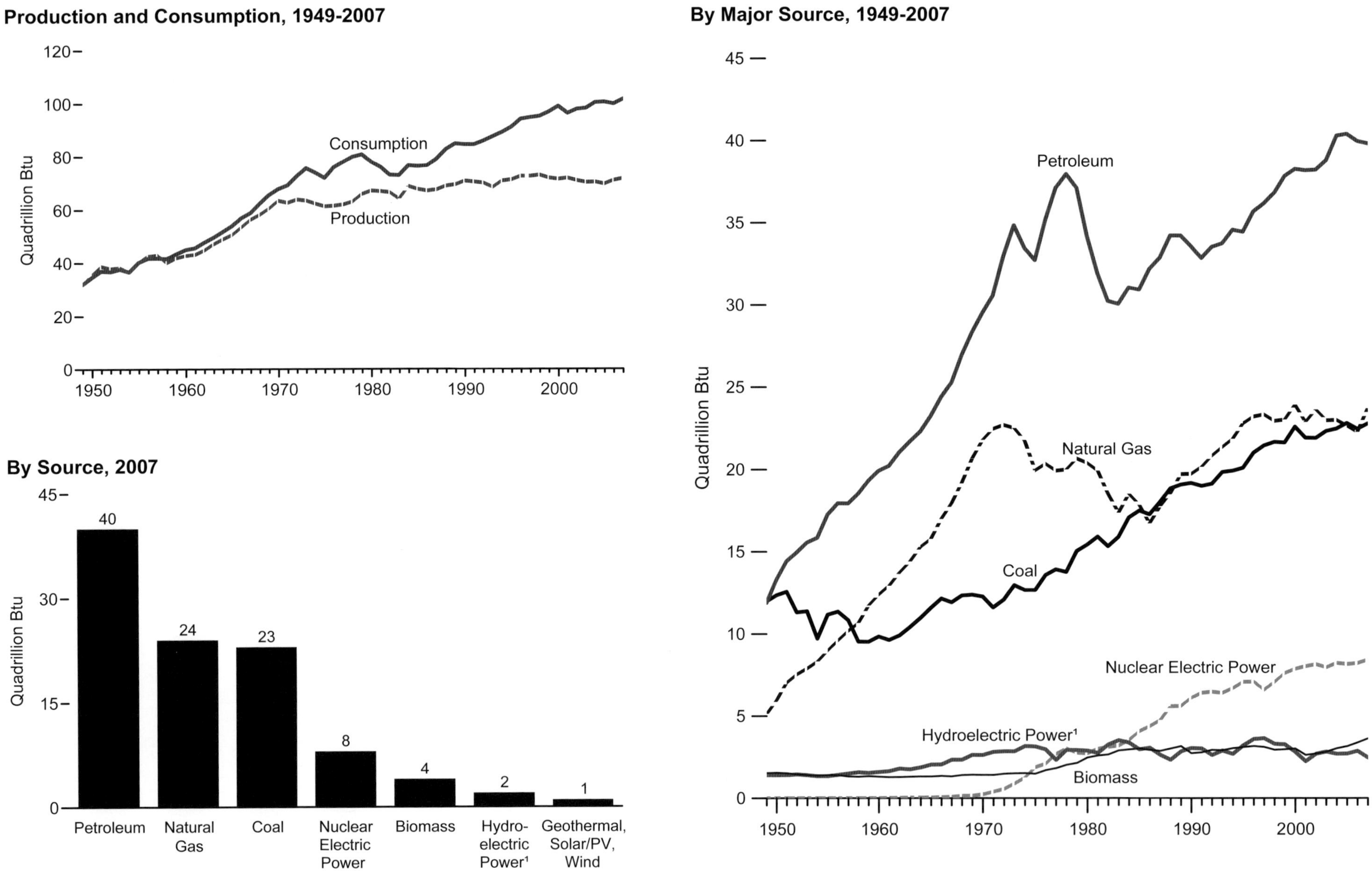

[1] Conventional hydroelectric power.

Note: Because vertical scales differ, graphs should not be compared.

Sources: Tables 1.2 and 1.3.

Table 1.3 Primary Energy Consumption by Source, Selected Years, 1949-2007

(Quadrillion Btu)

Year	Fossil Fuels					Nuclear Electric Power	Renewable Energy [1]						Electricity Net Imports [2]	Total
	Coal	Coal Coke Net Imports [2]	Natural Gas [3]	Petroleum [4]	Total		Hydro-electric Power [5]	Geothermal	Solar/PV	Wind	Biomass	Total		
1949	11.981	-0.007	5.145	11.883	29.002	0.000	1.425	NA	NA	NA	1.549	2.974	0.005	31.982
1950	12.347	.001	5.968	13.315	31.632	.000	1.415	NA	NA	NA	1.562	2.978	.006	34.616
1955	11.167	-.010	8.998	17.255	37.410	.000	1.360	NA	NA	NA	1.424	2.784	.014	40.208
1960	9.838	-.006	12.385	19.919	42.137	.006	1.608	.001	NA	NA	1.320	2.929	.015	45.087
1965	11.581	-.018	15.769	23.246	50.577	.043	2.059	.004	NA	NA	1.335	3.398	(s)	54.017
1970	12.265	-.058	21.795	29.521	63.522	.239	2.634	.011	NA	NA	1.431	4.076	.007	67.844
1971	11.598	-.033	22.469	30.561	64.596	.413	2.824	.012	NA	NA	1.432	4.268	.012	69.289
1972	12.077	-.026	22.698	32.947	67.696	.584	2.864	.031	NA	NA	1.503	4.398	.026	72.704
1973	12.971	-.007	22.512	34.840	70.316	.910	2.861	.043	NA	NA	1.529	4.433	.049	75.708
1974	12.663	.056	21.732	33.455	67.906	1.272	3.177	.053	NA	NA	1.540	4.769	.043	73.991
1975	12.663	.014	19.948	32.731	65.355	1.900	3.155	.070	NA	NA	1.499	4.723	.021	71.999
1976	13.584	(s)	20.345	35.175	69.104	2.111	2.976	.078	NA	NA	1.713	4.768	.029	76.012
1977	13.922	.015	19.931	37.122	70.989	2.702	2.333	.077	NA	NA	1.838	4.249	.059	78.000
1978	13.766	.125	20.000	37.965	71.856	3.024	2.937	.064	NA	NA	2.038	5.039	.067	79.986
1979	15.040	.063	20.666	37.123	72.892	2.776	2.931	.084	NA	NA	2.152	5.166	.069	80.903
1980	15.423	-.035	20.235	34.202	69.826	2.739	2.900	.110	NA	NA	2.476	5.485	.071	78.122
1981	15.908	-.016	19.747	31.931	67.570	3.008	2.758	.123	NA	NA	2.596	5.477	.113	76.168
1982	15.322	-.022	18.356	30.232	63.888	3.131	3.266	.105	NA	NA	2.664	6.034	.100	73.153
1983	15.894	-.016	17.221	30.054	63.154	3.203	3.527	.129	NA	(s)	2.904	6.561	.121	73.038
1984	17.071	-.011	18.394	31.051	66.504	3.553	3.386	.165	(s)	(s)	2.971	6.522	.135	76.714
1985	17.478	-.013	17.703	30.922	66.091	4.076	2.970	.198	(s)	(s)	3.016	6.185	.140	76.491
1986	17.260	-.017	16.591	32.196	66.031	4.380	3.071	.219	(s)	(s)	2.932	6.223	.122	76.756
1987	18.008	.009	17.640	32.865	68.522	4.754	2.635	.229	(s)	(s)	2.875	5.739	.158	79.173
1988	18.846	.040	18.448	34.222	71.556	5.587	2.334	.217	(s)	(s)	3.016	5.568	.108	82.819
1989	19.070	.030	19.602	34.211	72.913	5.602	2.837	.317	.055	.022	3.160	6.391	.037	84.944
1990	19.173	.005	19.603	33.553	72.333	6.104	3.046	.336	.060	.029	2.735	6.206	.008	84.652
1991	18.992	.010	20.033	32.845	71.880	6.422	3.016	.346	.063	.031	2.782	6.238	.067	84.607
1992	19.122	.035	20.714	33.527	73.397	6.479	2.617	.349	.064	.030	2.933	5.993	.087	85.956
1993	19.835	.027	21.229	33.744	74.836	6.410	2.892	.364	.066	.031	2.910	6.262	.095	87.603
1994	19.909	.058	21.728	34.562	76.258	6.694	2.683	.338	.069	.036	3.030	6.155	.153	89.260
1995	20.089	.061	22.671	34.437	77.258	7.075	3.205	.294	.070	.033	3.104	6.705	.134	91.173
1996	21.002	.023	23.085	35.673	79.783	7.087	3.590	.316	.071	.033	3.159	7.168	.137	94.175
1997	21.445	.046	23.223	36.160	80.874	6.597	3.640	.325	.070	.034	3.108	7.178	.116	94.765
1998	21.656	.067	22.830	36.817	81.370	7.068	3.297	.328	.070	.031	2.931	6.657	.088	95.183
1999	21.623	.058	22.909	37.838	82.428	7.610	3.268	.331	.069	.046	2.967	6.681	.099	96.817
2000	22.580	.065	23.824	38.264	84.733	7.862	2.811	.317	.066	.057	3.013	6.264	.115	98.975
2001	21.914	.029	22.773	38.186	82.903	8.033	2.242	.311	.065	.070	2.627	R5.316	.075	96.326
2002	21.904	.061	23.558	38.227	83.750	8.143	2.689	.328	.064	.105	2.706	5.893	.072	97.858
2003	22.321	.051	22.897	38.809	84.078	7.959	2.825	.331	.064	.115	2.817	R6.150	.022	98.209
2004	22.466	.138	22.931	40.294	85.830	8.222	2.690	.341	.065	.142	3.023	6.261	.039	100.351
2005	R22.797	.044	R22.583	40.393	R85.817	8.160	2.703	.343	.066	.178	R3.154	R6.444	.084	R100.506
2006	R22.447	.061	R22.191	R39.958	R84.658	R8.214	R2.869	R.343	R.072	R.264	R3.374	R6.922	R.063	R99.856
2007P	22.767	.025	23.638	39.818	86.248	8.415	2.463	.353	.080	.319	3.615	6.830	.107	101.600

[1] Most data are estimates. See Tables 10.1-10.2c for notes on series components and estimation.

[2] Net imports equal imports minus exports. Minus sign indicates exports are greater than imports.

[3] Natural gas only; excludes supplemental gaseous fuels. See Note 1, "Supplemental Gaseous Fuels," at end of Section 6.

[4] Petroleum products supplied, including natural gas plant liquids and crude oil burned as fuel. Does not include the fuel ethanol portion of motor gasoline—fuel ethanol is included in "Biomass."

[5] Conventional hydroelectric power.

R=Revised. P=Preliminary. NA=Not available. (s)=Less than 0.0005 and greater than -0.0005 quadrillion Btu.

Notes: • See Note 2, "Primary Energy Consumption," at end of section. • See Table E1 for estimated energy consumption for 1635-1945. • See Note 3, "Electricity Imports and Exports," at end of Section 8. • Totals may not equal sum of components due to independent rounding.

Web Page: For all data beginning in 1949, see http://www.eia.doe.gov/emeu/aer/overview.html.

Sources: Tables 5.12, 6.1, 7.1, 7.7, 8.1, 8.2a, 10.1, 10.3, A4, A5, and A6.

Figure 1.4 Primary Energy Trade by Source, 1949-2007

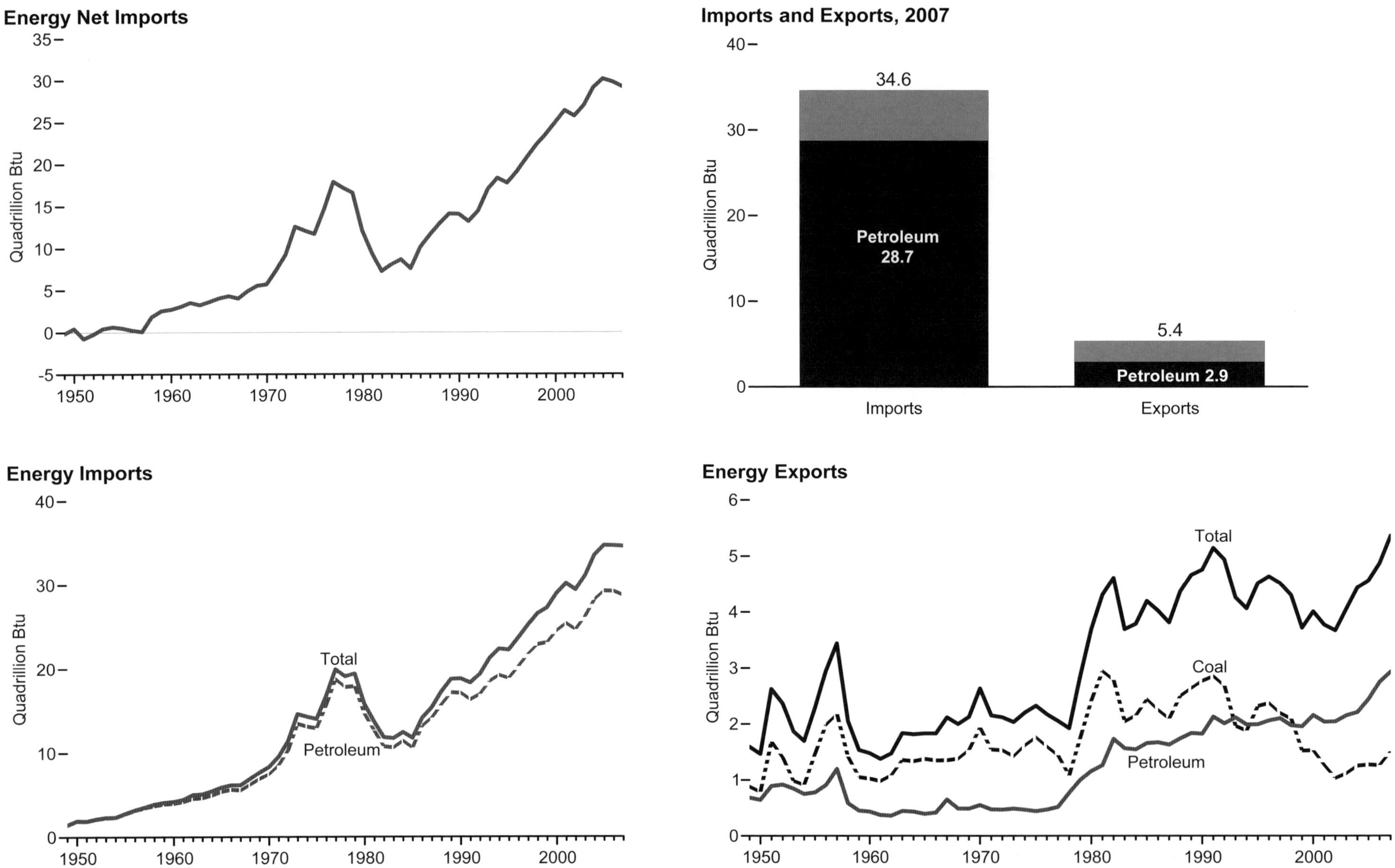

Notes: • Negative net imports are net exports. • Because vertical scales differ, graphs should not be compared.

Source: Table 1.4.

Table 1.4 Primary Energy Trade by Source, Selected Years, 1949-2007

(Quadrillion Btu)

Year	Imports									Exports								Net Imports [1]
				Petroleum									Petroleum					
	Coal	Coal Coke	Natural Gas	Crude Oil [2]	Petroleum Products [3]	Total	Fuel Ethanol	Elec-tricity	Total	Coal	Coal Coke	Natural Gas	Crude Oil	Petroleum Products [3]	Total	Elec-tricity	Total	Total
1949	0.008	0.007	0.000	0.915	0.513	1.427	NA	0.006	1.448	0.877	0.014	0.021	0.192	0.488	0.680	0.001	1.592	-0.144
1950	.009	.011	.000	1.056	.830	1.886	NA	.007	1.913	.786	.010	.027	.202	.440	.642	.001	1.465	.448
1955	.008	.003	.011	1.691	1.061	2.752	NA	.016	2.790	1.465	.013	.032	.067	.707	.774	.002	2.286	.504
1960	.007	.003	.161	2.196	1.802	3.999	NA	.018	4.188	1.023	.009	.012	.018	.413	.431	.003	1.477	2.710
1965	.005	.002	.471	2.654	2.748	5.402	NA	.012	5.892	1.376	.021	.027	.006	.386	.392	.013	1.829	4.063
1970	.001	.004	.846	2.814	4.656	7.470	NA	.021	8.342	1.936	.061	.072	.029	.520	.549	.014	2.632	5.709
1971	.003	.004	.964	3.573	4.968	8.540	NA	.024	9.535	1.546	.037	.083	.003	.470	.473	.012	2.151	7.384
1972	.001	.005	1.047	4.712	5.587	10.299	NA	.036	11.387	1.531	.031	.080	.001	.466	.467	.010	2.118	9.269
1973	.003	.027	1.060	6.887	6.578	13.466	NA	.057	14.613	1.425	.035	.079	.004	.482	.486	.009	2.033	12.580
1974	.052	.088	.985	7.395	5.731	13.127	NA	.053	14.304	1.620	.032	.078	.006	.458	.465	.009	2.203	12.101
1975	.024	.045	.978	8.721	4.227	12.948	NA	.038	14.032	1.761	.032	.074	.012	.427	.439	.017	2.323	11.709
1976	.030	.033	.988	11.239	4.434	15.672	NA	.037	16.760	1.597	.033	.066	.017	.452	.469	.008	2.172	14.588
1977	.041	.045	1.037	14.027	4.728	18.756	NA	.069	19.948	1.442	.031	.056	.106	.408	.514	.009	2.052	17.896
1978	.074	.142	.995	13.460	4.364	17.824	NA	.072	19.106	1.078	.017	.053	.335	.432	.767	.005	1.920	17.186
1979	.051	.099	1.300	13.825	4.108	17.933	NA	.077	19.460	1.753	.036	.056	.497	.505	1.002	.007	2.855	16.605
1980	.030	.016	1.006	11.195	3.463	14.658	NA	.085	15.796	2.421	.051	.049	.609	.551	1.160	.014	3.695	12.101
1981	.026	.013	.917	9.336	3.303	12.639	NA	.124	13.719	2.944	.029	.060	.482	.781	1.264	.010	4.307	9.412
1982	.019	.003	.950	7.418	3.360	10.777	NA	.112	11.861	2.787	.025	.052	.500	1.231	1.732	.012	4.608	7.253
1983	.032	.001	.940	7.079	3.568	10.647	NA	.132	11.752	2.045	.016	.055	.348	1.217	1.565	.011	3.693	8.059
1984	.032	.014	.847	7.302	4.131	11.433	NA	.144	12.471	2.151	.026	.055	.384	1.161	1.545	.009	3.786	8.685
1985	.049	.014	.952	6.814	3.796	10.609	NA	.157	11.781	2.438	.028	.056	.432	1.225	1.657	.017	4.196	7.584
1986	.055	.008	.748	9.002	4.199	13.201	NA	.139	14.151	2.248	.025	.062	.326	1.344	1.670	.016	4.021	10.130
1987	.044	.023	.992	10.067	4.095	14.162	NA	.178	15.398	2.093	.014	.055	.319	1.311	1.630	.020	3.812	11.586
1988	.053	.067	1.296	11.027	4.720	15.747	NA	.133	17.296	2.499	.027	.075	.329	1.412	1.741	.024	4.366	12.929
1989	.071	.057	1.387	12.596	4.565	17.162	NA	.089	18.766	2.637	.027	.109	.300	1.536	1.836	.052	4.661	14.105
1990	.067	.019	1.551	12.766	4.351	17.117	NA	.063	18.817	2.772	.014	.087	.230	1.594	1.824	.055	4.752	14.065
1991	.085	.029	1.798	12.553	3.794	16.348	NA	.075	18.335	2.854	.020	.132	.246	1.882	2.128	.008	5.141	13.194
1992	.095	.052	2.161	13.253	3.714	16.968	NA	.096	19.372	2.682	.017	.220	.188	1.819	2.008	.010	4.937	14.435
1993	.205	.053	2.397	14.749	R3.760	18.510	.001	.107	21.273	1.962	.026	.142	.208	1.907	2.115	.012	4.258	R17.014
1994	.222	.083	2.682	15.340	R3.904	R19.243	.001	.160	R22.390	1.879	.024	.164	.209	1.779	1.988	.007	4.061	R18.329
1995	.237	.095	2.901	15.669	R3.211	R18.881	.001	.146	R22.260	2.318	.034	.156	.200	1.791	1.991	.012	4.511	R17.750
1996	.203	.063	3.002	16.341	R3.943	R20.284	.001	.148	R23.702	2.368	.040	.155	.233	1.825	2.059	.011	4.633	R19.069
1997	.187	.078	3.063	17.876	R3.864	21.740	(s)	.147	R25.215	2.193	.031	.159	.228	1.872	2.100	.031	4.514	R20.701
1998	.218	.095	3.225	18.916	3.992	22.908	(s)	.135	26.581	2.092	.028	.161	.233	1.740	1.972	.047	4.299	R22.281
1999	.227	.080	3.664	18.935	4.198	23.133	(s)	.147	R27.252	1.525	.022	.164	.250	1.705	1.955	.049	3.715	R23.537
2000	.313	.094	3.869	19.783	4.749	R24.531	(s)	.166	28.973	1.528	.028	.245	.106	2.048	2.154	.051	4.006	24.967
2001	.495	.063	4.068	20.348	R5.051	R25.398	.001	.131	R30.157	1.265	.033	.377	.043	1.996	2.039	.056	3.770	R26.386
2002	.422	.080	4.104	19.920	R4.754	R24.674	.001	.125	R29.407	1.032	.020	.520	.019	2.023	2.042	.054	3.668	R25.739
2003	.626	.068	4.042	21.060	R5.159	R26.219	.001	.104	R31.061	1.117	.018	.686	.026	2.124	2.151	.082	4.054	R27.007
2004	.682	.170	4.365	22.082	R6.114	R28.196	.013	.117	R33.543	1.253	.033	.862	.057	2.151	2.208	.078	4.433	R29.110
2005	.762	.088	4.450	22.091	R7.157	R29.248	.011	.152	R34.710	1.273	.043	.735	.067	2.374	2.442	.068	4.561	R30.149
2006	.906	.101	R4.291	R22.085	R7.083	R29.168	R.062	R.146	R34.673	1.264	.040	R.730	.052	R2.699	R2.751	R.083	R4.868	R29.805
2007P	.909	.061	4.717	21.868	6.833	28.701	.037	.175	34.599	1.507	.036	.816	.058	2.876	2.934	.069	5.361	29.238

[1] Net imports equal imports minus exports. Minus sign indicates exports are greater than imports.

[2] Crude oil and lease condensate. Includes imports into the Strategic Petroleum Reserve, which began in 1977.

[3] Petroleum products, unfinished oils, pentanes plus, and gasoline blending components. Does not include fuel ethanol.

R=Revised. P=Preliminary. NA=Not available. (s)=Less than 0.0005 quadrillion Btu.

Notes: • Includes trade between the United States (50 States and the District of Columbia) and its territories and possessions. • See Note 3, "Electricity Imports and Exports," at end of Section 8. • Totals may not equal sum of components due to independent rounding.

Web Page: For all data beginning in 1949, see http://www.eia.doe.gov/emeu/aer/overview.html.

Sources: Tables 5.1, 5.3, 5.5, 6.1, 7.1, 7.7, 8.1, 10.3, A2, A4, A5, and A6.

Figure 1.5 Energy Consumption and Expenditures Indicators

Energy Consumption, 1949-2007

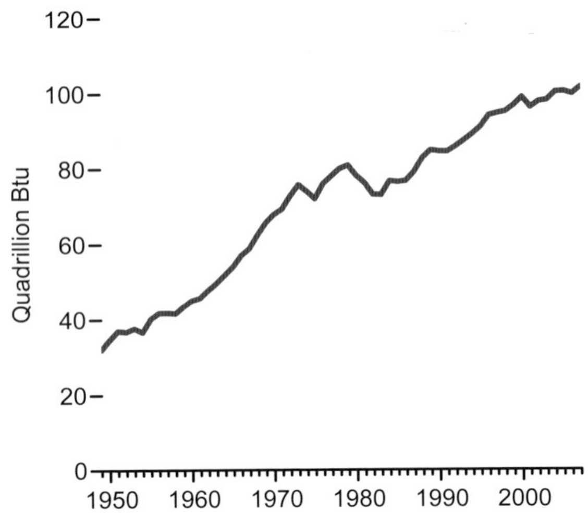

Energy Expenditures, 1970-2005

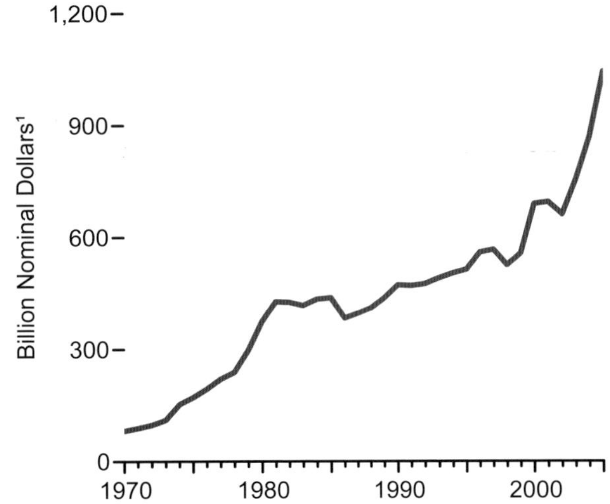

Energy Consumption per Real Dollar[2] of Gross Domestic Product, 1949-2007

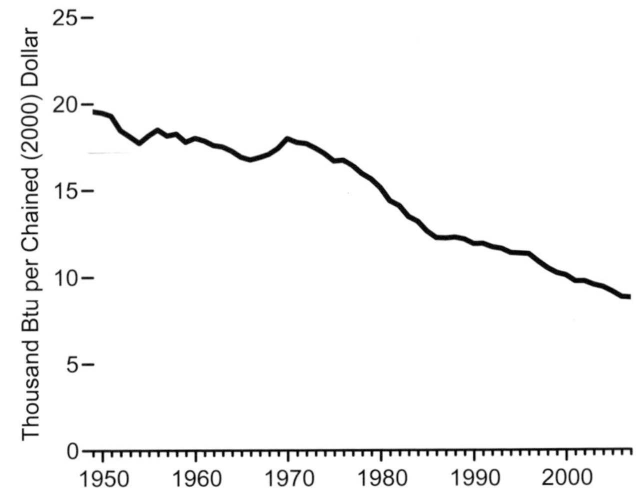

Energy Consumption per Person, 1949-2007

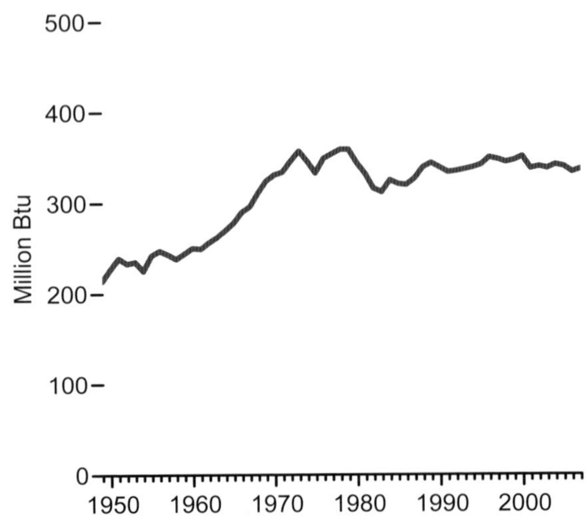

Energy Expenditures per Person, 1970-2005

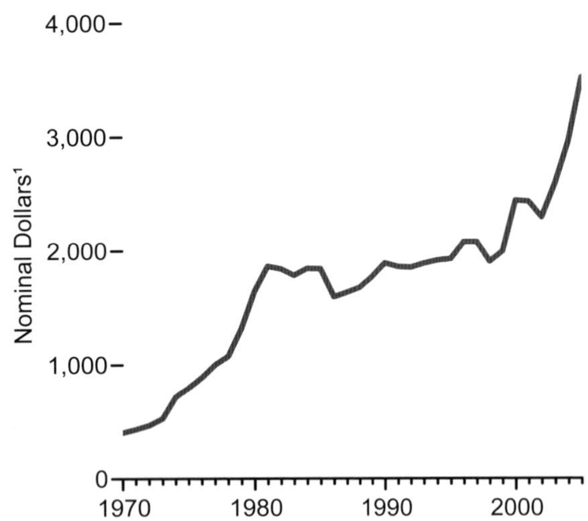

Energy Expenditures as Share of Gross Domestic Product, 1970-2005

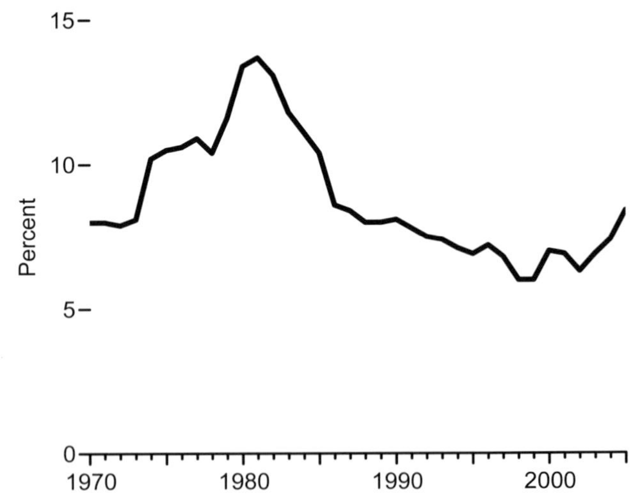

[1] See "Nominal Dollars" in Glossary

[2] In chained (2000) dollars, calculated by using gross domestic product implicit price deflators, See Appendix D1.

Source: Table 1.5.

Table 1.5 Energy Consumption, Expenditures, and Emissions Indicators, Selected Years, 1949-2007

Year	Energy Consumption (Quadrillion Btu)	Energy Consumption per Person (Million Btu)	Energy Expenditures [1] (Million Nominal Dollars [4])	Energy Expenditures [1] per Person (Nominal Dollars [4])	Gross Domestic Product (GDP) (Billion Nominal Dollars [4])	Energy Expenditures [1] as Share of GDP (Percent)	Gross Domestic Product (GDP) (Billion Chained (2000) Dollars [5])	Energy Consumption per Real Dollar of GDP (Thousand Btu per Chained (2000) Dollar [5])	Greenhouse Gas Emissions [2] per Real Dollar of GDP (Metric Tons Carbon Dioxide Equivalent per Million Chained (2000) Dollars [5])	Carbon Dioxide Emissions [3] per Real Dollar of GDP (Metric Tons Carbon Dioxide per Million Chained (2000) Dollars [5])
1949	31.98	[R]214	NA	NA	267.3	NA	1,634.6	19.57	NA	NA
1950	34.62	[R]227	NA	NA	293.8	NA	1,777.3	19.48	NA	NA
1955	40.21	[R]242	NA	NA	414.8	NA	2,212.8	18.17	NA	NA
1960	45.09	[R]250	NA	NA	526.4	NA	2,501.8	18.02	NA	NA
1965	54.02	[R]278	NA	NA	719.1	NA	3,191.1	16.93	NA	NA
1970	67.84	[R]331	82,911	[R]404	1,038.5	8.0	3,771.9	17.99	NA	NA
1971	69.29	[R]334	90,071	[R]434	1,127.1	8.0	3,898.6	17.77	NA	NA
1972	72.70	[R]346	98,108	[R]467	1,238.3	7.9	4,105.0	17.71	NA	NA
1973	75.71	[R]357	111,928	[R]528	1,382.7	8.1	4,341.5	17.44	NA	NA
1974	73.99	[R]346	153,370	[R]717	1,500.0	10.2	4,319.6	17.13	NA	NA
1975	72.00	[R]333	171,846	[R]796	1,638.3	10.5	4,311.2	16.70	NA	NA
1976	76.01	349	193,897	[R]889	1,825.3	10.6	4,540.9	16.74	NA	NA
1977	78.00	[R]354	220,461	[R]1,001	2,030.9	10.9	4,750.5	16.42	NA	NA
1978	79.99	[R]359	239,230	[R]1,075	2,294.7	10.4	5,015.0	15.95	NA	NA
1979	80.90	[R]359	297,543	[R]1,322	2,563.3	11.6	5,173.4	15.64	NA	NA
1980	78.12	[R]344	374,346	[R]1,647	2,789.5	13.4	5,161.7	15.13	1,131	917
1981	76.17	332	427,877	1,865	3,128.4	13.7	5,291.7	14.39	1,085	872
1982	73.15	316	426,437	1,841	3,255.0	13.1	5,189.3	14.10	1,053	843
1983	73.04	312	417,419	1,785	3,536.7	11.8	5,423.8	13.47	998	800
1984	76.71	325	434,982	1,845	3,933.2	11.1	5,813.6	13.20	982	788
1985	76.49	321	438,184	1,842	4,220.3	10.4	6,053.7	12.64	946	755
1986	76.76	320	383,409	1,597	4,462.8	8.6	6,263.6	12.25	913	731
1987	79.17	327	396,515	1,637	4,739.5	8.4	6,475.1	12.23	910	732
1988	82.82	339	410,426	1,679	5,103.8	8.0	6,742.7	12.28	907	735
1989	84.94	344	437,611	1,773	5,484.4	8.0	6,981.4	12.17	892	723
1990	84.65	[R]339	472,539	[R]1,893	5,803.1	8.1	7,112.5	11.90	[R]864	[R]705
1991	84.61	334	470,559	1,860	5,995.9	7.8	7,100.5	11.92	[R]859	[R]699
1992	85.96	335	475,587	1,854	6,337.7	7.5	7,336.6	11.72	[R]848	[R]690
1993	87.60	337	491,168	1,890	6,657.4	7.4	7,532.7	11.63	[R]839	[R]686
1994	89.26	339	504,204	1,916	7,072.2	7.1	7,835.5	11.39	[R]821	[R]670
1995	91.17	342	514,049	1,930	7,397.7	6.9	8,031.7	11.35	[R]806	[R]660
1996	94.17	350	559,954	2,079	7,816.9	7.2	8,328.9	11.31	[R]799	[R]659
1997	94.77	348	566,770	2,079	8,304.3	6.8	8,703.5	10.89	[R]773	[R]640
1998	95.18	345	525,737	1,906	8,747.0	6.0	9,066.9	10.50	[R]744	[R]618
1999	96.82	347	[R]556,533	1,994	9,268.4	6.0	9,470.3	10.22	[R]718	[R]599
2000	98.98	[R]351	[R]689,338	[R]2,443	9,817.0	7.0	9,817.0	10.08	[R]711	[R]596
2001	96.33	338	[R]694,054	[R]2,434	10,128.0	6.9	9,890.7	9.74	[R]694	[R]581
2002	97.86	340	[R]660,894	[R]2,296	10,469.6	6.3	10,048.8	9.74	[R]691	[R]578
2003	98.21	338	[R]754,131	[R]2,596	10,960.8	6.9	10,301.0	9.53	[R]681	[R]569
2004	100.35	342	[R]868,773	[R]2,963	[R]11,685.9	7.4	[R]10,675.8	[R]9.40	[R]668	[R]558
2005	[R]100.51	340	[R]1,042,934	[R]3,525	[R]12,433.9	[R]8.4	[R]11,003.4	[R]9.13	[R]653	[R]544
2006	[R]99.86	334	NA	NA	[R]13,194.7	NA	[R]11,319.4	[R]8.82	[R]625	[R]520
2007[P]	101.60	337	NA	NA	13,841.3	NA	11,566.8	8.78	NA	NA

[1] Expenditures include taxes where data are available.
[2] Greenhouse gas emissions from anthropogenic sources. See Table 12.1.
[3] Carbon dioxide emissions from energy consumption. See Table 12.2
[4] See "Nominal Dollars" in Glossary.
[5] See "Chained Dollars" in Glossary.

R=Revised. P=Preliminary. NA=Not available.
Web Page: For all data beginning in 1949, see http://www.eia.doe.gov/emeu/aer/overview.html.
Sources: **Energy Consumption:** Table 1.3. **Energy Expenditures:** Table 3.5. **Gross Domestic Product:** Table D1. **Population Data:** Table D1. **Greenhouse Gas Emissions:** Table 12.1. **Carbon Dioxide Emissions:** Table 12.2. **Other Columns:** Calculated by EIA.

Figure 1.6 State-Level Energy Consumption and Consumption per Person, 2005

Consumption

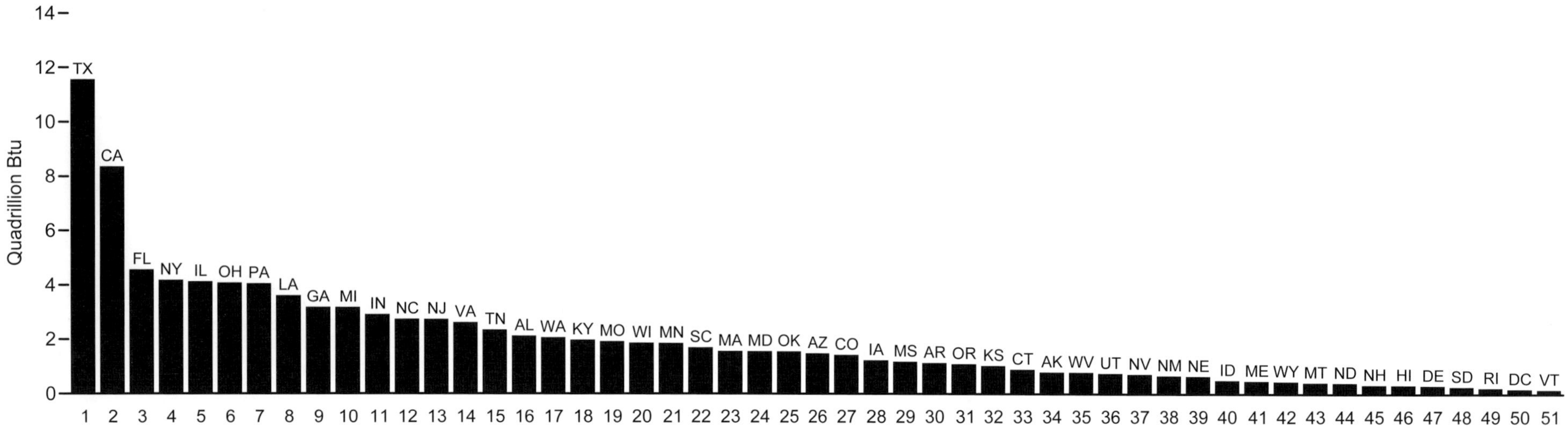

Consumption per Person

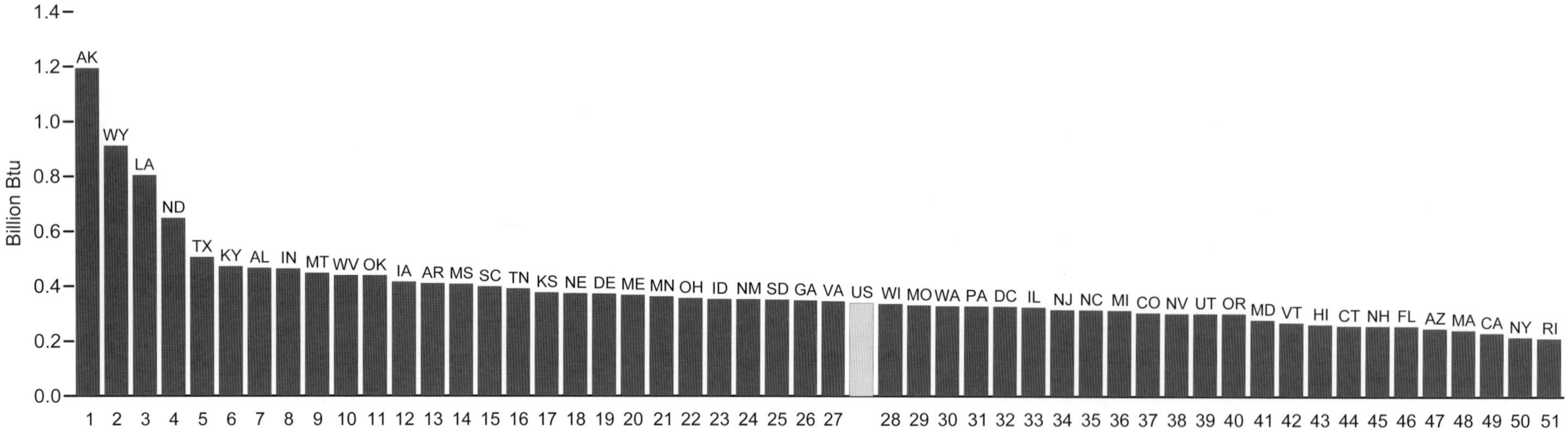

Note: Because vertical scales differ, graphs should not be compared.

Source: Table 1.6.

Table 1.6 State-Level Energy Consumption, Expenditures, and Prices, 2005

	Consumption		Consumption per Person		Expenditures [1]		Expenditures [1] per Person		Prices [1]	
Rank	Trillion Btu		Million Btu		Million Nominal Dollars [2]		Nominal Dollars [2]		Nominal Dollars [2] per Million Btu	
1	Texas	11,558.3	Alaska	1,193.9	Texas	114,410	Alaska	7,806	Hawaii	21.56
2	California	8,359.8	Wyoming	911.9	California	103,604	Wyoming	7,230	District of Columbia	20.21
3	Florida	4,563.3	Louisiana	803.7	New York	56,690	Louisiana	6,621	Connecticut	19.40
4	New York	4,179.5	North Dakota	648.1	Florida	49,756	North Dakota	5,183	Massachusetts	19.37
5	Illinois	4,121.5	Texas	506.0	Pennsylvania	42,786	Texas	5,008	New York	18.92
6	Ohio	4,081.6	Kentucky	472.3	Illinois	42,103	Montana	4,307	Rhode Island	18.79
7	Pennsylvania	4,050.2	Alabama	466.8	Ohio	41,895	Iowa	4,214	Vermont	18.78
8	Louisiana	3,613.0	Indiana	464.2	Michigan	32,404	Maine	4,157	New Hampshire	18.68
9	Georgia	3,173.0	Montana	448.2	Georgia	32,367	Kentucky	4,084	Florida	17.98
10	Michigan	3,166.5	West Virginia	439.7	New Jersey	32,213	Indiana	4,041	Nevada	17.94
11	Indiana	2,904.7	Oklahoma	438.6	Louisiana	29,766	Oklahoma	4,019	Arizona	17.78
12	North Carolina	2,732.0	Iowa	415.4	North Carolina	28,152	Alabama	3,953	California	17.37
13	New Jersey	2,728.6	Arkansas	409.5	Virginia	26,208	Hawaii	3,943	Maryland	17.11
14	Virginia	2,610.2	Mississippi	407.6	Indiana	25,287	Mississippi	3,889	New Mexico	16.68
15	Tennessee	2,338.8	South Carolina	398.2	Tennessee	21,833	West Virginia	3,848	North Carolina	16.53
16	Alabama	2,119.0	Tennessee	390.5	Massachusetts	21,704	Arkansas	3,827	Delaware	16.25
17	Washington	2,058.8	Kansas	376.4	Missouri	19,827	New Jersey	3,721	Maine	16.21
18	Kentucky	1,970.1	Nebraska	373.4	Wisconsin	19,190	South Carolina	3,717	New Jersey	16.19
19	Missouri	1,914.7	Delaware	371.9	Washington	18,550	South Dakota	3,710	Mississippi	15.81
20	Wisconsin	1,861.8	Maine	367.6	Minnesota	18,261	Delaware	3,663	Georgia	15.75
21	Minnesota	1,852.2	Minnesota	362.2	Alabama	17,946	Nebraska	3,657	Pennsylvania	15.69
22	South Carolina	1,694.3	Ohio	356.2	Maryland	17,571	Ohio	3,656	Missouri	15.34
23	Massachusetts	1,561.8	Idaho	352.9	Kentucky	17,032	Vermont	3,653	South Carolina	15.32
24	Maryland	1,555.2	New Mexico	352.3	Arizona	16,599	Tennessee	3,645	Ohio	15.27
25	Oklahoma	1,550.7	South Dakota	350.6	South Carolina	15,815	Minnesota	3,571	Virginia	15.22
26	Arizona	1,479.7	Georgia	348.4	Colorado	14,661	Connecticut	3,571	Wisconsin	15.22
27	Colorado	1,425.7	Virginia	345.4	Oklahoma	14,211	Georgia	3,554	Colorado	15.18
28	Iowa	1,227.8	Wisconsin	336.0	Iowa	12,456	Nevada	3,523	Arkansas	15.15
29	Mississippi	1,182.3	Missouri	330.8	Connecticut	12,449	New Hampshire	3,516	Oregon	15.10
30	Arkansas	1,135.1	Washington	328.3	Mississippi	11,280	District of Columbia	3,496	South Dakota	15.10
31	Oregon	1,095.7	Pennsylvania	327.5	Oregon	10,726	Kansas	3,487	Kansas	14.97
32	Kansas	1,031.8	District of Columbia	327.0	Arkansas	10,608	Virginia	3,468	Tennessee	14.88
33	Connecticut	900.2	Illinois	324.0	Kansas	9,559	Wisconsin	3,464	Montana	14.83
34	Alaska	799.2	New Jersey	315.2	Nevada	8,486	Pennsylvania	3,460	Oklahoma	14.80
35	West Virginia	793.9	North Carolina	314.8	Utah	6,978	Missouri	3,426	Texas	14.73
36	Utah	756.7	Michigan	313.3	West Virginia	6,948	Massachusetts	3,376	Illinois	14.72
37	Nevada	727.8	Colorado	305.1	New Mexico	6,462	New Mexico	3,372	Minnesota	14.58
38	New Mexico	675.0	Nevada	302.1	Nebraska	6,414	Illinois	3,310	Michigan	14.56
39	Nebraska	654.9	Utah	302.1	Maine	5,455	North Carolina	3,244	Nebraska	14.55
40	Idaho	503.2	Oregon	301.8	Alaska	5,226	Michigan	3,206	Washington	14.45
41	Maine	482.4	Maryland	279.1	Hawaii	4,997	Maryland	3,153	Iowa	14.31
42	Wyoming	461.9	Vermont	269.5	New Hampshire	4,582	Colorado	3,137	Alabama	14.28
43	Montana	419.4	Hawaii	263.0	Idaho	4,464	Idaho	3,131	Idaho	14.13
44	North Dakota	412.2	Connecticut	258.2	Montana	4,030	Washington	2,958	Alaska	13.98
45	New Hampshire	335.4	New Hampshire	257.4	Wyoming	3,662	Oregon	2,955	Kentucky	13.90
46	Hawaii	333.4	Florida	257.3	North Dakota	3,296	New York	2,943	Utah	13.69
47	Delaware	312.6	Arizona	248.6	Rhode Island	3,109	Rhode Island	2,915	West Virginia	13.50
48	South Dakota	273.5	Massachusetts	242.9	Delaware	3,079	California	2,879	Louisiana	13.23
49	Rhode Island	227.6	California	232.3	South Dakota	2,894	Florida	2,805	Wyoming	12.93
50	District of Columbia	190.4	New York	217.0	Vermont	2,264	Arizona	2,789	Indiana	12.67
51	Vermont	167.0	Rhode Island	213.3	District of Columbia	2,035	Utah	2,786	North Dakota	11.08
	United States	[3,4] **100,368.6**	**United States**	**339.2**	**United States**	[5] **1,042,934**	**United States**	**3,525**	**United States**	**15.66**

[1] Prices and expenditures include taxes where data are available.
[2] See "Nominal Dollars" in Glossary.
[3] Includes 44.2 trillion Btu of coal coke net imports, which are not allocated to the States.
[4] The U.S. consumption value in this table does not match those in Tables 1.1 and 1.3 because it: 1) does not include biofuels losses and co-products, and biodiesel; 2) includes supplemental gaseous fuels; 3) does not incorporate the latest data revisions; and 4) is the sum of State values, which use State average heat contents to convert physical units of coal and natural gas to Btu.
[5] Includes $633 million for coal coke net imports, which are not allocated to the States.

Note: Rankings based on unrounded data.
Web Page: For related information, see http://www.eia.doe.gov/emeu/states/_seds.html.
Sources: • **Consumption:** Energy Information Administration (EIA), "State Energy Data 2005: Consumption" (February 2008), Tables R1 and R2. • **Expenditures and Prices:** EIA, "State Energy Data 2005: Prices and Expenditures" (February 2008), Table R1. • "State Energy Data 2005" includes State-level data by end-use sector and type of energy. Consumption estimates are annual 1960 through 2005, and price and expenditure estimates are annual 1970 through 2005.

Figure 1.7 Heating Degree-Days by Month, 1949-2007

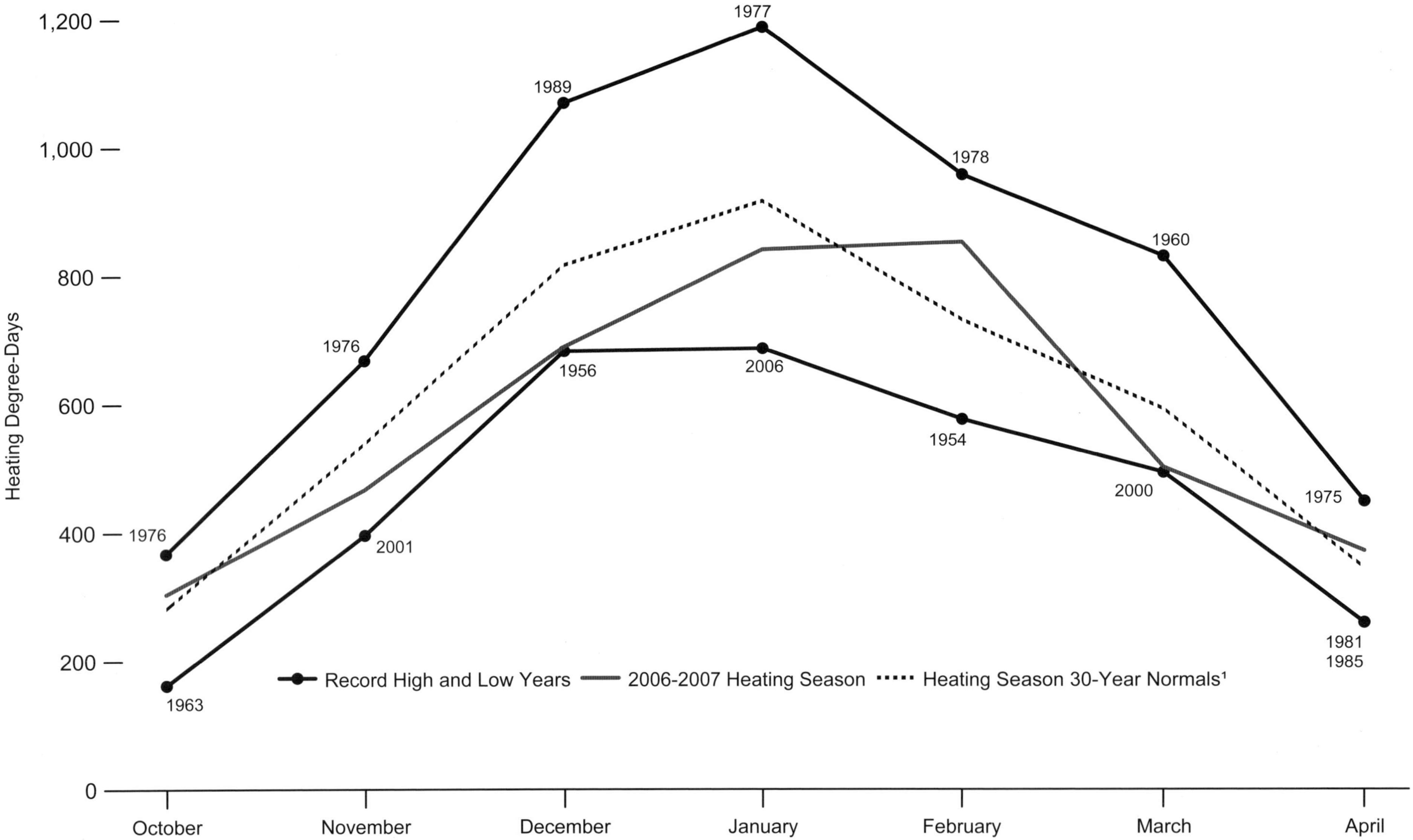

[1] Based on calculations of data from 1971 through 2000.

Source: Table 1.7.

Table 1.7 Heating Degree-Days by Month, Selected Years, 1949-2007

Year	January	February	March	April	May	June	July	August	September	October	November	December	Total
1949	858	701	611	330	128	21	7	9	94	209	503	763	4,234
1950	761	721	693	412	162	40	11	18	85	196	565	872	4,536
1955	927	759	600	272	121	48	9	6	56	237	600	886	4,521
1960	884	780	831	278	160	33	7	11	48	254	502	936	4,724
1965	907	780	738	355	114	48	11	14	78	271	494	739	4,549
1970	1,063	758	685	344	120	31	4	9	55	253	541	801	4,664
1971	976	760	681	375	194	29	10	12	47	187	553	723	4,547
1972	890	785	608	377	137	49	7	12	65	330	613	832	4,705
1973	893	772	504	356	182	22	6	9	61	212	497	799	4,313
1974	838	754	556	310	171	42	6	13	94	303	524	795	4,406
1975	821	742	686	449	117	37	5	13	100	235	462	805	4,472
1976	974	609	544	309	178	28	8	19	81	367	668	941	4,726
1977	1,188	751	529	270	119	38	6	13	59	295	493	844	4,605
1978	1,061	958	677	350	157	31	7	11	59	283	517	847	4,958
1979	1,079	950	575	364	148	37	6	15	58	271	528	750	4,781
1980	887	831	680	338	142	49	5	10	54	316	564	831	4,707
1981	984	689	620	260	165	25	6	11	76	327	504	845	4,512
1982	1,067	776	620	408	114	62	7	19	75	264	515	692	4,619
1983	874	706	588	421	189	35	6	5	53	251	509	990	4,627
1984	1,000	645	704	371	172	28	7	7	88	223	565	704	4,514
1985	1,057	807	557	260	123	47	5	17	69	243	506	951	4,642
1986	859	734	542	295	123	30	9	18	76	258	558	793	4,295
1987	920	714	573	309	107	20	8	13	61	345	491	773	4,334
1988	1,004	778	594	344	134	30	3	5	72	352	506	831	4,653
1989	789	832	603	344	163	32	5	14	73	259	542	1,070	4,726
1990	728	655	535	321	184	29	6	10	56	246	457	789	4,016
1991	921	639	564	287	98	30	6	7	69	242	586	751	4,200
1992	852	644	603	345	152	46	14	24	74	301	564	822	4,441
1993	860	827	664	368	128	38	11	9	89	302	580	824	4,700
1994	1,031	813	594	293	174	21	6	16	65	268	479	723	4,483
1995	847	750	556	375	174	31	4	7	77	233	605	872	4,531
1996	945	748	713	360	165	27	8	9	72	276	630	760	4,713
1997	932	672	552	406	198	31	7	16	63	273	592	800	4,542
1998	765	623	596	331	109	41	4	5	33	245	482	717	3,951
1999	861	647	645	319	139	31	5	12	62	275	413	760	4,169
2000	886	643	494	341	115	29	12	12	69	244	610	1,005	4,460
2001	935	725	669	302	115	29	8	6	69	260	396	689	4,203
2002	776	669	622	281	184	23	3	8	37	298	560	812	4,273
2003	944	801	572	344	165	41	4	5	62	260	477	784	4,459
2004	968	766	495	303	107	37	7	20	47	251	487	802	4,290
2005	859	676	648	305	186	25	3	6	39	236	466	866	4,315
2006	687	731	600	264	137	23	2	9	82	304	467	690	3,996
2007[P]	841	853	502	372	111	24	5	7	44	175	521	800	4,255
Normals[1]	917	732	593	345	159	39	9	15	77	282	539	817	4,524

[1] Based on calculations of data from 1971 through 2000.

P=Preliminary.

Notes: • This table excludes Alaska and Hawaii. • Degree-days are relative measurements of outdoor air temperature. Heating degree-days are deviations below the mean daily temperature of 65° F. For example, a weather station recording a mean daily temperature of 40° F would report 25 heating degree-days. • Temperature information recorded by weather stations is used to calculate State-wide degree-day averages based on resident State population. Beginning in July 2001, data are weighted by the 2000 population. The population-weighted State figures are aggregated into Census divisions and the national average.

Web Pages: • For all data beginning in 1949, see http://www.eia.doe.gov/emeu/aer/overview.html. • For current data, see http://www.eia.doe.gov/emeu/mer/overview.html.

Source: U.S. Department of Commerce, National Oceanic and Atmospheric Administration (NOAA), National Climatic Data Center, Asheville, North Carolina, Historical Climatology Series 5-1.

Figure 1.8 Cooling Degree-Days by Month, 1949-2007

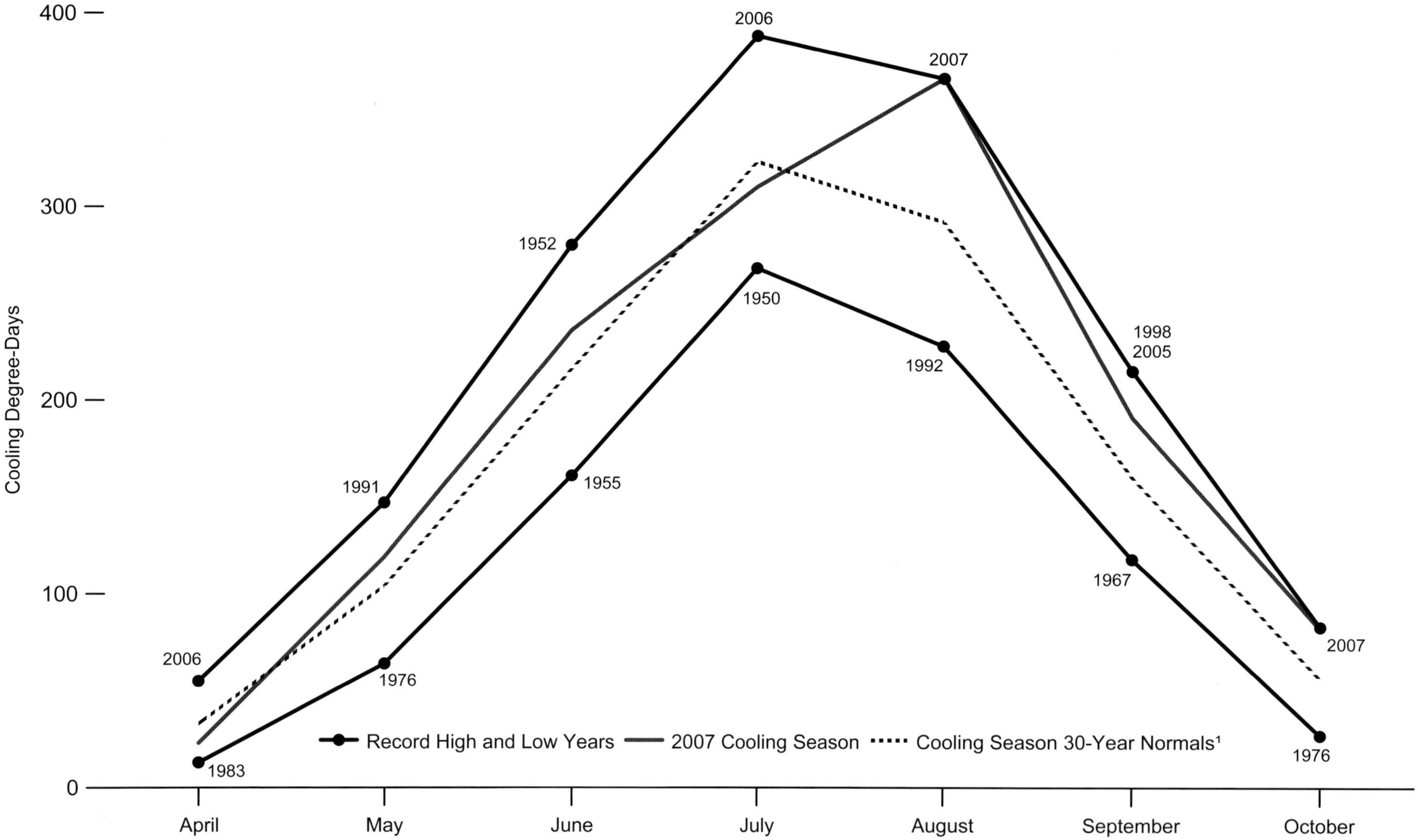

[1] Based on calculations of data from 1971 through 2000.

Source: Table 1.8.

Table 1.8 Cooling Degree-Days by Month, Selected Years, 1949-2007

Year	January	February	March	April	May	June	July	August	September	October	November	December	Total
1949	16	14	14	27	110	253	367	294	131	70	12	10	1,318
1950	27	12	13	21	105	201	268	244	128	78	9	4	1,110
1955	6	7	20	45	121	161	381	355	182	50	10	6	1,344
1960	7	4	6	37	76	215	301	302	181	59	15	3	1,206
1965	9	7	10	42	125	179	280	273	155	48	19	6	1,153
1970	3	4	10	36	104	201	323	313	185	48	6	9	1,242
1971	8	7	10	22	68	244	288	269	182	77	12	17	1,204
1972	15	6	22	36	88	174	299	276	169	44	9	8	1,146
1973	7	3	24	18	75	236	318	303	166	66	21	4	1,241
1974	21	6	28	29	101	173	317	267	120	40	10	5	1,117
1975	14	11	14	24	117	203	301	296	120	55	12	5	1,172
1976	5	11	23	27	64	208	282	243	127	27	8	4	1,029
1977	2	5	21	35	121	212	351	293	180	44	15	6	1,285
1978	3	1	10	31	93	218	310	300	180	52	19	9	1,226
1979	4	4	13	32	82	187	295	266	160	53	11	6	1,113
1980	9	4	13	23	95	199	374	347	192	42	10	5	1,313
1981	3	6	10	52	75	257	333	275	138	43	12	5	1,209
1982	6	10	21	26	115	165	318	262	140	47	15	11	1,136
1983	6	5	9	13	72	193	353	362	172	58	12	5	1,260
1984	5	6	14	24	92	233	291	312	143	70	9	15	1,214
1985	3	5	22	39	108	193	313	269	145	68	25	4	1,194
1986	8	10	17	33	106	231	340	259	161	52	23	9	1,249
1987	5	7	13	23	127	244	334	298	156	40	14	8	1,269
1988	5	5	13	28	89	218	359	348	149	45	18	6	1,283
1989	15	7	19	36	88	208	312	266	138	49	16	2	1,156
1990	15	14	21	29	86	234	316	291	172	57	16	9	1,260
1991	10	9	19	42	147	235	336	305	149	62	8	9	1,331
1992	6	10	15	29	77	170	286	228	150	49	13	7	1,040
1993	13	5	11	19	91	207	347	317	146	47	11	4	1,218
1994	7	9	18	37	76	262	328	263	141	50	20	9	1,220
1995	7	7	18	29	91	202	348	363	150	61	12	5	1,293
1996	7	6	8	26	116	226	299	287	139	45	14	7	1,180
1997	8	11	31	19	81	189	315	268	171	48	10	5	1,156
1998	12	7	10	23	135	228	350	337	215	62	20	11	1,410
1999	12	11	12	40	94	219	374	305	152	55	17	6	1,297
2000	10	10	25	28	131	221	284	302	156	50	8	4	1,229
2001	3	12	11	37	114	220	302	333	138	46	18	11	1,245
2002	8	6	17	53	92	243	370	332	202	57	11	5	1,396
2003	5	7	24	30	110	187	336	345	156	65	21	4	1,290
2004	6	6	28	29	138	208	299	252	177	67	17	5	1,232
2005	10	7	12	24	82	250	367	351	215	55	20	4	1,397
2006	13	5	18	53	109	236	388	337	138	46	14	11	1,368
2007P	10	5	29	23	119	236	310	366	191	82	16	12	1,399
Normals[1]	8	8	18	33	104	216	323	292	160	56	16	8	1,242

[1] Based on calculations of data from 1971 through 2000.
P=Preliminary.
Notes: • This table excludes Alaska and Hawaii. • Degree-days are relative measurements of outdoor air temperature. Cooling degree-days are deviations above the mean daily temperature of 65° F. For example, a weather station recording a mean daily temperature of 78° F would report 13 cooling degree-days. • Temperature information recorded by weather stations is used to calculate State-wide degree-day averages based on resident State population. Beginning in 2002, data are weighted by the 2000 population. The population-weighted State figures are aggregated into Census divisions and the national average.
Web Pages: • For all data beginning in 1949, see http://www.eia.doe.gov/emeu/aer/overview.html. • For current data, see http://www.eia.doe.gov/emeu/mer/overview.html.
Source: U.S. Department of Commerce, National Oceanic and Atmospheric Administration (NOAA), National Climatic Data Center, Asheville, North Carolina, Historical Climatology Series 5-2.

Figure 1.9 Heating Degree-Days by Census Division

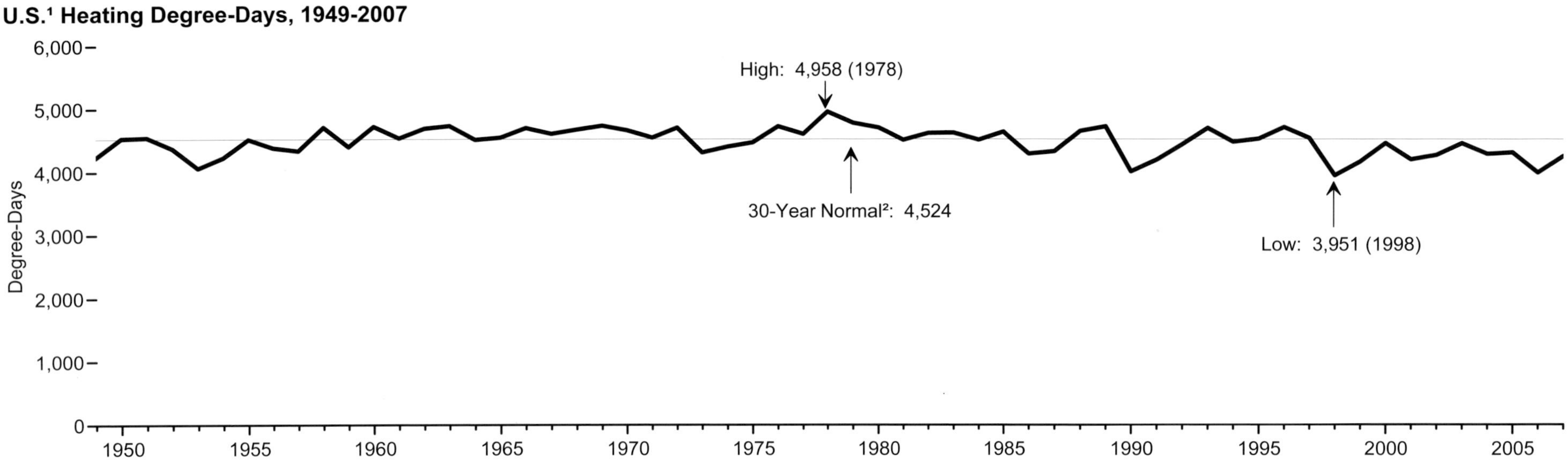

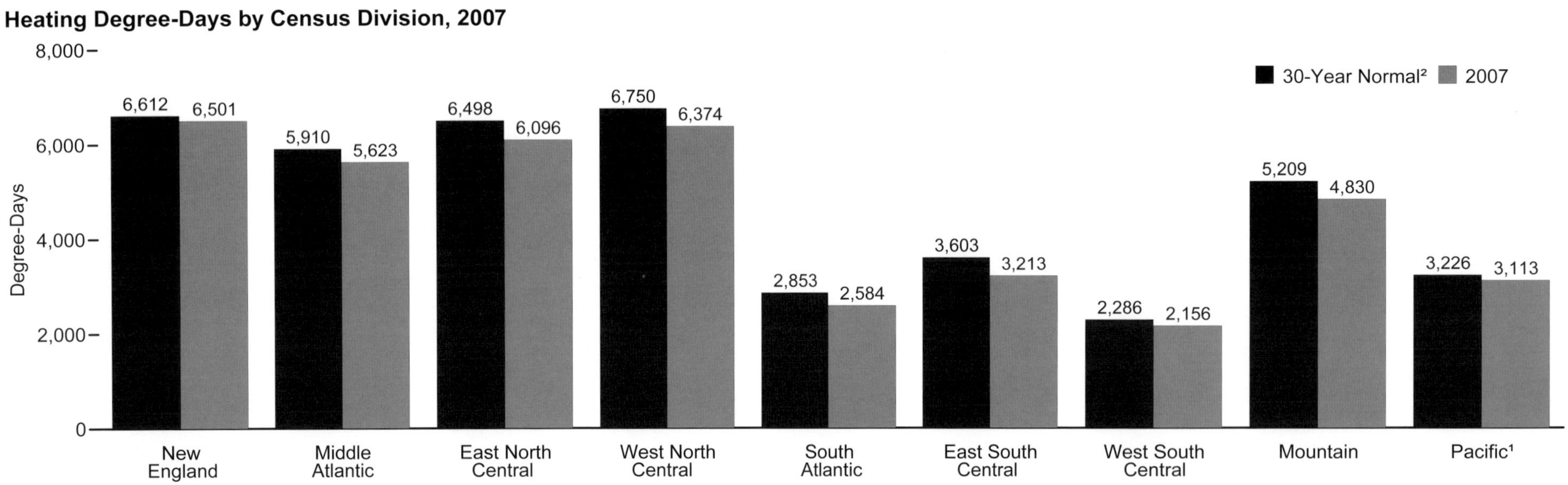

[1] Excludes Alaska and Hawaii.
[2] Normals are based on calculations of data from 1971 through 2000.

Note: See Appendix C for Census Divisions.
Source: Table 1.9.

Table 1.9 Heating Degree-Days by Census Division, Selected Years, 1949-2007

Year	New England	Middle Atlantic	East North Central	West North Central	South Atlantic	East South Central	West South Central	Mountain	Pacific [1]	United States [1]
1949	5,829	5,091	5,801	6,479	2,367	2,942	2,133	5,483	3,729	4,234
1950	6,470	5,765	6,619	7,136	2,713	3,315	1,974	4,930	3,355	4,536
1955	6,577	5,708	6,101	6,630	2,786	3,314	2,083	5,517	3,723	4,521
1960	6,561	5,901	6,544	6,884	3,147	3,958	2,551	5,328	3,309	4,724
1965	6,825	5,933	6,284	6,646	2,830	3,374	2,078	5,318	3,378	4,549
1970	6,839	5,943	6,455	6,835	2,997	3,685	2,396	5,436	3,257	4,664
1971	6,695	5,761	6,236	6,594	2,763	3,395	1,985	5,585	3,698	4,547
1972	7,001	6,064	6,772	7,094	2,759	3,438	2,259	5,352	3,376	4,705
1973	6,120	5,327	5,780	6,226	2,718	3,309	2,256	5,562	3,383	4,313
1974	6,621	5,670	6,259	6,478	2,551	3,171	2,080	5,281	3,294	4,406
1975	6,362	5,477	6,169	6,678	2,640	3,336	2,187	5,693	3,623	4,472
1976	6,839	6,097	6,768	6,670	3,040	3,881	2,446	5,303	3,115	4,726
1977	6,579	5,889	6,538	6,506	3,047	3,812	2,330	5,060	3,135	4,605
1978	7,061	6,330	7,095	7,324	3,187	4,062	2,764	5,370	3,168	4,958
1979	6,348	5,851	6,921	7,369	2,977	3,900	2,694	5,564	3,202	4,781
1980	6,900	6,143	6,792	6,652	3,099	3,855	2,378	5,052	2,986	4,707
1981	6,612	5,989	6,446	6,115	3,177	3,757	2,162	4,671	2,841	4,512
1982	6,697	5,866	6,542	7,000	2,721	3,357	2,227	5,544	3,449	4,619
1983	6,305	5,733	6,423	6,901	3,057	3,892	2,672	5,359	3,073	4,627
1984	6,442	5,777	6,418	6,582	2,791	3,451	2,194	5,592	3,149	4,514
1985	6,571	5,660	6,546	7,119	2,736	3,602	2,466	5,676	3,441	4,642
1986	6,517	5,665	6,150	6,231	2,686	3,294	2,058	4,870	2,807	4,295
1987	6,546	5,699	5,810	5,712	2,937	3,466	2,292	5,153	3,013	4,334
1988	6,715	6,088	6,590	6,634	3,122	3,800	2,346	5,148	2,975	4,653
1989	6,887	6,134	6,834	6,996	2,944	3,713	2,439	5,173	3,061	4,726
1990	5,848	4,998	5,681	6,011	2,230	2,929	1,944	5,146	3,148	4,016
1991	5,960	5,177	5,906	6,319	2,503	3,211	2,178	5,259	3,109	4,200
1992	6,844	5,964	6,297	6,262	2,852	3,498	2,145	5,054	2,763	4,441
1993	6,728	5,948	6,646	7,168	2,981	3,768	2,489	5,514	3,052	4,700
1994	6,672	5,934	6,378	6,509	2,724	3,394	2,108	5,002	3,155	4,483
1995	6,559	5,831	6,664	6,804	2,967	3,626	2,145	4,953	2,784	4,531
1996	6,679	5,986	6,947	7,345	3,106	3,782	2,285	5,011	2,860	4,713
1997	6,661	5,809	6,617	6,761	2,845	3,664	2,418	5,188	2,754	4,542
1998	5,680	4,812	5,278	5,774	2,429	3,025	2,021	5,059	3,255	3,951
1999	5,952	5,351	5,946	5,921	2,652	3,142	1,835	4,768	3,158	4,169
2000	6,489	5,774	6,284	6,456	2,959	3,548	2,194	4,881	3,012	4,460
2001	6,055	5,323	5,824	6,184	2,641	3,312	2,187	4,895	3,136	4,203
2002	6,099	5,372	6,122	6,465	2,671	3,420	2,307	5,018	3,132	4,273
2003	6,851	6,090	6,528	6,539	2,891	3,503	2,230	4,605	2,918	4,459
2004	6,612	5,749	6,199	6,290	2,748	3,289	2,088	4,844	2,925	4,290
2005	6,551	5,804	6,241	6,202	2,844	3,402	2,051	4,759	2,959	4,315
2006	5,809	5,050	5,712	5,799	2,535	3,239	1,863	4,778	3,116	3,996
2007P	6,501	5,623	6,096	6,374	2,584	3,213	2,156	4,830	3,113	4,255
Normals[2]	6,612	5,910	6,498	6,750	2,853	3,603	2,286	5,209	3,226	4,524

[1] Excludes Alaska and Hawaii.
[2] Based on calculations of data from 1971 through 2000.
P=Preliminary.
Notes: • Degree-days are relative measurements of outdoor air temperature. Heating degree-days are deviations below the mean daily temperature of 65° F. For example, a weather station recording a mean daily temperature of 40° F would report 25 heating degree-days. • Temperature information recorded by weather stations is used to calculate State-wide degree-day averages based on resident State population. Beginning in July 2001, data are weighted by the 2000 population. The population-weighted State figures are aggregated into Census divisions and the national average. • See Appendix C for Census divisions.

Web Pages: • For all data beginning in 1949, see http://www.eia.doe.gov/emeu/aer/overview.html. • For current data, see http://www.eia.doe.gov/emeu/mer/overview.html.

Source: U.S. Department of Commerce, National Oceanic and Atmospheric Administration (NOAA), National Climatic Data Center, Asheville, North Carolina, Historical Climatology Series 5-1.

Figure 1.10 Cooling Degree-Days by Census Division

U.S.[1] Cooling Degree-Days, 1949-2007

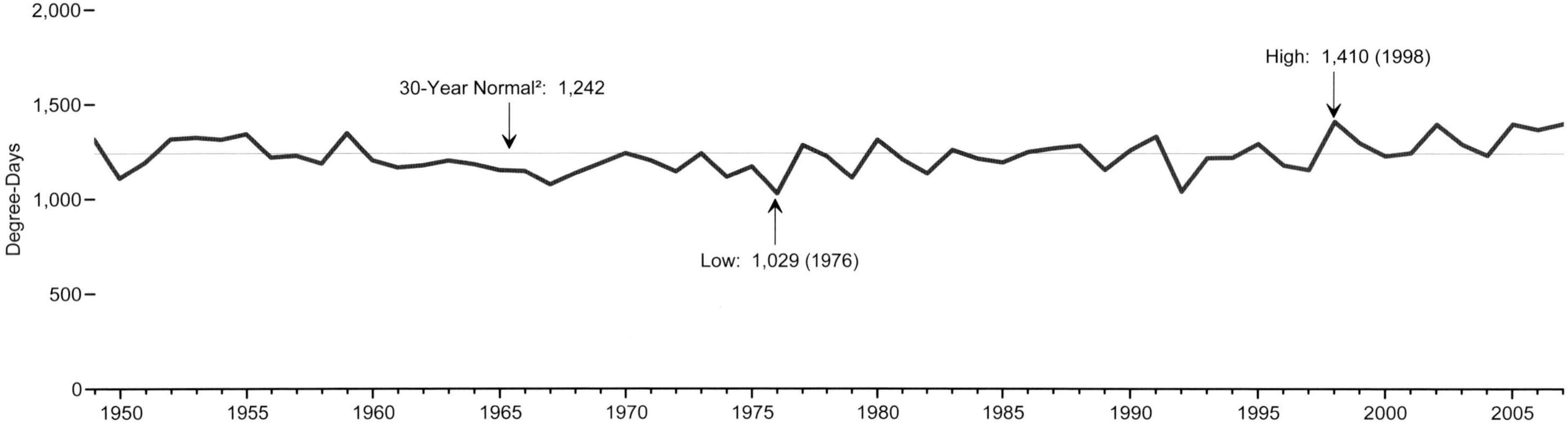

Cooling Degree-Days by Census Division, 2007

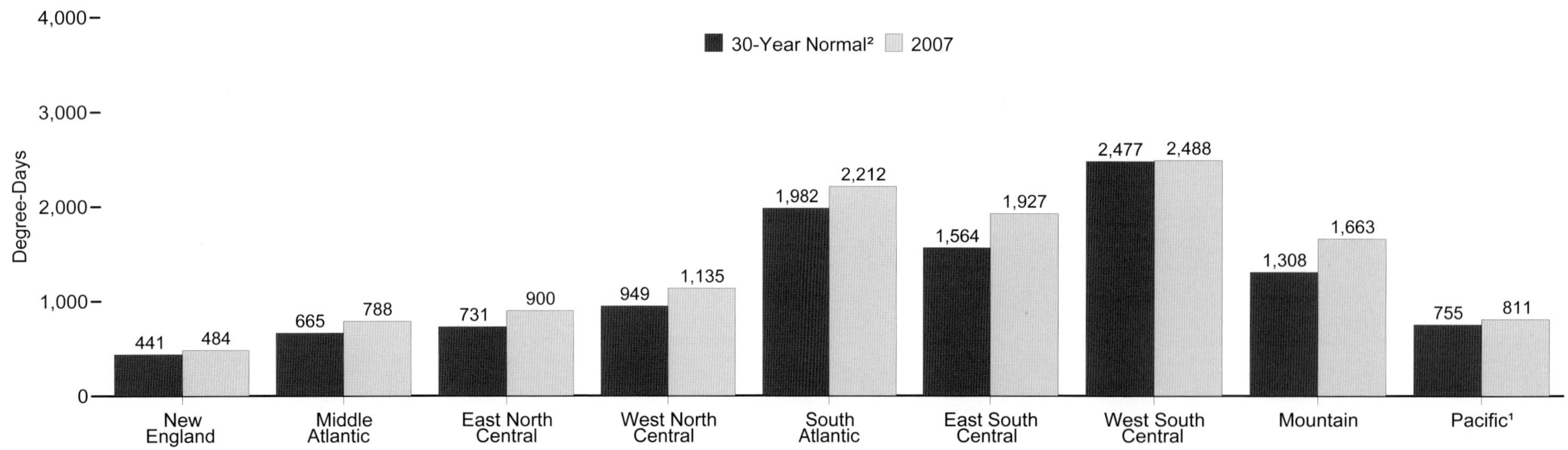

[1] Excludes Alaska and Hawaii.
[2] Normals are based on calculations of data from 1971 through 2000.

Note: See Appendix C for Census Divisions.
Source: Table 1.10.

Table 1.10 Cooling Degree-Days by Census Division, Selected Years, 1949-2007

Year	New England	Middle Atlantic	East North Central	West North Central	South Atlantic	East South Central	West South Central	Mountain	Pacific [1]	United States [1]
1949	654	901	949	1,038	2,128	1,776	2,510	1,198	593	1,318
1950	353	542	602	729	1,919	1,568	2,473	1,120	597	1,110
1955	602	934	1,043	1,238	2,045	1,791	2,643	1,124	560	1,344
1960	368	640	722	961	1,926	1,613	2,492	1,308	770	1,206
1965	352	638	688	914	1,931	1,634	2,579	961	542	1,153
1970	479	779	827	1,066	2,007	1,662	2,375	1,163	689	1,242
1971	465	730	783	960	1,932	1,577	2,448	1,074	685	1,204
1972	364	614	643	908	1,843	1,525	2,513	1,141	698	1,146
1973	551	830	864	1,009	2,000	1,665	2,359	1,123	624	1,241
1974	393	614	626	878	1,842	1,382	2,342	1,188	690	1,117
1975	467	708	788	1,003	2,011	1,520	2,261	1,031	547	1,172
1976	402	597	619	939	1,675	1,232	2,035	1,058	620	1,029
1977	407	689	823	1,122	2,020	1,808	2,720	1,256	715	1,285
1978	378	615	741	1,027	1,972	1,685	2,638	1,174	738	1,226
1979	434	588	618	871	1,833	1,412	2,242	1,164	770	1,113
1980	487	793	816	1,217	2,075	1,834	2,734	1,202	658	1,313
1981	436	657	658	924	1,889	1,576	2,498	1,331	876	1,209
1982	321	541	643	859	1,958	1,537	2,502	1,121	619	1,136
1983	538	799	934	1,178	1,925	1,579	2,288	1,174	776	1,260
1984	468	649	724	955	1,865	1,508	2,469	1,190	956	1,214
1985	372	627	643	830	2,004	1,596	2,599	1,210	737	1,194
1986	301	626	738	1,021	2,149	1,792	2,618	1,188	664	1,249
1987	406	729	918	1,115	2,067	1,718	2,368	1,196	706	1,269
1988	545	782	975	1,230	1,923	1,582	2,422	1,320	729	1,283
1989	426	658	652	864	1,977	1,417	2,295	1,330	685	1,156
1990	477	656	647	983	2,143	1,622	2,579	1,294	827	1,260
1991	511	854	959	1,125	2,197	1,758	2,499	1,182	672	1,331
1992	276	460	449	637	1,777	1,293	2,201	1,206	905	1,040
1993	486	764	735	817	2,092	1,622	2,369	1,113	708	1,218
1994	548	722	664	887	2,005	1,448	2,422	1,436	801	1,220
1995	507	803	921	985	2,081	1,671	2,448	1,234	754	1,293
1996	400	623	629	821	1,867	1,474	2,515	1,381	856	1,180
1997	395	586	574	873	1,886	1,393	2,361	1,335	921	1,156
1998	505	788	889	1,138	2,277	1,928	3,026	1,271	732	1,410
1999	631	882	855	970	2,024	1,733	2,645	1,242	635	1,297
2000	317	542	658	1,023	1,929	1,736	2,787	1,488	756	1,229
2001	519	722	744	1,028	1,891	1,535	2,565	1,498	794	1,245
2002	570	863	933	1,087	2,209	1,808	2,545	1,543	739	1,396
2003	522	685	645	946	2,007	1,494	2,522	1,639	941	1,290
2004	402	670	604	752	2,037	1,549	2,485	1,376	823	1,232
2005	642	990	960	1,094	2,081	1,696	2,636	1,457	728	1,397
2006	528	778	752	1,079	2,037	1,670	2,776	1,586	916	1,368
2007P	484	788	900	1,135	2,212	1,927	2,488	1,663	811	1,399
Normals[2]	441	665	731	949	1,982	1,564	2,477	1,308	755	1,242

[1] Excludes Alaska and Hawaii.
[2] Based on calculations of data from 1971 through 2000.
P=Preliminary.
Notes: • Degree-days are relative measurements of outdoor air temperature. Cooling degree-days are deviations above the mean daily temperature of 65° F. For example, a weather station recording a mean daily temperature of 78° F would report 13 cooling degree-days. • Temperature information recorded by weather stations is used to calculate State-wide degree-day averages based on resident State population. Beginning in 2002, data are weighted by the 2000 population. The population-weighted State figures are aggregated into Census divisions and the national average. • See Appendix C for Census divisions.
Web Pages: • For all data beginning in 1949, see http://www.eia.doe.gov/emeu/aer/overview.html. • For current data, see http://www.eia.doe.gov/emeu/mer/overview.html.
Source: U.S. Department of Commerce, National Oceanic and Atmospheric Administration (NOAA), National Climatic Data Center, Asheville, North Carolina, Historical Climatology Series 5-2.

Figure 1.11 U.S. Government Energy Consumption by Agency

Total and U.S. Department of Defense, Fiscal Years 1975-2007

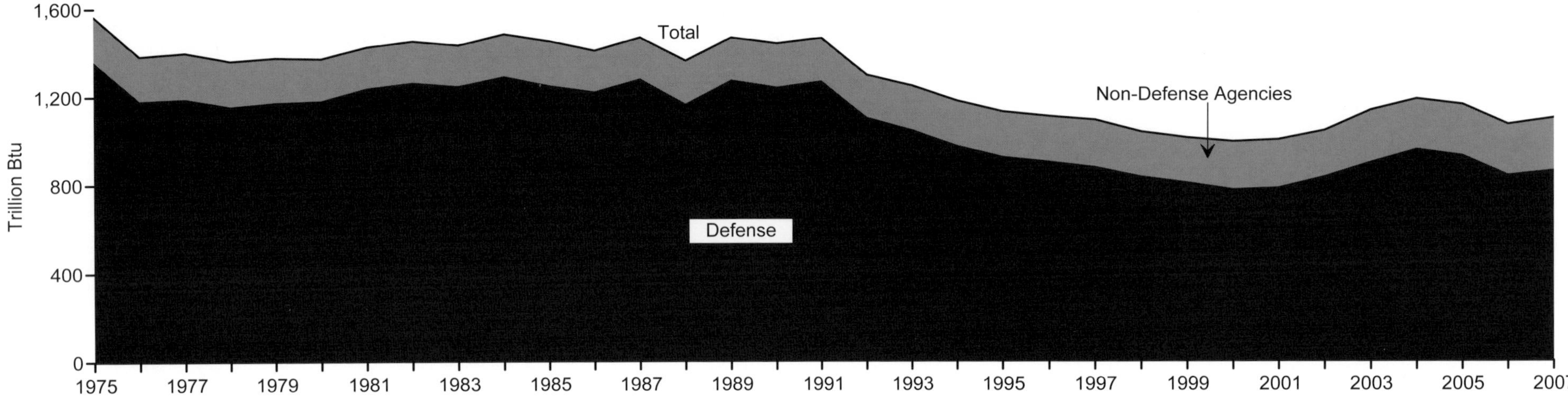

Non-Defense Agencies, Fiscal Year 2007

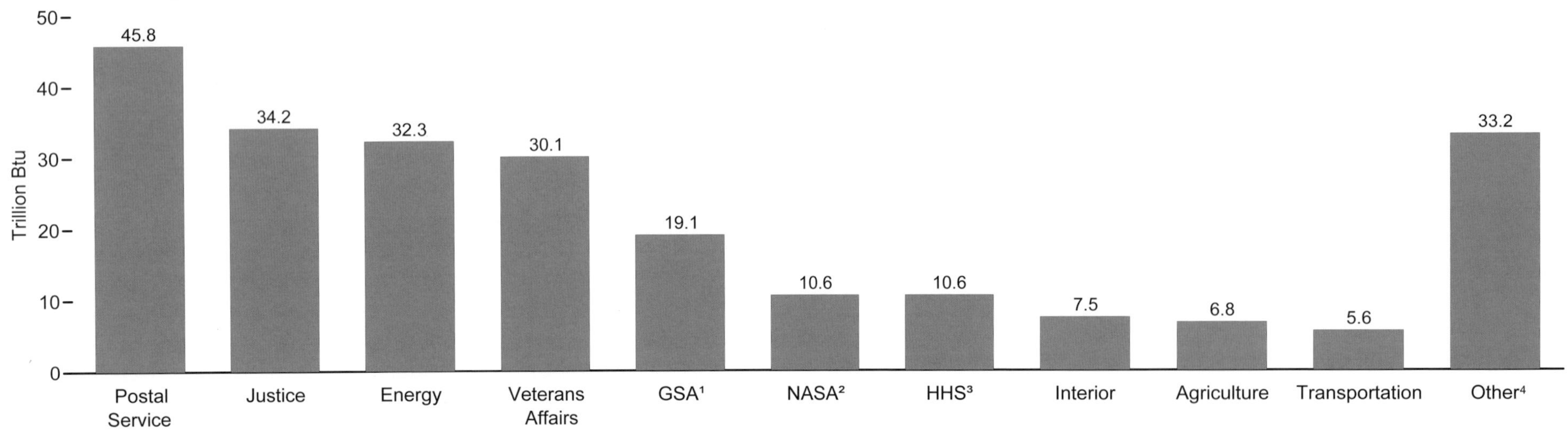

[1] General Services Administration.
[2] National Aeronautics and Space Administration.
[3] Health and Human Services.
[4] See Table 1.11 for list of agencies.

Notes: • The U.S. Government's fiscal year was October 1 through September 30, except in 1975 and 1976 when it was July 1 through June 30. • Because vertical scales differ, graphs should not be compared.

Source: Table 1.11.

Table 1.11 U.S. Government Energy Consumption by Agency, Fiscal Years 1975-2007

(Trillion Btu)

Year	Agriculture	Defense	Energy	GSA [1]	HHS [2]	Interior	Justice	NASA [3]	Postal Service	Transportation	Veterans Affairs	Other [4]	Total
1975	9.5	1,360.2	50.4	22.3	6.5	9.4	5.9	13.4	30.5	19.3	27.1	10.5	1,565.0
1976	9.3	1,183.3	50.3	20.6	6.7	9.4	5.7	12.4	30.0	19.5	25.0	11.2	1,383.4
1977	8.9	1,192.3	51.6	20.4	6.9	9.5	5.9	12.0	32.7	20.4	25.9	11.9	1,398.5
1978	9.1	1,157.8	50.1	20.4	6.5	9.2	5.9	11.2	30.9	20.6	26.8	12.4	1,360.9
1979	9.2	1,175.8	49.6	19.6	6.4	10.4	6.4	11.1	29.3	19.6	25.7	12.3	1,375.4
1980	8.6	1,183.1	47.4	18.1	6.0	8.5	5.7	10.4	27.2	19.2	24.8	12.3	1,371.2
1981	7.9	1,239.5	47.3	18.0	6.7	7.6	5.4	10.0	27.9	18.8	24.0	11.1	1,424.2
1982	7.6	1,264.5	49.0	18.1	6.4	7.4	5.8	10.1	27.5	19.1	24.2	11.6	1,451.4
1983	7.4	1,248.3	49.5	16.1	6.2	7.7	5.5	10.3	26.5	19.4	24.1	10.8	1,431.8
1984	7.9	1,292.1	51.6	16.2	6.4	8.4	6.4	10.6	27.7	19.8	24.6	10.7	1,482.5
1985	8.4	1,250.6	52.2	20.7	6.0	7.8	8.2	10.9	27.8	19.6	25.1	13.1	1,450.3
1986	6.8	1,222.8	46.9	14.0	6.2	6.9	8.6	11.2	28.0	19.4	25.0	10.8	1,406.7
1987	7.3	1,280.5	48.5	13.1	6.6	6.6	8.1	11.3	28.5	19.0	24.9	11.9	1,466.3
1988	7.8	1,165.8	49.9	12.4	6.4	7.0	9.4	11.3	29.6	18.7	26.3	15.8	1,360.3
1989	8.7	1,274.4	44.2	12.7	6.7	7.1	7.7	12.4	30.3	18.5	26.2	15.6	1,464.7
1990	9.6	1,241.7	43.5	R17.5	7.1	7.4	7.0	12.4	30.6	19.0	24.9	17.5	R1,438.0
1991	9.6	1,269.3	42.1	14.0	6.2	7.1	8.0	12.5	30.8	19.0	25.1	18.1	1,461.7
1992	9.1	1,104.0	44.3	13.8	6.8	7.0	7.5	12.6	31.7	17.0	25.3	15.7	1,294.8
1993	9.3	1,048.8	43.4	14.1	7.2	7.5	9.1	12.4	33.7	19.4	25.7	16.2	1,246.8
1994	9.4	977.0	42.1	14.0	7.5	7.9	10.3	12.6	35.0	19.8	25.6	17.1	1,178.2
1995	9.0	926.0	47.3	13.7	6.1	6.4	10.2	12.4	36.2	18.7	25.4	17.9	1,129.3
1996	9.1	904.5	44.6	14.5	6.6	4.3	12.1	11.5	36.4	19.6	26.8	18.5	1,108.5
1997	7.4	880.0	43.1	14.4	7.9	6.6	12.0	12.0	40.8	19.1	27.3	21.6	1,092.0
1998	7.9	837.1	31.5	14.1	7.4	6.4	15.8	11.7	39.5	18.5	27.6	20.3	1,037.9
1999	7.8	810.7	27.0	14.4	7.1	7.5	15.4	11.4	39.8	22.6	27.5	20.6	1,011.6
2000	7.4	779.1	30.5	17.6	8.0	7.8	19.7	11.1	43.3	21.2	27.0	21.0	993.8
2001	7.4	787.2	31.1	18.4	8.5	9.5	19.7	10.9	43.4	17.8	27.7	21.4	1,003.0
2002	7.2	837.5	30.7	17.5	8.0	8.2	17.7	10.7	41.6	18.3	27.7	19.8	1,044.8
2003	R7.7	R902.3	R31.6	R19.6	10.1	8.2	R22.7	R10.8	R50.9	5.6	30.5	R36.2	R1,136.3
2004	7.0	960.7	31.4	18.3	8.8	8.7	17.5	9.9	R50.5	5.2	29.9	39.2	R1,187.0
2005	7.5	933.2	29.6	18.4	9.6	8.6	18.8	10.3	R53.5	5.0	30.0	R37.2	R1,161.6
2006	6.8	843.7	R32.9	R18.2	9.3	8.1	23.5	R10.2	R51.8	4.6	29.3	R33.2	R1,071.5
2007P	6.8	864.6	32.3	19.1	10.6	7.5	34.2	10.6	45.8	5.6	30.1	33.2	1,100.4

[1] General Services Administration.

[2] Health and Human Services.

[3] National Aeronautics and Space Administration.

[4] Includes National Archives and Records Administration, U.S. Department of Commerce, Panama Canal Commission, Tennessee Valley Authority, U.S. Department of Labor, National Science Foundation, Federal Trade Commission, Federal Communications Commission, Environmental Protection Agency, U.S. Department of Homeland Security, U.S. Department of Housing and Urban Development, Railroad Retirement Board, Commodity Futures Trading Commission, Equal Employment Opportunity Commission, Nuclear Regulatory Commission, U.S. Department of State, U.S. Department of the Treasury, Small Business Administration, Office of Personnel Management, Federal Emergency Management Agency, Central Intelligence Agency, Consumer Product Safety Commission, Social Security Administration, and U.S. Information Agency (International Broadcasting Bureau).

R = Revised. P = Preliminary.

Notes: • For 1975 and 1976, the U.S. Government's fiscal year was July 1 through June 30. Beginning in 1977, the U.S. Government's fiscal year is October 1 through September 30 (for example, fiscal year 2007 is October 2006 through September 2007). • Data in this table are developed using the following conversion factors (which in most cases are different from those in Tables A1-A6)—coal: 24.580 million Btu/short ton; natural gas: 1,031 Btu/cubic foot; aviation gasoline: 5.250 million Btu/barrel; fuel oil: 5.8254 million Btu/barrel; jet fuel: 5.460 million Btu/barrel; liquefied petroleum gases: 4.011 million Btu/barrel; motor gasoline: 5.250 million Btu/barrel; electricity: 3,412 Btu/kilowatthour; and purchased steam: 1,000 Btu/pound. • Data include energy consumed at foreign installations and in foreign operations, including aviation and ocean bunkering, primarily by the U.S. Department of Defense. U.S. Government energy use for electricity generation and uranium enrichment is excluded. • Totals may not equal sum of components due to independent rounding.

Web Page: For related information, see http://www1.eere.energy.gov/femp/about/annual_report.html.

Source: U.S. Department of Energy, Office of Energy Efficiency and Renewable Energy, Federal Energy Management Program.

Figure 1.12 U.S. Government Energy Consumption by Source, Fiscal Years 1975-2007

Total U.S. Govenrment Energy Consumption

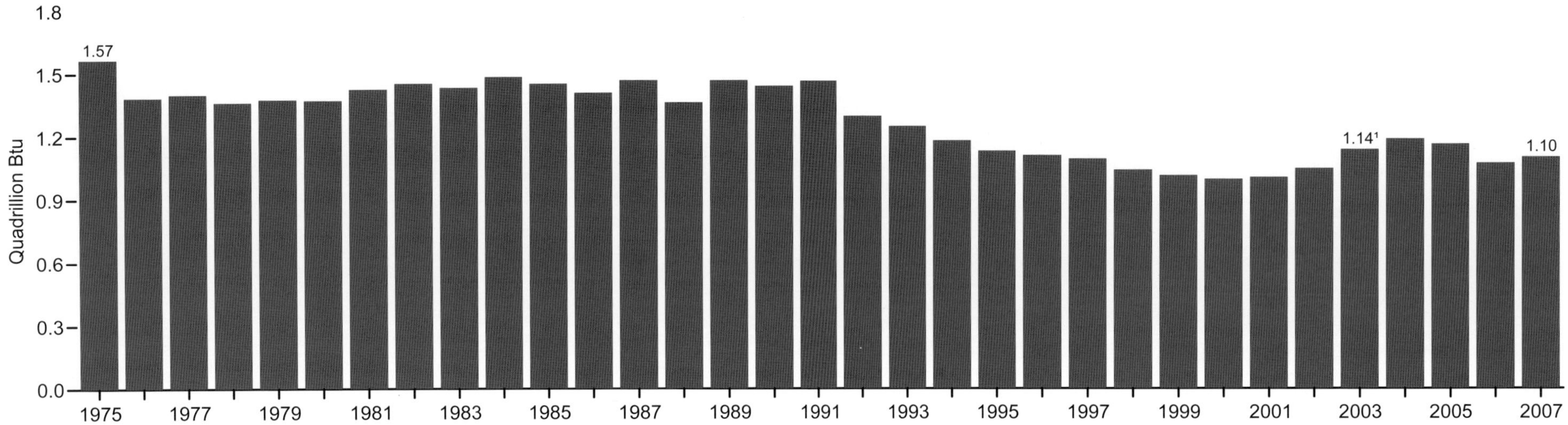

By Major Energy Source

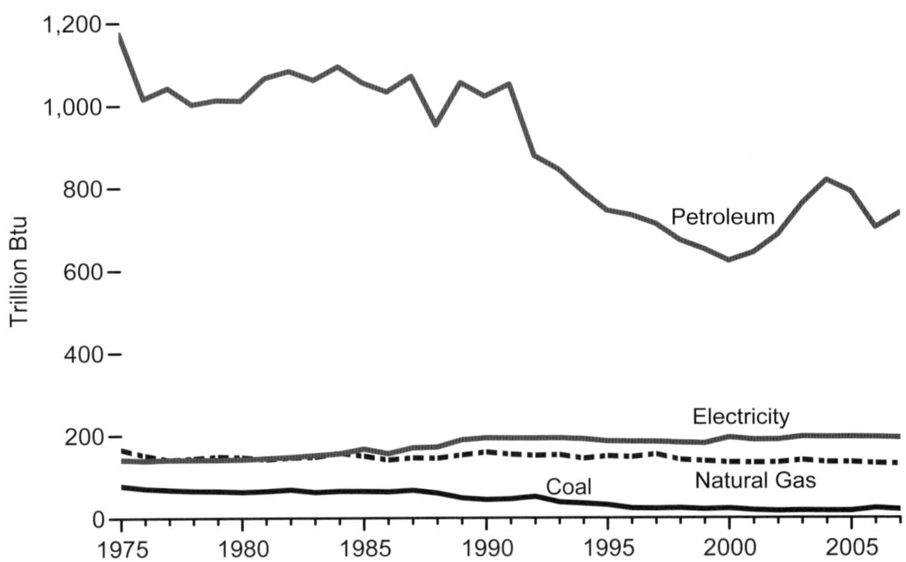

By Selected Petroleum Product

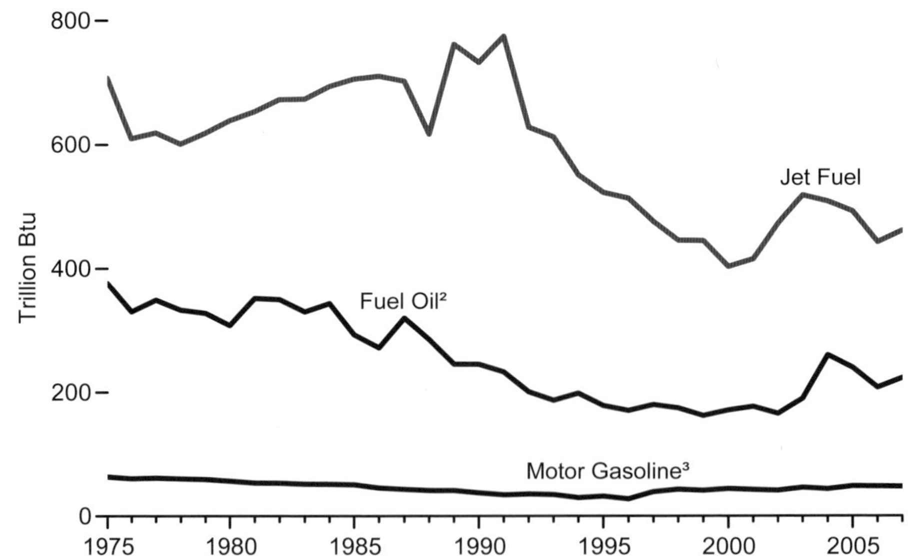

[1]Noted in reference to Executive Order 13423 (January 24, 2007), Sec. 2(a) that requires Federal agencies to "improve energy efficiency and reduce greenhouse gasses...relative to the baseline of the agency's energy use in fiscal year 2003."

[2]Distillate fuel oil and residual fuel oil.

[3]Includes ethanol blended into motor gasoline.

Notes: U.S. Government's fiscal year was October 1 through September 30, except in 1975 and 1976 when it was July 1 through June 30. • Because vertical scales differ, graphs should not be compared.

Source: Table 1.12.

Table 1.12 U.S. Government Energy Consumption by Source, Fiscal Years 1975-2007

(Trillion Btu)

Year	Coal	Natural Gas [1]	Petroleum: Aviation Gasoline	Petroleum: Fuel Oil [2]	Petroleum: Jet Fuel	Petroleum: LPG [3] and Other [4]	Petroleum: Motor Gasoline [5]	Petroleum: Total	Electricity	Purchased Steam and Other [6]	Total
1975	77.9	166.2	22.0	376.0	707.4	5.6	63.2	1,174.2	141.5	5.1	1,565.0
1976	71.3	151.8	11.6	329.7	610.0	4.7	60.4	1,016.4	139.3	4.6	1,383.4
1977	68.4	141.2	8.8	348.5	619.2	4.1	61.4	1,042.1	141.1	5.7	1,398.5
1978	66.0	144.7	6.2	332.3	601.1	3.0	60.1	1,002.9	141.0	6.4	1,360.9
1979	65.1	148.9	4.7	327.1	618.6	3.7	59.1	1,013.1	141.2	7.1	1,375.4
1980	63.5	147.3	4.9	307.7	638.7	4.0	56.5	1,011.8	141.9	6.8	1,371.2
1981	65.1	142.2	4.6	351.3	653.3	3.7	53.2	1,066.2	144.5	6.2	1,424.2
1982	68.6	146.2	3.6	349.4	672.7	3.9	53.1	1,082.8	147.5	6.2	1,451.4
1983	62.4	147.8	2.6	329.5	673.4	4.0	51.6	1,061.1	151.5	9.0	1,431.8
1984	65.3	157.4	1.9	342.9	693.7	4.1	51.2	1,093.8	155.9	10.1	1,482.5
1985	64.8	149.9	1.9	292.6	705.7	4.0	50.4	1,054.6	167.2	13.9	1,450.3
1986	63.8	140.9	1.4	271.6	710.2	3.9	45.3	1,032.4	155.8	13.7	1,406.7
1987	67.0	145.6	1.0	319.5	702.3	4.0	43.1	1,069.9	169.9	13.9	1,466.3
1988	60.2	144.6	6.0	284.8	617.2	3.2	41.2	952.4	171.2	32.0	1,360.3
1989	48.7	152.4	.8	245.3	761.7	5.7	41.1	1,054.5	188.6	20.6	1,464.7
1990	[R]44.3	[R]159.4	.5	[R]245.2	732.4	6.4	37.2	[R]1,021.7	[R]193.6	[R]19.1	[R]1,438.0
1991	45.9	154.1	.4	232.6	774.5	9.0	34.1	1,050.7	192.7	18.3	1,461.7
1992	51.7	151.2	1.0	200.6	628.2	11.4	35.6	876.8	192.5	22.5	1,294.8
1993	38.3	152.9	.7	187.0	612.4	9.3	34.5	843.9	193.1	18.6	1,246.8
1994	35.0	143.9	.6	198.5	550.7	10.9	29.5	790.2	190.9	18.2	1,178.2
1995	31.7	149.7	.3	178.5	522.3	11.4	31.9	744.4	185.3	18.2	1,129.3
1996	23.3	147.4	.2	170.6	513.0	21.7	27.6	733.2	184.5	20.1	1,108.5
1997	22.5	154.0	.3	180.1	475.7	17.2	39.0	712.2	184.0	19.2	1,092.0
1998	23.9	140.7	.2	174.6	445.5	9.4	43.1	672.8	181.8	18.8	1,037.9
1999	21.2	137.6	.1	162.2	444.7	2.9	41.1	650.9	180.4	21.5	1,011.6
2000	22.7	134.0	.2	171.4	403.1	4.3	43.9	622.9	194.0	20.2	993.8
2001	18.8	133.9	.2	177.0	415.2	7.9	42.5	642.9	188.8	18.6	1,003.0
2002	16.9	134.1	.2	165.7	472.9	6.0	41.3	686.1	189.1	18.5	1,044.8
2003	[R]17.7	[R]139.7	.3	[R]189.8	517.9	6.6	45.7	[R]760.3	[R]196.1	[R]22.5	[R]1,136.3
2004	17.4	[R]134.8	.2	[R]259.8	508.2	6.0	43.5	[R]817.8	[R]195.4	[R]21.6	[R]1,187.0
2005	17.1	[R]135.1	.4	[R]239.8	492.2	9.0	48.2	[R]789.6	[R]195.9	[R]23.9	[R]1,161.6
2006	[R]23.5	[R]132.0	.6	[R]207.8	[R]442.6	4.7	47.8	[R]703.5	[R]194.9	[R]17.7	[R]1,071.5
2007[P]	20.4	130.8	.4	223.2	461.1	6.6	47.4	738.8	193.8	16.6	1,100.4

[1] Natural gas, plus a small amount of supplemental gaseous fuels.
[2] Distillate fuel oil and residual fuel oil.
[3] Liquefied petroleum gases.
[4] Other types of fuel used in vehicles and equipment, primarily alternative fuels like methanol, ethanol, compressed natural gas, and biodiesel.
[5] Includes ethanol blended into motor gasoline.
[6] "Other" is chilled water, renewable energy, and other fuels reported as used in facilities.
R = Revised. P = Preliminary.

Notes: • For 1975 and 1976, the U.S. Government's fiscal year was July 1 through June 30. Beginning in 1977, the U.S. Government's fiscal year is October 1 through September 30 (for example, fiscal year 2007 is October 2006 through September 2007). • Data in this table are developed using the following conversion factors (which in most cases are different from those in Tables A1-A6)—coal: 24.580 million Btu/short ton; natural gas: 1,031 Btu/cubic foot; aviation gasoline: 5.250 million Btu/barrel; fuel oil: 5.8254 million Btu/barrel; jet fuel: 5.460 million Btu/barrel; liquefied petroleum gases: 4.011 million Btu/barrel; motor gasoline: 5.250 million Btu/barrel; electricity: 3,412 Btu/kilowatthour; and purchased steam: 1,000 Btu/pound. • Data include energy consumed at foreign installations and in foreign operations, including aviation and ocean bunkering, primarily by the U.S. Department of Defense. U.S. Government energy use for electricity generation and uranium enrichment is excluded. • Totals may not equal sum of components due to independent rounding.

Web Page: For related information, see http://www1.eere.energy.gov/femp/about/annual_report.html.

Source: U.S. Department of Energy, Office of Energy Efficiency and Renewable Energy, Federal Energy Management Program.

Figure 1.13 U.S. Government Energy Consumption by Agency and Source, Fiscal Years 2003, 2006, and 2007

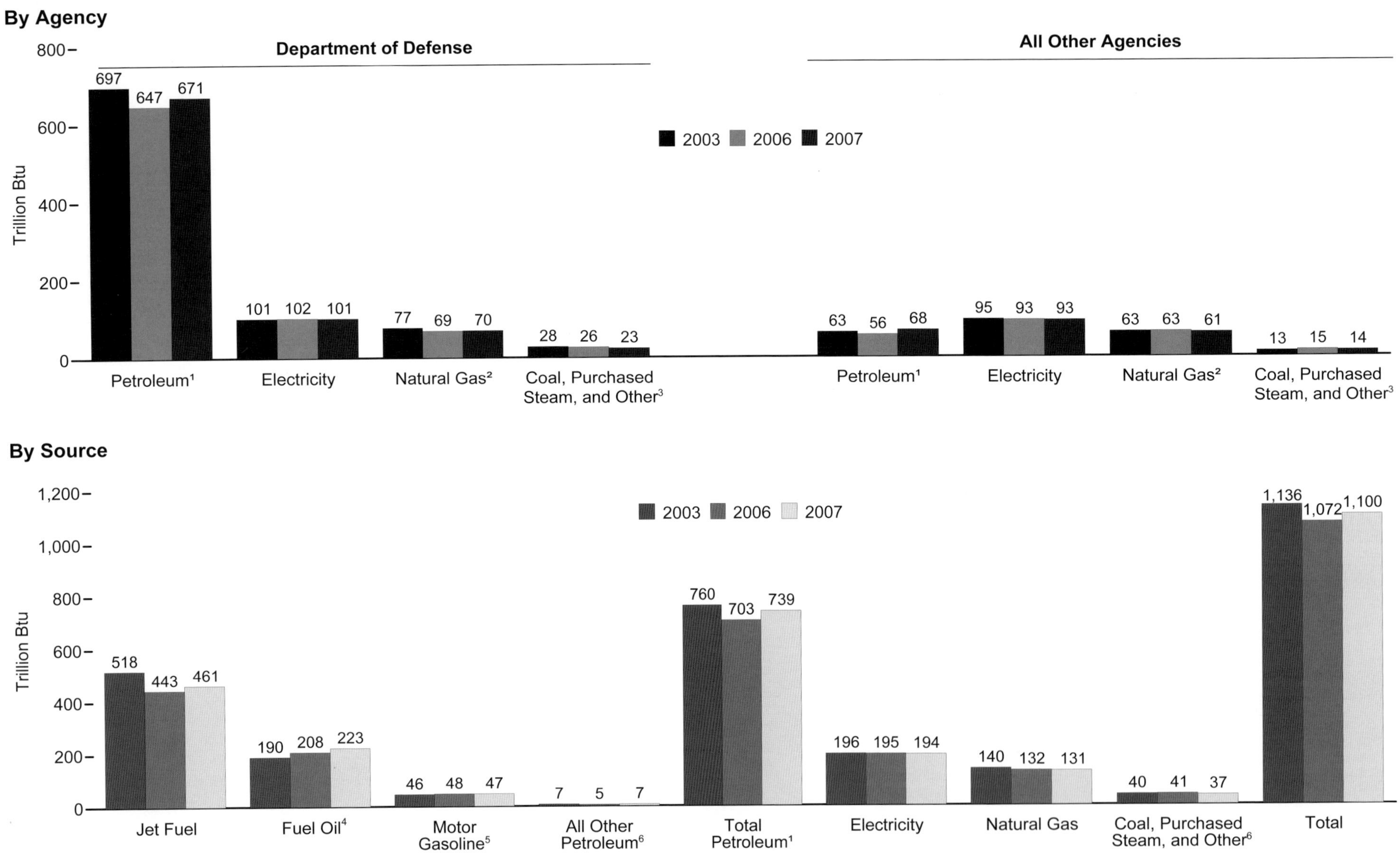

[1]Includes small amount of renewable energy; see Table 1.13, footnote 2.
[2]Natural gas, plus a small amount of supplemental gaseous fuels.
[3]Chilled water, renewable energy, and other fuels reported as used in facilities.
[4]Distillate fuel oil and residual fuel oil.
[5]Includes ethanol blended into motor gasoline.
[6]Aviation gasoline, liquefied petroleum gas, and other types of fuel used in vehicles and equipment, primarily alternative fuels like methanol, ethanol, compressed natural gas, and biodiesel.

Notes: • The U.S. Government's fiscal year runs from October 1 through September 30. • Because vertical scales differ, graphs should not be compared.

Source: Table 1.13.

Table 1.13 U.S. Government Energy Consumption by Agency and Source, Fiscal Years 2003, 2006, and 2007

(Trillion Btu)

Resource and Fiscal Years	Agriculture	Defense	Energy	GSA [1]	HHS [2]	Interior	Justice	NASA [3]	Postal Service	Transportation	Veterans Affairs	Other [4]	Total
Coal													
2003	0.0	15.4	2.0	0.0	0.0	0.0	0.0	0.0	0.0	0.0	0.2	0.0	17.7
2006	.0	17.0	R 6.2	.0	.0	.0	.0	.0	.0	.0	.2	.0	R 23.5
2007 P	.0	15.4	4.8	.0	.0	.0	.0	.0	.0	.0	.2	.0	20.4
Natural Gas [5]													
2003	1.4	76.6	7.0	7.6	3.7	1.3	8.6	2.9	10.4	.7	15.6	4.2	139.7
2006	1.7	68.6	R 6.0	6.4	5.1	1.3	10.7	R 2.9	R 9.8	.7	15.1	3.8	R 132.0
2007 P	1.6	69.5	6.2	7.2	5.1	1.2	10.5	2.7	6.5	.8	15.5	4.0	130.8
Petroleum													
2003	3.3	697.1	3.0	.2	1.5	4.4	6.5	1.4	18.2	1.6	2.8	20.3	760.3
2006	2.9	647.2	R 2.6	.2	.8	4.4	5.9	R 1.3	R 16.5	1.2	2.0	18.5	R 703.5
2007 P	3.0	670.6	2.7	.2	1.7	4.1	16.9	1.6	16.5	1.5	2.0	18.1	738.8
Aviation Gasoline													
2003	.0	.0	.0	.0	.0	.0	.1	.0	.0	.0	.0	.0	.3
2006	.0	.1	.0	.0	.0	.0	.1	.0	.0	.0	.0	.4	.6
2007 P	.1	.1	.0	.0	.0	.0	.1	.0	.0	.0	.0	.1	.4
Fuel Oil [6]													
2003	.4	166.5	2.0	.1	.9	1.2	.4	.4	5.1	.3	1.9	10.7	189.8
2006	.4	191.3	R 1.5	.1	.4	1.2	1.0	.4	R 2.8	.2	1.2	7.2	R 207.8
2007 P	.5	194.1	1.6	.1	.6	1.1	12.1	.4	3.1	.6	1.2	7.9	223.2
Jet Fuel													
2003	.0	509.9	.0	.0	.0	.1	1.5	.6	.0	.6	.0	5.2	517.9
2006	.0	436.3	.1	.0	.0	.1	.1	R .7	.0	.5	.0	4.8	R 442.6
2007 P	.0	455.6	.0	.0	.0	.0	.0	1.0	.0	.5	.0	4.0	461.1
LPG [7] and Other [8]													
2003	.7	4.2	.1	.0	.1	.7	.0	.1	.2	.1	.0	.3	6.6
2006	.2	2.2	.2	.0	.1	1.0	.1	.1	.2	.0	.0	.5	4.7
2007 P	.3	3.1	.4	.0	.1	.9	1.0	.1	.4	.0	.0	.2	6.6
Motor Gasoline [9]													
2003	2.2	16.5	.9	.1	.5	2.4	4.5	.2	12.9	.7	.9	4.1	45.7
2006	2.2	17.3	.8	.0	.3	2.1	4.7	R .1	13.4	.4	.8	5.6	47.8
2007 P	2.1	17.8	.7	.0	.9	2.0	3.7	.1	12.9	.4	.8	5.9	47.4
Electricity													
2003	2.6	101.1	18.0	10.0	3.6	2.4	7.0	5.8	21.7	3.2	10.2	10.5	196.1
2006	2.0	101.7	R 16.7	9.9	2.9	2.3	6.1	R 5.4	R 25.0	2.7	10.4	R 9.9	R 194.9
2007 P	1.9	101.2	17.3	10.0	3.3	2.1	6.0	5.5	22.5	3.2	10.7	10.1	193.8
Purchased Steam and Other [10]													
2003	.3	12.2	1.6	1.8	1.3	.1	.7	.8	.7	.0	1.7	1.2	22.5
2006	.3	9.2	R 1.4	1.7	.4	.1	.8	R .6	R .5	.1	1.6	1.0	R 17.7
2007 P	.2	7.9	1.4	1.7	.4	.1	.8	.8	.4	.0	1.7	1.0	16.6
Total Energy													
2003	7.7	902.3	31.6	19.6	10.1	8.2	22.7	10.8	50.9	5.6	30.5	36.2	1,136.3
2006	6.8	843.7	R 32.9	R 18.2	9.3	8.1	23.5	R 10.2	R 51.8	4.6	29.3	R 33.2	R 1,071.5
2007 P	6.8	864.6	32.3	19.1	10.6	7.5	34.2	10.6	45.8	5.6	30.1	33.2	1,100.4

[1] General Services Administration.
[2] Health and Human Services.
[3] National Aeronautics and Space Administration.
[4] Includes National Archives and Records Administration, U.S. Department of Commerce, Tennessee Valley Authority, U.S. Department of Labor, National Science Foundation, Federal Trade Commission, Federal Communications Commission, Environmental Protection Agency, U.S. Department of Homeland Security, U.S. Department of Housing and Urban Development, Railroad Retirement Board, Equal Employment Opportunity Commission, Nuclear Regulatory Commission, U.S. Department of State, U.S. Department of the Treasury, Office of Personnel Management, Consumer Product Safety Commission, Central Intelligence Agency, Social Security Administration, U.S. Information Agency (International Broadcasting Bureau).
[5] Natural gas, plus a small amount of supplemental gaseous fuels.
[6] Distillate fuel oil and residual fuel oil.
[7] Liquefied petroleum gases.
[8] Other types of fuel used in vehicles and equipment, primarily alternative fuels like methanol, ethanol, compressed natural gas, and biodiesel.
[9] Includes ethanol blended into motor gasoline.
[10] Chilled water, renewable energy, and other fuels reported as used in facilities.

R=Revised. P=Preliminary.

Notes: • For 1975 and 1976, the U.S. Government's fiscal year was July 1 through June 30. Beginning in 1977, the U.S. Government's fiscal year is October 1 through September 30 (for example, fiscal year 2007 is October 2006 through September 2007). • Data in this table are developed using the following conversion factors (which in most cases are different from those in Tables A1-A6)—coal: 24.580 million Btu/short ton; natural gas: 1,031 Btu/cubic foot; aviation gasoline: 5.250 million Btu/barrel; fuel oil: 5.8254 million Btu/barrel; jet fuel: 5.460 million Btu/barrel; liquefied petroleum gases: 4.011 million Btu/barrel; motor gasoline: 5.250 million Btu/barrel; electricity: 3,412 Btu/kilowatthour; and purchased steam: 1,000 Btu/pound. • Data include energy consumed at foreign installations and in foreign operations, including aviation and ocean bunkering, primarily by the U.S. Department of Defense. U.S. Government energy use for electricity generation and uranium enrichment is excluded. • Totals may not equal sum of components due to independent rounding.

Web Page: See http://www1.eere.energy.gov/femp/about/annual_report.html for related information.

Source: U.S. Department of Energy, Office of Energy Efficiency and Renewable Energy, Federal Energy Management Program.

Figure 1.14 Fossil Fuel Production on Federally Administered Lands

Total, 1949-2007

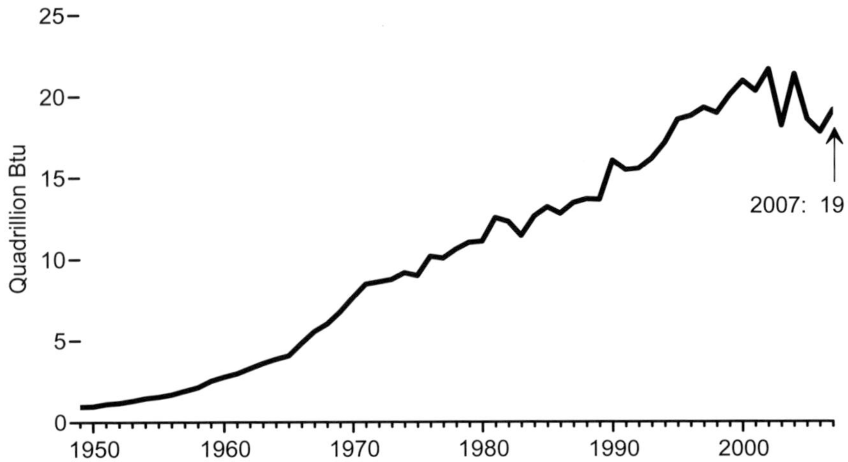

Federal Lands Fossil Fuel Production as a Share of U.S. Fossil Fuel Production, 1949-2007

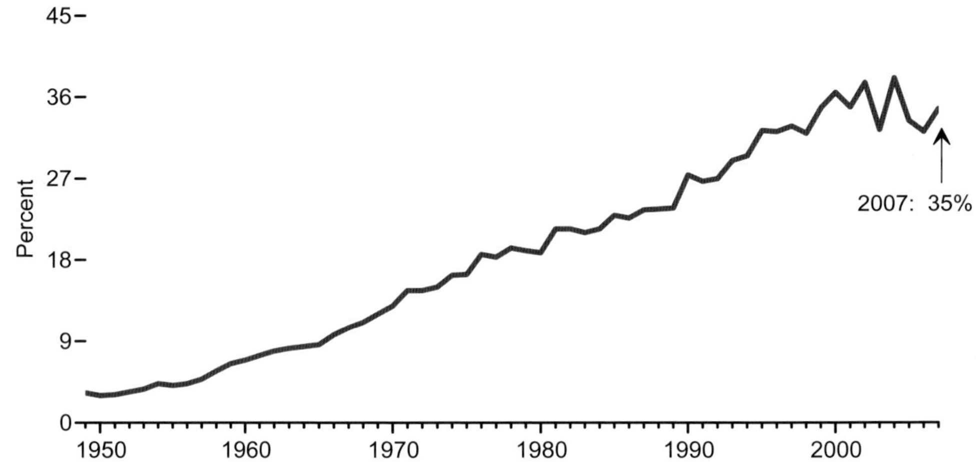

By Source, 1949-2007

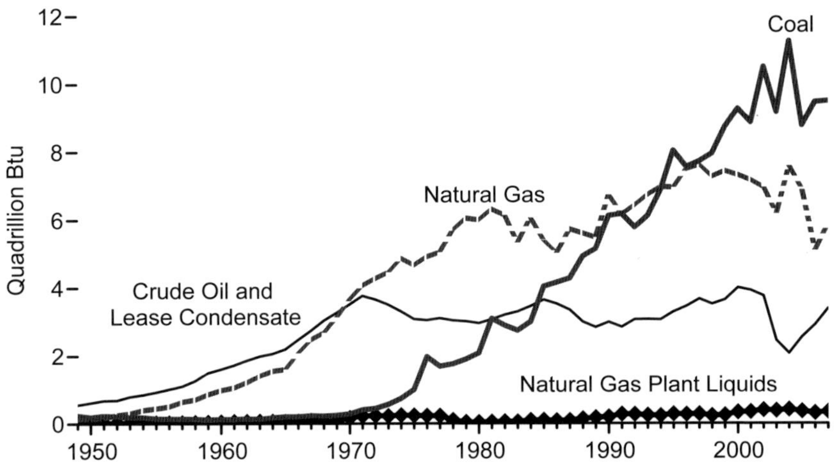

Federal Lands Fossil Fuel Production as a Share of U.S. Fossil Fuel Production, By Source, 2007

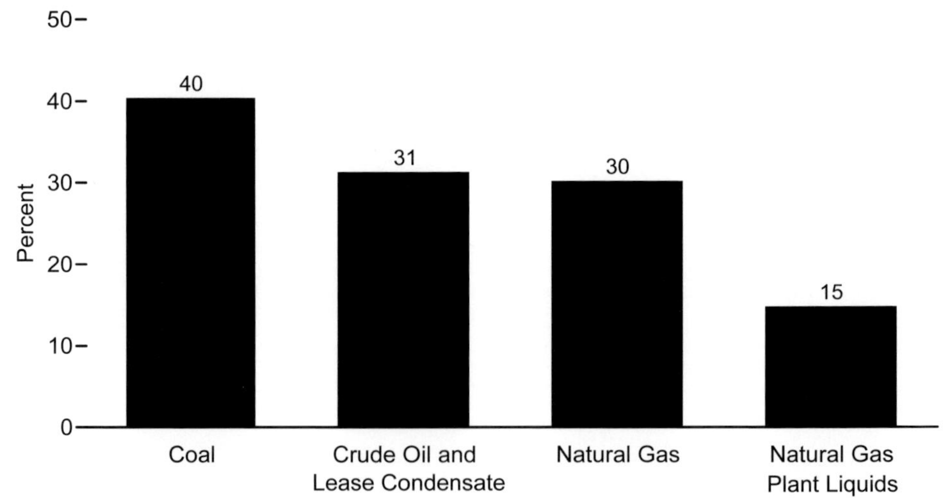

Notes: • Data through 2000 are on a calendar-year basis; data for 2001-2007 are on a fiscal-year basis (October–September). • "Federally Administered Lands" include all classes of land owned by the Federal Government, including acquired military, Outer Continental Shelf, and public lands. • Because vertical scales differ, graphs should not be compared.

Source: Table 1.14.

Table 1.14 Fossil Fuel Production on Federally Administered Lands, Selected Years, 1949-2007

Year	Crude Oil and Lease Condensate			Natural Gas Plant Liquids[1]			Natural Gas[2]			Coal			Fossil Fuels	
	Million Barrels[3]	Quadrillion Btu	Percent of U.S. Total	Million Barrels[3]	Quadrillion Btu	Percent of U.S. Total	Trillion Cubic Feet[3]	Quadrillion Btu	Percent of U.S. Total	Million Short Tons[3]	Quadrillion Btu	Percent of U.S. Total	Quadrillion Btu	Percent of U.S. Total
							Calendar-Year Data[4]							
1949	95.2	0.55	5.2	4.4	0.02	2.8	0.15	0.15	2.8	9.5	0.24	2.0	0.96	3.3
1950	105.9	.61	5.4	4.4	.02	2.4	.14	.15	2.4	7.7	.19	1.4	.98	3.0
1955	159.5	.92	6.4	6.0	.03	2.1	.43	.45	4.8	5.9	.15	1.2	1.55	4.1
1960	277.3	1.61	10.8	11.6	.05	3.4	.95	.98	7.8	5.2	.13	1.2	2.77	6.9
1965	378.6	2.20	13.3	14.3	.06	3.2	1.56	1.61	10.2	8.2	.20	1.6	4.07	8.6
1970	605.6	3.51	17.2	40.6	.17	6.7	3.56	3.67	16.9	12.0	.29	2.0	7.64	12.9
1971	648.9	3.76	18.8	54.0	.22	8.7	3.95	4.08	18.3	17.3	.41	3.1	8.47	14.6
1972	630.5	3.66	18.2	56.7	.23	8.9	4.17	4.28	19.3	19.0	.44	3.1	8.61	14.6
1973	604.3	3.51	18.0	54.9	.22	8.7	4.37	4.46	20.1	24.2	.57	4.1	8.75	15.0
1974	570.2	3.31	17.8	61.9	.25	10.1	4.75	4.87	22.9	32.1	.74	5.3	9.16	16.3
1975	531.5	3.08	17.4	59.7	.24	10.0	4.57	4.67	23.8	43.6	1.00	6.7	8.99	16.4
1976	525.7	3.05	17.7	57.2	.23	9.7	4.81	4.91	25.2	86.4	1.98	12.6	10.16	18.6
1977	535.0	3.10	17.8	57.4	.23	9.7	4.94	5.04	25.8	74.8	1.69	10.7	10.06	18.3
1978	523.6	3.04	16.5	25.9	.10	4.5	5.60	5.71	29.3	79.2	1.76	11.8	10.61	19.3
1979	519.8	3.01	16.7	11.9	.05	2.1	5.93	6.05	30.1	84.9	1.91	10.9	11.02	19.0
1980	510.4	2.96	16.2	10.5	.04	1.8	5.85	6.01	30.2	92.9	2.08	11.2	11.09	18.8
1981	529.3	3.07	16.9	12.3	.05	2.1	6.15	6.31	32.1	138.8	3.10	16.8	12.53	21.4
1982	552.3	3.20	17.5	15.0	.06	2.7	5.97	6.14	33.5	130.0	2.89	15.5	12.29	21.4
1983	568.8	3.30	17.9	14.0	.05	2.5	5.17	5.33	32.1	124.3	2.74	15.9	11.43	21.0
1984	595.8	3.46	18.3	25.4	.10	4.3	5.88	6.07	33.7	136.3	3.00	15.2	12.62	21.4
1985	628.3	3.64	19.2	26.6	.10	4.5	5.24	5.41	31.8	184.6	4.04	20.9	13.19	22.9
1986	608.4	3.53	19.2	23.3	.09	4.1	4.87	5.01	30.3	189.7	4.16	21.3	12.79	22.6
1987	577.3	3.35	18.9	23.7	.09	4.1	5.56	5.73	33.4	195.2	4.28	21.2	13.45	23.5
1988	516.3	2.99	17.3	37.0	.14	6.2	5.45	5.61	31.9	225.4	4.92	23.7	13.67	23.6
1989	488.9	2.84	17.6	45.1	.17	8.0	5.32	5.49	30.7	236.3	5.14	24.1	13.64	23.7
1990	515.9	2.99	19.2	50.9	.19	8.9	6.55	6.74	36.8	280.6	6.12	27.2	16.05	27.4
1991	491.0	2.85	18.1	72.7	.28	12.0	5.99	6.17	33.8	285.1	6.18	28.5	15.47	26.7
1992	529.1	3.07	20.2	70.7	.27	11.4	6.25	6.43	35.0	266.7	5.78	26.6	15.55	27.0
1993	529.3	3.07	21.2	64.4	.24	10.2	6.56	6.74	36.3	285.7	6.12	30.0	16.17	29.0
1994	527.7	3.06	21.7	60.0	.23	9.5	6.78	6.97	36.0	321.4	6.88	30.9	17.14	29.5
1995	567.4	3.29	23.7	74.0	.28	11.5	6.78	6.96	36.4	376.9	8.04	36.2	18.56	32.3
1996	596.5	3.46	25.2	71.2	.27	10.6	7.31	7.50	38.8	354.5	7.56	33.0	18.79	32.2
1997	632.8	3.67	26.9	74.7	.28	11.3	7.43	7.62	39.3	362.6	7.72	33.0	19.29	32.8
1998	[5]606.3	[5]3.52	[5]26.6	[5]60.3	[5].23	[5]9.4	[5]7.06	[5]7.27	[5]37.1	371.1	7.95	33.0	[5]18.97	[5]32.0
1999	[6]628.9	[6]3.65	[6]29.3	[6]66.5	[6].25	[6]9.9	[6]7.24	[6]7.44	[6]38.4	414.5	8.73	37.4	[6]20.07	[6]34.8
2000	689.2	4.00	32.3	88.9	.33	12.7	7.14	7.32	37.2	440.2	9.27	40.7	20.92	36.5
							Fiscal-Year Data[7]							
2001	676.5	3.92	32.0	93.0	0.35	14.0	6.98	7.17	35.7	425.4	8.87	38.1	20.31	34.9
2002	647.8	3.76	30.5	106.5	.40	15.2	6.78	6.96	35.4	507.8	10.51	45.7	21.63	37.6
2003	[8]422.6	[8]2.45	[8]20.4	101.0	.38	16.0	6.01	6.19	31.5	446.7	9.18	41.3	18.19	32.4
2004	356.4	2.07	17.7	110.7	.41	16.8	7.38	7.59	39.4	551.1	11.27	49.7	21.33	38.1
2005	439.9	2.55	22.7	96.6	.36	14.8	6.70	6.89	36.6	431.0	8.78	37.8	18.58	33.4
2006	502.1	2.91	27.4	84.1	.31	13.7	4.96	[R]5.10	[R]27.3	466.2	[R]9.47	[R]40.1	[R]17.80	[R]32.2
2007	584.7	3.39	31.3	94.5	.35	14.8	5.73	5.89	30.2	467.5	9.51	40.4	19.14	34.7

[1] Includes only those quantities for which the royalties were paid on the basis of the value of the natural gas plant liquids produced. Additional quantities of natural gas plant liquids were produced; however, the royalties paid were based on the value of natural gas processed. These latter quantities are included with natural gas.

[2] Includes some quantities of natural gas processed into liquids at natural gas processing plants and fractionators.

[3] Data from the U.S. Department of the Interior (DOI), U.S. Minerals Management Service (MMS), are for sales volumes.

[4] Through 2000, data are on a calendar-year (January through December) basis. The only exception is in 1949-1974 with production from Naval Petroleum Reserve No. 1, which is on a fiscal-year (July through June) basis.

[5] There is a discontinuity in this time series between 1997 and 1998 due to the sale of "Elk Hills," Naval Petroleum Reserve No. 1.

[6] There is a discontinuity in this time series between 1998 and 1999; beginning in 1999 Naval Petroleum Reserve data have become insignificant and are no longer included.

[7] Beginning in 2001, data are on a fiscal-year (October through September) basis; for example, fiscal-year 2006 data are for October 2005 through September 2006.

[8] A significant amount of Federal offshore crude oil was diverted to the Strategic Petroleum Reserve.

R=Revised.

Note: "Federally Administered Lands" include all classes of land owned by the Federal Government, including acquired military, Outer Continental Shelf, and public lands.

Web Pages: • For all data beginning in 1949, see http://www.eia.doe.gov/emeu/aer/overview.html. • For related information, see http://www.mrm.mms.gov.

Sources: See end of section.

Figure 1.15 Fossil Fuel Consumption for Nonfuel Use

Total, 1980-2007

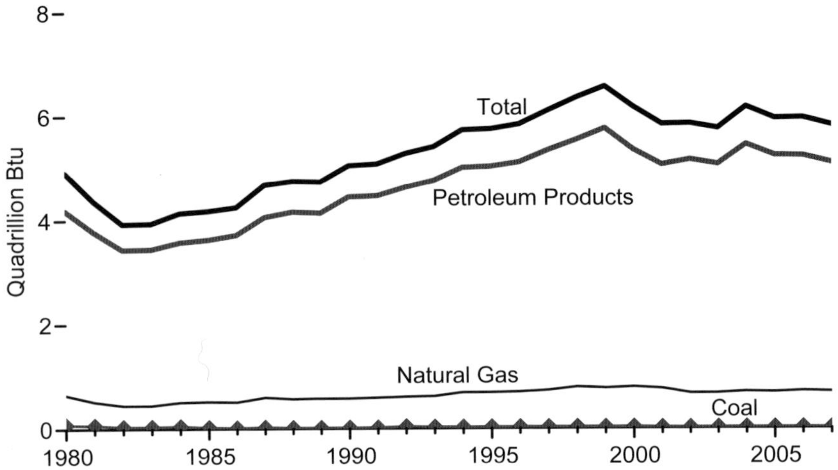

As Share of Total Energy Consumption, 1980-2007

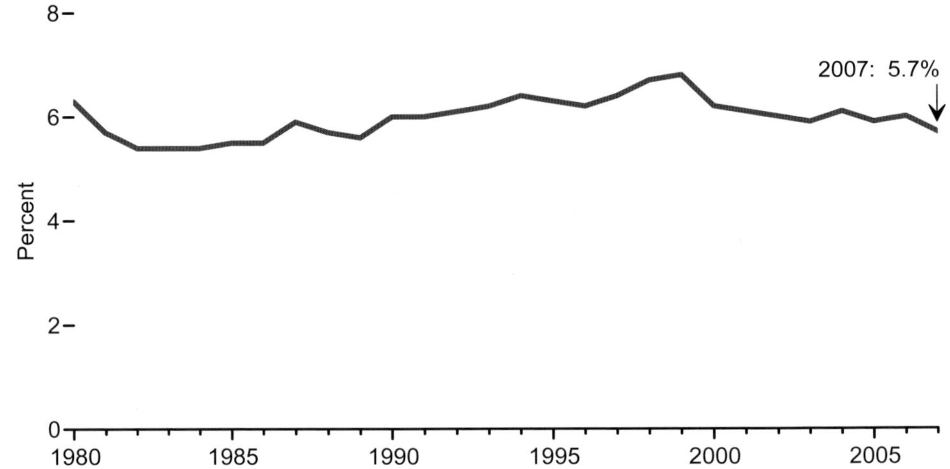

By Fuel, 2007

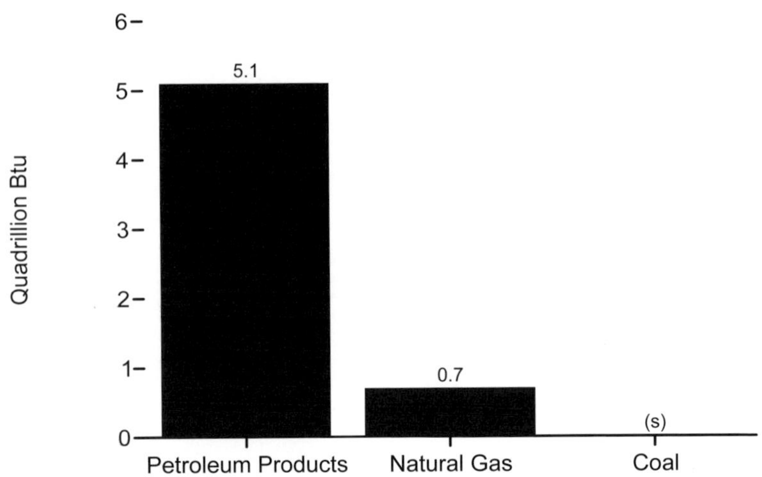

By Petroleum Product, 2007

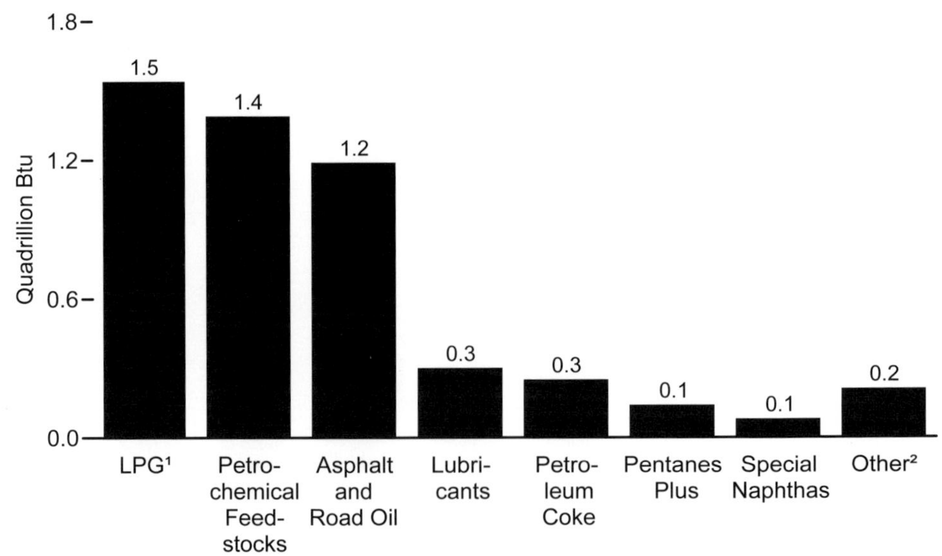

[1] Liquefied petroleum gases.
[2] Distillate fuel oil, residual fuel oil, waxes, and miscellaneous products.
(s)=Less than 0.05 quadrillion Btu.

Notes: • See Note 3, "Nonfuel Use of Fossil Fuels," at end of section. • Because vertical scales differ, graphs should not be compared.
Source: Table 1.15.

Table 1.15 Fossil Fuel Consumption for Nonfuel Use, 1980-2007

Year	Petroleum Products: Asphalt and Road Oil	Liquefied Petroleum Gases	Pentanes Plus	Lubricants	Petro-chemical Feedstocks	Petroleum Coke	Special Naphthas	Other [1]	Total	Natural Gas	Coal	Total	Percent of Total Energy Consumption
	Physical Units [2]												
1980	145	230	([3])	58	253	24	37	58	805	639	2.4	– –	– –
1981	125	229	([3])	56	216	29	27	54	736	507	2.1	– –	– –
1982	125	256	([3])	51	157	23	25	48	686	438	1.4	– –	– –
1983	136	264	([3])	53	151	10	30	45	689	441	1.2	– –	– –
1984	150	247	10	57	145	16	40	41	705	495	1.5	– –	– –
1985	156	265	13	53	144	15	30	41	718	500	1.1	– –	– –
1986	164	248	17	52	169	14	25	38	727	496	.7	– –	– –
1987	170	303	12	59	170	24	28	36	802	578	.8	– –	– –
1988	171	319	21	57	173	25	22	40	827	554	.7	– –	– –
1989	165	332	17	58	172	23	20	39	827	563	.6	– –	– –
1990	176	344	18	60	199	30	20	39	886	559	.6	– –	– –
1991	162	394	10	53	200	25	17	44	906	573	.6	– –	– –
1992	166	397	13	54	214	38	20	35	937	594	1.2	– –	– –
1993	174	389	60	55	216	21	20	33	969	607	.9	– –	– –
1994	176	437	56	58	222	23	15	35	1,022	673	.9	– –	– –
1995	178	450	66	57	215	22	13	34	1,035	668	.9	– –	– –
1996	177	470	69	55	217	25	14	34	1,061	681	.9	– –	– –
1997	184	473	65	58	250	20	14	35	1,100	706	.9	– –	– –
1998	190	494	44	61	252	35	20	39	1,137	762	.8	– –	– –
1999	200	520	57	62	238	47	28	37	1,188	753	.8	– –	– –
2000	192	479	51	61	243	23	19	38	1,106	767	.8	– –	– –
2001	189	445	44	56	214	34	15	39	1,036	732	.7	– –	– –
2002	187	465	37	55	229	32	20	38	1,063	657	.7	– –	– –
2003	184	441	37	51	247	27	15	36	1,038	R655	.7	– –	– –
2004	196	453	37	52	287	41	10	34	1,110	678	.7	– –	– –
2005	199	428	33	51	266	38	12	34	1,061	R677	.7	– –	– –
2006	185	439	23	42	R276	43	13	37	R1,059	688	.6	– –	– –
2007P	179	448	30	49	249	41	15	36	1,047	685	.6	– –	– –
	Quadrillion Btu												
1980	0.96	0.78	([3])	0.35	1.43	0.14	0.19	0.34	4.19	0.65	0.08	4.92	6.3
1981	.83	.77	([3])	.34	1.21	.17	.14	.31	3.78	.52	.07	4.37	5.7
1982	.83	.87	([3])	.31	.88	.14	.13	.28	3.44	.45	.04	3.93	5.4
1983	.90	.89	([3])	.32	.85	.06	.16	.26	3.45	.45	.04	3.94	5.4
1984	.99	.84	.05	.35	.82	.09	.21	.24	3.58	.51	.05	4.14	5.4
1985	1.03	.90	.06	.32	.82	.09	.16	.24	3.63	.52	.03	4.18	5.5
1986	1.09	.85	.08	.31	.95	.08	.13	.22	3.72	.51	.02	4.25	5.5
1987	1.13	1.06	.06	.36	.96	.14	.14	.21	4.06	.60	.03	4.68	5.9
1988	1.14	1.11	.10	.34	.97	.15	.11	.23	4.16	.57	.02	4.75	5.7
1989	1.10	1.18	.08	.35	.96	.14	.11	.23	4.14	.58	.02	4.74	5.6
1990	1.17	1.20	.08	.36	1.12	.18	.11	.23	4.45	.58	.02	5.05	6.0
1991	1.08	1.38	.04	.32	1.15	.15	.09	.26	4.47	.59	.02	5.08	6.0
1992	1.10	1.39	.06	.33	1.20	.23	.10	.20	4.63	.61	.04	5.28	6.1
1993	1.15	1.35	.28	.34	1.22	.12	.10	.20	4.76	.62	.03	5.41	6.2
1994	1.17	1.55	.26	.35	1.26	.14	.08	.20	5.01	.69	.03	5.73	6.4
1995	1.18	1.59	.30	.35	1.21	.13	.07	.20	5.03	.69	.03	5.75	6.3
1996	1.18	1.65	.32	.34	1.21	.15	.07	.20	5.11	.70	.03	5.84	6.2
1997	1.22	1.67	.30	.35	1.40	.12	.07	.21	5.34	.73	.03	6.10	6.4
1998	1.26	1.74	.20	.37	1.40	.21	.11	.23	5.54	.79	.03	6.35	6.7
1999	1.32	1.82	.26	.37	1.33	.28	.15	.22	5.76	.77	.03	6.56	6.8
2000	1.28	1.67	.24	.37	1.35	.14	.10	.22	5.35	.79	.03	6.17	6.2
2001	1.26	1.55	.20	.34	1.19	.21	.08	.23	5.06	.76	.02	5.84	6.1
2002	1.24	1.62	.17	.33	1.27	.19	.10	.22	5.16	.67	.02	5.85	6.0
2003	1.22	1.55	.17	.31	1.37	.16	.08	.21	5.07	R.67	.02	R5.76	5.9
2004	1.30	1.58	.17	.31	1.59	.25	.05	.20	5.45	.70	.02	6.17	6.1
2005	1.32	1.49	.15	.31	1.47	.23	.06	.20	5.24	R.69	.02	R5.95	5.9
2006	R1.26	1.52	.11	.25	R1.54	.26	.07	R.22	R5.23	.71	.02	R5.96	R6.0
2007P	1.19	1.54	.14	.30	1.39	.25	.08	.21	5.10	.70	.02	5.82	5.7

[1] Distillate fuel oil, residual fuel oil, waxes, and miscellaneous products.
[2] Petroleum—million barrels; natural gas—billion cubic feet; and coal—million short tons.
[3] Included in "Liquefied Petroleum Gases."
R=Revised. P=Preliminary. – – = Not applicable.
Notes: • Estimates of consumption for nonfuel use shown in this table are included in total energy consumption (see Table 1.3). • See Note 3, "Nonfuel Use of Fossil Fuels," at end of section. • Because of changes in methodology, data series may be revised annually. • Estimates of nonfuel use in this table are considered industrial uses with the exception of approximately half of the lubricants which are considered transportation use. • Totals may not equal sum of components due to independent rounding.
Web Page: For related information, see http://www.eia.doe.gov/environment.html.
Sources: See end of section.

Energy Overview

Note 1. Primary Energy Production. Primary energy production consists of coal production, waste coal supplied, and coal refuse recovery; crude oil and lease condensate production; natural gas plant liquids production; natural gas (dry) production; nuclear electricity net generation (converted to Btu using the nuclear plants heat rate); conventional hydroelectricity net generation (converted to Btu using the fossil-fueled plants heat rate); geothermal electricity net generation (converted to Btu using the geothermal plants heat rate), geothermal heat pump energy, and geothermal direct use energy; solar thermal and photovoltaic electricity net generation (converted to Btu using the fossil-fueled plants heat rate), and solar thermal direct use energy; wind electricity net generation (converted to Btu using the fossil-fueled plants heat rate); wood and wood-derived fuels consumption; biomass waste (municipal solid waste from biogenic sources, landfill gas, sludge waste, agricultural byproducts, and other biomass) consumption; and biofuels feedstock (biomass inputs to the production of fuel ethanol and biodiesel).

Note 2. Primary Energy Consumption. Primary energy consumption consists of coal consumption; coal coke net imports; petroleum consumption (petroleum products supplied, including natural gas plant liquids and crude oil burned as fuel, but excluding ethanol blended into motor gasoline); natural gas (excluding supplemental gaseous fuels) consumption; nuclear electricity net generation (converted to Btu using the nuclear plants heat rate); conventional hydroelectricity net generation (converted to Btu using the fossil-fueled plants heat rate); geothermal electricity net generation (converted to Btu using the geothermal plants heat rate), and geothermal heat pump and geothermal direct use energy; solar thermal and photovoltaic electricity net generation (converted to Btu using the fossil-fueled plants heat rate), and solar thermal direct use energy; wind electricity net generation (converted to Btu using the fossil-fueled plants heat rate); wood and wood-derived fuels consumption; biomass waste (municipal solid waste from biogenic sources, landfill gas, sludge waste, agricultural byproducts, and other biomass) consumption; fuel ethanol and biodiesel consumption; losses and co-products from the production of fuel ethanol and biodiesel; and electricity net imports (converted to Btu using the electricity heat content of 3,412 Btu per kilowatthour).

Note 3. Nonfuel Use of Fossil Fuels. Most fossil fuels consumed in the United States and elsewhere are combusted to produce heat and power. However, some are used directly for nonfuel use as construction materials, lubricants, chemical feedstocks, solvents, and waxes. For example, asphalt and road oil are used for roofing and paving; liquefied petroleum gases are used to create intermediate products that are used in making plastics; lubricants, including motor oil and greases, are used in vehicles and various industrial processes; petrochemical feedstocks are used to make plastics, synthetic fabrics, and related products; and natural gas is used to make nitrogenous fertilizers and as feedstock in the chemical industry. For more information, see Energy Information Administration, "Emissions of Greenhouse Gases in the United States" ("Nonfuel Use of Energy Inputs" section in Chapter 2), athttp://www.eia.doe.gov/environment.html.

Table 1.14 Sources; Physical Data (Columns 1, 4, 7, and 10): • 1949-1980—U.S. Geological Survey (USGS), *Federal and Indian Lands Oil and Gas Production, Royalty Income, and Related Statistics,* and *Federal and Indian Lands Coal, Phosphate, Potash, Sodium, and Other Mineral Production, Royalty Income, and Related Statistics* (June 1981). U.S. Department of Energy (DOE), Office of Naval Petroleum and Oil Shale Reserves (NPOSR), unpublished data; and USGS, National Petroleum Reserve in Alaska, unpublished data. • 1981-1983—DOI, Minerals Management Service (MMS), *Mineral Revenues Report on Receipts from Federal and Indian Leases*, annual reports; DOE, NPOSR, unpublished data; and USGS, National Petroleum Reserve in Alaska, unpublished data. • 1984-1998—DOI, MMS, *Mineral Revenues Report on Receipts from Federal and Indian Leases*, annual reports; and DOE, NPOSR, unpublished data. • 1999 and 2000—DOI, MMS. *Mineral Revenues Report on Receipts from Federal and American Indian Leases*, annual reports. • 2001 forward—DOI, MMS, "2001-Forward MRM Statistical Information." **Btu Data:** Data in columns 2, 5, 8, and 11 are calculated by multiplying the physical data by approximate heat contents for total U.S. production in Tables A2, A4, and A5. Data in column 13 are the sum of the other Btu columns. **Percent of U.S. Total:** Percentages are calculated by dividing production on federally administered lands by total U.S. production, then multiplying by 100. Calendar-year values for total U.S. production are from Tables 5.1, 6.1, and 7.1; fiscal-year values for total U.S. production are the sum of October-September values from the *Monthly Energy Review* (May 2008), Tables 3.1, 4.1, and 6.1.

Table 1.15 Sources; Petroleum Products: • 1980—Energy Information Administration (EIA), Energy Data Reports, *Petroleum Statement, Annual* and *Sales of Liquefied Petroleum Gases and Ethane in 1980.* • 1981 forward—EIA, *Petroleum Supply Annual*, annual reports, and unpublished data. **Natural Gas:** • 1980—Bureau of the Census, 1980 Survey of Manufactures, *Hydrocarbon, Coal, and Coke Materials Consumed.* • 1981 forward—U.S. Department of Commerce. **Coal:** • 1960-1995—U.S. International Trade Commission, *Synthetic Organic Chemicals, United States Production and Sales, 1995* (January 1997). • 1996 forward—EIA estimates. **Percent of Total Energy Consumption:** Derived by dividing total by total consumption on Table 1.3.

2 Energy Consumption by Sector

Office buildings, industries, residences, and transport systems, Baltimore, Maryland; east view from the inner harbor.
Source: U.S. Department of Energy.

Figure 2.1a Energy Consumption by Sector Overview

Total Consumption by End-Use Sector, 1949-2007

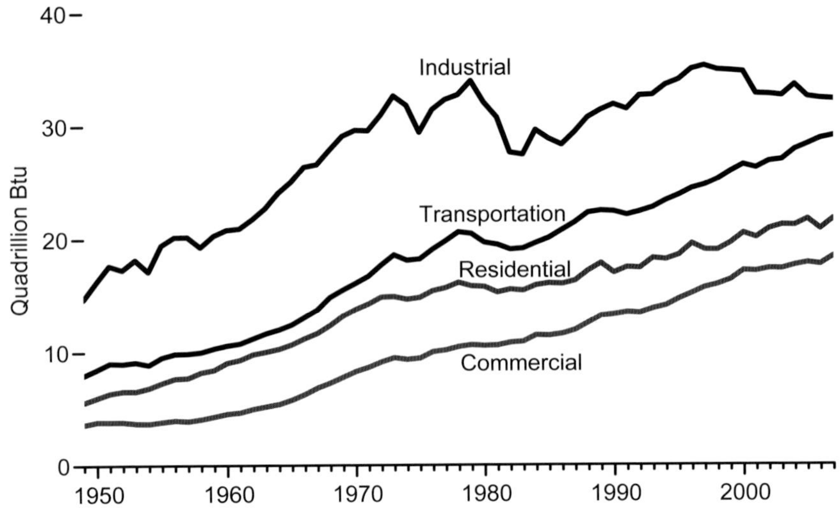

End-Use Sector Shares of Total Consumption, 2007

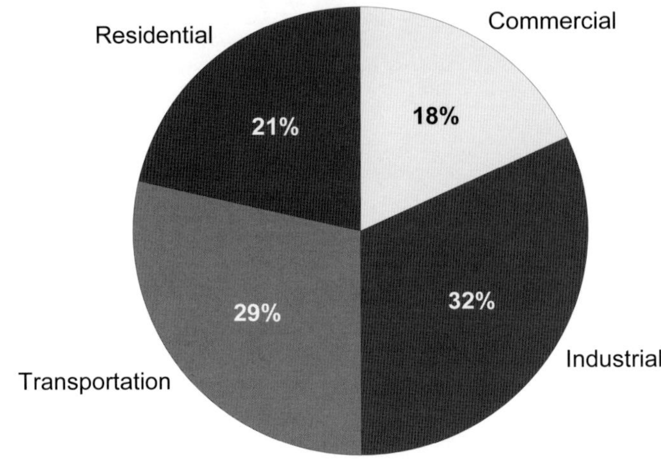

Primary and Total Consumption by Sector, 2007

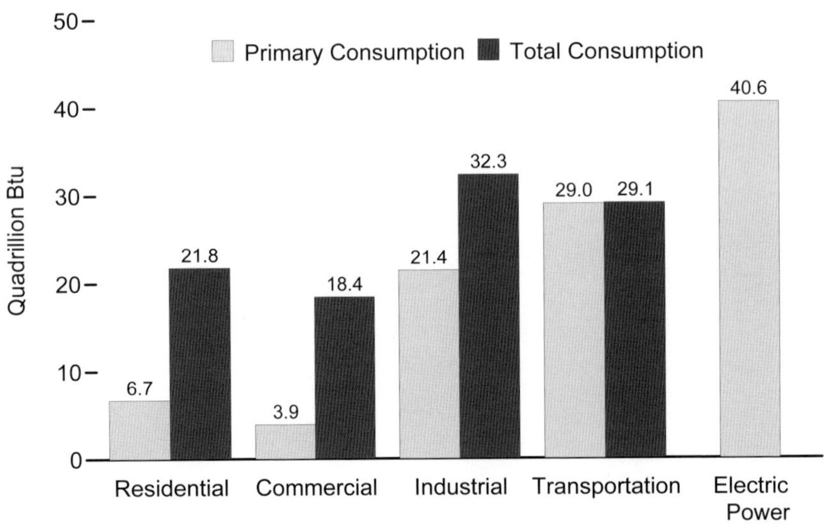

Electric Power Sector, 1949-2007

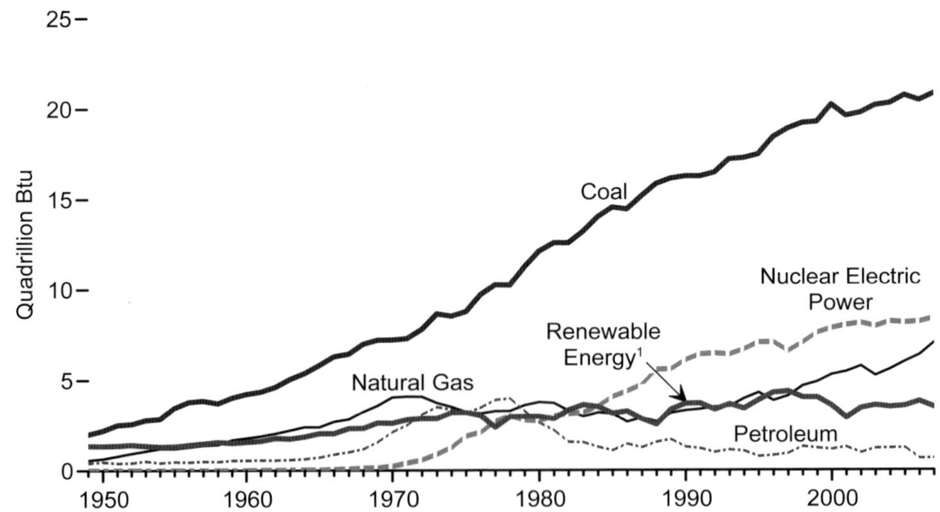

[1] Conventional hydroelectric power, geothermal, solar/photovoltaic, wind, and biomass.
Notes: • See Note 2, "Primary Energy Consumption," at end of Section 1. • Because vertical scales differ, graphs should not be compared.

Sources: Tables 2.1a and 2.1f.

Figure 2.1b Energy Consumption by End-Use Sector, 1949-2007

Residential, By Major Source

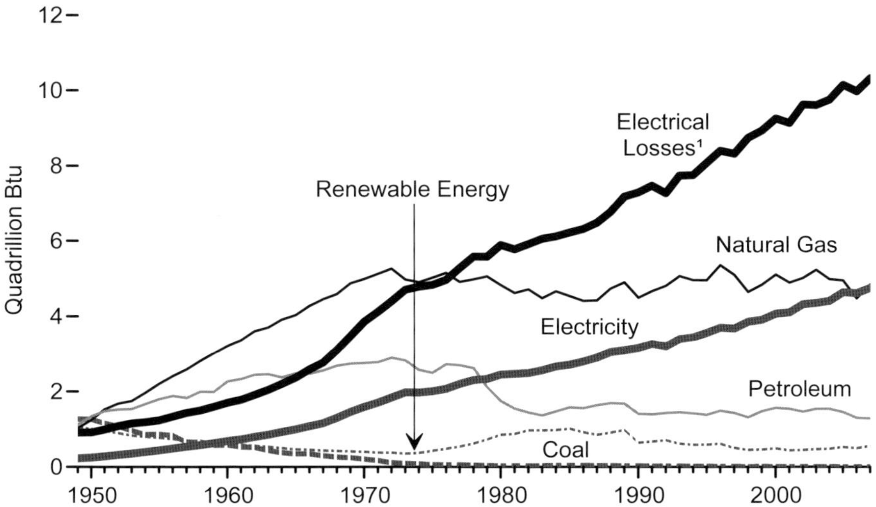

Commercial, By Major Source

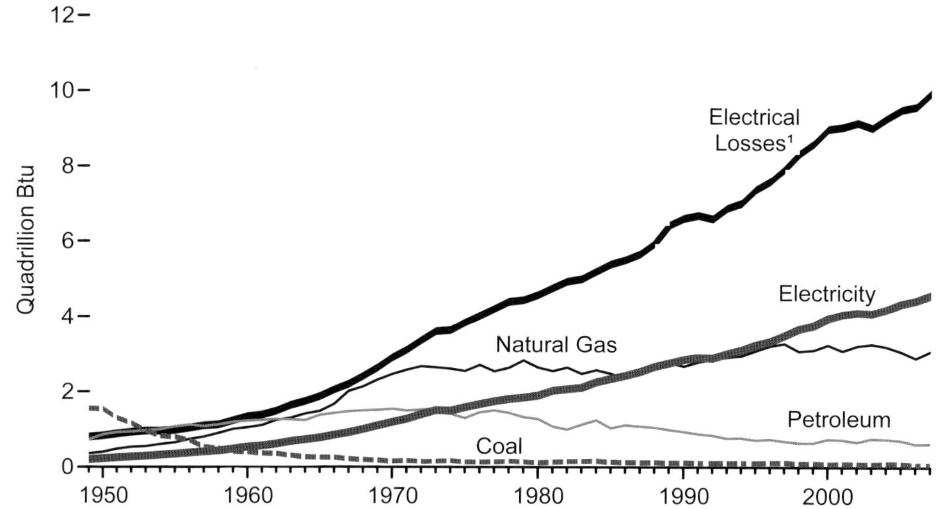

Industrial, By Major Source

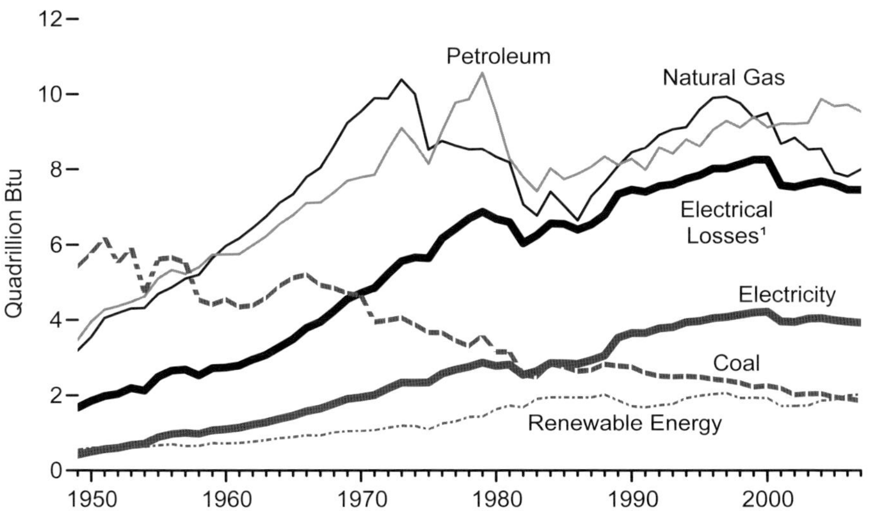

Transportation, By Major Source

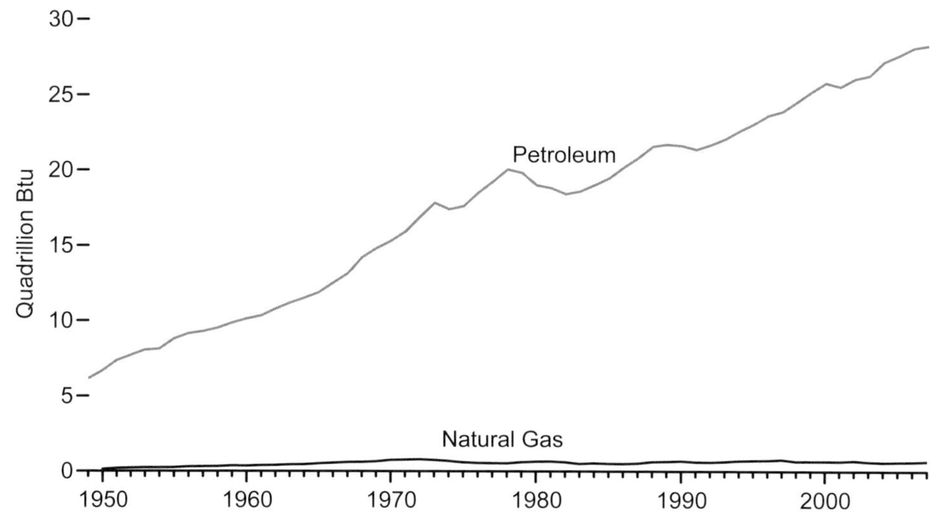

[1] Electrical system energy losses associated with the generation, transmission, and distribution of energy in the form of electricity.

Note: Because vertical scales differ, graphs should not be compared.
Sources: Tables 2.1b–2.1e.

Table 2.1a Energy Consumption by Sector, Selected Years, 1949-2007

(Trillion Btu)

Year	End-Use Sectors								Electric Power Sector [3,4]	Balancing Item [7]	Total [8]
	Residential		Commercial [1]		Industrial [2]		Transportation				
	Primary [5]	Total [6]	Primary [5]	Total [6]	Primary [5]	Total [6]	Primary [5]	Total [6]	Primary [5]		
1949	4,475	5,614	2,661	3,661	12,627	14,717	7,880	7,990	4,339	(s)	31,982
1950	4,848	6,007	2,824	3,883	13,881	16,233	8,384	8,493	4,679	(s)	34,616
1955	5,633	7,303	2,548	3,882	16,091	19,472	9,475	9,551	6,461	(s)	40,208
1960	6,689	9,078	2,702	4,589	16,977	20,823	10,560	10,597	8,158	(s)	45,087
1965	7,328	10,689	3,150	5,820	20,124	25,075	12,400	12,434	11,014	(s)	54,017
1970	8,353	13,798	4,196	8,307	22,975	29,641	16,061	16,098	16,259	(s)	67,844
1971	8,457	14,278	4,283	8,681	22,732	29,601	16,693	16,729	17,124	(s)	69,289
1972	8,655	14,891	4,369	9,145	23,532	30,953	17,681	17,716	18,466	(s)	72,704
1973	8,250	14,930	4,381	9,507	24,741	32,653	18,576	18,612	19,753	7	75,708
1974	7,928	14,683	4,221	9,363	23,816	31,819	18,086	18,119	19,933	7	73,991
1975	8,006	14,842	4,023	9,466	21,454	29,447	18,209	18,244	20,307	1	71,999
1976	8,408	15,441	4,333	10,035	22,685	31,430	19,065	19,099	21,513	8	76,012
1977	8,207	15,689	4,217	10,177	23,193	32,307	19,784	19,820	22,591	7	78,000
1978	8,272	16,156	4,269	10,481	23,276	32,733	20,580	20,615	23,587	2	79,986
1979	7,934	15,842	4,333	10,627	24,211	33,962	20,436	20,471	23,987	2	80,903
1980	7,453	15,787	4,074	10,563	22,610	32,077	19,658	19,696	24,327	-1	78,122
1981	7,058	15,295	3,805	10,602	21,338	30,756	19,476	19,513	24,488	3	76,168
1982	7,154	15,557	3,835	10,847	19,075	27,656	19,051	19,088	24,034	4	73,153
1983	6,841	15,457	3,806	10,923	18,577	27,480	19,132	19,175	24,679	3	73,038
1984	7,148	15,926	4,026	11,494	20,211	29,638	19,607	19,654	25,719	3	76,714
1985	7,161	16,088	3,695	11,444	19,466	28,875	20,041	20,087	26,132	-4	76,491
1986	6,922	16,029	3,657	11,604	19,096	28,331	20,740	20,789	26,338	3	76,756
1987	6,941	16,321	3,736	11,943	19,974	29,441	21,419	21,469	27,104	-3	79,173
1988	7,372	17,186	3,958	12,575	20,882	30,736	22,267	22,318	28,338	3	82,819
1989	7,586	17,858	4,004	13,203	20,895	31,395	22,425	22,479	[4]30,025	9	84,944
1990	6,570	17,015	3,858	13,333	21,206	31,894	22,366	22,420	30,660	-9	84,652
1991	6,758	17,490	3,906	13,512	20,852	31,485	22,065	22,118	31,025	1	84,607
1992	6,963	17,427	3,951	13,454	21,785	32,659	22,363	22,416	30,893	(s)	85,956
1993	7,156	18,289	3,934	13,836	21,783	32,719	22,716	22,770	32,025	-10	87,603
1994	6,991	18,181	3,979	14,111	22,420	33,606	23,312	23,367	32,563	-6	89,260
1995	6,946	18,578	4,063	14,698	22,746	34,045	23,793	23,849	33,621	3	91,173
1996	7,471	19,562	4,235	15,181	23,443	34,988	24,384	24,439	34,638	4	94,175
1997	7,040	19,026	4,257	15,694	23,721	35,288	24,697	24,752	35,045	6	94,765
1998	6,424	19,021	3,964	15,979	23,211	34,928	25,203	25,258	36,385	-3	95,183
1999	6,784	19,621	4,007	16,384	22,991	34,855	25,894	25,951	37,136	6	96,817
2000	7,169	20,488	4,227	17,176	22,871	34,757	26,492	26,552	38,214	2	98,975
2001	6,879	20,106	4,036	17,141	21,836	32,806	26,215	26,278	37,366	-6	96,326
2002	6,938	20,874	4,099	17,367	21,857	32,765	26,787	26,848	38,171	5	97,858
2003	7,252	21,208	4,239	17,351	21,576	32,650	26,928	R27,002	38,218	-3	98,209
2004	R7,019	R21,178	R4,180	R17,664	22,455	R33,609	27,820	27,899	38,876	(s)	100,351
2005	R6,941	R21,717	R4,014	R17,875	R21,467	R32,546	R28,280	R28,361	39,799	6	R100,506
2006	R6,276	R20,855	R3,716	R17,737	R21,495	R32,404	R28,781	R28,861	R39,589	R (s)	R99,856
2007P	6,688	21,753	3,898	18,430	21,435	32,321	29,012	29,096	40,567	(s)	101,600

[1] Commercial sector, including commercial combined-heat-and-power (CHP) and commercial electricity-only plants.

[2] Industrial sector, including industrial combined-heat-and-power (CHP) and industrial electricity-only plants.

[3] Electricity-only and combined-heat-and-power (CHP) plants within the NAICS 22 category whose primary business is to sell electricity, or electricity and heat, to the public.

[4] Through 1988, data are for electric utilities only; beginning in 1989, data are for electric utilities and independent power producers.

[5] See Note 2, "Primary Energy Consumption," at end of Section 1.

[6] Total energy consumption in the end-use sectors consists of primary energy consumption, electricity retail sales, and electrical system energy losses. See Note, "Electrical System Energy Losses," at end of section.

[7] A balancing item. The sum of primary consumption in the five energy-use sectors equals the sum of total consumption in the four end-use sectors. However, total energy consumption does not equal the sum of the sectoral components due to the use of sector-specific conversion factors for natural gas and coal.

[8] Primary energy consumption total. See Table 1.3.

R=Revised. P=Preliminary. (s)=Less than 0.5 trillion Btu.

Notes: • See Note 2, "Classification of Power Plants Into Energy-Use Sectors," at end of Section 8. • Totals may not equal sum of components due to independent rounding.

Web Page: For all data beginning in 1949, see http://www.eia.doe.gov/emeu/aer/consump.html.

Sources: Tables 1.3 and 2.1b-2.1f.

Table 2.1b Residential Sector Energy Consumption, Selected Years, 1949-2007

(Trillion Btu)

Year	Primary Consumption [1]										Electricity Retail Sales [4]	Electrical System Energy Losses [5]	Total
	Fossil Fuels				Renewable Energy [2]				Total Primary			
	Coal	Natural Gas [3]	Petroleum	Total	Geothermal	Solar/PV	Biomass	Total				
1949	1,272	1,027	1,121	3,420	NA	NA	1,055	1,055	4,475	228	911	5,614
1950	1,261	1,240	1,340	3,842	NA	NA	1,006	1,006	4,848	246	913	6,007
1955	867	2,198	1,792	4,858	NA	NA	775	775	5,633	438	1,232	7,303
1960	585	3,212	2,265	6,062	NA	NA	627	627	6,689	687	1,701	9,078
1965	352	4,028	2,481	6,860	NA	NA	468	468	7,328	993	2,368	10,689
1970	209	4,987	2,755	7,952	NA	NA	401	401	8,353	1,591	3,854	13,798
1971	172	5,126	2,777	8,075	NA	NA	382	382	8,457	1,704	4,116	14,278
1972	116	5,264	2,895	8,276	NA	NA	380	380	8,655	1,838	4,397	14,891
1973	94	4,977	2,825	7,896	NA	NA	354	354	8,250	1,976	4,703	14,930
1974	82	4,901	2,573	7,557	NA	NA	371	371	7,928	1,973	4,783	14,683
1975	63	5,023	2,495	7,580	NA	NA	425	425	8,006	2,007	4,829	14,842
1976	59	5,147	2,720	7,927	NA	NA	482	482	8,408	2,069	4,963	15,441
1977	57	4,913	2,695	7,666	NA	NA	542	542	8,207	2,202	5,280	15,689
1978	49	4,981	2,620	7,651	NA	NA	622	622	8,272	2,301	5,582	16,156
1979	37	5,055	2,114	7,206	NA	NA	728	728	7,934	2,330	5,578	15,842
1980	31	4,825	1,748	6,603	NA	NA	850	850	7,453	2,448	5,885	15,787
1981	30	4,614	1,543	6,188	NA	NA	870	870	7,058	2,464	5,773	15,295
1982	32	4,711	1,441	6,184	NA	NA	970	970	7,154	2,489	5,914	15,557
1983	31	4,478	1,362	5,871	NA	NA	970	970	6,841	2,562	6,054	15,457
1984	40	4,661	1,468	6,168	NA	NA	980	980	7,148	2,662	6,116	15,926
1985	39	4,534	1,578	6,151	NA	NA	1,010	1,010	7,161	2,709	6,219	16,088
1986	40	4,405	1,556	6,002	NA	NA	920	920	6,922	2,795	6,313	16,029
1987	37	4,420	1,634	6,091	NA	NA	850	850	6,941	2,902	6,479	16,321
1988	37	4,735	1,690	6,462	NA	NA	910	910	7,372	3,046	6,768	17,186
1989	31	4,899	1,679	6,608	5	53	920	978	7,586	3,090	7,182	17,858
1990	31	4,491	1,407	5,929	6	56	580	641	6,570	3,153	7,291	17,015
1991	25	4,667	1,392	6,085	6	58	610	674	6,758	3,260	7,472	17,490
1992	26	4,805	1,427	6,257	6	60	640	706	6,963	3,193	7,270	17,427
1993	26	5,063	1,448	6,537	7	62	550	618	7,156	3,394	7,739	18,289
1994	21	4,960	1,420	6,401	6	64	520	590	6,991	3,441	7,750	18,181
1995	17	4,954	1,383	6,355	7	65	520	591	6,946	3,557	8,075	18,578
1996	17	5,354	1,488	6,859	7	65	540	612	7,471	3,694	8,397	19,562
1997	16	5,093	1,428	6,537	8	65	430	503	7,040	3,671	8,315	19,026
1998	12	4,646	1,314	5,971	8	65	380	452	6,424	3,856	8,741	19,021
1999	14	4,835	1,473	6,322	8	64	390	462	6,784	3,906	8,931	19,621
2000	11	5,105	1,563	6,679	9	61	420	490	7,169	4,069	9,250	20,488
2001	12	4,889	1,539	6,440	9	60	370	439	6,879	4,100	9,127	20,106
2002	12	5,014	1,463	6,489	10	59	380	449	6,938	4,317	9,619	20,874
2003	12	5,230	1,539	6,781	13	58	400	471	7,252	4,353	9,603	21,208
2004	[R]11	4,986	1,539	[R]6,537	14	59	410	483	[R]7,019	4,408	9,750	[R]21,178
2005	8	[R]4,951	[R]1,455	[R]6,414	16	61	[R]450	[R]527	[R]6,941	4,638	10,139	[R]21,717
2006	[R]6	[R]4,476	[R]1,299	[R]5,780	18	[R]67	[R]410	[R]495	[R]6,276	[R]4,611	[R]9,968	[R]20,855
2007[P]	6	4,842	1,283	6,131	22	74	460	556	6,688	4,749	10,316	21,753

[1] See Note 2, "Primary Energy Consumption," at end of Section 1.

[2] Data are estimates. See Table 10.2a for notes on series components.

[3] Natural gas only; excludes the estimated portion of supplemental gaseous fuels. See Note 1, "Supplemental Gaseous Fuels," at end of Seciton 6.

[4] Electricity retail sales to ultimate customers reported by electric utilities and, beginning in 1996, other energy service providers.

[5] Total losses are calculated as the primary energy consumed by the electric power sector minus the energy content of electricity retail sales. Total losses are allocated to the end-use sectors in proportion to each sector's share of total electricity retail sales. See Note, "Electrical System Energy Losses," at end of section.

R=Revised. P=Preliminary. NA=Not available.

Note: Totals may not equal sum of components due to independent rounding.

Web Page: For all data beginning in 1949, see http://www.eia.doe.gov/emeu/aer/consump.html.

Sources: Tables 2.1f, 5.14a, 6.5, 7.3, 8.9, 10.2a, A4, A5, and A6.

Table 2.1c Commercial Sector Energy Consumption, Selected Years, 1949-2007

(Trillion Btu)

Year	Primary Consumption [1]										Electricity Retail Sales [6]	Electrical System Energy Losses [7]	Total
	Fossil Fuels				Renewable Energy [2]					Total Primary			
	Coal	Natural Gas [3]	Petroleum [4]	Total	Hydroelectric Power [5]	Geothermal	Biomass	Total					
1949	1,554	360	727	2,641	NA	NA	20	20		2,661	200	800	3,661
1950	1,542	401	862	2,805	NA	NA	19	19		2,824	225	834	3,883
1955	801	651	1,081	2,533	NA	NA	15	15		2,548	350	984	3,882
1960	407	1,056	1,228	2,690	NA	NA	12	12		2,702	543	1,344	4,589
1965	265	1,490	1,386	3,142	NA	NA	9	9		3,150	789	1,880	5,820
1970	165	2,473	1,551	4,189	NA	NA	8	8		4,196	1,201	2,910	8,307
1971	179	2,587	1,510	4,276	NA	NA	7	7		4,283	1,288	3,111	8,681
1972	153	2,678	1,530	4,362	NA	NA	7	7		4,369	1,408	3,368	9,145
1973	160	2,649	1,565	4,374	NA	NA	7	7		4,381	1,517	3,609	9,507
1974	175	2,617	1,423	4,214	NA	NA	7	7		4,221	1,501	3,640	9,363
1975	147	2,558	1,310	4,015	NA	NA	8	8		4,023	1,598	3,845	9,466
1976	144	2,718	1,461	4,323	NA	NA	9	9		4,333	1,678	4,025	10,035
1977	148	2,548	1,511	4,207	NA	NA	10	10		4,217	1,754	4,206	10,177
1978	165	2,643	1,450	4,257	NA	NA	12	12		4,269	1,813	4,398	10,481
1979	149	2,836	1,334	4,319	NA	NA	14	14		4,333	1,854	4,439	10,627
1980	115	2,651	1,287	4,053	NA	NA	21	21		4,074	1,906	4,582	10,563
1981	137	2,557	1,090	3,784	NA	NA	21	21		3,805	2,033	4,763	10,602
1982	155	2,650	1,008	3,813	NA	NA	22	22		3,835	2,077	4,935	10,847
1983	162	2,486	1,136	3,784	NA	NA	22	22		3,806	2,116	5,001	10,923
1984	169	2,582	1,252	4,004	NA	NA	22	22		4,026	2,264	5,203	11,494
1985	137	2,488	1,045	3,670	NA	NA	24	24		3,695	2,351	5,398	11,444
1986	135	2,367	1,126	3,629	NA	NA	27	27		3,657	2,439	5,508	11,604
1987	125	2,489	1,093	3,707	NA	NA	30	30		3,736	2,539	5,669	11,943
1988	131	2,731	1,063	3,925	NA	NA	33	33		3,958	2,675	5,943	12,575
1989	115	2,785	1,002	3,902	1	2	99	102		4,004	2,767	6,431	13,203
1990	124	2,682	953	3,760	1	3	94	98		3,858	2,860	6,615	13,333
1991	116	2,795	895	3,806	1	3	95	100		3,906	2,918	6,689	13,512
1992	117	2,871	854	3,842	1	3	105	109		3,951	2,900	6,603	13,454
1993	117	2,923	780	3,820	1	3	109	114		3,934	3,019	6,883	13,836
1994	118	2,962	787	3,867	1	4	106	112		3,979	3,116	7,017	14,111
1995	117	3,096	732	3,945	1	4	113	118		4,063	3,252	7,382	14,698
1996	122	3,226	751	4,099	1	5	129	135		4,235	3,344	7,603	15,181
1997	129	3,285	704	4,118	1	6	131	138		4,257	3,503	7,935	15,694
1998	93	3,083	661	3,837	1	7	118	127		3,964	3,678	8,338	15,979
1999	103	3,115	661	3,879	1	7	121	129		4,007	3,766	8,610	16,384
2000	92	3,252	756	4,099	1	8	119	128		4,227	3,956	8,993	17,176
2001	97	3,097	741	3,935	1	8	92	101		4,036	4,062	9,043	17,141
2002	90	3,225	[R]680	3,995	(s)	9	95	104		4,099	4,110	9,158	17,367
2003	82	3,274	770	4,126	1	11	101	113		4,239	4,090	9,023	17,351
2004	[R]103	3,204	755	[R]4,062	1	12	105	118		[R]4,180	4,198	9,286	[R]17,664
2005	[R]97	[R]3,076	[R]721	[R]3,894	1	14	105	119		[R]4,014	4,351	9,511	[R]17,875
2006	[R]66	[R]2,904	[R]629	[R]3,599	1	14	[R]102	[R]117		[R]3,716	[R]4,435	[R]9,586	[R]17,737
2007[P]	65	3,083	631	3,780	1	14	104	119		3,898	4,581	9,951	18,430

[1] See Note 2, "Primary Energy Consumption," at end of Section 1.
[2] Most data are estimates. See Table 10.2a for notes on series components and estimation.
[3] Natural gas only; excludes the estimated portion of supplemental gaseous fuels. See Note 1, "Supplemental Gaseous Fuels," at end of Section 6.
[4] Does not include the fuel ethanol portion of motor gasoline—fuel ethanol is included in "Biomass."
[5] Conventional hydroelectric power.
[6] Electricity retail sales to ultimate customers reported by electric utilities and, beginning in 1996, other energy service providers.
[7] Total losses are calculated as the primary energy consumed by the electric power sector minus the energy content of electricity retail sales. Total losses are allocated to the end-use sectors in proportion to each sector's share of total electricity retail sales. See Note, "Electrical System Energy Losses," at end of section.

R=Revised. P=Preliminary. NA=Not available. (s)=Less than 0.5 trillion Btu.

Notes: • The commercial sector includes commercial combined-heat-and-power (CHP) and commercial electricity-only plants. See Note 2, "Classification of Power Plants Into Energy-Use Sectors," at end of Section 8. • Totals may not equal sum of components due to independent rounding.

Web Page: For all data beginning in 1949, see http://www.eia.doe.gov/emeu/aer/consump.html.

Sources: Tables 2.1f, 5.14a, 6.5, 7.3, 8.9, 10.2a, A4, A5, and A6.

Table 2.1d Industrial Sector Energy Consumption, Selected Years, 1949-2007

(Trillion Btu)

Year	Primary Consumption [1]										Electricity Retail Sales [6]	Electrical System Energy Losses [7]	Total
	Fossil Fuels					Renewable Energy [2]				Total Primary			
	Coal	Coal Coke Net Imports	Natural Gas [3]	Petroleum [4]	Total	Hydroelectric Power [5]	Geothermal	Biomass	Total				
1949	5,433	-7	3,188	3,468	12,083	76	NA	468	544	12,627	418	1,672	14,717
1950	5,781	1	3,546	3,951	13,279	69	NA	532	602	13,881	500	1,852	16,233
1955	5,620	-10	4,701	5,111	15,421	38	NA	631	669	16,091	887	2,495	19,472
1960	4,543	-6	5,973	5,747	16,258	39	NA	680	719	16,977	1,107	2,739	20,823
1965	5,127	-18	7,339	6,789	19,236	33	NA	855	888	20,124	1,463	3,488	25,075
1970	4,656	-58	9,536	7,787	21,922	34	NA	1,019	1,053	22,975	1,948	4,719	29,641
1971	3,944	-33	9,892	7,856	21,659	34	NA	1,040	1,074	22,732	2,011	4,857	29,601
1972	3,993	-26	9,884	8,534	22,385	34	NA	1,113	1,147	23,532	2,187	5,233	30,953
1973	4,057	-7	10,388	9,104	23,541	35	NA	1,165	1,200	24,741	2,341	5,571	32,653
1974	3,870	56	10,004	8,694	22,624	33	NA	1,159	1,192	23,816	2,337	5,666	31,819
1975	3,667	14	8,532	8,146	20,359	32	NA	1,063	1,096	21,454	2,346	5,647	29,447
1976	3,661	(s)	8,762	9,010	21,432	33	NA	1,220	1,253	22,685	2,573	6,171	31,430
1977	3,454	15	8,635	9,774	21,879	33	NA	1,281	1,314	23,193	2,682	6,432	32,307
1978	3,314	125	8,539	9,867	21,844	32	NA	1,400	1,432	23,276	2,761	6,696	32,733
1979	3,593	63	8,549	10,568	22,773	34	NA	1,405	1,439	24,211	2,873	6,878	33,962
1980	3,155	-35	8,333	9,525	20,977	33	NA	1,600	1,633	22,610	2,781	6,686	32,077
1981	3,157	-16	8,185	8,285	19,610	33	NA	1,695	1,728	21,338	2,817	6,600	30,756
1982	2,552	-22	7,068	7,795	17,393	33	NA	1,649	1,682	19,075	2,542	6,039	27,656
1983	2,490	-16	6,776	7,420	16,670	33	NA	1,874	1,907	18,577	2,648	6,256	27,480
1984	2,842	-11	7,405	8,025	18,260	33	NA	1,917	1,950	20,211	2,859	6,568	29,638
1985	2,760	-13	7,032	7,738	17,516	33	NA	1,917	1,950	19,466	2,855	6,554	28,875
1986	2,641	-17	6,646	7,880	17,150	33	NA	1,914	1,947	19,096	2,834	6,401	28,331
1987	2,673	9	7,283	8,065	18,029	33	NA	1,912	1,945	19,974	2,928	6,538	29,441
1988	2,828	40	7,655	8,339	18,861	33	NA	1,988	2,020	20,882	3,059	6,795	30,736
1989	2,787	30	8,088	8,120	19,025	28	2	1,840	1,870	20,895	3,158	7,342	31,395
1990	2,756	5	8,451	8,278	19,490	31	2	1,683	1,716	21,206	3,226	7,461	31,894
1991	2,601	10	8,572	7,987	19,169	30	2	1,651	1,683	20,852	3,230	7,403	31,485
1992	2,515	35	8,918	8,581	20,048	31	2	1,704	1,737	21,785	3,319	7,556	32,659
1993	2,496	27	9,070	8,417	20,011	30	2	1,740	1,772	21,783	3,334	7,602	32,719
1994	2,510	58	9,126	8,799	20,493	62	3	1,862	1,927	22,420	3,439	7,746	33,606
1995	2,488	61	9,592	8,613	20,754	55	3	1,935	1,992	22,746	3,455	7,844	34,045
1996	2,434	23	9,901	9,052	21,410	61	3	1,970	2,033	23,443	3,527	8,018	34,988
1997	2,395	46	9,933	9,289	21,663	58	3	1,997	2,058	23,721	3,542	8,024	35,288
1998	2,335	67	9,763	9,114	21,280	55	3	1,873	1,931	23,211	3,587	8,131	34,928
1999	2,227	58	9,375	9,395	21,054	49	4	1,883	1,936	22,991	3,611	8,254	34,855
2000	2,256	65	9,500	9,119	20,941	42	4	1,884	1,930	22,871	3,631	8,255	34,757
2001	2,192	29	8,676	9,217	20,115	32	5	1,684	1,721	21,836	3,400	7,570	32,806
2002	2,019	61	8,845	9,209	20,135	39	5	1,679	1,723	21,857	3,379	7,528	32,765
2003	2,041	51	8,521	9,232	19,845	43	3	1,684	1,731	21,576	3,454	7,620	32,650
2004	2,047	138	8,544	R9,865	20,594	33	4	1,824	1,861	22,455	3,473	7,682	R33,609
2005	1,954	44	R7,911	R9,673	R19,583	32	4	R1,848	R1,884	R21,467	3,477	7,602	R32,546
2006	R1,914	61	R7,810	R9,711	R19,495	R29	4	R1,966	R1,999	R21,495	R3,451	R7,459	R32,404
2007P	1,861	25	7,999	9,523	19,409	23	5	1,998	2,025	21,435	3,432	7,454	32,321

[1] See Note 2, "Primary Energy Consumption," at end of Section 1.
[2] Most data are estimates. See Table 10.2b for notes on series components and estimation.
[3] Natural gas only; excludes the estimated portion of supplemental gaseous fuels. See Note 1, "Supplemental Gaseous Fuels," at end of Section 6.
[4] Does not include the fuel ethanol portion of motor gasoline—fuel ethanol is included in "Biomass."
[5] Conventional hydroelectric power.
[6] Electricity retail sales to ultimate customers reported by electric utilities and, beginning in 1996, other energy service providers.
[7] Total losses are calculated as the primary energy consumed by the electric power sector minus the energy content of electricity retail sales. Total losses are allocated to the end-use sectors in proportion to each sector's share of total electricity retail sales. See Note, "Electrical System Energy Losses," at end of section.

R=Revised. P=Preliminary. NA=Not available. (s)=Less than +0.5 trillion Btu and greater than -0.5 trillion Btu.

Notes: • The industrial sector includes industrial combined-heat-and-power (CHP) and industrial electricity-only plants. See Note 2, "Classification of Power Plants Into Energy-Use Sectors," at end of Section 8. • Totals may not equal sum of components due to independent rounding.

Web Page: For all data beginning in 1949, see http://www.eia.doe.gov/emeu/aer/consump.html.

Sources: Tables 2.1f, 5.14b, 6.5, 7.3, 7.7, 8.9, 10.2b, A4, A5, and A6.

Table 2.1e Transportation Sector Energy Consumption, Selected Years, 1949-2007

(Trillion Btu)

Year	Primary Consumption [1]						Electricity Retail Sales [5]	Electrical System Energy Losses [6]	Total
	Fossil Fuels				Renewable Energy [2]	Total Primary			
	Coal	Natural Gas [3]	Petroleum [4]	Total	Biomass				
1949	1,727	NA	6,152	7,880	NA	7,880	22	88	7,990
1950	1,564	130	6,690	8,384	NA	8,384	23	86	8,493
1955	421	254	8,800	9,475	NA	9,475	20	56	9,551
1960	75	359	10,126	10,560	NA	10,560	10	26	10,597
1965	16	517	11,868	12,400	NA	12,400	10	24	12,434
1970	7	745	15,310	16,061	NA	16,061	11	26	16,098
1971	5	766	15,923	16,693	NA	16,693	10	25	16,729
1972	4	787	16,891	17,681	NA	17,681	10	25	17,716
1973	3	743	17,831	18,576	NA	18,576	11	25	18,612
1974	2	685	17,399	18,086	NA	18,086	10	24	18,119
1975	1	595	17,614	18,209	NA	18,209	10	24	18,244
1976	(s)	559	18,506	19,065	NA	19,065	10	24	19,099
1977	(s)	543	19,241	19,784	NA	19,784	10	25	19,820
1978	([7])	539	20,041	20,580	NA	20,580	10	24	20,615
1979	([7])	612	19,825	20,436	NA	20,436	10	24	20,471
1980	([7])	650	19,008	19,658	NA	19,658	11	27	19,696
1981	([7])	658	18,811	19,469	7	19,476	11	25	19,513
1982	([7])	612	18,420	19,032	19	19,051	11	26	19,088
1983	([7])	505	18,593	19,098	34	19,132	13	30	19,175
1984	([7])	545	19,020	19,565	42	19,607	14	33	19,654
1985	([7])	519	19,471	19,990	51	20,041	14	32	20,087
1986	([7])	499	20,182	20,681	59	20,740	15	34	20,789
1987	([7])	535	20,816	21,352	67	21,419	16	35	21,469
1988	([7])	632	21,567	22,198	68	22,267	16	35	22,318
1989	([7])	649	21,706	22,355	69	22,425	16	38	22,479
1990	([7])	680	21,625	22,305	62	22,366	16	37	22,420
1991	([7])	620	21,373	21,994	72	22,065	16	37	22,118
1992	([7])	608	21,674	22,282	81	22,363	16	37	22,416
1993	([7])	645	21,976	22,621	96	22,716	16	37	22,770
1994	([7])	709	22,496	23,205	107	23,312	17	38	23,367
1995	([7])	724	22,954	23,678	115	23,793	17	39	23,849
1996	([7])	737	23,565	24,302	82	24,384	17	38	24,439
1997	([7])	780	23,813	24,593	104	24,697	17	38	24,752
1998	([7])	666	24,422	25,088	115	25,203	17	38	25,258
1999	([7])	675	25,098	25,774	120	25,894	17	40	25,951
2000	([7])	672	25,682	26,354	138	26,492	18	42	26,552
2001	([7])	658	R25,413	26,071	145	26,215	20	43	26,278
2002	([7])	702	25,913	26,615	172	26,787	19	42	26,848
2003	([7])	630	26,063	26,693	235	26,928	23	51	R27,002
2004	([7])	603	26,922	27,525	R296	27,820	25	55	27,899
2005	([7])	625	R27,309	R27,934	R346	R28,280	26	56	R28,361
2006	([7])	R626	R27,672	R28,298	R483	R28,781	R25	R54	R28,861
2007P	([7])	667	27,719	28,386	626	29,012	26	57	29,096

[1] See Note 2, "Primary Energy Consumption," at end of Section 1.

[2] Data are estimates. See Table 10.2b for notes on series components.

[3] Natural gas only; does not include supplemental gaseous fuels—see Note 1, "Supplemental Gaseous Fuels," at end of Section 6. Data are for natural gas consumed in the operation of pipelines (primarily in compressors) and small amounts consumed as vehicle fuel—see Table 6.5.

[4] Does not include the fuel ethanol portion of motor gasoline—fuel ethanol is included in "Biomass."

[5] Electricity retail sales to ultimate customers reported by electric utilities and, beginning in 1996, other energy service providers.

[6] Total losses are calculated as the primary energy consumed by the electric power sector minus the energy content of electricity retail sales. Total losses are allocated to the end-use sectors in proportion to each sector's share of total electricity retail sales. See Note, "Electrical System Energy Losses," at end of section.

[7] Beginning in 1978, the small amounts of coal consumed for transportation are reported as industrial sector consumption.

R=Revised. P=Preliminary. NA=Not available. (s)=Less than 0.5 trillion Btu.

Note: Totals may not equal sum of components due to independent rounding.

Web Page: For all data beginning in 1949, see http://www.eia.doe.gov/emeu/aer/consump.html.

Sources: Tables 2.1f, 5.14c, 6.5, 7.3, 8.9, 10.2b, A4, A5, and A6.

Table 2.1f Electric Power Sector Energy Consumption, Selected Years, 1949-2007

(Trillion Btu)

Year	Primary Consumption [1]												
	Fossil Fuels				Nuclear Electric Power	Renewable Energy [2]						Electricity Net Imports [6]	Total Primary
	Coal	Natural Gas [3]	Petroleum [4]	Total		Hydroelectric Power [5]	Geothermal	Solar/PV	Wind	Biomass	Total		
1949	1,995	569	415	2,979	0	1,349	NA	NA	NA	6	1,355	5	4,339
1950	2,199	651	472	3,322	0	1,346	NA	NA	NA	5	1,351	6	4,679
1955	3,458	1,194	471	5,123	0	1,322	NA	NA	NA	3	1,325	14	6,461
1960	4,228	1,785	553	6,565	6	1,569	1	NA	NA	2	1,571	15	8,158
1965	5,821	2,395	722	8,938	43	2,026	4	NA	NA	3	2,033	(s)	11,014
1970	7,227	4,054	2,117	13,399	239	2,600	11	NA	NA	4	2,615	7	16,259
1971	7,299	4,099	2,495	13,893	413	2,790	12	NA	NA	3	2,806	12	17,124
1972	7,811	4,084	3,097	14,992	584	2,829	31	NA	NA	3	2,864	26	18,466
1973	8,658	3,748	3,515	15,921	910	2,827	43	NA	NA	3	2,873	49	19,753
1974	8,534	3,519	3,365	15,418	1,272	3,143	53	NA	NA	3	3,199	43	19,933
1975	8,786	3,240	3,166	15,191	1,900	3,122	70	NA	NA	2	3,194	21	20,307
1976	9,720	3,152	3,477	16,349	2,111	2,943	78	NA	NA	3	3,024	29	21,513
1977	10,262	3,284	3,901	17,446	2,702	2,301	77	NA	NA	5	2,383	59	22,591
1978	10,238	3,297	3,987	17,522	3,024	2,905	64	NA	NA	3	2,973	67	23,587
1979	11,260	3,613	3,283	18,156	2,776	2,897	84	NA	NA	5	2,986	69	23,987
1980	12,123	3,778	2,634	18,534	2,739	2,867	110	NA	NA	4	2,982	71	24,327
1981	12,583	3,730	2,202	18,516	3,008	2,725	123	NA	NA	4	2,852	113	24,488
1982	12,582	3,312	1,568	17,462	3,131	3,233	105	NA	NA	3	3,341	100	24,034
1983	13,213	2,972	1,544	17,729	3,203	3,494	129	NA	(s)	4	3,627	121	24,679
1984	14,019	3,199	1,286	18,504	3,553	3,353	165	(s)	(s)	9	3,527	135	25,719
1985	14,542	3,135	1,090	18,767	4,076	2,937	198	(s)	(s)	14	3,150	140	26,132
1986	14,444	2,670	1,452	18,566	4,380	3,038	219	(s)	(s)	12	3,270	122	26,338
1987	15,173	2,916	1,257	19,346	4,754	2,602	229	(s)	(s)	15	2,846	158	27,104
1988	15,850	2,693	1,563	20,106	5,587	2,302	217	(s)	(s)	17	2,536	108	28,338
1989[7]	16,137	3,173	1,703	21,013	5,602	2,808	308	3	22	232	3,372	37	30,025
1990	16,261	3,309	1,289	20,859	6,104	3,014	326	4	29	317	3,689	8	30,660
1991	16,250	3,377	1,198	20,825	6,422	2,985	335	5	31	354	3,710	67	31,025
1992	16,466	3,512	991	20,968	6,479	2,586	338	4	30	402	3,360	87	30,893
1993	17,196	3,538	1,124	21,857	6,410	2,861	351	5	31	415	3,662	95	32,025
1994	17,261	3,977	1,059	22,297	6,694	2,620	325	5	36	434	3,420	153	32,563
1995	17,466	4,302	755	22,523	7,075	3,149	280	5	33	422	3,889	134	33,621
1996	18,429	3,862	817	23,109	7,087	3,528	300	5	33	438	4,305	137	34,638
1997	18,905	4,126	927	23,957	6,597	3,581	309	5	34	446	4,375	116	35,045
1998	19,216	4,675	1,306	25,197	7,068	3,241	311	5	31	444	4,032	88	36,385
1999	19,279	4,902	1,211	25,393	7,610	3,218	312	5	46	453	4,034	99	37,136
2000	20,220	5,293	1,144	26,658	7,862	2,768	296	5	57	453	3,579	115	38,214
2001	19,614	5,458	1,277	26,348	8,033	2,209	289	6	70	337	2,910	75	37,366
2002	19,783	5,767	961	26,511	8,143	2,650	305	6	105	380	3,445	72	38,171
2003	20,185	5,246	1,205	26,636	7,959	2,781	303	5	115	397	3,601	22	38,218
2004	20,305	5,595	1,212	27,112	8,222	2,656	311	6	142	388	3,503	39	38,876
2005	20,737	6,015	1,235	[R]27,986	8,160	2,670	309	6	178	406	3,568	84	39,799
2006	[R]20,462	[R]6,375	[R]648	[R]27,485	[R]8,214	[R]2,839	[R]306	5	[R]264	[R]412	[R]3,827	[R]63	[R]39,589
2007[P]	20,835	7,046	660	28,542	8,415	2,440	312	6	319	427	3,503	107	40,567

[1] See Note 2, "Primary Energy Consumption," at end of Section 1.
[2] See Table 10.2c for notes on series components.
[3] Natural gas only; excludes the estimated portion of supplemental gaseous fuels. See Note 1, "Supplemental Gaseous Fuels," at end of Section 6.
[4] See Table 5.14c for series components.
[5] Conventional hydroelectric power.
[6] Net imports equal imports minus exports.
[7] Through 1988, data are for electric utilities only. Beginning in 1989, data are for electric utilities and independent power producers.

R=Revised. P=Preliminary. NA=Not available. (s)=Less than 0.5 trillion Btu.

Notes: • Data are for fuels consumed to produce electricity and useful thermal output. • The electric power sector comprises electricity-only and combined-heat-and-power (CHP) plants within the NAICS 22 category whose primary business is to sell electricity, or electricity and heat, to the public. • See Note 3, "Electricity Imports and Exports," at end of Section 8. • Totals may not equal sum of components due to independent rounding.

Web Page: For all data beginning in 1949, see http://www.eia.doe.gov/emeu/aer/consump.html.

Sources: Tables 5.14c, 6.5, 7.3, 8.1, 8.2b, 10.2c, A4, A5, and A6.

Figure 2.2 Manufacturing Energy Consumption for All Purposes, 2002

By Energy Source

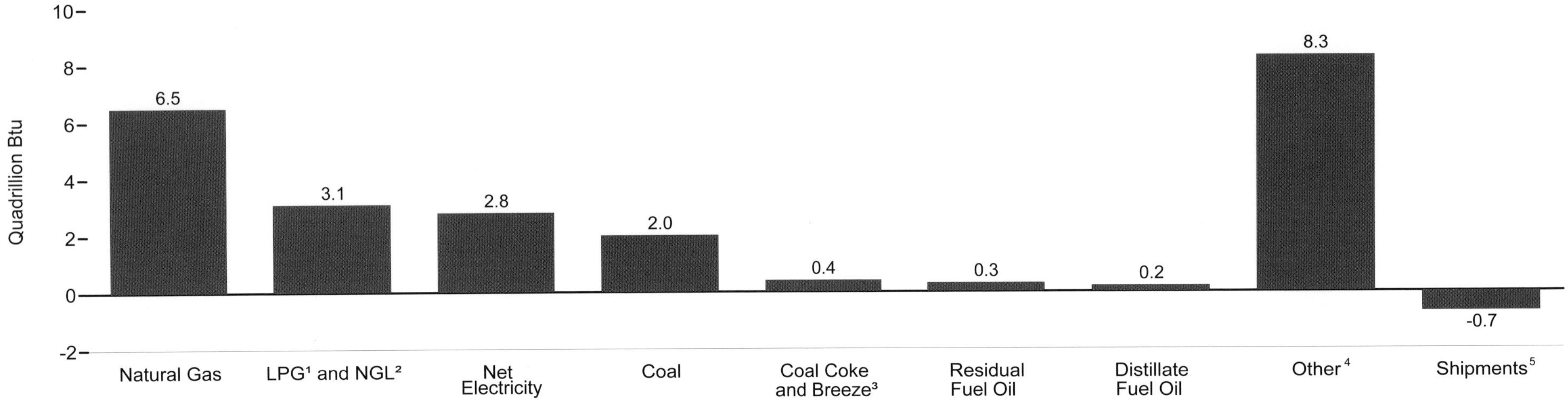

By North American Industry Classification System (NAICS) Code[6]

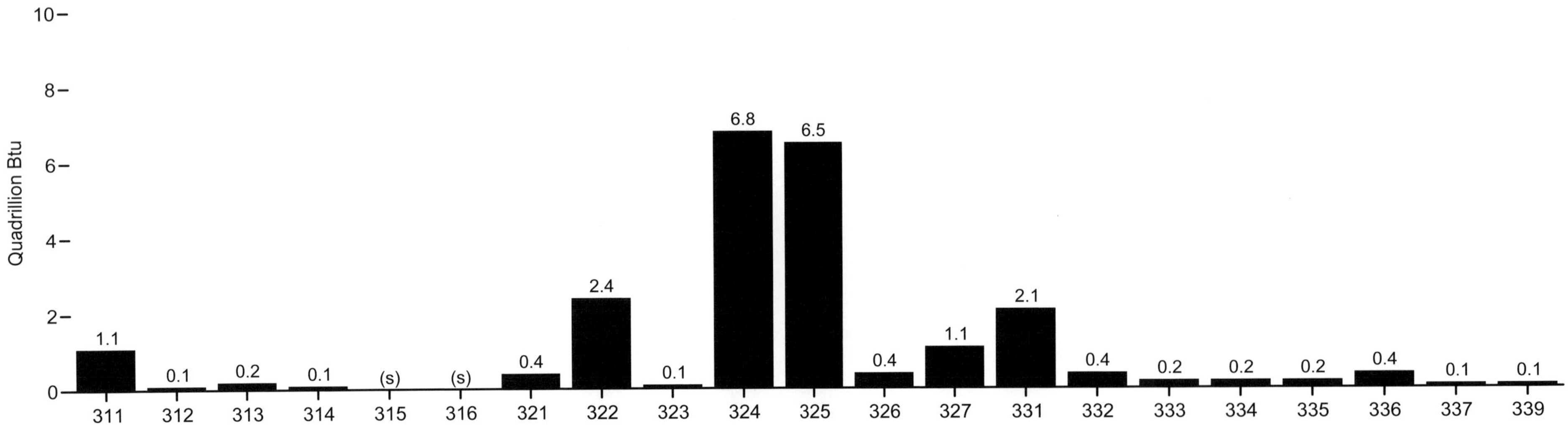

[1] Liquefied petroleum gases.
[2] Natural gas liquids.
[3] See "Breeze" in Glossary.
[4] Includes all other types of energy that respondents indicated were consumed or allocated.
[5] Energy sources produced onsite from the use of other energy sources but sold or transferred to another entity.

[6] See Table 2.2 for Manufacturing Group titles of industries that correspond to the 3-digit NAICS codes.
(s)=Less than 0.05 quadrillion Btu.
Source: Table 2.2.

Table 2.2 Manufacturing Energy Consumption for All Purposes, 2002

(Trillion Btu)

NAICS [1] Code	Manufacturing Group	Coal	Coal Coke and Breeze [2]	Natural Gas	Distillate Fuel Oil	LPG [3] and NGL [4]	Residual Fuel Oil	Net Electricity [5]	Other [6]	Shipments of Energy Sources [7]	Total [8]
311	Food	184	1	582	19	5	13	230	89	-0	1,123
312	Beverage and Tobacco Products	17	0	46	2	1	2	26	11	-0	105
313	Textile Mills	22	0	75	2	2	4	86	15	-0	207
314	Textile Product Mills	Q	0	29	Q	1	2	17	Q	-0	60
315	Apparel	0	0	16	1	(s)	(s)	12	(s)	-0	30
316	Leather and Allied Products	0	0	4	(s)	(s)	(s)	2	(s)	-0	7
321	Wood Products	1	0	57	13	5	1	72	228	-0	377
322	Paper	236	4	504	13	6	100	223	1,276	-0	2,363
323	Printing and Related Support	0	0	46	(s)	1	(s)	50	1	-0	98
324	Petroleum and Coal Products	Q	2	878	19	24	25	127	5,520	-83	6,799
325	Chemicals	344	6	2,307	14	3,001	87	522	687	-504	6,465
326	Plastics and Rubber Products	Q	0	128	2	6	7	181	5	-0	351
327	Nonmetallic Mineral Products	309	11	422	34	3	3	142	136	-0	1,059
331	Primary Metals	515	355	704	15	3	1	493	178	-143	2,120
332	Fabricated Metal Products	1	Q	210	6	3	Q	161	3	-0	388
333	Machinery	1	0	82	3	3	(s)	84	4	-0	177
334	Computer and Electronic Products	(s)	0	65	1	(s)	1	131	3	-0	201
335	Electrical Equipment, Appliances, and Components	(s)	(s)	53	1	1	(s)	47	70	-0	172
336	Transportation Equipment	8	Q	203	4	4	6	172	30	-0	429
337	Furniture and Related Products	1	0	25	1	1	(s)	24	11	-0	64
339	Miscellaneous	0	0	32	1	1	(s)	35	2	-0	71
—	Total Manufacturing	1,958	385	6,468	152	3,070	255	2,840	8,271	-730	22,666

[1] North American Industry Classification System (NAICS).

[2] See "Breeze" in Glossary.

[3] Liquefied petroleum gases.

[4] Natural gas liquids.

[5] "Net Electricity" is the sum of purchases, transfers in, and onsite generation from noncombustible renewable energy sources, minus quantities sold and transferred out; it excludes onsite generation from combustible fuels.

[6] Includes all other types of energy that respondents indicated were consumed or allocated, such as asphalt and road oil, lubricants, naphtha less than 401 degrees Fahrenheit, other oils greater than or equal to 401 degrees Fahrenheit, special naphthas, waxes, and miscellaneous nonfuel products, which are nonfuel products assigned to the petroleum refining industry group (NAICS Code 324110).

[7] Energy sources produced onsite from the use of other energy sources but sold or transferred to another entity. Note that shipments of energy sources are subtracted from consumption.

[8] The sum of coal, coal coke and breeze, natural gas, distillate fuel oil, liquefied petroleum gases, natural gas liquids, residual fuel oil, net electricity, and other, minus shipments of energy sources.

(s)=Less than 0.5 trillion Btu. Q=Data withheld because the relative standard error was greater than 50 percent.

Notes: • Data are estimates for the first use of energy for heat and power and as feedstocks or raw material inputs. "First use" is the consumption of energy that was originally produced offsite or was produced onsite from input materials not classified as energy. • Totals may not equal sum of components due to independent rounding.

Web Page: For related information, see http://www.eia.doe.gov/emeu/mecs.

Source: Energy Information Administration, Form EIA-846, "2002 Manufacturing Energy Consumption Survey."

Figure 2.3 Manufacturing Energy Consumption for Heat, Power, and Electricity Generation, 2002

By Selected End Use[1]

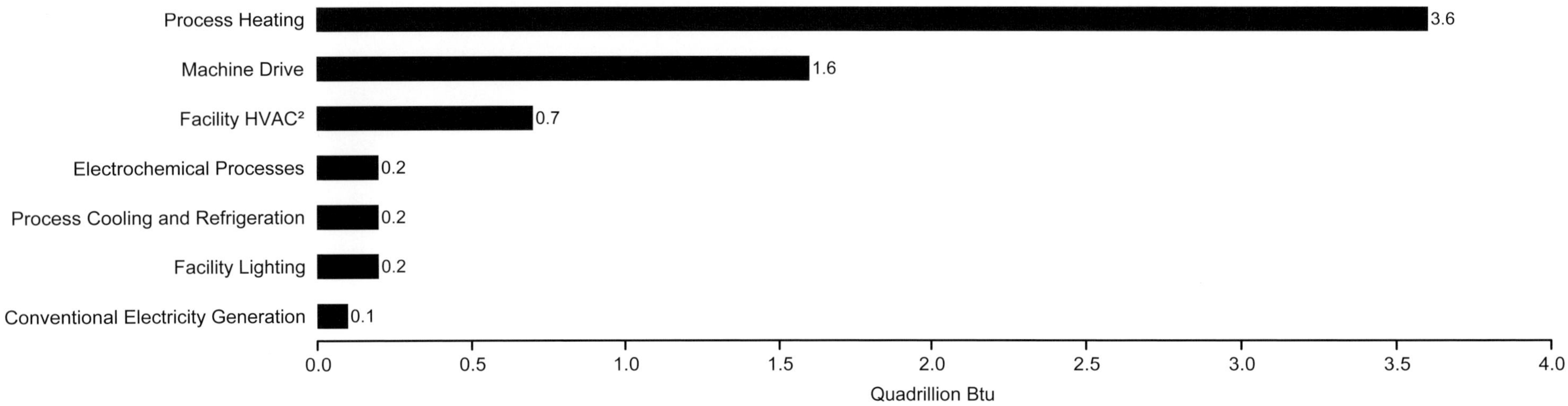

By Energy Source

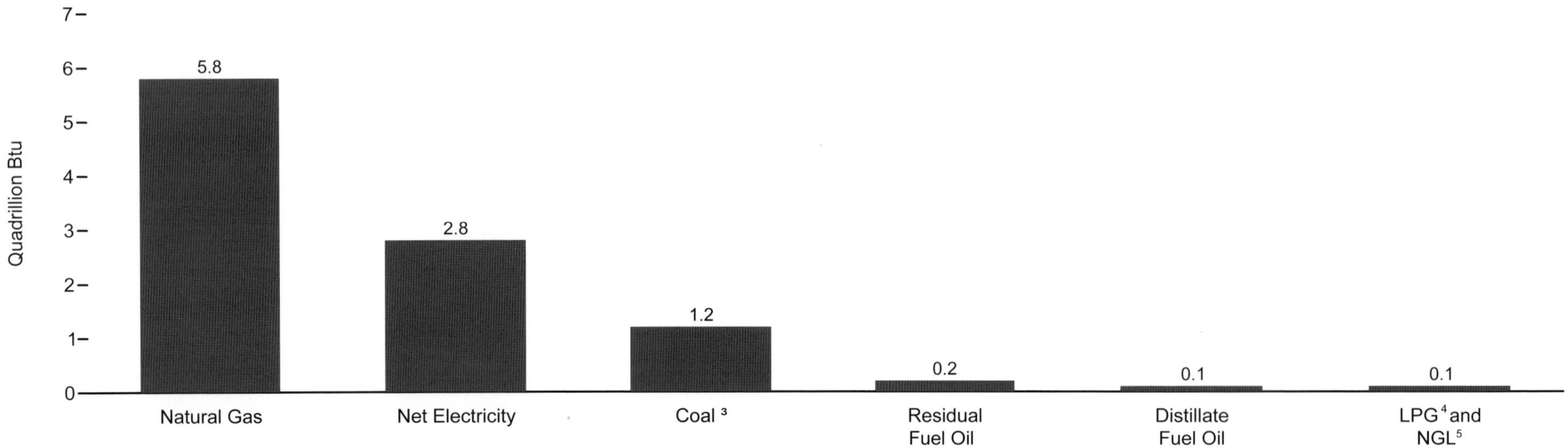

[1] Excludes inputs of unallocated energy sources (6,006 trillion Btu).
[2] Heating, ventilation, and air conditioning. Excludes steam and hot water.
[3] Excludes coal coke and breeze.
[4] Liquefied petroleum gases.
[5] Natural gas liquids.
Source: Table 2.3.

Table 2.3 Manufacturing Energy Consumption for Heat, Power, and Electricity Generation by End Use, 2002

End-Use Category	Net Electricity [1]	Residual Fuel Oil	Distillate Fuel Oil	LPG [2] and NGL [3]	Natural Gas	Coal [4]	Total [5]
	Million Kilowatthours	Million Barrels			Billion Cubic Feet	Million Short Tons	
Indirect End Use (Boiler Fuel)	**3,540**	**20**	**6**	**2**	**2,105**	**35**	– –
Conventional Boiler Use	2,496	12	4	2	1,271	11	– –
CHP [6] and/or Cogeneration Process	1,043	8	2	(s)	834	23	– –
Direct End Use							
All Process Uses	**650,100**	**10**	**7**	**16**	**2,878**	**17**	– –
Process Heating	100,541	9	4	15	2,670	17	– –
Process Cooling and Refrigeration	56,723	(s)	(s)	(s)	44	(s)	– –
Machine Drive	417,998	(s)	3	1	106	(s)	– –
Electrochemical Processes	71,045	– –	– –	– –	– –	– –	– –
Other Process Uses	3,793	(s)	(s)	(s)	58	(s)	– –
All Non-Process Uses	**150,530**	**1**	**9**	**6**	**500**	**1**	– –
Facility Heating, Ventilation, and Air Conditioning [7]	76,840	1	1	1	406	(s)	– –
Facility Lighting	57,460	– –	– –	– –	– –	– –	– –
Other Facility Support	14,087	(s)	(s)	(s)	29	(s)	– –
Onsite Transportation	1,212	– –	6	5	2	– –	– –
Conventional Electricity Generation	– –	(s)	Q	(s)	54	1	– –
Other Non-Process Use	931	(s)	Q	(s)	10	0	– –
End Use Not Reported	**28,087**	**3**	**2**	**2**	**157**	**(s)**	– –
Total	**832,257**	**33**	**24**	**26**	**5,641**	**53**	– –
	Trillion Btu						
Indirect End Use (Boiler Fuel)	**12**	**127**	**35**	**8**	**2,162**	**776**	**3,120**
Conventional Boiler Use	9	76	25	8	1,306	255	1,679
CHP [6] and/or Cogeneration Process	4	51	10	(s)	857	521	1,443
Direct End Use							
All Process Uses	**2,218**	**60**	**43**	**64**	**2,956**	**381**	**5,722**
Process Heating	343	58	24	60	2,742	368	3,595
Process Cooling and Refrigeration	194	(s)	2	(s)	45	(s)	241
Machine Drive	1,426	2	16	4	109	5	1,562
Electrochemical Processes	242	– –	– –	– –	– –	– –	242
Other Process Uses	13	(s)	1	(s)	60	7	81
All Non-Process Uses	**514**	**4**	**50**	**24**	**513**	**19**	**1,124**
Facility Heating, Ventilation, and Air Conditioning [7]	262	3	5	5	417	5	697
Facility Lighting	196	– –	– –	– –	– –	– –	196
Other Facility Support	48	(s)	1	(s)	30	(s)	79
Onsite Transportation	4	– –	35	18	2	– –	59
Conventional Electricity Generation	– –	1	Q	(s)	55	14	70
Other Non-Process Use	3	(s)	Q	(s)	10	0	13
End Use Not Reported	**96**	**17**	**12**	**6**	**162**	**6**	**299**
Total	**2,840**	**208**	**141**	**103**	**5,794**	**1,182**	**10,268**

[1] "Net Electricity" is the sum of purchases, transfers in, and onsite generation from noncombustible renewable energy sources, minus quantities sold and transferred out; it excludes onsite generation from combustible fuels.
[2] Liquefied petroleum gases.
[3] Natural gas liquids.
[4] Excludes coal coke and breeze.
[5] Total of listed energy sources. Excludes inputs of unallocated energy sources (6,006 trillion Btu).
[6] Combined-heat-and-power plants.
[7] Excludes steam and hot water.
– – = Not applicable. (s)=Estimate less than 0.5. Q=Withheld because relative standard error is greater than 50 percent.

Notes: • Data are estimates for the total consumption of energy for the production of heat, power, and electricity generation, regardless of where the energy was produced. Specifically, the estimates include the quantities of energy that were originally produced offsite and purchased by or transferred to the establishment, plus those that were produced onsite from other energy or input materials not classified as energy, or were extracted from captive (onsite) mines or wells. • Allocations to end uses are made on the basis of reasonable approximations by respondents. • Totals may not equal sum of components due to independent rounding, the presence of estimates that round to zero, and the presence of estimates that are withheld because the relative standard error is greater than 50 percent.

Web Page: For related information, see http://www.eia.doe.gov/emeu/mecs.

Source: Energy Information Administration, Form EIA-846, "2002 Manufacturing Energy Consumption Survey."

Figure 2.4 Household Energy Consumption

Consumption by All Households, Selected Years, 1978-2001

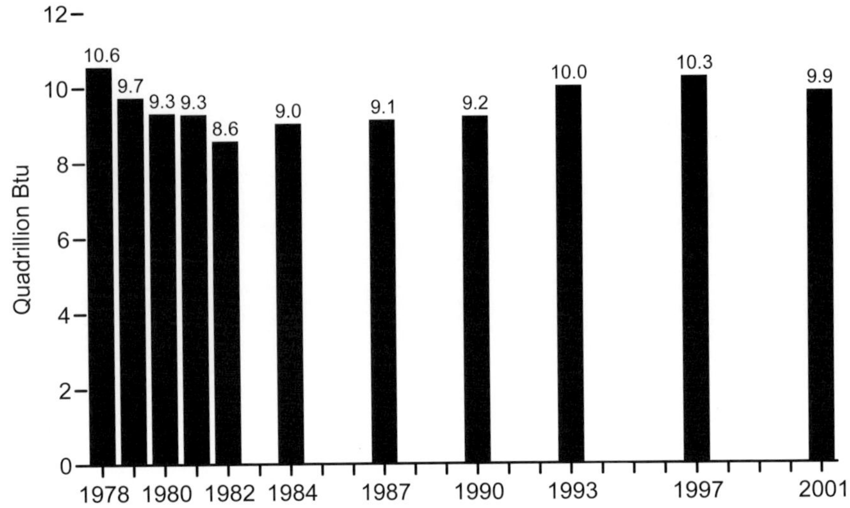

Consumption by All Households, by Census Region, 2001

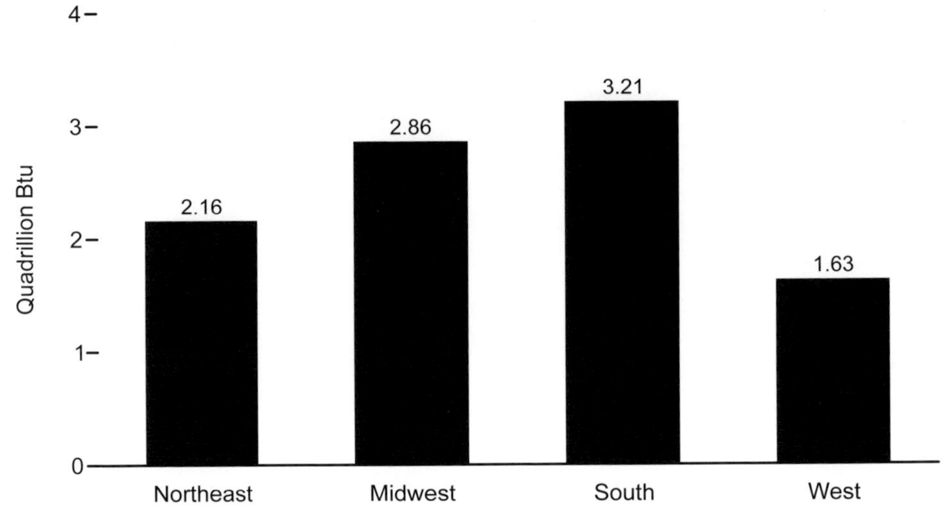

Consumption per Household, Selected Years, 1978-2001

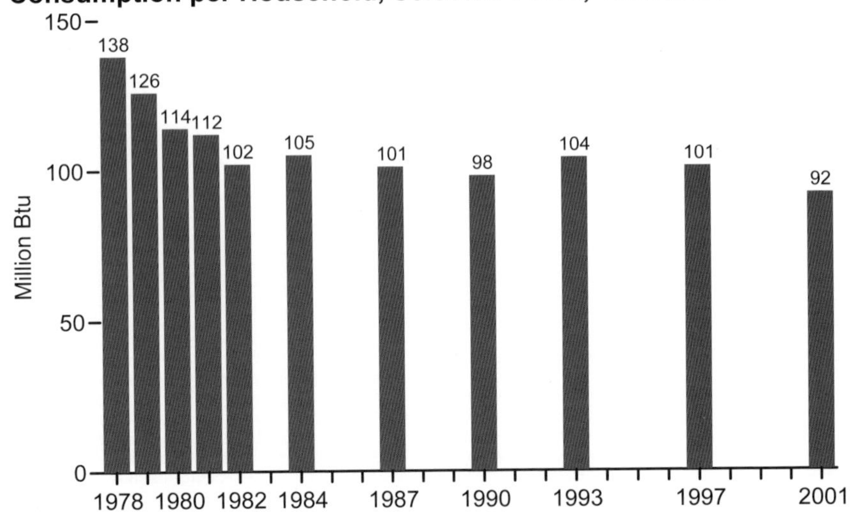

Consumption per Household, by Census Region, 2001

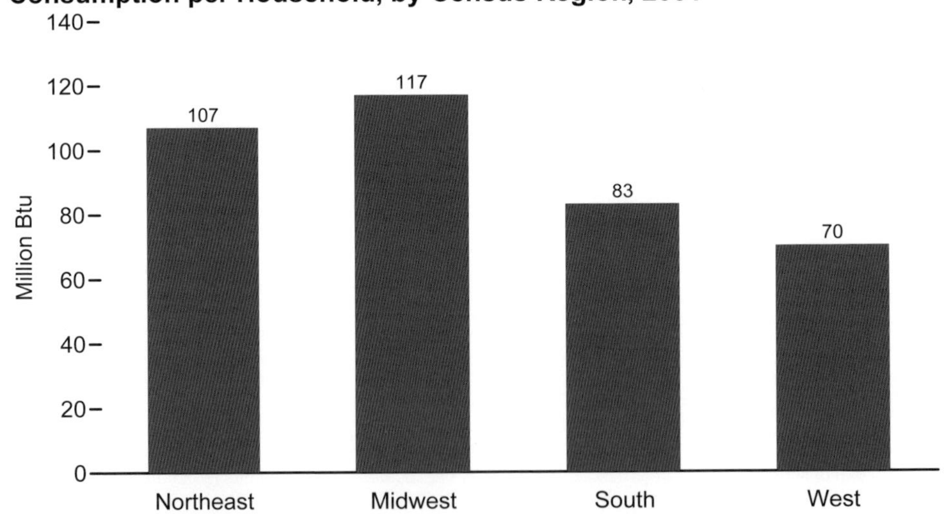

Notes: • Data include natural gas, electricity, distillate fuel oil, kerosene, and liquefied petroleum gases; data do not include wood. • For years not shown, there are no data available. Data for 1978-1984 are for April of the year shown through March of following year; data for 1987 forward are for the calendar year. • Because vertical scales differ, graphs should not be compared. • See Appendix C for Census regions.

Source: Table 2.4.

Table 2.4 Household Energy Consumption by Census Region, Selected Years, 1978-2001
(Quadrillion Btu, Except as Noted)

Census Region [1]	1978	1979	1980	1981	1982	1984	1987	1990	1993	1997	2001
Northeast (total does not include wood)	**2.89**	**2.50**	**2.44**	**2.36**	**2.19**	**2.29**	**2.37**	**2.30**	**2.38**	**2.38**	**2.16**
Natural Gas	1.14	1.05	.94	1.01	.96	.93	1.03	1.03	1.11	1.03	.98
Electricity [2]	.39	.39	.41	.40	.37	.41	.44	.47	.47	.49	.53
Distillate Fuel Oil and Kerosene	1.32	1.03	1.07	.93	.83	.93	.87	.78	.78	.84	.60
Liquefied Petroleum Gases	.03	.03	.03	.03	.02	.03	.02	.02	.03	.03	.05
Wood [3]	NA	NA	.26	.27	.24	.21	.17	.12	.14	.14	.10
Consumption per Household (million Btu) [3]	166	145	138	132	122	125	124	120	122	121	107
Midwest (total does not include wood)	**3.70**	**3.48**	**2.96**	**3.09**	**2.61**	**2.80**	**2.73**	**2.81**	**3.13**	**3.22**	**2.86**
Natural Gas	2.53	2.48	2.05	2.22	1.78	1.99	1.83	1.88	2.07	2.20	1.84
Electricity [2]	.60	.59	.60	.56	.56	.55	.61	.66	.74	.75	.81
Distillate Fuel Oil and Kerosene	.46	.31	.17	.19	.16	.13	.16	.13	.13	.11	.06
Liquefied Petroleum Gases	.12	.10	.15	.13	.11	.13	.13	.13	.19	.17	.15
Wood [3]	NA	NA	.25	.25	.27	.27	.25	.17	.11	.08	.09
Consumption per Household (million Btu) [3]	180	168	141	146	122	129	123	122	134	134	117
South (total does not include wood)	**2.43**	**2.30**	**2.57**	**2.41**	**2.45**	**2.50**	**2.61**	**2.60**	**2.95**	**3.01**	**3.21**
Natural Gas	.96	.91	1.12	1.15	1.14	1.15	1.09	1.03	1.18	1.13	1.13
Electricity [2]	1.00	.97	1.06	1.01	1.01	1.06	1.22	1.36	1.51	1.67	1.89
Distillate Fuel Oil and Kerosene	.32	.28	.25	.14	.18	.16	.17	.11	.13	.10	.08
Liquefied Petroleum Gases	.15	.14	.14	.12	.12	.12	.12	.10	.13	.12	.12
Wood [3]	NA	NA	.23	.21	.33	.33	.26	.17	.17	.11	.09
Consumption per Household (million Btu) [3]	99	92	95	87	87	85	84	81	88	84	83
West (total does not include wood)	**1.54**	**1.47**	**1.34**	**1.42**	**1.33**	**1.45**	**1.42**	**1.51**	**1.55**	**1.63**	**1.63**
Natural Gas	.95	.88	.86	.90	.85	.91	.88	.92	.91	.93	.90
Electricity [2]	.48	.47	.41	.46	.41	.47	.48	.54	.56	.64	.66
Distillate Fuel Oil and Kerosene	.09	.09	.04	.03	.03	.04	.02	.02	.03	.03	.02
Liquefied Petroleum Gases	.03	.04	.04	.04	.04	.03	.05	.03	.04	.04	.06
Wood [3]	NA	NA	.11	.13	.13	.17	.17	.12	.12	.10	.10
Consumption per Household (million Btu) [3]	110	100	84	87	81	85	78	78	76	75	70
United States (total does not include wood)	**10.56**	**9.74**	**9.32**	**9.29**	**8.58**	**9.04**	**9.13**	**9.22**	**10.01**	**10.25**	**9.86**
Natural Gas	5.58	5.31	4.97	5.27	4.74	4.98	4.83	4.86	5.27	5.28	4.84
Electricity [2]	2.47	2.42	2.48	2.42	2.35	2.48	2.76	3.03	3.28	3.54	3.89
Distillate Fuel Oil and Kerosene	2.19	1.71	1.52	1.28	1.20	1.26	1.22	1.04	1.07	1.07	.75
Liquefied Petroleum Gases	.33	.31	.35	.31	.29	.31	.32	.28	.38	.36	.38
Wood [3]	NA	NA	.85	.87	.97	.98	.85	.58	.55	.43	.37
Consumption per Household (million Btu) [3]	138	126	114	112	102	105	101	98	104	101	92

[1] See Appendix C for Census regions.
[2] Retail electricity. One kilowatthour = 3,412 Btu.
[3] Wood is not included in the region and U.S. totals, or in the consumption-per-household data.
NA=Not available.
Notes: • Data are estimates, and are for major energy sources only. • For years not shown, there are no data available. • Data for 1978-1984 are for April of year shown through March of following year; data for 1987 forward are for the calendar year. • Totals may not equal sum of components due to independent rounding.
Web Page: For related information, see http://www.eia.doe.gov/emeu/recs.
Sources: • 1978 and 1979—Energy Information Administration (EIA), Form EIA-84, "Residential Energy Consumption Survey." • 1980 forward—EIA, Form EIA-457, "Residential Energy Consumption Survey."

Figure 2.5 Household Energy Consumption and Expenditures

Consumption by Energy Source, 2001

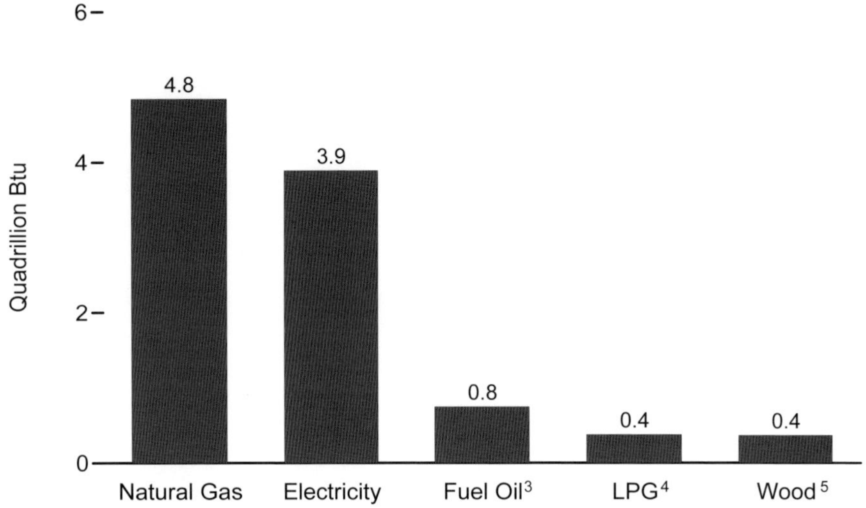

Expenditures[1], Selected Years, 1978-2001

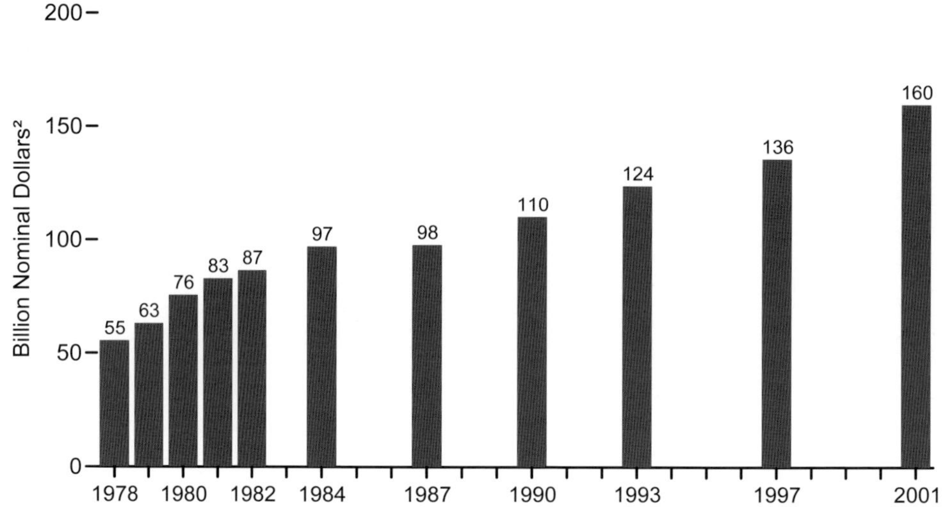

Consumption[1] by End Use, 2001

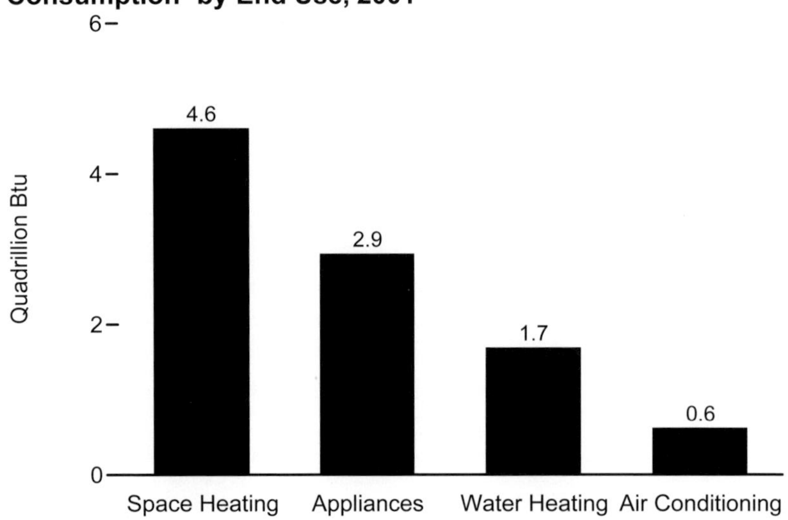

Consumption[1] for Space Heating, 2001

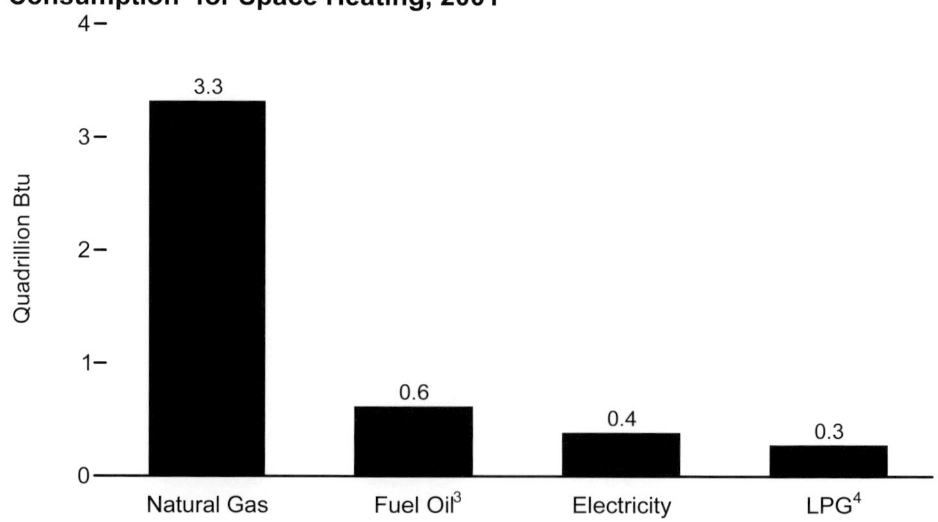

[1] Does not include wood, which is used for both space heating and ambiance.
[2] See "Nominal Dollars" in Glossary.
[3] Distillate fuel oil and kerosene.
[4] Liquefied petroleum gases.
[5] Used for both space heating and ambiance.

Notes: • For years not shown, there are no data available. • Because vertical scales differ, graphs should not be compared.

Source: Table 2.5.

Table 2.5 Household Energy Consumption and Expenditures by End Use and Energy Source, Selected Years, 1978-2001

Year	Space Heating [1]				Air Conditioning [2]	Water Heating				Appliances [3,4]			Total				
	Natural Gas	Elec-tricity [5]	Fuel Oil [6]	LPG [7]	Electricity [5]	Natural Gas	Elec-tricity [5]	Fuel Oil [6]	LPG [7]	Natural Gas	Elec-tricity [5]	LPG [7]	Natural Gas [2]	Elec-tricity [5]	Fuel Oil [4,6]	LPG [7]	Wood [8]
	Consumption (quadrillion Btu)																
1978	4.26	0.40	2.05	0.23	0.32	1.04	0.29	0.14	0.06	0.28	1.45	0.03	5.58	2.47	2.19	0.33	NA
1979	NA	NA	NA	NA	NA	NA	NA	NA	NA	NA	NA	NA	5.31	2.42	1.71	.31	NA
1980	3.41	.27	1.30	.23	.36	1.15	.30	.22	.07	.36	1.54	.05	4.97	2.48	1.52	.35	.85
1981	3.69	.26	1.06	.21	.34	1.13	.30	.22	.06	.43	1.52	.05	5.27	2.42	1.28	.31	.87
1982	3.14	.25	1.04	.19	.31	1.15	.28	.15	.06	.43	1.50	.05	4.74	2.35	1.20	.29	.97
1984	3.51	.25	1.11	.21	.32	1.10	.32	.15	.06	.35	1.59	.04	4.98	2.48	1.26	.31	.98
1987	3.38	.28	1.05	.22	.44	1.10	.31	.17	.06	.34	1.72	.04	4.83	2.76	1.22	.32	.85
1990	3.37	.30	.93	.19	.48	1.16	.34	.11	.06	.33	1.91	.03	4.86	3.03	1.04	.28	.58
1993	3.67	.41	.95	.30	.46	1.31	.34	.12	.05	.29	2.08	.03	5.27	3.28	1.07	.38	.55
1997	3.61	.40	.91	.26	.42	1.29	.39	.16	.08	.37	2.33	.02	5.28	3.54	1.07	.36	.43
2001	3.32	.39	.62	.28	.62	1.15	.36	.13	.05	.37	2.52	.05	4.84	3.89	.75	.38	.37
	Expenditures (billion nominal dollars [9])																
1978	11.49	3.53	8.06	1.05	4.12	2.88	3.14	0.56	0.36	0.93	19.10	0.25	15.30	29.89	8.62	1.66	NA
1979	NA	NA	NA	NA	NA	NA	NA	NA	NA	NA	NA	NA	17.84	32.56	10.73	2.06	NA
1980	13.22	3.78	10.48	1.78	5.84	4.51	4.45	1.76	.57	1.91	26.74	.44	19.77	40.81	12.24	2.80	NA
1981	16.62	3.93	9.44	1.78	6.23	5.13	4.94	1.94	.51	2.17	29.70	.52	24.03	44.80	11.29	2.81	NA
1982	17.74	4.21	8.80	1.69	6.23	6.51	5.00	1.28	.54	2.58	31.29	.52	26.96	46.74	10.07	2.75	NA
1984	20.66	4.62	8.51	2.00	7.06	6.63	6.44	1.09	.58	2.31	36.36	.54	29.78	54.48	9.60	3.12	NA
1987	18.05	5.53	6.25	1.85	9.77	6.02	6.45	.94	.50	2.02	39.83	.46	26.15	61.58	7.21	2.81	NA
1990	18.59	6.16	7.42	2.01	11.23	6.59	7.21	.83	.65	2.03	46.95	.48	27.26	71.54	8.25	3.14	NA
1993	21.95	8.66	6.24	2.81	11.31	8.08	7.58	.74	.58	1.98	53.52	.42	32.04	81.08	6.98	3.81	NA
1997	24.11	8.56	6.57	2.79	10.20	8.84	8.99	1.04	.89	2.86	60.57	.36	35.81	88.33	7.61	4.04	NA
2001	31.84	8.98	5.66	4.04	15.94	11.31	8.47	1.15	.69	3.83	66.94	.86	46.98	100.34	6.83	5.60	NA

[1] Wood used for space heating is included in "Total Wood."
[2] A small amount of natural gas used for air conditioning is included in "Total Natural Gas."
[3] Includes refrigerators.
[4] A small amount of distillate fuel oil and kerosene used for appliances is included in "Fuel Oil" under "Total."
[5] Retail electricity. One kilowatthour=3,412 Btu.
[6] Distillate fuel oil and kerosene.
[7] Liquefied petroleum gases.
[8] Wood used for both space heating and ambience.
[9] See "Nominal Dollars" in Glossary.
NA=Not available.
Notes: • Data are estimates. • For years not shown, there are no data available. • Totals may not equal sum of components due to independent rounding.
Web Page: For related information, see http://www.eia.doe.gov/emeu/recs.
Sources: • 1978 and 1979—Energy Information Administration (EIA), Form EIA-84, "Residential Energy Consumption Survey." • 1980 forward—EIA, Form EIA-457, "Residential Energy Consumption Survey."

Figure 2.6 Households With Selected Appliances and Types of Main Heating Fuel

Households With Selected Electric Appliances, 1980 and 2005

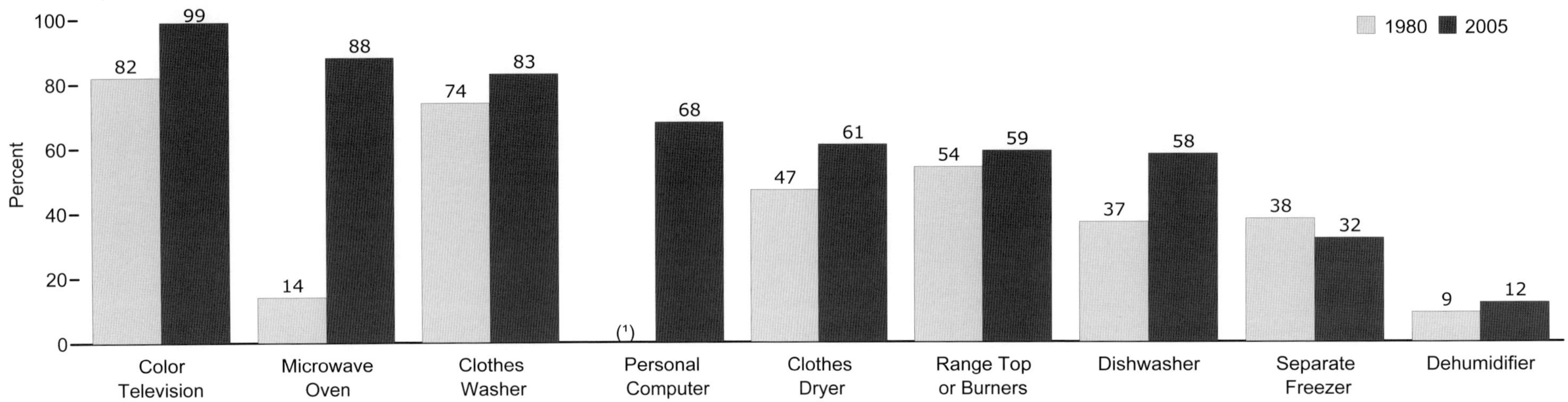

Households With Other Selected Appliances, 1980 and 2005

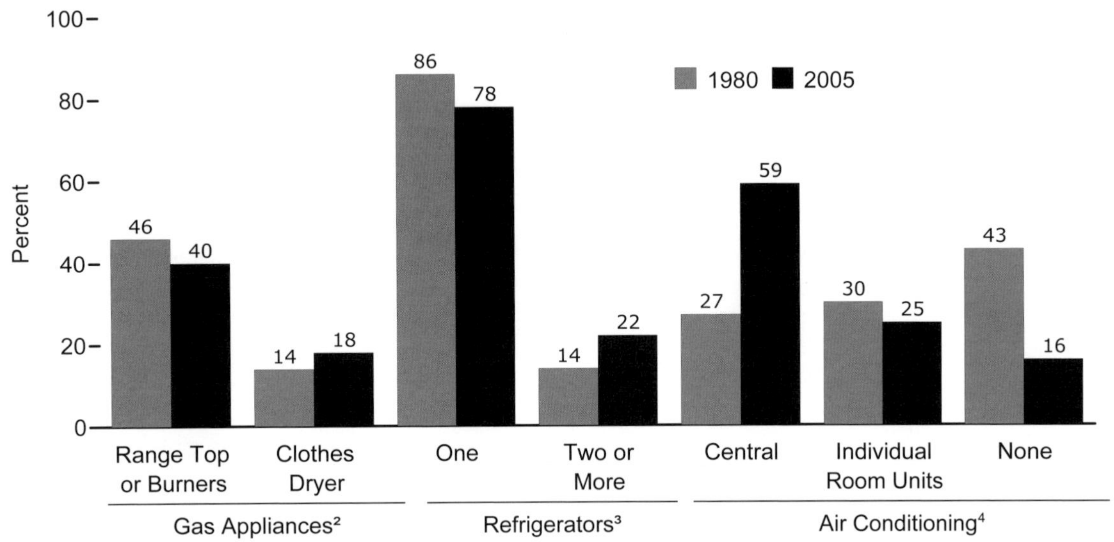

Type of Main Heating Fuel, 2005

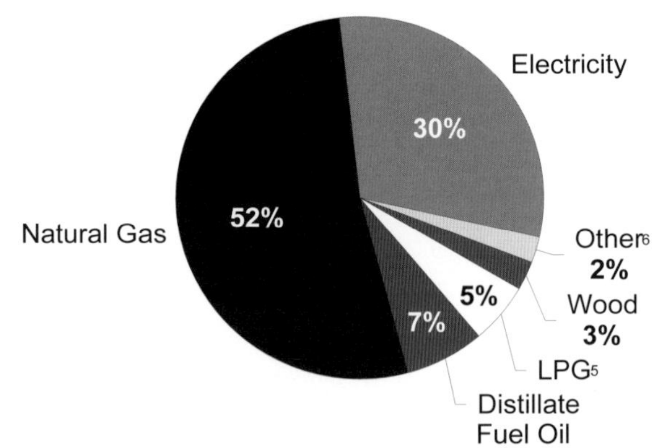

[1] Not collected in 1980.
[2] Natural gas or liquefied petroleum gases.
[3] Fewer than 0.5 percent of the households do not have a refrigerator.
[4] Households with both central and individual room units are counted only under "Central."
[5] Liquefied petroleum gases.
[6] Kerosene, district steam, coal, solar, other, and no heat.
Source: Table 2.6.

Table 2.6 Households With Selected Appliances and Types of Main Heating Fuel, Selected Years, 1978-2005

Appliance	Year												Change 1980-2005
	1978	1979	1980	1981	1982	1984	1987	1990	1993	1997	2001	2005	
Total Households (millions)	77	78	82	83	84	86	91	94	97	101	107	111	29
	Percent of Households												
Type of Main Heating Fuel													
Natural Gas	55	55	55	56	57	55	55	55	53	53	55	52	-3
Electricity [1]	16	17	18	17	16	17	20	23	26	29	29	30	12
Liquefied Petroleum Gases	4	5	5	4	5	5	5	5	5	5	5	5	0
Distillate Fuel Oil	20	17	15	14	13	12	12	11	11	9	R 8	7	-8
Wood	2	4	6	6	7	7	6	4	3	2	2	3	-3
Other [2]	3	2	2	3	3	3	3	2	2	2	R 1	1	-1
Type of Appliances													
Electric Appliances													
Television Set (Color)	NA	NA	82	83	85	88	93	96	98	99	99	99	17
Television Set (B/W)	NA	NA	51	48	46	43	36	31	20	NA	NA	NA	NA
Television Set (Any)	NA	NA	98	98	98	98	98	99	99	NA	NA	NA	NA
Clothes Washer	74	NA	74	73	71	73	75	76	77	77	79	83	9
Range Top or Burners	53	NA	54	54	53	54	57	58	61	60	60	59	5
Oven, Microwave	8	NA	14	17	21	34	61	79	84	83	86	88	74
Clothes Dryer	45	NA	47	45	45	46	51	53	57	55	57	61	14
Separate Freezer	35	NA	38	38	37	37	34	34	35	33	32	32	-6
Dishwasher	35	NA	37	37	36	38	43	45	45	50	53	58	21
Dehumidifier	NA	NA	9	9	9	9	10	12	9	NA	11	12	3
Waterbed Heaters	NA	NA	NA	NA	NA	10	14	15	12	8	5	2	NA
Window or Ceiling Fan	NA	NA	NA	NA	28	35	46	51	60	NA	NA	NA	NA
Whole House Fan	NA	NA	NA	NA	8	8	9	10	4	NA	NA	NA	NA
Evaporative Cooler	NA	NA	4	4	4	4	3	4	3	NA	3	3	-1
Personal Computer	NA	NA	NA	NA	NA	NA	NA	16	23	35	56	68	NA
Pump for Well Water	NA	NA	NA	NA	NA	NA	NA	15	13	14	13	13	NA
Swimming-Pool Pump [3]	NA	NA	3	4	3	NA	NA	5	5	5	6	7	4
Gas [4] Appliances													
Range Top or Burners	48	NA	46	46	47	45	43	42	38	39	39	40	-6
Clothes Dryer	14	NA	14	16	15	16	15	16	15	16	17	18	4
Outdoor Gas Grill	6	NA	9	9	11	13	20	26	29	NA	NA	8	-1
Outdoor Gas Light	2	NA	2	2	2	1	1	1	1	1	R 6	1	-1
Swimming Pool Heater [5]	NA	NA	(s)	(s)	(s)	1	1	1	1	1	1	1	NA
Refrigerators [6]													
One	86	NA	86	87	86	88	86	84	85	85	83	78	-8
Two or More	14	NA	14	13	13	12	14	15	15	15	17	22	8
Air Conditioning (A/C)													
Central [7]	23	24	27	27	28	30	34	39	44	47	55	59	32
Individual Room Units [7]	33	31	30	31	30	30	30	29	25	25	23	25	-5
None	44	45	43	42	42	40	36	32	32	28	23	16	-27
Portable Kerosene Heaters	(s)	NA	(s)	1	3	6	6	5	3	2	2	1	NA

[1] Retail electricity.
[2] Kerosene, district steam, coal, solar, other, or no heat.
[3] Through 1990, data are for all reported swimming pools, which were assumed to have an electric pump for filtering and circulating the water. Beginning in 1993, data are explicitly for pools with filters.
[4] Natural gas or liquefied petroleum gases.
[5] In 1984 and 1987, also includes heaters for hot tubs.
[6] Fewer than 0.5 percent of the households do not have a refrigerator.
[7] Households with both central and individual room units are counted only under "Central."

R=Revised. NA=Not available. (s)=Less than 0.5 percent.
Notes: • Data are estimates. • For years not shown, there are no data available.
Web Page: For related information, see http://www.eia.doe.gov/emeu/recs.
Sources: • 1978 and 1979—Energy Information Administration (EIA), Form EIA-84, "Residential Energy Consumption Survey." • 1980 forward—EIA, Form EIA-457, "Residential Energy Consumption Survey."

Figure 2.7 Type of Heating in Occupied Housing Units, 1950 and 2005

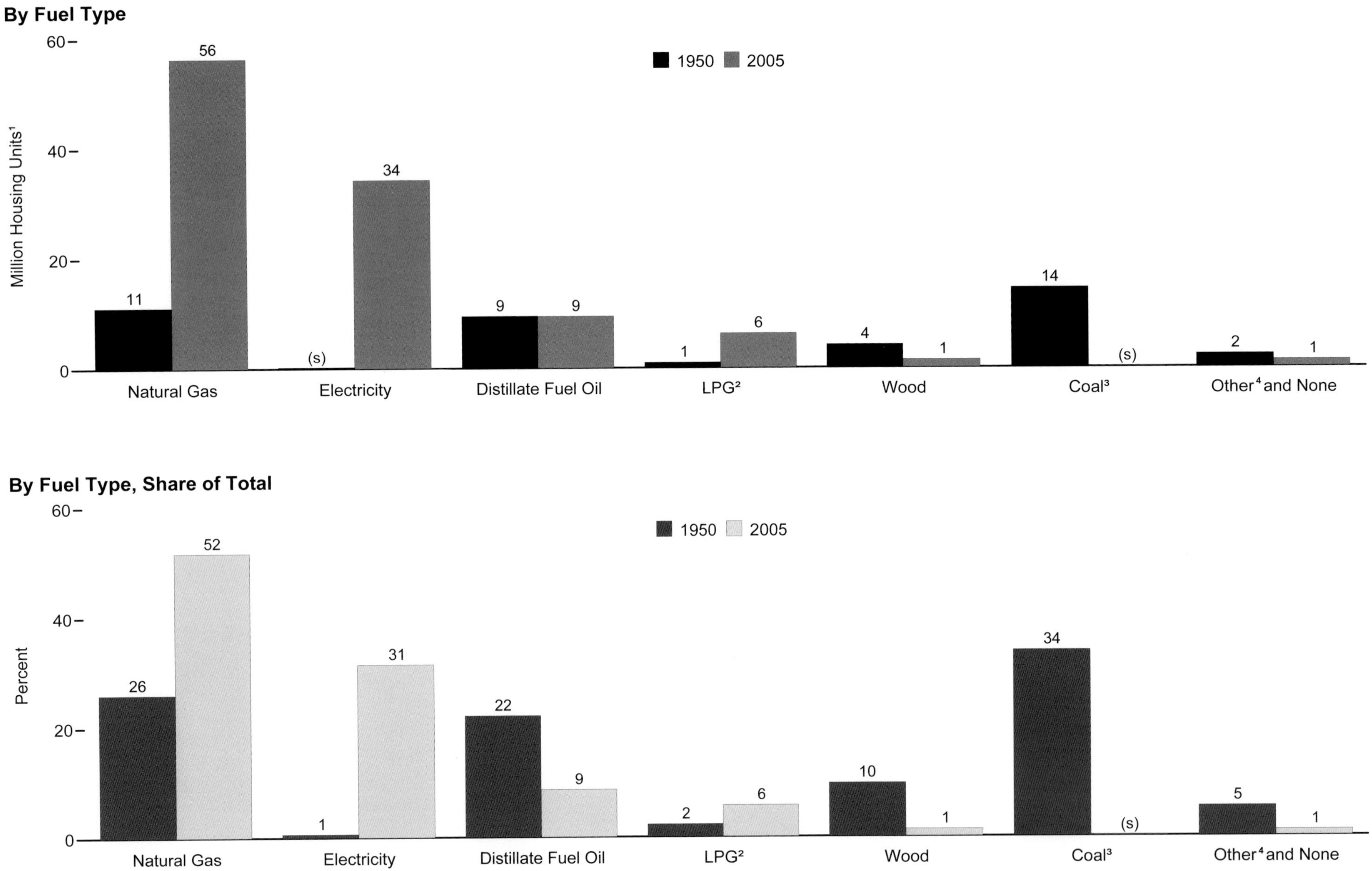

[1] Sum of components do not equal total due to independent rounding.
[2] Liquefied petroleum gases.
[3] Includes coal coke.
[4] Kerosene, solar, and other.
(s)=Less than 0.5.
Source: Table 2.7.

Table 2.7 Type of Heating in Occupied Housing Units, Selected Years, 1950-2005

Year	Coal [1]	Distillate Fuel Oil	Kerosene	Liquefied Petroleum Gases	Natural Gas	Electricity	Wood	Solar	Other [2]	None [3]	Total
						Millions					
1950	14.48	9.46	([4])	0.98	11.12	0.28	4.17	NA	0.77	1.57	42.83
1960	6.46	17.16	([4])	2.69	22.85	.93	2.24	NA	.22	.48	53.02
1970	1.82	16.47	([4])	3.81	35.01	4.88	.79	NA	.27	.40	63.45
1973	.80	17.24	([4])	4.42	38.46	7.21	.60	NA	.15	.45	69.34
1974	.74	16.84	([4])	4.14	39.47	8.41	.66	NA	.09	.48	70.83
1975	.57	16.30	([4])	4.15	40.93	9.17	.85	NA	.08	.47	72.52
1976	.48	16.45	([4])	4.24	41.22	10.15	.91	NA	.09	.46	74.01
1977	.45	15.62	.44	4.18	41.54	11.15	1.24	NA	.15	.51	75.28
1978	.40	15.65	.42	4.13	42.52	12.26	1.07	NA	.12	.60	77.17
1979	.36	15.30	.41	4.13	43.32	13.24	1.14	NA	.10	.57	78.57
1980	.33	14.50	.37	4.17	44.40	14.21	1.38	NA	.11	.61	80.07
1981	.36	14.13	.37	4.17	46.08	15.49	1.89	NA	.10	.59	83.18
1983 [5]	.43	12.59	.45	3.87	46.70	15.68	4.09	NA	.16	.68	84.64
1985	.45	12.44	1.06	3.58	45.33	18.36	6.25	.05	.37	.53	88.43
1987	.41	12.74	1.08	3.66	45.96	20.61	5.45	.05	.28	.66	90.89
1989	.34	12.47	1.07	3.66	47.40	23.06	4.59	.04	.40	.66	93.68
1991	.32	11.47	.99	3.88	47.02	23.71	4.44	.03	.41	.86	93.15
1993	.30	11.17	1.02	3.92	47.67	25.11	4.10	.03	.50	.91	94.73
1995	.21	10.98	1.06	4.25	49.20	26.77	3.53	.02	.64	1.04	97.69
1997	.18	10.10	.75	5.40	51.05	29.20	1.79	.03	.36	.62	99.49
1999	.17	10.03	.72	5.91	52.37	31.14	1.70	.02	.21	.54	102.80
2001 [6]	.13	9.81	.65	6.04	54.13	32.41	1.67	.02	.19	.39	105.44
2003	.13	9.50	.64	6.13	54.93	32.34	1.56	.02	.16	.44	105.84
2005	.10	9.38	.55	6.23	56.32	34.26	1.41	.02	.21	.40	108.87
						Percent					
1950	33.8	22.1	([4])	2.3	26.0	0.6	9.7	NA	1.8	3.7	100.0
1960	12.2	32.4	([4])	5.1	43.1	1.8	4.2	NA	.4	.9	100.0
1970	2.9	26.0	([4])	6.0	55.2	7.7	1.3	NA	.4	.6	100.0
1973	1.2	24.9	([4])	6.4	55.5	10.4	.9	NA	.2	.7	100.0
1974	1.0	23.8	([4])	5.8	55.7	11.9	.9	NA	.1	.7	100.0
1975	.8	22.5	([4])	5.7	56.4	12.6	1.2	NA	.1	.6	100.0
1976	.7	22.2	([4])	5.7	55.7	13.7	1.2	NA	.1	.6	100.0
1977	.6	20.7	.6	5.6	55.2	14.8	1.6	NA	.2	.7	100.0
1978	.5	20.3	.5	5.4	55.1	15.9	1.4	NA	.2	.8	100.0
1979	.5	19.5	.5	5.3	55.1	16.9	1.4	NA	.1	.7	100.0
1980	.4	18.1	.5	5.2	55.4	17.7	1.7	NA	.1	.8	100.0
1981	.4	17.0	.4	5.0	55.4	18.6	2.3	NA	.1	.7	100.0
1983 [5]	.5	14.9	.5	4.6	55.2	18.5	4.8	NA	.2	.8	100.0
1985	.5	14.1	1.2	4.1	51.3	20.8	7.1	.1	.4	.6	100.0
1987	.4	14.0	1.2	4.0	50.6	22.7	6.0	.1	.3	.7	100.0
1989	.4	13.3	1.1	3.9	50.6	24.6	4.9	(s)	.4	.7	100.0
1991	.3	12.3	1.1	4.2	50.5	25.5	4.8	(s)	.4	.9	100.0
1993	.3	11.8	1.1	4.1	50.3	26.5	4.3	(s)	.5	1.0	100.0
1995	.2	11.2	1.1	4.4	50.4	27.4	3.6	(s)	.7	1.1	100.0
1997	.2	10.2	.8	5.4	51.3	29.4	1.8	(s)	.4	.6	100.0
1999	.2	9.8	.7	5.7	50.9	30.3	1.7	(s)	.2	.5	100.0
2001 [6]	.1	9.3	.6	5.7	51.3	30.7	1.6	(s)	.2	.4	100.0
2003	.1	9.0	.6	5.8	51.9	30.6	1.5	(s)	.1	.4	100.0
2005	.1	8.6	.5	5.7	51.7	31.5	1.3	(s)	.2	.4	100.0

[1] Includes coal coke.
[2] Includes briquettes (made of pitch and sawdust), coal dust, waste material (such as corncobs), purchased steam, and other fuels not separately displayed.
[3] In 1950 and 1960, also includes nonreporting units, which totaled 997 and 2,000 units, respectively.
[4] Included in "Distillate Fuel Oil."
[5] Beginning in 1983, the *American Housing Survey for the United States* has been a biennial survey.
[6] Beginning in 2001, data are consistent with the 2000 Census. For 2001 data consistent with the 1990 Census, see *American Housing Survey for the United States: 2001.*

NA=Not available. (s)=Less than 0.05 percent.

Notes: • Includes mobile homes and individual housing units in apartment buildings. Housing units with more than one type of heating system are classified according to the principal type of heating system. • Totals may not equal sum of components due to independent rounding.

Web Page: For related information, see http://www.census.gov/hhes/www/ahs.html.

Sources: • 1950, 1960, and 1970—Bureau of the Census, *Census of Population and Housing.* • 1973 forward—Bureau of the Census, *American Housing Survey for the United States,* biennial surveys, Table 2-5.

Figure 2.8 Motor Vehicle Mileage, Fuel Consumption, and Fuel Rates

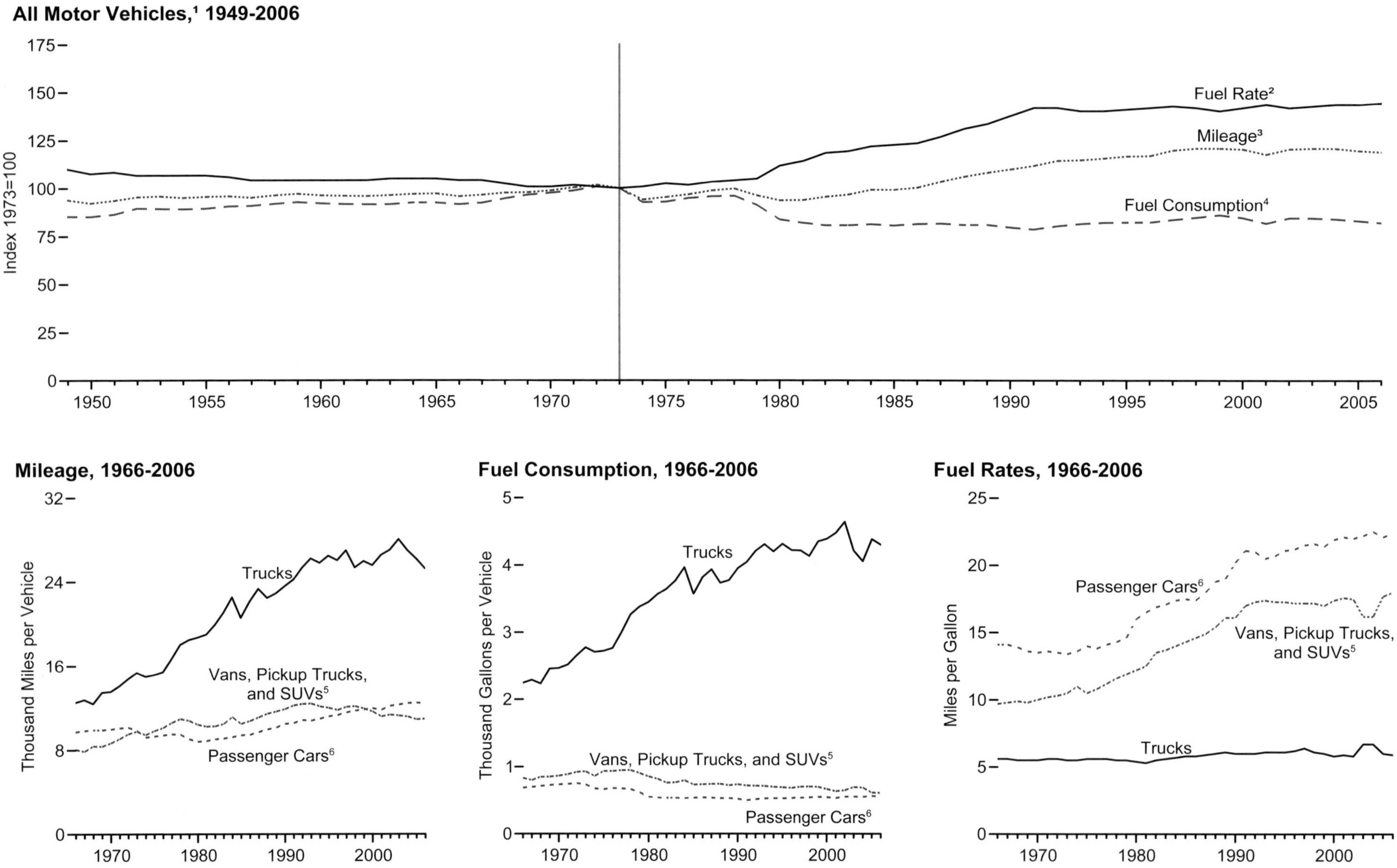

[1] Passenger cars, motorcycles, vans, pickup trucks, sport utility vehicles, trucks, and buses.
[2] Miles per gallon.
[3] Miles per vehicle.
[4] Gallons per vehicle.
[5] Sport utility vehicle.
[6] Through 1989, includes motorcycles.
Source: Table 2.8.

Table 2.8 Motor Vehicle Mileage, Fuel Consumption, and Fuel Rates, Selected Years, 1949-2006

	Passenger Cars [1]			Vans, Pickup Trucks, and Sport Utility Vehicles [2]			Trucks [3]			All Motor Vehicles [4]		
	Mileage	Fuel Consumption	Fuel Rate	Mileage	Fuel Consumption	Fuel Rate	Mileage	Fuel Consumption	Fuel Rate	Mileage	Fuel Consumption	Fuel Rate
Year	Miles per Vehicle	Gallons per Vehicle	Miles per Gallon	Miles per Vehicle	Gallons per Vehicle	Miles per Gallon	Miles per vehicle	Gallons per vehicle	Miles per Gallon	Miles per Vehicle	Gallons per Vehicle	Miles per Gallon
1949	9,388	627	15.0	(5)	(5)	(5)	9,712	1,080	9.0	9,498	726	13.1
1950	9,060	603	15.0	(5)	(5)	(5)	10,316	1,229	8.4	9,321	725	12.8
1955	9,447	645	14.6	(5)	(5)	(5)	10,576	1,293	8.2	9,661	761	12.7
1960	9,518	668	14.3	(5)	(5)	(5)	10,693	1,333	8.0	9,732	784	12.4
1965	9,603	661	14.5	(5)	(5)	(5)	10,851	1,387	7.8	9,826	787	12.5
1970	9,989	737	13.5	8,676	866	10.0	13,565	2,467	5.5	9,976	830	12.0
1971	10,097	743	13.6	9,082	888	10.2	14,117	2,519	5.6	10,133	839	12.1
1972	10,171	754	13.5	9,534	922	10.3	14,780	2,657	5.6	10,279	857	12.0
1973	9,884	737	13.4	9,779	931	10.5	15,370	2,775	5.5	10,099	850	11.9
1974	9,221	677	13.6	9,452	862	11.0	14,995	2,708	5.5	9,493	788	12.0
1975	9,309	665	14.0	9,829	934	10.5	15,167	2,722	5.6	9,627	790	12.2
1976	9,418	681	13.8	10,127	934	10.8	15,438	2,764	5.6	9,774	806	12.1
1977	9,517	676	14.1	10,607	947	11.2	16,700	3,002	5.6	9,978	814	12.3
1978	9,500	665	14.3	10,968	948	11.6	18,045	3,263	5.5	10,077	816	12.4
1979	9,062	620	14.6	10,802	905	11.9	18,502	3,380	5.5	9,722	776	12.5
1980	8,813	551	16.0	10,437	854	12.2	18,736	3,447	5.4	9,458	712	13.3
1981	8,873	538	16.5	10,244	819	12.5	19,016	3,565	5.3	9,477	697	13.6
1982	9,050	535	16.9	10,276	762	13.5	19,931	3,647	5.5	9,644	686	14.1
1983	9,118	534	17.1	10,497	767	13.7	21,083	3,769	5.6	9,760	686	14.2
1984	9,248	530	17.4	11,151	797	14.0	22,550	3,967	5.7	10,017	691	14.5
1985	9,419	538	17.5	10,506	735	14.3	20,597	3,570	5.8	10,020	685	14.6
1986	9,464	543	17.4	10,764	738	14.6	22,143	3,821	5.8	10,143	692	14.7
1987	9,720	539	18.0	11,114	744	14.9	23,349	3,937	5.9	10,453	694	15.1
1988	9,972	531	18.8	11,465	745	15.4	22,485	3,736	6.0	10,721	688	15.6
1989	[1]10,157	[1]533	[1]19.0	11,676	724	16.1	22,926	3,776	6.1	10,932	688	15.9
1990	10,504	520	20.2	11,902	738	16.1	23,603	3,953	6.0	11,107	677	16.4
1991	10,571	501	21.1	12,245	721	17.0	24,229	4,047	6.0	11,294	669	16.9
1992	10,857	517	21.0	12,381	717	17.3	25,373	4,210	6.0	11,558	683	16.9
1993	10,804	527	20.5	12,430	714	17.4	26,262	4,309	6.1	11,595	693	16.7
1994	10,992	531	20.7	12,156	701	17.3	25,838	4,202	6.1	11,683	698	16.7
1995	11,203	530	21.1	12,018	694	17.3	26,514	4,315	6.1	11,793	700	16.8
1996	11,330	534	21.2	11,811	685	17.2	26,092	4,221	6.2	11,813	700	16.9
1997	11,581	539	21.5	12,115	703	17.2	27,032	4,218	6.4	12,107	711	17.0
1998	11,754	544	21.6	12,173	707	17.2	25,397	4,135	6.1	12,211	721	16.9
1999	11,848	553	21.4	11,957	701	17.0	26,014	4,352	6.0	12,206	732	16.7
2000	11,976	547	21.9	11,672	669	17.4	25,617	4,391	5.8	12,164	720	16.9
2001	11,831	534	22.1	11,204	636	17.6	26,602	4,477	5.9	11,887	695	17.1
2002	12,202	555	22.0	11,364	650	17.5	27,071	4,642	5.8	12,171	719	16.9
2003	12,325	556	22.2	11,287	697	16.2	28,093	4,215	6.7	12,208	718	17.0
2004	12,460	553	22.5	11,184	690	16.2	27,023	4,057	6.7	12,200	714	17.1
2005	[R]12,510	[R]567	[R]22.1	[R]10,920	[R]617	[R]17.7	[R]26,235	[R]4,385	[R]6.0	[R]12,082	[R]706	[R]17.1
2006[P]	12,427	554	22.4	10,986	612	18.0	25,290	4,300	5.9	12,016	697	17.2

[1] Through 1989, includes motorcycles.
[2] Includes a small number of trucks with 2 axles and 4 tires, such as step vans.
[3] Single-unit trucks with 2 axles and 6 or more tires, and combination trucks.
[4] Includes buses and motorcycles, which are not separately displayed.
[5] Included in "Trucks."
R=Revised. P=Preliminary.

Web Pages: • For all data beginning in 1949, see http://www.eia.doe.gov/aer/consump.html. • For related information, see http://www.fhwa.dot.gov/policy/ohpi/hss/index.htm.

Sources: **Passenger Cars, 1990-1994:** U.S. Department of Transportation, Bureau of Transportation Statistics, *National Transportation Statistics 1998,* Table 4-13. **All Other Data:** • 1949-1994—Federal Highway Administration (FHWA), *Highway Statistics Summary to 1995,* Table VM-201A. • 1995 forward—FHWA, *Highway Statistics,* annual reports, Table VM-1.

Figure 2.9 Commercial Buildings Consumption by Energy Source

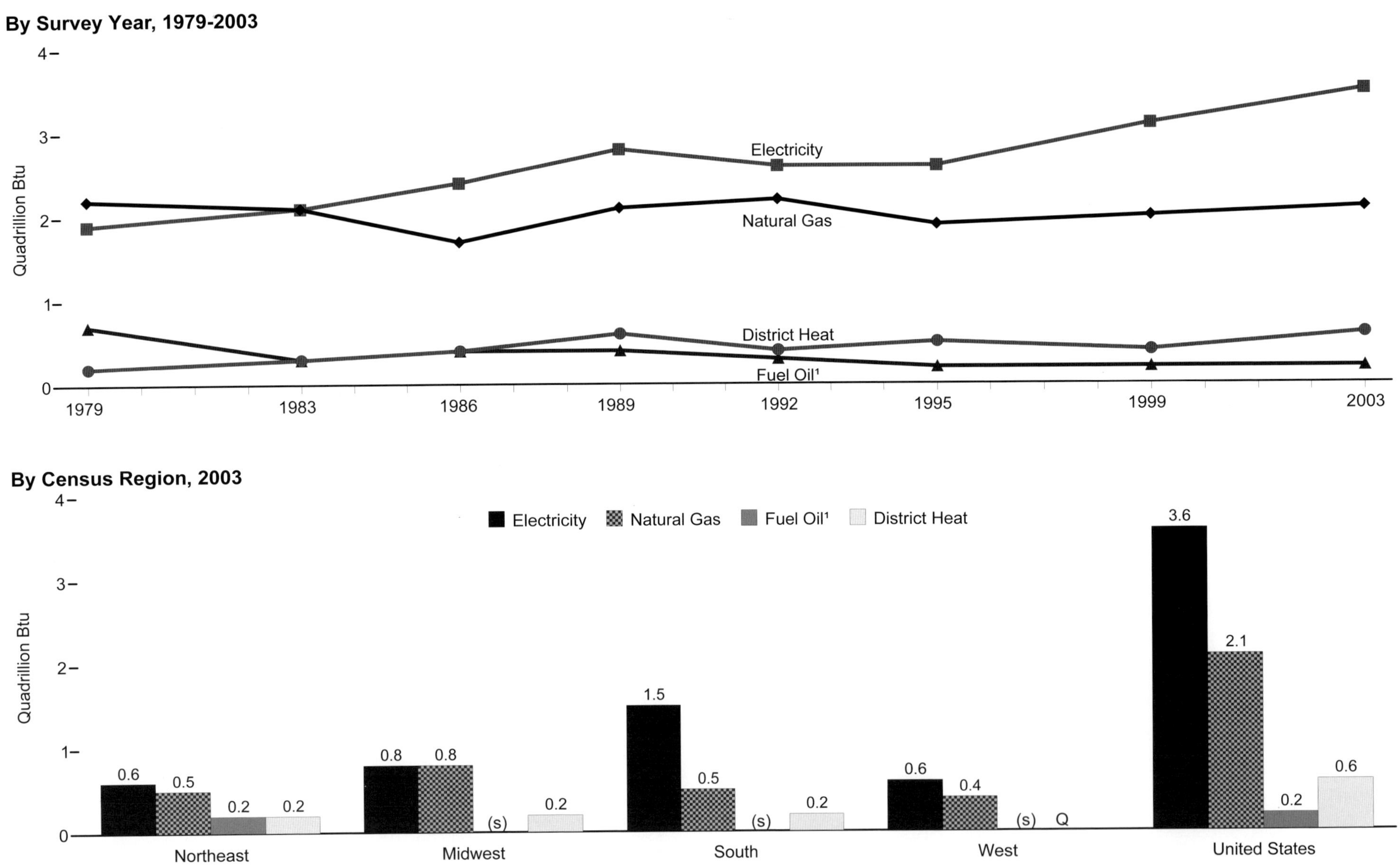

[1] Distillate fuel oil, residual fuel oil, and kerosene.
(s)=Less than 0.05 quadrillion Btu.
Q=Data withheld because either the relative standard error was greater than 50 percent or fewer than 20 buildings were sampled.

Note: See Appendix C for Census regions.
Source: Table 2.9.

Table 2.9 Commercial Buildings Consumption by Energy Source, Selected Years, 1979-2003

(Trillion Btu)

Energy Source and Year	Square Footage Category			Principal Building Activity								Census Region [1]				All Buildings
	1,001 to 10,000	10,001 to 100,000	Over 100,000	Education	Food Sales	Food Service	Health Care	Lodging	Mercantile and Service	Office	All Other	Northeast	Midwest	South	West	
Major Sources [2]																
1979	1,255	2,202	1,508	511	([3])	336	469	278	894	861	1,616	1,217	1,826	1,395	526	4,965
1983	1,242	1,935	1,646	480	([3])	414	463	362	812	1,018	1,274	858	1,821	1,462	682	4,823
1986	1,273	2,008	1,696	633	147	247	456	299	985	1,008	1,202	1,037	1,585	1,459	896	4,977
1989	1,259	2,402	2,127	704	139	255	449	425	1,048	1,230	1,538	1,354	1,659	1,648	1,126	5,788
1992	1,258	2,301	1,932	637	137	307	403	463	892	1,247	1,404	1,090	1,578	1,825	998	5,490
1995 [4]	1,332	2,152	1,838	614	137	332	561	461	973	1,019	1,225	1,035	1,497	1,684	1,106	5,321
1999	1,381	2,300	2,053	649	201	447	515	450	1,145	1,089	1,237	1,116	1,509	1,961	1,147	5,733
2003	1,248	2,553	2,721	820	251	427	594	510	1,333	1,134	1,455	1,396	1,799	2,265	1,063	6,523
Electricity																
1979	429	872	608	163	([3])	171	129	119	361	424	543	425	593	662	227	1,908
1983	469	903	758	152	([3])	212	147	151	426	509	532	324	673	801	331	2,129
1986	654	927	809	179	99	121	132	120	536	641	563	430	584	867	510	2,390
1989	572	1,145	1,056	217	105	113	154	138	550	781	715	586	609	975	604	2,773
1992	586	991	1,033	235	113	138	138	189	444	704	649	419	622	1,002	566	2,609
1995 [4]	618	1,064	926	221	119	166	211	187	508	676	521	436	558	1,027	587	2,608
1999	698	1,235	1,164	257	165	216	232	196	659	767	606	543	662	1,247	645	3,098
2003	685	1,405	1,469	371	208	217	248	235	883	719	679	587	799	1,542	631	3,559
Natural Gas																
1979	646	996	532	214	([3])	145	221	115	422	272	784	443	1,007	470	255	2,174
1983	684	809	597	246	([3])	188	218	170	327	365	576	278	978	523	311	2,091
1986	485	715	523	254	45	114	205	105	332	258	409	244	742	426	311	1,723
1989	568	836	670	323	27	128	186	187	417	238	566	353	831	498	391	2,073
1992	572	1,017	586	291	24	157	189	193	381	388	552	354	747	697	376	2,174
1995 [4]	535	830	580	245	18	158	258	213	395	239	420	297	750	528	371	1,946
1999	604	803	616	227	31	216	217	181	446	219	486	299	709	618	396	2,023
2003	482	909	709	268	39	203	243	215	403	269	460	462	751	527	360	2,100
Fuel Oil [5]																
1979	177	272	231	107	([3])	15	97	20	103	107	232	285	133	237	26	681
1983	85	140	90	61	([3])	Q	28	18	43	75	79	172	28	104	Q	314
1986	114	206	121	103	Q	Q	Q	20	105	39	130	270	63	86	23	442
1989	101	170	86	71	Q	Q	17	10	76	43	122	237	61	50	Q	357
1992	86	111	75	62	Q	Q	21	16	55	47	67	194	26	48	Q	272
1995 [4]	71	104	60	57	Q	Q	21	Q	49	28	70	168	16	45	7	235
1999	29	73	60	48	Q	Q	19	Q	18	29	65	138	5	29	8	179
2003	71	74	83	47	Q	Q	11	35	41	18	68	181	24	15	9	228
District Heat [6]																
1979	Q	61	136	27	([3])	Q	22	24	Q	58	57	64	93	Q	Q	201
1983	Q	83	202	21	([3])	Q	70	22	Q	68	87	84	141	34	30	289
1986	Q	159	243	97	Q	Q	80	Q	12	71	99	94	196	81	51	422
1989	19	252	315	Q	Q	Q	92	Q	Q	167	134	179	159	126	121	585
1992	Q	182	238	49	NC	Q	55	65	Q	109	135	123	183	78	51	435
1995 [4]	Q	154	271	91	Q	Q	70	57	Q	75	214	135	173	83	Q	533
1999	Q	158	213	117	Q	Q	46	68	Q	74	126	136	132	67	98	433
2003	Q	165	460	134	NC	Q	Q	Q	Q	128	247	166	225	182	Q	636

[1] See Appendix C for Census regions.
[2] Includes electricity, natural gas, fuel oil, and district heat.
[3] Included in "Food Service."
[4] Beginning in 1995, excludes commercial buildings at multi-building manufacturing facilities, and parking garages.
[5] Distillate fuel oil, residual fuel oil, and kerosene.
[6] Through 1983, includes purchased steam only. Beginning in 1986, includes purchased and non-purchased steam and hot water.

Q=Data withheld because either the relative standard error was greater than 50 percent or fewer than 20 buildings were sampled. NC=No cases in the sample.

Note: Data are estimates. Statistics for individual fuels are for all buildings using each fuel. Statistics for "Major Sources" are for the sum of "Electricity," "Natural Gas," "Fuel Oil," and "District Heat," across all buildings using any of those fuels.

Web Page: For related information, see http://www.eia.doe.gov/emeu/cbecs.

Sources: • 1979—Energy Information Administration (EIA), Form EIA-143, "Nonresidential Buildings Energy Consumption Survey." • 1983—EIA, Form EIA-788, "Nonresidential Buildings Energy Consumption Survey." • 1986—EIA, Form EIA-871, "Nonresidential Buildings Energy Consumption Survey." • 1989 forward—EIA, Form EIA-871A-F, "Commercial Buildings Energy Consumption Survey."

Figure 2.10 Commercial Buildings Energy Consumption and Expenditure Indicators, Selected Years, 1979-2003

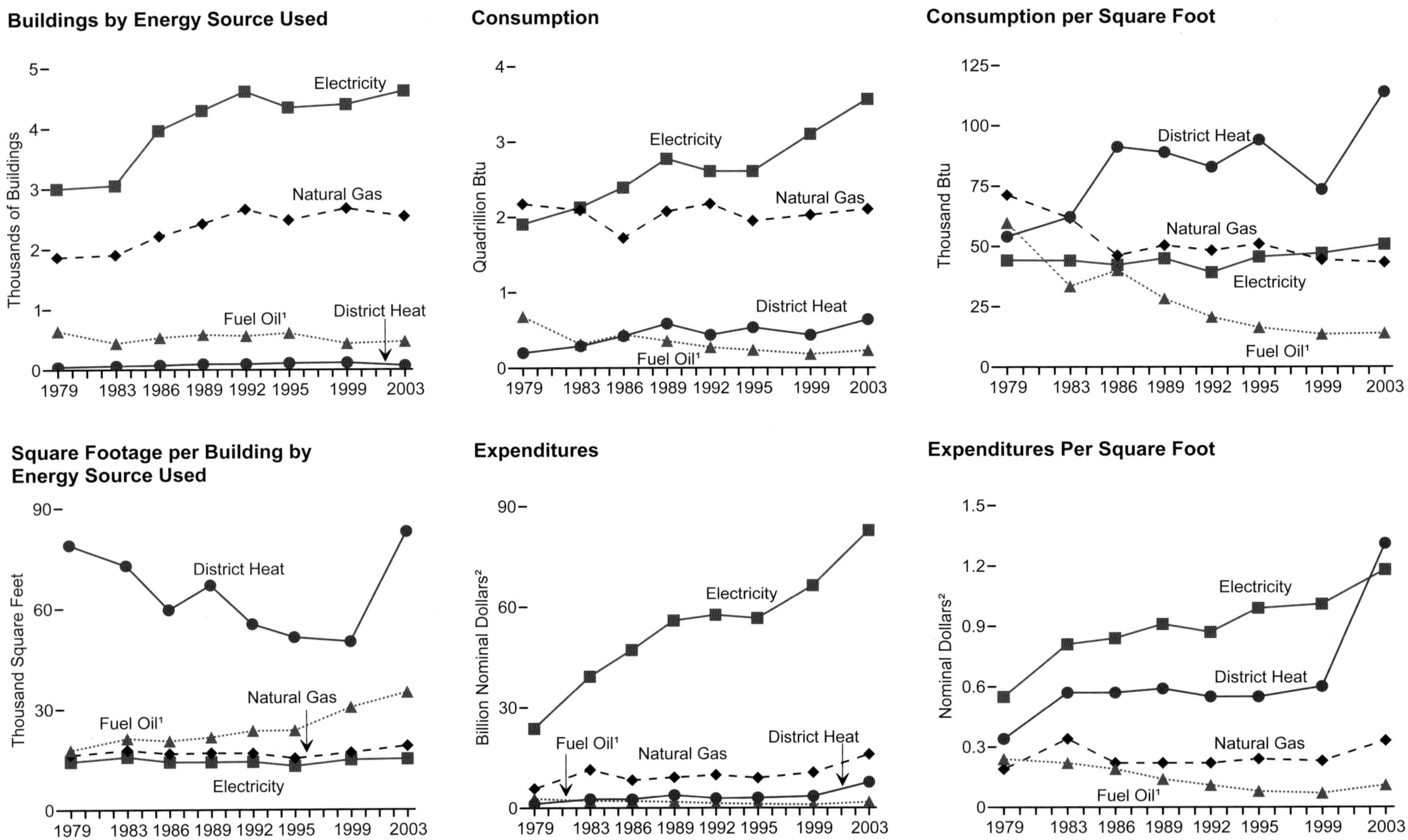

¹ Distillate fuel oil, residual fuel oil, and kerosene.
² See "Nominal Dollars" in Glossary.

Notes: • For years not shown, there are no data available. • Because vertical scales differ, graphs should not be compared.
Source: Table 2.10.

Table 2.10 Commercial Buildings Energy Consumption and Expenditure Indicators, Selected Years, 1979-2003

Energy Source and Year	Building Characteristics: Number of Buildings (Thousands)	Building Characteristics: Total Square Feet (Millions)	Building Characteristics: Square Feet per Building (Thousands)	Energy Consumption: Total (Trillion Btu)	Energy Consumption: Per Building (Million Btu)	Energy Consumption: Per Square Foot (Thousand Btu)	Energy Consumption: Per Employee (Million Btu)	Energy Expenditures: Total (Million Nominal Dollars [1])	Energy Expenditures: Per Building (Thousand Nominal Dollars [1])	Energy Expenditures: Per Square Foot (Nominal Dollars [1])	Energy Expenditures: Per Million Btu (Nominal Dollars [1])
Major Sources [2]											
1979	3,073	43,546	14.2	5,008	1,630	115.0	85.0	33,821	11.0	0.78	6.75
1983	3,185	49,471	15.5	4,856	1,525	98.2	65.7	55,764	17.5	1.13	11.48
1986	4,154	58,199	14.0	5,040	1,213	86.6	68.6	60,762	14.6	1.04	12.06
1989	4,528	63,184	14.0	5,788	1,278	91.6	81.9	70,826	15.6	1.12	12.24
1992	4,806	67,876	14.1	5,490	1,142	80.9	77.1	71,821	14.9	1.06	13.08
1995 [3]	4,579	58,772	12.8	5,321	1,162	90.5	69.3	69,918	15.3	1.19	13.14
1999	4,657	67,338	14.5	5,733	1,231	85.1	70.0	81,552	17.5	1.21	14.22
2003	4,859	71,658	14.7	6,523	1,342	91.0	([4])	107,897	22.2	1.51	16.54
Electricity											
1979	3,001	43,153	14.4	1,908	636	44.2	32.4	23,751	7.9	.55	12.45
1983	3,052	48,327	15.8	2,129	697	44.1	28.9	39,279	12.9	.81	18.45
1986	3,965	56,508	14.3	2,390	603	42.3	32.7	47,186	11.9	.84	19.74
1989	4,294	61,563	14.3	2,773	646	45.0	39.3	55,943	13.0	.91	20.17
1992	4,611	66,525	14.4	2,609	566	39.2	36.6	57,619	12.5	.87	22.09
1995 [3]	4,343	57,076	13.1	2,608	600	45.7	34.1	56,621	13.0	.99	21.71
1999	4,395	65,716	15.0	3,098	706	47.1	37.9	66,424	15.1	1.01	21.44
2003	4,617	70,181	15.2	3,559	771	50.7	([4])	82,783	17.9	1.18	23.26
Natural Gas											
1979	1,864	30,477	16.4	2,174	1,167	71.3	52.5	5,814	3.1	.19	2.67
1983	1,904	33,935	17.8	2,091	1,098	61.6	40.6	11,443	6.0	.34	5.47
1986	2,214	37,263	16.8	1,723	778	46.2	35.2	8,355	3.8	.22	4.85
1989	2,420	41,143	17.0	2,073	857	50.4	43.2	9,204	3.8	.22	4.44
1992	2,657	44,994	16.9	2,174	818	48.3	42.5	9,901	3.7	.22	4.55
1995 [3]	2,478	38,145	15.4	1,946	785	51.0	38.7	9,018	3.6	.24	4.63
1999	2,670	45,525	17.1	2,023	758	44.4	36.0	10,609	4.0	.23	5.24
2003	2,538	48,473	19.1	2,100	828	43.3	([4])	16,010	6.3	.33	7.62
Fuel Oil [5]											
1979	641	11,397	17.8	681	1,063	59.7	40.5	2,765	4.3	.24	4.06
1983	441	9,409	21.3	314	714	33.4	19.8	2,102	4.8	.22	6.68
1986	534	11,005	20.6	442	827	40.1	27.7	2,059	3.9	.19	4.66
1989	581	12,600	21.7	357	614	28.3	21.0	1,822	3.1	.14	5.11
1992	560	13,215	23.6	272	487	20.6	15.1	1,400	2.5	.11	5.14
1995 [3]	607	14,421	23.7	235	387	16.3	10.2	1,175	1.9	.08	5.00
1999	434	13,285	30.6	179	412	13.5	9.1	956	2.2	.07	5.35
2003	465	16,265	35.0	228	490	14.0	([4])	1,826	3.9	.11	8.01
District Heat [6]											
1979	47	3,722	79.0	201	4,267	54.0	26.5	1,267	26.9	.34	6.30
1983	64	4,643	72.9	289	4,530	62.1	34.4	2,627	41.2	.57	9.10
1986	77	4,625	59.7	422	5,446	91.2	52.4	2,620	33.8	.57	6.21
1989	98	6,578	67.0	585	5,964	89.0	56.5	3,857	39.3	.59	6.59
1992	95	5,245	55.4	435	4,596	82.9	60.9	2,901	30.7	.55	6.67
1995 [3]	110	5,658	51.5	533	4,849	94.1	51.2	3,103	28.3	.55	5.83
1999	117	5,891	50.2	433	3,692	73.6	50.1	3,564	30.4	.60	8.23
2003	67	5,576	83.0	636	9,470	114.0	([4])	7,279	108.4	1.31	11.45

[1] See "Nominal Dollars" in Glossary.
[2] Includes electricity, natural gas, fuel oil, and district heat.
[3] Beginning in 1995, excludes commercial buildings at multi-building manufacturing facilities, and parking garages.
[4] Total number of employees not collected in 2003.
[5] Distillate fuel oil, residual fuel oil, and kerosene.
[6] Through 1983, includes purchased steam only. Beginning in 1986, includes purchased and non-purchased steam and hot water.

Note: Data are estimates. Statistics for individual fuels are for all buildings using each fuel. Statistics for major sources are for all buildings, even buildings using no major fuel.

Web Page: For related information, see http://www.eia.doe.gov/emeu/cbecs.

Sources: • 1979—Energy Information Administration (EIA), Form EIA-143, "Nonresidential Buildings Energy Consumption Survey." • 1983—EIA, Form EIA-788, "Nonresidential Buildings Energy Consumption Survey." • 1986—EIA, Form EIA-871, "Nonresidential Buildings Energy Consumption Survey." • 1989 forward—EIA, Form EIA-871A-F, "Commercial Buildings Energy Consumption Survey."

Energy Consumption by Sector

Note. Electrical System Energy Losses. Electrical system energy losses are calculated as the difference between total primary consumption by the electric power sector—see Table 2.1f—and the total energy content of electricity retail sales—see Tables 8.9 and A6. Most of these losses occur at steam-electric power plants (conventional and nuclear) in the conversion of heat energy into mechanical energy to turn electric generators. The loss is a thermodynamically necessary feature of the steam-electric cycle. Part of the energy input-to-output losses is a result of imputing fossil energy equivalent inputs for hydroelectric, solar, and wind energy sources, since there is no generally accepted practice for measuring those thermal conversion rates. In addition to conversion losses, other losses include power plant use of electricity, transmission and distribution of electricity from power plants to end-use consumers (also called "line losses"), and unaccounted for electricity. Total losses are allocated to the end-use sectors in proportion to each sector's share of total electricity sales. Overall, approximately 67 percent of total energy input is lost in conversion; of electricity generated, approximately 5 percent is lost in plant use and 9 percent is lost in transmission and distribution.

3 Financial Indicators

Figure 3.1 Fossil Fuel Production Prices

Prices, 1949-2007

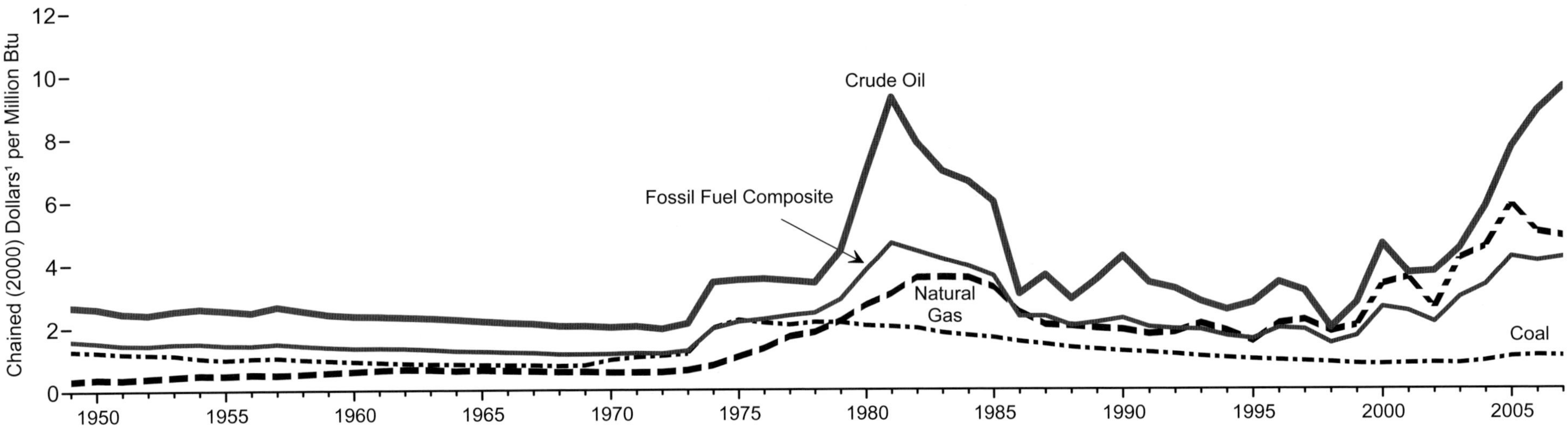

Fossil Fuel Composite Price[2], Change From Previous Year, 1950-2007

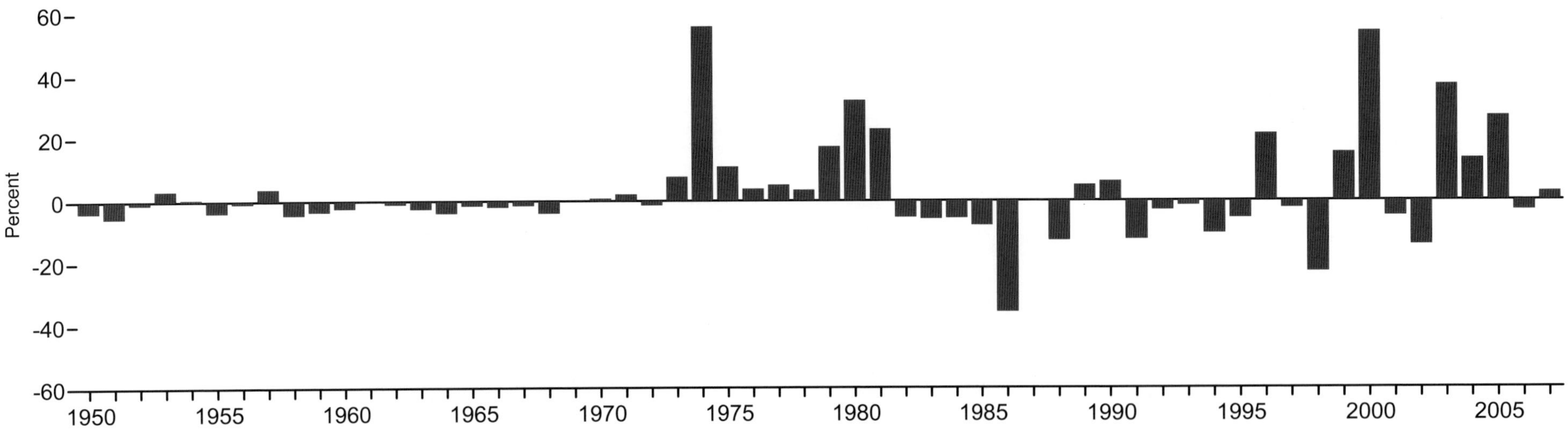

[1] Calculated by using gross domestic product implicit price deflators in Table D1. See "Chained Dollars" in Glossary.

[2] Based on real prices.

Source: Table 3.1.

Table 3.1 Fossil Fuel Production Prices, Selected Years, 1949-2007

(Dollars per Million Btu)

Year	Coal [1]		Natural Gas [2]		Crude Oil [3]		Fossil Fuel Composite [4]		
	Nominal [5]	Real [6]	Nominal [5]	Real [6]	Nominal [5]	Real [6]	Nominal [5]	Real [6]	Percent Change [7]
1949	0.21	1.29	0.05	0.33	0.44	2.68	0.26	1.60	– –
1950	.21	1.25	.06	.38	.43	2.62	.26	1.54	-3.6
1955	.19	.99	.09	.48	.48	2.55	.27	1.45	-3.7
1960	.19	.92	.13	.60	.50	2.36	.28	1.35	-2.3
1965	.18	.82	.15	.65	.49	2.19	.28	1.23	-1.5
1970	.27	.97	.15	.56	.55	1.99	.32	1.15	.8
1971	.30	1.05	.16	.56	.58	2.02	.34	1.18	2.1
1972	.33	1.09	.17	.57	.58	1.94	.35	1.16	-1.4
1973	.37	1.15	.20	.63	.67	2.11	.40	1.25	7.7
1974	.69	1.98	.27	.79	1.18	3.41	.68	1.95	55.8
1975	.85	2.22	.40	1.06	1.32	3.48	.82	2.16	10.9
1976	.86	2.13	.53	1.32	1.41	3.51	.90	2.24	3.8
1977	.88	2.07	.72	1.69	1.48	3.46	1.01	2.36	5.1
1978	.98	2.15	.84	1.83	1.55	3.39	1.12	2.44	3.4
1979	1.06	2.14	1.08	2.18	2.18	4.40	1.42	2.86	17.3
1980	1.10	2.04	1.45	2.68	3.72	6.89	2.04	3.78	32.1
1981	1.18	2.00	1.80	3.04	5.48	9.27	2.75	4.64	22.9
1982	1.23	1.95	2.22	3.54	4.92	7.84	2.76	4.40	-5.3
1983	1.18	1.81	2.32	3.56	4.52	6.93	2.70	4.14	-5.8
1984	1.16	1.72	2.40	3.55	4.46	6.60	2.65	3.91	-5.6
1985	1.15	1.65	2.26	3.24	4.15	5.96	2.51	3.60	-7.8
1986	1.09	1.52	1.75	2.45	2.16	3.03	1.65	2.32	-35.6
1987	1.05	1.44	1.50	2.05	2.66	3.63	1.70	2.32	.1
1988	1.01	1.34	1.52	2.01	2.17	2.87	1.53	2.03	-12.8
1989	1.00	1.28	1.53	1.94	2.73	3.48	1.67	2.13	5.0
1990	1.00	1.22	1.55	1.90	3.45	4.23	1.84	2.26	6.2
1991	.99	1.17	1.48	1.75	2.85	3.38	1.67	1.98	-12.5
1992	.97	1.12	1.57	1.82	2.76	3.19	1.66	1.92	-3.0
1993	.93	1.05	1.84	2.09	2.46	2.78	1.67	1.89	-1.5
1994	.91	1.01	1.67	1.86	2.27	2.52	1.53	1.69	-10.4
1995	.88	.96	1.40	1.52	2.52	2.74	1.47	1.60	-5.5
1996	.87	.92	1.96	2.09	3.18	3.39	1.82	1.94	21.3
1997	.85	.89	2.10	2.20	2.97	3.11	1.81	1.89	-2.4
1998	.83	.86	1.77	1.83	1.87	1.94	1.41	1.46	-22.8
1999	.79	.81	1.98	2.02	2.68	2.74	1.65	1.69	15.4
2000	.80	.80	3.32	3.32	4.61	4.61	2.60	2.60	54.2
2001	.84	.82	3.62	3.54	3.77	3.68	2.53	2.47	-4.9
2002	.87	.84	2.67	2.56	3.88	3.73	2.21	2.12	-14.2
2003	.87	.82	4.41	4.15	4.75	4.47	3.09	2.91	37.1
2004	.98	.89	4.94	R4.51	6.34	5.79	3.61	3.30	13.4
2005	1.16	1.03	R6.63	R5.87	8.67	R7.67	R4.73	R4.19	R27.1
2006	R1.24	R1.06	R5.80	R4.98	10.29	R8.83	R4.73	4.06	R-3.2
2007P	1.25	1.04	5.79	4.84	11.47	9.58	4.99	4.17	2.8

[1] Free-on-board (f.o.b.) rail/barge prices, which are the f.o.b. prices of coal at the point of first sale, excluding freight or shipping and insurance costs. See "Free on Board (F.O.B.)" in Glossary.

[2] Wellhead prices (converted to dollars per million Btu using marketed production heat contents). See "Natural Gas Wellhead Price" in Glossary.

[3] Domestic first purchase prices. See "Crude Oil Domestic First Purchase Price" in Glossary.

[4] Derived by multiplying the price per Btu of each fossil fuel by the total Btu content of the production of each fossil fuel and dividing this accumulated value of total fossil fuel production by the accumulated Btu content of total fossil fuel production.

[5] See "Nominal Dollars" in Glossary.

[6] In chained (2000) dollars, calculated by using gross domestic product implicit price deflators in Table D1. See "Chained Dollars" in Glossary.

[7] Based on real values.

R=Revised. P=Preliminary. – – = Not applicable.

Web Page: For all data beginning in 1949, see http://www.eia.doe.gov/emeu/aer/finan.html.

Sources: Tables 5.18, 6.7, 7.8, A2, A4, and A5.

Figure 3.2 Value of Fossil Fuel Production

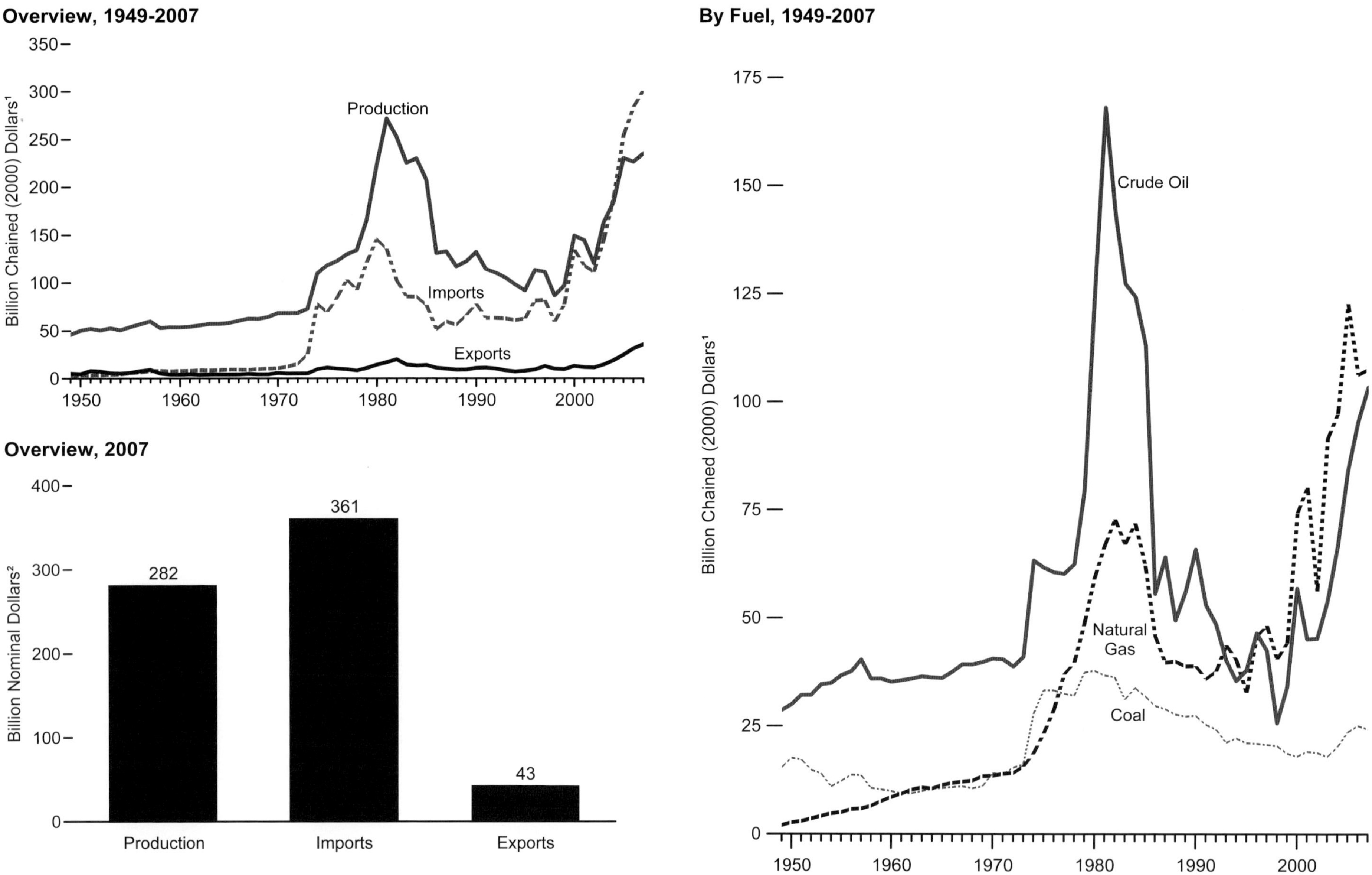

[1] In chained (2000) dollars, calculated by using gross domestic product implicit price deflators in Table D1. See "Chained Dollars" in Glossary.

[2] See "Nominal Dollars" in Glossary.

Note: Because vertical scales differ, graphs should not be compared.

Sources: Tables 3.2, 3.7, and 3.8.

Table 3.2 Value of Fossil Fuel Production, Selected Years, 1949-2007
(Billion Dollars)

Year	Coal [1]		Natural Gas [2]		Crude Oil [3,4]		Total	
	Nominal [5]	Real [6]	Nominal [5]	Real [6]	Nominal [5]	Real [6]	Nominal [5]	Real [6]
1949	2.52	15.40	0.33	1.99	4.68	28.61	7.52	46.00
1950	2.91	17.59	.44	2.66	4.95	29.97	8.30	50.22
1955	2.30	12.28	.94	5.02	6.88	36.72	10.12	54.02
1960	2.10	9.97	1.79	8.50	7.42	35.24	11.30	53.71
1965	2.40	10.64	2.57	11.39	8.15	36.15	13.11	58.18
1970	3.88	14.11	3.73	13.53	11.19	40.62	18.80	68.27
1971	4.01	13.87	4.05	14.00	11.71	40.50	19.77	68.38
1972	4.65	15.42	4.28	14.19	11.71	38.83	20.65	68.44
1973	5.14	16.14	4.98	15.64	13.07	41.05	23.20	72.84
1974	9.65	27.79	6.48	18.66	22.00	63.36	38.13	109.81
1975	12.67	33.33	8.85	23.28	23.45	61.70	44.96	118.31
1976	13.40	33.33	11.57	28.79	24.37	60.64	49.34	122.76
1977	13.91	32.53	15.82	37.00	25.79	60.32	55.52	129.86
1978	14.65	32.02	18.18	39.72	28.60	62.51	61.43	134.25
1979	18.55	37.44	24.16	48.75	39.45	79.63	82.16	165.82
1980	20.45	37.84	32.09	59.37	67.93	125.70	120.47	222.91
1981	21.75	36.79	39.51	66.84	99.40	168.13	160.66	271.75
1982	22.84	36.41	45.71	72.88	90.03	143.53	158.58	252.81
1983	20.32	31.16	43.73	67.06	83.05	127.36	147.10	225.58
1984	22.94	33.91	48.69	71.97	84.10	124.31	155.74	230.19
1985	22.27	31.94	43.35	62.18	78.88	113.16	144.50	207.28
1986	21.18	29.73	32.71	45.90	39.63	55.63	93.52	131.26
1987	21.20	28.96	29.11	39.77	46.93	64.12	97.24	132.85
1988	20.97	27.71	30.28	40.01	37.48	49.51	88.73	117.22
1989	21.40	27.24	30.58	38.93	44.07	56.10	96.05	122.27
1990	22.39	27.45	31.80	38.97	53.77	65.91	107.96	132.32
1991	21.40	25.35	30.39	35.99	44.77	53.02	96.57	114.36
1992	20.98	24.28	32.56	37.69	41.97	48.58	95.50	110.56
1993	18.77	21.23	38.72	43.81	35.61	40.29	93.10	105.34
1994	20.06	22.23	36.46	40.40	32.07	35.53	88.59	98.16
1995	19.45	21.12	30.24	32.83	35.00	38.00	84.69	91.95
1996	19.68	20.97	42.99	45.81	43.68	46.54	106.35	113.32
1997	19.77	20.72	46.09	48.30	40.57	42.52	106.43	111.55
1998	19.75	20.47	39.12	40.56	24.80	25.71	83.68	86.74
1999	18.30	18.70	43.37	44.32	33.40	34.13	95.08	97.15
2000	18.02	18.02	74.33	74.33	56.93	56.93	149.27	149.27
2001	19.60	19.14	82.28	80.35	46.25	45.16	148.13	144.66
2002	19.68	18.88	58.66	56.30	47.21	45.31	125.54	120.50
2003	19.13	17.98	97.47	91.61	57.14	53.71	173.75	163.29
2004	22.16	20.25	106.57	R97.35	72.93	R66.62	201.66	R184.22
2005	26.69	R23.62	R138.74	R122.77	95.03	R84.10	R260.46	R230.50
2006	R29.25	R25.10	R124.04	R106.41	R111.16	R95.36	R264.46	R226.87
2007P	29.10	24.32	128.77	107.61	123.89	103.53	281.75	235.45

[1] Coal values are based on free-on-board (f.o.b.) rail/barge prices, which are the f.o.b. prices of coal at the point of first sale, excluding freight or shipping and insurance costs. See "Free on Board (F.O.B.)" in Glossary.
[2] Natural gas values are for marketed production based on wellhead prices. See "Natural Gas Marketed Production" and "Natural Gas Wellhead Price" in Glossary.
[3] Includes lease condensate.
[4] Crude oil values are based on domestic first purchase prices. See "Crude Oil Domestic First Purchase Price" in Glossary.
[5] See "Nominal Dollars" in Glossary.
[6] In chained (2000) dollars, calculated by using gross domestic product implicit price deflators in Table D1. See "Chained Dollars" in Glossary.
R=Revised. P=Preliminary.
Note: Totals may not equal sum of components due to independent rounding.
Web Page: For all data beginning in 1949, see http://www.eia.doe.gov/emeu/aer/finan.html.
Sources: Tables 5.1, 5.18, 6.2, 6.7, 7.2, and 7.8.

Figure 3.3 Consumer Price Estimates for Energy by Source

Total Energy, 1970-2005

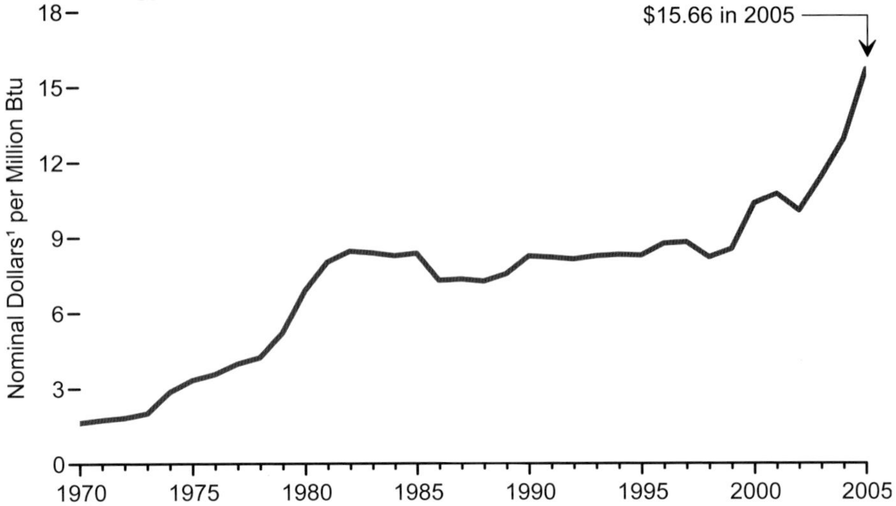

By Energy Type, 2005

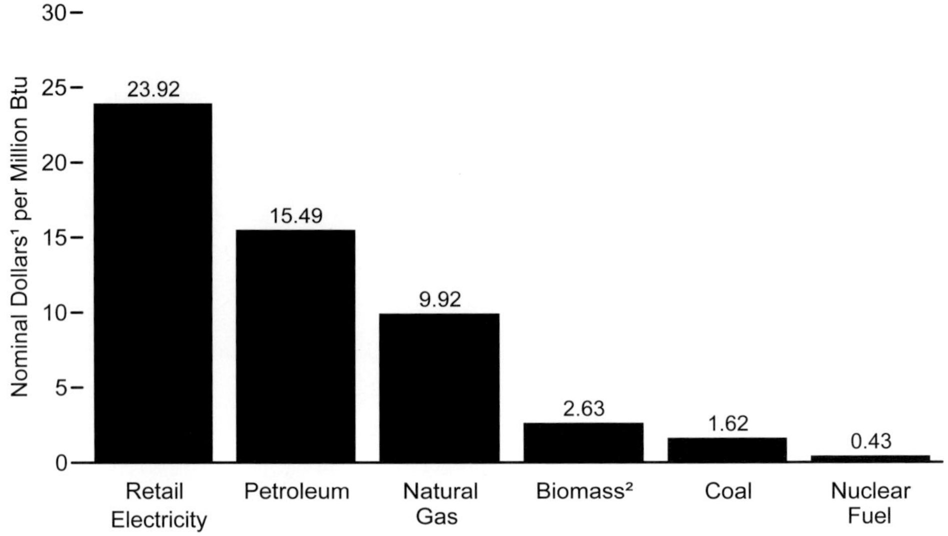

Prices[3] by Energy Type, Indexed, 1970-2005

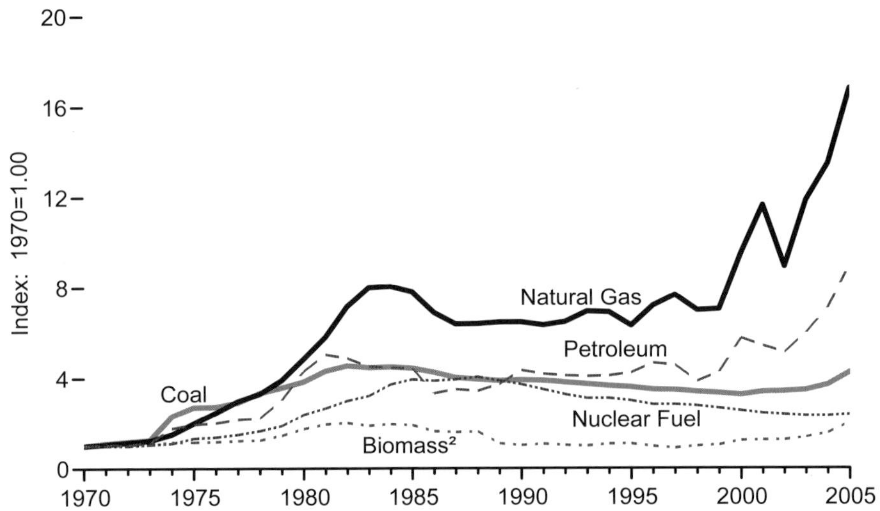

By Petroleum Product, 2005

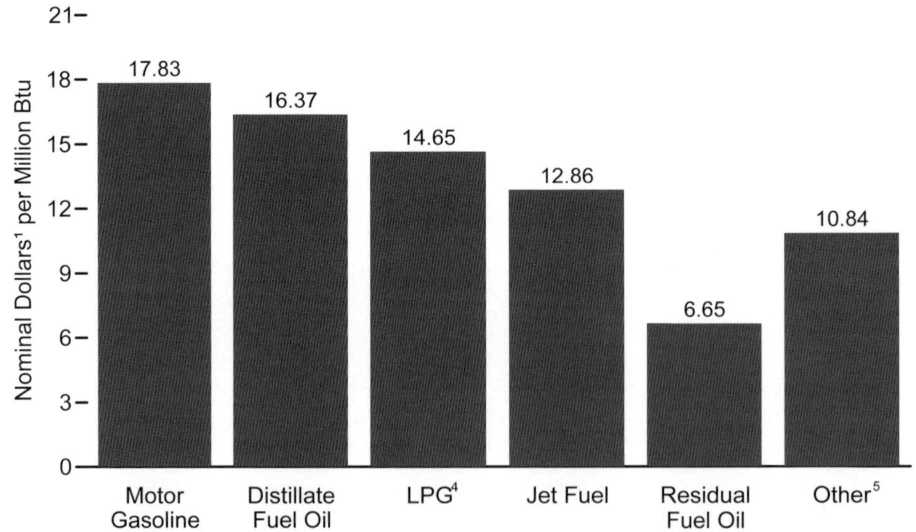

[1] See "Nominal Dollars" in Glossary.
[2] Wood and waste; excludes ethanol and biodiesel.
[3] Based on nominal dollars.
[4] Liquefied petroleum gases.
[5] Consumption-weighted average price for asphalt and road oil, aviation gasoline, kerosene, lubricants, petrochemical feedstocks, petroleum coke, special naphthas, waxes, and miscellaneous petroleum products.

Note: Because vertical scales differ, graphs should not be compared.
Source: Table 3.3.

Table 3.3 Consumer Price Estimates for Energy by Source, 1970-2005

(Nominal Dollars [1] per Million Btu)

Year	Primary Energy [2]: Coal	Primary Energy: Natural Gas [3]	Primary Energy: Petroleum: Distillate Fuel Oil	Petroleum: Jet Fuel [4]	Petroleum: LPG [5]	Petroleum: Motor Gasoline [6]	Petroleum: Residual Fuel Oil	Petroleum: Other [7]	Petroleum: Total	Primary Energy: Nuclear Fuel	Primary Energy: Biomass [8]	Primary Energy: Total [9,10]	Electric Power Sector [11,12]	Retail Electricity [13]	Total Energy [9,10,14]
1970	0.38	0.59	1.16	0.73	1.46	2.85	0.42	1.38	1.72	0.18	1.29	1.08	0.32	4.98	1.65
1971	.42	.63	1.22	.77	1.49	2.90	.58	1.45	1.79	.18	1.31	1.15	.38	5.30	1.76
1972	.45	.68	1.22	.79	1.52	2.88	.62	1.49	1.78	.18	1.33	1.18	.42	5.54	1.84
1973	.48	.73	1.46	.92	2.02	3.10	.75	1.58	1.97	.19	1.39	1.29	.47	5.86	2.02
1974	.88	.89	2.44	1.58	2.81	4.32	1.82	2.60	3.06	.20	1.50	1.94	.87	7.42	2.87
1975	1.03	1.18	2.60	2.05	2.97	4.65	1.93	2.94	3.35	.24	1.50	2.19	R.97	8.61	3.33
1976	1.04	1.46	2.77	2.25	3.21	4.84	1.90	3.08	3.47	.25	1.53	2.34	1.03	9.13	3.57
1977	1.11	1.76	3.11	2.59	3.65	5.13	2.14	3.27	3.73	.27	1.58	2.58	1.17	10.11	3.98
1978	1.27	1.95	3.26	2.87	3.60	5.24	2.08	3.45	3.84	.30	1.61	2.71	R1.27	10.92	4.23
1979	1.36	2.31	4.69	3.90	4.50	7.11	2.83	4.70	5.23	.34	1.88	3.47	R1.50	11.78	5.21
1980	1.46	2.86	6.70	6.36	5.64	9.84	3.88	7.04	7.40	.43	2.26	4.57	1.77	13.95	6.89
1981	1.64	3.43	8.03	7.57	6.18	10.94	4.91	8.67	8.68	.48	2.52	R5.25	R2.04	16.14	8.03
1982	1.73	4.23	7.78	7.23	6.66	10.39	4.65	7.87	8.40	.54	2.60	R5.33	R2.05	18.16	8.46
1983	1.70	4.72	7.32	6.53	7.17	9.12	4.50	7.60	7.77	.58	2.44	R5.12	R2.02	18.62	8.39
1984	1.71	4.75	7.37	6.25	6.93	8.89	4.75	7.67	7.68	.67	2.53	5.03	R2.02	18.50	8.28
1985	1.69	4.61	7.22	5.91	6.55	9.01	4.30	7.55	7.63	.71	2.47	4.92	R1.91	19.05	8.37
1986	1.62	4.07	5.68	3.92	6.44	6.79	2.37	5.80	5.73	.70	2.12	3.97	R1.60	19.05	7.30
1987	1.53	3.77	5.97	4.03	6.07	7.23	2.86	5.63	6.04	.71	2.07	R4.00	R1.57	18.74	7.34
1988	1.50	3.78	5.83	3.80	5.88	7.33	2.35	5.26	5.91	.73	2.09	3.89	R1.49	18.68	7.26
1989	1.48	3.82	6.43	4.39	5.54	8.02	2.72	5.50	6.43	.70	1.42	4.07	1.51	18.98	7.55
1990	1.49	3.82	7.68	5.68	6.77	9.12	3.17	5.82	7.47	.67	1.32	R4.46	1.48	19.32	8.25
1991	1.48	3.74	7.29	4.83	6.81	8.93	2.62	5.74	7.20	.63	1.39	R4.29	R1.40	19.84	8.20
1992	1.45	3.83	7.09	4.52	6.21	8.96	2.28	5.52	7.07	.59	1.32	4.24	R1.38	20.06	8.13
1993	1.42	4.10	7.08	4.29	6.23	8.83	2.26	5.50	7.01	.56	1.28	R4.26	R1.40	20.38	8.25
1994	1.39	4.08	6.99	3.95	6.66	8.96	2.32	5.47	7.06	.56	1.39	4.27	R1.36	20.33	8.30
1995	1.37	3.73	6.98	4.00	6.56	9.22	2.46	5.74	7.29	.54	1.40	4.23	R1.29	20.29	8.28
1996	1.33	4.25	7.87	4.82	8.03	9.85	2.80	6.19	8.02	.51	1.25	4.63	R1.35	20.16	8.75
1997	1.32	4.53	7.66	4.53	7.43	9.81	2.93	5.88	7.86	.51	1.15	R4.66	R1.38	20.13	8.80
1998	1.29	4.13	6.57	3.35	6.01	8.45	2.15	5.04	6.64	.50	1.27	R4.08	R1.32	19.80	8.20
1999	1.27	4.16	7.19	4.01	6.65	9.31	2.51	5.30	7.33	.48	1.34	R4.37	R1.33	19.52	8.53
2000	1.24	5.62	9.86	R6.64	10.20	12.01	4.32	6.97	9.91	.46	1.58	R5.73	R1.71	20.03	R10.34
2001	1.29	6.87	9.17	5.72	9.61	11.35	3.99	R6.36	9.32	.44	1.61	R5.81	R1.84	R21.41	R10.72
2002	R1.30	5.27	8.63	5.33	8.15	10.67	3.91	R6.54	8.82	.43	1.62	5.23	R1.53	R21.15	R10.05
2003	1.32	7.00	10.04	6.46	10.41	12.34	4.75	R7.55	R10.31	.42	1.78	R6.27	R1.83	R21.85	11.41
2004	1.41	7.95	12.22	8.93	R12.33	14.57	4.92	R8.48	R12.23	.42	2.03	R7.36	R1.99	R22.38	R12.90
2005	1.62	9.92	16.37	[4]12.86	14.65	17.83	6.65	10.84	15.49	.43	2.63	9.25	2.60	23.92	15.66

[1] See "Nominal Dollars" in Glossary.
[2] Consumption-weighted average prices for all sectors, including the electric power sector.
[3] Natural gas, plus a small amount of supplemental gaseous fuels.
[4] Through 2004, includes kerosene-type and naphtha-type jet fuel. Beginning in 2005, includes kerosene-type jet fuel only.
[5] Liquefied petroleum gases.
[6] Beginning in 1993, includes ethanol blended into motor gasoline.
[7] Consumption-weighted average price for asphalt and road oil, aviation gasoline, kerosene, lubricants, petrochemical feedstocks, petroleum coke, special naphthas, waxes, and miscellaneous petroleum products.
[8] Wood and waste; excludes ethanol and biodiesel.
[9] Includes coal coke imports and exports, which are not separately displayed. In 2005, coal coke imports averaged 8.92 dollars per million Btu, and coal coke exports averaged 3.39 dollars per million Btu.
[10] Includes electricity imports, which are not separately displayed. Also, in 1981-1992, includes ethanol blended into motor gasoline that is not included in the motor gasoline data for those years.
[11] Electricity-only and combined-heat-and-power (CHP) plants within the NAICS (North American Industry Classification System) 22 category whose primary business is to sell electricity, or electricity and heat, to the public. Through 1988, data are for electric utilities only; beginning in 1989, data are for electric utilities and independent power producers.
[12] Consumption-weighted average electric power sector price for coal, natural gas, petroleum, nuclear fuel, wood, waste, and electricity imports.
[13] Retail electricity prices paid by ultimate customers, reported by electric utilities and, beginning in 1996, other energy service providers.
[14] Consumption-weighted average price for primary energy and retail electricity in the four end-use sectors (residential, commercial, industrial, and transportation); excludes energy in the electric power sector.

R=Revised.

Notes: • Prices include taxes where data are available. • There are no direct fuel costs for hydroelectric, geothermal, wind, or solar energy.

Web Page: For related information, see http://www.eia.doe.gov/emeu/states/_seds.html.

Source: Energy Information Administration, "State Energy Data 2005: Prices and Expenditures" (February 2008), U.S. Table 1.

Figure 3.4 Consumer Price Estimates for Energy by End-Use Sector, 2005

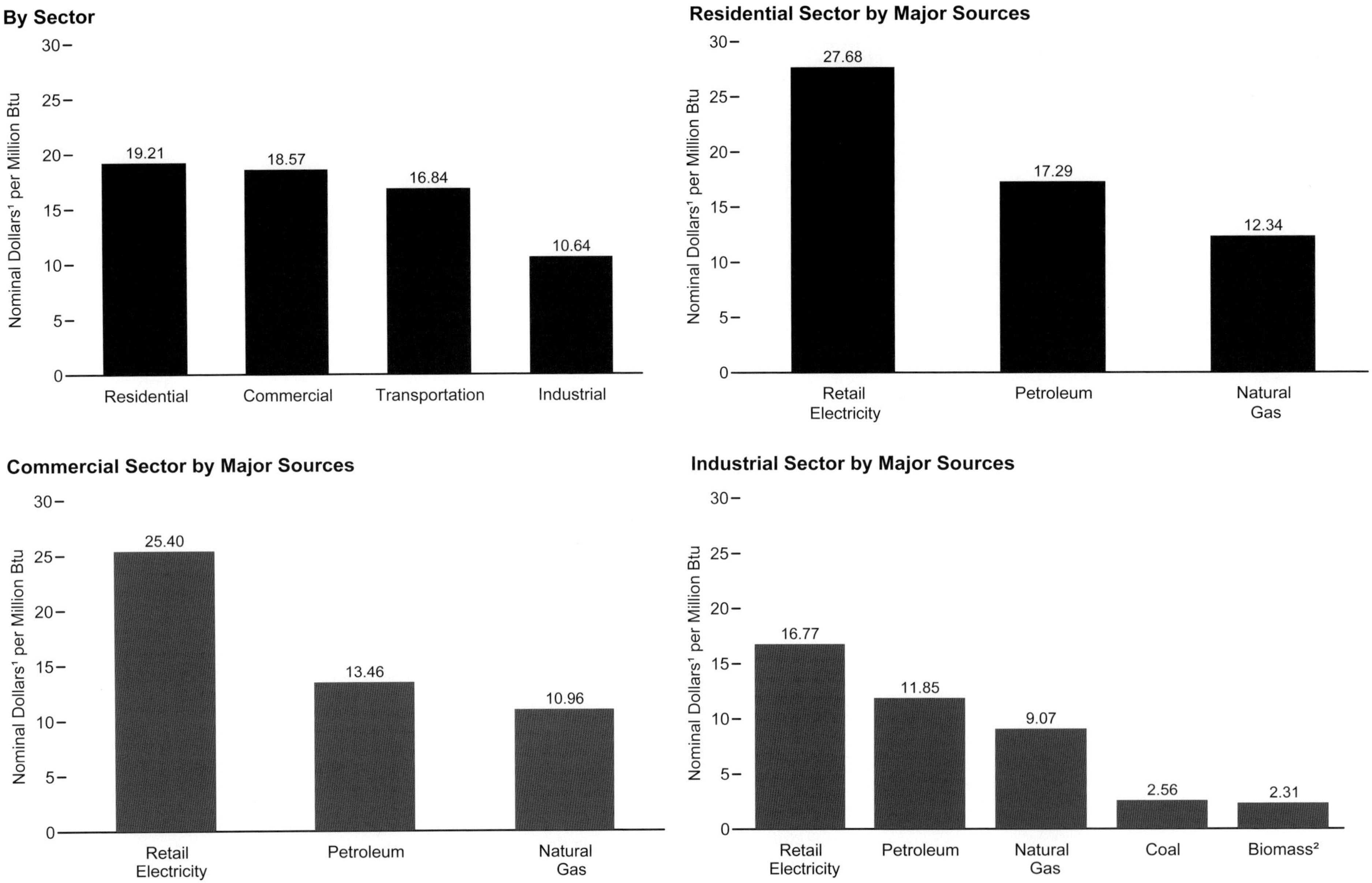

[1] See "Nominal Dollars" in Glossary.
[2] Wood and waste; excludes ethanol and biodiesel.

Notes: • Consumer prices are intended to represent prices paid by consumers. As such they include taxes where data are available. • There are no direct fuel costs for hydroelectric, geothermal, wind, or solar energy.
Source: Table 3.4.

Table 3.4 Consumer Price Estimates for Energy by End-Use Sector, 1970-2005

(Nominal Dollars [1] per Million Btu)

Year	Residential				Commercial				Industrial						Transportation	
	Natural Gas [2]	Petroleum	Retail Electricity [3]	Total [4]	Natural Gas [2]	Petroleum [5]	Retail Electricity [3]	Total [6]	Coal	Natural Gas [2]	Petroleum [5]	Biomass [7]	Retail Electricity [3]	Total [8]	Petroleum [5]	Total [9]
1970	1.06	1.56	6.51	2.10	0.75	0.90	6.09	1.98	0.45	0.38	0.98	1.59	2.99	0.84	2.31	2.31
1971	1.12	1.61	6.80	2.24	.80	1.01	6.44	2.16	.50	.41	1.05	1.59	3.22	.92	2.37	2.37
1972	1.18	1.63	7.09	2.37	.86	1.04	6.71	2.33	.55	.46	1.05	1.59	3.40	.99	2.38	2.38
1973	1.26	2.11	7.44	2.72	.91	1.20	7.06	2.56	.63	.50	1.18	1.60	3.66	1.10	2.57	2.57
1974	1.42	2.87	9.09	3.38	1.05	2.25	8.91	3.41	1.22	.67	2.24	1.60	4.95	1.78	3.70	3.70
1975	1.67	3.04	10.29	3.81	1.32	2.39	10.11	4.08	1.50	.95	2.46	1.60	6.07	2.20	4.02	4.02
1976	1.94	3.26	10.93	4.13	1.61	2.49	10.82	4.39	1.50	1.21	2.57	1.60	6.48	2.43	4.20	4.21
1977	2.30	3.66	11.87	4.77	2.00	2.84	11.99	5.13	1.56	1.48	2.84	1.59	7.33	2.78	4.47	4.48
1978	2.52	3.79	12.63	5.13	2.20	2.92	12.78	5.51	1.73	1.66	2.96	1.60	8.18	3.03	4.59	4.59
1979	2.92	5.33	13.60	6.00	2.69	4.15	13.72	6.28	1.75	1.96	3.99	1.60	8.94	3.63	6.19	6.19
1980	3.60	7.26	15.71	7.46	3.32	5.64	16.06	7.85	1.87	2.52	5.75	1.67	10.81	4.71	8.60	8.61
1981	4.19	8.68	18.17	8.82	3.91	7.00	18.44	9.49	2.06	3.07	6.84	1.67	12.57	5.52	9.83	9.84
1982	5.05	8.69	20.11	9.78	4.70	6.65	20.11	10.37	2.09	3.80	6.51	1.67	14.51	6.05	9.42	9.43
1983	5.88	8.43	21.04	10.66	5.43	6.51	20.57	10.94	1.91	4.10	6.57	1.67	14.54	6.21	8.44	8.45
1984	5.95	8.47	20.96	10.68	5.40	6.49	20.89	11.10	1.91	4.13	6.56	1.67	14.16	6.12	8.25	8.26
1985	5.94	8.17	21.66	10.91	5.34	6.38	21.30	11.65	1.90	3.87	6.29	1.67	14.57	6.03	8.26	8.27
1986	5.67	6.82	21.75	10.75	4.94	4.32	21.10	11.22	1.80	3.20	4.92	1.65	14.45	5.36	6.21	6.22
1987	5.39	6.66	21.82	10.71	4.64	4.65	20.44	10.98	1.67	2.88	4.96	1.65	13.98	5.17	6.57	6.59
1988	5.32	6.63	21.92	10.66	4.51	4.38	20.34	10.82	1.68	2.90	4.62	1.65	13.78	5.00	6.56	6.57
1989	5.47	7.59	22.41	11.02	4.61	4.99	20.77	11.27	1.68	2.93	4.69	1.20	13.85	4.92	7.17	7.18
1990	5.63	8.75	22.96	11.88	4.70	5.95	21.20	11.89	1.69	2.95	5.48	.99	13.92	5.23	8.27	8.28
1991	5.66	8.56	23.57	12.08	4.69	5.44	21.73	12.07	1.67	2.80	5.31	1.14	14.18	5.18	7.98	7.99
1992	5.73	7.89	24.06	11.98	4.75	5.23	22.15	12.17	1.69	2.91	5.00	1.13	14.18	5.13	7.91	7.93
1993	5.99	7.73	24.40	12.28	5.08	5.00	22.40	12.58	1.63	3.12	4.93	1.12	14.22	5.16	7.87	7.88
1994	6.23	7.81	24.57	12.63	5.35	4.89	22.35	12.74	1.62	3.09	5.04	1.15	14.00	5.15	7.91	7.92
1995	5.89	7.75	24.63	12.63	4.94	4.97	22.29	12.64	1.63	2.80	5.20	1.21	13.68	4.97	8.08	8.09
1996	6.16	8.92	24.50	12.73	5.26	6.01	22.17	12.78	1.62	3.30	6.04	1.01	13.49	5.40	8.76	8.77
1997	6.75	8.90	24.71	13.29	5.67	5.92	22.03	13.05	1.62	3.53	5.68	1.01	13.29	5.34	8.69	8.70
1998	6.61	7.88	24.21	13.48	5.38	4.88	21.48	13.07	1.58	3.16	4.54	1.24	13.13	4.91	7.47	7.48
1999	6.50	8.12	23.93	13.19	5.22	5.35	21.01	12.87	1.58	3.21	5.07	1.38	12.98	5.12	8.23	8.23
2000	7.64	11.55	24.14	14.27	6.56	8.09	21.52	13.93	1.55	4.61	7.50	1.43	13.60	6.49	10.78	R10.79
2001	R9.42	11.65	R25.16	R15.68	8.32	7.64	R22.99	R15.55	1.63	5.71	6.75	1.54	R14.78	R6.85	10.21	10.21
2002	7.71	10.40	R24.75	R14.72	6.49	6.96	R22.81	R14.68	1.75	4.37	R6.43	1.55	R14.30	R6.26	9.63	9.63
2003	9.23	12.30	R25.56	R15.85	8.02	8.38	R23.54	R15.61	1.74	6.03	R7.78	1.52	R14.97	R7.46	11.20	11.21
2004	R10.52	R13.86	R26.22	R17.12	R9.24	R9.89	R23.95	R16.58	1.99	R7.08	R9.32	1.75	R15.38	R8.53	13.36	13.37
2005	12.34	17.29	27.68	19.21	10.96	13.46	25.40	18.57	2.56	9.07	11.85	2.31	16.77	10.64	16.84	16.84

[1] See "Nominal Dollars" in Glossary.
[2] Natural gas, plus a small amount of supplemental gaseous fuels.
[3] Retail electricity prices paid by ultimate customers, reported by electric utilities and, beginning in 1996, other energy service providers.
[4] Includes coal and wood, which are not separately displayed.
[5] Beginning in 1993, includes ethanol blended into motor gasoline.
[6] Includes coal, wood, and waste, which are not separately displayed.
[7] Wood and waste; excludes ethanol and biodiesel.
[8] Includes coal coke imports and exports, which are not separately displayed.
[9] Includes coal, natural gas, and retail electricity, which are not separately displayed. Also, in 1981-1992, includes ethanol blended into motor gasoline that is not included in the petroleum data for those years.

R=Revised.

Notes: • Prices include taxes where data are available. • There are no direct fuel costs for hydroelectric, geothermal, wind, or solar energy.

Web Page: For related information, see http://www.eia.doe.gov/emeu/states/_seds.html.

Source: Energy Information Administration, "State Energy Data 2005: Prices and Expenditures" (February 2008), U.S. Tables 2-5.

Figure 3.5 Consumer Expenditure Estimates for Energy by Source

Total Energy, 1970-2005

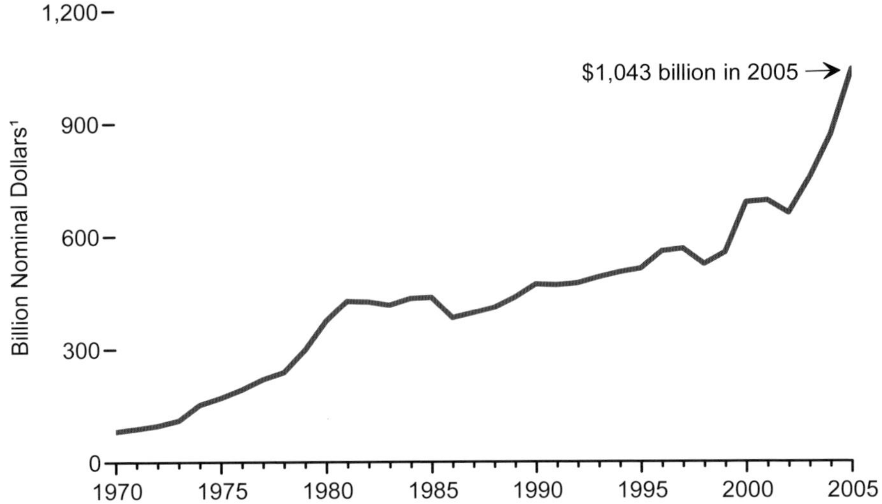

By Energy Type, 2005

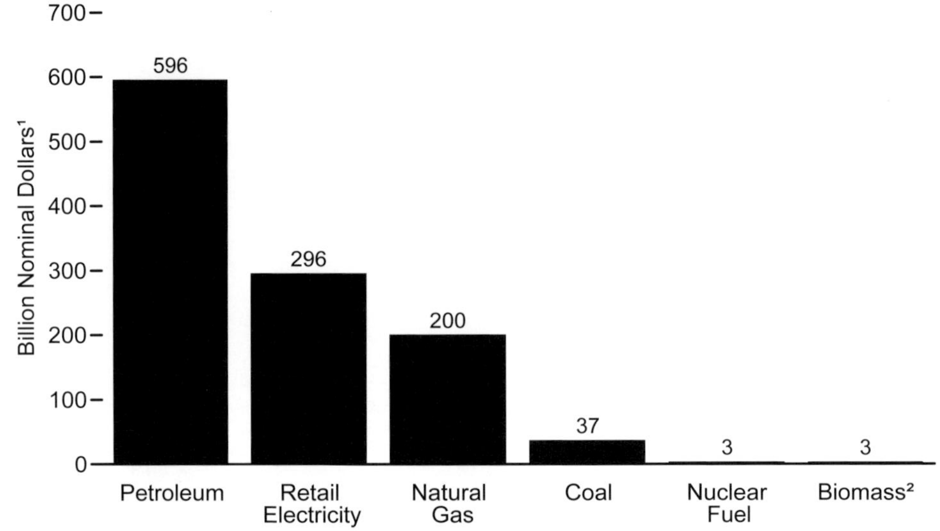

Expenditures[3] by Energy Type, Indexed, 1970-2005

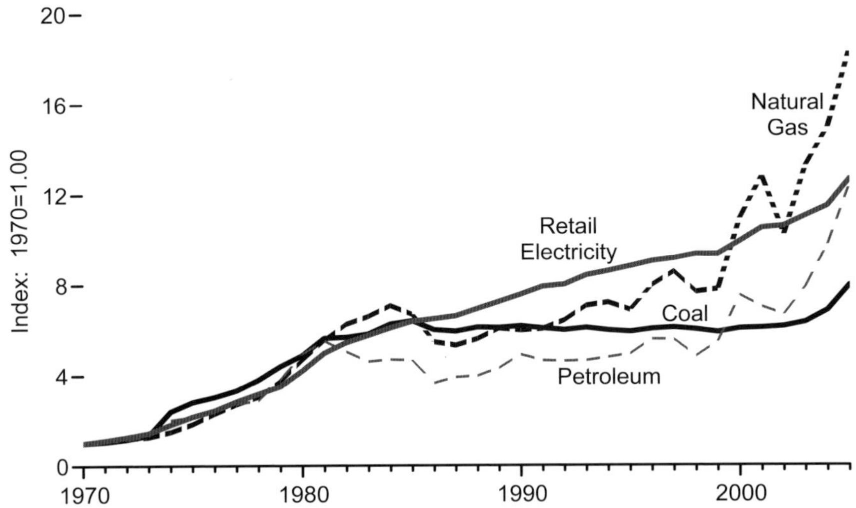

By Petroleum Product, 2005

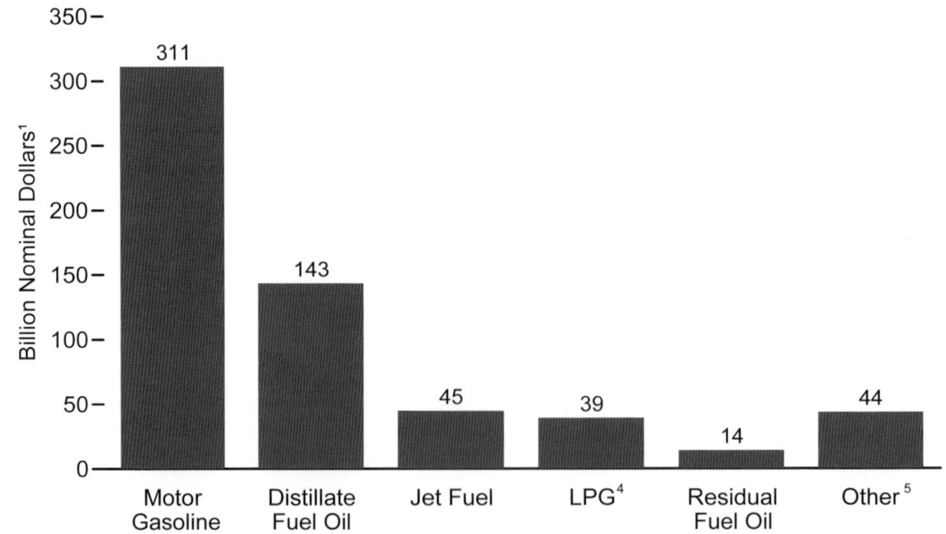

[1] See "Nominal Dollars" in Glossary.
[2] Wood and waste; excludes ethanol and biodiesel.
[3] Based on nominal dollars.
[4] Liquefied petroleum gases.

[5] Asphalt and road oil, aviation gasoline, kerosene, lubricants, petrochemical feedstocks, petroleum coke, special naphthas, waxes, and miscellaneous petroleum products.

Note: Because vertical scales differ, graphs should not be compared.

Source: Table 3.5.

Table 3.5 Consumer Expenditure Estimates for Energy by Source, 1970-2005

(Million Nominal Dollars [1])

Year	Primary Energy [2]															
				Petroleum												
	Coal	Coal Coke Net Imports [3]	Natural Gas [4]	Distillate Fuel Oil	Jet Fuel [5]	LPG [6]	Motor Gasoline [7]	Residual Fuel Oil	Other [8]	Total	Nuclear Fuel	Biomass [9]	Total [10,11]	Electric Power Sector [11,12,13]	Retail Electricity [14]	Total Energy [10,15]
1970	4,630	-75	10,891	6,253	1,441	2,446	31,596	2,046	4,172	47,955	44	438	R63,923	R-4,357	23,345	82,911
1971	4,902	-40	12,065	6,890	1,582	2,531	33,478	2,933	4,449	51,864	73	446	R69,360	R-5,491	26,202	90,071
1972	5,415	-26	13,198	7,552	1,682	2,889	35,346	3,458	4,777	55,702	104	476	R74,947	R-6,551	29,712	98,108
1973	6,243	7	13,933	9,524	2,001	3,933	39,667	4,667	5,318	65,109	177	502	R86,106	R-7,952	33,774	111,928
1974	11,118	150	16,380	15,217	3,208	5,273	54,194	10,547	8,284	96,723	259	544	R125,341	R-14,558	42,586	153,370
1975	13,021	82	20,061	15,680	4,193	5,231	59,446	10,374	8,493	103,416	448	534	R137,712	R-16,545	50,680	171,846
1976	14,051	44	25,097	18,402	4,567	5,993	64,977	11,648	9,925	115,513	520	622	R156,003	R-19,078	56,972	193,897
1977	15,416	67	29,602	22,004	5,517	6,824	70,591	14,381	11,790	131,106	743	694	R177,952	R-23,715	66,225	220,461
1978	17,551	362	33,185	23,587	6,205	6,621	74,513	13,747	13,348	138,021	915	782	R191,196	R-26,125	74,159	239,230
1979	20,376	259	40,785	32,854	8,603	9,383	95,916	17,656	18,785	183,197	941	964	R246,963	R-31,472	82,051	297,543
1980	22,607	-78	51,061	40,797	13,923	10,926	124,408	21,573	26,049	237,676	1,189	1,231	R314,278	R-38,027	98,095	374,346
1981	26,159	-31	60,544	48,200	15,607	11,900	138,138	22,668	28,571	265,084	1,436	1,429	R355,696	R-44,274	116,455	427,877
1982	26,349	-52	68,292	44,087	14,974	12,925	130,305	17,632	22,447	242,372	1,684	1,510	R341,398	R-42,354	127,393	426,437
1983	26,987	-44	72,000	41,846	13,979	14,083	115,803	14,099	21,573	221,382	1,859	1,517	R325,254	R-42,566	134,731	417,419
1984	29,025	-22	77,169	44,668	15,097	14,143	114,429	14,410	22,646	225,392	2,384	1,608	R337,248	R-44,686	142,420	434,982
1985	29,678	-34	72,938	43,972	14,747	13,579	118,048	11,493	22,088	223,928	2,878	1,597	R332,921	R-43,970	149,233	438,184
1986	27,855	-40	59,702	35,113	10,505	12,726	91,529	7,486	17,647	175,005	3,061	1,352	R268,633	R-37,016	151,793	383,409
1987	27,532	7	58,019	37,729	11,448	12,887	99,864	8,062	17,687	187,677	3,378	1,300	R280,012	R-38,182	154,685	396,515
1988	28,333	116	61,089	38,776	11,318	12,806	103,323	7,259	16,779	190,260	4,057	1,378	R286,915	R-38,553	162,063	410,426
1989	28,284	137	66,198	43,159	13,434	12,187	112,720	8,357	17,060	206,917	3,939	2,270	R309,108	R-40,829	169,332	437,611
1990	28,602	22	65,278	49,335	17,784	13,715	126,558	8,721	19,255	235,368	4,104	1,997	R336,475	R-40,626	176,691	472,539
1991	28,129	44	65,956	45,269	14,609	14,976	123,118	6,784	18,231	222,987	4,073	2,165	R324,544	R-38,752	184,767	470,559
1992	27,776	126	70,086	45,019	13,559	14,213	125,249	5,585	18,363	221,988	3,802	2,194	R327,344	R-38,663	186,906	475,587
1993	28,229	96	77,052	45,732	13,002	14,018	126,560	5,449	18,318	223,079	3,597	2,193	R334,953	R-40,317	196,532	491,168
1994	27,715	214	78,581	47,002	12,474	16,361	130,068	5,296	18,701	229,901	3,777	2,521	R343,724	R-40,352	200,831	504,204
1995	27,431	234	75,020	47,533	12,525	16,306	136,647	4,676	19,218	236,905	3,810	2,938	R347,245	R-39,073	205,876	514,049
1996	28,028	156	86,904	56,455	15,770	21,208	148,344	5,313	21,086	268,176	3,624	2,668	R390,501	R-41,652	211,105	559,954
1997	28,277	170	93,382	55,910	15,000	19,905	149,668	5,206	21,578	267,266	3,369	2,423	R395,874	R-42,947	213,843	566,770
1998	27,888	188	83,620	48,350	11,239	15,388	132,730	4,280	19,912	231,898	3,555	2,477	R350,687	R-43,311	218,361	525,737
1999	27,310	140	84,960	54,586	13,878	19,184	149,260	4,686	R21,221	R262,815	3,643	2,661	R382,809	R-44,689	218,413	R556,533
2000	28,080	146	R119,094	78,207	R23,777	29,879	193,947	8,870	R26,208	R360,889	3,628	3,196	R517,815	R-60,054	231,577	R689,338
2001	28,202	82	R139,388	74,924	19,602	25,734	185,892	7,266	R22,911	R336,329	3,524	2,912	R513,126	R-64,555	R245,483	R694,054
2002	R28,511	180	R111,368	69,271	17,802	23,148	179,511	6,156	R23,996	R319,884	3,504	2,792	R467,363	R-54,067	R247,598	R660,894
2003	29,402	169	R144,489	83,767	21,096	28,401	R209,592	8,325	R27,786	R378,967	3,362	2,978	R560,737	R-64,600	R257,995	R754,131
2004	R31,765	1,125	R162,702	105,662	30,219	R34,663	R253,228	9,717	R34,911	R468,400	3,445	3,301	R672,353	R-71,716	R268,136	R868,773
2005	36,933	633	200,303	143,261	[5]44,679	39,073	311,035	13,951	43,593	595,594	3,477	3,396	842,849	-95,704	295,789	1,042,934

[1] See "Nominal Dollars" in Glossary.
[2] Expenditures by all sectors, including the electric power sector.
[3] Values derive from U.S. Department of Commerce, Bureau of the Census, "Monthly Report IM-145" and "Monthly Report EM-545," and may differ slightly from those shown on Table 3.9, which derive from Bureau of the Census, *U.S. International Trade in Goods and Services,* FT600 series.
[4] Natural gas, plus a small amount of supplemental gaseous fuels.
[5] Through 2004, includes kerosene-type and naphtha-type jet fuel. Beginning in 2005, includes kerosene-type jet fuel only.
[6] Liquefied petroleum gases.
[7] Beginning in 1993, includes ethanol blended into motor gasoline.
[8] Asphalt and road oil, aviation gasoline, kerosene, lubricants, petrochemical feedstocks, petroleum coke, special naphthas, waxes, and miscellaneous petroleum products.
[9] Wood and waste; excludes ethanol and biodiesel.
[10] Includes electricity imports, which are not separately displayed. Also, in 1981-1992, includes ethanol blended into motor gasoline that is not included in the motor gasoline data for those years.
[11] Revisions are due to the removal of electricity exports.
[12] Electricity-only and combined-heat-and-power (CHP) plants within the NAICS (North American Industry Classification System) 22 category whose primary business is to sell electricity, or electricity and heat, to the public. Through 1988, data are for electric utilities only; beginning in 1989, data are for electric utilities and independent power producers.
[13] Expenditures by the electric power sector for coal, natural gas, petroleum, nuclear fuel, wood, waste, and electricity imports. Values are negative so the columns will sum to the "Total Energy" column.
[14] Retail electricity expenditures by ultimate customers, reported by electric utilities and, beginning in 1996, other energy service providers.
[15] Expenditures for primary energy and retail electricity by the four end-use sectors (residential, commercial, industrial, and transportation); excludes expenditures for energy by the electric power sector.

R=Revised.

Notes: • Expenditures include taxes where data are available. • There are no direct fuel costs for hydroelectric, geothermal, wind, or solar energy. • Totals may not equal the sum of components due to independent rounding.

Web Page: For related information, see http://www.eia.doe.gov/emeu/states/_seds.html.

Source: Energy Information Administration, "State Energy Data 2005: Prices and Expenditures" (February 2008), U.S. Table 1.

Figure 3.6 Consumer Expenditure Estimates for Energy by End-Use Sector, 2005

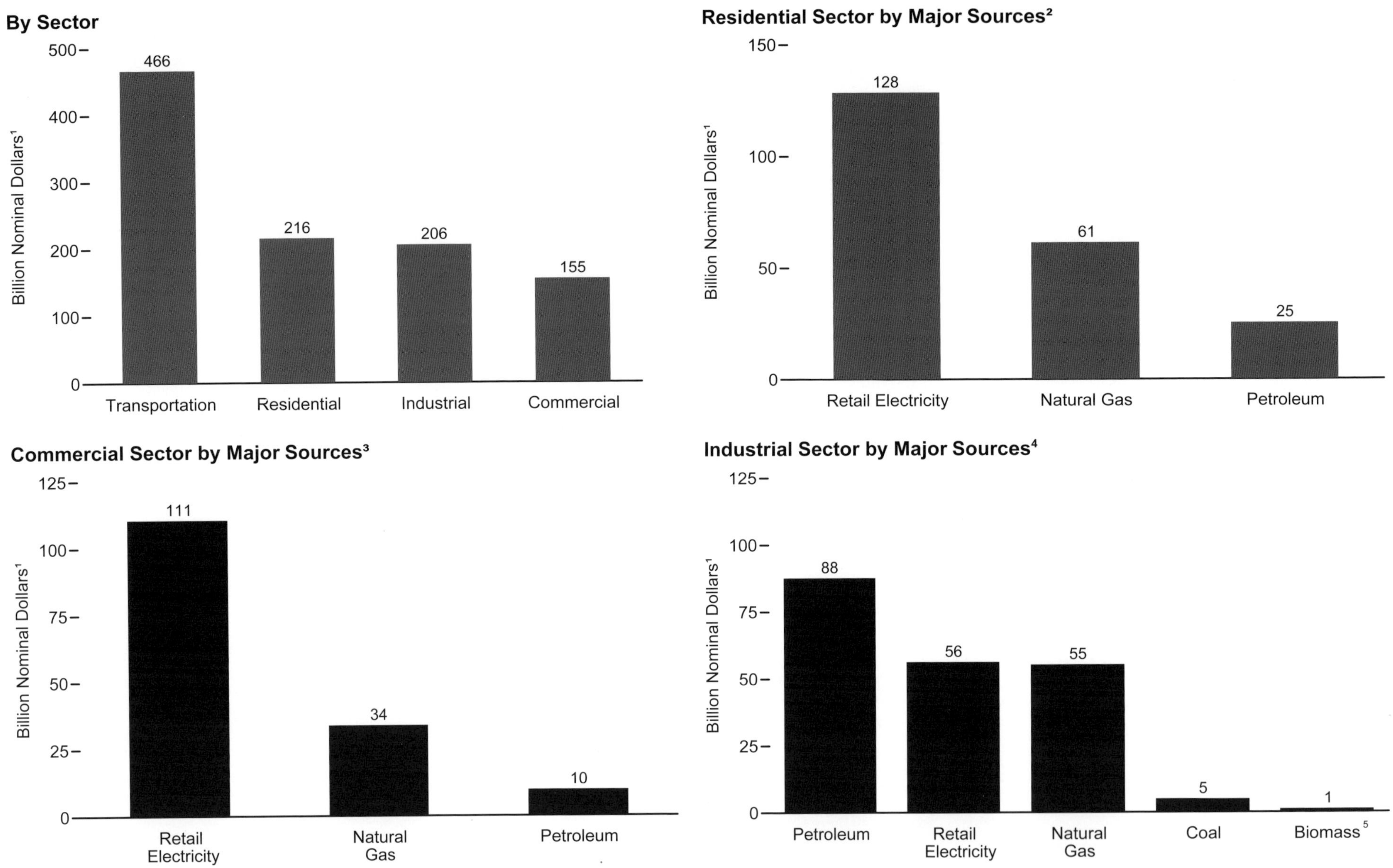

[1] See "Nominal Dollars" in Glossary.
[2] Expenditures for coal and wood are not displayed.
[3] Expenditures for coal, wood, and waste are not displayed.
[4] Expenditures for imports and exports of coal coke are not displayed.
[5] Wood and waste; excludes ethanol and biodiesel.

Notes: • Petroleum accounts for nearly all transportation sector expenditures. • There are no direct fuel costs for hydroelectric, geothermal, wind, or solar energy. • Totals may not equal the sum of components due to independent rounding. • Because vertical scales differ, graphs should not be compared.
Source: Table 3.6.

Table 3.6 Consumer Expenditure Estimates for Energy by End-Use Sector, 1970-2005

(Million Nominal Dollars [1])

	Residential				Commercial				Industrial						Transportation	
Year	Natural Gas [2]	Petroleum	Retail Electricity [3]	Total [4]	Natural Gas [2]	Petroleum [5]	Retail Electricity [3]	Total [6]	Coal	Natural Gas [2]	Petroleum [5]	Biomass [7]	Retail Electricity [3]	Total [8]	Petroleum [5]	Total [9]
1970	5,272	4,286	10,352	20,213	1,844	1,391	7,319	10,628	2,082	2,625	6,069	366	5,624	16,691	35,327	35,379
1971	5,702	4,466	11,589	22,033	2,060	1,523	8,301	11,970	1,971	3,019	6,663	374	6,256	18,244	37,766	37,824
1972	6,223	4,731	13,034	24,203	2,289	1,599	9,443	13,410	2,212	3,335	7,180	404	7,173	20,278	40,154	40,218
1973	6,282	5,957	14,712	27,137	2,421	1,881	10,707	15,094	2,527	3,936	8,600	425	8,284	23,779	45,846	45,918
1974	6,949	7,392	17,924	32,563	2,741	3,205	13,373	19,494	4,704	4,971	15,408	421	11,184	36,837	64,368	64,476
1975	8,410	7,582	20,644	36,932	3,385	3,133	16,157	22,869	5,498	5,844	15,544	386	13,760	41,113	70,813	70,933
1976	9,992	8,857	22,621	41,785	4,379	3,638	18,148	26,345	5,448	7,484	18,384	443	16,083	47,887	77,759	77,880
1977	11,324	9,872	26,132	47,694	5,094	4,297	21,023	30,611	5,360	8,958	22,190	464	18,956	55,996	86,047	86,160
1978	12,565	9,926	29,069	51,949	5,812	4,230	23,166	33,443	5,722	10,114	23,203	511	21,798	61,710	92,003	92,128
1979	14,772	11,276	31,683	58,267	7,623	5,534	25,433	38,819	6,247	12,110	33,705	512	24,797	77,630	122,688	122,826
1980	17,497	12,695	38,458	69,418	8,858	7,267	30,611	46,932	5,888	16,350	42,765	529	28,863	94,316	163,517	163,680
1981	19,502	13,394	44,780	78,630	10,085	7,628	37,484	55,462	6,441	20,432	47,171	558	34,007	108,578	184,946	185,207
1982	23,987	12,526	50,045	87,618	12,565	6,706	41,759	61,343	5,301	20,504	41,841	540	35,364	103,498	173,553	173,978
1983	26,564	11,486	53,918	92,948	13,602	7,392	43,529	64,825	4,735	21,461	38,437	610	37,017	102,216	156,841	157,430
1984	27,873	12,433	55,777	97,170	14,012	8,121	47,304	69,763	5,420	23,763	41,563	622	39,050	110,395	156,979	157,654
1985	27,136	12,894	58,672	99,772	13,368	6,671	50,092	70,396	5,252	21,615	38,876	619	40,190	106,518	160,745	161,498
1986	25,147	10,613	60,776	97,353	11,770	4,869	51,449	68,337	4,745	16,479	30,567	639	39,271	91,661	125,353	126,058
1987	23,926	10,875	63,318	98,851	11,601	5,078	51,900	68,795	4,448	15,909	31,092	636	39,109	91,201	136,807	137,667
1988	25,332	11,204	66,793	104,104	12,377	4,655	54,411	71,669	4,744	17,257	29,123	662	40,507	92,410	141,382	142,244
1989	26,951	12,739	69,243	109,770	12,908	4,998	57,460	75,610	4,650	18,770	28,561	1,323	42,255	95,695	155,591	156,536
1990	25,439	12,308	72,378	111,097	12,681	5,669	60,627	79,284	4,636	19,348	34,132	906	43,358	102,402	178,852	179,757
1991	26,508	11,916	76,828	116,212	13,175	4,871	63,407	81,740	4,332	18,912	32,511	1,034	44,201	101,034	170,589	171,572
1992	27,599	11,264	76,848	116,631	13,685	4,469	64,233	82,676	4,245	20,553	32,225	1,079	45,474	103,703	171,482	172,577
1993	30,533	11,200	82,814	125,351	14,967	3,903	67,626	86,790	4,060	22,367	31,550	1,146	45,726	104,946	173,704	174,082
1994	31,028	11,089	84,552	127,395	15,927	3,847	69,637	89,694	4,060	22,556	33,626	1,279	46,257	107,992	178,724	179,123
1995	29,362	10,715	87,610	128,388	15,383	3,638	72,481	91,788	4,068	21,487	34,170	1,699	45,402	107,060	186,411	186,813
1996	33,219	13,278	90,503	137,822	17,106	4,518	74,121	96,053	3,943	26,167	40,796	1,432	46,102	118,596	207,078	207,483
1997	34,590	12,712	90,704	138,674	18,755	4,168	77,153	100,396	3,887	28,411	39,833	1,435	45,610	119,347	207,940	208,353
1998	30,875	10,352	93,360	135,101	16,667	3,222	78,999	99,138	3,566	24,515	32,220	1,600	45,634	107,723	183,368	183,775
1999	31,577	11,956	93,482	137,573	16,351	3,540	79,141	99,290	3,457	24,079	[R]36,937	1,786	45,429	[R]111,828	207,433	207,843
2000	38,959	18,051	98,209	156,089	21,339	6,121	85,129	112,870	3,507	34,624	[R]53,509	1,888	47,859	[R]141,533	[R]278,398	[R]278,846
2001	46,189	17,924	[R]103,158	[R]167,997	25,879	5,666	[R]93,402	[R]125,231	3,572	38,597	[R]46,983	1,751	[R]48,519	[R]139,503	260,813	[R]261,323
2002	38,490	15,208	[R]106,834	[R]161,205	20,926	4,741	[R]93,763	[R]119,719	3,526	31,031	[R]45,505	1,537	[R]46,606	[R]128,386	[R]251,107	[R]251,585
2003	48,278	18,923	[R]111,249	[R]179,287	26,411	6,464	[R]96,263	[R]129,456	3,552	41,168	[R]53,953	1,400	[R]49,962	[R]150,205	[R]294,538	[R]295,183
2004	[R]52,265	[R]21,350	[R]115,577	[R]190,174	[R]29,518	[R]7,442	[R]100,546	[R]137,920	4,064	[R]47,322	[R]70,718	1,509	[R]51,491	[R]176,229	[R]363,764	[R]364,450
2005	61,196	25,150	128,393	216,046	33,838	9,724	110,522	154,576	5,004	55,247	87,679	1,183	56,229	205,975	465,476	466,337

[1] See "Nominal Dollars" in Glossary.
[2] Natural gas, plus a small amount of supplemental gaseous fuels.
[3] Retail electricity expenditures by ultimate customers, reported by electric utilities and, beginning in 1996, other energy service providers.
[4] Includes coal and wood, which are not separately displayed.
[5] Beginning in 1993, includes ethanol blended into motor gasoline.
[6] Includes coal, wood, and waste, which are not separately displayed.
[7] Wood and waste; excludes ethanol and biodiesel.
[8] Includes coal coke imports and exports, which are not separately displayed.
[9] Includes coal, natural gas, and retail electricity, which are not separately displayed. Also, in 1981-1992, includes ethanol blended into motor gasoline that is not included in the petroleum data for those years.

R=Revised.

Notes: • Expenditures include taxes where data are available. • There are no direct fuel costs for hydroelectric, geothermal, wind, or solar energy. • Totals may not equal the sum of components due to independent rounding.

Web Page: For related information, see http://www.eia.doe.gov/emeu/states/_seds.html.

Source: Energy Information Administration, "State Energy Data 2005: Prices and Expenditures" (February 2008), U.S. Tables 2-5.

Figure 3.7 Value of Fossil Fuel Imports

Total, 1949-2007

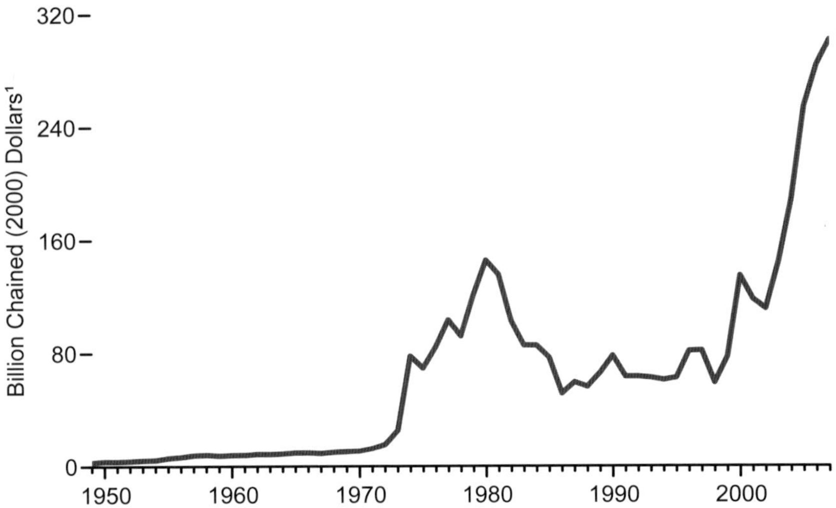

By Fuel, 2007

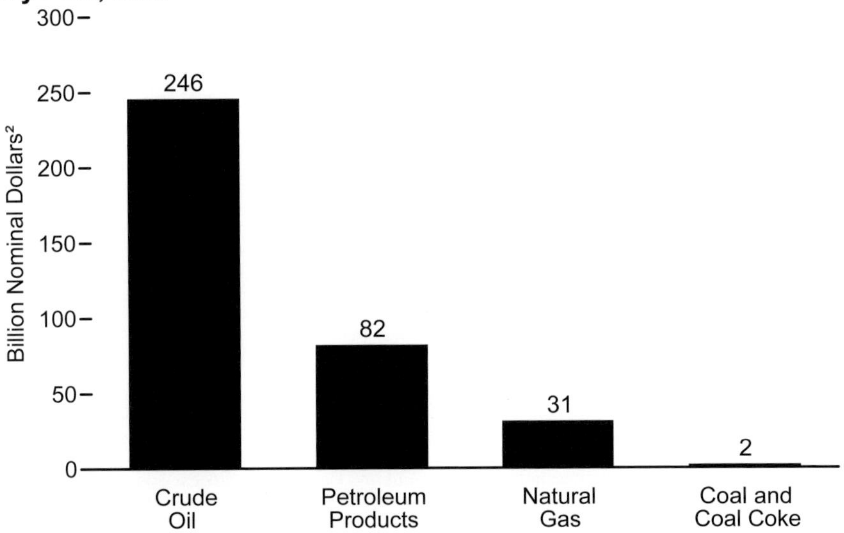

By Fuel, 1949-2007

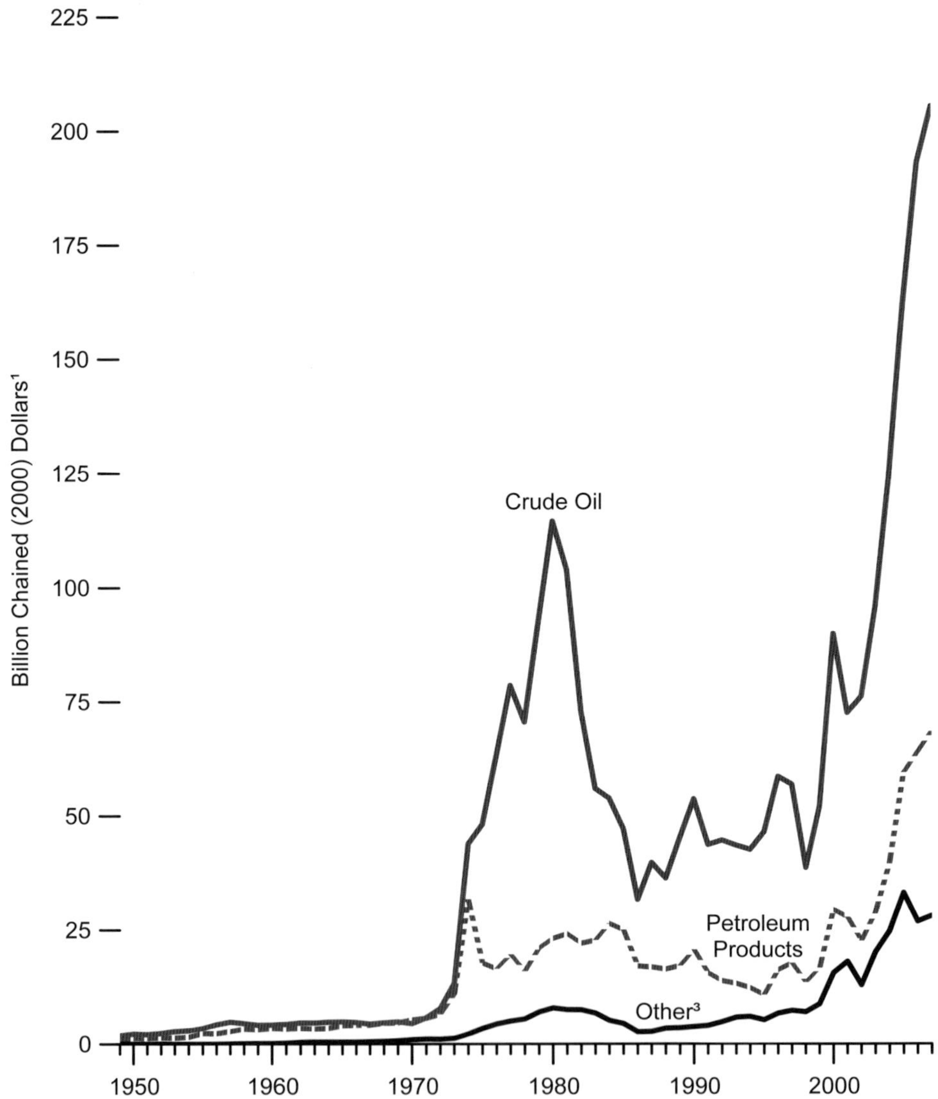

[1] Calculated by using gross domestic product implicit price deflators in Table D1. See "Chained Dollars" in Glossary.

[2] See "Nominal Dollars" in Glossary.

[3] Natural gas, coal, and coal coke.

Note: Because vertical scales differ, graphs should not be compared.

Source: Table 3.7.

Table 3.7 Value of Fossil Fuel Imports, Selected Years, 1949-2007

(Billion Dollars)

Year	Coal		Coal Coke		Natural Gas		Crude Oil [1]		Petroleum Products [2]		Total	
	Nominal [3]	Real [4]	Nominal [3]	Real [4]	Nominal [3]	Real [4]	Nominal [3]	Real [4]	Nominal [3]	Real [4]	Nominal [3]	Real [4]
1949	(s)	0.01	(s)	0.02	0.00	0.00	0.30	1.86	0.14	0.84	0.45	2.74
1950	(s)	.02	.01	.03	.00	.00	.37	2.23	.21	1.30	.59	3.58
1955	(s)	.01	(s)	.01	(s)	.01	.65	3.49	.44	2.36	1.10	5.88
1960	(s)	.01	(s)	.01	.03	.13	.90	4.25	.73	3.48	1.66	7.88
1965	(s)	.01	(s)	.01	.11	.47	1.12	4.97	.92	4.10	2.15	9.55
1970	(s)	(s)	(s)	.01	.26	.94	1.26	4.58	1.48	5.38	3.00	10.91
1971	(s)	.01	.01	.02	.31	1.08	1.69	5.84	1.66	5.73	3.66	12.67
1972	(s)	(s)	(s)	.02	.31	1.04	2.37	7.85	1.99	6.59	4.68	15.51
1973	(s)	.01	.04	.12	.36	1.14	4.24	13.31	3.50	10.98	8.14	25.56
1974	.06	.17	.19	.56	.53	1.53	15.25	43.92	11.01	31.71	27.05	77.89
1975	.02	.06	.16	.41	1.15	3.03	18.29	48.13	6.77	17.81	26.39	69.44
1976	.02	.04	.11	.28	1.66	4.13	25.46	63.33	6.65	16.55	33.90	84.33
1977	.04	.09	.13	.31	2.00	4.68	33.59	78.57	8.42	19.69	44.18	103.34
1978	.07	.16	.41	.89	2.06	4.50	32.30	70.59	7.30	15.96	42.15	92.11
1979	.05	.10	.34	.69	3.13	6.31	46.06	92.96	10.45	21.09	60.03	121.15
1980	.03	.06	.05	.10	4.21	7.80	61.90	114.54	12.54	23.21	78.74	145.69
1981	.03	.05	.04	.07	4.41	7.46	61.46	103.96	14.30	24.18	80.24	135.73
1982	.02	.04	.01	.01	4.69	7.48	45.72	72.89	13.86	22.10	64.31	102.53
1983	.04	.07	(s)	(s)	4.39	6.73	36.49	55.96	14.84	22.76	55.77	85.52
1984	.05	.07	.05	.07	3.44	5.08	36.44	53.87	17.87	26.41	57.84	85.49
1985	.07	.10	.04	.06	3.05	4.37	32.90	47.20	17.47	25.05	53.53	76.79
1986	.08	.11	.03	.04	1.82	2.56	22.61	31.73	12.18	17.10	36.72	51.53
1987	.06	.08	.05	.08	1.93	2.64	29.13	39.80	12.37	16.89	43.54	59.48
1988	.06	.08	.19	.26	2.38	3.14	27.55	36.39	12.43	16.43	42.62	56.30
1989	.10	.12	.22	.28	2.51	3.19	35.53	45.23	13.50	17.18	51.85	66.00
1990	.09	.11	.07	.09	2.97	3.64	43.78	53.66	16.90	20.72	63.83	78.23
1991	.11	.13	.09	.11	3.24	3.83	36.90	43.70	13.17	15.60	53.51	63.37
1992	.13	.15	.14	.17	3.96	4.58	38.55	44.63	11.98	13.87	54.77	63.40
1993	.25	.29	.17	.19	4.77	5.40	38.47	43.53	11.74	13.28	55.40	62.68
1994	.27	.30	.27	.30	4.90	5.43	38.48	42.63	11.14	12.35	55.07	61.01
1995	.32	.35	.33	.35	4.23	4.59	42.81	46.48	9.95	10.80	57.64	62.58
1996	.27	.29	.24	.26	5.79	6.17	54.93	58.53	15.27	16.27	76.51	81.52
1997	.26	.27	.25	.27	6.50	6.81	54.23	56.83	[5]16.93	[5]17.74	78.16	81.91
1998	.28	.29	.29	.30	6.21	6.44	37.25	38.61	13.01	13.49	57.05	59.13
1999	.28	.29	.23	.23	8.03	8.21	50.89	52.00	16.28	16.64	75.71	77.36
2000	.38	.38	.25	.25	14.94	14.94	89.88	89.88	29.38	29.38	134.81	134.81
2001	.67	.66	.19	.19	17.62	17.21	74.29	72.55	28.45	27.79	121.23	118.39
2002	.60	.58	.24	.23	12.61	12.10	79.25	76.07	23.52	22.57	116.22	111.55
2003	.79	.74	.24	.22	20.39	19.16	101.80	95.67	30.64	28.79	153.85	144.59
2004	1.02	.94	1.23	1.13	24.74	R22.60	136.03	R124.27	43.24	R39.50	206.26	R188.43
2005	1.42	1.26	.78	.69	35.25	R31.19	182.94	R161.90	67.12	R59.40	287.52	R254.44
2006	1.78	1.53	.64	.55	R28.80	R24.71	R225.16	R193.16	R74.56	R63.96	R330.93	R283.90
2007	P1.73	P1.45	P.48	P.40	E31.32	E26.17	P245.78	P205.39	P81.91	P68.45	P361.21	P301.86

[1] Beginning in 1977, includes imports into the Strategic Petroleum Reserve.

[2] Includes petroleum preparations, liquefied propane and butane, and, beginning in 1997, other mineral fuels.

[3] See "Nominal Dollars" in Glossary.

[4] In chained (2000) dollars, calculated by using gross domestic product implicit price deflators in Table D1. See "Chained Dollars" in Glossary.

[5] There is a discontinuity in this time series between 1996 and 1997 due to the addition of the commodity category "Other Mineral Fuels."

R=Revised. P=Preliminary. E=Estimate. (s)=Less than 0.005 billion.

Notes: • Includes value of imports into Puerto Rico from foreign countries; excludes receipts into the 50 States and the District of Columbia from the Virgin Islands and Puerto Rico. • Totals may not equal sum of components due to independent rounding.

Web Page: For all data beginning in 1949, see http://www.eia.doe.gov/emeu/aer/finan.html.

Sources: **Coal** and **Coal Coke:** Bureau of the Census, Foreign Trade Division, upublished data. **Natural Gas:** • 1949-1962—Bureau of the Census, *U.S. Imports of Merchandise for Consumption,* FT110. • 1963—Bureau of the Census, *U.S. Imports of Merchandise for Consumption,* FT125. • 1964-1971—Bureau of the Census, *U.S. Imports for Consumption and General Imports,* FT246. • 1972 and 1973—Federal Power Commission, *Pipeline Imports and Exports of Natural Gas - Imports and Exports of LNG.* • 1974-1977—Federal Power Commission, *United States Imports and Exports of Natural Gas,* annual reports. • 1978-1981—Energy Information Administration (EIA), *U.S. Imports and Exports of Natural Gas,* annual reports. • 1982-2005—EIA, *Natural Gas Monthly (NGM),* monthly reports. • 2006—EIA, *NGM* (March 2008), Table 4. • 2007—EIA estimate. **Crude Oil** and **Petroleum Products:** • 1949-1962—Bureau of the Census, *U.S. Imports of Merchandise for Consumption,* FT110. • 1963—Bureau of the Census, *U.S. Imports of Merchandise for Consumption,* FT125. • 1964-1988—Bureau of the Census, *U.S. Imports for Consumption,* FT135. • 1989 forward—Bureau of the Census, Foreign Trade Division, *U.S. Merchandise Trade,* FT900, "Exhibit 15. Exports and Imports of Goods by Principal SITC Commodity Groupings," December issues.

Figure 3.8 Value of Fossil Fuel Exports

Total, 1949-2007

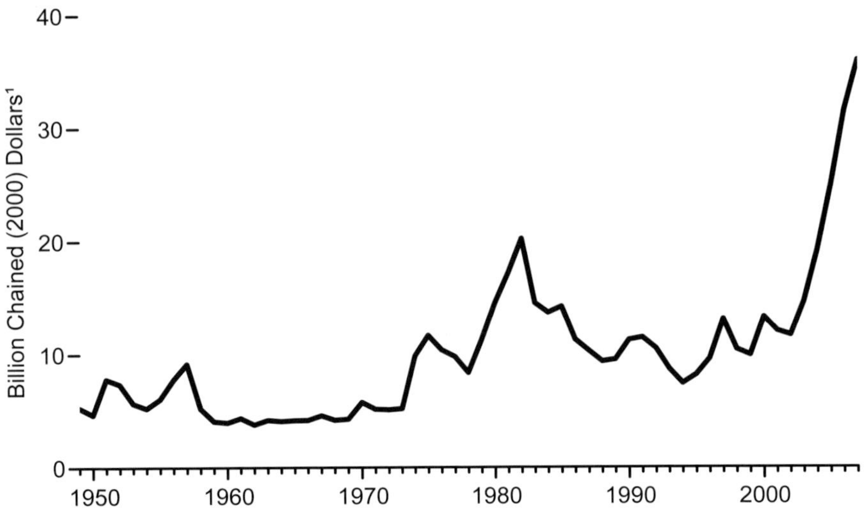

By Fuel, 2007

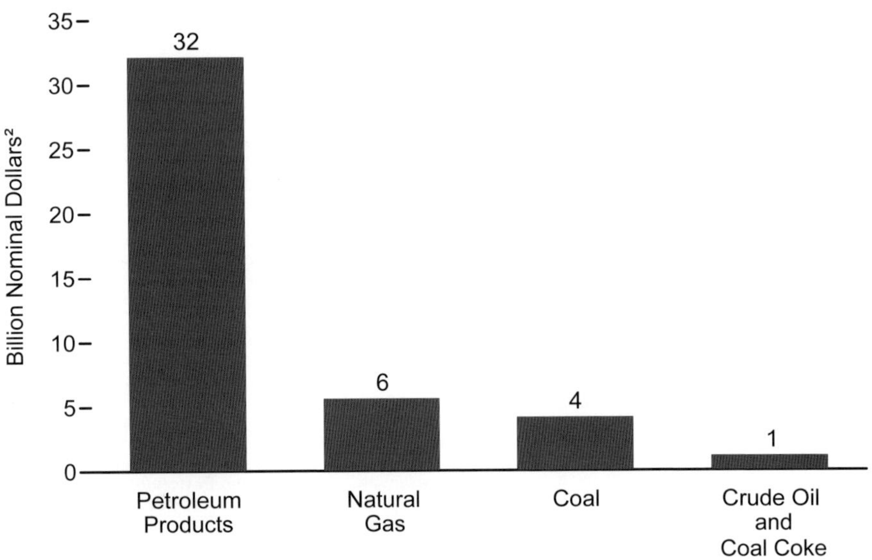

By Fuel, 1949-2007

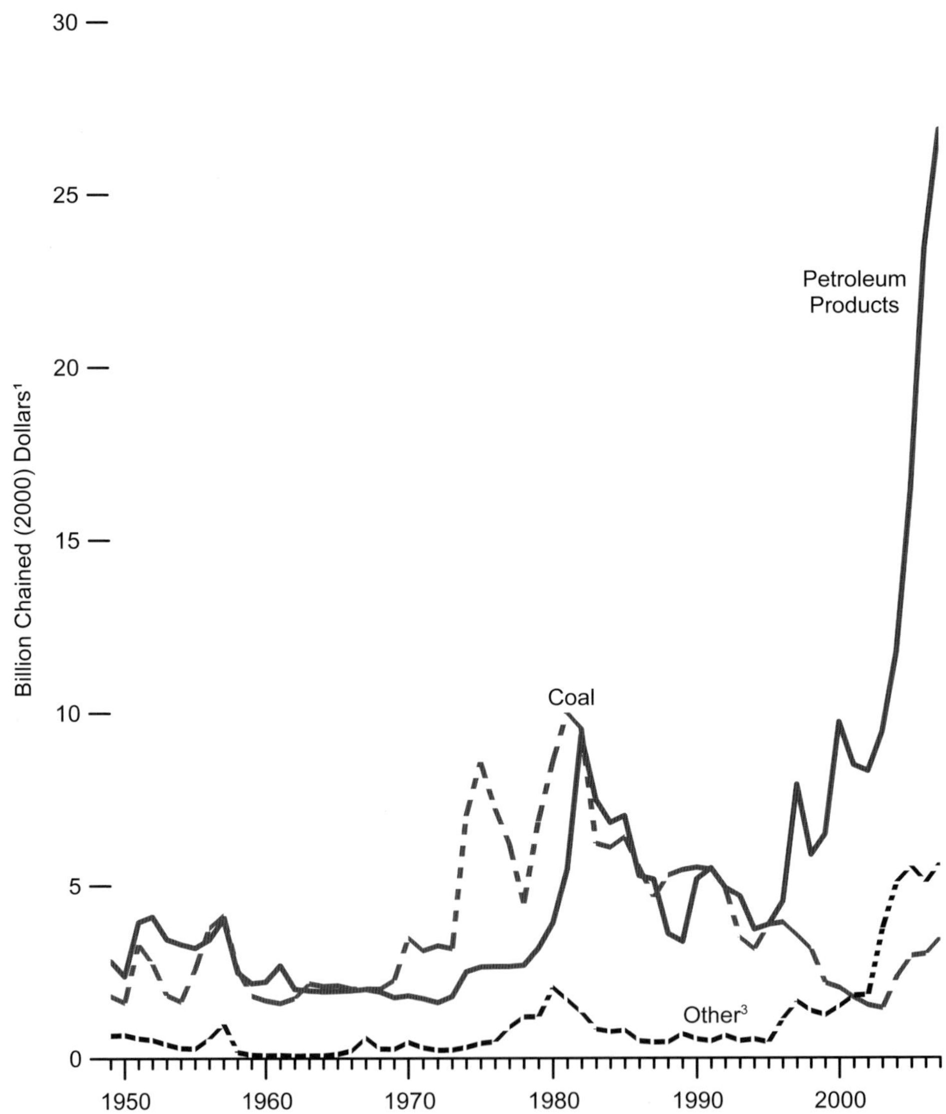

[1] Calculated by using gross domestic product implicit price deflators in Table D1. See "Chained Dollars" in Glossary.

[2] See "Nominal Dollars" in Glossary.

[3] Natural gas, crude oil, and coal coke.

Note: Because vertical scales differ, graphs should not be compared.

Source: Table 3.8.

Table 3.8 Value of Fossil Fuel Exports, Selected Years, 1949-2007

(Billion Dollars)

Year	Coal		Coal Coke		Natural Gas		Crude Oil		Petroleum Products [1]		Total	
	Nominal [2]	Real [3]	Nominal [2]	Real [3]	Nominal [2]	Real [3]	Nominal [2]	Real [3]	Nominal [2]	Real [3]	Nominal [2]	Real [3]
1949	0.30	1.82	0.01	0.05	(s)	0.01	0.10	0.60	0.46	2.82	0.87	5.30
1950	.27	1.63	.01	.04	(s)	.02	.10	.62	.39	2.39	.78	4.69
1955	.48	2.59	.01	.04	.01	.03	.04	.21	.60	3.20	1.14	6.07
1960	.35	1.68	.01	.03	(s)	.02	.01	.04	.47	2.22	.84	3.99
1965	.48	2.12	.02	.07	.01	.03	(s)	.02	.44	1.95	.95	4.19
1970	.96	3.49	.08	.29	.03	.11	.02	.07	.50	1.82	1.59	5.78
1971	.90	3.12	.04	.16	.04	.13	.01	.02	.50	1.74	1.49	5.16
1972	.98	3.26	.03	.10	.04	.13	(s)	.01	.49	1.62	1.55	5.12
1973	1.01	3.18	.03	.10	.04	.13	(s)	.01	.57	1.80	1.66	5.22
1974	2.44	7.02	.04	.13	.05	.16	.01	.04	.87	2.51	3.42	9.85
1975	3.26	8.58	.07	.20	.09	.24	(s)	(s)	1.01	2.65	4.43	11.66
1976	2.91	7.24	.07	.17	.10	.25	.03	.07	1.07	2.66	4.17	10.39
1977	2.66	6.21	.07	.17	.11	.25	.21	.49	1.14	2.66	4.18	9.78
1978	2.05	4.48	.05	.11	.11	.24	.39	.85	1.23	2.69	3.83	8.38
1979	3.40	6.86	.08	.16	.13	.26	.39	.80	1.58	3.20	5.58	11.27
1980	4.63	8.56	.13	.24	.23	.42	.75	1.39	2.12	3.92	7.86	14.54
1981	5.92	10.01	.07	.13	.35	.59	.58	.98	3.24	5.48	10.16	17.18
1982	5.99	9.55	.06	.10	.30	.48	.47	.75	5.86	9.34	12.68	20.22
1983	4.06	6.22	.05	.07	.28	.43	.22	.34	4.88	7.48	9.48	14.54
1984	4.13	6.11	.07	.10	.27	.40	.19	.27	4.62	6.82	9.27	13.71
1985	4.47	6.41	.08	.11	.26	.38	.23	.32	4.90	7.02	9.93	14.24
1986	3.93	5.52	.07	.09	.17	.24	.12	.16	3.77	5.29	8.05	11.30
1987	3.40	4.65	.05	.07	.17	.23	.13	.17	3.80	5.19	7.54	10.30
1988	4.01	5.30	.08	.10	.20	.27	.08	.10	2.72	3.60	7.09	9.37
1989	4.29	5.46	.08	.10	.27	.34	.21	.26	2.65	3.38	7.49	9.54
1990	4.51	5.53	.05	.06	.27	.32	.14	.17	4.23	5.19	9.20	11.27
1991	4.62	5.47	.05	.06	.33	.40	.03	.04	4.65	5.51	9.69	11.48
1992	4.24	4.91	.04	.05	.49	.56	.03	.04	4.27	4.94	9.07	10.50
1993	3.09	3.49	.06	.07	.36	.41	.02	.02	4.15	4.69	7.68	8.69
1994	2.85	3.16	.04	.04	.40	.45	.05	.05	3.36	3.72	6.71	7.43
1995	3.57	3.87	.05	.05	.37	.40	.01	.01	3.56	3.87	7.55	8.20
1996	3.69	3.93	.06	.07	.46	.49	.56	.60	4.25	4.53	9.02	9.61
1997	3.39	3.55	.05	.06	.47	.50	1.04	1.09	[4]7.55	[4]7.91	12.51	13.11
1998	3.04	3.15	.04	.05	.39	.40	.90	.93	5.68	5.89	10.04	10.41
1999	2.13	2.18	.03	.03	.43	.44	.77	.79	6.35	6.48	9.71	9.92
2000	2.04	2.04	.05	.05	1.00	1.00	.46	.46	9.73	9.73	13.28	13.28
2001	1.80	1.76	.11	.11	1.56	1.53	.19	.18	8.68	8.48	12.34	12.05
2002	1.60	1.54	.06	.06	1.76	1.69	.09	.09	8.65	8.30	12.17	11.68
2003	1.55	1.45	.07	.07	3.77	3.54	.16	.15	10.05	9.45	15.59	14.65
2004	2.60	2.37	.11	.10	5.20	4.75	.28	.25	12.85	R11.74	21.04	19.22
2005	3.35	2.97	.15	.13	5.53	R4.89	.60	.53	18.56	R16.42	28.18	R24.94
2006	3.52	R3.02	.13	.11	R4.94	R4.24	.85	.73	R27.32	R23.44	R36.77	R31.54
2007	P4.16	P3.47	P.13	P.11	E5.62	E4.70	P.99	P.83	P32.13	P26.85	P43.03	P35.96

[1] Includes petroleum preparations, liquefied propane and butane, and, beginning in 1997, other mineral fuels.

[2] See "Nominal Dollars" in Glossary.

[3] In chained (2000) dollars, calculated by using gross domestic product implicit price deflators in Table D1. See "Chained Dollars" in Glossary.

[4] There is a discontinuity in this time series between 1996 and 1997 due to the addition of the commodity category "Other Mineral Fuels."

R=Revised. P=Preliminary. E=Estimate. (s)=Less than 0.005 billion.

Notes: • Includes value of exports from Puerto Rico to foreign countries; excludes shipments from the 50 States and the District of Columbia to the Virgin Islands and Puerto Rico. • Totals may not equal sum of components due to independent rounding.

Web Page: For all data beginning in 1949, see http://www.eia.doe.gov/emeu/aer/finan.html.

Sources: **Coal** and **Coal Coke:** Bureau of the Census, Foreign Trade Division, unpublished data. **Natural Gas:** • 1949-1971—Bureau of the Census, *U.S. Exports,* FT410. • 1972 and 1973—Federal Power Commission, *Pipeline Imports and Exports of Natural Gas - Imports and Exports of LNG.* • 1974-1977—Federal Power Commission, *United States Imports and Exports of Natural Gas,* annual reports. • 1978-1981—Energy Information Administration (EIA), *U.S. Imports and Exports of Natural Gas,* annual reports. • 1982-2005—EIA, *Natural Gas Monthly (NGM),* monthly reports. • 2006—EIA, *NGM* (March 2008), Table 4. • 2007—EIA estimate. **Crude Oil** and **Petroleum Products:** • 1949-1988—Bureau of the Census, *U.S. Exports,* FT410. • 1989 forward—Bureau of the Census, Foreign Trade Division, *U.S. Merchandise Trade,* FT900, "Exhibit 15. Exports and Imports of Goods by Principal SITC Commodity Groupings," December issues.

Figure 3.9 Value of Fossil Fuel Net Imports, 1949-2007

Value of Fossil Fuel Net Imports

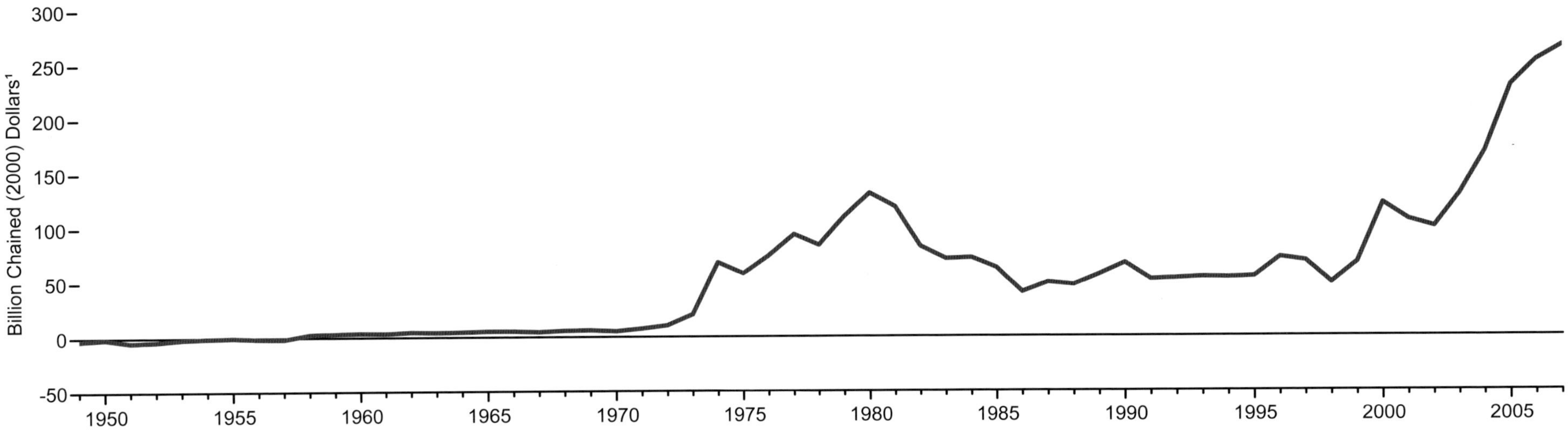

Value of Fossil Fuel Net Imports by Fuel

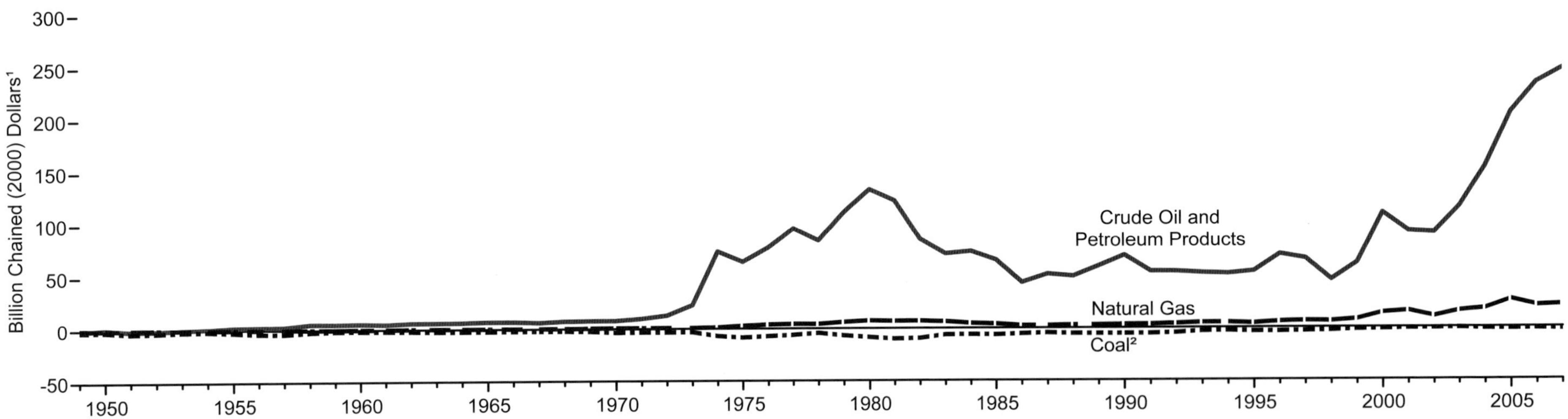

[1] Calculated by using gross domestic product implicit price deflators in Table D1. See "Chained Dollars" in Glossary.

[2] Includes small amounts of coal coke.

Note: Negative imports indicate that the value of exports is greater than the value of imports.

Source: Table 3.9.

Table 3.9 Value of Fossil Fuel Net Imports, Selected Years, 1949-2007

(Billion Dollars)

Year	Coal		Coal Coke		Natural Gas		Crude Oil		Petroleum Products [1]		Total	
	Nominal [2]	Real [3]	Nominal [2]	Real [3]	Nominal [2]	Real [3]	Nominal [2]	Real [3]	Nominal [2]	Real [3]	Nominal [2]	Real [3]
1949	-0.29	-1.80	(s)	-0.03	(s)	-0.01	0.21	1.26	-0.32	-1.98	-0.42	-2.56
1950	-.27	-1.61	(s)	-.01	(s)	-.02	.27	1.61	-.18	-1.09	-.18	-1.11
1955	-.48	-2.57	-.01	-.04	-.01	-.03	.62	3.29	-.16	-.84	-.04	-.19
1960	-.35	-1.67	-.01	-.03	.02	.12	.89	4.22	.26	1.26	.82	3.89
1965	-.48	-2.11	-.01	-.07	.10	.44	1.11	4.95	.48	2.15	1.21	5.36
1970	-.96	-3.49	-.08	-.27	.23	.83	1.24	4.51	.98	3.56	1.41	5.14
1971	-.90	-3.11	-.04	-.14	.27	.95	1.68	5.82	1.15	3.99	2.17	7.50
1972	-.98	-3.26	-.03	-.09	.28	.91	2.37	7.85	1.50	4.97	3.13	10.39
1973	-1.01	-3.18	.01	.02	.32	1.01	4.24	13.31	2.93	9.19	6.48	20.34
1974	-2.38	-6.85	.15	.43	.48	1.37	15.24	43.89	10.14	29.20	23.63	68.04
1975	-3.24	-8.52	.08	.22	1.06	2.79	18.29	48.13	5.76	15.16	21.96	57.78
1976	-2.89	-7.20	.04	.11	1.56	3.88	25.43	63.26	5.58	13.89	29.72	73.94
1977	-2.62	-6.12	.06	.14	1.89	4.43	33.38	78.08	7.28	17.03	40.00	93.56
1978	-1.98	-4.32	.36	.79	1.95	4.26	31.91	69.73	6.07	13.27	38.31	83.73
1979	-3.35	-6.75	.26	.52	3.00	6.05	45.66	92.16	8.87	17.89	54.44	109.88
1980	-4.60	-8.51	-.08	-.14	3.98	7.37	61.15	113.15	10.42	19.28	70.88	131.15
1981	-5.89	-9.96	-.03	-.05	4.06	6.87	60.88	102.98	11.06	18.71	70.09	118.55
1982	-5.97	-9.52	-.05	-.08	4.39	7.00	45.25	72.15	8.00	12.76	51.63	82.31
1983	-4.01	-6.16	-.04	-.07	4.11	6.30	36.27	55.62	9.96	15.28	46.28	70.98
1984	-4.09	-6.04	-.02	-.03	3.17	4.68	36.26	53.59	13.25	19.58	48.57	71.79
1985	-4.39	-6.30	-.03	-.05	2.79	4.00	32.68	46.87	12.57	18.03	43.60	62.55
1986	-3.85	-5.40	-.04	-.06	1.65	2.32	22.49	31.57	8.42	11.81	28.67	40.23
1987	-3.35	-4.58	.01	.01	1.76	2.41	29.00	39.63	8.57	11.71	36.00	49.18
1988	-3.95	-5.22	.12	.15	2.18	2.88	27.47	36.29	9.71	12.83	35.53	46.93
1989	-4.19	-5.33	.14	.17	2.24	2.85	35.32	44.97	10.85	13.81	44.35	56.46
1990	-4.42	-5.41	.02	.03	2.71	3.32	43.65	53.50	12.67	15.53	54.63	66.96
1991	-4.51	-5.34	.04	.05	2.90	3.43	36.87	43.66	8.52	10.09	43.82	51.90
1992	-4.11	-4.76	.10	.11	3.47	4.02	38.52	44.59	7.72	8.93	45.70	52.90
1993	-2.83	-3.21	.11	.12	4.41	4.99	38.45	43.50	7.59	8.59	47.72	54.00
1994	-2.58	-2.86	.23	.26	4.50	4.98	38.43	42.58	7.78	8.62	48.37	53.58
1995	-3.24	-3.52	.27	.30	3.86	4.19	42.81	46.48	6.39	6.94	50.09	54.38
1996	-3.41	-3.64	.18	.19	5.33	5.68	54.37	57.93	11.01	11.74	67.49	71.91
1997	-3.13	-3.28	.20	.21	6.02	6.31	53.19	55.74	[4]9.37	[4]9.82	65.65	68.80
1998	-2.75	-2.86	.25	.26	5.82	6.03	36.36	37.69	7.33	7.60	47.00	48.72
1999	-1.85	-1.90	.20	.20	7.61	7.77	50.12	51.21	9.94	10.15	66.00	67.44
2000	-1.66	-1.66	.20	.20	13.94	13.94	89.41	89.41	19.65	19.65	121.53	121.53
2001	-1.13	-1.10	.08	.08	16.05	15.68	74.11	72.37	19.77	19.31	108.89	106.34
2002	-1.00	-.96	.18	.17	10.85	10.41	79.16	75.98	14.87	14.27	104.06	99.87
2003	-.76	-.71	.17	.16	16.62	15.62	101.64	95.52	20.59	19.35	138.26	129.93
2004	-1.57	-1.44	1.12	1.03	19.54	R17.85	135.75	R124.02	30.38	R27.76	185.23	R169.22
2005	-1.93	-1.71	.63	.56	29.72	R26.30	182.35	R161.37	48.56	R42.98	259.34	R229.50
2006	-1.74	R-1.49	.51	.44	R23.86	R20.47	R224.30	R192.42	R47.24	R40.53	R294.17	R252.36
2007	P-2.42	P-2.03	P.35	P.29	E25.70	E21.48	P244.78	P204.56	P49.77	P41.59	P318.18	P265.89

[1] Includes petroleum preparations, liquefied propane and butane, and, beginning in 1997, other mineral fuels.

[2] See "Nominal Dollars" in Glossary.

[3] In chained (2000) dollars, calculated by using gross domestic product implicit price deflators in Table D1. See "Chained Dollars" in Glossary.

[4] There is a discontinuity in this time series between 1996 and 1997 due to the addition of the commodity category "Other Mineral Fuels."

R=Revised. P=Preliminary. E=Estimate. (s)=Less than 0.005 billion.

Notes: • Net imports equal imports minus exports. Minus sign indicates that the value of exports is greater than the value of imports. • Totals may not equal sum of components due to independent rounding. • Data on this table may not equal data on Table 3.7 minus data on Table 3.8 due to independent rounding.

Web Page: For all data beginning in 1949, see http://www.eia.doe.gov/emeu/aer/finan.html.

Sources: Tables 3.7 and 3.8.

Figure 3.10 Major U.S. Energy Companies' Domestic Production and Refining

Crude Oil and Natural Gas Liquids Production by Major Energy Companies, 1974-2006

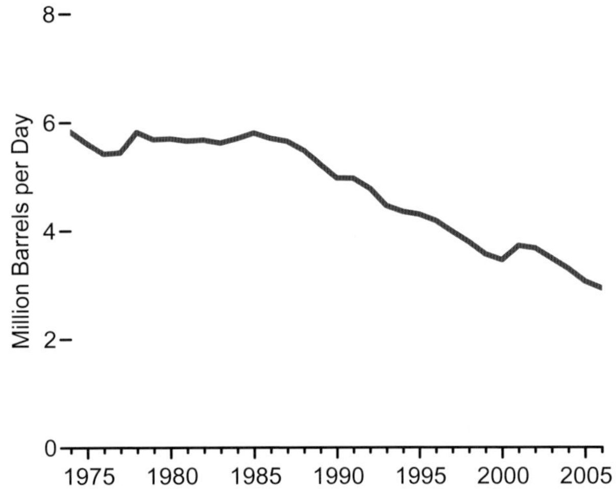

Dry Natural Gas Production by Major Energy Companies, 1974-2006

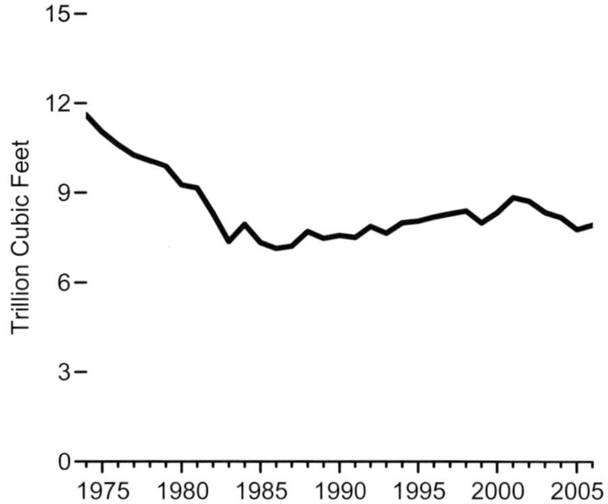

Coal Production by Major Energy Companies, 1974-2005

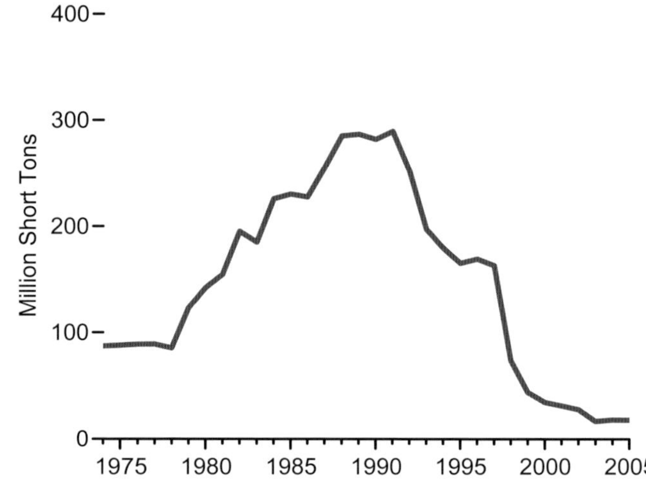

Major Energy Companies' Shares of U.S. Total Production, 1974-2006

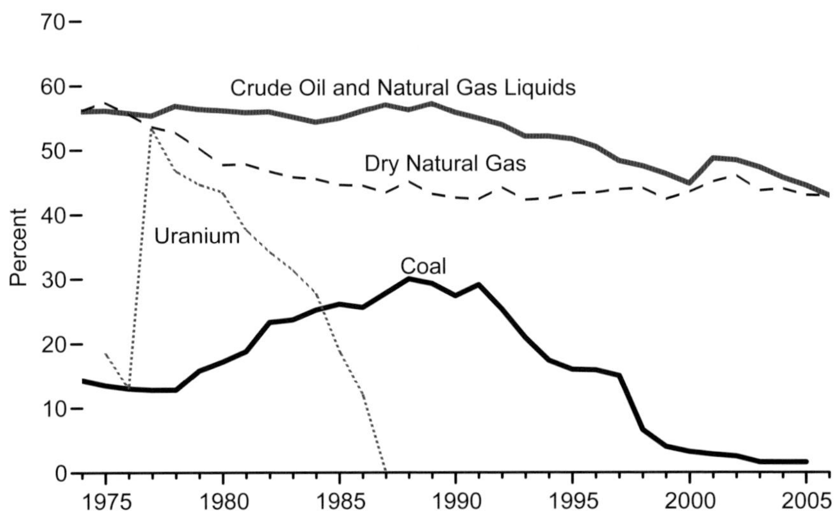

Major Energy Companies' Shares of U.S. Refining Capacity and Output, 1974-2006

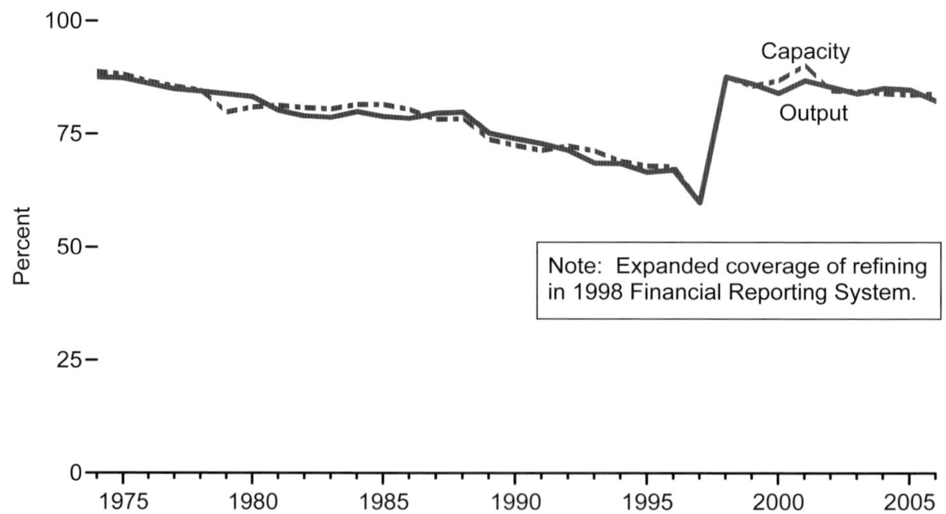

Notes: • "Major U.S. Energy Companies" are the top publicly-owned, U.S.-based crude oil and natural gas producers and petroleum refiners that form the Financial Reporting System (FRS). See Table 3.14. • Because vertical scales differ, graphs should not be compared.

Source: Table 3.10.

Table 3.10 Major U.S. Energy Companies' Domestic Production and Refining, 1974-2006

Year	Production								Refining			
	Crude Oil and Natural Gas Liquids [1]		Dry Natural Gas [1]		Coal [2]		Uranium		Capacity [3]		Output [4]	
	Million Barrels per Day	Percent of U.S. Total	Trillion Cubic Feet	Percent of U.S. Total	Million Short Tons	Percent of U.S. Total	Million Pounds [5]	Percent of U.S. Total [6]	Million Barrels per Day	Percent of U.S. Total [7]	Million Barrels per Day	Percent of U.S. Total [7]
1974	5.9	56.0	11.6	56.1	87.4	14.3	NA	NA	13.3	88.8	11.8	87.6
1975	5.6	56.1	11.0	57.4	88.1	13.5	4.3	18.6	13.4	88.2	12.0	87.4
1976	5.4	55.7	10.6	55.6	89.0	13.0	3.3	13.0	14.2	86.7	12.6	86.2
1977	5.5	55.3	10.3	53.6	89.1	12.8	16.0	53.4	14.6	85.6	13.5	85.0
1978	5.8	56.8	10.1	52.7	85.5	12.8	17.3	46.8	14.8	84.7	13.5	84.5
1979	5.7	56.3	9.9	50.3	123.3	15.8	16.7	44.7	14.4	79.9	13.2	83.9
1980	5.7	56.1	9.3	47.7	142.3	17.2	19.0	43.5	15.1	81.0	12.2	83.3
1981	5.7	55.8	9.2	47.8	154.8	18.8	14.5	37.7	14.6	81.4	11.2	80.3
1982	5.7	55.9	8.3	46.7	195.2	23.3	9.2	34.2	13.6	80.9	10.6	79.0
1983	5.6	55.1	7.4	45.8	185.2	23.7	6.6	31.4	13.0	80.6	10.3	78.7
1984	5.7	54.3	7.9	45.5	226.0	25.2	4.1	27.8	12.8	81.6	10.9	80.0
1985	5.8	54.9	7.3	44.6	230.4	26.1	2.1	18.9	12.6	81.6	10.8	78.9
1986	5.7	56.0	7.1	44.5	227.6	25.6	1.6	12.1	12.5	80.5	11.4	78.5
1987	5.7	57.0	7.2	43.4	255.3	27.8	NA	NA	12.5	78.3	11.7	79.7
1988	5.5	56.2	7.7	45.1	285.3	30.0	NA	NA	12.3	78.4	12.0	79.9
1989	5.2	57.2	7.5	43.2	286.9	29.3	NA	NA	11.5	73.8	11.4	75.2
1990	5.0	55.8	7.6	42.6	282.0	27.4	NA	NA	11.4	72.5	11.3	74.0
1991	5.0	54.9	7.5	42.4	289.6	29.1	NA	NA	11.2	71.4	11.1	72.9
1992	4.8	53.9	7.9	44.2	251.9	25.3	.0	.0	11.0	72.4	11.0	71.4
1993	4.5	52.1	7.7	42.3	197.3	20.9	.0	.0	10.7	71.3	10.8	68.6
1994	4.4	52.1	8.0	42.5	179.7	17.4	.0	.0	10.6	69.0	10.8	68.5
1995	4.3	51.7	8.1	43.3	165.4	16.0	.0	.0	10.4	68.0	10.7	66.6
1996	4.2	50.5	8.2	43.4	169.4	15.9	.0	.0	10.5	67.8	11.0	67.1
1997	4.0	48.3	8.3	43.9	163.3	15.0	.0	.0	9.4	59.9	10.0	59.9
1998	3.8	47.5	8.4	44.1	73.9	6.6	.0	.0	[8]14.3	[8]87.8	[8]14.9	[8]87.7
1999	3.6	46.3	8.0	42.4	44.0	4.0	.0	.0	14.2	85.7	14.6	86.2
2000	3.5	44.8	8.3	43.5	34.6	3.2	.0	.0	14.4	86.9	14.5	84.1
2001	3.7	48.7	8.8	45.1	31.3	2.8	.0	.0	15.2	90.3	15.0	86.9
2002	3.7	48.4	8.7	46.0	27.8	2.5	.0	.0	14.2	84.7	14.8	85.5
2003	3.5	47.3	8.3	43.7	16.8	1.6	.0	.0	14.3	84.5	14.7	84.0
2004	3.3	45.7	8.2	44.0	18.1	1.6	.0	.0	14.4	84.1	15.2	85.2
2005	3.1	44.5	7.8	R43.1	18.0	1.6	.0	.0	14.5	83.8	15.1	84.9
2006	2.9	43.1	7.9	42.9	NA	NA	.0	.0	14.7	84.0	14.8	82.4

[1] Production is on a net ownership basis. "Net ownership" is all reserve quantities owned, regardless of type of ownership (e.g., working interest or royalty).

[2] Bituminous coal, subbituminous coal, and lignite.

[3] Operable capacity as of January 1 of the following year.

[4] Includes refinery output at own refineries for own account and at others' refineries for own account.

[5] Production of uranium oxide (U_3O_8). See "Uranium Oxide" in Glossary.

[6] Percent of U.S. total uranium concentrate production. See "Uranium Concentrate" in Glossary.

[7] The Financial Reporting System (FRS) data include Puerto Rico and the Virgin Islands; U.S. Totals do not include Puerto Rico and the Virgin Islands.

[8] There is a discontinuity in this time series between 1997 and 1998 due to the expanded coverage of the FRS.

NA=Not available.

Note: "Major U.S. Energy Companies" are the top publicly-owned, U.S.-based crude oil and natural gas producers and petroleum refiners that form the FRS. See Table 3.14.

Web Page: For related information, see http://www.eia.doe.gov/emeu/finance.

Sources: **Production and Refining:** • 1974-1976—Energy Information Administration (EIA), Form EIA-28, "Financial Reporting System" database, November 1998. • 1977 forward—EIA, *Performance Profiles of Major Energy Producers,* annual reports. **Percent of U.S. Total:** Tables 5.1, 5.8, 5.9, 6.1, 7.1, and 9.3.

Figure 3.11 Major U.S. Energy Companies' Net Income

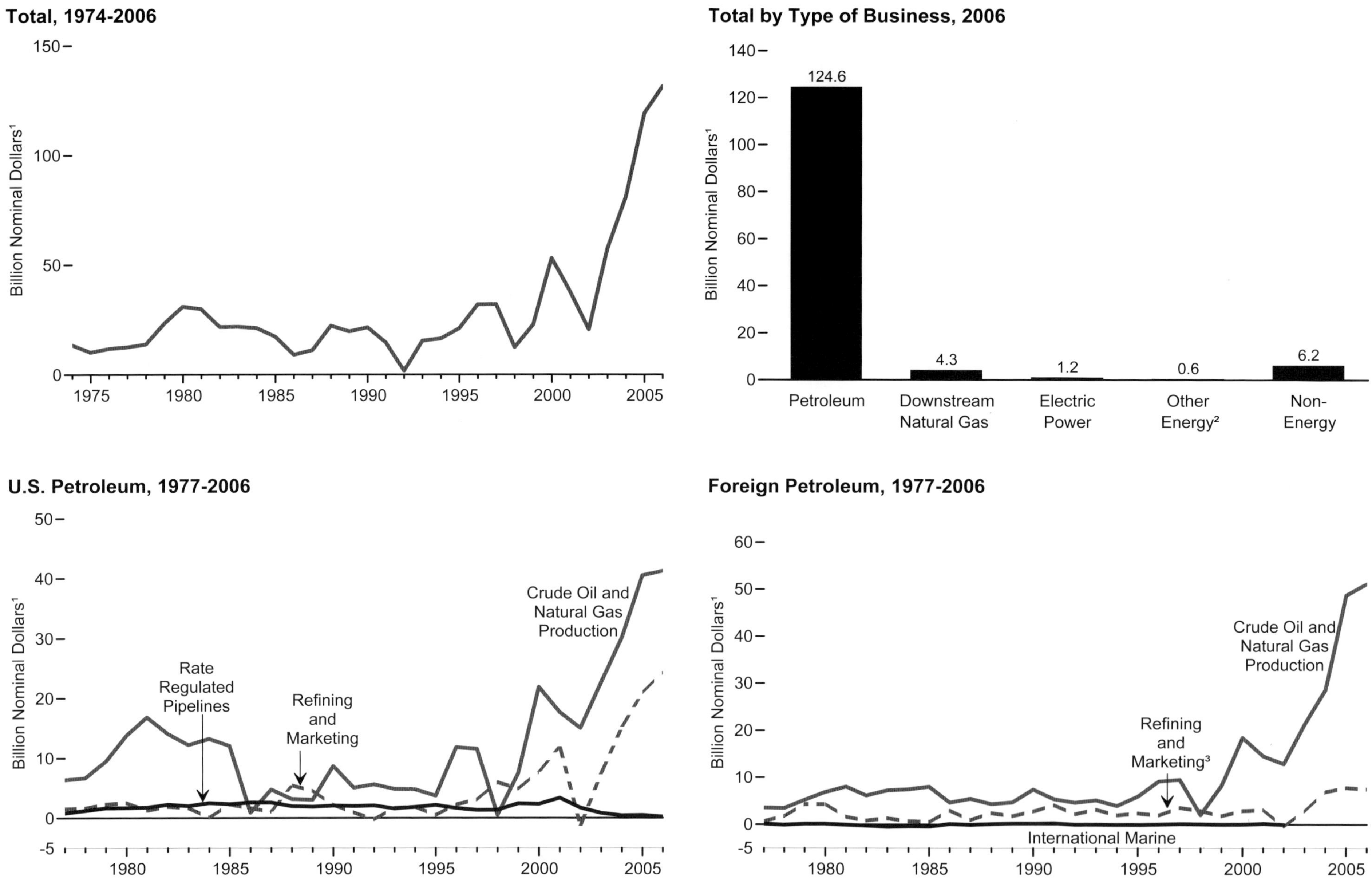

[1] See "Nominal Dollars" in Glossary.
[2] Coal, nuclear, renewable fuels, and nonconventional energy.
[3] Beginning in 2003, includes International Marine.

Notes: • "Major U.S. Energy Companies" are the top publicly-owned, U.S.-based crude oil and natural gas producers and petroleum refiners that form the Financial Reporting System (FRS). See Table 3.14. • Because vertical scales differ, graphs should not be compared.

Source: Table 3.11.

Table 3.11 Major U.S. Energy Companies' Net Income, 1974-2006

(Billion Nominal Dollars [1])

Year	Petroleum [2]									Down-stream Natural Gas [2,3]	Electric Power	Coal	Other Energy [4]	Non-Energy	Total [5]
	United States				Foreign										
	Crude Oil and Natural Gas Production	Refining and Marketing	Rate Regulated Pipelines	Total	Crude Oil and Natural Gas Production	Refining and Marketing	Inter-national Marine	Total	Total Petroleum						
1974	NA	NA	NA	NA	NA	NA	NA	NA	NA	NA	NA	NA	NA	NA	13.6
1975	NA	NA	NA	NA	NA	NA	NA	NA	NA	NA	NA	NA	NA	NA	10.3
1976	NA	NA	NA	NA	NA	NA	NA	NA	NA	NA	NA	NA	NA	NA	12.0
1977	6.4	1.5	.8	8.6	3.6	.7	.1	4.4	13.0	([6])	([7])	.2	(s)	1.7	12.7
1978	6.7	1.6	1.2	9.5	3.5	1.8	-.1	5.2	14.7	([6])	([7])	.1	-.1	1.8	13.9
1979	9.4	2.3	1.7	13.4	5.2	4.3	.1	9.7	23.0	([6])	([7])	.3	-.1	2.8	23.5
1980	13.8	2.5	1.7	17.9	6.9	4.3	.1	11.2	29.1	([6])	([7])	.3	(s)	2.3	31.0
1981	16.8	1.3	1.8	19.9	8.0	1.6	-.1	9.6	29.5	([6])	([7])	.4	-.3	1.6	30.0
1982	14.1	1.9	2.3	18.3	6.1	.8	-.3	6.7	25.0	([6])	([7])	.4	-.3	.4	21.8
1983	12.2	1.6	2.0	15.9	7.2	1.3	-.5	8.2	24.0	([6])	([7])	.5	(s)	1.8	21.9
1984	13.3	.1	2.5	15.8	7.5	.7	-.4	7.8	23.6	([6])	([7])	.6	-.1	2.9	21.3
1985	12.1	2.3	2.3	16.7	8.0	.5	-.4	8.1	24.8	([6])	([7])	.4	-.3	2.5	17.4
1986	.9	1.6	2.6	5.2	4.7	2.9	.1	7.7	12.9	([6])	([7])	.2	(s)	2.8	9.2
1987	4.7	1.1	2.6	8.4	5.4	1.0	-.1	6.4	14.8	([6])	([7])	.4	(s)	7.1	11.3
1988	3.2	5.4	2.0	10.6	4.3	2.4	.1	6.9	17.5	([6])	([7])	.6	-.1	10.8	22.3
1989	3.1	4.5	1.9	9.5	4.7	1.8	.2	6.7	16.2	([6])	([7])	.4	-.1	8.7	19.8
1990	8.7	2.2	2.1	12.9	7.4	2.8	.2	10.5	23.4	([6])	([7])	.3	.1	4.3	21.6
1991	5.1	.9	2.0	7.9	5.4	4.1	.3	9.8	17.7	([6])	([7])	.6	.1	1.6	14.7
1992	5.6	-.2	2.1	7.5	4.7	2.2	(s)	6.9	14.4	([6])	([7])	-.5	.1	1.2	1.8
1993	4.8	1.7	1.6	8.1	5.2	3.2	(s)	8.4	16.5	([6])	([7])	.4	.1	2.7	15.5
1994	4.8	1.8	1.8	8.5	4.0	2.0	(s)	5.9	14.4	([6])	([7])	.2	.2	6.2	16.5
1995	3.7	.5	2.2	6.4	5.9	2.4	(s)	8.3	14.7	([6])	([7])	.3	.2	12.6	21.1
1996	11.8	2.3	1.6	15.7	9.2	2.0	(s)	11.2	26.9	([6])	([7])	.5	.2	8.0	32.0
1997	11.6	3.1	1.3	16.0	9.6	3.6	.1	13.3	29.3	([6])	([7])	.3	.3	6.3	32.1
1998	.5	5.9	1.4	7.8	2.0	2.9	.1	5.1	12.8	([6])	([7])	.5	.9	1.8	12.5
1999	7.4	4.9	2.4	14.8	8.2	1.9	(s)	10.1	24.8	([6])	([7])	.2	.7	2.8	22.9
2000	21.9	7.7	2.3	31.8	18.5	2.9	(s)	21.4	53.3	([6])	([7])	(s)	2.7	3.6	53.2
2001	17.6	12.0	3.3	32.9	14.6	3.1	.2	17.8	50.8	([6])	([7])	.1	2.0	-2.7	37.7
2002	15.0	-1.4	1.7	15.4	12.9	-.4	(s)	12.5	27.9	([6])	([7])	(s)	-1.5	1.8	20.6
2003	[2]22.6	[2]7.4	[2].8	[2]30.9	[2]21.3	[2,8]2.9	([8])	[2]24.3	[2]55.1	3.6	1.0	([7])	.1	.9	57.4
2004	30.1	15.2	.4	45.8	28.6	[8]7.0	([8])	35.5	81.3	3.3	.6	([7])	1.1	4.2	81.1
2005	40.5	21.0	.5	61.9	48.7	[8]7.8	([8])	56.5	118.5	2.2	.4	([7])	1.0	4.2	119.2
2006	41.3	24.3	.2	65.8	51.2	[8]7.5	([8])	58.7	124.6	4.3	1.2	([7])	.6	6.2	131.5

[1] See "Nominal Dollars" in Glossary.

[2] Through 2002, natural gas operations are included in the "Petroleum" line of business. Beginning in 2003, downstream natural gas operations are included in their own line of business.

[3] "Downstream Natural Gas" is a line of business that begins with the procurement of natural gas, processes and gathers natural gas, produces natural gas liquids, imports liquefied natural gas, markets and trades natural gas and natural gas liquids, and delivers wholesale and retail volumes of natural gas and natural gas liquids.

[4] Through 2002, includes electric power, nuclear, renewable fuels, and nonconventional energy (including oil shale, tar sands, coal liquefaction and gasification, geothermal, and solar). Beginning in 2003, includes coal, nuclear, renewable fuels, and nonconventional energy.

[5] Total is sum of components shown, minus eliminations and nontraceables (see Notes).

[6] Included in "Petroleum."

[7] Included in "Other Energy."

[8] "International Marine" is included with "Foreign Refining and Marketing" to prevent disclosure.

NA=Not available. (s)=Less than 0.05 billion and greater than -0.05 billion.

Notes: • "Major U.S. Energy Companies" are the top publicly-owned, U.S.-based crude oil and natural gas producers and petroleum refiners that form the Financial Reporting System. See Table 3.14. • "Net income" is operating income plus other income and extraordinary income less operating expenses, taxes, interest charges, other deductions, and extraordinary deductions. • "Eliminations" are revenues and expenses resulting from transactions between segments of the energy industry. Consolidated company accounts do not include intersegment revenues and expenses. Therefore, such intersegment transactions must be eliminated. • "Nontraceables" are energy companies' revenues, costs, assays, and liabilities that cannot be directly attributed to a type of business by use of a reasonable allocation method developed on the basis of operating-level utilities.

Web Page: For related information, see http://www.eia.doe.gov/emeu/finance.

Sources: • 1974-1976—Energy Information Administration (EIA), Form EIA-28, "Financial Reporting System" database, November 1997. • 1977 forward—EIA, *Performance Profiles of Major Energy Producers,* annual reports.

Figure 3.12 Major U.S. Energy Companies' Profitability

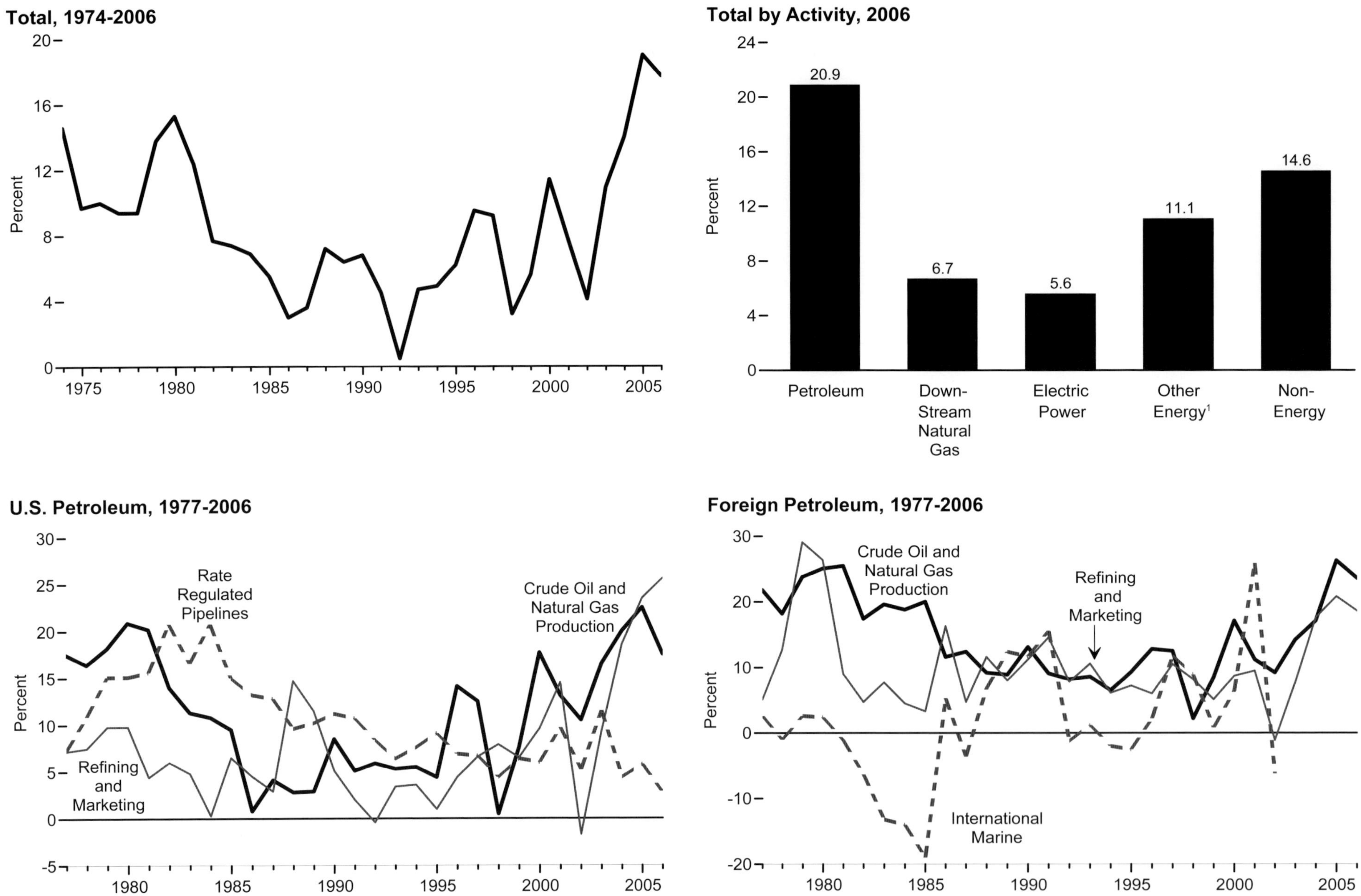

[1] Coal, nuclear, renewable fuels, and nonconventional energy.
Notes: • "Major U.S. Energy Companies" are the top publicly-owned crude oil and natural gas producers and petroleum refiners that form the Financial Reporting System (FRS). See Table 3.14. • Because vertical scales differ, graphs should not be compared.
Source: Table 3.12.

Table 3.12 Major U.S. Energy Companies' Profitability, 1974-2006

(Percent)

Year	Petroleum [1]									Downstream Natural Gas [1,2]	Electric Power	Coal	Other Energy [3]	Non-Energy	Total
	United States				Foreign				Total Petroleum						
	Crude Oil and Natural Gas Production	Refining and Marketing	Rate Regulated Pipelines	Total	Crude Oil and Natural Gas Production	Refining and Marketing	International Marine	Total							
1974	NA	NA	NA	NA	NA	NA	NA	NA	NA	NA	NA	NA	NA	NA	14.6
1975	NA	NA	NA	NA	NA	NA	NA	NA	NA	NA	NA	NA	NA	NA	9.7
1976	NA	NA	NA	NA	NA	NA	NA	NA	NA	NA	NA	NA	NA	NA	10.0
1977	17.5	7.2	7.3	12.5	21.8	5.1	2.6	12.4	12.5	([4])	([5])	8.8	-2.6	7.1	9.4
1978	16.4	7.5	10.9	12.8	18.2	12.7	-1.0	13.6	13.1	([4])	([5])	4.1	-4.2	6.5	9.4
1979	18.2	9.8	15.1	15.5	23.8	29.1	2.6	23.1	18.0	([4])	([5])	6.3	-3.7	8.8	13.8
1980	20.9	9.8	15.1	17.5	25.1	26.4	2.4	23.0	19.2	([4])	([5])	5.6	-.7	5.9	15.3
1981	20.2	4.4	15.6	16.1	25.5	9.0	-1.1	17.7	16.6	([4])	([5])	6.1	-6.8	3.5	12.4
1982	14.0	6.0	20.8	12.7	17.4	4.7	-6.3	11.8	12.5	([4])	([5])	4.4	-5.2	.6	7.7
1983	11.3	4.8	16.6	10.3	19.6	7.7	-13.2	14.1	11.3	([4])	([5])	5.0	.5	2.9	7.4
1984	10.8	.3	20.8	9.4	18.8	4.5	-14.0	13.3	10.4	([4])	([5])	6.2	-1.8	4.8	6.9
1985	9.5	6.5	15.0	9.4	20.0	3.3	-19.0	13.8	10.5	([4])	([5])	4.6	-8.4	4.2	5.5
1986	.8	4.5	13.2	3.0	11.6	16.3	5.3	12.8	5.5	([4])	([5])	2.7	-.8	5.1	3.0
1987	4.1	2.9	12.8	4.9	12.4	4.7	-3.6	9.5	6.2	([4])	([5])	5.1	.5	12.2	3.6
1988	2.8	14.7	9.6	6.3	9.2	11.6	6.8	9.9	7.3	([4])	([5])	6.7	-2.5	20.3	7.2
1989	2.9	11.5	10.2	5.8	8.9	8.0	12.4	8.7	6.7	([4])	([5])	5.0	-2.3	17.3	6.4
1990	8.5	5.1	11.2	7.9	13.1	11.2	11.7	12.5	9.5	([4])	([5])	3.3	2.6	7.8	6.8
1991	5.1	2.0	10.7	4.9	9.1	14.6	15.6	11.0	7.0	([4])	([5])	8.7	2.8	2.9	4.5
1992	5.9	-.4	8.4	4.4	8.2	7.8	-1.2	7.9	5.6	([4])	([5])	-9.3	1.8	2.1	.5
1993	5.3	3.4	6.4	4.9	8.6	10.6	1.2	9.2	6.4	([4])	([5])	7.6	4.1	4.7	4.7
1994	5.5	3.6	7.6	5.2	6.5	6.1	-2.0	6.2	5.6	([4])	([5])	4.0	4.8	10.5	4.9
1995	4.4	1.0	9.1	4.0	9.3	7.2	-2.5	8.4	5.7	([4])	([5])	6.9	6.1	19.4	6.2
1996	14.1	4.4	6.9	9.9	12.8	6.0	2.2	10.6	10.1	([4])	([5])	9.9	7.9	15.0	9.5
1997	12.5	6.6	6.7	10.0	12.5	10.5	11.8	11.9	10.8	([4])	([5])	7.2	7.0	10.9	9.2
1998	.5	7.9	4.4	3.8	2.2	8.2	8.9	4.0	3.9	([4])	([5])	26.4	13.2	4.5	3.2
1999	7.6	6.5	6.4	7.0	8.5	5.1	.8	7.6	7.2	([4])	([5])	9.5	7.6	5.8	5.6
2000	17.7	9.6	6.0	13.2	17.1	8.7	6.4	15.1	13.9	([4])	([5])	1.7	11.0	7.3	11.4
2001	13.1	14.5	9.7	13.1	11.2	9.5	25.9	10.9	12.2	([4])	([5])	9.0	9.0	-6.6	7.7
2002	10.5	-1.7	5.2	6.0	9.2	-1.1	-6.2	7.2	6.5	([4])	([5])	-8.5	-6.8	4.7	4.1
2003	[1]16.5	[1]9.3	[1]11.5	[1]13.7	[1]14.2	[1]7.7	W	[1]13.0	[1]13.4	8.8	5.2	([5])	2.8	2.4	10.9
2004	20.0	18.6	4.4	18.9	17.1	17.6	W	17.3	18.2	5.9	3.1	([5])	24.7	10.7	14.0
2005	22.5	23.5	5.8	R22.4	26.3	20.8	W	25.5	23.7	4.0	1.7	([5])	22.2	R11.6	19.0
2006	17.5	25.6	2.7	19.4	23.6	18.6	W	22.8	20.9	6.7	5.6	([5])	11.1	14.6	17.7

[1] Through 2002, natural gas operations are included in the "Petroleum" line of business. Beginning in 2003, downstream natural gas operations are included in their own line of business.

[2] "Downstream Natural Gas" is a line of business that begins with the procurement of natural gas, processes and gathers natural gas, produces natural gas liquids, imports liquefied natural gas, markets and trades natural gas and natural gas liquids, and delivers wholesale and retail volumes of natural gas and natural gas liquids.

[3] Through 2002, includes electric power, nuclear, renewable fuels, and nonconventional energy (including oil shale, tar sands, coal liquefaction and gasification, geothermal, and solar). Beginning in 2003, includes coal, nuclear, renewable fuels, and nonconventional energy.

[4] Included in "Petroleum."

[5] Included in "Other Energy."

R=Revised. NA=Not available. W=Value withheld to avoid disclosure of individual company data.

Notes: • "Major U.S. Energy Companies" are the top publicly-owned, U.S.-based crude oil and natural gas producers and petroleum refiners that form the Financial Reporting System. See Table 3.14. • Data are for return on investment, measured as net income divided by net investment in place. "Net income" is operating income plus other income and extraordinary income less operating expenses, taxes, interest charges, other deductions, and extraordinary deductions. "Net investment in place" is net property, plant, and equipment plus investments and advances to unconsolidated affiliates.

Web Page: For related information, see http://www.eia.doe.gov/emeu/finance.

Sources: • 1974-1976—Energy Information Administration (EIA), Form EIA-28, "Financial Reporting System" database, October 1996. • 1977 forward—EIA, *Performance Profiles of Major Energy Producers,* annual reports.

Figure 3.13 U.S. Energy Activities by Foreign-Affiliated Companies, 1978-2005

Petroleum Activities

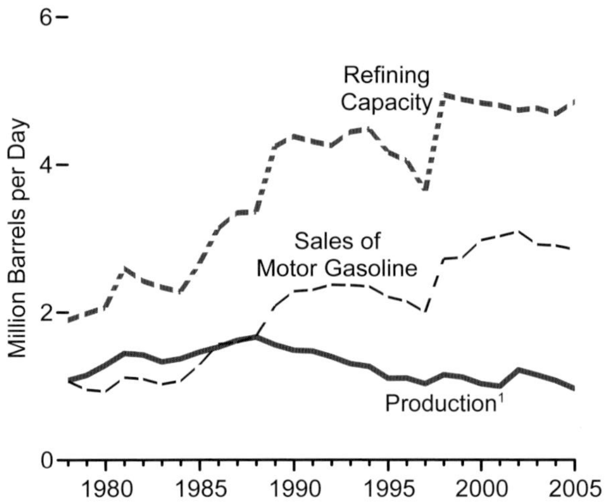

Petroleum Activities Share of U.S. Total

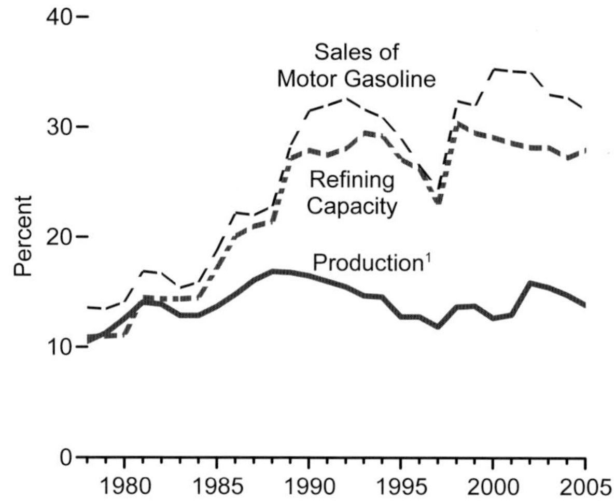

Share of U.S. Total Production by Fuel Type

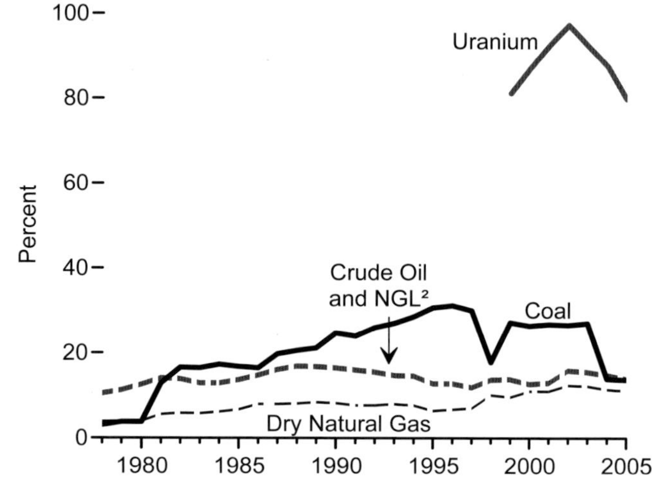

Dry Natural Gas Production

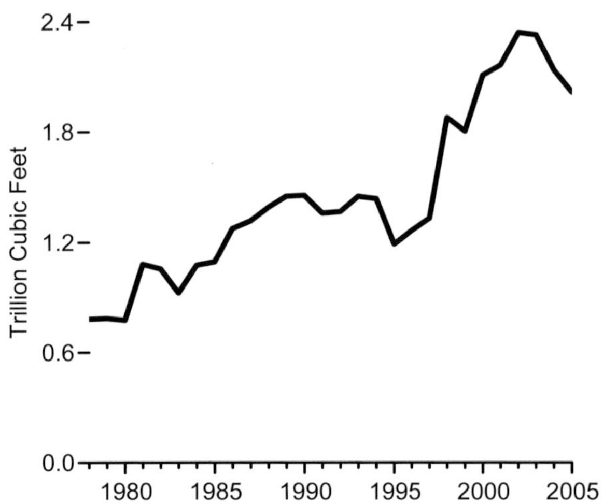

Coal Production

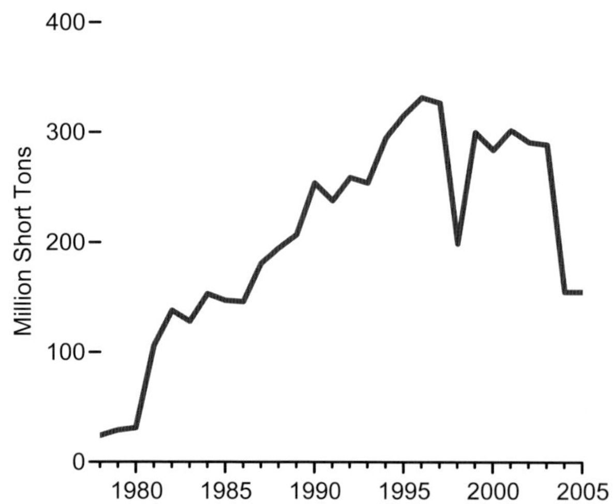

Expenditures for Exploration and Development of Uranium

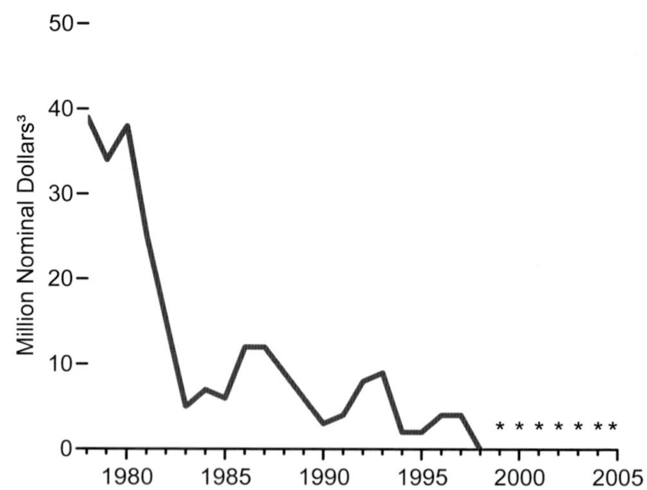

[1] Crude oil and natural gas liquids.
[2] Natural gas liquids.
[3] See "Nominal Dollars" in Glossary.

* 1999-2002 uranium values are withheld to avoid disclosure of individual company data; 2003-2005 values are not available.

Note: Because vertical scales differ, graphs should not be compared.

Source: Table 3.13.

Table 3.13 U.S. Energy Activities by Foreign-Affiliated Companies, 1978-2005

Year	Production: Crude Oil and Natural Gas Liquids: Thousand Barrels per Day	Production: Crude Oil and Natural Gas Liquids: Percent of U.S. Total	Production: Dry Natural Gas: Billion Cubic Feet	Production: Dry Natural Gas: Percent of U.S. Total	Production: Coal: Million Short Tons	Production: Coal: Percent of U.S. Total	Production: Uranium: Thousand Pounds [2]	Production: Uranium: Percent of U.S. Total [3]	Refining Capacity [1]: Thousand Barrels per Day	Refining Capacity [1]: Percent of U.S. Total	Sales of Motor Gasoline: Thousand Barrels per Day	Sales of Motor Gasoline: Percent of U.S. Total	Expenditures for Exploration and Development of Uranium: Million Nominal Dollars [4]	Expenditures for Exploration and Development of Uranium: Percent of U.S. Total
1978	1,076	10.5	783	3.9	24	3.1	NA	NA	1,895	10.9	1,066	13.6	39	12.5
1979	1,145	11.3	786	4.0	29	3.8	NA	NA	1,984	11.0	948	13.5	34	10.8
1980	1,280	12.6	776	4.0	31	3.8	NA	NA	2,066	11.1	926	14.1	38	14.1
1981	1,438	14.1	1,080	5.6	106	12.9	NA	NA	2,595	14.5	1,114	16.9	25	17.0
1982	1,421	13.9	1,055	5.9	138	16.6	NA	NA	2,423	14.4	1,092	16.7	15	19.8
1983	1,325	12.9	924	5.8	128	16.5	NA	NA	2,337	14.4	1,022	15.4	5	13.0
1984	1,365	12.9	1,075	6.2	153	17.3	NA	NA	2,276	14.5	1,066	15.9	7	24.9
1985	1,455	13.7	1,093	6.7	147	16.8	NA	NA	2,656	17.2	1,285	18.8	6	27.9
1986	1,523	14.8	1,276	8.0	146	16.5	NA	NA	3,133	20.1	1,565	22.2	12	54.3
1987	1,614	16.1	1,318	8.0	181	19.8	NA	NA	3,342	21.0	1,586	22.0	12	60.4
1988	1,659	16.9	1,392	8.1	195	20.6	NA	NA	3,356	21.4	1,673	22.8	9	44.2
1989	1,553	16.8	1,452	8.4	207	21.2	NA	NA	4,243	27.2	2,084	28.4	6	41.2
1990	1,481	16.5	1,457	8.2	254	24.7	NA	NA	4,379	27.9	2,282	31.5	3	14.6
1991	1,469	16.0	1,360	7.7	238	24.0	NA	NA	4,312	27.5	2,299	32.0	4	19.7
1992	1,392	15.5	1,368	7.7	259	26.0	NA	NA	4,256	28.1	2,369	32.6	8	55.2
1993	1,299	14.7	1,451	8.0	254	27.0	NA	NA	4,440	29.5	2,362	31.6	9	76.0
1994	1,261	14.6	1,439	7.7	295	28.6	NA	NA	4,479	29.2	2,346	30.9	2	51.0
1995	1,103	12.8	1,191	6.4	316	30.7	NA	NA	4,164	27.1	2,204	29.0	2	35.0
1996	1,105	12.8	1,265	6.7	332	31.2	NA	NA	4,050	26.2	2,145	26.5	4	44.0
1997	1,028	11.9	1,332	7.0	327	30.0	NA	NA	3,637	23.0	1,998	24.4	4	14.0
1998	1,149	13.7	1,881	10.1	199	17.8	NA	NA	4,940	30.4	2,721	32.4	(s)	1.0
1999	1,118	13.8	1,805	9.6	300	27.2	[5]3,745	[5]81.2	4,877	29.5	2,737	32.0	W	W
2000	1,027	12.7	2,112	11.0	284	26.4	3,443	87.0	4,831	29.1	2,971	35.3	W	W
2001	994	13.0	2,167	11.0	302	26.8	2,440	92.5	4,797	28.6	3,027	35.1	W	W
2002	1,214	15.9	2,344	12.4	291	26.6	2,280	97.3	4,733	28.2	3,090	35.0	W	W
2003	1,147	15.5	2,331	12.2	289	27.0	2,024	[6]NM	4,761	28.2	2,914	33.0	NA	NA
2004	1,073	14.8	2,140	R11.4	155	13.9	2,000	87.6	4,683	27.3	2,900	32.7	NA	NA
2005	962	13.9	2,019	11.1	155	13.7	2,147	79.8	4,848	28.0	2,845	31.6	NA	NA

[1] Operable capacity as of January 1 of the following year.
[2] Production of uranium oxide (U_3O_8). See "Uranium Oxide" in Glossary.
[3] Percent of U.S. total uranium concentrate production. See "Uranium Concentrate" in Glossary.
[4] See "Nominal Dollars" in Glossary.
[5] Includes a small amount produced by a U.S. company, which left the industry by the close of 1999.
[6] Total U.S. uranium production is slightly below that of the foreign-affiliated companies shown in this table. The U.S. data were rounded to avoid disclosure of individual company data.

R=Revised. NA=Not available. NM=Not meaningful. (s)=Less than 0.5 million dollars. W=Value withheld to avoid disclosure of individual company data.

Web Page: For related information, see http://www.eia.doe.gov/emeu/finance.

Sources: • 1978—U.S. Department of Energy, *Secretary's Annual Report to Congress*, (September 1983). • 1979-1992—Energy Information Administration (EIA), *Profiles of Foreign Direct Investment in U.S. Energy*, annual reports. • 1993-1996—EIA, *Performance Profiles of Major Energy Producers*, annual reports. • 1997 forward—EIA, *Foreign Direct Investment in U.S. Energy*, annual reports.

Table 3.14 Companies Reporting to the Financial Reporting System, 1974-2006

Company	1974-81	1982	1983-84	1985-86	1987	1988	1989-90	1991	1992-93	1994-96	1997	1998	1999	2000	2001	2002	2003-06
Amerada Hess Corporation	X	X	X	X	X	X	X	X	X	X	X	X	X	X	X	X	X
American Petrofina, Inc. [1]	X	X	X	X	X	X	X										
Anadarko Petroleum Corporation [2]								X	X	X	X	X	X	X	X	X	X
Apache Corporation														X	X	X	X
Ashland Inc. [3]	X	X	X	X	X	X	X	X	X	X	X						
Atlantic Richfield Co. (ARCO) [4]	X	X	X	X	X	X	X	X	X	X	X	X	X				
BP America, Inc. [5,6]					X	X	X	X	X	X	X	X	X			X	X
BP Amoco Corporation [4,5,7]	X	X	X	X	X	X	X	X	X	X	X	X	X	X	X		
Burlington Northern Inc. [8]	X	X	X	X	X												
Burlington Resources Inc. [8,9]						X	X	X	X	X	X	X	X	X	X	X	X
Chesapeake Energy Corporation																	X
Chevron Corporation [10,11,12,13]	X	X	X	X	X	X	X	X	X	X	X	X	X	X	X	X	X
Citgo Petroleum Corporation												X	X	X	X	X	X
Cities Service [14]	X	X															
ConocoPhillips, Inc. [9,15,16,17]	X											X	X	X	X	X	X
Devon Energy Corporaton														X	X	X	X
Dominion Resources														X	X	X	X
E.I. du Pont de Nemours and Co. [15,16]		X	X	X	X	X	X	X	X	X	X						
El Paso Energy Corporation [18]													X	X	X	X	X
Enron Corporation									X	X	X	X	X	X			
EOG Resources, Inc.														X	X	X	X
Equilon Enterprises, LLC [19]												X	X	X	X		
Equitable Resources, Inc.																	X
Exxon Mobil Corporation [20]	X	X	X	X	X	X	X	X	X	X	X	X	X	X	X	X	X
Getty Oil [21]	X	X	X														
Gulf Oil [11]	X	X	X														
Kerr-McGee Corporation [2,22]	X	X	X	X	X	X	X	X	X	X	X	X	X	X	X	X	X
LYONDELL-CITGO Refining, LP [23]												X	X	X	X	X	X
Marathon Oil Corporation [24]	X															X	X
Mobil Corporation [20,25]	X	X	X	X	X	X	X	X	X	X	X	X					
Motiva Enterprises, LLC [26]												X	X	X	X	X	X
Nerco, Inc. [27]									X								
Occidental Petroleum Corporation [14]	X	X	X	X	X	X	X	X	X	X	X	X	X	X	X	X	X
Oryx Energy Company [22,28]						X	X	X	X	X	X						
Phillips Petroleum Company [17,29]	X	X	X	X	X	X	X	X	X	X	X	X	X	X	X	X	
Premcor, Inc. [30,31]												X	X	X	X	X	X
Shell Oil Company	X	X	X	X	X	X	X	X	X	X	X	X	X	X	X	X	X
Sonat Inc.											X	X					
Standard Oil Co. (Ohio) (Sohio) [6]	X	X	X	X													
Sunoco, Inc. [28,32]	X	X	X	X	X	X	X	X	X	X		X	X	X	X	X	X
Superior Oil [25]	X	X	X														
Tenneco Inc. [33]	X	X	X	X	X	X											
Tesoro Petroleum Corporation												X	X	X	X	X	X
Texaco Inc. [12,21]	X	X	X	X	X	X	X	X	X	X	X	X	X	X			
The Coastal Corporation [18]	X	X	X	X	X	X	X	X	X	X	X	X	X	X			
The Williams Companies, Inc												X	X	X	X	X	X
Tosco Corporation [29]												X	X	X	X		
Total Holdings, USA, Inc. [1,34,35]								X	X	X	X	X	X	X	X	X	X
Total Petroleum (North America) Ltd. [36]							X	X									
Ultramar Diamond Shamrock Corporation [37]												X	X	X	X		
Union Pacific Resources Group, Inc. [38,39]	X	X	X	X	X	X	X	X	X	X	X	X	X				
Unocal Corporation [13]	X	X	X	X	X	X	X	X	X	X	X	X	X	X	X	X	X
USX Corporation	X	X	X	X	X	X	X	X	X	X	X	X	X	X	X		
Valero Energy Corporation [31,37]												X	X	X	X	X	X
XTO Energy, Inc.																X	X

Footnotes: See the following two pages.

Note: "X" indicates that the company was included in the Financial Reporting System for the year indicated.

Web Page: For related information, see http://www.eia.doe.gov/emeu/finance.

Source: Energy Information Administration, Form EIA-28, "Financial Reporting System."

Footnotes for Table 3.14

[1]American Petrofina, Inc. changed its name to Fina, Inc., effective April 17, 1991.
[2]Kerr-McGee merged with Anadarko on August 10, 2006. Anadrako's 2005 submission includes data for Kerr-McGee after August 10, 2006. Data for Kerr-McGee were submitted separately for the period from January 1, 2005, to August 10, 2006.
[3]Ashland was dropped from the FRS system for 1998 after spinning off downstream and coal operations and disposing of upstream operations.
[4]BP Amoco acquired Atlantic Richfield Company (ARCO) in April of 2000. The reporting was consolidated under BP Amoco for 2000. Data for ARCO is not included in the database for the period from January 1, 2000, to April 14, 2000.
[5]Amoco merged with British Petroleum plc and became BP Amoco plc on December 31, 1998. BP America was renamed BP Amoco, Inc. The companies reported separately for 1998 and 1999.
[6]In 1987, British Petroleum acquired all shares in Standard Oil Company (Ohio) that it did not already control and renamed its U.S. affiliate, BP America, Inc.
[7]Formerly Standard Oil Company (Indiana).
[8]Burlington Resources was added to the FRS system and Burlington Northern was dropped for 1988. Data for Burlington Resources cover the full year 1988 even though that company was not created until May of that year.
[9]Burlington Resources merged with ConocoPhillips on March 30, 2006. ConocoPhillip's 2006 submission includes data for Burlington Resources after March 30, 2006. Data for Burlington Resources were submitted separately for the period from January 1, 2006, to March 30, 2006.
[10]Formerly Standard Oil Company of California.
[11]Chevron acquired Gulf Oil in 1984, but separate data for Gulf continued to be available for the full 1984 year.
[12]In October 2000, Chevron and Texaco agreed to merge. Both companies reported separately for 2000. In May 2005, Chevron Texaco renamed itself Chevron.
[13]Unocal merged with Chevron on August 10, 2005. Chevron's 2005 submission includes data for Unocal after August 10, 2005. Data for Unocal were submitted separately for the period from January 1, 2005, to August 10, 2005.
[14]Occidental acquired Cities Service in 1982. Separate financial reports were available for 1982, so each company continued to be treated separately until 1983.
[15]DuPont acquired Conoco in 1981. Separate data for Conoco were available for 1981; DuPont was included in the FRS system in 1982.
[16]Dupont was dropped from the FRS system when Conoco was spun-off in 1998. Conoco began reporting separately again in 1998.
[17]In November 2001, Phillips and Conoco agreed to merge forming ConocoPhillips in 2002. Both companies reported separately in 2001. The companies reported separately in 2002 until the time of the merger.
[18]In January 2001, Coastal merged with a wholly owned subsidiary of El Paso Energy Corporation. The name was changed to El Paso CGP Company. Data were reported separately in 2000 under the name The Coastal Company.
[19]Equilon is a joint venture combining Shell's and Texaco's western and midwestern U.S. refining and marketing businesses and nationwide trading transportation and lubricants businesses. Net income is duplicated in the FRS system since Shell and Texaco account for this investment using the equity method.
[20]In December 1998, Exxon and Mobil agreed to merge. Both companies reported separately for 1998.
[21]Texaco acquired Getty in 1984; however, Getty was treated as a separate FRS company for that year.
[22]In 1998, Kerr-McGee and Oryx merged. The financial reporting for both was consolidated under Kerr-McGee for 1998.
[23]LYONDELL-CITGO is a limited partnership owned by Lyondell Chemical Company and Citgo. There will be some duplication of net income since Citgo accounts for its investment using the equity method.
[24]U.S. Steel (now USX) acquired Marathon in 1982.
[25]Mobil acquired Superior in 1984, but both companies were treated separately for that year.
[26]Motiva is a joint venture approximately equally owned by Shell, Texaco, and Saudi Refining, Inc. The joint venture combines the company's Gulf and east coast refining and marketing businesses. Duplication exists for the net income related to Shell and Texaco's interests, which are accounted for under the equity method.
[27]RTZ America acquired the common stock of Nerco, Inc., on February 17, 1994. In September 1993, Nerco, Inc. sold Nerco Oil & Gas, Inc., its subsidiary. Nerco's 1993 submission includes operations of Nerco Oil & Gas, Inc., through September 28, 1993.
[28]Sun Company spun off Sun Exploration and Development Company (later renamed Oryx Energy Company) during 1988. Both companies were included in the FRS system for 1988; therefore, some degree of duplication exists for that year.
[29]In September 2001, Phillips acquired Tosco. Both companies reported separately in 2001.
[30]In May 2000, Clark Refining & Marketing changed its name to Premcor Refining Group. At a later date the name was changed to Premcor, Inc.
[31]Premcor merged with Valero on September 1, 2005. Valero's 2005 submission includes data for Premcor after September 1, 2005. Data for Premcor were submitted separately for the period from January 1, 2005, to September 1, 2005.
[32]Sun company withdrew from oil and gas exploration and production in 1996. Sun's 1996 submission includes oil and gas exploration and production activities through September 30, 1996. Refining/marketing activities are included for the entire 1996 calendar year. In 1998 the company changed its name to Sunoco, Inc.
[33]Tenneco sold its worldwide oil and gas assets and its refining and marketing assets in 1988. Other FRS companies purchased approximately 70 percent of Tenneco's assets.
[34]Prior submissions were reported at the FINA, Inc. level. FINA, Inc. was the parent of Fina Oil and Chemical Company, which is now ATOFINA Petrochemicals. Due to a series of mergers and acquisitions, beginning in 2000, the submission is

reported at the American Petrofina Holding Company level, which is the holding company of ATOFINA.
[35]In 2002, the name was changed to Total Fina Elf and changed to Total Holdings, USA in 2003.
[36]Effective June 1, 1991, Total's exploration, production, and marketing operations in Canada were spun off to Total Oil & Gas, a new public entity.
[37]In December 2001, Valero and Ultramar Diamond Shamrock agreed to merge. Both companies reported separately in 2001.
[38]Effective October 15, 1996, Union Pacific Corporation distributed its ownership in the Union Pacific Resources Group, Inc. to its shareholders. Prior to 1996, the FRS system included Union Pacific Corporation. The FRS system includes only Union Pacific Resources Group, Inc. for 1996.
[39]Union Pacific merged with Anadarko on July 14, 2000. Anadarko's 2000 submission includes data for Union Pacific after July 14, 2000. Data for Union Pacific were not submitted for the period from January 1, 2000, to July 14, 2000.

4 Energy Resources

Semisubmersible drilling rig in the Gulf of Mexico. Source: U.S. Department of Energy.

Figure 4.1 Technically Recoverable Crude Oil, Natural Gas, and Natural Gas Liquids Resource Estimates, 2006

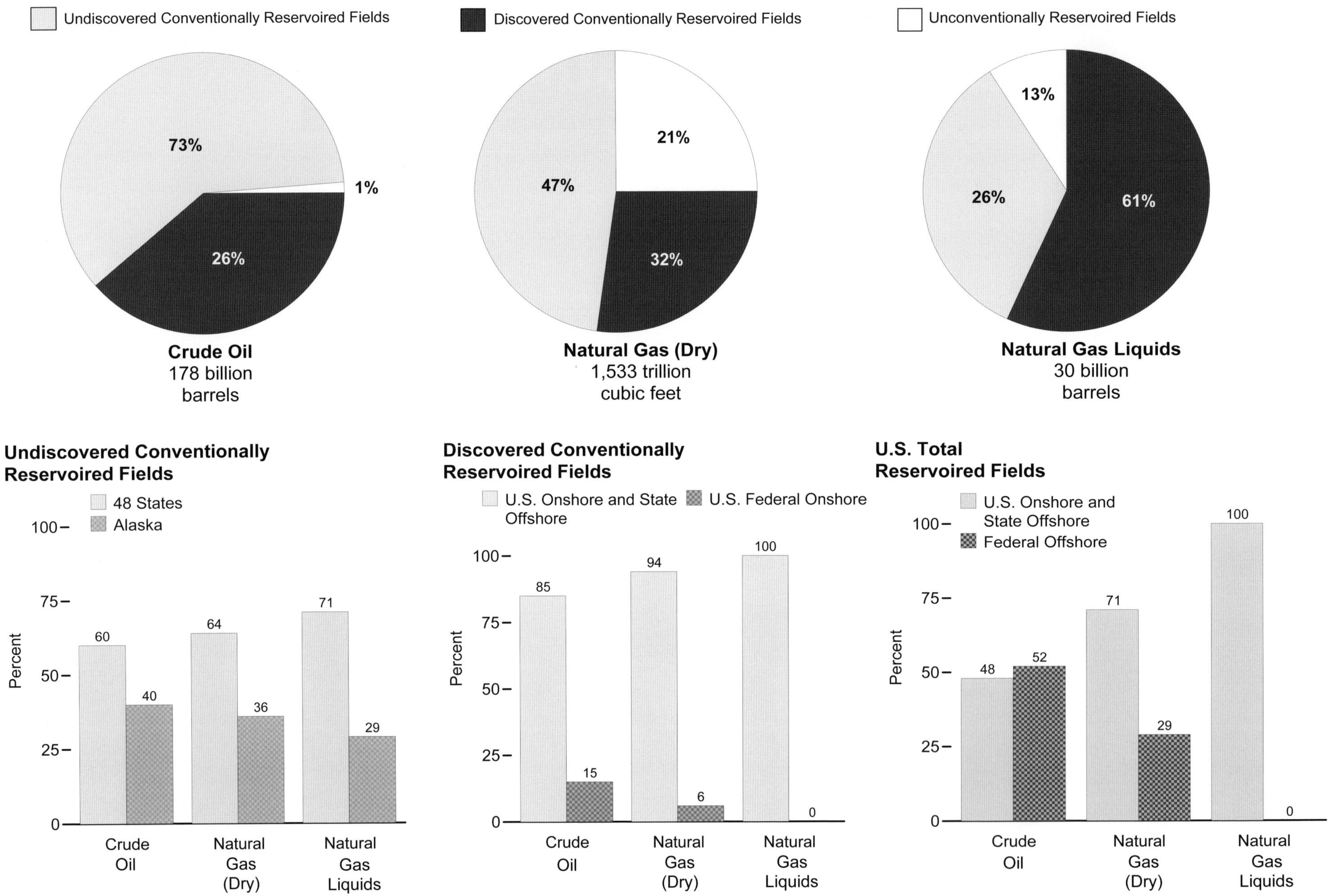

Source: Table 4.1.

Table 4.1 Technically Recoverable Crude Oil, Natural Gas, and Natural Gas Liquids Resource Estimates, 2006

Region	Crude Oil [1]	Natural Gas (Dry)	Natural Gas Liquids [1]
	Billion Barrels	Trillion Cubic Feet	Billion Barrels
Undiscovered Conventionally Reservoired Fields [2]	**130.16**	**724.84**	**7.79**
Alaska Onshore and State Offshore [3]	26.04	126.75	2.23
Alaska Federal Offshore [4]	26.61	132.06	.00
48 States Onshore and State Offshore [3]	18.24	178.21	5.56
48 States Federal Offshore [4]	59.27	287.82	.00
Discovered Conventionally Reservoired Fields [2] **Ultimate Recovery Appreciation** [5]	**45.54**	**485.71**	**18.26**
U.S. Onshore and State Offshore [3]	38.66	454.80	18.26
U.S. Federal Offshore [4]	6.88	30.91	.00
Unconventionally Reservoired Fields [6] **(Continuous-Type Deposits (all onshore))**	**2.13**	**322.27**	**3.80**
U.S. Total	**177.83**	**1,532.82**	**29.85**
U.S. Onshore and State Offshore [3]	85.07	1,082.03	29.85
Federal Offshore [4]	92.76	450.79	.00

[1] To the extent that lease condensate is measured or estimated it is included in "Natural Gas Liquids"; otherwise, lease condensate is included in "Crude Oil."

[2] Conventionally reservoired deposits are discrete subsurface accumulations of crude oil or natural gas usually defined, controlled, or limited by hydrocarbon/water contacts.

[3] Onshore plus State offshore waters (near-shore, shallow-water areas under State jurisdiction).

[4] Federal offshore jurisdictions (Outer Continental Shelf and deeper water areas seaward of State offshore).

[5] Proved reserves (see Table 4.2) are not included in these estimates. Ultimate recovery appreciation (reserve growth) is the volume by which the estimate of total recovery from a known crude oil or natural gas reservoir or aggregation of such reservoirs is expected to increase during the time between discovery and permanent abandonment.

[6] Unconventionally reservoired deposits (continuous-type accumulations) are geographically extensive subsurface accumulations of crude oil or natural gas that generally lack well-defined hydrocarbon/water contacts. Examples include coalbed methane, "tight gas," and self-sourced oil- and gas-shale reservoirs.

Notes: • "Technically recoverable" resources are those that are producible using current technology without reference to the economic viability thereof. • For purposes of comparison, the Potential Gas Committee, an industry-sponsored group of experts, biennially provides another geologically-based estimate of the Nation's natural gas resources. The latest mean estimate, published in "Potential Supply of Natural Gas in the United States," December 31, 2006, is 1,321 trillion cubic feet. This volume includes undiscovered conventionally reservoired deposits, expected ultimate recovery appreciation, coalbed methane, and tight gas where it is believed to be technically recoverable and marketable at reasonable costs. • A value of zero indicates either that none exists in this area or that no estimate of this resource has been made for this area. • "48 States" is the United States excluding Alaska and Hawaii.

Sources: • National Oil and Gas Resource Assessment Team, 2007 Assessment Updates United States Geological Survey, Washington, D.C., December 2007 at http://energy.cr.usgs.gov/oilgas/noga/ass_updates.html. • Resource Evaluation Division, Assessment of Undiscovered Technically Recoverable Oil and Gas Resources of the Nation's Outer Continental Shelf, 2006 MMS Fact Sheet RED-2006-01b, Minerals Management Service, Washington, D.C., February 2006, at http://www.mms.gov/revaldiv/PDFs/2006NationalAssessmentBrochure.pdf. • The ultimate recovery appreciation estimates for Alaska and the Lower 48 States Onshore Plus State Waters were developed by the Energy Information Administration, Reserves and Production Division, Office of Oil and Gas, based on data available as of year-end 2006.

Figure 4.2 Crude Oil and Natural Gas Cumulative Production, Proved Reserves, and Proved Ultimate Recovery, 1977-2006

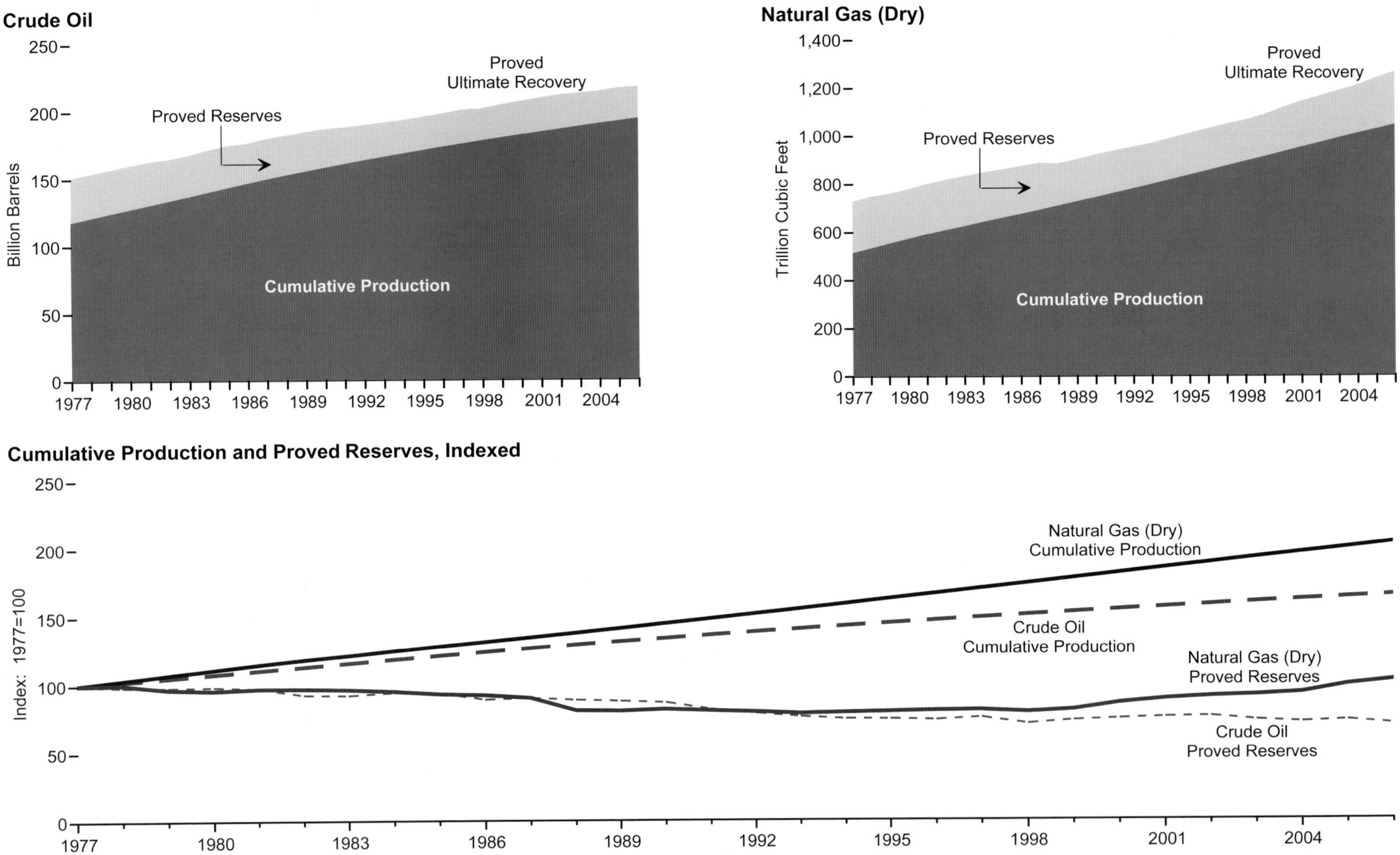

Notes: • Data are at end of year. • Crude oil includes lease condensate.

Source: Table 4.2.

Table 4.2 Crude Oil and Natural Gas Cumulative Production, Proved Reserves, and Proved Ultimate Recovery, 1977-2006

Year	Crude Oil and Lease Condensate [1]			Natural Gas (Dry)		
	Cumulative Production	Proved Reserves	Proved Ultimate Recovery	Cumulative Production	Proved Reserves	Proved Ultimate Recovery
	Billion Barrels			Trillion Cubic Feet		
1977	118.1	31.8	149.9	514.4	207.4	721.9
1978	121.3	31.4	152.6	533.6	208.0	741.6
1979	124.4	31.2	155.6	553.2	201.0	754.2
1980	127.5	31.3	158.9	572.6	199.0	771.6
1981	130.7	31.0	161.7	591.8	201.7	793.5
1982	133.8	29.5	163.3	609.6	201.5	811.1
1983	137.0	29.3	166.3	625.7	200.2	826.0
1984	140.2	30.0	170.2	643.2	197.5	840.7
1985	143.5	29.9	173.4	659.6	193.4	853.0
1986	146.7	28.3	175.0	675.7	191.6	867.3
1987	149.7	28.7	178.4	692.3	187.2	879.5
1988	152.7	28.2	180.9	709.4	168.0	877.4
1989	155.5	27.9	183.4	726.7	167.1	893.9
1990	158.2	27.6	185.7	744.5	169.3	913.9
1991	160.9	25.9	186.8	762.2	167.1	929.3
1992	163.5	25.0	188.5	780.1	165.0	945.1
1993	166.0	24.1	190.2	798.2	162.4	960.6
1994	168.4	23.6	192.0	817.0	163.8	980.8
1995	170.8	23.5	194.4	835.6	165.1	1,000.7
1996	173.2	23.3	196.5	854.5	166.5	1,020.9
1997	175.6	23.9	199.4	873.4	167.2	1,040.6
1998	177.8	22.4	200.2	892.4	164.0	1,056.4
1999	180.0	23.2	203.1	911.2	167.4	1,078.6
2000	182.1	23.5	205.6	930.4	177.4	1,107.8
2001	184.2	23.8	208.1	950.0	183.5	1,133.5
2002	186.3	24.0	210.4	968.9	186.9	1,155.9
2003	188.4	23.1	211.5	988.0	189.0	1,177.1
2004	190.4	22.6	213.0	1,006.6	192.5	1,199.1
2005	192.3	23.0	215.3	1,024.6	204.4	1,229.0
2006	194.1	22.1	216.3	1,043.1	211.1	1,254.2

[1] Lease condensate is the portion of natural gas liquids that is separated from the wellhead gas stream at a lease or field separation facility.

Notes: • Data are at end of year. • See "Proved Reserves, Crude Oil," "Proved Reserves, Lease Condensate," "Proved Reserves, Natural Gas," and "Proved Reserves, Natural Gas Liquids" in Glossary.

Web Pages: See http://www.eia.doe.gov/oil_gas/petroleum/info_glance/petroleum.html and http://www.eia.doe.gov/oil_gas/natural_gas/info_glance/natural_gas.html for related information.

Sources: **Cumulative Production:** Calculated from Energy Information Administration (EIA), *Petroleum Supply Annual,* annual reports and *Natural Gas Annual,* annual reports. **Proved Reserves:** • 1977-2005—EIA, *U.S. Crude Oil, Natural Gas, and Natural Gas Liquids Reserves,* annual reports. • 2006—EIA, *U.S. Crude Oil, Natural Gas, and Natural Gas Liquids Reserves 2006 Annual Report* (December 2007), Tables 6, 8, and 15. **Proved Ultimate Recovery:** Calculated as the sum of cumulative production and proved reserves.

Figure 4.3 Crude Oil, Natural Gas, and Natural Gas Liquids Proved Reserves

Total, 1949-2006

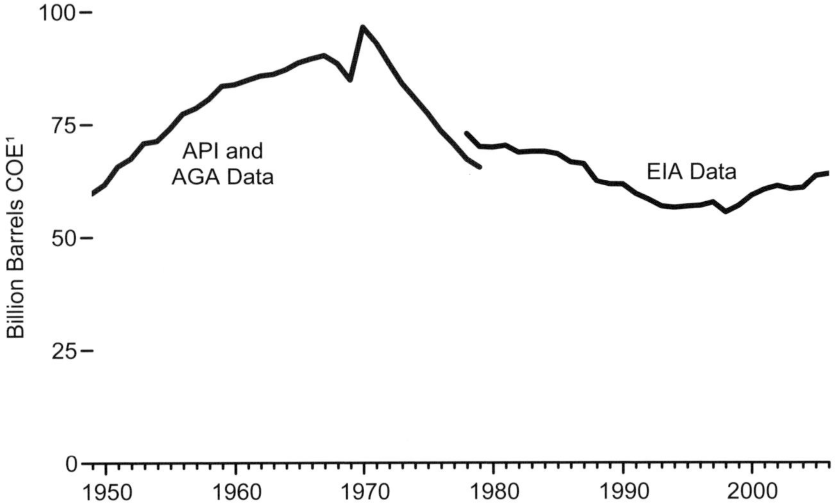

By Type, 2006

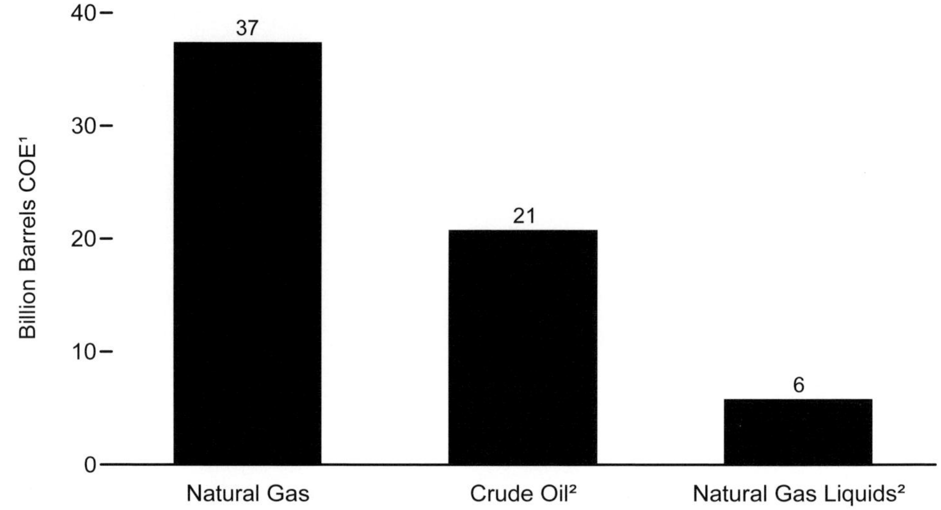

By Type, 1949-2006

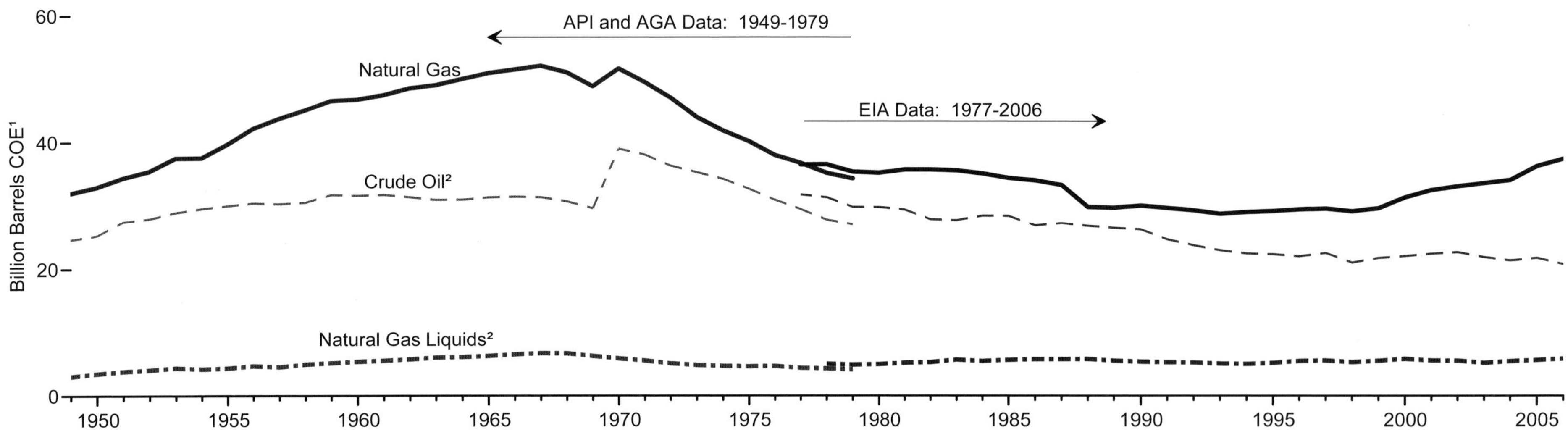

[1] COE=crude oil equivalent.

[2] To the extent that lease condensate is measured or estimated it is included in "Natural Gas Liquids"; otherwise, lease condensate is included in "Crude Oil."

Notes: • Data are at end of year. • API=American Petroleum Institute. AGA=American Gas Association. EIA=Energy Information Administration. • Because vertical scales differ, graphs should not be compared.

Source: Table 4.3.

Table 4.3 Crude Oil, Natural Gas, and Natural Gas Liquids Proved Reserves, Selected Years, 1949-2006

Year	Crude Oil [1]	Natural Gas (Dry)		Natural Gas Liquids [1]		Total
	Billion Barrels	Trillion Cubic Feet [2]	Billion Barrels COE [3]	Billion Barrels	Billion Barrels COE [3]	Billion Barrels COE [3]
	American Petroleum Institute and American Gas Association Data					
1949	24.6	179.4	32.0	3.7	3.1	59.7
1950	25.3	184.6	32.9	4.3	3.5	61.7
1955	30.0	222.5	39.7	5.4	4.4	74.1
1960	31.6	262.3	46.8	6.8	5.4	83.8
1965	31.4	286.5	51.0	8.0	6.3	88.6
1970	39.0	290.7	51.7	7.7	5.9	96.6
1971	38.1	278.8	49.6	7.3	5.5	93.2
1972	36.3	266.1	47.1	6.8	5.1	88.5
1973	35.3	250.0	44.0	6.5	4.8	84.1
1974	34.2	237.1	41.9	6.4	4.7	80.8
1975	32.7	228.2	40.2	6.3	4.6	77.5
1976	30.9	216.0	38.0	6.4	4.7	73.6
1977	29.5	208.9	36.8	6.0	4.4	70.6
1978	27.8	200.3	35.2	5.9	4.3	67.3
1979	27.1	194.9	34.3	5.7	4.1	65.5
	Energy Information Administration Data					
1977	31.8	207.4	36.5	NA	NA	NA
1978	31.4	208.0	36.5	6.8	5.0	73.0
1979	29.8	201.0	35.4	6.6	4.9	70.1
1980	29.8	199.0	35.2	6.7	5.0	70.0
1981	29.4	201.7	35.7	7.1	5.2	70.4
1982	27.9	201.5	35.7	7.2	5.3	68.8
1983	27.7	200.2	35.6	7.9	5.7	69.0
1984	28.4	197.5	35.1	7.6	5.5	69.0
1985	28.4	193.4	34.4	7.9	5.6	68.5
1986	26.9	191.6	34.0	8.2	5.8	66.7
1987	27.3	187.2	33.3	8.1	5.8	66.3
1988	26.8	168.0	29.8	8.2	5.8	62.4
1989	26.5	167.1	29.7	7.8	5.5	61.7
1990	26.3	169.3	30.0	7.6	5.4	61.7
1991	24.7	167.1	29.7	7.5	5.3	59.6
1992	23.7	165.0	29.3	7.5	5.2	58.3
1993	23.0	162.4	28.8	7.2	5.1	56.8
1994	22.5	163.8	29.0	7.2	5.0	56.5
1995	22.4	165.1	29.2	7.4	5.2	56.8
1996	22.0	166.5	29.4	7.8	5.5	56.9
1997	22.5	167.2	29.6	8.0	5.6	57.7
1998	21.0	164.0	29.2	7.5	5.3	55.5
1999	21.8	167.4	29.6	7.9	5.5	56.9
2000	22.0	177.4	31.4	8.3	5.8	59.2
2001	22.4	183.5	32.5	8.0	5.6	60.5
2002	22.7	186.9	33.1	8.0	5.6	61.3
2003	21.9	189.0	33.6	7.5	5.2	60.7
2004	21.4	192.5	34.1	7.9	5.5	60.9
2005	21.8	204.4	36.3	8.2	5.6	63.6
2006	20.8	211.1	37.4	8.5	5.8	64.0

[1] To the extent that lease condensate is measured or estimated it is included in "Natural Gas Liquids"; otherwise, lease condensate is included in "Crude Oil."

[2] The American Gas Association estimates of natural gas proved reserves include volumes of natural gas held in underground storage. In 1979, this volume amounted to 4.9 trillion cubic feet. Energy Information Administration (EIA) data do not include natural gas in underground storage.

[3] Natural gas is converted to crude oil equivalent (COE) by multiplying by the natural gas dry production approximate heat content (see Table A4) and then dividing by the crude oil production approximate heat content (see Table A2). The lease condensate portion of natural gas liquids is converted to COE by multiplying by the lease condensate production approximate heat content (5.5 million Btu per barrel) and then dividing by the crude oil production approximate heat content. Other natural gas liquids are converted to COE by multiplying by the natural gas plant liquids production approximate heat content (see Table A2) and then dividing by the crude oil production approximate heat content.

NA=Not available.

Notes: • Data are at end of year. • See "Proved Reserves, Crude Oil," "Proved Reserves, Natural Gas," and "Proved Reserves, Natural Gas Liquids" in Glossary.

Web Pages: • For all data beginning in 1949, see http://www.eia.doe.gov/emeu/aer/resource.html. • For related information, see http://www.eia.doe.gov/oil_gas/petroleum/info_glance/petroleum.html

Sources: **American Petroleum Institute and American Gas Association Data:** American Petroleum Institute, American Gas Association, and Canadian Petroleum Association (published jointly), *Reserves of Crude Oil, Natural Gas Liquids and Natural Gas in the United States and Canada as of December 31, 1979,* Volume 34 (June 1980). **Energy Information Administraton Data:** • 1977-1995—EIA, *U.S. Crude Oil, Natural Gas, and Natural Gas Liquids Reserves,* annual reports. • 1996 forward—EIA, *U.S. Crude Oil, Natural Gas, and Natural Gas Liquids Reserves 2006 Annual Report* (December 2007), Table 1.

Figure 4.4 Crude Oil and Natural Gas Rotary Rigs in Operation

By Site, 1949-2007

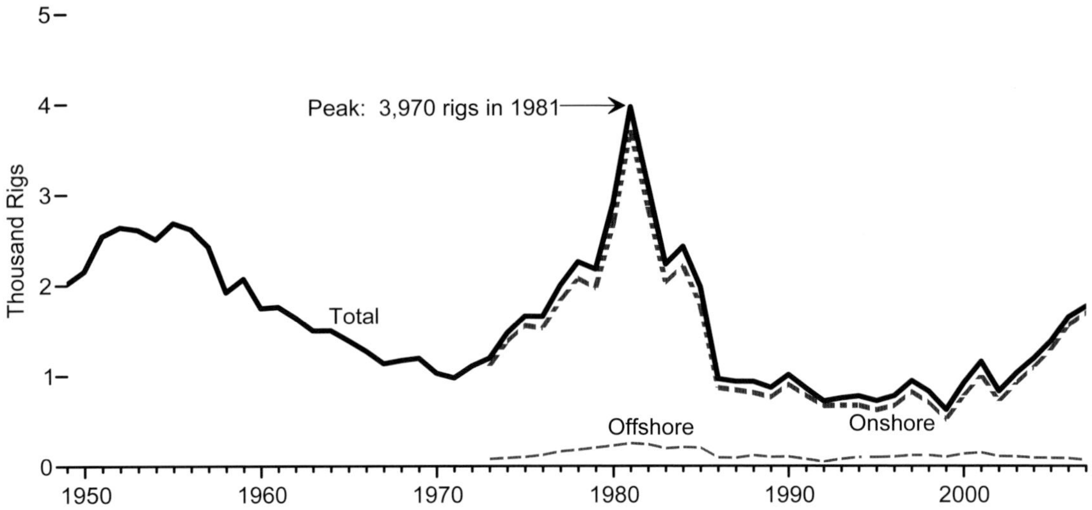

By Site, 2007

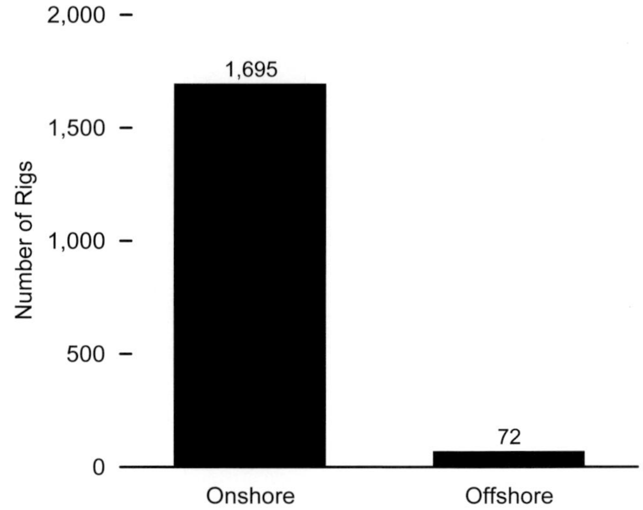

By Type, 1949-2007

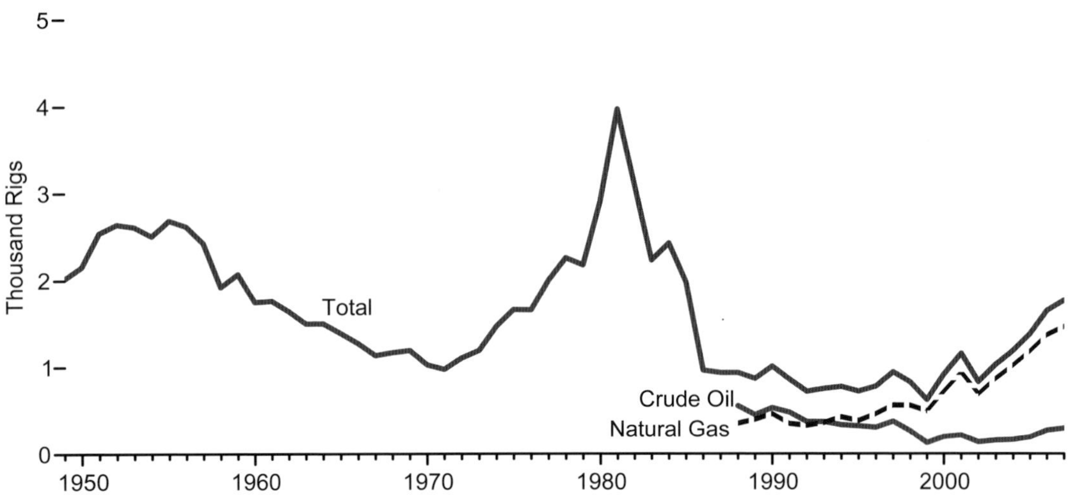

By Type, 2007

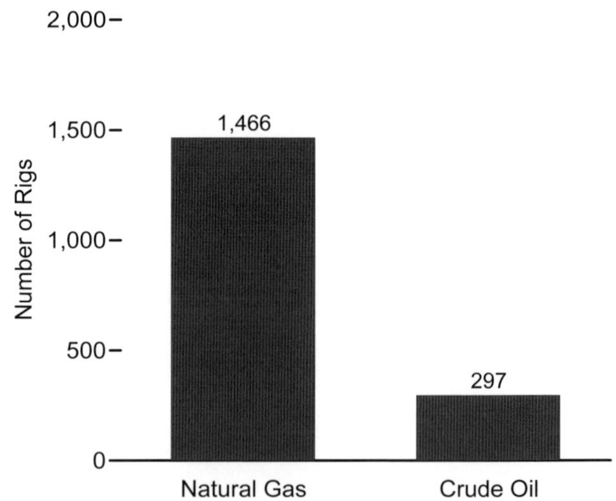

Source: Table 4.4.

Table 4.4 Crude Oil and Natural Gas Rotary Rigs in Operation, Selected Years, 1949-2007

(Number of Rigs)

Year	By Site		By Type		Total [1]
	Onshore	Offshore	Crude Oil	Natural Gas	
1949	NA	NA	NA	NA	2,017
1950	NA	NA	NA	NA	2,154
1955	NA	NA	NA	NA	2,686
1960	NA	NA	NA	NA	1,748
1965	NA	NA	NA	NA	1,388
1970	NA	NA	NA	NA	1,028
1971	NA	NA	NA	NA	976
1972	NA	NA	NA	NA	1,107
1973	1,110	84	NA	NA	1,194
1974	1,378	94	NA	NA	1,472
1975	1,554	106	NA	NA	1,660
1976	1,529	129	NA	NA	1,658
1977	1,834	167	NA	NA	2,001
1978	2,074	185	NA	NA	2,259
1979	1,970	207	NA	NA	2,177
1980	2,678	231	NA	NA	2,909
1981	3,714	256	NA	NA	3,970
1982	2,862	243	NA	NA	3,105
1983	2,033	199	NA	NA	2,232
1984	2,215	213	NA	NA	2,428
1985	1,774	206	NA	NA	1,980
1986	865	99	NA	NA	964
1987	841	95	NA	NA	936
1988	813	123	554	354	936
1989	764	105	453	401	869
1990	902	108	532	464	1,010
1991	779	81	482	351	860
1992	669	52	373	331	721
1993	672	82	373	364	754
1994	673	102	335	427	775
1995	622	101	323	385	723
1996	671	108	306	464	779
1997	821	122	376	564	943
1998	703	123	264	560	827
1999	519	106	128	496	625
2000	778	140	197	720	918
2001	1,003	153	217	939	1,156
2002	717	113	137	691	830
2003	924	108	157	872	1,032
2004	1,095	97	165	1,025	1,192
2005	1,290	93	194	1,186	1,383
2006	1,559	90	274	1,372	1,649
2007	1,695	72	297	1,466	1,768

[1] Sum of rigs drilling for crude oil, rigs drilling for natural gas, and other rigs (not shown) drilling for miscellaneous purposes, such as service wells, injection wells, and stratigraphic tests.

NA=Not available.

Notes: • Data are not for the exact calendar year but are an average for the 52 or 53 consecutive whole weeks that most nearly coincide with the calendar year. • Geographic coverage is the 50 States and the District of Columbia. • Totals may not equal sum of components due to independent rounding.

Web Page: For all data beginning in 1949, see http://www.eia.doe.gov/emeu/aer/resource.html.

Source: Baker Hughes, Inc., Houston, Texas, *Rotary Rigs Running—By State.*

Figure 4.5 Crude Oil and Natural Gas Exploratory and Development Wells

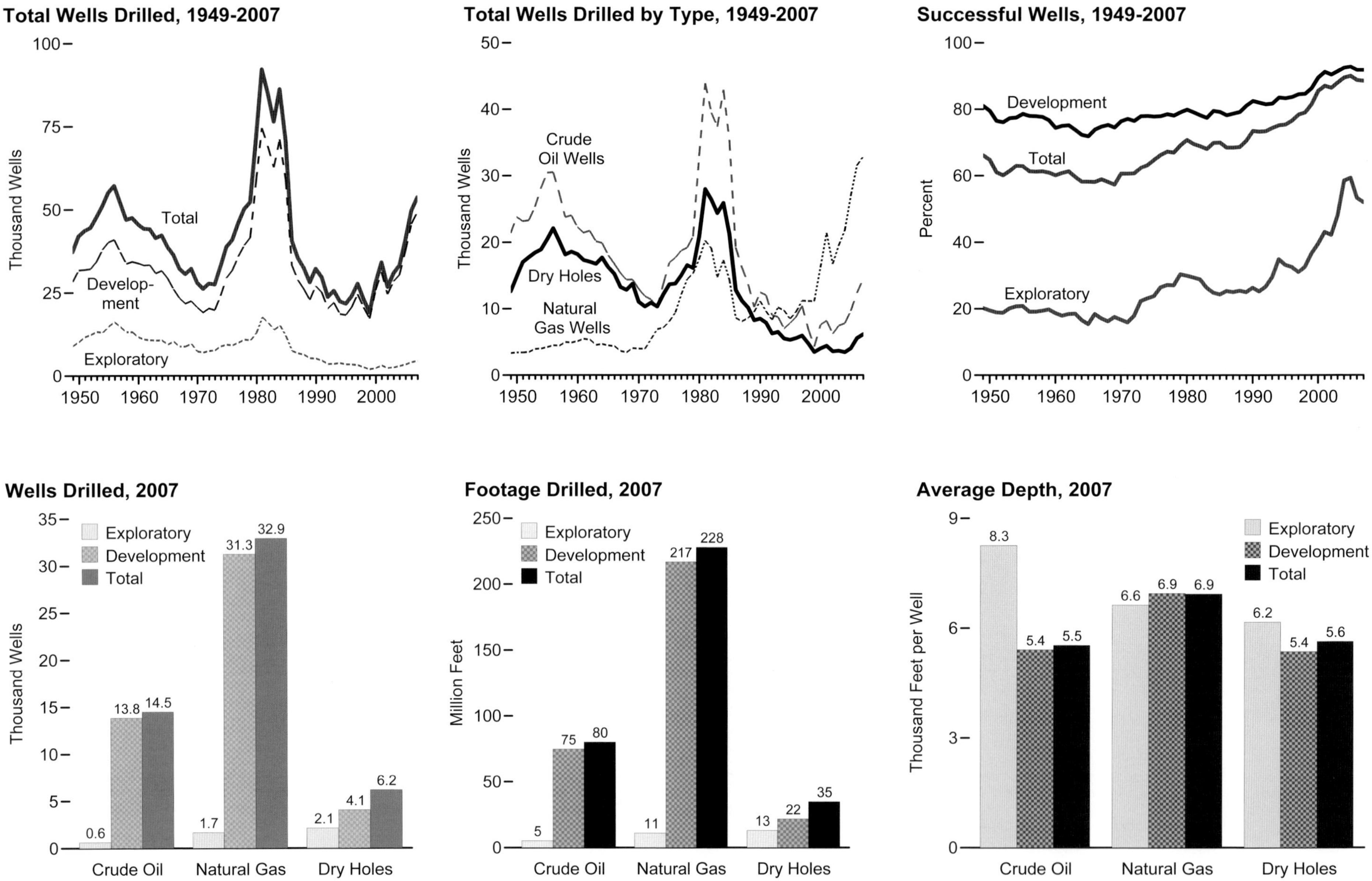

Sources: Tables 4.5–4.7.

Table 4.5 Crude Oil and Natural Gas Exploratory and Development Wells, Selected Years, 1949-2007

Year	Wells Drilled: Crude Oil [2]	Wells Drilled: Natural Gas [3]	Wells Drilled: Dry Holes [4]	Wells Drilled: Total	Successful Wells	Footage Drilled [1]: Crude Oil [2]	Footage Drilled [1]: Natural Gas [3]	Footage Drilled [1]: Dry Holes [4]	Footage Drilled [1]: Total	Average Depth: Crude Oil [2]	Average Depth: Natural Gas [3]	Average Depth: Dry Holes [4]	Average Depth: Total
	Number				Percent	Thousand Feet				Feet per Well			
1949	21,352	3,363	12,597	37,312	66.2	79,428	12,437	43,754	135,619	3,720	3,698	3,473	3,635
1950	23,812	3,439	14,799	42,050	64.8	92,695	13,685	50,977	157,358	3,893	3,979	3,445	3,742
1955	30,432	4,266	20,452	55,150	62.9	121,148	19,930	85,103	226,182	3,981	4,672	4,161	4,101
1960	22,258	5,149	18,212	45,619	60.1	86,568	28,246	77,361	192,176	3,889	5,486	4,248	4,213
1965	18,065	4,482	16,226	38,773	58.2	73,322	24,931	76,629	174,882	4,059	5,562	4,723	4,510
1970	12,968	4,011	11,031	28,010	60.6	56,859	23,623	58,074	138,556	4,385	5,860	5,265	4,943
1971	11,853	3,971	10,309	26,133	60.6	49,109	23,460	54,685	127,253	4,126	5,890	5,305	4,858
1972	11,378	5,440	10,891	27,709	60.7	49,269	30,006	58,556	137,831	4,330	5,516	5,377	4,974
1973	10,167	6,933	10,320	27,420	62.4	44,416	38,045	55,761	138,223	4,369	5,488	5,403	5,041
1974	13,647	7,138	12,116	32,901	63.2	52,025	38,449	62,899	153,374	3,812	5,387	5,191	4,662
1975	16,948	8,127	13,646	38,721	64.8	66,819	44,454	69,220	180,494	3,943	5,470	5,073	4,661
1976	17,688	9,409	13,758	40,855	66.3	68,892	49,113	68,977	186,982	3,895	5,220	5,014	4,577
1977	18,745	12,122	14,985	45,852	67.3	75,451	63,686	76,728	215,866	4,025	5,254	5,120	4,708
1978	19,181	14,413	16,551	50,145	67.0	77,041	75,841	85,788	238,669	4,017	5,262	5,183	4,760
1979	20,851	15,254	16,099	52,204	69.2	82,688	80,468	81,642	244,798	3,966	5,275	5,071	4,689
1980	32,959	17,461	20,785	71,205	70.8	125,262	92,106	99,575	316,943	3,801	5,275	4,791	4,451
1981	43,887	20,250	27,953	92,090	69.6	172,167	108,353	134,934	415,454	3,923	5,351	4,827	4,511
1982	39,459	19,076	26,379	84,914	68.9	149,674	107,149	123,746	380,569	3,793	5,617	4,691	4,482
1983	37,366	14,684	24,355	76,405	68.1	136,849	78,108	105,222	320,179	3,662	5,319	4,320	4,191
1984	42,906	17,338	25,884	86,128	69.9	162,653	91,480	119,860	373,993	3,791	5,276	4,631	4,342
1985	35,261	14,324	21,211	70,796	70.0	137,728	76,293	100,388	314,409	3,906	5,326	4,733	4,441
1986	19,213	8,599	12,799	40,611	68.5	76,825	45,039	60,961	182,825	3,999	5,238	4,763	4,502
1987	16,210	8,096	11,167	35,473	68.5	66,358	42,584	53,588	162,530	4,094	5,260	4,799	4,582
1988	13,646	8,578	10,119	32,343	68.7	58,639	45,363	52,517	156,519	4,297	5,288	5,190	4,839
1989	10,230	9,522	8,236	27,988	70.6	43,266	49,081	42,099	134,446	4,229	5,154	5,112	4,804
1990	12,445	11,126	8,496	32,067	73.5	55,269	56,775	44,160	156,204	4,441	5,103	5,198	4,871
1991	12,035	9,611	7,882	29,528	73.3	55,268	50,757	40,307	146,332	4,592	5,281	5,114	4,956
1992	9,019	8,305	6,284	23,608	73.4	44,851	46,615	31,814	123,280	4,973	5,613	5,063	5,222
1993	8,764	10,174	6,513	25,451	74.4	43,922	61,186	33,323	138,431	5,012	6,014	5,116	5,439
1994	7,001	9,739	5,515	22,255	75.2	37,270	61,576	30,293	129,139	5,324	6,323	5,493	5,803
1995	7,827	8,454	5,319	21,600	75.4	39,125	52,872	29,312	121,309	4,999	6,254	5,511	5,616
1996	8,760	9,539	5,587	23,886	76.6	42,196	59,800	31,366	133,362	4,817	6,269	5,614	5,583
1997	10,445	11,186	5,955	27,586	78.4	51,466	69,593	34,233	155,292	4,927	6,221	5,749	5,629
1998	6,979	11,127	4,805	22,911	79.0	34,340	67,789	29,008	131,137	4,920	6,092	6,037	5,724
1999	4,314	11,121	3,504	18,939	81.5	18,860	55,331	20,404	94,595	4,372	4,975	5,823	4,995
2000	7,585	16,242	4,046	27,873	85.5	33,777	79,605	23,193	136,575	4,453	4,901	5,732	4,900
2001	8,186	21,403	4,432	34,021	87.0	38,716	108,482	25,047	172,245	4,730	5,069	5,651	5,063
2002	6,226	16,728	3,610	26,564	86.4	27,869	91,788	20,316	139,973	4,476	5,487	5,628	5,269
2003	7,465	19,522	3,688	30,675	88.0	35,220	112,990	20,968	169,178	4,718	5,788	5,685	5,515
2004E	R7,806	R21,816	R3,474	R33,096	R89.5	R36,691	R127,953	R20,056	R184,701	R4,700	R5,865	R5,773	R5,581
2005E	R9,668	R27,014	R4,063	R40,745	R90.0	R46,011	R159,778	R21,795	R227,584	R4,759	R5,915	R5,364	R5,586
2006E	12,339	31,587	5,581	49,507	88.7	60,494	194,504	29,533	284,531	4,903	6,158	5,292	5,747
2007E	14,477	32,910	6,171	53,558	88.5	80,086	227,968	34,753	342,807	5,532	6,927	5,632	6,401

[1] See "Footage Drilled" in Glossary.
[2] See "Crude Oil Well" in Glossary.
[3] See "Natural Gas Well" in Glossary.
[4] See "Dry Hole" in Glossary.
R=Revised. E=Estimate.

Notes: • Data are for exploratory and development wells combined; see Table 4.6 for exploratory wells only, and Table 4.7 for development wells only. • Service wells, stratigraphic tests, and core tests are excluded. • For 1949-1959, data represent wells completed in a given year. For 1960-1969, data are for well completion reports received by the American Petroleum Institute during the reporting year. For 1970 forward, the data represent wells completed in a given year. The as-received well completion data for recent years are incomplete due to delays in the reporting of wells drilled. The Energy Information Administration (EIA) therefore statistically imputes the missing data. • Totals may not equal sum of components due to independent rounding. Average depth may not equal average of components due to independent rounding.

Web Pages: • For all data beginning in 1949, see http://www.eia.doe.gov/emeu/aer/resource.html. • For related information, see http://www.eia.doe.gov/oil_gas/petroleum/info_glance/petroleum.html.

Sources: • 1949-1965—Gulf Publishing Company, *World Oil,* "Forecast-Review" issue. • 1966-1969—American Petroleum Institute (API), *Quarterly Review of Drilling Statistics for the United States,* annual summaries and monthly reports. • 1970-1994—EIA computations based on well reports submitted to the API. • 1995 forward—EIA computations based on well reports submitted to the Information Handling Services Energy Group, Inc. For current data see the EIA, *Monthly Energy Review,* Table 5.2.

Figure 4.6 Crude Oil and Natural Gas Exploratory Wells, 1949-2007

Exploratory Wells Drilled by Well Type

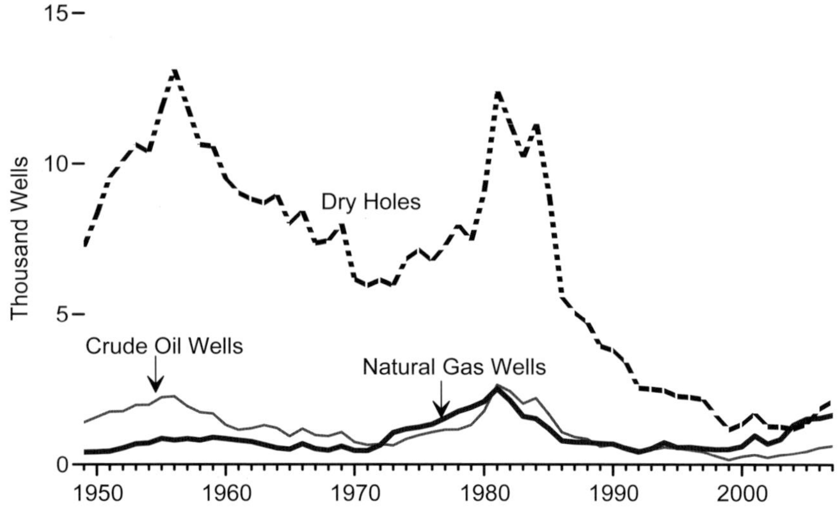

Exploratory Footage Drilled by Well Type

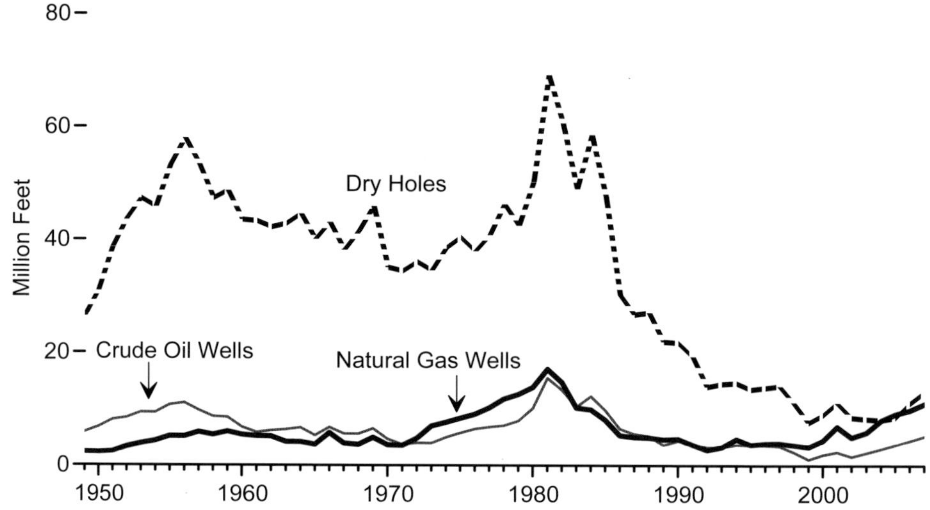

Exploratory Wells Average Depth, All Wells

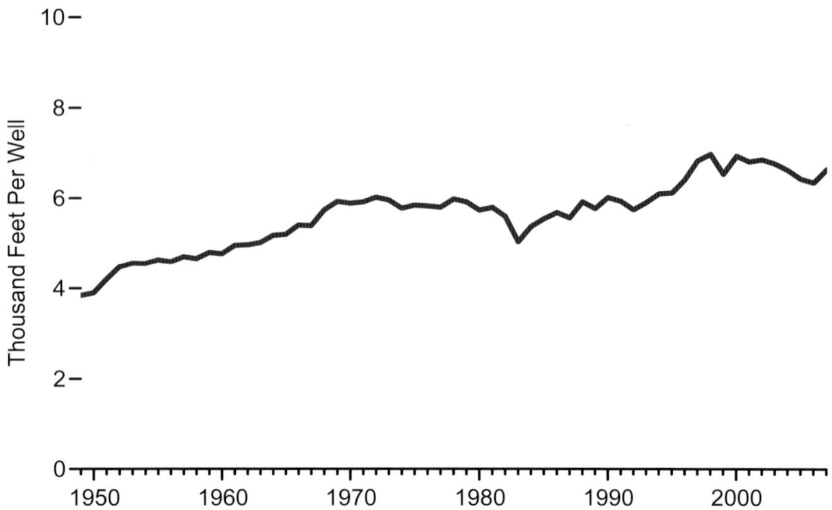

Exploratory Wells Average Depth by Well Type

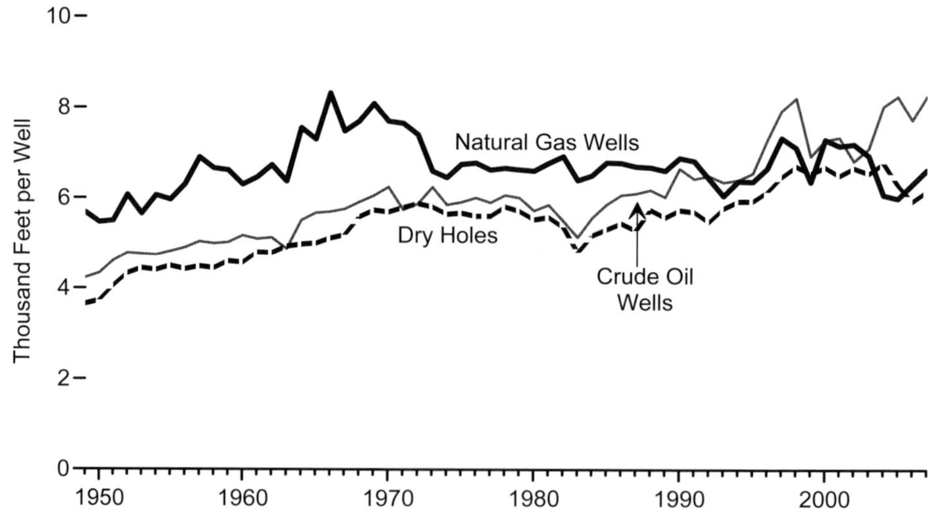

Note: These graphs depict exploratory wells only; see Figure 4.5 for all wells and Figure 4.7 for development wells only.

Source: Table 4.6.

Table 4.6 Crude Oil and Natural Gas Exploratory Wells, Selected Years, 1949-2007

Year	Wells Drilled				Successful Wells	Footage Drilled [1]				Average Depth			
	Crude Oil [2]	Natural Gas [3]	Dry Holes [4]	Total		Crude Oil [2]	Natural Gas [3]	Dry Holes [4]	Total	Crude Oil [2]	Natural Gas [3]	Dry Holes [4]	Total
	Number				Percent	Thousand Feet				Feet per Well			
1949	1,406	424	7,228	9,058	20.2	5,950	2,409	26,439	34,798	4,232	5,682	3,658	3,842
1950	1,583	431	8,292	10,306	19.5	6,862	2,356	30,957	40,175	4,335	5,466	3,733	3,898
1955	2,236	874	11,832	14,942	20.8	10,774	5,212	53,220	69,206	4,819	5,964	4,498	4,632
1960	1,321	868	9,515	11,704	18.7	6,829	5,466	43,535	55,831	5,170	6,298	4,575	4,770
1965	946	515	8,005	9,466	15.4	5,366	3,757	40,081	49,204	5,672	7,295	5,007	5,198
1970	757	477	6,162	7,396	16.7	4,729	3,678	35,123	43,530	6,247	7,695	5,700	5,885
1971	659	470	5,952	7,081	15.9	3,786	3,610	34,499	41,895	5,745	7,649	5,796	5,915
1972	685	656	6,134	7,475	17.9	4,028	4,847	36,081	44,956	5,880	7,400	5,882	6,015
1973	642	1,067	5,952	7,661	22.3	4,008	7,038	34,571	45,618	6,243	6,596	5,808	5,955
1974	859	1,190	6,833	8,882	23.1	5,029	7,683	38,603	51,315	5,855	6,456	5,649	5,777
1975	982	1,248	7,129	9,359	23.8	5,806	8,422	40,448	54,677	5,913	6,748	5,674	5,842
1976	1,086	1,346	6,772	9,204	26.4	6,527	9,121	37,969	53,617	6,010	6,777	5,607	5,825
1977	1,164	1,548	7,283	9,995	27.1	6,870	10,255	40,823	57,949	5,902	6,625	5,605	5,798
1978	1,171	1,771	7,965	10,907	27.0	7,105	11,798	46,295	65,197	6,067	6,662	5,812	5,978
1979	1,321	1,907	7,437	10,665	30.3	7,941	12,643	42,512	63,096	6,011	6,630	5,716	5,916
1980	1,777	2,099	9,081	12,957	29.9	10,177	13,862	50,249	74,288	5,727	6,604	5,533	5,733
1981	2,651	2,522	12,400	17,573	29.4	15,515	17,079	69,214	101,808	5,853	6,772	5,582	5,793
1982	2,437	2,133	11,307	15,877	28.8	13,413	14,763	60,680	88,856	5,504	6,921	5,367	5,597
1983	2,030	1,605	10,206	13,841	26.3	10,437	10,264	48,989	69,690	5,141	6,395	4,800	5,035
1984	2,209	1,528	11,321	15,058	24.8	12,294	9,935	58,624	80,853	5,565	6,502	5,178	5,369
1985	1,680	1,200	8,954	11,834	24.3	9,854	8,144	47,604	65,602	5,865	6,787	5,317	5,544
1986	1,084	797	5,567	7,448	25.3	6,579	5,401	30,325	42,305	6,069	6,777	5,447	5,680
1987	926	756	5,052	6,734	25.0	5,652	5,064	26,746	37,462	6,104	6,698	5,294	5,563
1988	855	747	4,711	6,313	25.4	5,286	4,992	27,079	37,357	6,182	6,683	5,748	5,917
1989	607	706	3,934	5,247	25.0	3,659	4,664	21,947	30,270	6,028	6,606	5,579	5,769
1990	664	693	3,793	5,150	26.3	4,420	4,774	21,777	30,971	6,657	6,889	5,741	6,014
1991	601	544	3,390	4,535	25.2	3,865	3,712	19,330	26,907	6,431	6,824	5,702	5,933
1992	498	427	2,550	3,475	26.6	3,236	2,749	13,983	19,968	6,498	6,438	5,484	5,746
1993	509	541	2,509	3,559	29.5	3,235	3,277	14,504	21,016	6,356	6,057	5,781	5,905
1994	579	740	2,465	3,784	34.9	3,708	4,720	14,632	23,060	6,404	6,378	5,936	6,094
1995	549	583	2,279	3,411	33.2	3,601	3,713	13,545	20,859	6,559	6,369	5,943	6,115
1996	496	591	2,246	3,333	32.6	3,615	3,938	13,805	21,358	7,288	6,663	6,146	6,408
1997	434	543	2,178	3,155	31.0	3,446	3,981	14,105	21,532	7,940	7,331	6,476	6,825
1998	286	510	1,649	2,445	32.6	2,351	3,629	11,062	17,042	8,220	7,116	6,708	6,970
1999	156	519	1,167	1,842	36.6	1,081	3,300	7,648	12,029	6,929	6,358	6,554	6,530
2000	267	615	1,349	2,231	39.5	1,945	4,488	9,024	15,457	7,285	7,298	6,689	6,928
2001	330	972	1,716	3,018	43.1	2,423	6,954	11,165	20,542	7,342	7,154	6,506	6,806
2002	236	701	1,283	2,220	42.2	1,609	5,042	8,559	15,210	6,818	7,193	6,671	6,851
2003	321	853	1,266	2,440	48.1	2,280	5,920	8,291	16,491	7,103	6,940	6,549	6,759
2004E	R368	R1,323	R1,200	R2,891	R58.5	R2,957	R8,036	R8,134	R19,127	R8,036	R6,074	R6,778	R6,616
2005E	R448	R1,532	R1,358	R3,338	R59.3	R3,700	R9,191	R8,551	R21,442	R8,258	R5,999	R6,297	R6,424
2006E	576	1,559	1,870	4,005	53.3	4,452	9,845	11,107	25,404	7,729	6,315	5,940	6,343
2007E	634	1,658	2,119	4,411	52.0	5,236	10,990	13,054	29,280	8,259	6,628	6,161	6,638

[1] See "Footage Drilled" in Glossary.
[2] See "Crude Oil Well" in Glossary.
[3] See "Natural Gas Well" in Glossary.
[4] See "Dry Hole" in Glossary.
R=Revised. E=Estimate.

Notes: • Data are for exploratory wells only; see Table 4.5 for exploratory and development wells combined, and Table 4.7 for development wells only. • For 1949-1959, data represent wells completed in a given year. For 1960-1969, data are for well completion reports received by the American Petroleum Institute (API) during the reporting year. For 1970 forward, the data represent wells completed in a given year. The as-received well completion data for recent years are incomplete due to delays in the reporting of wells drilled. The Energy Information Administration (EIA) therefore statistically imputes the missing data. • Totals may not equal sum of components due to independent rounding. Average depth may not equal average of components due to independent rounding.

Web Pages: • For all data beginning in 1949, see http://www.eia.doe.gov/emeu/aer/resource.html. • For related information, see http://www.eia.doe.gov/oil_gas/petroleum/info_glance/petroleum.html.

Sources: • 1949-1960—American Association of Petroleum Geologists, *Statistics on Exploratory Drilling in the United States, 1940 through 1960* (1962), pp. 4-19. • 1961-1965—*Bulletin of the American Association of Petroleum Geologists,* "North American Developments" issue. • 1966-1969—API, *Quarterly Review of Drilling Statistics for the United States,* annual summaries and monthly reports. • 1970-1994—EIA computations based on well reports submitted to the API. • 1995 forward—EIA computations based on well reports submitted to the Information Handling Services Energy Group, Inc. For current data see the EIA, *Monthly Energy Review,* Table 5.2.

Figure 4.7 Crude Oil and Natural Gas Development Wells, 1949-2007

Development Wells Drilled by Well Type

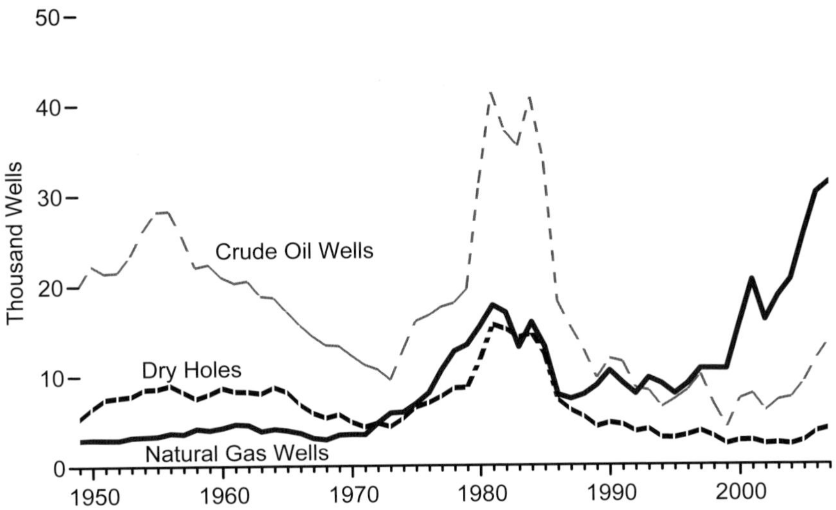

Development Footage Drilled by Well Type

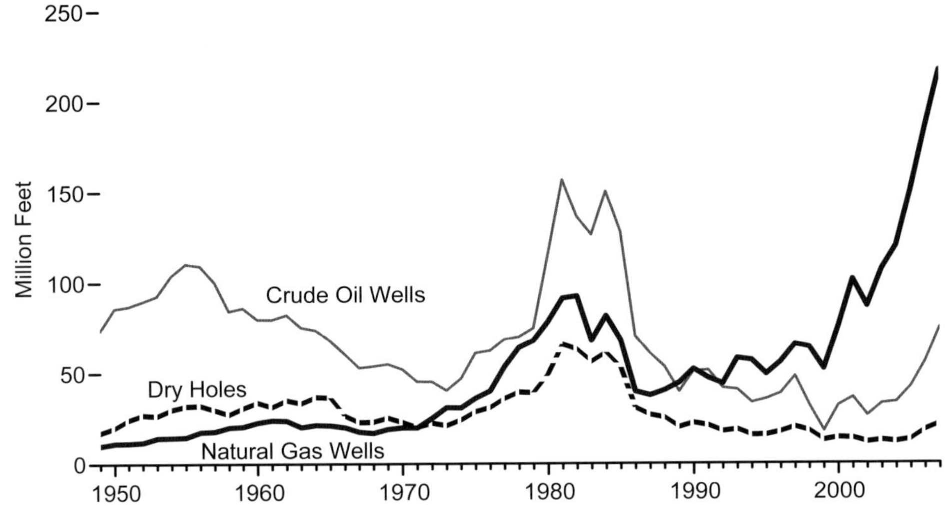

Development Wells Average Depth, All Wells

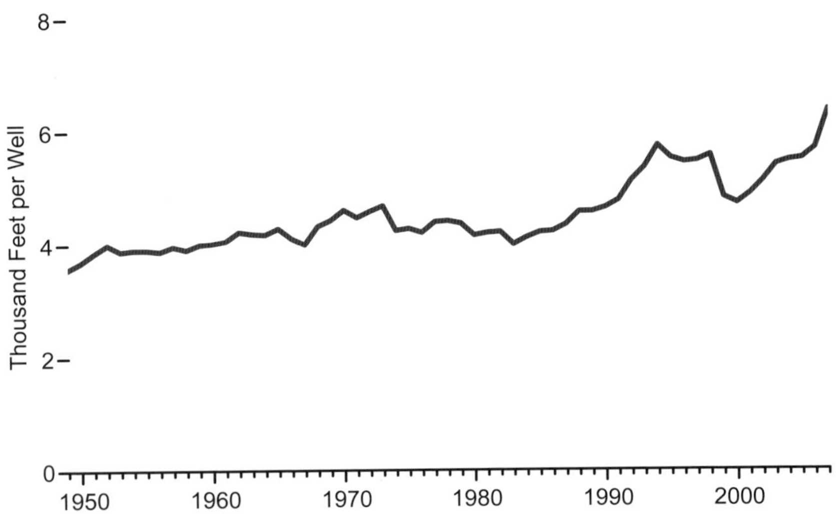

Development Wells Average Depth by Well Type

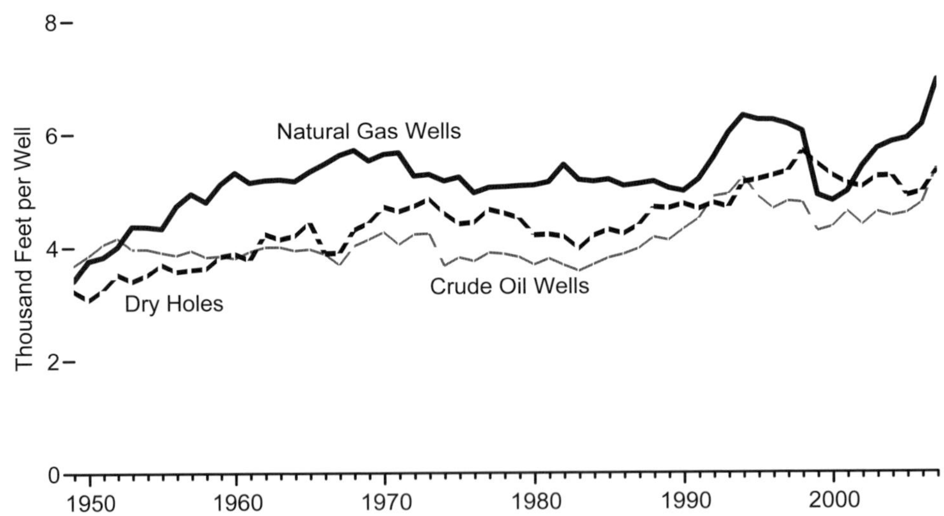

Note: These graphs depict development wells only; see Figure 4.5 for all wells and Figure 4.6 for exploratory wells only.

Source: Table 4.7.

Table 4.7 Crude Oil and Natural Gas Development Wells, Selected Years, 1949-2007

Year	Wells Drilled				Successful Wells	Footage Drilled [1]				Average Depth			
	Crude Oil [2]	Natural Gas [3]	Dry Holes [4]	Total		Crude Oil [2]	Natural Gas [3]	Dry Holes [4]	Total	Crude Oil [2]	Natural Gas [3]	Dry Holes [4]	Total
	Number				Percent	Thousand Feet				Feet per Well			
1949	19,946	2,939	5,369	28,254	81.0	73,478	10,028	17,315	100,821	3,684	3,412	3,225	3,568
1950	22,229	3,008	6,507	31,744	79.5	85,833	11,329	20,020	117,183	3,861	3,766	3,077	3,691
1955	28,196	3,392	8,620	40,208	78.6	110,374	14,718	31,883	156,976	3,915	4,339	3,699	3,904
1960	20,937	4,281	8,697	33,915	74.4	79,739	22,780	33,826	136,345	3,809	5,321	3,889	4,020
1965	17,119	3,967	8,221	29,307	71.9	67,956	21,174	36,548	125,678	3,970	5,337	4,446	4,288
1970	12,211	3,534	4,869	20,614	76.4	52,130	19,945	22,951	95,026	4,269	5,644	4,714	4,610
1971	11,194	3,501	4,357	19,052	77.1	45,323	19,850	20,186	85,358	4,049	5,670	4,633	4,480
1972	10,693	4,784	4,757	20,234	76.5	45,241	25,159	22,475	92,875	4,231	5,259	4,725	4,590
1973	9,525	5,866	4,368	19,759	77.9	40,408	31,007	21,190	92,605	4,242	5,286	4,851	4,687
1974	12,788	5,948	5,283	24,019	78.0	46,996	30,766	24,296	102,059	3,675	5,173	4,599	4,249
1975	15,966	6,879	6,517	29,362	77.8	61,013	36,032	28,772	125,817	3,821	5,238	4,415	4,285
1976	16,602	8,063	6,986	31,651	77.9	62,365	39,992	31,008	133,365	3,756	4,960	4,439	4,214
1977	17,581	10,574	7,702	35,857	78.5	68,581	53,431	35,905	157,917	3,901	5,053	4,662	4,404
1978	18,010	12,642	8,586	39,238	78.1	69,936	64,043	39,493	173,472	3,883	5,066	4,600	4,421
1979	19,530	13,347	8,662	41,539	79.1	74,747	67,825	39,130	181,702	3,827	5,082	4,517	4,374
1980	31,182	15,362	11,704	58,248	79.9	115,085	78,244	49,326	242,655	3,691	5,093	4,214	4,166
1981	41,236	17,728	15,553	74,517	79.1	156,652	91,274	65,720	313,646	3,799	5,149	4,226	4,209
1982	37,022	16,943	15,072	69,037	78.2	136,261	92,386	63,066	291,713	3,681	5,453	4,184	4,225
1983	35,336	13,079	14,149	62,564	77.4	126,412	67,844	56,233	250,489	3,577	5,187	3,974	4,004
1984	40,697	15,810	14,563	71,070	79.5	150,359	81,545	61,236	293,140	3,695	5,158	4,205	4,125
1985	33,581	13,124	12,257	58,962	79.2	127,874	68,149	52,784	248,807	3,808	5,193	4,306	4,220
1986	18,129	7,802	7,232	33,163	78.2	70,246	39,638	30,636	140,520	3,875	5,080	4,236	4,237
1987	15,284	7,340	6,115	28,739	78.7	60,706	37,520	26,842	125,068	3,972	5,112	4,390	4,352
1988	12,791	7,831	5,408	26,030	79.2	53,353	40,371	25,438	119,162	4,171	5,155	4,704	4,578
1989	9,623	8,816	4,302	22,741	81.1	39,607	44,417	20,152	104,176	4,116	5,038	4,684	4,581
1990	11,781	10,433	4,703	26,917	82.5	50,849	52,001	22,383	125,233	4,316	4,984	4,759	4,653
1991	11,434	9,067	4,492	24,993	82.0	51,403	47,045	20,977	119,425	4,496	5,189	4,670	4,778
1992	8,521	7,878	3,734	20,133	81.5	41,615	43,866	17,831	103,312	4,884	5,568	4,775	5,131
1993	8,255	9,633	4,004	21,892	81.7	40,687	57,909	18,819	117,415	4,929	6,012	4,700	5,363
1994	6,422	8,999	3,050	18,471	83.5	33,562	56,856	15,661	106,079	5,226	6,318	5,135	5,743
1995	7,278	7,871	3,040	18,189	83.3	35,524	49,159	15,767	100,450	4,881	6,246	5,187	5,523
1996	8,264	8,948	3,341	20,553	83.7	38,581	55,862	17,561	112,004	4,669	6,243	5,256	5,450
1997	10,011	10,643	3,777	24,431	84.5	48,020	65,612	20,128	133,760	4,797	6,165	5,329	5,475
1998	6,693	10,617	3,156	20,466	84.6	31,989	64,160	17,946	114,095	4,779	6,043	5,686	5,575
1999	4,158	10,602	2,337	17,097	86.3	17,779	52,031	12,756	82,566	4,276	4,908	5,458	4,829
2000	7,318	15,627	2,697	25,642	89.5	31,832	75,117	14,169	121,118	4,350	4,807	5,254	4,723
2001	7,856	20,431	2,716	31,003	91.2	36,293	101,528	13,882	151,703	4,620	4,969	5,111	4,893
2002	5,990	16,027	2,327	24,344	90.4	26,260	86,746	11,757	124,763	4,384	5,412	5,052	5,125
2003	7,144	18,669	2,422	28,235	91.4	32,940	107,070	12,677	152,687	4,611	5,735	5,234	5,408
2004E	R7,438	R20,493	R2,274	R30,205	R92.5	R33,734	R119,917	R11,922	R165,573	R4,535	R5,852	R5,243	R5,482
2005E	R9,220	R25,482	R2,705	R37,407	R92.8	R42,312	R150,587	R13,243	R206,142	R4,589	R5,910	R4,896	R5,511
2006E	11,763	30,028	3,711	45,502	91.8	56,042	184,659	18,425	259,127	4,764	6,150	4,965	5,695
2007E	13,843	31,252	4,052	49,147	91.8	74,850	216,978	21,699	313,527	5,407	6,943	5,355	6,379

[1] See "Footage Drilled" in Glossary.
[2] See "Crude Oil Well" in Glossary.
[3] See "Natural Gas Well" in Glossary.
[4] See "Dry Hole" in Glossary.
R=Revised. E=Estimate.

Notes: • Data are for development wells only; see Table 4.5 for exploratory and development wells combined, and Table 4.6 for exploratory wells only. • Service wells, stratigraphic tests, and core tests are excluded. • For 1949-1959, data represent wells completed in a given year. For 1960-1969, data are for well completion reports received by the American Petroleum Institute during the reporting year. For 1970 forward, the data represent wells completed in a given year. The as-received well completion data for recent years are incomplete due to delays in the reporting of wells drilled. The Energy Information Administration (EIA) therefore statistically imputes the missing data. • Totals may not equal sum of components due to independent rounding. Average depth may not equal average of components due to independent rounding.

Web Pages: • For all data beginning in 1949, see http://www.eia.doe.gov/emeu/aer/resource.html. • For related information, see http://www.eia.doe.gov/oil_gas/petroleum/info_glance/petroleum.html.

Sources: • 1949-1965—Gulf Publishing Company, *World Oil*, "Forecast-Review" issue. • 1966-1969—American Petroleum Institute, *Quarterly Review of Drilling Statistics for the United States*, annual summaries and monthly reports. • 1970-1994—EIA computations based on well reports submitted to the American Petroleum Institute. • 1995 forward—EIA computations based on well reports submitted to the Information Handling Services Energy Group, Inc. For current data see the EIA, *Monthly Energy Review*, Table 5.2.

Figure 4.8 Costs of Crude Oil and Natural Gas Wells Drilled

Costs per Well, All Wells, 1960-2006

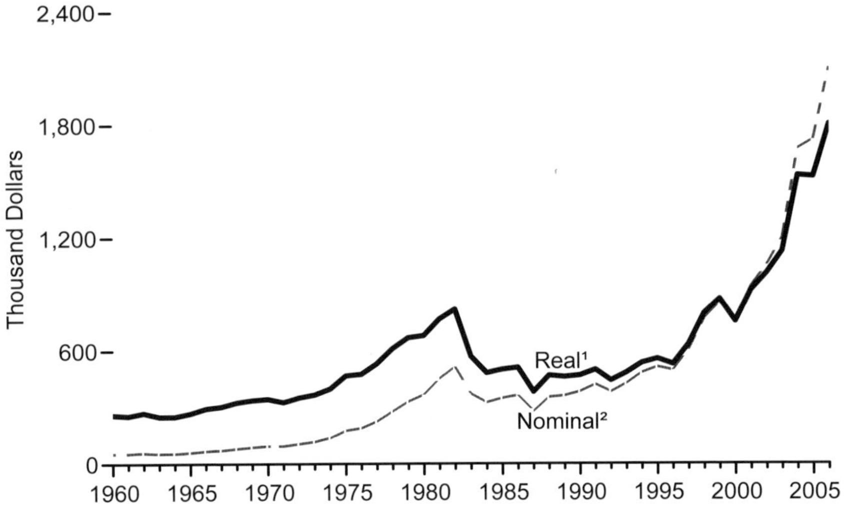

Costs per Foot, All Wells, 1960-2006

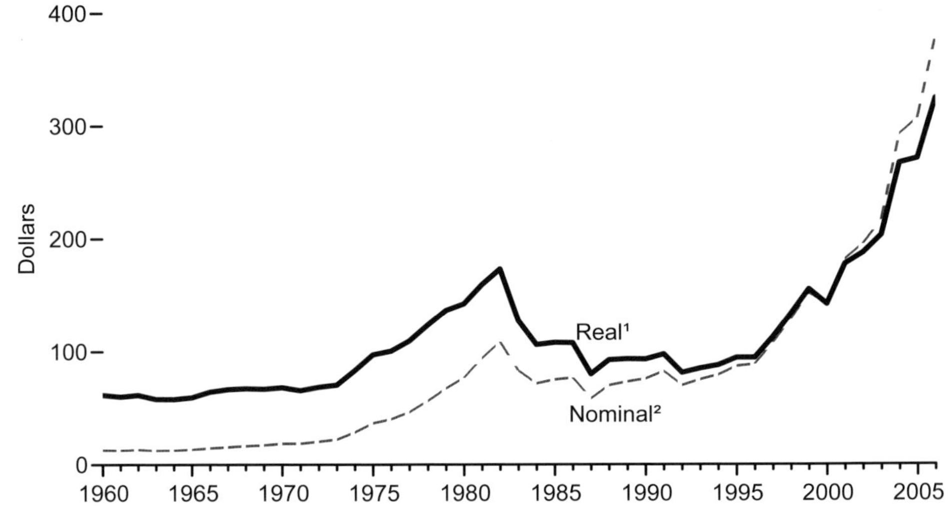

Costs per Well by Well Type, 2006

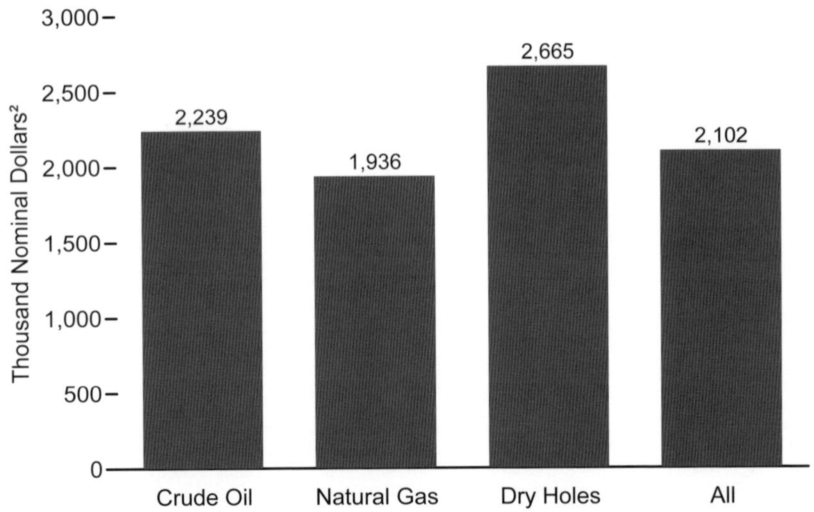

Costs per Foot by Well Type, 2006

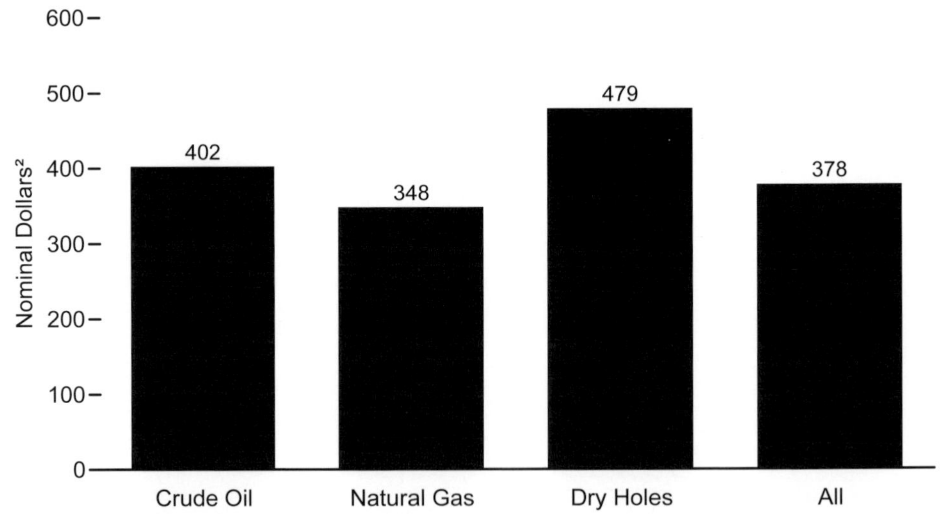

[1] In chained (2000) dollars, calculated by using gross domestic product implicit price deflators. See Table D1.

[2] See "Nominal Dollars" in Glossary.

Note: Because vertical scales differ, graphs should not be compared.
Source: Table 4.8.

Table 4.8 Costs of Crude Oil and Natural Gas Wells Drilled, 1960-2006

	Thousand Dollars per Well					Dollars per Foot				
	Crude Oil [1]	Natural Gas [2]	Dry Holes [3]	All		Crude Oil [1]	Natural Gas [2]	Dry Holes [3]	All	
Year	Nominal [4]	Nominal [4]	Nominal [4]	Nominal [4]	Real [5]	Nominal [4]	Nominal [4]	Nominal [4]	Nominal [4]	Real [5]
1960	52.2	102.7	44.0	54.9	261.1	13.22	18.57	10.56	13.01	61.83
1961	51.3	94.7	45.2	54.5	256.2	13.11	17.65	10.56	12.85	60.39
1962	54.2	97.1	50.8	58.6	271.8	13.41	18.10	11.20	13.31	61.71
1963	51.8	92.4	48.2	55.0	252.4	13.20	17.19	10.58	12.69	58.22
1964	50.6	104.8	48.5	55.8	252.2	13.12	18.57	10.64	12.86	58.11
1965	56.6	101.9	53.1	60.6	269.1	13.94	18.35	11.21	13.44	59.64
1966	62.2	133.8	56.9	68.4	295.1	15.04	21.75	12.34	14.95	64.51
1967	66.6	141.0	61.5	72.9	305.1	16.61	23.05	12.87	15.97	66.84
1968	79.1	148.5	66.2	81.5	327.0	18.63	24.05	12.88	16.83	67.56
1969	86.5	154.3	70.2	88.6	338.7	19.28	25.58	13.23	17.56	67.15
1970	86.7	160.7	80.9	94.9	344.6	19.29	26.75	15.21	18.84	68.42
1971	78.4	166.6	86.8	94.7	327.6	18.41	27.70	16.02	19.03	65.82
1972	93.5	157.8	94.9	106.4	352.8	20.77	27.78	17.28	20.76	68.82
1973	103.8	155.3	105.8	117.2	367.8	22.54	27.46	19.22	22.50	70.65
1974	110.2	189.2	141.7	138.7	399.5	27.82	34.11	26.76	28.93	83.31
1975	138.6	262.0	177.2	177.8	467.9	34.17	46.23	33.86	36.99	97.34
1976	151.1	270.4	190.3	191.6	476.7	37.35	49.78	36.94	40.46	100.66
1977	170.0	313.5	230.2	227.2	531.4	41.16	57.57	43.49	46.81	109.49
1978	208.0	374.2	281.7	280.0	611.8	49.72	68.37	52.55	56.63	123.76
1979	243.1	443.1	339.6	331.4	668.8	58.29	80.66	64.60	67.70	136.64
1980	272.1	536.4	376.5	367.7	680.4	66.36	95.16	73.70	77.02	142.52
1981	336.3	698.6	464.0	453.7	767.4	80.40	122.17	90.03	94.30	159.51
1982	347.4	864.3	515.4	514.4	820.0	86.34	146.20	104.09	108.73	173.34
1983	283.8	608.1	366.5	371.7	570.1	72.65	108.37	79.10	83.34	127.81
1984	262.1	489.8	329.2	326.5	482.5	66.32	88.80	67.18	71.90	106.27
1985	270.4	508.7	372.3	349.4	501.2	66.78	93.09	73.69	75.35	108.09
1986	284.9	522.9	389.2	364.6	511.7	68.35	93.02	76.53	76.88	107.90
1987	246.0	380.4	259.1	279.6	382.0	58.35	69.55	51.05	58.71	80.21
1988	279.4	460.3	366.4	354.7	468.6	62.28	84.65	66.96	70.23	92.78
1989	282.3	457.8	355.4	362.2	461.1	64.92	86.86	67.61	73.55	93.63
1990	321.8	471.3	367.5	383.6	470.2	69.17	90.73	67.49	76.07	93.23
1991	346.9	506.6	441.2	421.5	499.1	73.75	93.10	83.05	82.64	97.86
1992	362.3	426.1	357.6	382.6	442.9	69.50	72.83	67.82	70.27	81.35
1993	356.6	521.2	387.7	426.8	482.9	67.52	83.15	72.56	75.30	85.20
1994	409.5	535.1	491.5	483.2	535.4	70.57	81.90	86.60	79.49	88.07
1995	415.8	629.7	481.2	513.4	557.4	78.09	95.97	84.60	87.22	94.70
1996	341.0	616.0	541.0	496.1	528.6	70.60	98.67	95.74	88.92	94.74
1997	445.6	728.6	655.6	603.9	632.9	90.48	117.55	115.09	107.83	113.01
1998	566.0	815.6	973.2	769.1	797.2	108.88	127.94	157.79	128.97	133.69
1999	783.0	798.4	1,115.5	856.1	874.8	156.45	138.42	182.99	152.02	155.33
2000	593.4	756.9	1,075.4	754.6	754.6	125.96	138.39	181.83	142.16	142.16
2001	729.1	896.5	1,620.4	943.2	921.1	153.72	172.05	271.63	181.94	177.68
2002	882.8	991.9	1,673.4	1,054.2	1,011.9	194.55	175.78	284.17	195.31	187.46
2003	1,037.3	1,106.0	2,065.1	1,199.5	1,127.4	221.13	189.95	345.94	216.27	203.25
2004	1,441.8	1,716.4	1,977.3	1,673.1	R1,528.5	298.45	284.78	327.91	292.57	R267.28
2005	1,920.4	1,497.6	2,392.9	1,720.7	R1,522.7	314.36	280.03	429.92	306.50	R271.24
2006	2,238.6	1,936.2	2,664.6	2,101.7	1,803.0	402.45	348.36	479.33	378.03	324.30

[1] See "Crude Oil Well" in Glossary.
[2] See "Natural Gas Well" in Glossary.
[3] See "Dry Hole" in Glossary.
[4] See "Nominal Dollars" in Glossary.
[5] In chained (2000) dollars, calculated by using gross domestic product implicit price deflators in Table D1. See "Chained Dollars" in Glossary.
R=Revised.

Notes: • The information reported for 1965 and prior years is not strictly comparable to that in more recent surveys. • Average cost is the arithmetic mean and includes all costs for drilling and equipping wells and for surface-producing facilities. Wells drilled include exploratory and development wells; excludes service wells, stratigraphic tests, and core tests. See "Development Well" and "Exploratory Well" in Glossary.

Web Page: For related information, see http://www.api.org/statistics/accessapi/api-reports.cfm.

Source: American Petroleum Institute, *2006 Joint Association Survey on Drilling Costs* (May 2008).

Figure 4.9 Crude Oil, Natural Gas, and Natural Gas Liquids Gross Additions to Proved Reserves, and Exploration and Development Expenditures

Gross Additions to Proved Reserves of Crude Oil, Natural Gas, and Natural Gas Liquids, 1974-2006

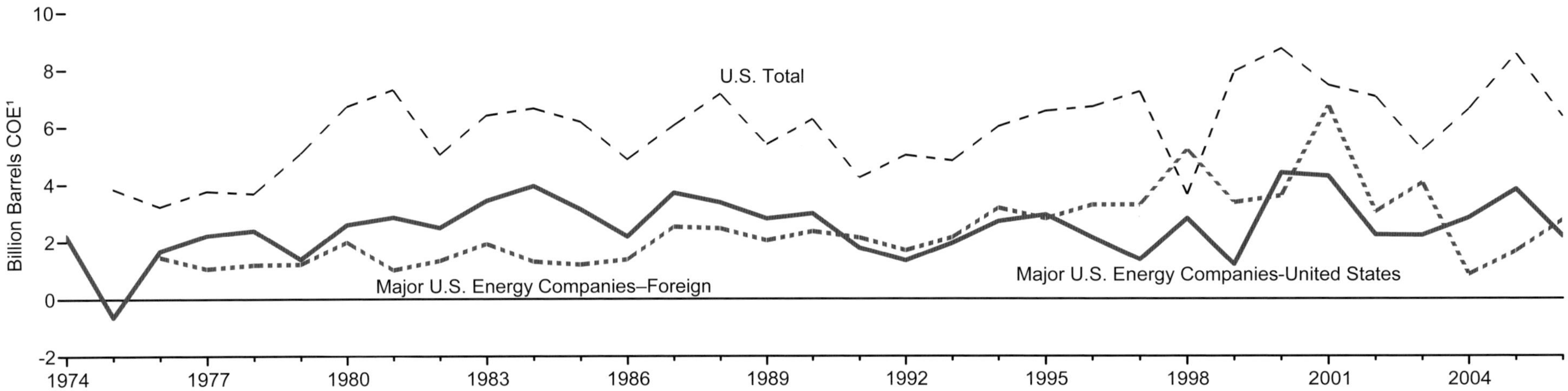

Crude Oil and Natural Gas Exploration and Development Expenditures, 1974-2006

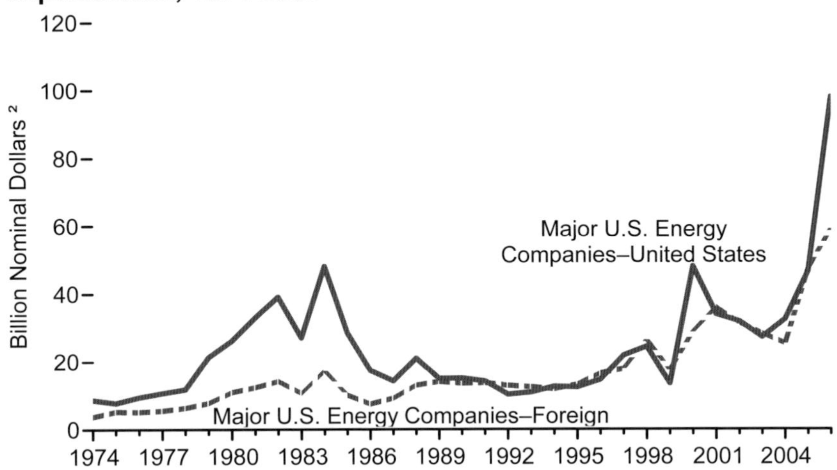

Expenditures per Barrel of Reserve Additions, 1975-2005 Three-Year Moving Average

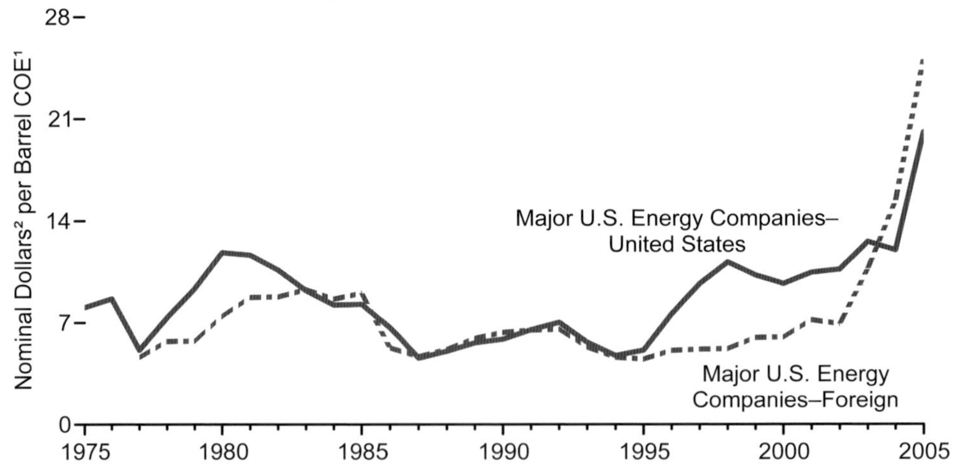

[1] Crude oil equivalent.
[2] See "Nominal Price" in Glossary.

Note: "Major U.S. Energy Companies" are the top publicly-owned crude oil and natural gas producers and petroleum refiners that form the Financial Reporting System (FRS). See Table 3.14.

Source: Table 4.9.

Table 4.9 Crude Oil, Natural Gas, and Natural Gas Liquids Gross Additions to Proved Reserves, and Exploration and Development Expenditures, 1974-2006

Year	Gross Additions to Proved Reserves [1] of Crude Oil, Natural Gas, and Natural Gas Liquids			Crude Oil and Natural Gas Exploration and Development Expenditures		Expenditures per Barrel of Reserve Additions, Three-Year Moving Average	
		Major U.S. Energy Companies [2]		Major U.S. Energy Companies [2]		Major U.S. Energy Companies [2]	
	U.S. Total	United States	Foreign	United States	Foreign	United States	Foreign
	Million Barrels COE [3]			Billion Nominal Dollars [4]		Nominal Dollars [4] per Barrel COE [3]	
1974	NA	2,205	NA	8.7	3.8	NA	NA
1975	3,846	-634	NA	7.8	5.3	8.05	NA
1976	3,224	1,663	1,459	9.5	5.2	8.64	NA
1977	3,765	2,210	1,055	10.7	5.6	5.12	4.64
1978	3,679	2,383	1,191	11.8	6.4	7.34	5.73
1979	5,071	1,378	[5]1,208	21.3	7.8	9.34	[5]5.75
1980	6,723	2,590	1,977	26.2	11.0	11.80	7.45
1981	7,304	2,848	1,006	33.0	12.4	11.63	8.74
1982	5,030	2,482	1,332	39.1	14.2	[6]10.62	[6]8.78
1983	6,412	3,427	1,918	27.1	10.7	9.20	9.28
1984	6,653	3,941	1,298	48.1	17.3	[6]8.21	[6]8.63
1985	6,190	[7]3,129	1,192	28.5	10.1	[7]8.27	9.03
1986	4,866	2,178	[5]1,375	17.4	7.5	6.67	[5]5.28
1987	6,059	[7]3,698	2,516	14.3	9.2	[7]4.58	4.69
1988	7,156	3,359	2,460	21.0	13.0	5.05	5.18
1989	5,385	2,798	2,043	15.0	14.1	5.62	5.94
1990	6,275	2,979	2,355	15.1	13.6	5.87	6.34
1991	4,227	1,772	2,135	14.2	13.7	6.52	6.50
1992	5,006	1,332	1,694	10.3	12.9	7.02	6.55
1993	4,814	1,945	2,147	10.9	12.5	5.66	5.33
1994	6,021	2,703	3,173	12.6	11.9	4.74	4.63
1995	6,558	2,929	2,799	12.4	13.2	5.11	4.51
1996	6,707	2,131	3,280	14.6	16.6	7.61	5.10
1997	7,233	1,367	3,279	21.8	17.9	9.67	5.18
1998	3,628	2,798	5,206	24.4	26.4	11.15	5.22
1999	7,929	1,197	3,360	13.5	17.5	10.25	5.98
2000	8,725	4,392	3,593	48.0	28.8	9.67	6.01
2001	7,449	4,271	6,744	33.9	35.9	10.44	7.19
2002	7,056	2,232	3,036	31.8	31.4	10.65	6.91
2003	5,189	2,216	4,047	27.2	28.2	12.57	10.71
2004	6,624	2,825	841	32.4	25.3	R11.99	15.38
2005	8,543	3,818	1,664	R46.6	47.3	R20.08	R25.07
2006	6,346	2,174	2,747	98.0	59.1	NA	NA

[1] Gross additions to proved reserves equal annual change in proved reserves plus annual production. See "Proved Reserves, Crude Oil," "Proved Reserves, Natural Gas," and "Proved Reserves, Natural Gas Liquids" in Glossary.
[2] "Major U.S. Energy Companies" are the top publicly-owned, U.S.-based crude oil and natural gas producers and petroleum refiners that form the Financial Reporting System (FRS) (see Table 3.14).
[3] Crude oil equivalent: converted to Btu on the basis of annual average conversion factors. See Appendix A.
[4] See "Nominal Dollars" in Glossary.
[5] Data for 1979 exclude downward revisions of 1,225 million barrels COE due to Iranian policies. Data for 1986 exclude downward revisions due to Libyan sanctions.
[6] Data for 1982 and 1984 are adjusted to exclude purchases of proved reserves associated with mergers among the FRS companies.
[7] Data for 1985 and 1987 exclude downward revisions of 1,477 million barrels COE and 2,396 million barrels COE, respectively, of Alaska North Slope natural gas reserves.

R=Revised. NA=Not available.

Web Page: For related information, see http://www.eia.doe.gov/emeu/finance.

Sources: **Major U.S. Energy Companies:** • 1974-1976—Energy Information Administration (EIA), Form EIA-28, "Financial Reporting System" database, November 1997. • 1977 forward—EIA, *Performance Profiles of Major Energy Producers,* annual reports. **U.S. Total, Gross Additions to Proved Reserves:** • 1975-1979—American Gas Association, American Petroleum Institute, and Canadian Petroleum Association (published jointly), *Reserves of Crude Oil, Natural Gas Liquids, and Natural Gas in the United States and Canada as of December 31, 1979,* Volume 34 (June 1980). • 1980 forward—EIA, *U.S. Crude Oil, Natural Gas, and Natural Gas Liquids Reserves,* annual reports.

Figure 4.10 Major U.S. Energy Companies' Expenditures for Crude Oil and Natural Gas Exploration and Development by Region

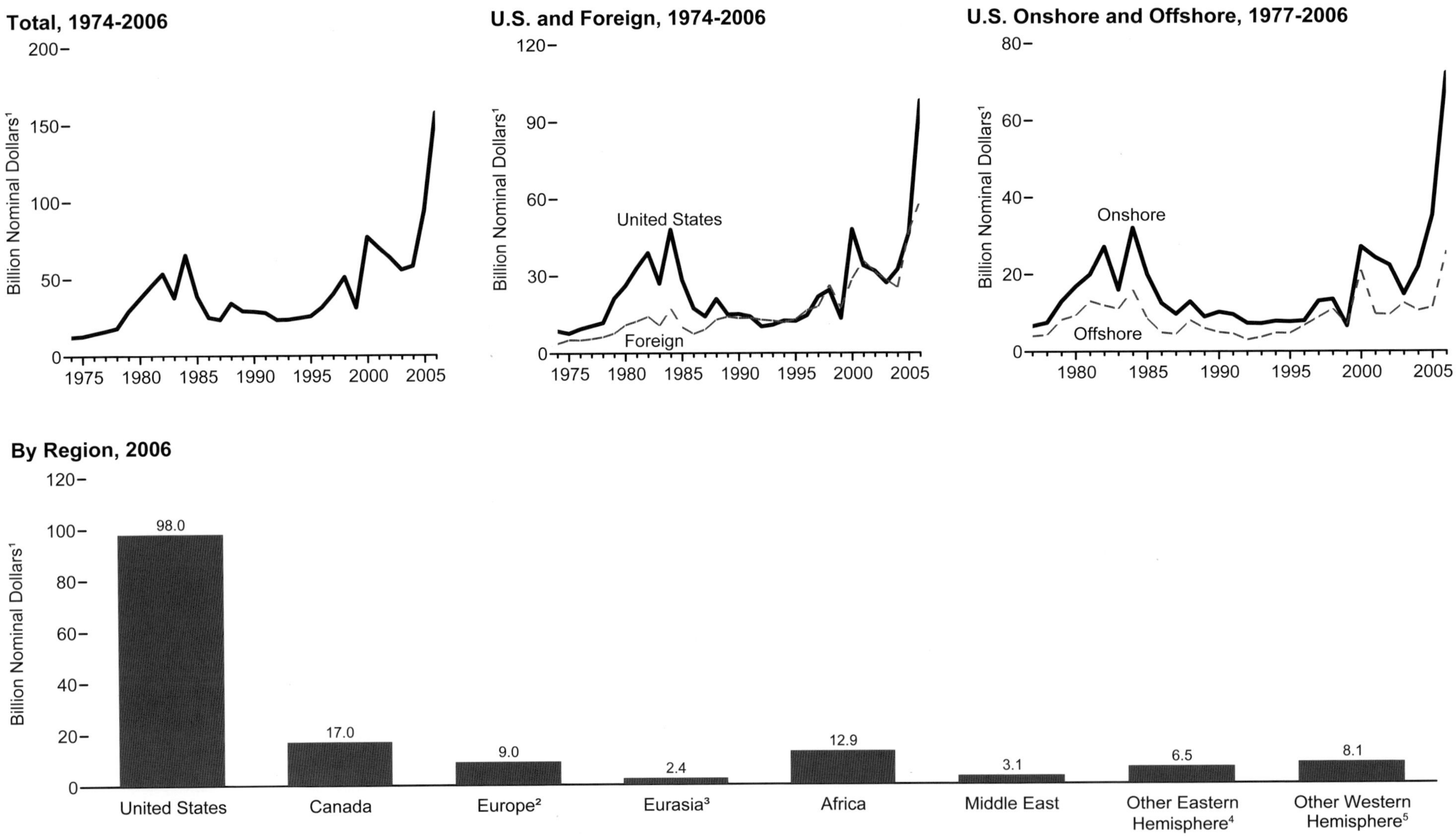

[1] See "Nominal Dollars" in Glossary.

[2] Includes all Europe except countries that were part of the former U.S.S.R. See "U.S.S.R" in Glossary.

[3] Includes only countries that were part of the former U.S.S.R. See "U.S.S.R" in Glossary.

[4] This region includes areas that are eastward of the Greenwich prime meridian to 180° longitude and that are not included in other domestic or foreign classifications.

[5] This region includes areas that are westward of the Greenwich prime meridian to 180° longitude and that are not included in other domestic or foreign classifications.

Notes: • "Major U.S. Energy Companies" are the top publicly-owned, U.S.-based crude oil and natural gas producers and petroleum refiners that form the Financial Reporting System (FRS). See Table 3.14. • Because vertical scales differ, graphs should not be compared.

Source: Table 4.10.

Table 4.10 Major U.S. Energy Companies' Expenditures for Crude Oil and Natural Gas Exploration and Development by Region, 1974-2006 (Billion Nominal Dollars [1])

Year	United States			Foreign								Total
	Onshore	Offshore	Total	Canada	Europe [2]	Eurasia [3]	Africa	Middle East	Other Eastern Hemisphere [4]	Other Western Hemisphere [5]	Total	
1974	NA	NA	8.7	NA	NA	– –	NA	NA	NA	NA	3.8	12.5
1975	NA	NA	7.8	NA	NA	– –	NA	NA	NA	NA	5.3	13.1
1976	NA	NA	9.5	NA	NA	– –	NA	NA	NA	NA	5.2	14.7
1977	6.7	4.0	10.7	1.5	2.5	– –	.7	.2	.3	.4	5.6	16.3
1978	7.5	4.3	11.8	1.6	2.6	– –	.8	.3	.4	.6	6.4	18.2
1979	13.0	8.3	21.3	2.3	3.0	– –	.8	.2	.5	.8	7.8	29.1
1980	16.8	9.4	26.2	3.1	4.3	– –	1.4	.2	.8	1.0	11.0	37.2
1981	19.9	13.0	33.0	1.8	5.0	– –	2.1	.3	1.9	1.3	12.4	45.4
1982	27.2	11.9	39.1	1.9	6.3	– –	2.1	.4	2.4	1.1	14.2	53.3
1983	16.0	11.1	27.1	1.6	4.3	– –	1.7	.5	2.0	.6	10.7	37.7
1984	32.1	16.0	48.1	5.4	5.5	– –	3.4	.5	2.0	.5	17.3	65.3
1985	20.0	8.5	28.5	1.9	3.7	– –	1.6	.9	1.3	.7	10.1	38.6
1986	12.5	4.9	17.4	1.1	3.2	– –	1.1	.3	1.2	.6	7.5	24.9
1987	9.7	4.5	14.3	1.9	3.0	– –	.8	.4	2.8	.5	9.2	23.5
1988	12.9	8.1	21.0	5.4	4.3	– –	.8	.4	1.4	.7	13.0	34.1
1989	9.0	6.0	15.0	6.3	3.5	– –	1.0	.4	2.3	.6	14.1	29.1
1990	10.2	4.9	15.1	1.8	6.6	– –	1.4	.6	2.4	.7	13.6	28.7
1991	9.6	4.6	14.2	1.7	6.8	– –	1.5	.5	2.4	.7	13.7	27.9
1992	7.3	3.0	10.3	1.1	6.8	– –	1.4	.6	2.4	.6	12.9	23.2
1993	7.2	3.7	10.9	1.6	5.5	.3	1.5	.7	2.5	.6	12.5	23.5
1994	7.8	4.8	12.6	1.8	4.4	.3	1.4	.4	2.8	.7	11.9	24.5
1995	7.7	4.7	12.4	1.9	5.2	.4	2.0	.4	2.4	.9	13.2	25.6
1996	7.9	6.7	14.6	1.6	5.6	.5	2.8	.5	4.1	1.6	16.6	31.3
1997	13.0	8.8	21.8	2.0	7.1	.6	3.0	.6	3.0	1.6	17.9	39.8
1998	13.5	11.0	24.4	4.8	8.6	1.3	3.1	.9	3.9	3.7	26.4	50.8
1999	6.6	6.9	13.5	2.1	4.1	.6	3.1	.4	3.4	3.8	17.5	31.0
2000	27.1	21.0	48.0	4.9	7.5	.9	2.7	.6	6.8	5.4	28.8	76.8
2001	24.2	9.6	33.9	15.3	5.4	.9	5.5	.7	5.0	3.1	35.9	69.8
2002	22.3	9.5	31.8	6.7	9.8	1.3	5.1	.8	6.2	1.6	31.4	63.2
2003	14.7	12.5	27.2	4.9	5.7	2.1	9.2	1.0	4.2	1.1	28.2	55.4
2004	21.9	10.5	32.4	5.3	4.4	2.0	6.9	1.3	3.8	1.6	25.3	57.7
2005	35.2	R11.3	R46.6	9.1	6.1	6.3	10.7	1.5	12.0	1.7	47.3	R93.8
2006	72.1	25.9	98.0	17.0	[2]9.0	[3]2.4	12.9	3.1	6.5	8.1	59.1	157.1

[1] See "Nominal Dollars" in Glossary.

[2] Through 2005, includes Austria, Belgium, Denmark, Finland, France, Germany (the Federal Republic of), Greece, Ireland, Italy, Luxembourg, Netherlands, Norway, Portugal, Spain, Sweden, Switzerland, Turkey, and the United Kingdom. Beginning in 2006, includes all Europe except countries that were part of the former U.S.S.R. See "U.S.S.R" in Glossary.

[3] Through 2005, includes countries that were part of the former U.S.S.R. as well as Albania, Bosnia and Herzegovina, Bulgaria, Croatia, Czech Republic, Macedonia, Serbia and Montenegro, Slovakia, and Slovenia. Beginning in 2006, includes only countries that were part of the former U.S.S.R. See "U.S.S.R." in Glossary.

[4] This region includes areas that are eastward of the Greenwich prime meridian to 180° longitude and that are not included in other domestic or foreign classifications.

[5] This region includes areas that are westward of the Greenwich prime meridian to 180° longitude and that are not included in other domestic or foreign classifications.

R=Revised. NA=Not available. – – = Not applicable.

Notes: • "Major U.S. Energy Companies" are the top publicly-owned, U.S.-based crude oil and natural gas producers and petroleum refiners that form the Financial Reporting System (FRS). See Table 3.14. • Totals may not equal sum of components due to independent rounding.

Web Page: For related information, see http://www.eia.doe.gov/emeu/finance.

Sources: • 1974-1976—Energy Information Administration (EIA), Office of Energy Markets and End Use, FRS Database, November 1997. • 1977 forward—EIA, *Performance Profiles of Major Energy Producers,* annual reports.

Figure 4.11 Coal Demonstrated Reserve Base, January 1, 2007

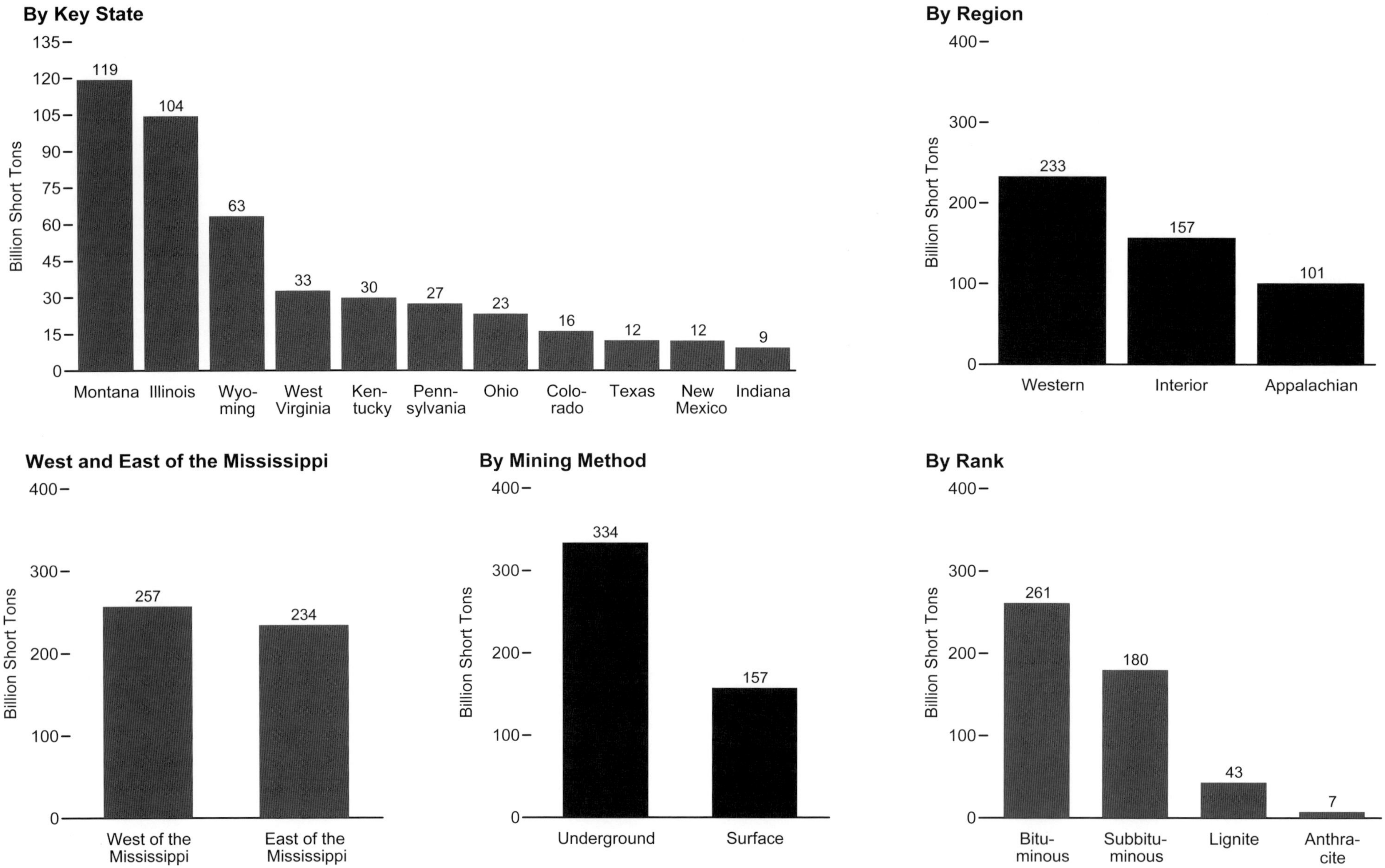

Note: Because vertical scales differ, graphs should not be compared.

Source: Table 4.11.

Table 4.11 Coal Demonstrated Reserve Base, January 1, 2007

(Billion Short Tons)

Region and State	Anthracite	Bituminous Coal		Subbituminous Coal		Lignite	Total		
		Underground	Surface	Underground	Surface	Surface [1]	Underground	Surface	Total
Appalachian	**7.3**	**70.0**	**22.5**	**0.0**	**0.0**	**1.1**	**73.9**	**27.0**	**100.9**
Alabama	.0	1.0	2.1	.0	.0	1.1	1.0	3.2	4.2
Kentucky, Eastern	.0	1.1	9.3	.0	.0	.0	1.1	9.3	10.4
Ohio	.0	17.5	5.7	.0	.0	.0	17.5	5.7	23.2
Pennsylvania	7.2	19.3	.9	.0	.0	.0	23.1	4.3	27.4
Virginia	.1	1.0	.5	.0	.0	.0	1.1	.5	1.6
West Virginia	.0	29.0	3.7	.0	.0	.0	29.0	3.7	32.7
Other [2]	.0	1.1	.3	.0	.0	.0	1.1	.3	1.4
Interior	**.1**	**117.1**	**27.2**	**.0**	**.0**	**12.7**	**117.2**	**39.9**	**157.1**
Illinois	.0	87.9	16.5	.0	.0	.0	87.9	16.5	104.4
Indiana	.0	8.7	.7	.0	.0	.0	8.7	.7	9.4
Iowa	.0	1.7	.5	.0	.0	.0	1.7	.5	2.2
Kentucky, Western	.0	15.8	3.6	.0	.0	.0	15.8	3.6	19.4
Missouri	.0	1.5	4.5	.0	.0	.0	1.5	4.5	6.0
Oklahoma	.0	1.2	.3	.0	.0	.0	1.2	.3	1.5
Texas	.0	.0	.0	.0	.0	12.3	.0	12.3	12.3
Other [3]	.1	.3	1.1	.0	.0	0.4	.4	1.5	1.9
Western	**(s)**	**21.5**	**2.4**	**121.3**	**58.5**	**29.4**	**142.8**	**90.3**	**233.1**
Alaska	.0	.6	.1	4.8	.6	(s)	5.4	.7	6.1
Colorado	(s)	7.6	.6	3.7	.0	4.2	11.3	4.8	16.1
Montana	.0	1.4	.0	69.6	32.5	15.8	71.0	48.3	119.3
New Mexico	(s)	2.7	.9	3.5	5.1	.0	6.2	6.0	12.2
North Dakota	.0	.0	.0	.0	.0	9.0	.0	9.0	9.0
Utah	.0	5.1	.3	(s)	.0	.0	5.1	.3	5.4
Washington	.0	.3	.0	1.0	.0	(s)	1.3	.0	1.3
Wyoming	.0	3.8	.5	38.7	20.3	.0	42.5	20.8	63.3
Other [4]	.0	.0	.0	(s)	(s)	.4	.0	.4	.4
U.S. Total	**7.4**	**208.6**	**52.1**	**121.3**	**58.5**	**43.2**	**333.9**	**157.2**	**491.1**
States East of the Mississippi River	7.3	182.4	43.3	.0	.0	1.1	186.3	47.8	234.1
States West of the Mississippi River	.1	26.2	8.8	121.3	58.5	42.1	147.6	109.4	257.0

[1] Lignite resources are not mined underground in the United States.
[2] Georgia, Maryland, North Carolina, and Tennessee.
[3] Arkansas, Kansas, Louisiana, and Michigan.
[4] Arizona, Idaho, Oregon, and South Dakota.
(s)=Less than 0.05 billion short tons.

Notes: • See *U.S. Coal Reserves: 1997 Update* on the Web Page for a description of the methodology used to produce these data. • Data represent remaining measured and indicated coal resources, analyzed and on file, meeting minimum seam and depth criteria, and in the ground as of January 1, 2007. These coal resources are not totally recoverable. Net recoverability with current mining technologies ranges from 0 percent (in far northern Alaska) to more than 90 percent. Fifty-four percent of the demonstrated reserve base of coal in the United States is estimated to be recoverable. • Totals may not equal sum of components due to independent rounding.

Web Page: For related information, see http://www.eia.doe.gov/fuelcoal.html.

Source: Energy Information Administration, Coal Reserves Database.

Figure 4.12 Uranium Exploration and Development Drilling

Total Holes Drilled, 1958-2001 and 2004-2007

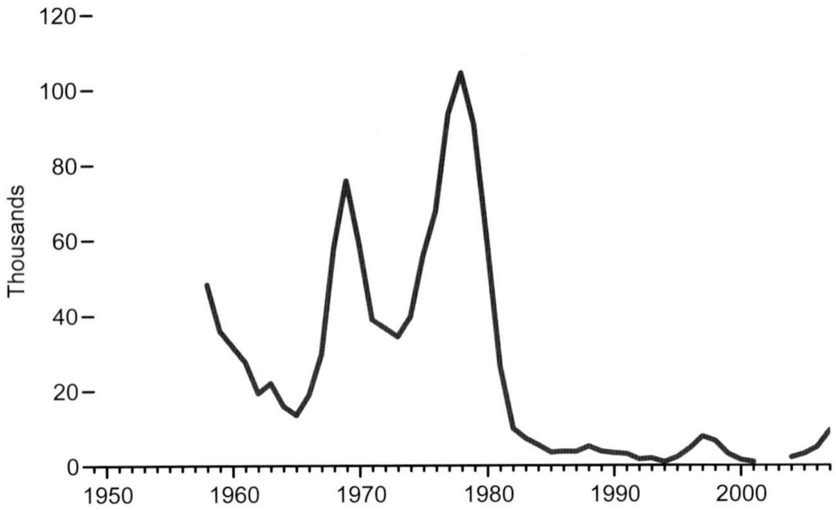

Holes Drilled, Selected Years

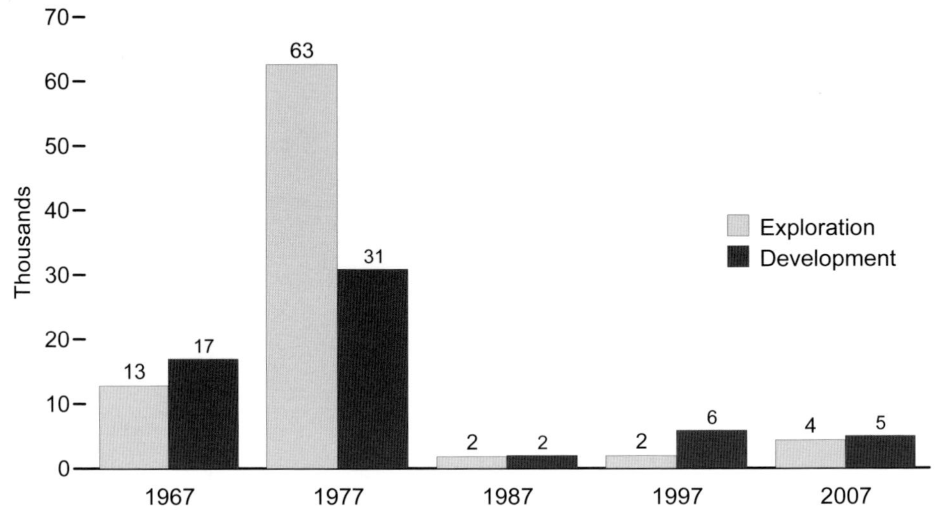

Total Footage Drilled, 1949-2001 and 2004-2007

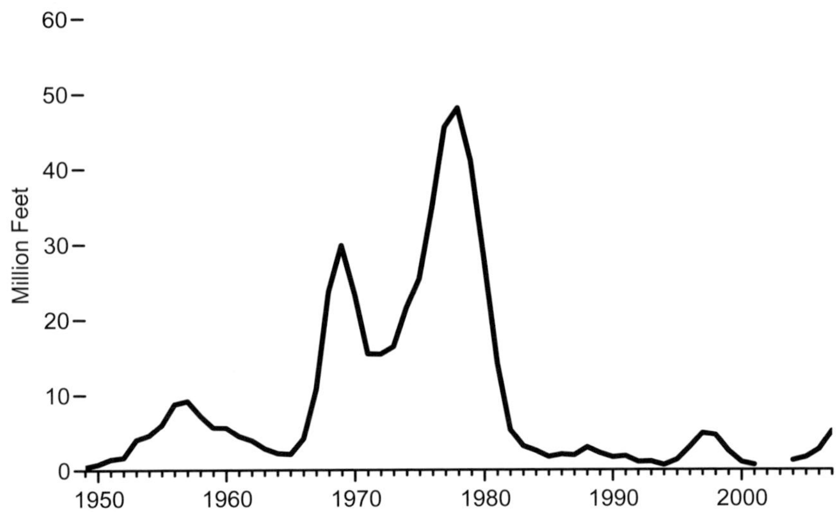

Footage Drilled, Selected Years

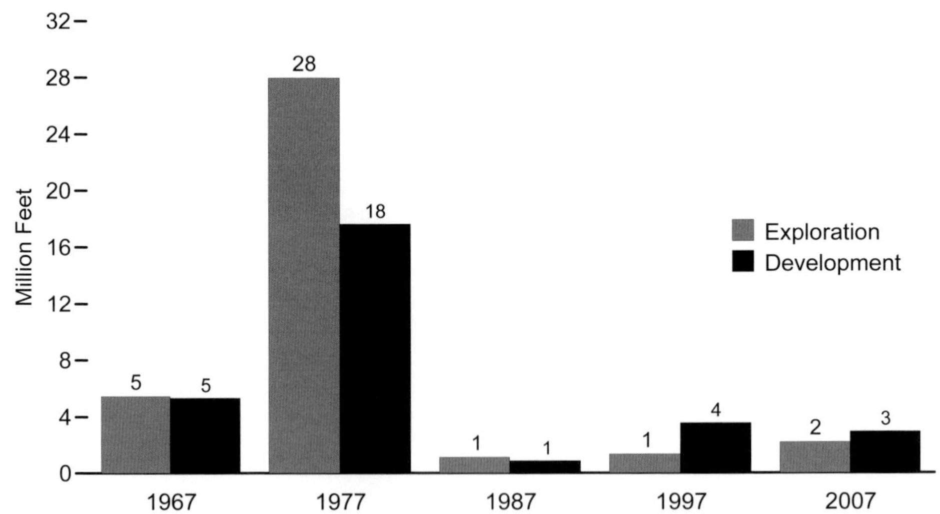

Source: Table 4.12.

Table 4.12 Uranium Exploration and Development Drilling, Selected Years, 1949-2007

Year	Exploration [1]		Development [2]		Total	
	Holes Drilled	Footage Drilled	Holes Drilled	Footage Drilled	Holes Drilled	Footage Drilled
	Thousands	Million Feet	Thousands	Million Feet	Thousands	Million Feet
1949	NA	0.36	NA	0.05	NA	0.41
1950	NA	.57	NA	.21	NA	.78
1955	NA	5.27	NA	.76	NA	6.03
1960	7.34	1.40	24.40	4.21	31.73	5.61
1965	6.23	1.16	7.33	.95	13.56	2.11
1970	43.98	17.98	14.87	5.55	58.85	23.53
1971	28.42	11.40	10.44	4.05	38.86	15.45
1972	26.91	11.82	9.71	3.61	36.62	15.42
1973	22.56	10.83	11.70	5.59	34.26	16.42
1974	27.40	14.72	12.30	6.84	39.70	21.56
1975	34.29	15.69	21.60	9.73	55.89	25.42
1976	40.41	20.36	27.23	14.44	67.64	34.80
1977	62.60	27.96	30.86	17.62	93.45	45.58
1978	75.07	28.95	29.29	19.15	104.35	48.10
1979	60.46	28.07	30.19	13.01	90.65	41.08
1980	39.61	19.60	20.19	8.59	59.80	28.19
1981	17.75	10.87	8.67	3.35	26.42	14.22
1982	6.97	4.23	3.00	1.13	9.97	5.36
1983	4.29	2.09	3.01	1.08	7.30	3.17
1984	4.80	2.26	.72	.29	5.52	2.55
1985	2.88	1.42	.77	.34	3.65	1.76
1986	1.99	1.10	1.85	.97	3.83	2.07
1987	1.82	1.11	1.99	.86	3.81	1.97
1988	2.03	1.28	3.18	1.73	5.21	3.01
1989	2.09	1.43	1.75	.80	3.84	2.23
1990	1.51	.87	1.91	.81	3.42	1.68
1991	1.62	.97	1.57	.87	3.20	1.84
1992	.94	.56	.83	.50	1.77	1.06
1993	.36	.22	1.67	.89	2.02	1.11
1994	.52	.34	.48	.32	1.00	.66
1995	.58	.40	1.73	.95	2.31	1.35
1996	1.12	.88	3.58	2.16	4.70	3.05
1997	1.94	1.33	5.86	3.56	7.79	4.88
1998	1.37	.89	5.23	3.75	6.60	4.64
1999	.27	.18	2.91	2.33	3.18	2.50
2000	W	W	W	W	1.55	1.02
2001	.00	.00	1.02	.66	1.02	.66
2002	W	W	W	W	W	W
2003	NA	NA	NA	NA	W	W
2004	W	W	W	W	2.19	1.25
2005	W	W	W	W	3.14	1.67
2006	1.47	.82	3.43	1.89	4.90	2.71
2007	4.35	2.20	5.00	2.95	9.35	5.15

[1] Includes surface drilling in search of new ore deposits or extensions of known deposits and drilling at the location of a discovery up to the time the company decides sufficient ore reserves are present to justify commercial exploitation.

[2] Includes all surface drilling on an ore deposit to determine more precisely size, grade, and configuration subsequent to the time that commercial exploitation is deemed feasible.

NA=Not available. W=Value withheld to avoid disclosure of individual company data.

Note: Totals may not equal sum of components due to independent rounding.

Web Pages: • For all data beginning in 1949, see http://www.eia.doe.gov/emeu/aer/resource.html. • For related information, see http://www.eia.doe.gov/fuelnuclear.html.

Sources: • 1949-1981—U.S. Department of Energy, Grand Junction Office, *Statistical Data of the Uranium Industry, January 1, 1983,* Report No. GJO-100 (1983), Table VIII-5. • 1982-2002—Energy Information Administration (EIA), *Uranium Industry Annual,* annual reports. • 2003 forward—EIA, "Domestic Uranium Production Report" (May 2008).

Figure 4.13 Uranium Reserves and Resources, 2003

Reserves
(Million Pounds Uranium Oxide)

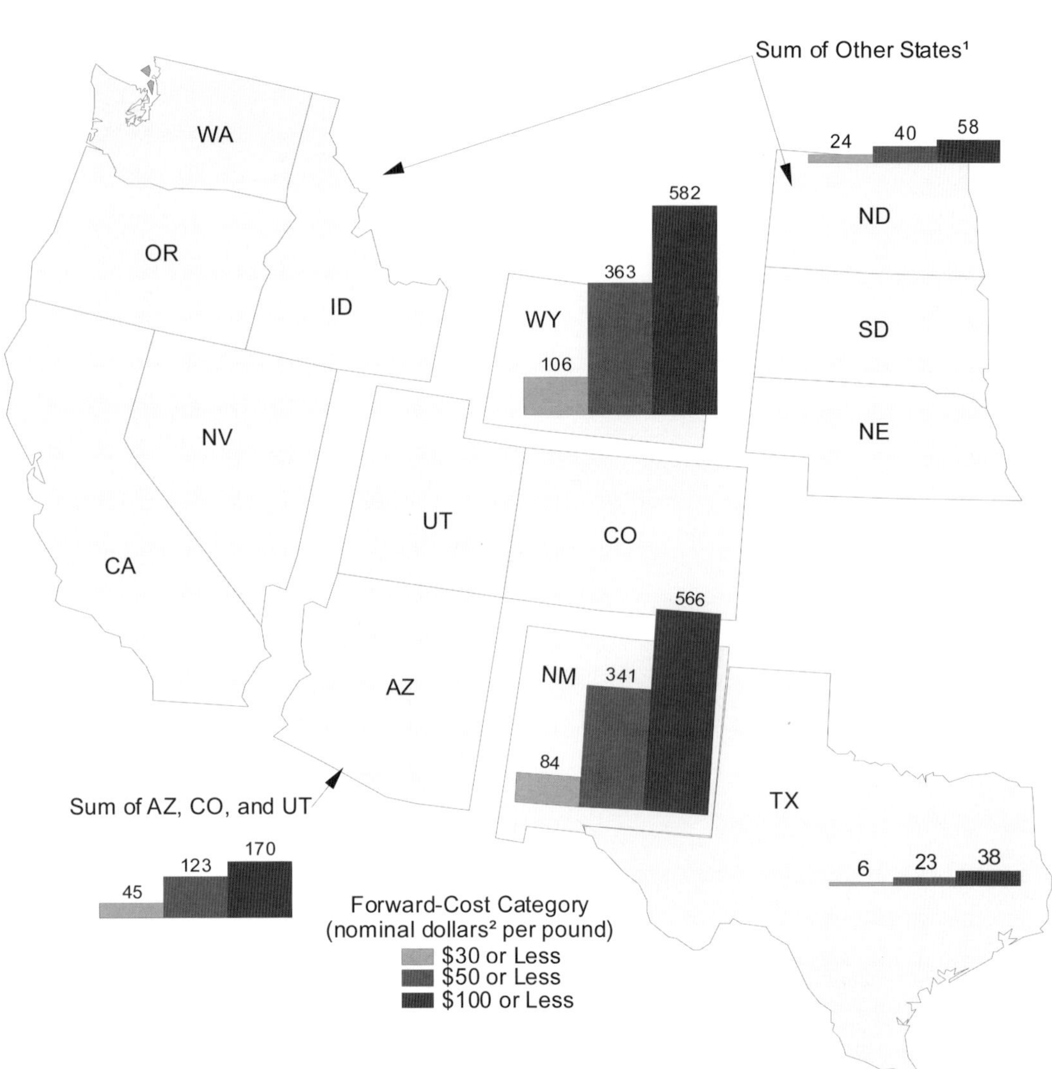

Reserves and Resources

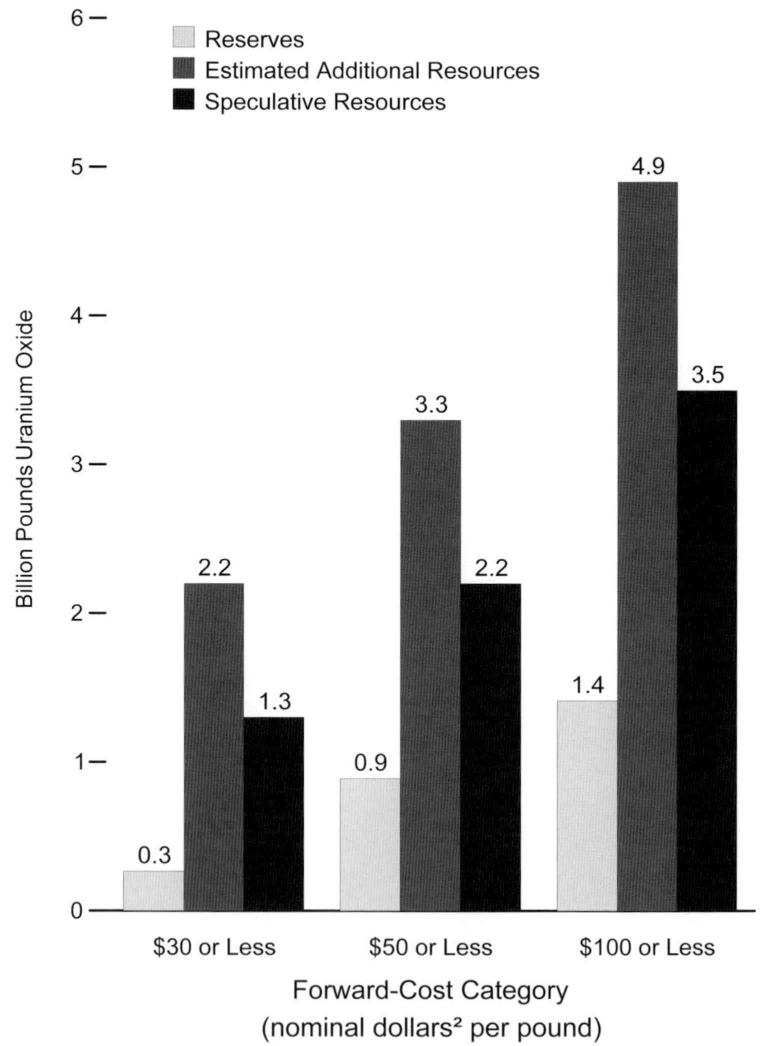

[1]California, Idaho, Nebraska, Nevada, North Dakota, Oregon, South Dakota, and Washington.
[2]See "Nominal Dollars" in Glossary.

Notes: • See "Uranium Oxide" in Glossary. • Data are at end of year.
Source: Table 4.13.

Table 4.13 Uranium Reserves and Resources, 2003

(Million Pounds Uranium Oxide)

Resource Category and State	Forward-Cost [1] Category (nominal dollars [2] per pound)		
	$30 or Less	$50 or Less	$100 or Less
Reserves [3]	**265**	**890**	**1,414**
New Mexico	84	341	566
Wyoming	106	363	582
Texas	6	23	38
Arizona, Colorado, Utah	45	123	170
Others [4]	24	40	58
Potential Resources [5]			
Estimated Additional Resources	2,180	3,310	4,850
Speculative Resources	1,310	2,230	3,480

[1] Forward costs are all operating and capital costs yet to be incurred in the production of uranium from estimated resources. Excluded are previous expenditures (such as exploration and land acquisitions), taxes, profit, and the cost of money. Generally, forward costs are lower than market prices. Resource values in forward-cost categories are cumulative; that is, the quantity at each level of forward cost includes all reserves/resources at the lower cost in that category.

[2] See "Nominal Dollars" in Glossary.

[3] The Energy Information Administration (EIA) category of uranium reserves is equivalent to the internationally reported category of "Reasonably Assured Resources" (RAR).

[4] California, Idaho, Nebraska, Nevada, North Dakota, Oregon, South Dakota, and Washington.

[5] Shown are the mean values for the distribution of estimates for each forward-cost category, rounded to the nearest million pounds uranium oxide.

Notes: • Data are at end of year. • Until further notice, these estimates will not be updated annually. • See "Uranium Oxide" in Glossary.

Web Page: For related information, see http://www.eia.doe.gov/fuelnuclear.html.

Sources: • **Forward Costs $30 or Less and $50 or Less:** EIA, "U.S. Uranium Reserves Estimates" (June 2004). • **Forward Costs $100 or Less:** EIA, Office of Coal, Nuclear, Electric and Alternate Fuels database as of June 2004.

5

Petroleum

Oil pumping unit and drilling rig, Texas. Source: U.S. Department of Energy.

Diagram 2. Petroleum Flow, 2007
(Million Barrels per Day)

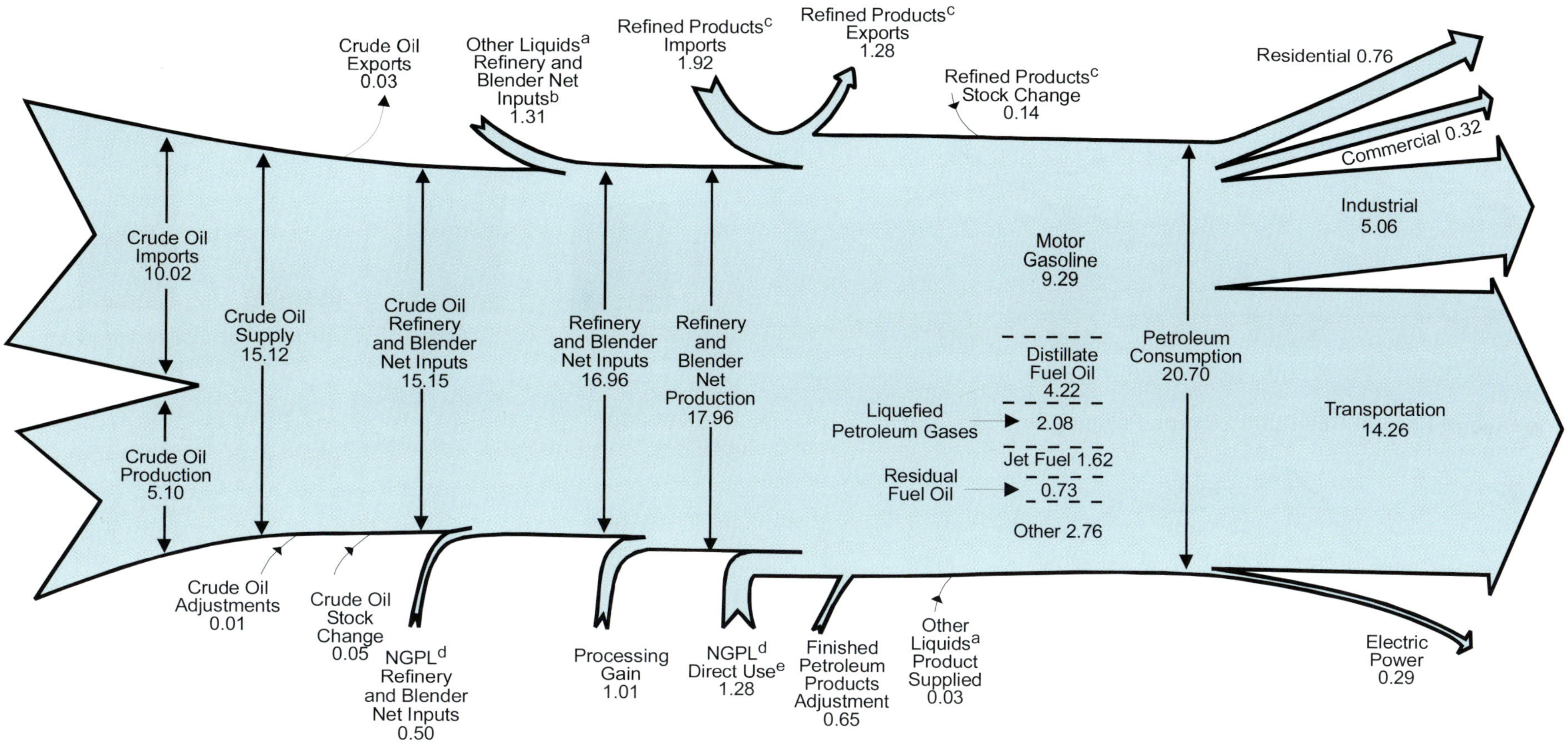

[a] Unfinished oils, other hydrocarbons/hydrogen, and motor gasoline and aviation gasoline blending components.
[b] Net imports (1.41) and adjustments (-0.05) minus stock change (0.02) and product supplied (0.03).
[c] Finished petroleum products, liquefied petroleum gases, and pentanes plus.
[d] Natural gas plant liquids.
[e] Production minus refinery input.

Notes: • Data are preliminary. • Values are derived from source data prior to rounding for publication. • Totals may not equal sum of components due to independent rounding.

Sources: Tables 5.1, 5.3, 5.5, 5.8, 5.11, 5.13a-5.13d, 5.16, and *Petroleum Supply Monthly*, February 2008, Table 4.

Figure 5.1 Petroleum Overview

Overview, 1949-2007

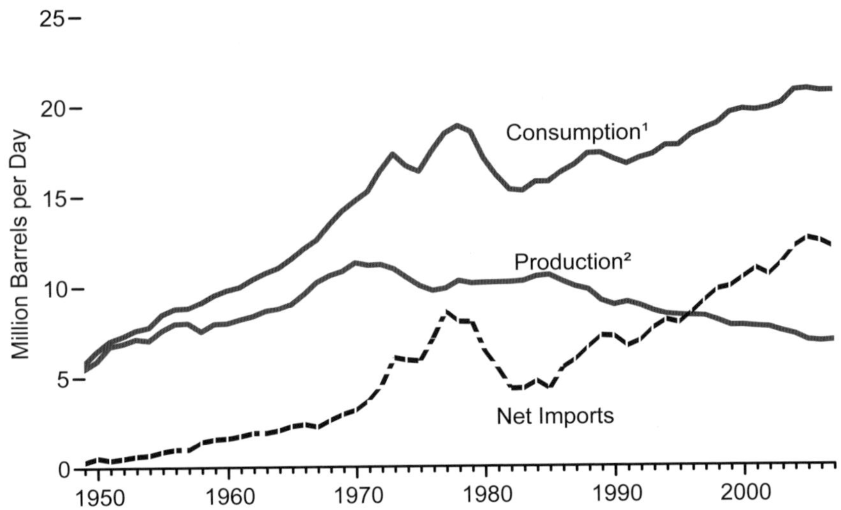

Overview, 2007

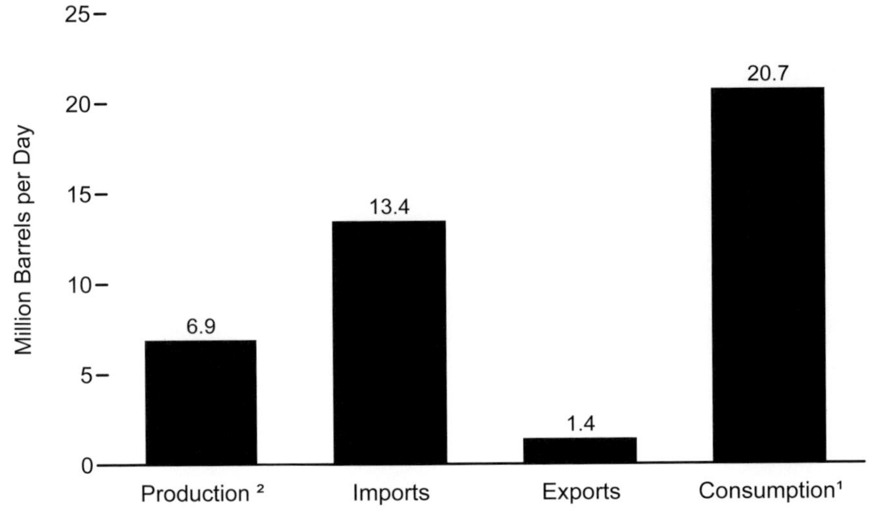

Crude Oil and Natural Gas Plant Liquids Production, 1949-2007

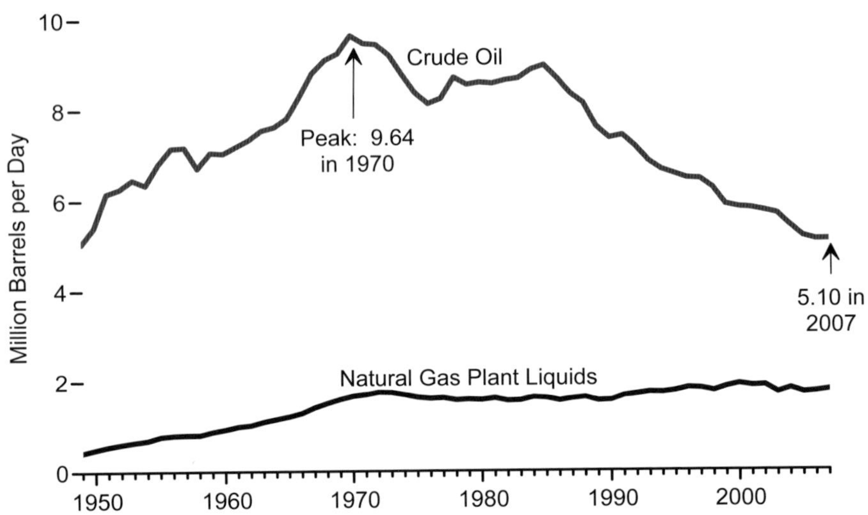

Trade, 1949-2007

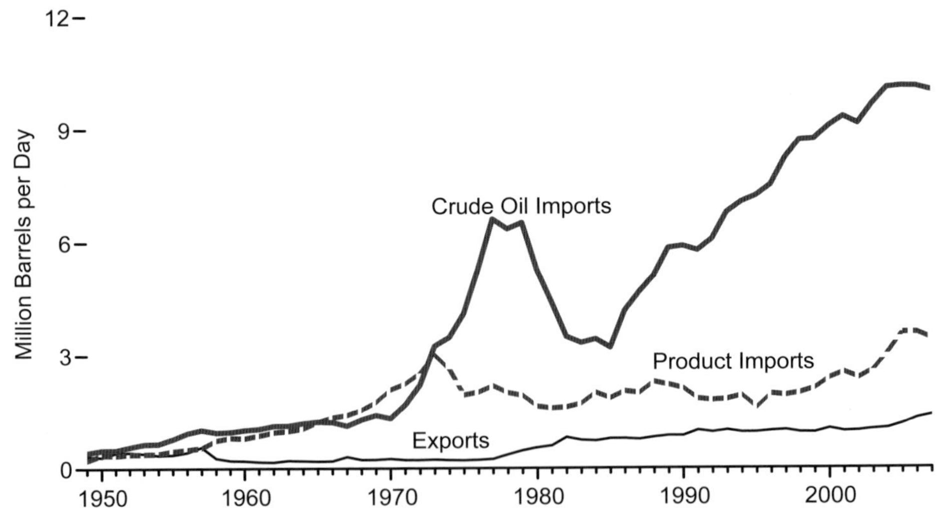

[1] Petroleum products supplied is used as an approximation for consumption.
[2] Crude oil and natural gas plant liquids production.

Note: Because vertical scales differ, graphs should not be compared.
Sources: Tables 5.1 and 5.3.

Table 5.1 Petroleum Overview, Selected Years, 1949-2007

(Thousand Barrels per Day)

Year	Field Production [1]					Processing Gain [5]	Trade			Stock Change [8]	Adjust-ments [9]	Petroleum Products Supplied
	Crude Oil [2]			Natural Gas Plant Liquids [4]	Total		Imports [6]	Exports	Net Imports [7]			
	48 States [3]	Alaska	Total									
1949	5,046	0	5,046	430	5,477	-2	645	327	318	-8	-38	5,763
1950	5,407	0	5,407	499	5,906	2	850	305	545	-56	-51	6,458
1955	6,807	0	6,807	771	7,578	34	1,248	368	880	(s)	-37	8,455
1960	7,034	2	7,035	929	7,965	146	1,815	202	1,613	-83	-8	9,797
1965	7,774	30	7,804	1,210	9,014	220	2,468	187	2,281	-8	-10	11,512
1970	9,408	229	9,637	1,660	11,297	359	3,419	259	3,161	103	-16	14,697
1971	9,245	218	9,463	1,693	11,155	382	3,926	224	3,701	71	45	15,212
1972	9,242	199	9,441	1,744	11,185	388	4,741	222	4,519	-232	43	16,367
1973	9,010	198	9,208	1,738	10,946	453	6,256	231	6,025	135	18	17,308
1974	8,581	193	8,774	1,688	10,462	480	6,112	221	5,892	179	-2	16,653
1975	8,183	191	8,375	1,633	10,007	460	6,056	209	5,846	32	41	16,322
1976	7,958	173	8,132	1,604	9,736	477	7,313	223	7,090	-58	101	17,461
1977	7,781	464	8,245	1,618	9,862	524	8,807	243	8,565	548	28	18,431
1978	7,478	1,229	8,707	1,567	10,275	496	8,363	362	8,002	-94	-20	18,847
1979	7,151	1,401	8,552	1,584	10,135	527	8,456	471	7,985	173	38	18,513
1980	6,980	1,617	8,597	1,573	10,170	597	6,909	544	6,365	140	64	17,056
1981	6,962	1,609	8,572	1,609	10,180	508	5,996	595	5,401	160	129	16,058
1982	6,953	1,696	8,649	1,550	10,199	531	5,113	815	4,298	-147	121	15,296
1983	6,974	1,714	8,688	1,559	10,246	488	5,051	739	4,312	-20	165	15,231
1984	7,157	1,722	8,879	1,630	10,509	553	5,437	722	4,715	280	228	15,726
1985	7,146	1,825	8,971	1,609	10,581	557	5,067	781	4,286	-103	200	15,726
1986	6,814	1,867	8,680	1,551	10,231	616	6,224	785	5,439	202	197	16,281
1987	6,387	1,962	8,349	1,595	9,944	639	6,678	764	5,914	41	209	16,665
1988	6,123	2,017	8,140	1,625	9,765	655	7,402	815	6,587	-28	249	17,283
1989	5,739	1,874	7,613	1,546	9,159	661	8,061	859	7,202	-43	260	17,325
1990	5,582	1,773	7,355	1,559	8,914	683	8,018	857	7,161	107	338	16,988
1991	5,618	1,798	7,417	1,659	9,076	715	7,627	1,001	6,626	-10	287	16,714
1992	5,457	1,714	7,171	1,697	8,868	772	7,888	950	6,938	-68	386	17,033
1993	5,264	1,582	6,847	1,736	8,582	766	8,620	1,003	7,618	151	422	17,237
1994	5,103	1,559	6,662	1,727	8,388	768	8,996	942	8,054	15	523	17,718
1995	5,076	1,484	6,560	1,762	8,322	774	8,835	949	7,886	-246	496	17,725
1996	5,071	1,393	6,465	1,830	8,295	837	9,478	981	8,498	-151	528	18,309
1997	5,156	1,296	6,452	1,817	8,269	850	10,162	1,003	9,158	143	487	18,620
1998	5,077	1,175	6,252	1,759	8,011	886	10,708	945	9,764	239	495	18,917
1999	4,832	1,050	5,881	1,850	7,731	886	10,852	940	9,912	-422	567	19,519
2000	4,851	970	5,822	1,911	7,733	948	11,459	1,040	10,419	-69	532	19,701
2001	4,839	963	5,801	1,868	7,670	903	11,871	971	10,900	325	501	19,649
2002	4,761	984	5,746	1,880	7,626	957	11,530	984	10,546	-105	527	19,761
2003	4,706	974	5,681	1,719	7,400	974	12,264	1,027	11,238	56	478	20,034
2004	4,510	908	5,419	1,809	7,228	1,051	13,145	1,048	12,097	209	564	20,731
2005	4,314	864	5,178	1,717	6,895	989	13,714	1,165	12,549	145	513	20,802
2006	R4,361	741	R5,102	R1,739	R6,841	R994	R13,707	R1,317	R12,390	R60	R522	R20,687
2007P	4,384	719	5,103	1,776	6,879	1,005	13,439	1,399	12,040	-162	611	20,698

[1] Crude oil production on leases, and natural gas liquids (liquefied petroleum gases, pentanes plus, and a small amount of finished petroleum products) production at natural gas processing plants. Excludes what was previously classified as "Field Production" of finished motor gasoline, motor gasoline blending components, and other hydrocarbons and oxygenates; these are now included in "Adjustments."

[2] Includes lease condensate.

[3] United States excluding Alaska and Hawaii.

[4] See Table 5.10.

[5] Refinery and blender net production minus refinery and blender net inputs. See Table 5.8.

[6] Includes crude oil imports for the Strategic Petroleum Reserve, which began in 1977. See Table 5.17.

[7] Net imports equal imports minus exports.

[8] A negative value indicates a decrease in stocks and a positive value indicates an increase. Includes crude oil stocks in the Strategic Petroleum Reserve, but excludes distillate fuel oil stocks in the Northeast Heating Oil Reserve. See Table 5.16.

[9] An adjustment for crude oil, finished motor gasoline, motor gasoline blending components, fuel ethanol, and distillate fuel oil. See EIA, *Petroleum Supply Monthly,* Appendix B, Note 3.

R=Revised. P=Preliminary. (s)=Less than 500 barrels per day.

Notes: • See Note 1, "Petroleum Products Supplied and Petroleum Consumption," Note 2, "Adjustment to Total Petroleum Products Supplied," and Note 3, "Changes Affecting Petroleum Production and Product Supplied Statistics," at end of section. • Totals may not equal sum of components due to independent rounding.

Web Pages: • For all data beginning in 1949, see http://www.eia.doe.gov/emeu/aer/petro.html. • For related information, see http://www.eia.doe.gov/oil_gas/petroleum/info_glance/petroleum.html.

Sources: • 1949-1975—Bureau of Mines, Mineral Industry Surveys, *Petroleum Statement, Annual,* annual reports. • 1976-1980—Energy Information Administration (EIA), Energy Data Reports, *Petroleum Statement, Annual,* annual reports. • 1981-2006—EIA, *Petroleum Supply Annual,* annual reports. • 2007—EIA, *Petroleum Supply Monthly* (February 2008).

Figure 5.2 Crude Oil Production and Crude Oil Well Productivity, 1954-2007

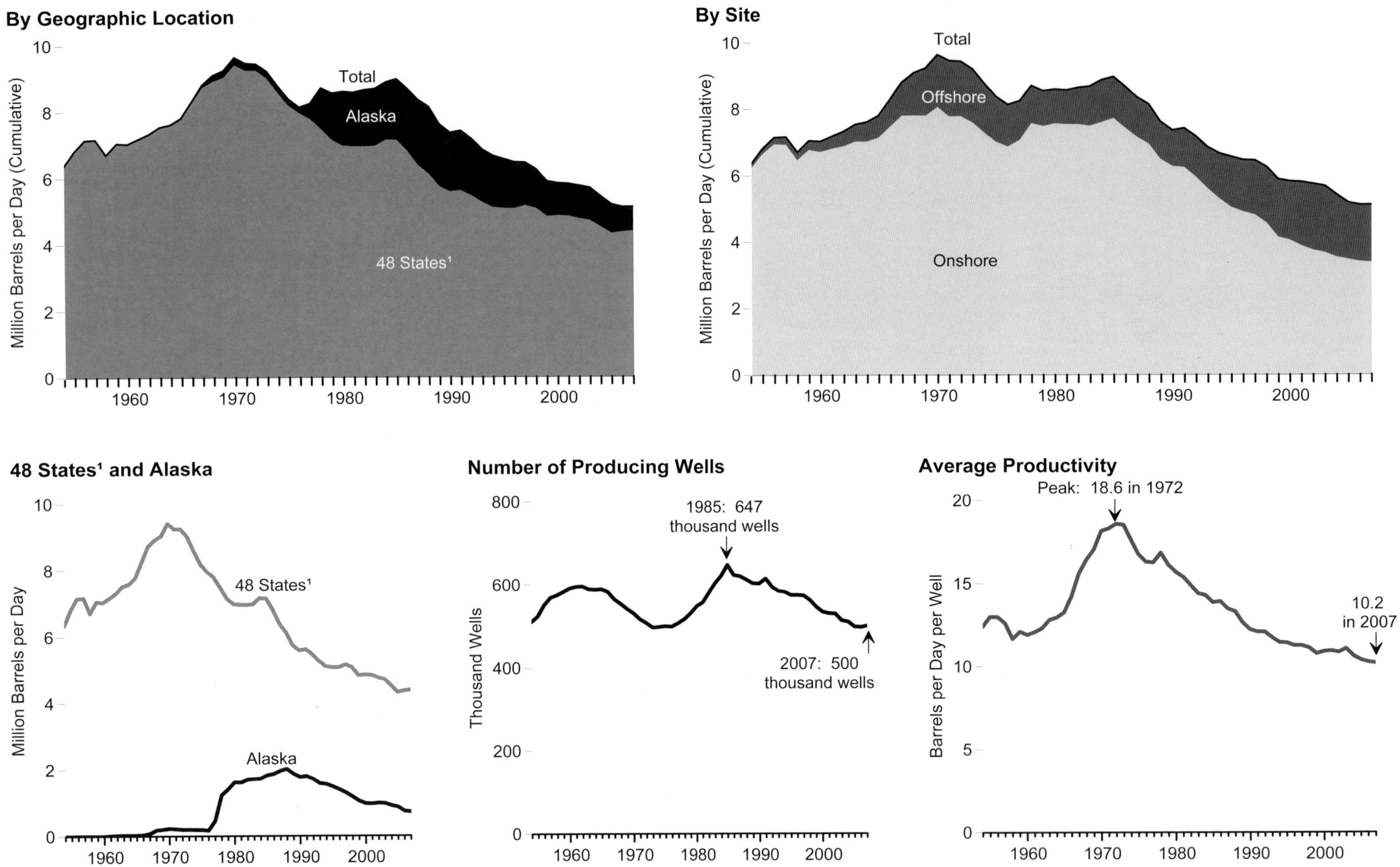

¹ United States excluding Alaska and Hawaii.
Note: Crude oil includes lease condensate.

Source: Table 5.2.

Table 5.2 Crude Oil Production and Crude Oil Well Productivity, 1954-2007

Year	Geographic Location		Site		Type		Total Production	Crude Oil Well [1] Productivity	
	48 States [2]	Alaska	Onshore	Offshore	Crude Oil	Lease Condensate		Producing Wells [3]	Average Productivity [4]
	Thousand Barrels per Day							Thousands	Barrels per Day per Well
1954	6,342	0	6,209	133	6,342	(5)	6,342	511	12.4
1955	6,807	0	6,645	162	6,807	(5)	6,807	524	13.0
1956	7,151	0	6,951	201	7,151	(5)	7,151	551	13.0
1957	7,170	0	6,940	229	7,170	(5)	7,170	569	12.6
1958	6,710	0	6,473	236	6,710	(5)	6,710	575	11.7
1959	7,053	1	6,779	274	7,054	(5)	7,054	583	12.1
1960	7,034	2	6,716	319	7,035	(5)	7,035	591	11.9
1961	7,166	17	6,817	365	7,183	(5)	7,183	595	12.1
1962	7,304	28	6,888	444	7,332	(5)	7,332	596	12.3
1963	7,512	29	7,026	515	7,542	(5)	7,542	589	12.8
1964	7,584	30	7,027	587	7,614	(5)	7,614	588	12.9
1965	7,774	30	7,140	665	7,804	(5)	7,804	589	13.2
1966	8,256	39	7,473	823	8,295	(5)	8,295	583	14.2
1967	8,730	80	7,802	1,009	8,810	(5)	8,810	565	15.6
1968	8,915	181	7,808	1,287	8,660	436	9,096	554	16.4
1969	9,035	203	7,797	1,441	8,778	460	9,238	542	17.0
1970	9,408	229	8,060	1,577	9,180	457	9,637	531	18.1
1971	9,245	218	7,779	1,684	9,032	431	9,463	517	18.3
1972	9,242	199	7,780	1,660	8,998	443	9,441	508	18.6
1973	9,010	198	7,592	1,616	8,784	424	9,208	497	18.5
1974	8,581	193	7,285	1,489	8,375	399	8,774	498	17.6
1975	8,183	191	7,012	1,362	8,007	367	8,375	500	16.8
1976	7,958	173	6,868	1,264	7,776	356	8,132	499	16.3
1977	7,781	464	7,069	1,176	7,875	370	8,245	507	16.3
1978	7,478	1,229	7,571	1,136	8,353	355	8,707	517	16.8
1979	7,151	1,401	7,485	1,067	8,181	371	8,552	531	16.1
1980	6,980	1,617	7,562	1,034	8,210	386	8,597	548	15.7
1981	6,962	1,609	7,537	1,034	8,176	395	8,572	557	15.4
1982	6,953	1,696	7,538	1,110	8,261	387	8,649	580	14.9
1983	6,974	1,714	7,492	1,196	8,688	(5)	8,688	603	14.4
1984	7,157	1,722	7,596	1,283	8,879	(5)	8,879	621	14.3
1985	7,146	1,825	7,722	1,250	8,971	(5)	8,971	647	13.9
1986	6,814	1,867	7,426	1,254	8,680	(5)	8,680	623	13.9
1987	6,387	1,962	7,153	1,196	8,349	(5)	8,349	620	13.5
1988	6,123	2,017	6,949	1,191	8,140	(5)	8,140	612	13.3
1989	5,739	1,874	6,486	1,127	7,613	(5)	7,613	603	12.6
1990	5,582	1,773	6,273	1,082	7,355	(5)	7,355	602	12.2
1991	5,618	1,798	6,245	1,172	7,417	(5)	7,417	614	12.1
1992	5,457	1,714	5,953	1,218	7,171	(5)	7,171	594	12.1
1993	5,264	1,582	5,606	1,241	6,847	(5)	6,847	584	11.7
1994	5,103	1,559	5,291	1,370	6,662	(5)	6,662	582	11.4
1995	5,076	1,484	5,035	1,525	6,560	(5)	6,560	574	11.4
1996	5,071	1,393	4,902	1,562	6,465	(5)	6,465	574	11.3
1997	5,156	1,296	4,803	1,648	6,452	(5)	6,452	573	11.3
1998	5,077	1,175	4,560	1,692	6,252	(5)	6,252	562	11.1
1999	4,832	1,050	4,132	1,750	5,881	(5)	5,881	546	10.8
2000	4,851	970	4,049	1,773	5,822	(5)	5,822	534	10.9
2001	4,839	963	3,879	1,923	5,801	(5)	5,801	530	10.9
2002	4,761	984	3,743	2,003	5,746	(5)	5,746	529	10.9
2003	4,706	974	3,668	2,012	5,681	(5)	5,681	513	11.1
2004	4,510	908	3,536	1,883	5,419	(5)	5,419	510	10.6
2005	4,314	864	3,466	1,712	5,178	(5)	5,178	498	10.4
2006	R4,361	741	R3,401	R1,701	R5,102	(5)	R5,102	R497	10.3
2007	P4,384	P719	E3,379	E1,724	P5,103	(5)	P5,103	P500	P10.2

[1] See "Crude Oil Well" in Glossary.
[2] United States excluding Alaska and Hawaii.
[3] As of December 31.
[4] Through 1976, average productivity is based on the average number of producing wells. Beginning in 1977, average productivity is based on the number of wells producing at end of year.
[5] Included in "Crude Oil."
R=Revised. P=Preliminary. E=Estimate.
Note: Totals may not equal sum of components due to independent rounding.
Web Page: See http://www.eia.doe.gov/oil_gas/petroleum/info_glance/petroleum.html for related infomation.
Sources: **Onshore:** • 1954-1975—Bureau of Mines, Mineral Industry Surveys, *Petroleum Statement (PS), Annual,* annual reports. • 1976-1980—Energy Information Administration (EIA), Energy Data Reports, *PS, Annual,* annual reports. • 1981-2006—EIA, *Petroleum Supply Annual (PSA),* annual reports. • 2007—EIA estimate. **Offshore:** • 1954-1969—U.S. Geological Survey, *Outer Continental Shelf Statistics* (June 1979). • 1970-1975—Bureau of Mines, Mineral Industry Surveys, *PS, Annual,* annual reports. • 1976-1980—EIA, Energy Data Reports, *PS, Annual,* annual reports. • 1981-2006—EIA, *PSA,* annual reports. • 2007—EIA estimate. **Producing Wells:** • 1954-1975—Bureau of Mines, *Minerals Yearbook,* "Crude Petroleum and Petroleum Products" chapter. • 1976-1980—EIA, Energy Data Reports, *PS, Annual,* annual reports. • 1981-1994—Independent Petroleum Association of America, *The Oil Producing Industry in Your State.* • 1995 forward—Gulf Publishing Co., *World Oil,* February issues. **All Other Data:** • 1954-1975—Bureau of Mines, Mineral Industry Surveys, *PS, Annual,* annual reports. • 1976-1980—EIA, Energy Data Reports, *PS, Annual,* annual reports. • 1981-2006—EIA, *PSA,* annual reports. • 2007—EIA, *Petroleum Supply Monthly* (February 2008).

Figure 5.3 Petroleum Imports by Type

Total, 1949-2007

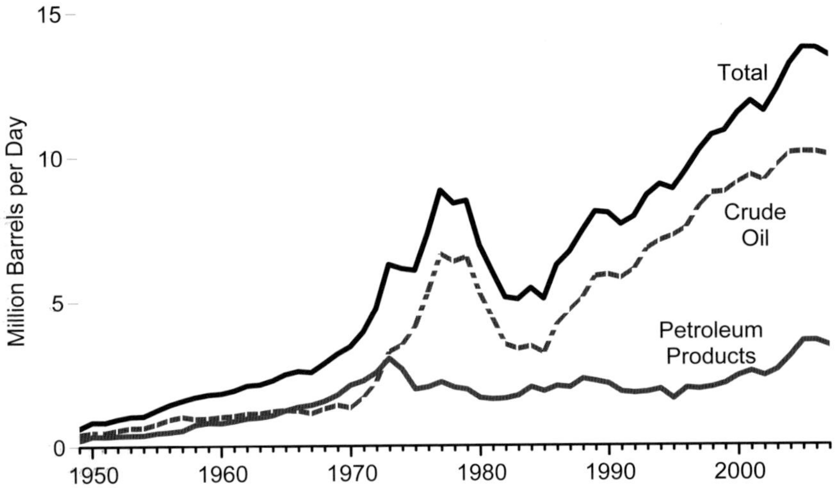

By Selected Product, 1949-2007

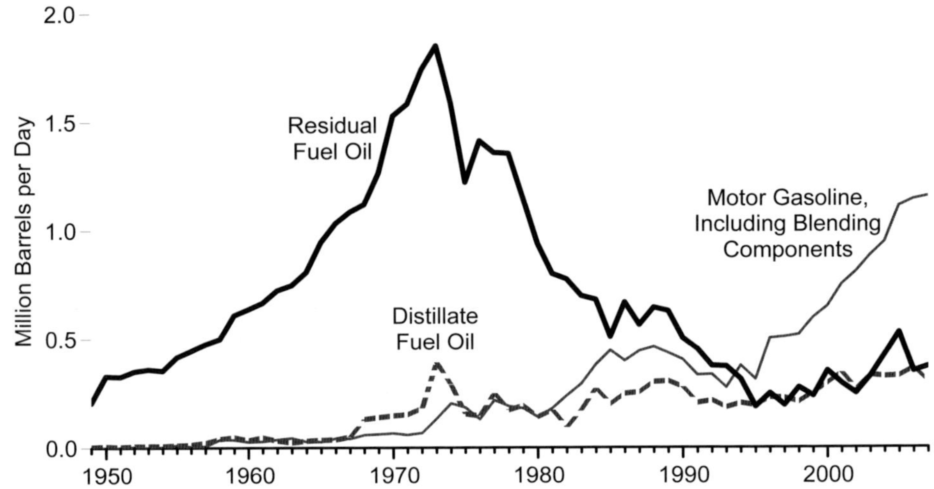

By Product, 2007

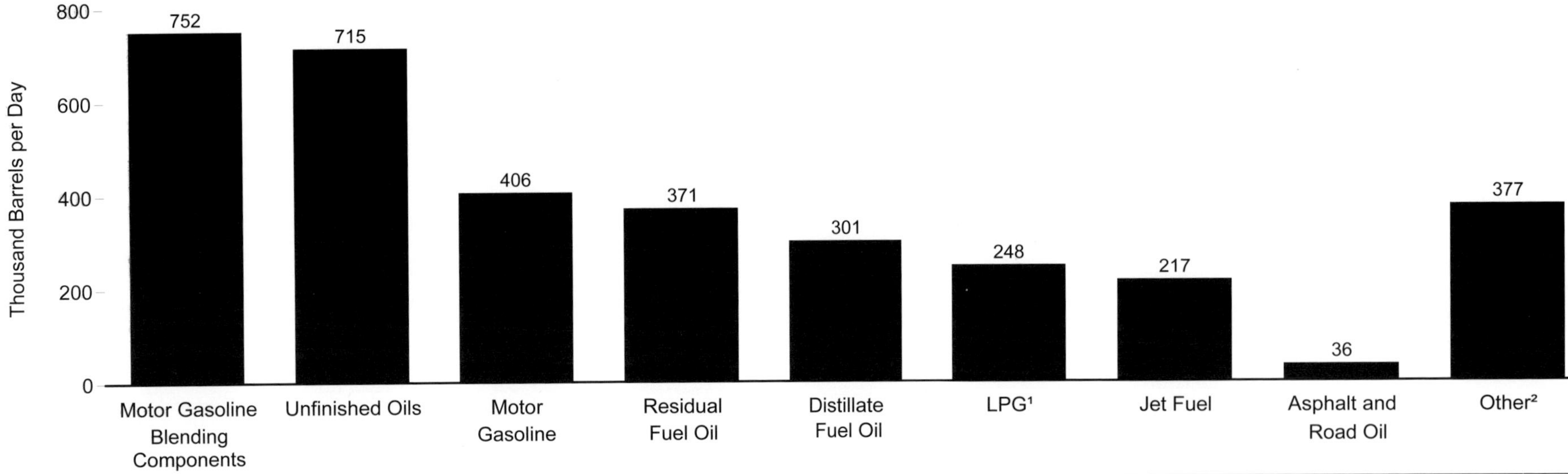

[1] Liquefied petroleum gases.

[2] Aviation gasoline and blending components, kerosene, lubricants, pentanes plus, petrochemical feedstocks, petroleum coke, special naphthas, waxes, other hydrocarbons and oxygenates, and miscellaneous products.

Note: Because vertical scales differ, graphs should not be compared.
Source: Table 5.3.

Table 5.3 Petroleum Imports by Type, Selected Years, 1949-2007

(Thousand Barrels per Day)

Year	Crude Oil [1,2]	Petroleum Products											Total Petroleum
		Asphalt and Road Oil	Distillate Fuel Oil	Jet Fuel [3]	Liquefied Petroleum Gases		Motor Gasoline [5]	Motor Gasoline Blending Components	Residual Fuel Oil	Unfinished Oils	Other Products [6]	Total	
					Propane [4]	Total							
1949	421	3	5	([3])	0	0	0	0	206	10	0	224	645
1950	487	5	7	([3])	0	0	(s)	([7])	329	21	1	363	850
1955	782	9	12	([3])	0	0	13	([7])	417	15	0	466	1,248
1960	1,015	17	35	34	NA	4	27	([7])	637	45	(s)	799	1,815
1965	1,238	17	36	81	NA	21	28	([7])	946	92	10	1,229	2,468
1970	1,324	17	147	144	26	52	67	([7])	1,528	108	32	2,095	3,419
1971	1,681	20	153	180	32	70	59	([7])	1,583	124	56	2,245	3,926
1972	2,216	25	182	194	43	89	68	([7])	1,742	125	101	2,525	4,741
1973	3,244	23	392	212	71	132	134	([7])	1,853	137	129	3,012	6,256
1974	3,477	31	289	163	59	123	204	([7])	1,587	121	117	2,635	6,112
1975	4,105	14	155	133	60	112	184	([7])	1,223	36	95	1,951	6,056
1976	5,287	11	146	76	68	130	131	([7])	1,413	32	87	2,026	7,313
1977	6,615	4	250	75	86	161	217	([7])	1,359	31	95	2,193	8,807
1978	6,356	2	173	86	57	123	190	([7])	1,355	27	50	2,008	8,363
1979	6,519	4	193	78	88	217	181	([7])	1,151	59	54	1,937	8,456
1980	5,263	4	142	80	69	216	140	([7])	939	55	72	1,646	6,909
1981	4,396	4	173	38	70	244	157	24	800	112	48	1,599	5,996
1982	3,488	5	93	29	63	226	197	42	776	174	84	1,625	5,113
1983	3,329	7	174	29	44	190	247	47	699	234	94	1,722	5,051
1984	3,426	18	272	62	67	195	299	83	681	231	171	2,011	5,437
1985	3,201	35	200	39	67	187	381	67	510	318	130	1,866	5,067
1986	4,178	29	247	57	110	242	326	72	669	250	153	2,045	6,224
1987	4,674	36	255	67	88	190	384	60	565	299	146	2,004	6,678
1988	5,107	31	302	90	106	209	405	57	644	360	196	2,295	7,402
1989	5,843	31	306	106	111	181	369	66	629	348	183	2,217	8,061
1990	5,894	32	278	108	115	188	342	62	504	413	198	2,123	8,018
1991	5,782	28	205	67	91	147	297	36	453	413	198	1,844	7,627
1992	6,083	27	216	82	85	131	294	41	375	443	195	1,805	7,888
1993	6,787	32	184	100	103	160	247	27	373	491	219	1,833	8,620
1994	7,063	37	203	117	124	183	356	20	314	413	291	1,933	8,996
1995	7,230	36	193	106	102	146	265	48	187	349	276	1,605	8,835
1996	7,508	27	230	111	119	166	336	166	248	367	319	1,971	9,478
1997	8,225	32	228	91	113	169	309	200	194	353	360	1,936	10,162
1998	8,706	28	210	124	137	194	311	209	275	302	350	2,002	10,708
1999	8,731	34	250	128	122	182	382	217	237	317	375	2,122	10,852
2000	9,071	28	295	162	161	215	427	223	352	274	414	2,389	11,459
2001	9,328	26	344	148	140	206	454	298	295	378	393	2,543	11,871
2002	9,140	27	267	107	145	183	498	311	249	410	337	2,390	11,530
2003	9,665	12	333	109	168	225	518	367	327	335	373	2,599	12,264
2004	10,088	43	325	127	209	263	496	451	426	490	436	3,057	13,145
2005	10,126	43	329	190	233	328	603	510	530	582	473	3,588	13,714
2006	R10,118	R50	R365	R186	R228	R332	R475	669	R350	R689	R473	R3,589	R13,707
2007P	10,017	36	301	217	180	248	406	752	371	715	377	3,422	13,439

[1] Includes lease condensate.
[2] Includes imports for the Strategic Petroleum Reserve, which began in 1977. See Table 5.17.
[3] Through 1955, naphtha-type jet fuel is included in "Motor Gasoline." Through 1964, kerosene-type jet fuel is included with kerosene in "Other Products." Beginning in 2005, naphtha-type jet fuel is included in "Other Products."
[4] Includes propylene.
[5] Finished motor gasoline. Through 1955, also includes naphtha-type jet fuel. Through 1963, also includes aviation gasoline and special naphthas. Through 1980, also includes motor gasoline blending components.
[6] Aviation gasoline blending components, kerosene, lubricants, pentanes plus, petrochemical feedstocks, petroleum coke, waxes, other hydrocarbons and oxygenates, and miscellaneous products. Through 1964, also includes kerosene-type jet fuel. Beginning in 1964, also includes aviation gasoline and special naphthas. Beginning in 2005, also includes naphtha-type jet fuel.
[7] Included in "Motor Gasoline."

R=Revised. P=Preliminary. NA=Not available. (s)=Less than 500 barrels per day.

Notes: • Includes imports from U.S. possessions and territories. • Totals may not equal sum of components due to independent rounding.

Web Pages: • For all data beginning in 1949, see http://www.eia.doe.gov/emeu/aer/petro.html. • For related information, see http://www.eia.doe.gov/oil_gas/petroleum/info_glance/petroleum.html.

Sources: • 1949-1975—Bureau of Mines, Mineral Industry Surveys, *Petroleum Statement, Annual,* annual reports. • 1976-1980—Energy Information Administration (EIA), Energy Data Reports, *Petroleum Statement, Annual,* annual reports. • 1981-2006—EIA, *Petroleum Supply Annual,* annual reports. • 2007—EIA, *Petroleum Supply Monthly* (February 2008).

Figure 5.4 Petroleum Imports by Country of Origin

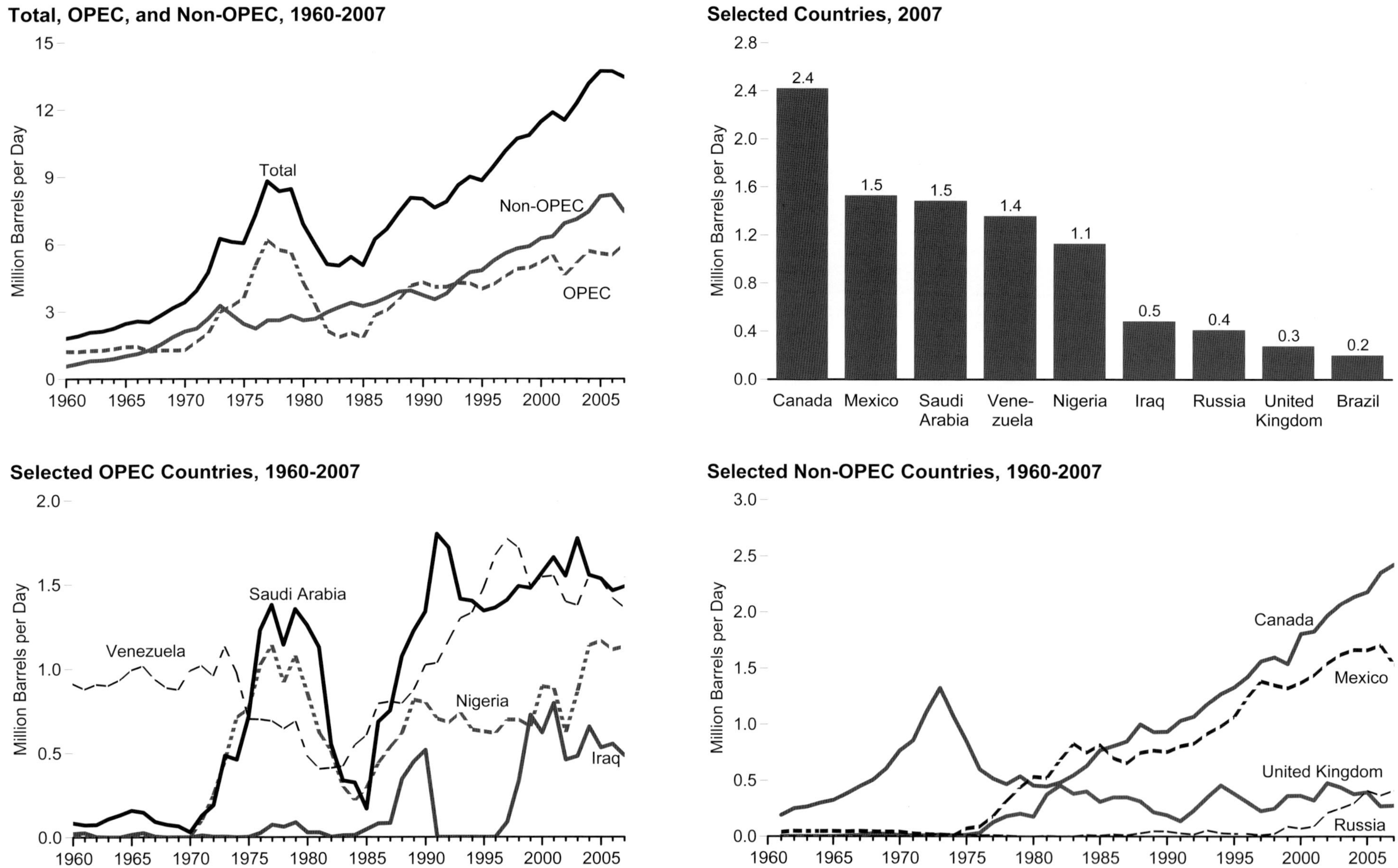

Notes: • OPEC=Organization of the Petroleum Exporting Countries. • Because vertical scales differ, graphs should not be compared.

Source: Table 5.4.

Table 5.4 Petroleum Imports by Country of Origin, 1960-2007

Year	Persian Gulf [2]	Selected OPEC [1] Countries: Iraq	Nigeria	Saudi Arabia [3]	Venezuela	Total OPEC [4]	Selected Non-OPEC [1] Countries: Brazil	Canada	Mexico	Russia [5]	United Kingdom	Total Non-OPEC [4]	Total Imports	Imports From Persian Gulf [2] as Share of Total Imports	Imports From OPEC [1] as Share of Total Imports
	Thousand Barrels per Day													Percent	
1960	NA	R22	R(6)	84	911	R1,233	1	120	16	0	(s)	R581	1,815	NA	R68.0
1961	346	25	R(6)	73	879	R1,224	7	190	40	0	1	R693	1,917	18.0	R63.8
1962	272	2	R(6)	74	906	1,265	5	250	49	0	2	816	2,082	13.0	60.8
1963	303	1	R(6)	108	900	R1,282	6	265	48	0	3	R840	2,123	14.3	R60.4
1964	315	R0	R(6)	131	933	R1,352	1	299	47	0	(s)	R907	2,259	13.9	R59.8
1965	345	16	R(6)	158	994	R1,439	0	323	48	0	(s)	R1,029	2,468	14.0	R58.3
1966	306	26	R(6)	147	1,018	R1,444	0	384	45	0	6	R1,129	2,573	11.9	R56.1
1967	198	5	R(6)	92	938	R1,247	2	450	49	0	11	R1,290	2,537	7.8	R49.2
1968	202	R0	R(6)	74	886	R1,287	(s)	506	45	0	28	R1,553	2,840	7.1	R45.3
1969	179	R0	R(6)	65	875	R1,286	0	608	43	2	20	R1,879	3,166	5.7	R40.6
1970	121	R0	R(6)	30	989	R1,294	2	766	42	3	11	R2,126	3,419	3.5	R37.8
1971	299	11	102	128	1,020	1,673	3	857	27	0	10	2,253	3,926	7.6	42.6
1972	471	4	251	190	959	R2,046	5	1,108	21	8	9	R2,695	4,741	9.9	R43.2
1973	848	4	459	486	1,135	2,993	9	1,325	16	26	15	3,263	6,256	13.6	47.8
1974	1,039	0	713	461	979	R3,256	2	1,070	8	20	8	R2,856	6,112	17.0	R53.3
1975	1,165	2	762	715	702	3,601	5	846	71	14	14	2,454	6,056	19.2	59.5
1976	1,840	26	1,025	1,230	700	5,066	0	599	87	11	31	2,247	7,313	25.2	69.3
1977	2,448	74	1,143	1,380	690	6,193	0	517	179	12	126	2,614	8,807	27.8	70.3
1978	2,219	62	919	1,144	646	5,751	0	467	318	8	180	2,612	8,363	26.5	68.8
1979	2,069	88	1,080	1,356	690	5,637	1	538	439	1	202	2,819	8,456	24.5	66.7
1980	1,519	28	857	1,261	481	4,300	3	455	533	1	176	2,609	6,909	22.0	62.2
1981	1,219	(s)	620	1,129	406	3,323	23	447	522	5	375	2,672	5,996	20.3	55.4
1982	696	3	514	552	412	2,146	47	482	685	1	456	2,968	5,113	13.6	42.0
1983	442	10	302	337	422	1,862	41	547	826	1	382	3,189	5,051	8.8	36.9
1984	506	12	216	325	548	2,049	60	630	748	13	402	3,388	5,437	9.3	37.7
1985	311	46	293	168	605	1,830	61	770	816	8	310	3,237	5,067	6.1	36.1
1986	912	81	440	685	793	2,837	50	807	699	18	350	3,387	6,224	14.7	45.6
1987	1,077	83	535	751	804	3,060	84	848	655	11	352	3,617	6,678	16.1	45.8
1988	1,541	345	618	1,073	794	3,520	98	999	747	29	315	3,882	7,402	20.8	47.6
1989	1,861	449	815	1,224	873	4,140	82	931	767	48	215	3,921	8,061	23.1	51.4
1990	1,966	518	800	1,339	1,025	4,296	49	934	755	45	189	3,721	8,018	24.5	53.6
1991	1,845	0	703	1,802	1,035	4,092	22	1,033	807	29	138	3,535	7,627	24.2	53.7
1992	1,778	0	681	1,720	1,170	4,092	20	1,069	830	18	230	3,796	7,888	22.5	51.9
1993	1,782	0	740	1,414	1,300	4,273	33	1,181	919	55	350	4,347	8,620	20.7	49.6
1994	1,728	0	637	1,402	1,334	4,247	31	1,272	984	30	458	4,749	8,996	19.2	47.2
1995	1,573	0	627	1,344	1,480	4,002	8	1,332	1,068	25	383	4,833	8,835	17.8	45.3
1996	1,604	1	617	1,363	1,676	4,211	9	1,424	1,244	25	308	5,267	9,478	16.9	44.4
1997	1,755	89	698	1,407	1,773	4,569	5	1,563	1,385	13	226	5,593	10,162	17.3	45.0
1998	2,136	336	696	1,491	1,719	4,905	26	1,598	1,351	24	250	5,803	10,708	19.9	45.8
1999	2,464	725	657	1,478	1,493	4,953	26	1,539	1,324	89	365	5,899	10,852	22.7	45.6
2000	2,488	620	896	1,572	1,546	5,203	51	1,807	1,373	72	366	6,257	11,459	21.7	45.4
2001	2,761	795	885	1,662	1,553	5,528	82	1,828	1,440	90	324	6,343	11,871	23.3	46.6
2002	2,269	459	621	1,552	1,398	4,605	116	1,971	1,547	210	478	6,925	11,530	19.7	39.9
2003	2,501	481	867	1,774	1,376	5,162	108	2,072	1,623	254	440	7,103	12,264	20.4	42.1
2004	2,493	656	1,140	1,558	1,554	5,701	104	2,138	1,665	298	380	7,444	13,145	19.0	43.4
2005	2,334	531	1,166	1,537	1,529	5,587	156	2,181	1,662	410	396	8,127	13,714	17.0	40.7
2006	R2,211	553	R1,114	R1,463	R1,419	R5,517	193	R2,353	R1,705	R369	R272	R8,190	R13,707	R16.1	R40.2
2007P	2,170	485	1,132	1,489	1,362	5,983	202	2,426	1,533	413	278	7,456	13,439	16.1	44.5

[1] See "Organization of the Petroleum Exporting Countries (OPEC)" in Glossary.

[2] Bahrain, Iran, Iraq, Kuwait, Qatar, Saudi Arabia, United Arab Emirates, and the Neutral Zone (between Kuwait and Saudi Arabia).

[3] Through 1970, includes half the imports from the Neutral Zone. Beginning in 1971, includes imports from the Neutral Zone that are reported to U.S. Customs as originating in Saudi Arabia.

[4] On this table, "Total OPEC" for all years includes Iran, Iraq, Kuwait, Saudi Arabia, Venezuela, and the Neutral Zone (between Kuwait and Saudi Arabia); beginning in 1961, also includes Qatar; beginning in 1962, also includes Indonesia and Libya; beginning in 1967, also includes United Arab Emirates; beginning in 1969, also includes Algeria; beginning in 1971, also includes Nigeria; for 1973-1992, also includes Ecuador (although Ecuador rejoined OPEC in November 2007, on this table Ecuador is included in "Total Non-OPEC" for 2007); for 1975-1994, also includes Gabon; and beginning in 2007, also includes Angola. Data for all countries not included in "Total OPEC" are included in "Total Non-OPEC."

[5] Through 1992, may include imports from republics other than Russia in the former U.S.S.R. See "U.S.S.R." in Glossary.

[6] Nigeria joined OPEC in 1971. For 1960-1970, Nigeria is included in "Total Non-OPEC."

R=Revised. P=Preliminary. NA=Not available. (s)=Less than 500 barrels per day.

Notes: • The country of origin for refined petroleum products may not be the country of origin for the crude oil from which the refined products were produced. For example, refined products imported from refineries in the Caribbean may have been produced from Middle East crude oil. • Data include any imports for the Strategic Petroleum Reserve, which began in 1977. • Totals may not equal sum of components due to independent rounding.

Web Page: See http://www.eia.doe.gov/oil_gas/petroleum/info_glance/petroleum.html for related information.

Sources: • 1960-1975—Bureau of Mines, *Minerals Yearbook,* "Crude Petroleum and Petroleum Products" chapter. • 1976-1980—Energy Information Administration (EIA), Energy Data Reports, *P.A.D. Districts Supply/Demand, Annual,* annual reports. • 1981-2006—EIA, *Petroleum Supply Annual,* annual reports. • 2007—EIA, *Petroleum Supply Monthly* (February 2008).

Figure 5.5 Petroleum Exports by Type

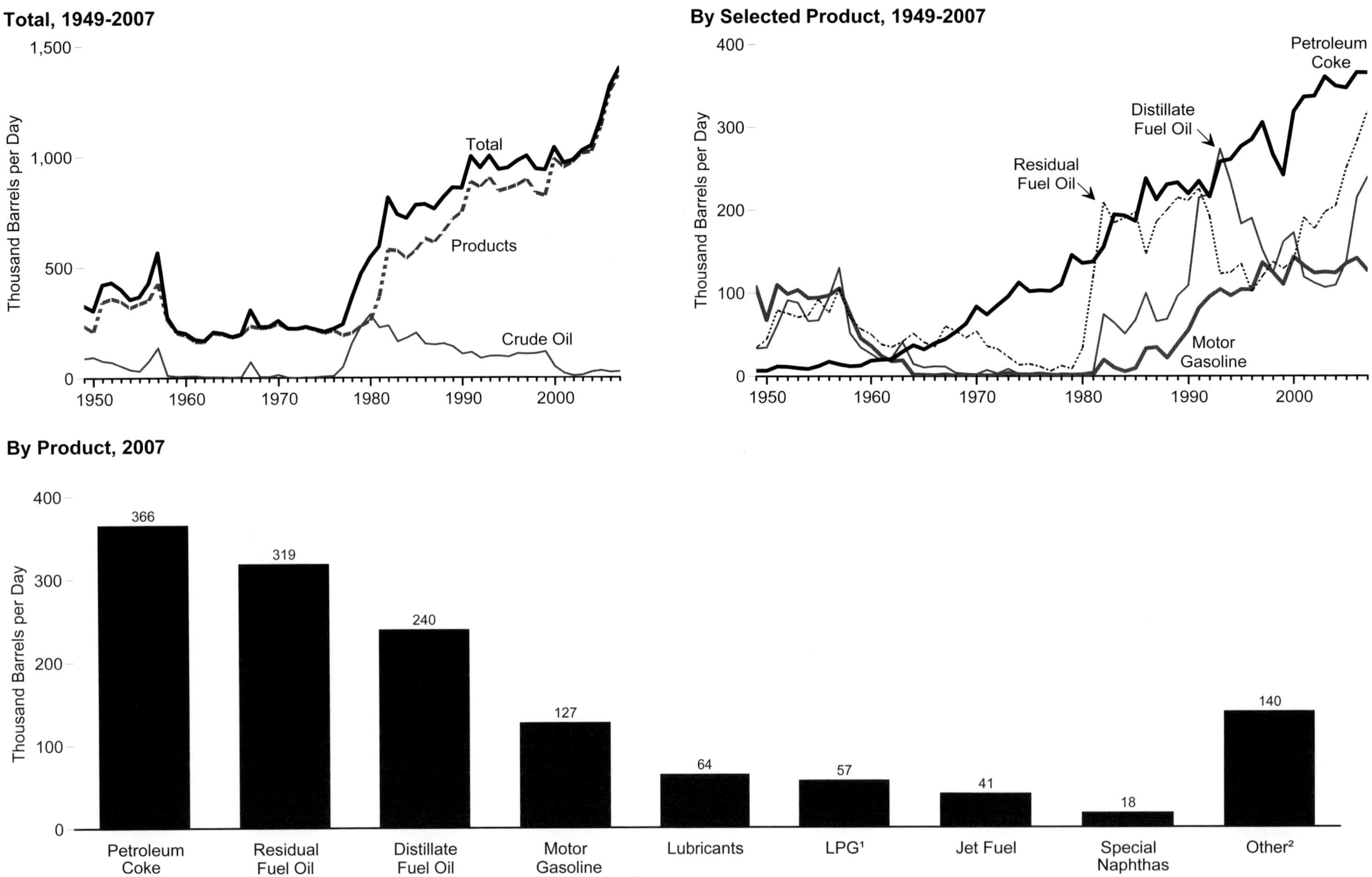

[1] Liquefied petroleum gases.

[2] Asphalt and road oil, aviation gasoline, kerosene, motor gasoline blending components, pentanes plus, waxes, other hydrocarbons and oxygenates, and miscellaneous products.

Note: Because vertical scales differ, graphs should not be compared.
Source: Table 5.5.

Table 5.5 Petroleum Exports by Type, Selected Years, 1949-2007

(Thousand Barrels per Day)

		Petroleum Products												
				Liquefied Petroleum Gases										
Year	Crude Oil [1]	Distillate Fuel Oil	Jet Fuel [2]	Propane [3]	Total	Lubricants	Motor Gasoline [4]	Petroleum Coke	Petrochemical Feedstocks	Residual Fuel Oil	Special Naphthas	Other Products [5]	Total	Total Petroleum
1949	91	34	([2])	NA	4	35	108	7	0	35	NA	15	236	327
1950	95	35	([2])	NA	4	39	68	7	0	44	NA	12	210	305
1955	32	67	(s)	NA	12	39	95	12	0	93	NA	18	336	368
1960	8	27	(s)	NA	8	43	37	19	0	51	NA	9	193	202
1965	3	10	3	NA	21	45	2	32	5	41	4	20	184	187
1970	14	2	6	6	27	44	1	84	10	54	4	12	245	259
1971	1	8	4	13	26	43	1	74	14	36	4	12	223	224
1972	1	3	3	18	31	41	1	85	13	33	4	8	222	222
1973	2	9	4	15	27	35	4	96	19	23	5	8	229	231
1974	3	2	3	14	25	33	2	113	15	14	4	7	218	221
1975	6	1	2	13	26	25	2	102	22	15	3	6	204	209
1976	8	1	2	13	25	26	3	103	30	12	7	6	215	223
1977	50	1	2	10	18	26	2	102	24	6	4	7	193	243
1978	158	3	1	9	20	27	1	111	23	13	2	2	204	362
1979	235	3	1	8	15	23	(s)	146	31	9	5	3	236	471
1980	287	3	1	10	21	23	1	136	29	33	5	4	258	544
1981	228	5	2	18	42	19	2	138	26	118	11	4	367	595
1982	236	74	6	31	65	16	20	156	24	209	5	4	579	815
1983	164	64	6	43	73	16	10	195	20	185	3	3	575	739
1984	181	51	9	30	48	15	6	193	21	190	2	6	541	722
1985	204	67	13	48	62	15	10	187	19	197	1	4	577	781
1986	154	100	18	28	42	23	33	238	22	147	1	8	631	785
1987	151	66	24	24	38	23	35	213	20	186	2	7	613	764
1988	155	69	28	31	49	26	22	231	23	200	7	6	661	815
1989	142	97	27	24	35	19	39	233	26	215	12	15	717	859
1990	109	109	43	28	40	20	55	220	26	211	11	13	748	857
1991	116	215	43	28	41	18	82	235	0	226	15	9	885	1,001
1992	89	219	43	33	49	16	96	216	0	193	14	16	861	950
1993	98	274	59	26	43	19	105	258	0	123	4	20	904	1,003
1994	99	234	20	24	38	22	97	261	0	125	20	26	843	942
1995	95	183	26	38	58	25	104	277	0	136	21	25	855	949
1996	110	190	48	28	51	34	104	285	0	102	21	36	871	981
1997	108	152	35	32	50	31	137	306	0	120	22	44	896	1,003
1998	110	124	26	25	42	25	125	267	0	138	18	70	835	945
1999	118	162	32	33	50	28	111	242	0	129	16	52	822	940
2000	50	173	32	53	74	26	144	319	0	139	20	64	990	1,040
2001	20	119	29	31	44	26	133	336	0	191	23	50	951	971
2002	9	112	15	55	67	33	124	337	0	177	15	94	975	984
2003	12	107	20	37	56	37	125	361	0	197	22	89	1,014	1,027
2004	27	110	40	28	43	41	124	350	0	205	27	82	1,021	1,048
2005	32	138	53	37	53	40	136	347	0	251	21	94	1,133	1,165
2006	25	215	41	45	56	[R]55	142	366	0	283	14	121	[R]1,292	[R]1,317
2007[P]	27	240	41	42	57	64	127	366	0	319	18	140	1,371	1,399

[1] Includes lease condensate.

[2] Through 1952, naphtha-type jet fuel is included in the products from which it was blended: gasoline, kerosene, and distillate fuel oil. Through 1964, kerosene-type jet fuel is included with kerosene in "Other Products." Beginning in 2005, naphtha-type jet fuel is included in "Other Products."

[3] Includes propylene.

[4] Finished motor gasoline. Through 1963, also includes aviation gasoline.

[5] Asphalt and road oil, kerosene, motor gasoline blending components, pentanes plus, waxes, other hydrocarbons and oxygenates, and miscellaneous products. Through 1964, also includes kerosene-type jet fuel. Beginning in 1964, also includes aviation gasoline. Beginning in 2005, also includes naphtha-type jet fuel.

R=Revised. P=Preliminary. NA=Not available. (s)=Less than 500 barrels per day.

Notes: • Includes exports to U.S. possessions and territories. • Totals may not equal sum of components due to independent rounding.

Web Pages: • For all data beginning in 1949, see http://www.eia.doe.gov/emeu/aer/petro.html. • For related information, see http://www.eia.doe.gov/oil_gas/petroleum/info_glance/petroleum.html.

Sources: • 1949-1975—Bureau of Mines, Mineral Industry Surveys, *Petroleum Statement, Annual,* annual reports. • 1976-1980—Energy Information Administration (EIA), Energy Data Reports, *Petroleum Statement, Annual,* annual reports. • 1981-2006—EIA, *Petroleum Supply Annual,* annual reports. • 2007—EIA, *Petroleum Supply Monthly* (February 2008).

Figure 5.6 Petroleum Exports by Country of Destination

Total Exports and Exports to Canada and Mexico, 1960-2007

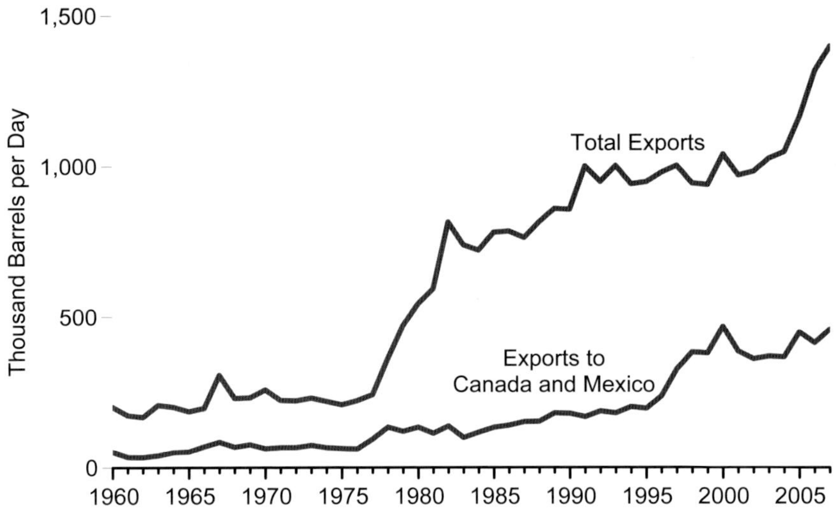

By Selected Country, 1960-2007

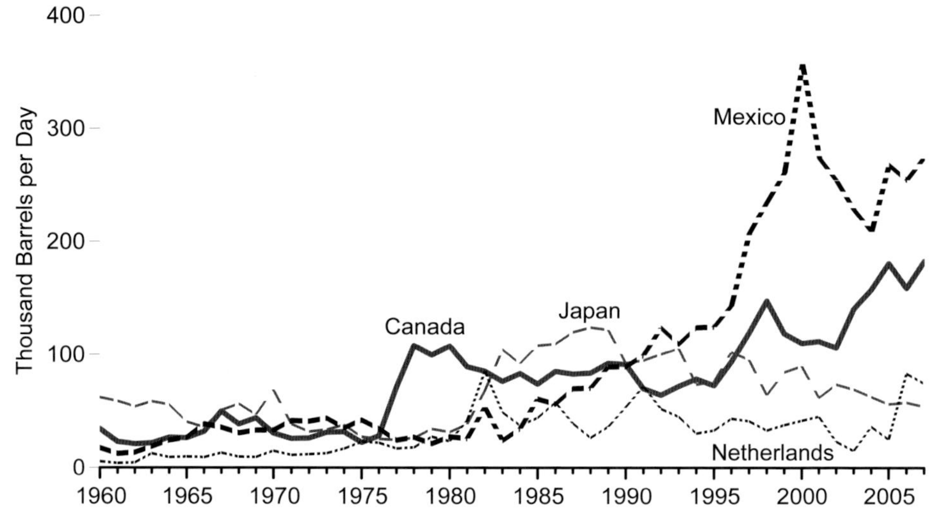

By Selected Country, 2007

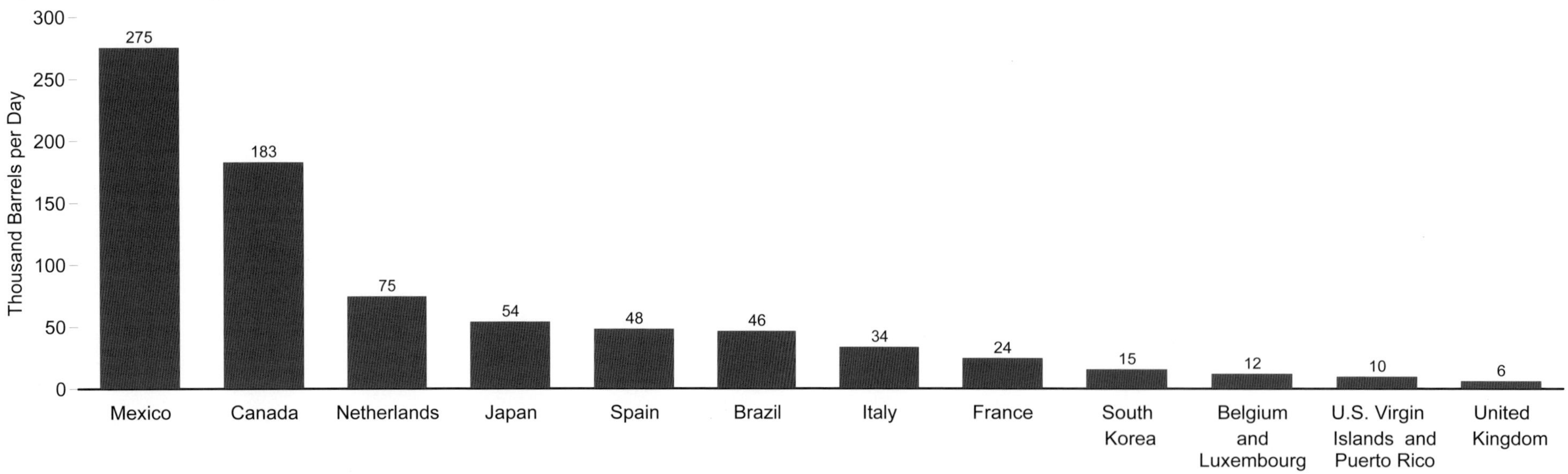

Note: Because vertical scales differ, graphs should not be compared.

Source: Table 5.6.

Table 5.6 Petroleum Exports by Country of Destination, 1960-2007

(Thousand Barrels per Day)

Year	Belgium and Luxembourg	Brazil	Canada	France	Italy	Japan	Mexico	Nether-lands	South Korea	Spain	United Kingdom	U.S. Virgin Islands and Puerto Rico	Other	Total
1960	3	4	34	4	6	62	18	6	NA	NA	12	1	52	202
1961	4	4	23	4	5	59	12	4	NA	NA	10	1	48	174
1962	3	5	21	3	5	54	14	5	NA	NA	8	1	50	168
1963	9	4	22	4	8	58	19	13	NA	NA	11	1	59	208
1964	4	4	27	4	8	56	24	9	NA	NA	10	2	55	202
1965	3	3	26	3	7	40	27	10	NA	NA	12	1	54	187
1966	3	4	32	4	7	36	39	9	NA	NA	12	3	49	198
1967	5	6	50	3	9	51	36	13	NA	NA	62	7	65	307
1968	4	8	39	4	8	56	31	10	NA	NA	14	2	55	231
1969	4	7	44	4	9	47	33	9	NA	NA	13	2	59	233
1970	5	7	31	5	10	69	33	15	NA	NA	12	2	71	259
1971	7	9	26	5	8	39	42	11	NA	NA	9	3	67	224
1972	13	9	26	5	9	32	41	12	NA	4	10	4	59	222
1973	15	8	31	5	9	34	44	13	NA	4	9	3	56	231
1974	13	9	32	4	9	38	35	17	NA	4	6	6	48	221
1975	9	6	22	6	10	27	42	23	NA	4	7	12	40	209
1976	12	7	28	6	10	25	35	22	NA	4	13	22	39	223
1977	16	6	71	9	10	25	24	17	NA	5	9	11	39	243
1978	15	8	108	9	10	26	27	18	NA	5	7	86	42	362
1979	19	7	100	13	15	34	21	28	2	9	7	170	45	471
1980	20	4	108	11	14	32	28	23	2	8	7	220	70	544
1981	12	1	89	15	22	38	26	42	10	18	5	220	97	595
1982	17	8	85	24	32	68	53	85	28	24	14	212	165	815
1983	22	2	76	23	35	104	24	49	15	34	8	144	202	739
1984	21	1	83	18	39	92	35	37	17	29	14	152	182	722
1985	26	3	74	11	30	108	61	44	27	28	14	162	193	781
1986	30	3	85	11	39	110	56	58	12	39	8	113	222	785
1987	17	2	83	12	42	120	70	39	25	31	6	136	179	764
1988	25	3	84	12	29	124	70	26	24	36	9	147	226	815
1989	23	5	92	11	37	122	89	36	17	28	9	141	249	859
1990	20	2	91	17	48	92	89	54	60	33	11	101	240	857
1991	22	13	70	27	55	95	99	72	66	23	13	117	330	1,001
1992	22	20	64	9	38	100	124	52	80	21	12	95	315	950
1993	21	16	72	8	34	105	110	45	74	30	10	108	370	1,003
1994	26	15	78	11	35	74	124	30	66	30	10	104	338	942
1995	21	16	73	11	46	76	125	33	57	38	14	123	317	949
1996	27	29	94	18	32	102	143	43	60	34	9	72	318	981
1997	21	15	119	11	30	95	207	41	50	42	12	18	340	1,003
1998	14	18	148	8	30	64	235	33	33	30	11	4	317	945
1999	11	27	119	7	25	84	261	38	49	26	9	8	276	940
2000	14	28	110	10	34	90	358	42	20	40	10	10	277	1,040
2001	16	23	112	13	33	62	274	45	14	51	13	4	312	971
2002	19	26	106	12	29	74	254	23	11	54	12	9	354	984
2003	13	27	141	9	39	69	228	15	10	39	6	9	421	1,027
2004	20	27	158	18	32	63	209	36	12	42	14	10	408	1,048
2005	21	39	181	14	28	56	268	25	16	35	21	11	449	1,165
2006	R23	42	159	13	39	58	255	R83	21	42	28	R10	R543	R1,317
2007P	12	46	183	24	34	54	275	75	15	48	6	10	616	1,399

R=Revised. P=Preliminary. NA=Not available.

Note: Totals may not equal sum of components due to independent rounding.

Web Page: See http://www.eia.doe.gov/oil_gas/petroleum/info_glance/petroleum.html for related information.

Sources: • 1960-1975—Bureau of Mines, Mineral Industry Surveys, *Petroleum Statement, Annual,* annual reports. • 1976-1980—Energy Information Administration (EIA), Energy Data Reports, *Petroleum Statement, Annual,* annual reports. • 1981-2006—EIA, *Petroleum Supply Annual,* annual reports. • 2007—EIA, *Petroleum Supply Monthly* (February 2008).

Figure 5.7 Petroleum Net Imports by Country of Origin, 1960-2007

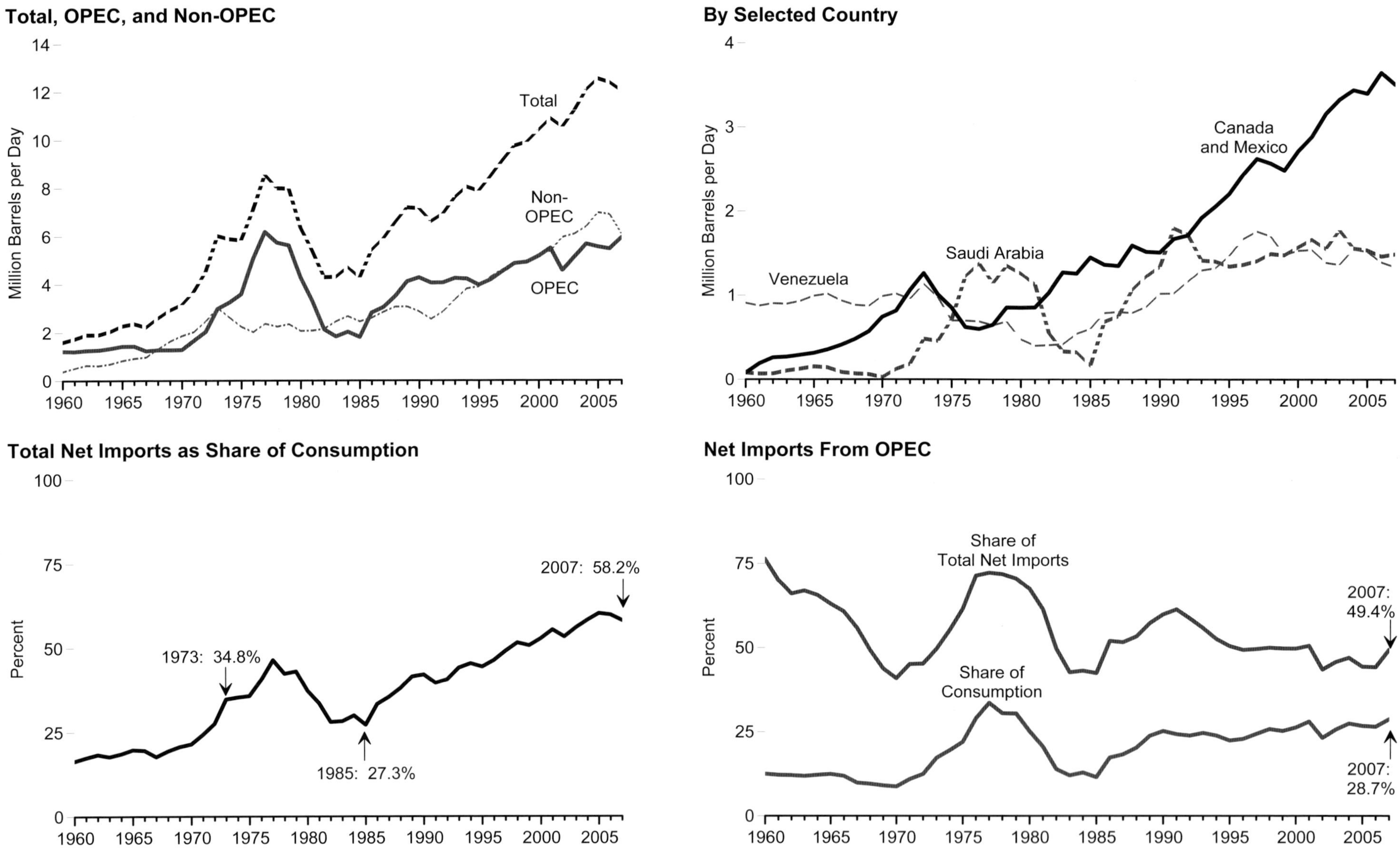

Notes: • OPEC=Organization of the Petroleum Exporting Countries. • Because vertical scales differ, graphs should not be compared.

Source: Table 5.7.

Table 5.7 Petroleum Net Imports by Country of Origin, 1960-2007

Year	Persian Gulf [2]	Selected OPEC [1] Countries					Selected Non-OPEC [1] Countries					Total Net Imports	Total Net Imports as Share of Consumption [5]	Net Imports From OPEC [1]	
		Algeria	Nigeria	Saudi Arabia [3]	Venezuela	Total OPEC [4]	Canada	Mexico	United Kingdom	U.S. Virgin Islands and Puerto Rico	Total Non-OPEC [4]			Share of Total Net Imports [6]	Share of Consumption [7]
	Thousand Barrels per Day												Percent		
1960	NA	R(8)	R(9)	84	910	R1,232	86	-2	-12	34	R381	1,613	16.5	R76.4	R12.6
1961	NA	R(8)	R(9)	73	878	R1,223	167	27	-10	42	R520	1,743	17.5	R70.2	R12.3
1962	NA	R(8)	R(9)	74	905	R1,263	229	35	-6	40	R650	1,913	18.4	R66.0	R12.2
1963	NA	R(8)	R(9)	108	899	R1,281	243	29	-7	43	R633	1,915	17.8	R66.9	11.9
1964	NA	R(8)	R(9)	131	932	R1,350	272	23	-9	45	R707	2,057	18.7	R65.6	R12.2
1965	NA	R(8)	R(9)	158	994	R1,438	297	21	-11	45	R843	2,281	19.8	R63.0	R12.5
1966	NA	R(8)	R(9)	147	1,018	R1,443	352	6	-6	58	R932	2,375	19.7	R60.7	R11.9
1967	NA	R(8)	R(9)	92	937	R1,246	400	13	-51	89	R984	2,230	17.8	R55.9	R9.9
1968	NA	R(8)	R(9)	74	886	R1,286	468	15	13	143	R1,322	2,609	19.5	R49.3	R9.6
1969	NA	R2	R(9)	65	875	R1,286	564	10	7	186	R1,647	2,933	20.8	R43.8	R9.1
1970	NA	R8	R(9)	30	989	R1,294	736	9	-1	270	R1,867	3,161	21.5	R40.9	R8.8
1971	NA	R15	102	128	1,019	1,671	831	-14	1	365	2,030	3,701	24.3	R45.1	11.0
1972	NA	R92	251	189	959	R2,044	1,082	-20	-1	428	R2,475	4,519	27.6	R45.2	R12.5
1973	NA	R136	459	485	1,134	2,991	1,294	-28	6	426	3,034	6,025	34.8	49.6	17.3
1974	NA	R190	713	461	978	R3,254	1,038	-27	1	475	R2,638	5,892	35.4	R55.2	R19.5
1975	NA	R282	762	714	702	3,599	824	29	7	484	2,248	5,846	35.8	61.6	22.0
1976	NA	R432	1,025	1,229	699	5,063	571	53	19	488	2,027	7,090	40.6	71.4	29.0
1977	NA	R559	1,143	1,379	689	6,190	446	155	117	560	2,375	8,565	46.5	72.3	33.6
1978	NA	R649	919	1,142	644	5,747	359	291	173	436	2,255	8,002	42.5	71.8	30.5
1979	NA	R636	1,080	1,354	688	5,633	438	418	196	353	2,352	7,985	43.1	70.5	30.4
1980	NA	R488	857	1,259	478	4,293	347	506	169	256	2,071	6,365	37.3	67.5	25.2
1981	1,215	311	620	1,128	403	3,315	358	497	370	169	2,086	5,401	33.6	61.4	20.6
1982	692	170	512	551	409	2,136	397	632	442	154	2,163	4,298	28.1	49.7	14.0
1983	439	240	299	336	420	1,843	471	802	374	178	2,469	4,312	28.3	42.7	12.1
1984	502	323	215	324	544	2,037	547	714	388	184	2,679	4,715	30.0	43.2	13.0
1985	309	187	293	167	602	1,821	696	755	295	114	2,465	4,286	27.3	42.5	11.6
1986	909	271	440	685	788	2,828	721	642	342	152	2,611	5,439	33.4	52.0	17.4
1987	1,074	295	535	751	801	3,055	765	585	346	158	2,859	5,914	35.5	51.7	18.3
1988	1,529	300	618	1,064	790	3,513	916	677	306	117	3,074	6,587	38.1	53.3	20.3
1989	1,858	269	815	1,224	861	4,124	839	678	206	212	3,078	7,202	41.6	57.3	23.8
1990	1,962	280	800	1,339	1,016	4,285	843	666	179	213	2,876	7,161	42.2	59.8	25.2
1991	1,833	253	703	1,796	1,020	4,065	963	707	125	153	2,561	6,626	39.6	61.3	24.3
1992	1,773	196	680	1,720	1,161	4,071	1,005	706	219	180	2,867	6,938	40.7	58.7	23.9
1993	1,774	219	736	1,413	1,296	4,253	1,109	809	340	175	3,365	7,618	44.2	55.8	24.7
1994	1,723	243	637	1,402	1,322	4,233	1,194	860	448	246	3,822	8,054	45.5	52.6	23.9
1995	1,563	234	626	1,343	1,468	3,980	1,260	943	369	170	3,906	7,886	44.5	50.5	22.4
1996	1,596	256	616	1,362	1,667	4,193	1,330	1,101	299	262	4,305	8,498	46.4	49.3	22.9
1997	1,747	285	693	1,407	1,758	4,542	1,444	1,178	214	298	4,616	9,158	49.2	49.6	24.4
1998	2,132	290	693	1,491	1,700	4,880	1,451	1,116	239	305	4,884	9,764	51.6	50.0	25.8
1999	2,459	259	655	1,478	1,480	4,934	1,421	1,063	356	284	4,978	9,912	50.8	49.8	25.3
2000	2,483	225	896	1,571	1,530	5,181	1,697	1,015	356	297	5,238	10,419	52.9	49.7	26.3
2001	2,758	278	884	1,662	1,540	5,510	1,717	1,166	311	268	5,390	10,900	55.5	50.5	28.0
2002	2,265	264	620	1,551	1,387	4,589	1,864	1,292	467	224	5,958	10,546	53.4	43.5	23.2
2003	2,497	381	866	1,774	1,364	5,144	1,932	1,395	434	279	6,094	11,238	56.1	45.8	25.7
2004	2,489	452	1,139	1,557	1,548	5,688	1,980	1,456	366	321	6,409	12,097	58.4	47.0	27.4
2005	2,330	478	1,165	1,536	1,515	5,567	2,001	1,394	375	317	6,982	12,549	60.3	44.4	26.8
2006	R2,208	R657	R1,111	R1,462	R1,392	R5,480	R2,194	R1,450	R244	R318	R6,910	R12,390	R59.9	R44.2	26.5
2007P	2,166	663	1,131	1,487	1,336	5,946	2,243	1,258	272	336	6,094	12,040	58.2	49.4	28.7

[1] See "Organization of the Petroleum Exporting Countries (OPEC)" in Glossary.
[2] Bahrain, Iran, Iraq, Kuwait, Qatar, Saudi Arabia, United Arab Emirates, and the Neutral Zone (between Kuwait and Saudi Arabia).
[3] Through 1970, includes half the imports from the Neutral Zone. Beginning in 1971, includes imports from the Neutral Zone that are reported to U.S. Customs as originating in Saudi Arabia.
[4] On this table, "Total OPEC" for all years includes Iran, Iraq, Kuwait, Saudi Arabia, Venezuela, and the Neutral Zone (between Kuwait and Saudi Arabia); beginning in 1961, also includes Qatar; beginning in 1962, also includes Indonesia and Libya; beginning in 1967, also includes United Arab Emirates; beginning in 1969, also includes Algeria; beginning in 1971, also includes Nigeria; for 1973-1992, also includes Ecuador (although Ecuador rejoined OPEC in November 2007, on this table Ecuador is included in "Total Non-OPEC" for 2007); for 1975-1994, also includes Gabon; and beginning in 2007, also includes Angola. Data for all countries not included in "Total OPEC" are included in "Total Non-OPEC."
[5] Calculated by dividing total net petroleum imports by total U.S. petroleum products supplied (consumption).
[6] Calculated by dividing net petroleum imports from OPEC countries by total net petroleum imports.
[7] Calculated by dividing net petroleum imports from OPEC countries by total U.S. petroleum product supplied (consumption).
[8] Algeria joined OPEC in 1969. For 1960-1968, Algeria is included in "Total Non-OPEC."
[9] Nigeria joined OPEC in 1971. For 1960-1970, Nigeria is included in "Total Non-OPEC."
R=Revised. P=Preliminary. NA=Not available.

Notes: • The country of origin for refined petroleum products may not be the country of origin for the crude oil from which the refined products were produced. For example, refined products imported from refineries in the Caribbean may have been produced from Middle East crude oil. • Net imports equal imports minus exports. Minus sign indicates exports are greater than imports. • Data include any imports for the Strategic Petroleum Reserve, which began in 1977. • Totals may not equal sum of components due to independent rounding.

Web Page: See http://www.eia.doe.gov/oil_gas/petroleum/info_glance/petroleum.html for related information.

Sources: • 1960-1975—Bureau of Mines, *Minerals Yearbook*, "Crude Petroleum and Petroleum Products" chapter. • 1976-1980—Energy Information Administration (EIA), Energy Data Reports, *P.A.D. Districts Supply/Demand, Annual*, annual reports. • 1981-2006—EIA, *Petroleum Supply Annual*, annual reports. • 2007—EIA, *Petroleum Supply Monthly* (February 2008).

Figure 5.8 Refinery and Blender Net Inputs and Net Production, 1949-2007

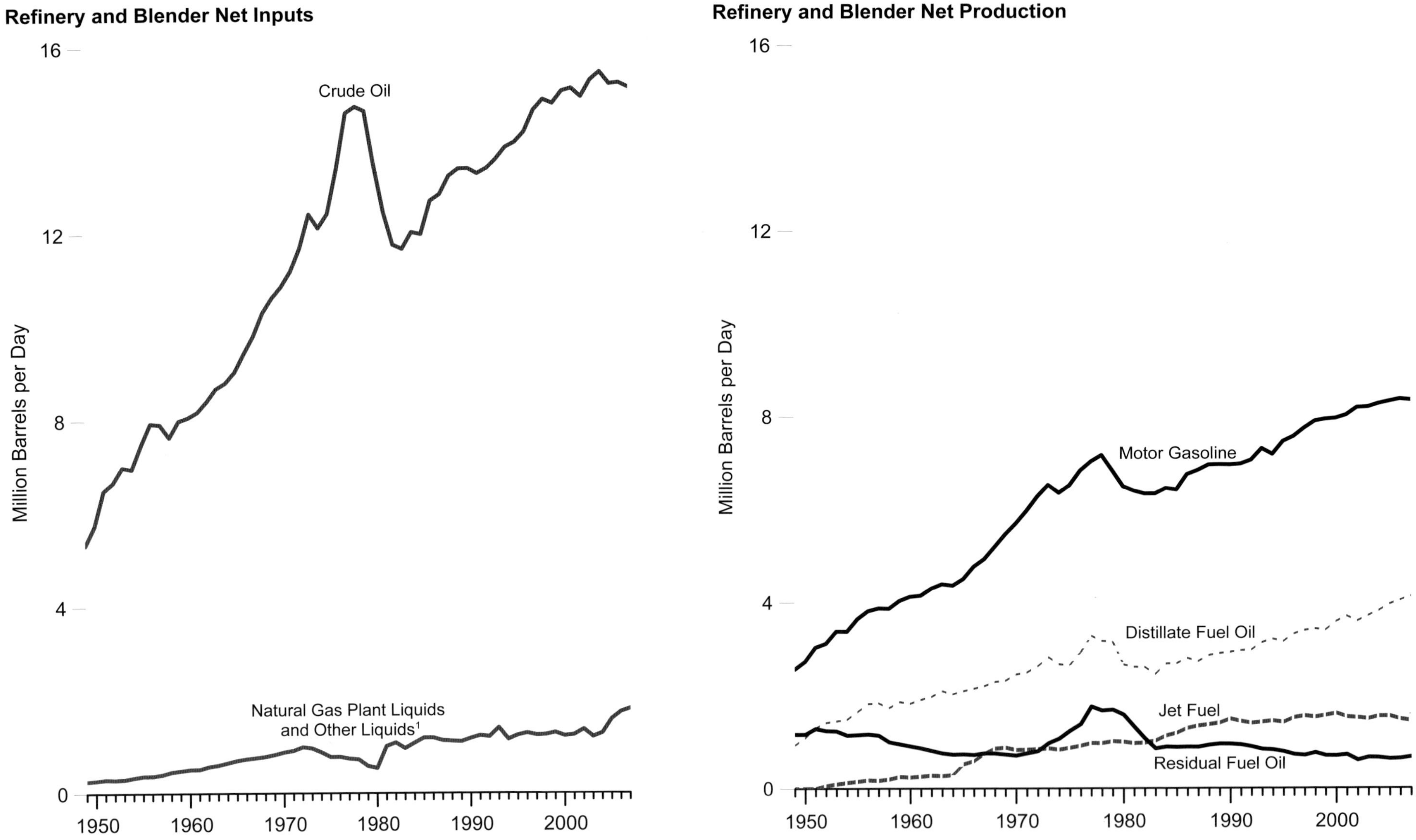

[1] See Table 5.8, footnote 4.

Source: Table 5.8.

Table 5.8 Refinery and Blender Net Inputs and Net Production, Selected Years, 1949-2007

(Thousand Barrels per Day)

Year	Refinery and Blender Net Inputs [1]				Refinery and Blender Net Production [2]										Processing Gain
	Crude Oil [3]	Natural Gas Plant Liquids	Other Liquids [4]	Total	Asphalt and Road Oil	Distillate Fuel Oil	Jet Fuel [5]	Liquefied Petroleum Gases	Motor Gasoline [6]	Petroleum Coke	Residual Fuel Oil	Still Gas	Other Products [7]	Total	
1949	5,327	234	28	5,588	155	934	([5])	64	2,572	46	1,164	226	425	5,587	-2
1950	5,739	259	19	6,018	179	1,093	([5])	80	2,735	47	1,165	229	492	6,019	2
1955	7,480	345	32	7,857	251	1,651	155	119	3,648	78	1,152	319	518	7,891	34
1960	8,067	455	61	8,583	286	1,823	241	212	4,126	164	908	354	616	8,729	146
1965	9,043	618	88	9,750	357	2,096	523	293	4,507	236	736	395	827	9,970	220
1970	10,870	763	121	11,754	428	2,454	827	345	5,699	296	706	483	876	12,113	359
1971	11,199	781	136	12,116	454	2,495	835	357	5,970	299	753	474	861	12,498	382
1972	11,696	826	168	12,691	446	2,630	847	356	6,281	327	799	507	886	13,080	388
1973	12,431	815	155	13,401	480	2,820	859	375	6,527	362	971	518	940	13,854	453
1974	12,133	746	138	13,018	470	2,668	836	338	6,358	339	1,070	521	900	13,498	480
1975	12,442	710	72	13,225	408	2,653	871	311	6,518	354	1,235	523	811	13,685	460
1976	13,416	725	59	14,200	391	2,924	918	340	6,838	356	1,377	541	993	14,677	477
1977	14,602	673	74	15,349	431	3,277	973	352	7,031	369	1,754	572	1,114	15,874	524
1978	14,739	639	92	15,470	482	3,167	970	355	7,167	369	1,667	603	1,186	15,966	496
1979	14,648	510	78	15,236	467	3,152	1,012	340	6,837	376	1,687	598	1,296	15,763	527
1980	13,481	462	81	14,025	393	2,661	999	330	6,492	370	1,580	581	1,215	14,622	597
1981	12,470	524	488	13,482	340	2,613	968	315	6,400	390	1,321	565	1,078	13,990	508
1982	11,774	515	572	12,861	329	2,606	978	270	6,336	410	1,070	554	839	13,391	531
1983	11,685	460	505	12,650	372	2,456	1,022	328	6,338	420	852	550	801	13,138	488
1984	12,044	500	581	13,126	386	2,680	1,132	363	6,453	439	891	559	776	13,679	553
1985	12,002	509	681	13,192	401	2,686	1,189	391	6,419	455	882	584	743	13,750	557
1986	12,716	479	711	13,906	410	2,796	1,293	417	6,752	506	889	641	818	14,522	616
1987	12,854	466	667	13,987	434	2,729	1,343	449	6,841	512	885	643	791	14,626	639
1988	13,246	511	610	14,367	443	2,857	1,370	499	6,956	544	926	670	758	15,022	655
1989	13,401	499	613	14,513	424	2,899	1,403	554	6,963	542	954	681	755	15,175	661
1990	13,409	467	713	14,589	449	2,925	1,488	499	6,959	552	950	673	778	15,272	683
1991	13,301	472	768	14,541	430	2,962	1,438	536	6,975	568	934	651	761	15,256	715
1992	13,411	469	745	14,626	419	2,974	1,399	607	7,058	596	892	659	796	15,398	772
1993	13,613	491	917	15,021	451	3,132	1,422	592	7,304	619	835	653	780	15,787	766
1994	13,866	465	691	15,023	451	3,205	1,448	611	7,181	622	826	657	790	15,791	768
1995	13,973	471	775	15,220	467	3,155	1,416	654	7,459	630	788	647	778	15,994	774
1996	14,195	450	843	15,487	459	3,316	1,515	662	7,565	664	726	654	764	16,324	837
1997	14,662	416	832	15,909	485	3,392	1,554	691	7,743	689	708	661	836	16,759	850
1998	14,889	403	853	16,144	498	3,424	1,526	674	7,892	712	762	656	886	17,030	886
1999	14,804	372	927	16,103	505	3,399	1,565	684	7,934	713	698	656	835	16,989	886
2000	15,067	380	849	16,295	525	3,580	1,606	705	7,951	727	696	659	793	17,243	948
2001	15,128	429	825	16,382	485	3,695	1,530	667	8,022	767	721	670	729	17,285	903
2002	14,947	429	941	16,316	492	3,592	1,514	671	8,183	781	601	667	771	17,273	957
2003	15,304	419	791	16,513	496	3,707	1,488	658	8,194	798	660	702	784	17,487	974
2004	15,475	422	866	16,762	508	3,814	1,547	645	8,265	836	655	704	838	17,814	1,051
2005	15,220	441	1,149	16,811	512	3,954	1,546	573	8,318	835	628	684	752	17,800	989
2006	R15,242	R501	R1,238	R16,981	506	R4,040	1,481	R627	R8,364	R848	R635	709	R764	R17,975	R994
2007P	15,148	497	1,312	16,957	455	4,131	1,448	647	8,344	824	670	693	752	17,963	1,005

[1] See "Refinery and Blender Net Inputs" in Glossary.
[2] See "Refinery and Blender Net Production" in Glossary.
[3] Includes lease condensate.
[4] Unfinished oils (net), other hydrocarbons, and hydrogen. Beginning in 1981, also includes aviation and motor gasoline blending components (net). Beginning in 1993, also includes oxygenates (net). See Note 2, "Adjustment to Total Petroleum Products Supplied," at end of section.
[5] Through 1951, naphtha-type jet fuel is included in the products from which it was blended: in 1952, 71 percent gasoline, 17 percent kerosene, and 12 percent distillate fuel oil. Through 1964, kerosene-type jet fuel is included with kerosene in "Other Products." Beginning in 2005, naphtha-type jet fuel is included in "Other Products."
[6] Finished motor gasoline. Through 1963, also includes aviation gasoline and special naphthas. Beginning in 1993, also includes ethanol blended into motor gasoline.
[7] Kerosene, lubricants, petrochemical feedstocks, waxes, and miscellaneous products. Through 1964, also includes kerosene-type jet fuel. Beginning in 1964, also includes aviation gasoline and special naphthas. Beginning in 2005, also includes naphtha-type jet fuel.

R=Revised. P=Preliminary.

Note: Totals may not equal sum of components due to independent rounding.

Web Pages: • For all data beginning in 1949, see http://www.eia.doe.gov/emeu/aer/petro.html. • For related information, see http://www.eia.doe.gov/oil_gas/petroleum/info_glance/petroleum.html.

Sources: • 1949-1975—Bureau of Mines, Mineral Industry Surveys, *Petroleum Statement, Annual,* annual reports. • 1976-1980—Energy Information Administration (EIA), Energy Data Reports, *Petroleum Statement, Annual,* annual reports. • 1981-2006—EIA, *Petroleum Supply Annual,* annual reports. • 2007—EIA, *Petroleum Supply Monthly* (February 2008).

Figure 5.9 Refinery Capacity and Utilization, 1949-2007

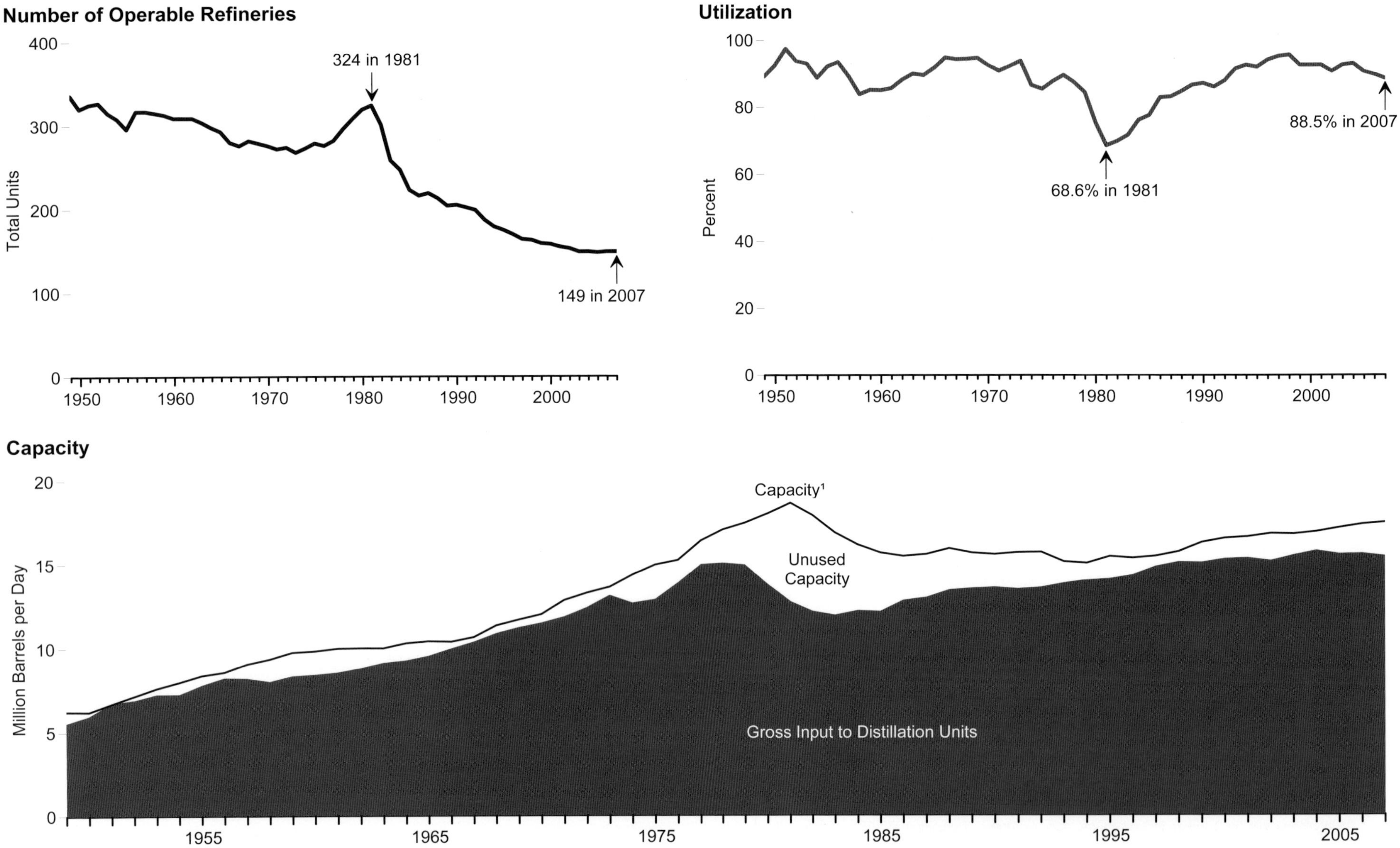

[1] Operable refineries capacity on January 1.

Source: Table 5.9.

Table 5.9 Refinery Capacity and Utilization, Selected Years, 1949-2007

Year	Operable Refineries [1]	Operable Refineries Capacity: On January 1	Operable Refineries Capacity: Annual Average [2]	Gross Input to Distillation Units [3]	Utilization [4]
	Number	Thousand Barrels per Day	Thousand Barrels per Day	Thousand Barrels per Day	Percent
1949	336	6,231	NA	5,556	89.2
1950	320	6,223	NA	5,980	92.5
1955	296	8,386	NA	7,820	92.2
1960	309	9,843	NA	8,439	85.1
1965	293	10,420	NA	9,557	91.8
1970	276	12,021	NA	11,517	92.6
1971	272	12,860	NA	11,881	90.9
1972	274	13,292	NA	12,431	92.3
1973	268	13,642	NA	13,151	93.9
1974	273	14,362	NA	12,689	86.6
1975	279	14,961	NA	12,902	85.5
1976	276	15,237	NA	13,884	87.8
1977	282	16,398	NA	14,982	89.6
1978	296	17,048	NA	15,071	87.4
1979	308	17,441	NA	14,955	84.4
1980	319	17,988	NA	13,796	75.4
1981	324	18,621	18,603	12,752	68.6
1982	301	17,890	17,432	12,172	69.9
1983	258	16,859	16,668	11,947	71.7
1984	247	16,137	16,035	12,216	76.2
1985	223	15,659	15,671	12,165	77.6
1986	216	15,459	15,459	12,826	82.9
1987	219	15,566	15,642	13,003	83.1
1988	213	15,915	15,927	13,447	84.7
1989	204	15,655	15,701	13,551	86.6
1990	205	15,572	15,623	13,610	87.1
1991	202	15,676	15,707	13,508	86.0
1992	199	15,696	15,460	13,600	87.9
1993	187	15,121	15,143	13,851	91.5
1994	179	15,034	15,150	14,032	92.6
1995	175	15,434	15,346	14,119	92.0
1996	170	15,333	15,239	14,337	94.1
1997	164	15,452	15,594	14,838	95.2
1998	163	15,711	15,802	15,113	95.6
1999	159	16,261	16,282	15,080	92.6
2000	158	16,512	16,525	15,299	92.6
2001	155	16,595	16,582	15,352	92.6
2002	153	16,785	16,744	15,180	90.7
2003	149	16,757	16,748	15,508	92.6
2004	149	16,894	16,974	15,783	93.0
2005	148	17,125	17,196	15,578	90.6
2006	149	17,339	R17,385	R15,602	89.7
2007P	149	17,443	17,447	15,449	88.5

[1] Through 1956, includes only those refineries in operation on January 1; beginning in 1957, includes all "operable" refineries on January 1. See "Operable Refineries" in Glossary.
[2] Weighted average of monthly capacity data.
[3] See Note 4, "Gross Input to Distillation Units," at end of section.
[4] Through 1980, utilization is calculated by dividing gross input to distillation units by one-half of the sum of the current year's January 1 capacity and the following year's January 1 capacity. Beginning in 1981, utilization is calculated by dividing gross input to distillation units by the annual average capacity.
R=Revised. P=Preliminary. NA=Not available.
Web Pages: • For all data beginning in 1949, see http://www.eia.doe.gov/emeu/aer/petro.html. • For related information, see http://www.eia.doe.gov/oil_gas/petroleum/info_glance/petroleum.html.
Sources: **Operable Refineries** and **Operable Refineries Capacity:** • 1949-1961—Bureau of Mines Information Circular, "Petroleum Refineries, Including Cracking Plants in the United States." • 1962-1977—Bureau of Mines, Mineral Industry Surveys, *Petroleum Refineries, Annual,* annual reports. • 1978-1981—Energy Information Administration (EIA), Energy Data Reports, *Petroleum Refineries in the United States.* • 1982-2006—EIA, *Petroleum Supply Annual,* annual reports. • 2007—EIA, *Refinery Capacity Report* (June 2007), Table 1. **Gross Input to Distillation Units:** • 1949-1966—Bureau of Mines, *Minerals Yearbook,* "Natural Gas Liquids" and "Crude Petroleum and Petroleum Products" chapters. • 1967-1977—Bureau of Mines, Mineral Industry Surveys, *Petroleum Refineries, Annual,* annual reports. • 1978-1980—EIA, Energy Data Reports, *Petroleum Refineries in the United States and U.S. Territories.* • 1981-2006—EIA, *Petroleum Supply Annual,* annual reports. • 2007—EIA, *Petroleum Supply Monthly* (January-December 2007 issues). **Utilization:** • 1949-1980—Calculated. • 1981-2006—EIA, *Petroleum Supply Annual,* annual reports. • 2007—Calculated.

Figure 5.10 Natural Gas Plant Liquids Production

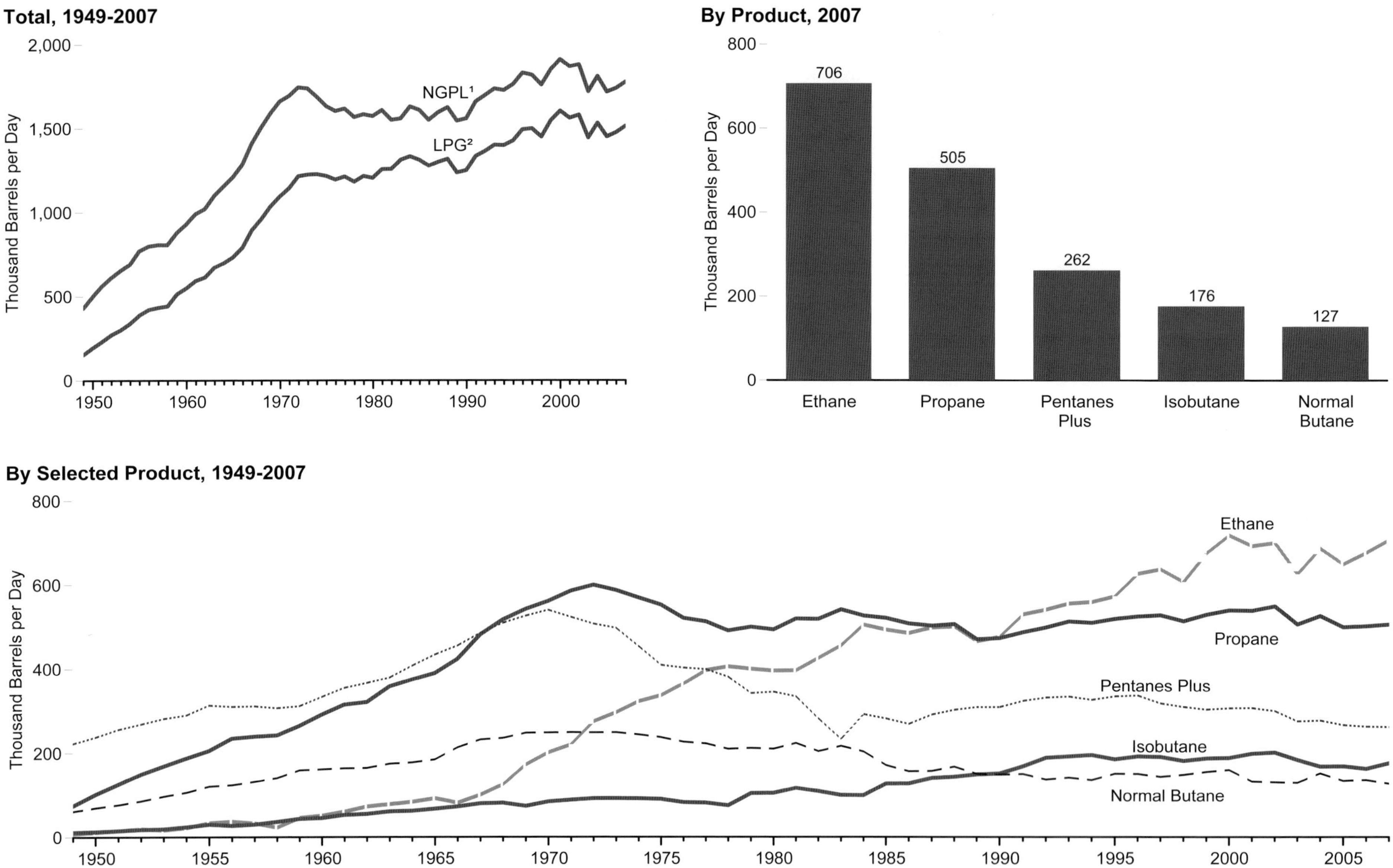

[1] Natural gas plant liquids.
[2] Liquefied petroleum gases.

Note: Because vertical scales differ, graphs should not be compared.
Source: Table 5.10.

Table 5.10 Natural Gas Plant Liquids Production, Selected Years, 1949-2007

(Thousand Barrels per Day)

Year	Finished Petroleum Products [1]	Liquefied Petroleum Gases					Pentanes Plus [4]	Total
		Ethane [2]	Isobutane	Normal Butane [3]	Propane [2,3]	Total		
1949	53	8	11	61	74	155	223	430
1950	66	12	13	69	101	195	238	499
1955	68	34	30	120	205	390	313	771
1960	47	51	45	161	291	549	333	929
1965	41	92	67	185	390	734	434	1,210
1970	25	201	84	248	561	1,095	540	1,660
1971	25	221	88	249	586	1,144	523	1,693
1972	21	275	92	249	600	1,215	507	1,744
1973	16	296	92	249	587	1,225	497	1,738
1974	7	323	92	244	569	1,227	454	1,688
1975	7	337	90	237	552	1,217	409	1,633
1976	6	365	82	227	521	1,195	403	1,604
1977	5	397	81	223	513	1,214	399	1,618
1978	3	406	75	210	491	1,182	382	1,567
1979	26	400	104	212	500	1,216	342	1,584
1980	23	396	105	210	494	1,205	345	1,573
1981	18	397	117	224	519	1,256	334	1,609
1982	11	426	109	204	519	1,258	282	1,550
1983	12	456	100	217	541	1,314	233	1,559
1984	4	505	99	203	527	1,334	292	1,630
1985	14	493	127	171	521	1,313	282	1,609
1986	4	485	128	157	508	1,277	269	1,551
1987	4	499	141	157	503	1,300	291	1,595
1988	4	501	144	167	506	1,319	302	1,625
1989	NA	466	149	151	471	1,237	309	1,546
1990	NA	477	151	149	474	1,250	309	1,559
1991	NA	530	169	150	487	1,336	324	1,659
1992	NA	541	189	137	499	1,365	332	1,697
1993	NA	556	192	142	513	1,402	334	1,736
1994	NA	559	195	136	510	1,400	326	1,727
1995	NA	573	185	151	519	1,428	335	1,762
1996	NA	627	192	150	525	1,494	336	1,830
1997	NA	637	191	144	528	1,499	318	1,817
1998	NA	607	181	148	513	1,450	309	1,759
1999	NA	675	187	155	529	1,547	303	1,850
2000	NA	717	188	160	539	1,605	306	1,911
2001	NA	692	198	133	538	1,562	307	1,868
2002	NA	700	201	131	549	1,581	300	1,880
2003	NA	625	183	129	506	1,444	275	1,719
2004	NA	686	168	152	526	1,532	277	1,809
2005	NA	649	168	134	499	1,451	266	1,717
2006	NA	R676	163	R136	R501	R1,476	R263	R1,739
2007P	NA	706	176	127	505	1,515	262	1,776

[1] Motor gasoline, aviation gasoline, special naphthas, distillate fuel oil, and miscellaneous products.
[2] Reported production of ethane-propane mixtures has been allocated 70 percent ethane and 30 percent propane.
[3] Reported production of butane-propane mixtures has been allocated 60 percent butane and 40 percent propane.
[4] Through 1983, "Pentanes Plus" was reported separately as natural gasoline, isopentane, and plant condensate.
R=Revised. P=Preliminary. NA=Not available.

Note: Totals may not equal sum of components due to independent rounding.

Web Pages: • For all data beginning in 1949, see http://www.eia.doe.gov/emeu/aer/petro.html. • For related information, see http://www.eia.doe.gov/oil_gas/petroleum/info_glance/petroleum.html.

Sources: • 1949-1968—Bureau of Mines, *Minerals Yearbook,* "Crude Petroleum and Petroleum Products" chapter. • 1969-1975—Bureau of Mines, Mineral Industry Surveys, *Petroleum Statement, Annual,* annual reports. • 1976-1980—Energy Information Administration (EIA), Energy Data Reports, *Petroleum Statement, Annual,* annual reports. • 1981-2006—EIA, *Petroleum Supply Annual,* annual reports. • 2007—EIA, *Petroleum Supply Monthly* (February 2008).

Figure 5.11 Petroleum Products Supplied by Type

By Selected Product, 1949-2007

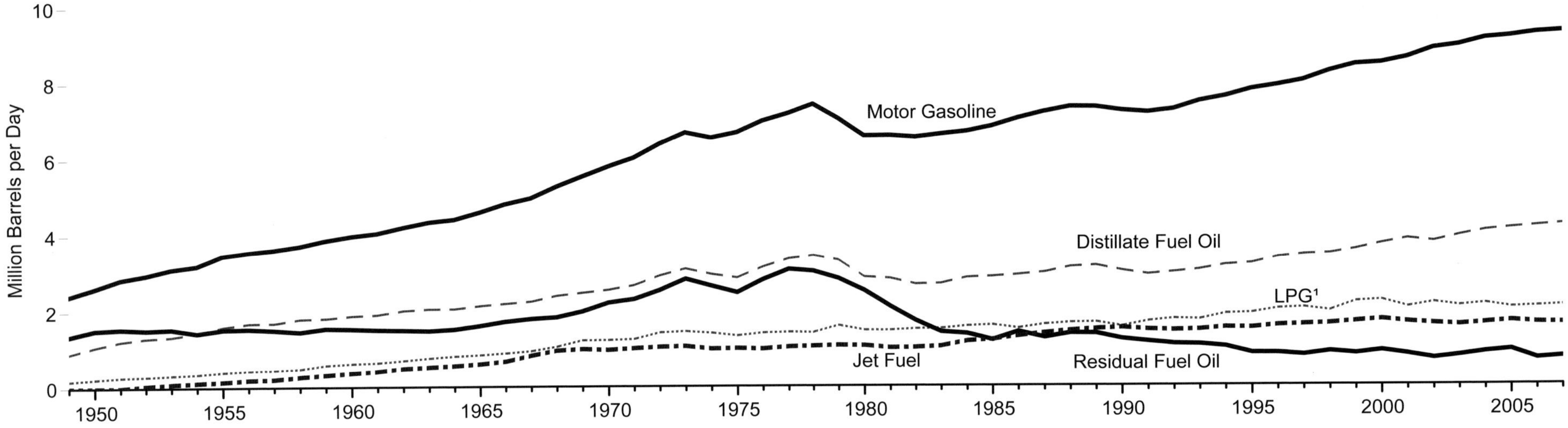

By Product, 2007

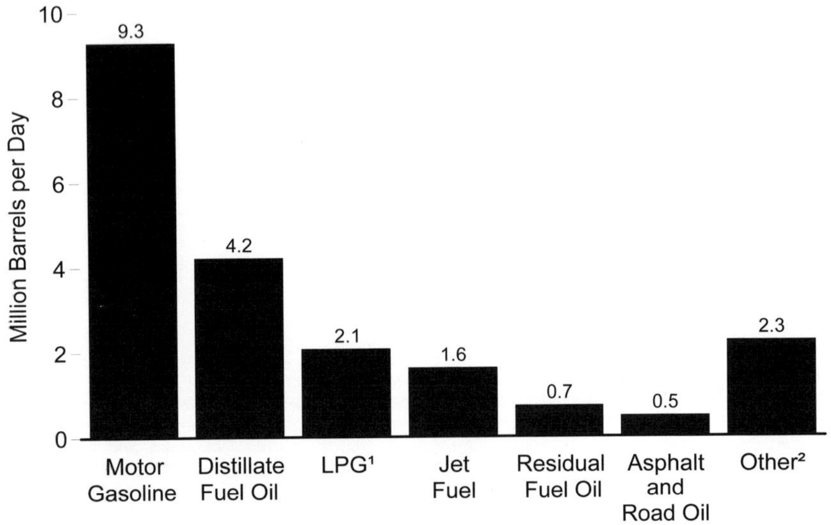

Motor Gasoline's Share of Total Petroleum Products Supplied, 1949-2007

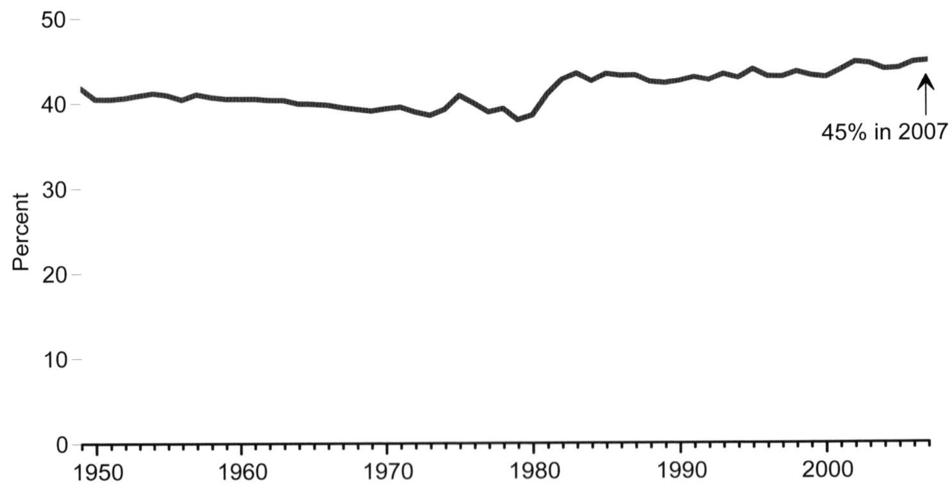

[1] Liquefied petroleum gases.

[2] Aviation gasoline, kerosene, lubricants, naphtha-type jet fuel, pentanes plus, petrochemical-feedstocks, petroleum coke, special naphthas, still gas (refinery gas), waxes, miscellaneous products, and crude oil burned as fuel.

Source: 5.11.

Table 5.11 Petroleum Products Supplied by Type, Selected Years, 1949-2007

(Thousand Barrels per Day)

Year	Asphalt and Road Oil	Aviation Gasoline	Distillate Fuel Oil	Jet Fuel [1]	Kerosene	Liquefied Petroleum Gases		Lubricants	Motor Gasoline [3]	Petroleum Coke	Residual Fuel Oil	Other [4]	Total	Percentage Change From Previous Year [5]
						Propane [2]	Total							
1949	157	93	902	([1])	281	NA	187	91	2,410	40	1,359	243	5,763	– –
1950	180	108	1,082	([1])	323	NA	234	106	2,616	41	1,517	250	6,458	12.1
1955	254	192	1,592	154	320	NA	404	116	3,463	67	1,526	366	8,455	9.0
1960	302	161	1,872	371	271	NA	621	117	3,969	149	1,529	435	9,797	3.1
1965	368	120	2,126	602	267	NA	841	129	4,593	202	1,608	657	11,512	4.2
1970	447	55	2,540	967	263	776	1,224	136	5,785	212	2,204	866	14,697	4.0
1971	458	49	2,661	1,010	249	794	1,251	135	6,014	219	2,296	870	15,212	3.5
1972	468	46	2,913	1,045	235	893	1,420	144	6,376	241	2,529	949	16,367	7.9
1973	522	45	3,092	1,059	216	872	1,449	162	6,674	261	2,822	1,005	17,308	5.5
1974	481	44	2,948	993	176	830	1,406	155	6,537	239	2,639	1,034	16,653	-3.8
1975	419	39	2,851	1,001	159	783	1,333	137	6,675	247	2,462	1,001	16,322	-2.0
1976	411	37	3,133	987	169	830	1,404	152	6,978	243	2,801	1,145	17,461	7.3
1977	436	38	3,352	1,039	175	821	1,422	160	7,177	268	3,071	1,294	18,431	5.3
1978	479	39	3,432	1,057	175	778	1,413	172	7,412	256	3,023	1,391	18,847	2.3
1979	476	38	3,311	1,076	188	849	1,592	180	7,034	246	2,826	1,546	18,513	-1.8
1980	396	35	2,866	1,068	158	754	1,469	159	6,579	237	2,508	1,581	17,056	-7.6
1981	342	31	2,829	1,007	127	773	1,466	153	6,588	252	2,088	1,176	16,058	-6.1
1982	342	25	2,671	1,013	129	798	1,499	140	6,539	248	1,716	973	15,296	-4.7
1983	373	26	2,690	1,046	127	751	1,509	146	6,622	229	1,421	1,042	15,231	-.4
1984	408	24	2,845	1,175	115	833	1,572	156	6,693	247	1,369	1,120	15,726	3.5
1985	425	27	2,868	1,218	114	883	1,599	145	6,831	264	1,202	1,032	15,726	-.3
1986	448	32	2,914	1,307	98	831	1,512	142	7,034	268	1,418	1,105	16,281	3.5
1987	467	25	2,976	1,385	95	924	1,612	161	7,206	299	1,264	1,176	16,665	2.4
1988	468	27	3,122	1,449	96	923	1,656	155	7,336	312	1,378	1,286	17,283	4.0
1989	453	26	3,157	1,489	84	990	1,668	159	7,328	307	1,370	1,284	17,325	(s)
1990	483	24	3,021	1,522	43	917	1,556	164	7,235	339	1,229	1,373	16,988	-1.9
1991	444	23	2,921	1,471	46	982	1,689	146	7,188	328	1,158	1,299	16,714	-1.6
1992	454	22	2,979	1,454	41	1,032	1,755	149	7,268	382	1,094	1,434	17,033	2.2
1993	474	21	3,041	1,469	50	1,006	1,734	152	7,476	366	1,080	1,373	17,237	.9
1994	484	21	3,162	1,527	49	1,082	1,880	159	7,601	361	1,021	1,454	17,718	2.8
1995	486	21	3,207	1,514	54	1,096	1,899	156	7,789	365	852	1,381	17,725	(s)
1996	484	20	3,365	1,578	62	1,136	2,012	151	7,891	379	848	1,518	18,309	3.6
1997	505	22	3,435	1,599	66	1,170	2,038	160	8,017	377	797	1,605	18,620	1.4
1998	521	19	3,461	1,622	78	1,120	1,952	168	8,253	447	887	1,508	18,917	1.6
1999	547	21	3,572	1,673	73	1,246	2,195	169	8,431	477	830	1,532	19,519	3.2
2000	525	20	3,722	1,725	67	1,235	2,231	166	8,472	406	909	1,458	19,701	1.2
2001	519	19	3,847	1,655	72	1,142	2,044	153	8,610	437	811	1,481	19,649	-.5
2002	512	18	3,776	1,614	43	1,248	2,163	151	8,848	463	700	1,474	19,761	.6
2003	503	16	3,927	1,578	55	1,215	2,074	140	8,935	455	772	1,579	20,034	1.4
2004	537	17	4,058	1,630	64	1,276	2,132	141	9,105	524	865	1,657	20,731	3.8
2005	546	19	4,118	1,679	70	1,229	2,030	141	9,159	515	920	1,605	20,802	.1
2006	R521	18	R4,169	R1,633	54	R1,215	R2,052	R137	R9,253	R522	R689	R1,640	R20,687	R-.6
2007P	490	17	4,220	1,623	31	1,231	2,081	135	9,290	490	732	1,587	20,698	(s)

[1] Through 1951, naphtha-type jet fuel is included in the products from which it was blended: in 1952, 71 percent gasoline, 17 percent kerosene, and 12 percent distillate fuel oil. Beginning in 1952, includes naphtha-type jet fuel. Beginning in 1957, also includes kerosene-type jet fuel. Beginning in 2005, naphtha-type jet fuel is included in "Other."

[2] Includes propylene.

[3] Finished motor gasoline. Through 1963, also includes special naphthas. Beginning in 1993, also includes ethanol blended into motor gasoline.

[4] Pentanes plus, petrochemical feedstocks, still gas (refinery gas), waxes, and miscellaneous products. Beginning in 1964, also includes special naphthas. Beginning in 1981, also includes negative barrels per day of distillate and residual fuel oil reclassified as unfinished oils, and other products (from both primary and secondary supply) reclassified as gasoline blending components. Beginning in 1983, also includes crude oil burned as fuel. Beginning in 2005, also includes naphtha-type jet fuel.

[5] Percent change from previous year calculated from data in thousand barrels per year.

R=Revised. P=Preliminary. NA=Not available. – – = Not applicable. (s)=Less than 0.05 percent and greater than -0.05 percent.

Notes: • Petroleum products supplied is an approximation of petroleum consumption and is synonymous with the term "petroleum consumption" in Tables 5.13a-d and 5.14a-c. • See Note 1, "Petroleum Products Supplied and Petroleum Consumption," Note 2, "Adjustment to Total Petroleum Products Supplied," and Note 3, "Changes Affecting Petroleum Production and Product Supplied Statistics," at end of section. • Totals may not equal sum of components due to independent rounding.

Web Pages: • For all data beginning in 1949, see http://www.eia.doe.gov/emeu/aer/petro.html. • For related information, see http://www.eia.doe.gov/oil_gas/petroleum/info_glance/petroleum.html.

Sources: • 1949-1975—Bureau of Mines, Mineral Industry Surveys, *Petroleum Statement, Annual,* annual reports. • 1976-1980—Energy Information Administration (EIA), Energy Data Reports, *Petroleum Statement, Annual,* annual reports. • 1981-2006—EIA, *Petroleum Supply Annual,* annual reports. • 2007—EIA, *Petroleum Supply Monthly* (February 2008).

Figure 5.12 Heat Content of Petroleum Products Supplied

Petroleum Products Supplied as Share of Total Energy Consumption, 1949-2007

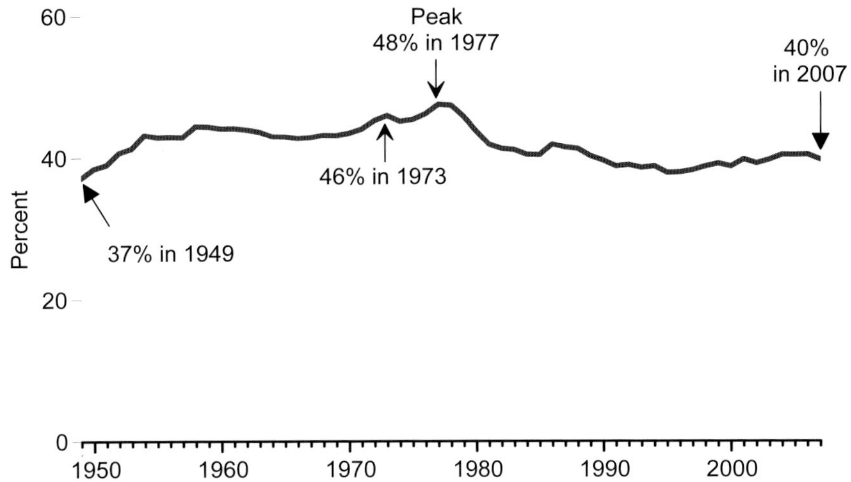

Total Petroleum and Motor Gasoline Product Supplied, 1949-2007

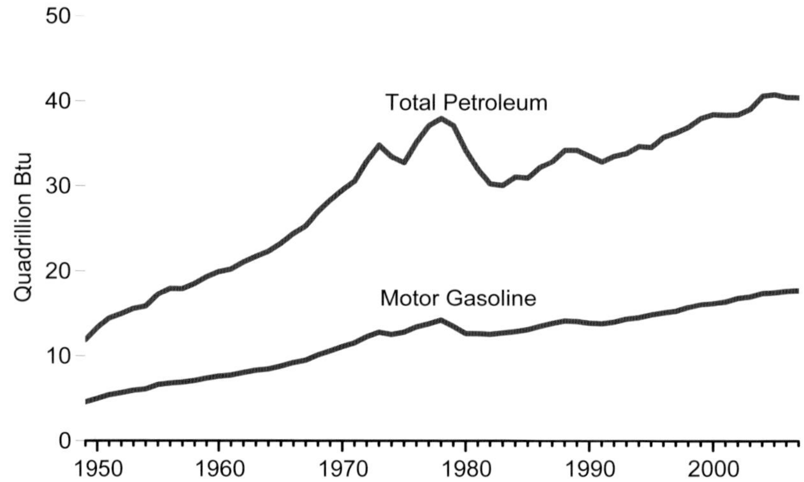

By Product, 2007

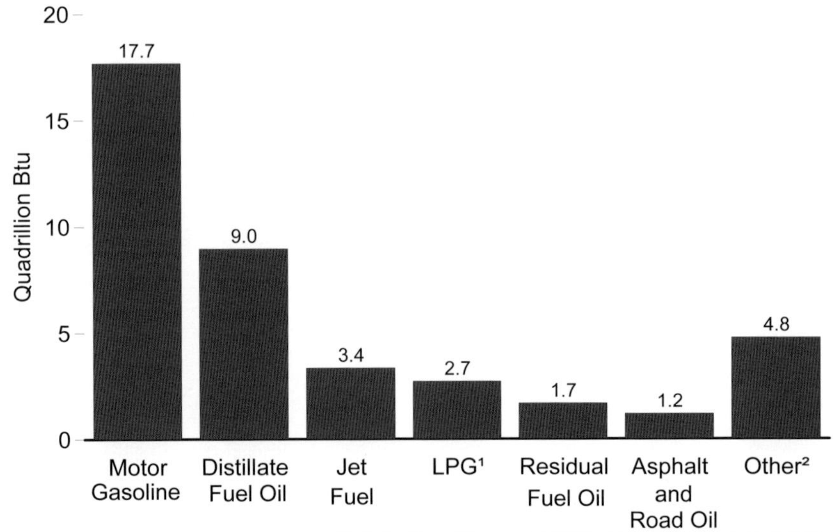

By Selected Product, 1949-2007

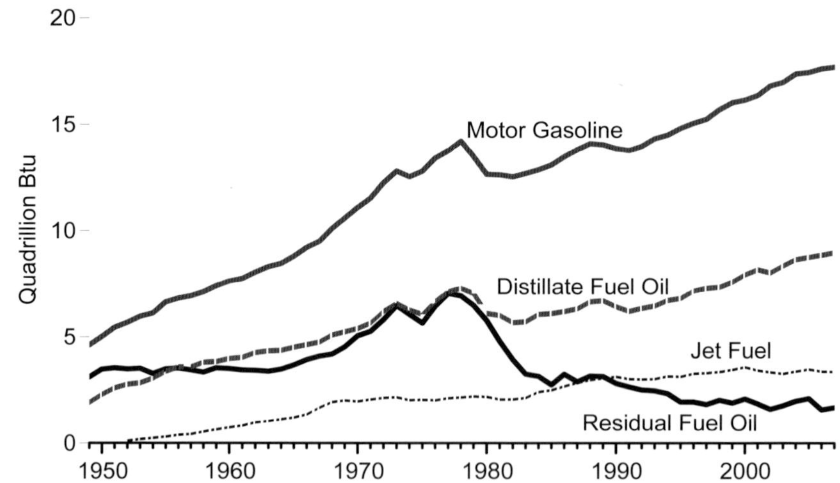

[1] Liquefied petroleum gases.

[2] Aviation gasoline, kerosene, lubricants, naphtha-type jet fuel, pentanes plus, petrochemical feedstocks, petroleum coke, special naphthas, still gas (refinery gas), waxes, miscellaneous products, and crude burned as fuel.

Sources: Tables 1.3 and 5.12.

Table 5.12 Heat Content of Petroleum Products Supplied, Selected Years, 1949-2007

(Trillion Btu)

Year	Asphalt and Road Oil	Aviation Gasoline	Distillate Fuel Oil	Jet Fuel [1]	Kerosene	Liquefied Petroleum Gases: Propane [2]	Liquefied Petroleum Gases: Total	Lubricants	Motor Gasoline [3]	Petroleum Coke	Residual Fuel Oil	Other [4]	Total	Percentage Change From Previous Year
1949	380	172	1,918	([1])	582	NA	274	201	4,621	87	3,118	530	11,883	– –
1950	435	199	2,300	([1])	668	NA	343	236	5,015	90	3,482	546	13,315	12.1
1955	615	354	3,385	301	662	NA	592	258	6,640	147	3,502	798	17,255	8.9
1960	734	298	3,992	739	563	NA	912	259	7,631	328	3,517	947	19,919	3.1
1965	890	222	4,519	1,215	553	NA	1,232	286	8,806	444	3,691	1,390	23,246	4.2
1970	1,082	100	5,401	1,973	544	1,086	1,689	301	11,091	465	5,057	1,817	29,521	4.2
1971	1,108	90	5,658	2,061	515	1,111	1,723	299	11,532	481	5,269	1,825	30,561	3.5
1972	1,137	85	6,210	2,141	487	1,254	1,955	320	12,259	532	5,820	2,001	32,947	7.8
1973	1,264	83	6,575	2,167	447	1,221	1,981	359	12,797	573	6,477	2,117	34,840	5.7
1974	1,165	82	6,267	2,030	365	1,163	1,914	344	12,535	524	6,056	2,173	33,455	-4.0
1975	1,014	71	6,061	2,047	329	1,097	1,807	304	12,798	542	5,649	2,107	32,731	-2.2
1976	998	67	6,679	2,026	351	1,166	1,907	338	13,415	537	6,445	2,410	35,175	7.5
1977	1,056	70	7,126	2,126	363	1,150	1,908	354	13,760	589	7,047	2,722	37,122	5.5
1978	1,160	71	7,296	2,164	363	1,089	1,892	380	14,211	562	6,936	2,930	37,965	2.3
1979	1,153	70	7,039	2,204	389	1,189	2,138	397	13,487	541	6,485	3,219	37,123	-2.2
1980	962	64	6,110	2,190	329	1,059	1,976	354	12,648	522	5,772	3,275	34,202	-7.9
1981	828	56	6,014	2,062	263	1,082	1,949	339	12,631	553	4,791	2,445	31,931	-6.6
1982	829	47	5,679	2,072	266	1,117	1,978	309	12,538	545	3,939	2,029	30,232	-5.3
1983	904	48	5,720	2,141	263	1,051	1,990	324	12,697	503	3,260	2,204	30,054	-.6
1984	992	44	6,065	2,414	239	1,170	2,071	346	12,867	545	3,151	2,317	31,051	3.3
1985	1,029	50	6,098	2,497	236	1,236	2,103	322	13,098	582	2,759	2,149	30,922	-.4
1986	1,086	59	6,196	2,682	203	1,163	2,009	315	13,487	590	3,255	2,313	32,196	4.1
1987	1,130	46	6,328	2,843	196	1,294	2,153	356	13,816	657	2,901	2,440	32,865	2.1
1988	1,136	49	6,655	2,982	200	1,296	2,213	343	14,105	687	3,170	2,681	34,222	4.1
1989	1,096	48	6,712	3,059	174	1,387	2,243	352	14,050	676	3,144	2,658	34,211	(s)
1990	1,170	45	6,422	3,129	88	1,284	2,059	362	13,872	745	2,820	2,840	33,553	-1.9
1991	1,077	42	6,210	3,025	96	1,374	2,227	324	13,781	722	2,657	2,685	32,845	-2.1
1992	1,102	41	6,351	3,001	86	1,449	2,328	330	13,973	843	2,518	2,953	33,527	2.1
1993	1,149	38	6,466	3,028	103	1,409	2,282	337	14,335	804	2,479	2,821	33,841	.9
1994	1,173	38	6,723	3,154	101	1,515	2,494	352	14,511	793	2,342	2,988	34,670	2.4
1995	1,178	40	6,818	3,132	112	1,534	2,512	346	14,825	802	1,955	2,834	34,553	-.3
1996	1,176	37	7,175	3,274	128	1,594	2,660	335	15,064	837	1,952	3,119	35,757	3.5
1997	1,224	40	7,304	3,308	136	1,638	2,690	354	15,254	829	1,828	3,298	36,266	1.4
1998	1,263	35	7,359	3,357	162	1,568	2,575	371	15,701	982	2,036	3,093	36,934	1.8
1999	1,324	39	7,595	3,462	151	1,745	2,897	375	16,036	1,048	1,905	3,128	37,960	2.8
2000	1,276	36	7,935	3,580	140	1,734	2,945	369	16,155	895	2,091	2,981	38,404	1.2
2001	1,257	35	8,179	3,426	150	1,598	2,697	338	16,373	961	1,861	3,056	38,333	-.2
2002	1,240	34	8,028	3,340	90	1,747	2,852	334	16,819	1,018	1,605	3,041	38,401	.2
2003	1,220	30	8,349	3,265	113	1,701	2,747	309	16,981	1,000	1,772	3,260	39,047	1.7
2004	1,304	31	8,652	3,383	133	1,791	2,824	313	17,379	1,156	1,990	3,429	40,594	4.0
2005	1,323	35	8,755	3,475	144	1,721	2,682	312	17,444	1,133	2,111	3,320	40,735	.3
2006	[R]1,261	[R]33	[R]8,864	[R]3,379	111	[R]1,701	[R]2,701	[R]303	[R]17,622	[R]1,148	[R]1,581	[R]3,416	[R]40,420	[R]-.8
2007[P]	1,188	32	8,973	3,358	64	1,724	2,728	299	17,698	1,078	1,681	3,296	40,394	-.1

[1] Through 1951, naphtha-type jet fuel is included in the products from which it was blended: in 1952, 71 percent gasoline, 17 percent kerosene, and 12 percent distillate fuel oil. Beginning in 1952, includes naphtha-type jet fuel. Beginning in 1957, also includes kerosene-type jet fuel. Beginning in 2005, naphtha-type jet fuel is included in "Other."

[2] Includes propylene.

[3] Finished motor gasoline. Through 1963, also includes special naphthas. Beginning in 1993, also includes ethanol blended into motor gasoline.

[4] Pentanes plus, petrochemical feedstocks, still gas (refinery gas), waxes, and miscellaneous products. Beginning in 1964, also includes special naphthas. Beginning in 1981, also includes negative barrels per day of distillate and residual fuel oil reclassified as unfinished oils, and other products (from both primary and secondary supply) reclassified as gasoline blending components. Beginning in 1983, also includes crude oil burned as fuel. Beginning in 2005, also includes naphtha-type jet fuel.

R=Revised. P=Preliminary. NA=Not available. – – = Not applicable. (s)=Less than 0.05 percent and greater than -0.05 percent.

Notes: • Petroleum products supplied is an approximation of petroleum consumption and is synonymous with the term "petroleum consumption" in Tables 5.13a-d and 5.14a-c. • See Note 1, "Petroleum Products Supplied and Petroleum Consumption," Note 2, "Adjustment to Total Petroleum Products Supplied," and Note 3, "Changes Affecting Petroleum Production and Product Supplied Statistics," at end of section. • Totals may not equal sum of components due to independent rounding.

Web Pages: • For all data beginning in 1949, see http://www.eia.doe.gov/emeu/aer/petro.html. • For related information, see http://www.eia.doe.gov/oil_gas/petroleum/info_glance/petroleum.html

Sources: Tables 5.11, A1, and A3.

Figure 5.13a Estimated Petroleum Consumption by Sector

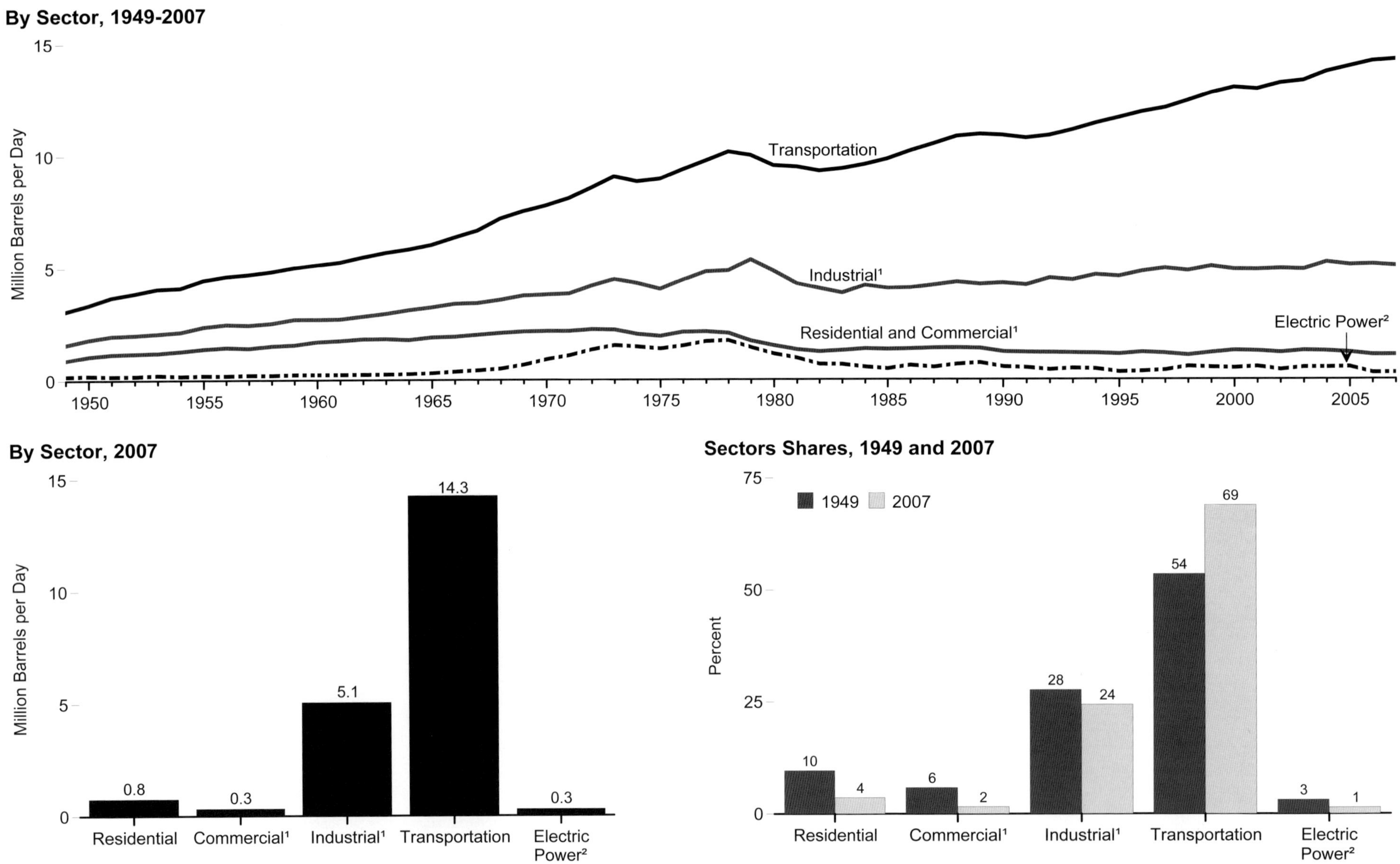

[1] Includes combined-heat-and-power plants and a small number of electricity-only plants.
[2] Electricity-only and combined-heat-and-power plants whose primary business is to sell electricity, or electricity and heat, to the public.

Note: See related Figure 5.13b.
Sources: Tables 5.13a–5.13d.

Figure 5.13b Estimated Petroleum Consumption by Product by Sector, 1949-2007

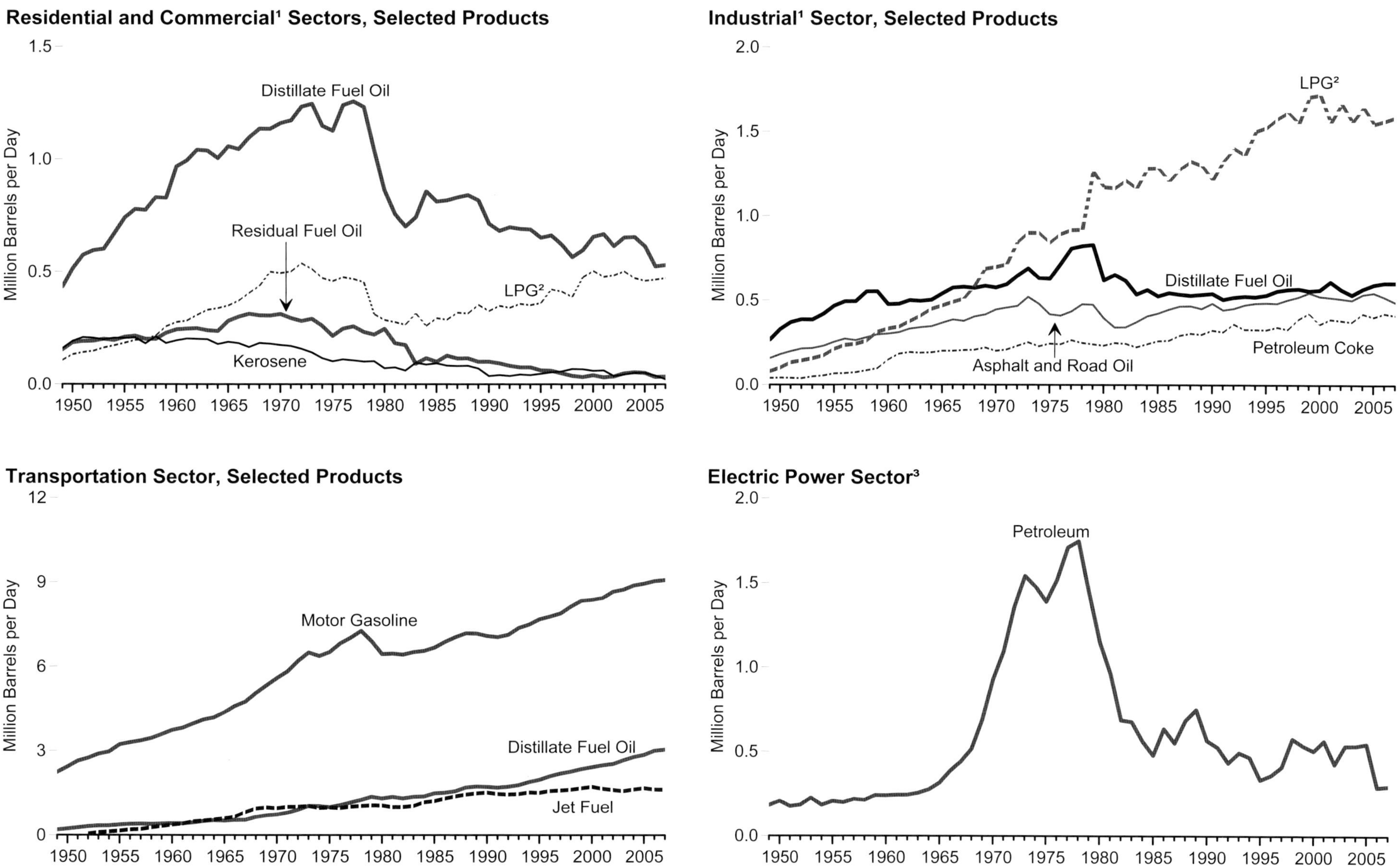

[1] Includes combined-heat-and-power plants and a small number of electricity-only plants.
[2] Liquefied petroleum gases.
[3] Electricity-only and combined-heat-and-power plants whose primary business is to sell electricity, or electricity and heat, to the public.

Notes: • See related Figure 5.13a. • Because vertical scales differ, graphs should not be compared.
Sources: Tables 5.13a–5.13d.

Table 5.13a Estimated Petroleum Consumption: Residential and Commercial Sectors, Selected Years, 1949-2007

(Thousand Barrels per Day)

Year	Residential Sector: Distillate Fuel Oil	Residential Sector: Kerosene	Residential Sector: Liquefied Petroleum Gases	Residential Sector: Total	Commercial Sector: Distillate Fuel Oil: CHP [1]	Commercial Sector: Distillate Fuel Oil: Other [2]	Commercial Sector: Distillate Fuel Oil: Total	Commercial Sector: Kerosene	Commercial Sector: Liquefied Petroleum Gases	Commercial Sector: Motor Gasoline [3]	Commercial Sector: Petroleum Coke	Commercial Sector: Residual Fuel Oil: CHP [1]	Commercial Sector: Residual Fuel Oil: Other [2]	Commercial Sector: Residual Fuel Oil: Total	Commercial Sector: Total
1949	329	140	90	559	(4)	104	104	19	16	48	NA	(4)	153	153	340
1950	390	168	112	670	(4)	123	123	23	20	52	NA	(4)	185	185	403
1955	562	179	155	896	(4)	177	177	24	27	69	NA	(4)	209	209	508
1960	736	171	234	1,140	(4)	232	232	23	41	35	NA	(4)	243	243	573
1965	805	161	296	1,263	(4)	251	251	26	52	40	NA	(4)	281	281	651
1970	883	144	420	1,447	(4)	276	276	30	74	45	NA	(4)	311	311	736
1971	892	143	425	1,460	(4)	280	280	27	75	44	NA	(4)	293	293	718
1972	936	131	456	1,523	(4)	296	296	27	81	45	NA	(4)	280	280	729
1973	942	110	435	1,487	(4)	303	303	31	77	45	NA	(4)	290	290	746
1974	867	89	401	1,357	(4)	280	280	26	71	43	NA	(4)	259	259	679
1975	850	78	389	1,316	(4)	276	276	24	69	46	NA	(4)	214	214	629
1976	932	89	404	1,425	(4)	308	308	21	71	50	NA	(4)	247	247	697
1977	938	81	397	1,416	(4)	318	318	25	70	52	NA	(4)	256	256	722
1978	917	74	386	1,377	(4)	313	313	26	68	56	NA	(4)	232	232	695
1979	765	64	264	1,093	(4)	274	274	38	47	54	NA	(4)	220	220	634
1980	617	51	242	911	(4)	243	243	20	43	56	NA	(4)	245	245	606
1981	540	41	234	815	(4)	215	215	34	41	48	NA	(4)	182	182	519
1982	494	46	224	764	(4)	207	207	15	40	46	NA	(4)	174	174	480
1983	435	41	267	743	(4)	306	306	54	47	53	NA	(4)	91	91	552
1984	512	42	220	774	(4)	345	345	45	39	56	NA	(4)	115	115	600
1985	514	77	249	839	(4)	297	297	16	44	50	NA	(4)	99	99	506
1986	523	59	243	825	(4)	293	293	24	43	55	NA	(4)	126	126	542
1987	544	57	269	870	(4)	286	286	24	48	58	NA	(4)	114	114	529
1988	558	69	267	894	(4)	281	281	13	47	57	NA	(4)	115	115	513
1989	546	57	299	901	3	267	270	13	53	53	0	2	97	99	488
1990	460	31	276	767	3	249	252	6	49	58	0	3	97	100	465
1991	438	35	295	768	2	241	243	6	52	44	0	2	91	92	438
1992	460	31	288	779	1	236	238	5	51	41	(s)	2	80	82	418
1993	458	37	303	797	2	230	232	7	53	15	(s)	2	73	75	383
1994	451	31	298	781	3	233	236	9	53	13	(s)	2	73	75	386
1995	426	36	306	767	2	223	225	11	54	10	(s)	1	61	62	361
1996	434	43	358	835	2	225	227	10	63	14	(s)	1	58	60	373
1997	411	45	349	805	3	206	209	12	62	22	(s)	1	47	48	353
1998	363	52	329	744	2	199	202	15	58	20	(s)	3	35	37	332
1999	389	54	404	847	2	204	206	13	71	15	(s)	2	30	32	338
2000	424	46	427	897	2	228	230	14	75	23	(s)	2	38	40	383
2001	427	46	406	879	3	236	239	15	72	20	(s)	2	28	30	376
2002	404	29	412	845	2	207	209	8	73	24	(s)	1	34	35	348
2003	425	34	426	885	2	225	226	9	75	32	(s)	2	46	48	391
2004	433	41	401	875	3	218	221	10	71	25	(s)	2	51	53	380
2005	402	R40	R391	R833	2	208	210	R10	R69	R24	(s)	2	48	50	365
2006	R335	R32	RP395	RP762	1	R188	R189	R7	RP70	RP24	(s)	1	R31	R33	RP323
2007	P338	P19	P401	P758	P1	P190	P191	P4	P71	P24	P (s)	P1	P33	P34	P325

[1] Commercial combined-heat-and-power (CHP) and a small number of commercial electricity-only plants. See Note 2, "Classification of Power Plants Into Energy-Use Sectors," at end of Section 8.

[2] All commercial sector fuel use other than that in "CHP."

[3] Finished motor gasoline. Through 1963, also includes special naphthas. Beginning in 1993, also includes ethanol blended into motor gasoline.

[4] Included in "Other."

R=Revised. P=Preliminary. NA=Not available. (s)=Less than 500 barrels per day.

Notes: • For total petroleum consumption by all sectors, see petroleum products supplied data in Table 5.11. Petroleum products supplied is an approximation of petroleum consumption and is synonymous with the term "petroleum consumption" in Tables 5.13a-d and 5.14a-c. • See Note 1, "Petroleum Products Supplied and Petroleum Consumption," at end of section. • Totals may not equal sum of components due to independent rounding.

Web Pages: • For all data beginning in 1949, see http://www.eia.doe.gov/emeu/aer/petro.html. • For related information, see http://www.eia.doe.gov/emeu/states/_seds.html.

Sources: **CHP** and **Petroleum Coke:** Table 8.7c. **All Other Data:** • 1949-1959—Bureau of Mines, Mineral Industry Surveys, *Petroleum Statement, Annual,* annual reports, and Energy Information Administration (EIA) estimates. • 1960-2005—EIA, "State Energy Data 2005: Consumption" (February 2008), U.S. Tables 8 and 9. • 2006 and 2007—EIA estimates.

Table 5.13b Estimated Petroleum Consumption: Industrial Sector, Selected Years, 1949-2007

(Thousand Barrels per Day)

Year	Industrial Sector															
	Asphalt and Road Oil	Distillate Fuel Oil			Kerosene	Liquefied Petroleum Gases	Lubricants	Motor Gasoline [3]	Petroleum Coke			Residual Fuel Oil			Other Petroleum [4]	Total
		CHP [1]	Other [2]	Total					CHP [1]	Other [2]	Total	CHP [1]	Other [2]	Total		
1949	157	([5])	265	265	123	80	36	121	([5])	40	40	([5])	534	534	243	1,598
1950	180	([5])	328	328	132	100	43	131	([5])	41	41	([5])	617	617	250	1,822
1955	254	([5])	466	466	116	212	47	173	([5])	67	67	([5])	686	686	366	2,387
1960	302	([5])	476	476	78	333	48	198	([5])	149	149	([5])	689	689	435	2,708
1965	368	([5])	541	541	80	470	62	179	([5])	202	202	([5])	689	689	657	3,247
1970	447	([5])	577	577	89	699	70	150	([5])	203	203	([5])	708	708	866	3,808
1971	458	([5])	596	596	80	715	69	143	([5])	211	211	([5])	705	705	870	3,845
1972	468	([5])	648	648	77	846	73	132	([5])	233	233	([5])	765	765	949	4,191
1973	522	([5])	691	691	75	902	88	133	([5])	254	254	([5])	809	809	1,005	4,479
1974	481	([5])	633	633	61	901	85	123	([5])	230	230	([5])	753	753	1,034	4,301
1975	419	([5])	630	630	58	844	68	116	([5])	246	246	([5])	658	658	1,001	4,038
1976	411	([5])	717	717	59	895	75	110	([5])	242	242	([5])	792	792	1,145	4,447
1977	436	([5])	809	809	69	918	82	102	([5])	266	266	([5])	844	844	1,294	4,821
1978	479	([5])	823	823	75	921	88	93	([5])	250	250	([5])	748	748	1,391	4,867
1979	476	([5])	830	830	86	1,266	92	84	([5])	243	243	([5])	721	721	1,546	5,343
1980	396	([5])	621	621	87	1,172	82	82	([5])	234	234	([5])	586	586	1,581	4,842
1981	342	([5])	653	653	52	1,166	79	83	([5])	250	250	([5])	471	471	1,176	4,273
1982	342	([5])	617	617	68	1,211	72	72	([5])	246	246	([5])	456	456	973	4,058
1983	373	([5])	537	537	32	1,166	75	59	([5])	225	225	([5])	345	345	1,042	3,854
1984	408	([5])	564	564	28	1,283	80	83	([5])	244	244	([5])	386	386	1,120	4,198
1985	425	([5])	526	526	21	1,285	75	114	([5])	261	261	([5])	326	326	1,032	4,065
1986	448	([5])	546	546	16	1,207	73	108	([5])	264	264	([5])	321	321	1,105	4,087
1987	467	([5])	537	537	14	1,279	83	107	([5])	294	294	([5])	253	253	1,176	4,210
1988	468	([5])	530	530	14	1,326	80	100	([5])	306	306	([5])	237	237	1,286	4,347
1989	453	5	531	536	14	1,300	82	104	5	295	300	58	121	178	1,284	4,251
1990	483	7	534	541	6	1,215	84	97	25	300	325	64	115	179	1,373	4,304
1991	444	12	495	507	5	1,326	75	101	22	293	315	55	91	146	1,299	4,219
1992	454	10	509	519	5	1,402	77	101	26	336	362	59	109	168	1,434	4,522
1993	474	10	515	525	6	1,363	78	94	22	308	330	65	129	194	1,373	4,438
1994	484	10	513	522	8	1,505	82	101	25	304	329	69	113	183	1,454	4,667
1995	486	6	526	532	7	1,527	80	105	26	302	328	60	87	147	1,381	4,594
1996	484	8	549	557	9	1,580	78	105	27	317	343	66	80	146	1,518	4,819
1997	505	8	558	566	9	1,617	82	111	37	294	331	56	71	127	1,605	4,953
1998	521	16	554	570	11	1,553	86	105	29	362	390	60	40	100	1,508	4,844
1999	547	16	542	558	6	1,709	87	80	31	395	426	52	38	90	1,532	5,035
2000	525	10	553	563	8	1,720	86	79	19	342	361	48	57	105	1,458	4,903
2001	519	9	602	611	11	1,557	79	155	15	375	390	46	42	89	1,481	4,892
2002	512	6	561	566	7	1,668	78	163	21	362	383	37	46	83	1,474	4,934
2003	503	10	525	534	12	1,561	72	171	17	358	375	38	58	96	1,579	4,903
2004	537	9	561	570	14	1,647	73	195	18	405	423	46	62	108	1,657	5,223
2005	546	11	583	594	[R]19	[R]1,549	72	[R]187	14	390	404	46	77	123	1,605	[R]5,100
2006	[R]521	[R]5	[R]589	[R]594	[R]14	[RP]1,567	[RP]71	[RP]189	[R]21	[R]404	[R]425	[R]33	[R]71	[R]104	[RP]1,640	[RP]5,124
2007	[P]490	[P]6	[P]594	[P]600	[P]8	[P]1,589	[P]70	[P]190	[P]23	[P]391	[P]414	[P]28	[P]81	[P]109	[P]1,587	[P]5,056

[1] Industrial combined-heat-and-power (CHP) and a small number of industrial electricity-only plants. See Note 2, "Classification of Power Plants Into Energy-Use Sectors," at end of Section 8.

[2] All industrial sector fuel use other than that in "CHP."

[3] Finished motor gasoline. Through 1963, also includes special naphthas. Beginning in 1993, also includes ethanol blended into motor gasoline.

[4] Pentanes plus, petrochemical feedstocks, still gas (refinery gas), waxes, and miscellaneous products. Beginning in 1964, also includes special naphthas. Beginning in 1981, also includes negative barrels per day of distillate and residual fuel oil reclassified as unfinished oils, and other products (from both primary and secondary supply) reclassified as gasoline blending components. Beginning in 1983, also includes crude oil burned as fuel. Beginning in 2005, also includes naphtha-type jet fuel.

[5] Included in "Other."

R=Revised. P=Preliminary.

Notes: • For total petroleum consumption by all sectors, see petroleum products supplied data in Table 5.11. Petroleum products supplied is an approximation of petroleum consumption and is synonymous with the term "petroleum consumption" in Tables 5.13a-d and 5.14a-c. • See Note 1, "Petroleum Products Supplied and Petroleum Consumption," at end of section. • Totals may not equal sum of components due to independent rounding.

Web Pages: • For all data beginning in 1949, see http://www.eia.doe.gov/emeu/aer/petro.html. • For related information, see http://www.eia.doe.gov/emeu/states/_seds.html.

Sources: **CHP:** Table 8.7c. **All Other Data:** • 1949-1959—Bureau of Mines, Mineral Industry Surveys, *Petroleum Statement, Annual,* annual reports, and Energy Information Administration (EIA) estimates. • 1960-2005—EIA, "State Energy Data 2005: Consumption" (February 2008), U.S. Table 10. • 2006 and 2007—EIA estimates.

Table 5.13c Estimated Petroleum Consumption: Transportation Sector, Selected Years, 1949-2007

(Thousand Barrels per Day)

Year	Transportation Sector								
	Aviation Gasoline	Distillate Fuel Oil	Jet Fuel		Liquefied Petroleum Gases	Lubricants	Motor Gasoline [2]	Residual Fuel Oil	Total
			Kerosene Type	Total [1]					
1949	93	190	0	([1])	1	54	2,241	504	3,084
1950	108	226	0	([1])	2	64	2,433	524	3,356
1955	192	372	0	154	9	70	3,221	440	4,458
1960	161	418	91	371	13	68	3,736	367	5,135
1965	120	514	334	602	23	67	4,374	336	6,036
1970	55	738	718	967	32	66	5,589	332	7,778
1971	49	800	751	1,010	37	67	5,827	305	8,095
1972	46	910	779	1,021	38	71	6,199	280	8,566
1973	45	1,045	825	1,042	35	74	6,496	317	9,054
1974	44	1,036	757	979	33	71	6,372	304	8,838
1975	39	998	782	992	31	70	6,512	310	8,951
1976	37	1,073	777	976	33	77	6,817	358	9,372
1977	38	1,171	814	1,022	36	78	7,022	396	9,761
1978	39	1,260	845	1,044	38	83	7,264	431	10,160
1979	38	1,366	867	1,067	16	87	6,896	535	10,005
1980	35	1,311	845	1,062	13	77	6,441	608	9,546
1981	31	1,365	808	1,005	24	74	6,456	531	9,487
1982	25	1,312	803	1,011	24	68	6,421	444	9,307
1983	26	1,367	839	1,046	29	71	6,510	358	9,406
1984	24	1,383	953	1,175	30	76	6,554	351	9,592
1985	27	1,491	1,005	1,218	21	71	6,667	342	9,838
1986	32	1,514	1,105	1,307	19	69	6,871	379	10,191
1987	25	1,568	1,181	1,385	15	78	7,041	392	10,505
1988	27	1,701	1,236	1,449	17	75	7,179	399	10,846
1989	26	1,734	1,284	1,489	16	77	7,171	423	10,937
1990	24	1,722	1,340	1,522	16	80	7,080	443	10,888
1991	23	1,694	1,296	1,471	15	71	7,042	447	10,763
1992	22	1,728	1,310	1,454	14	72	7,125	465	10,881
1993	21	1,785	1,357	1,469	14	74	7,367	393	11,124
1994	21	1,896	1,480	1,527	24	77	7,487	385	11,417
1995	21	1,973	1,497	1,514	13	76	7,674	397	11,668
1996	20	2,096	1,575	1,578	11	73	7,772	370	11,921
1997	22	2,198	1,598	1,599	10	78	7,883	310	12,099
1998	19	2,263	1,623	1,622	13	81	8,128	294	12,420
1999	21	2,352	1,675	1,673	10	82	8,336	290	12,765
2000	20	2,422	1,725	1,725	8	81	8,370	386	13,012
2001	19	2,489	1,656	1,655	10	74	8,435	255	12,938
2002	18	2,536	1,621	1,614	10	73	8,662	295	13,208
2003	16	2,665	1,578	1,578	12	68	8,733	249	13,321
2004	17	2,783	1,630	1,630	14	69	8,885	321	13,718
2005	19	2,858	1,679	1,679	R20	68	R8,948	365	R13,957
2006	18	R3,017	R1,633	R1,633	RP20	RP67	RP9,039	R395	RP14,189
2007	P17	P3,048	P1,623	P1,623	P21	P66	P9,076	P414	P14,265

[1] Through 1951, naphtha-type jet fuel is included in the products from which jet fuel was blended: in 1952, 71 percent gasoline, 17 percent kerosene, and 12 percent distillate fuel oil. Beginning in 1952, includes naphtha-type jet fuel. Beginning in 1957, also includes kerosene-type jet fuel. Beginning in 2005, includes kerosene-type jet fuel only.

[2] Finished motor gasoline. Through 1963, also includes special naphthas. Beginning in 1993, also includes ethanol blended into motor gasoline.

R=Revised. P=Preliminary.

Notes: • For total petroleum consumption by all sectors, see petroleum products supplied data in Table 5.11. Petroleum products supplied is an approximation of petroleum consumption and is synonymous with the term "petroleum consumption" in Tables 5.13a-d and 5.14a-c. • See Note 1, "Petroleum Products Supplied and Petroleum Consumption," at end of section. • Totals may not equal sum of components due to independent rounding.

Web Pages: • For all data beginning in 1949, see http://www.eia.doe.gov/emeu/aer/petro.html. • For related information, see http://www.eia.doe.gov/emeu/states/_seds.html.

Sources: • 1949-1959—Bureau of Mines, Mineral Industry Surveys, *Petroleum Statement, Annual,* annual reports, and Energy Information Administration (EIA) estimates. • 1960-2005—EIA, "State Energy Data 2005: Consumption" (February 2008), U.S. Table 11. • 2006 and 2007—EIA estimates.

Table 5.13d Petroleum Consumption: Electric Power Sector, Selected Years, 1949-2007

(Thousand Barrels per Day)

	Electric Power Sector [1]											
	Electricity Only				Combined Heat and Power (CHP)				Total			
Year	Distillate Fuel Oil [2]	Petroleum Coke	Residual Fuel Oil [3]	Total	Distillate Fuel Oil [2]	Petroleum Coke	Residual Fuel Oil [3]	Total	Distillate Fuel Oil [2]	Petroleum Coke	Residual Fuel Oil [3]	Total
1949	13	NA	169	182	NA	NA	NA	NA	13	NA	169	182
1950	15	NA	192	207	NA	NA	NA	NA	15	NA	192	207
1955	15	NA	191	206	NA	NA	NA	NA	15	NA	191	206
1960	10	NA	231	241	NA	NA	NA	NA	10	NA	231	241
1965	14	NA	302	316	NA	NA	NA	NA	14	NA	302	316
1970	66	9	853	928	NA	NA	NA	NA	66	9	853	928
1971	94	8	992	1,095	NA	NA	NA	NA	94	8	992	1,095
1972	146	9	1,203	1,358	NA	NA	NA	NA	146	9	1,203	1,358
1973	129	7	1,406	1,542	NA	NA	NA	NA	129	7	1,406	1,542
1974	146	9	1,324	1,478	NA	NA	NA	NA	146	9	1,324	1,478
1975	107	1	1,280	1,388	NA	NA	NA	NA	107	1	1,280	1,388
1976	114	1	1,405	1,520	NA	NA	NA	NA	114	1	1,405	1,520
1977	134	1	1,575	1,710	NA	NA	NA	NA	134	1	1,575	1,710
1978	130	5	1,612	1,747	NA	NA	NA	NA	130	5	1,612	1,747
1979	84	4	1,350	1,437	NA	NA	NA	NA	84	4	1,350	1,437
1980	79	2	1,069	1,151	NA	NA	NA	NA	79	2	1,069	1,151
1981	58	2	904	964	NA	NA	NA	NA	58	2	904	964
1982	42	2	642	686	NA	NA	NA	NA	42	2	642	686
1983	45	4	627	676	NA	NA	NA	NA	45	4	627	676
1984	42	3	517	562	NA	NA	NA	NA	42	3	517	562
1985	40	3	435	478	NA	NA	NA	NA	40	3	435	478
1986	39	4	592	636	NA	NA	NA	NA	39	4	592	636
1987	42	5	504	551	NA	NA	NA	NA	42	5	504	551
1988	51	6	627	683	NA	NA	NA	NA	51	6	627	683
1989[4]	70	7	663	740	2	0	6	8	72	7	669	748
1990	41	14	497	551	4	0	10	15	45	14	507	566
1991	38	13	469	520	1	0	4	5	39	13	473	526
1992	33	18	371	422	2	2	8	12	34	20	379	434
1993	37	21	409	467	4	15	9	27	41	36	418	494
1994	46	16	369	431	11	15	10	36	56	32	379	467
1995	44	15	237	296	7	22	9	38	51	37	247	334
1996	47	14	263	325	4	22	10	36	51	36	273	360
1997	48	23	301	373	4	23	10	37	52	46	311	410
1998	61	30	448	539	3	26	8	37	64	56	456	576
1999	63	26	409	497	3	25	9	38	66	51	418	535
2000	77	20	370	466	6	25	8	39	82	45	378	505
2001	76	25	430	531	4	22	7	33	80	47	437	564
2002	59	54	281	394	1	26	6	33	60	80	287	427
2003	71	66	373	510	5	14	6	24	76	79	379	534
2004	49	83	376	509	3	17	6	26	52	101	382	535
2005	51	94	376	521	3	17	6	26	54	111	382	547
2006	34	[R]82	[R]151	[R]267	1	15	6	22	35	[R]97	[R]157	[R]289
2007[P]	41	63	168	273	2	13	6	22	43	77	174	294

[1] Electricity-only and combined-heat-and-power (CHP) plants within the NAICS 22 category whose primary business is to sell electricity, or electricity and heat, to the public. Electric utility CHP plants are included in "Electricity Only."

[2] Fuel oil nos. 1, 2, and 4. For 1949-1979, data are for gas turbine and internal combustion plant use of petroleum. For 1980-2000, electric utility data also include small amounts of kerosene and jet fuel.

[3] Fuel oil nos. 5 and 6. For 1949-1979, data are for steam plant use of petroleum. For 1980-2000, electric utility data also include a small amount of fuel oil no. 4.

[4] Through 1988, data are for electric utilities only. Beginning in 1989, data are for electric utilities and independent power producers.

R=Revised. P=Preliminary. NA=Not available.

Notes: • For total petroleum consumption by all sectors, see petroleum products supplied data in Table 5.11. Petroleum products supplied is an approximation of petroleum consumption and is synonymous with the term "petroleum consumption" in Tables 5.13a-d and 5.14a-c. • See Tables 8.5a-8.5d for the amount of petroleum used to produce electricity and Tables 8.6a-8.6c for the amount of petroleum used to produce useful thermal output. • See Note 1, "Petroleum Products Supplied and Petroleum Consumption," at end of section. • Totals may not equal sum of components due to independent rounding.

Web Pages: • For all data beginning in 1949, see http://www.eia.doe.gov/emeu/aer/petro.html. • For related information, see http://www.eia.doe.gov/fuelelectric.html.

Sources: Tables 8.5b, 8.5c, 8.6b, and 8.7b.

Figure 5.14 Heat Content of Petroleum Consumption by Product by Sector, 1949-2007

Residential and Commercial[1] Sectors, Selected Products

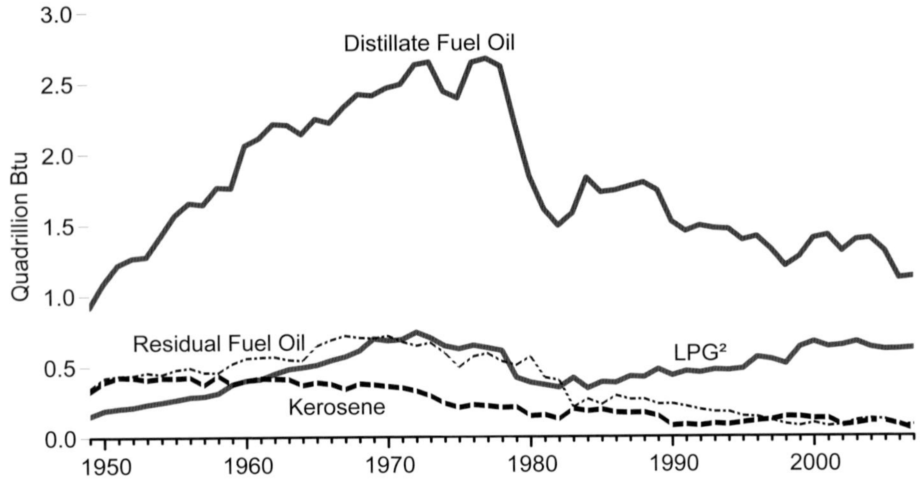

Industrial[1] Sector, Selected Products

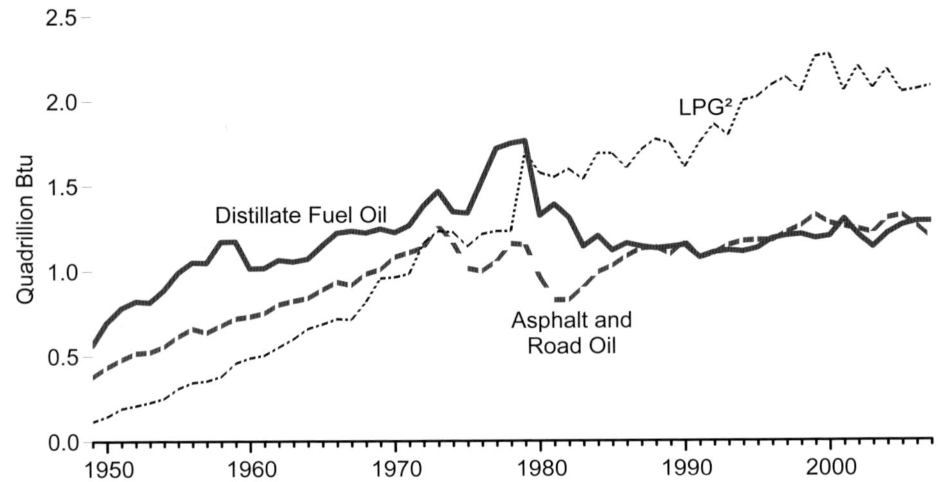

Transportation Sector, Selected Products

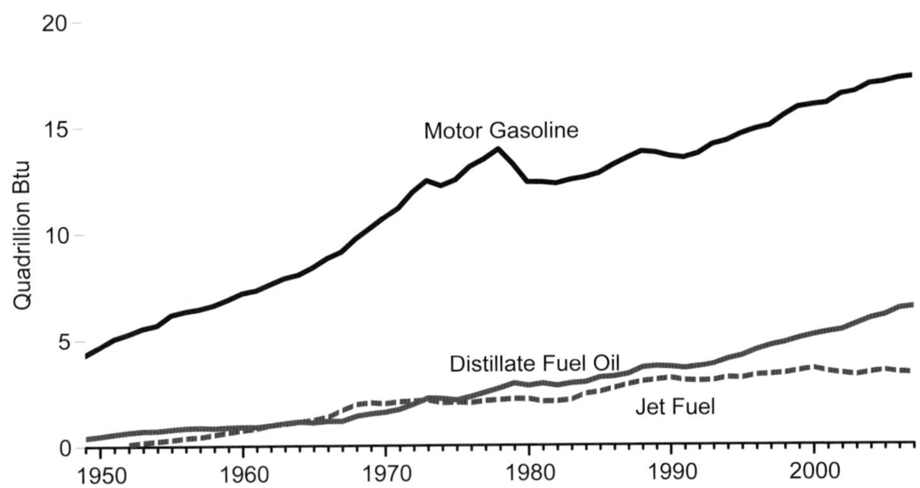

Electric Power Sector[3]

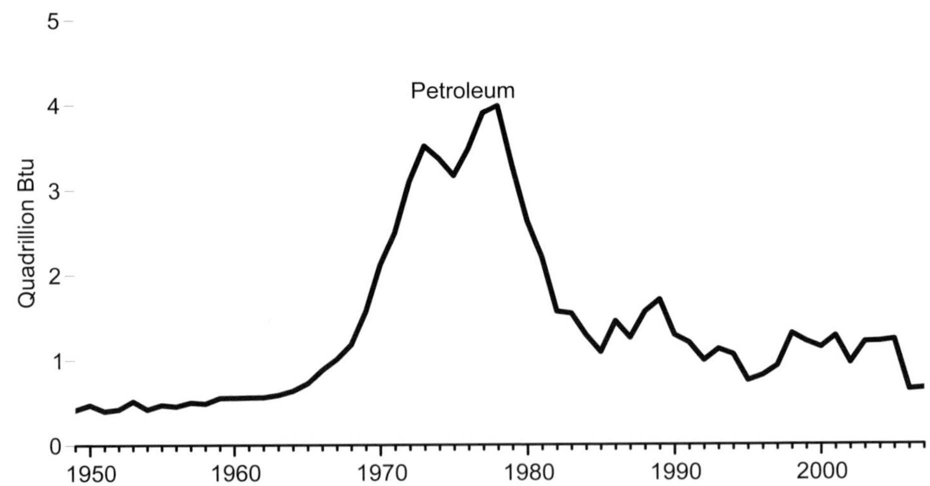

[1] Includes combined-heat-and-power plants and a small number of electricity-only plants.
[2] Liquefied petroleum gases.
[3] Electricity-only and combined-heat-and-power plants whose primary business is to sell electricity, or electricity and heat, to the public.

Note: Because vertical scales differ, graphs should not be compared.
Sources: Tables 5.14a–5.14c.

Table 5.14a Heat Content of Petroleum Consumption: Residential and Commercial Sectors, Selected Years, 1949-2007

(Trillion Btu)

Year	Residential Sector: Distillate Fuel Oil	Residential Sector: Kerosene	Residential Sector: Liquefied Petroleum Gases	Residential Sector: Total	Commercial Sector: Distillate Fuel Oil	Commercial Sector: Kerosene	Commercial Sector: Liquefied Petroleum Gases	Commercial Sector: Motor Gasoline[1]	Commercial Sector: Petroleum Coke	Commercial Sector: Residual Fuel Oil	Commercial Sector: Total
1949	700	289	132	1,121	221	39	23	92	NA	351	727
1950	829	347	164	1,340	262	47	29	100	NA	424	862
1955	1,194	371	227	1,792	377	51	40	133	NA	480	1,081
1960	1,568	354	343	2,265	494	48	61	67	NA	559	1,228
1965	1,713	334	434	2,481	534	54	77	77	NA	645	1,386
1970	1,878	298	579	2,755	587	61	102	86	NA	714	1,551
1971	1,897	295	585	2,777	595	55	103	84	NA	672	1,510
1972	1,996	271	628	2,895	632	55	111	87	NA	645	1,530
1973	2,003	227	595	2,825	644	65	105	87	NA	665	1,565
1974	1,844	184	546	2,573	596	55	96	83	NA	593	1,423
1975	1,807	161	528	2,495	587	49	93	89	NA	492	1,310
1976	1,987	184	549	2,720	656	44	97	97	NA	567	1,461
1977	1,994	167	533	2,695	676	52	94	101	NA	588	1,511
1978	1,951	153	516	2,620	666	55	91	107	NA	532	1,450
1979	1,626	133	355	2,114	584	78	63	104	NA	505	1,334
1980	1,316	107	325	1,748	518	41	57	107	NA	565	1,287
1981	1,147	85	311	1,543	457	69	55	92	NA	417	1,090
1982	1,050	95	296	1,441	440	30	52	88	NA	399	1,008
1983	924	85	352	1,362	651	111	62	102	NA	208	1,136
1984	1,091	88	290	1,468	735	93	51	107	NA	266	1,252
1985	1,092	159	327	1,578	631	33	58	96	NA	228	1,045
1986	1,111	121	323	1,556	623	50	57	106	NA	290	1,126
1987	1,156	119	360	1,634	607	49	63	111	NA	263	1,093
1988	1,190	144	356	1,690	600	26	63	110	NA	264	1,063
1989	1,160	117	402	1,679	574	28	71	102	0	228	1,002
1990	978	64	365	1,407	536	12	64	111	0	230	953
1991	930	72	389	1,392	517	12	69	85	0	212	895
1992	980	65	382	1,427	507	11	68	80	(s)	189	854
1993	974	76	399	1,448	493	14	70	30	(s)	173	780
1994	960	65	395	1,420	501	19	70	25	(s)	172	787
1995	905	74	404	1,383	479	22	71	18	(s)	141	732
1996	926	89	473	1,488	483	21	84	27	(s)	137	751
1997	874	93	461	1,428	444	25	81	43	(s)	111	704
1998	772	108	434	1,314	429	31	77	39	(s)	85	661
1999	828	111	534	1,473	438	27	94	28	(s)	73	661
2000	905	95	564	1,563	491	30	99	45	(s)	92	756
2001	908	95	535	1,539	508	31	94	37	(s)	70	742
2002	860	60	543	1,463	444	16	96	45	(s)	80	681
2003	905	70	564	1,539	481	19	100	60	(s)	111	771
2004	924	85	531	1,539	470	20	94	49	(s)	122	756
2005	854	R84	R517	R1,455	447	22	R91	R46	(s)	116	R722
2006	R712	R66	RP520	RP1,299	R401	R15	RP92	RP46	(s)	R75	RP630
2007	P719	P38	P526	P1,283	P405	P9	P93	P46	P(s)	P79	P633

[1] Finished motor gasoline. Through 1963, also includes special naphthas. Beginning in 1993, also includes ethanol blended into motor gasoline.

R=Revised. P=Preliminary. NA=Not available. (s)=Less than 0.5 trillion Btu.

Notes: • Data are estimates. • For total heat content of petroleum consumption by all sectors, see data for heat content of petroleum products supplied in Table 5.12. Petroleum products supplied is an approximation of petroleum consumption and is synonymous with the term "petroleum consumption" in Tables 5.13a-d and 5.14a-c. • Totals may not equal sum of components due to independent rounding.

Web Pages: • For all data beginning in 1949, see http://www.eia.doe.gov/emeu/aer/petro.html. • For related information, see http://www.eia.doe.gov/emeu/sedr/contents.html.

Sources: Tables 5.13a, A1, and A3.

Table 5.14b Heat Content of Petroleum Consumption: Industrial Sector, Selected Years, 1949-2007

(Trillion Btu)

Year	Industrial Sector									
	Asphalt and Road Oil	Distillate Fuel Oil	Kerosene	Liquefied Petroleum Gases	Lubricants	Motor Gasoline [1]	Petroleum Coke	Residual Fuel Oil	Other Petroleum [2]	Total
1949	380	564	254	117	80	231	87	1,225	530	3,468
1950	435	698	274	147	94	251	90	1,416	546	3,951
1955	615	991	241	310	103	332	147	1,573	798	5,111
1960	734	1,016	161	489	107	381	328	1,584	947	5,747
1965	890	1,150	165	688	137	342	444	1,582	1,390	6,789
1970	1,082	1,226	185	964	155	288	446	1,624	1,817	7,787
1971	1,108	1,266	165	984	152	275	463	1,618	1,825	7,856
1972	1,137	1,381	160	1,164	163	254	513	1,761	2,001	8,534
1973	1,264	1,469	156	1,233	195	255	558	1,858	2,117	9,104
1974	1,165	1,346	126	1,227	187	235	506	1,728	2,173	8,694
1975	1,014	1,339	119	1,144	149	223	540	1,509	2,107	8,146
1976	998	1,530	123	1,216	166	211	535	1,822	2,410	9,010
1977	1,056	1,719	143	1,232	182	196	586	1,937	2,722	9,774
1978	1,160	1,750	156	1,233	195	178	550	1,716	2,930	9,867
1979	1,153	1,764	177	1,700	204	162	533	1,655	3,219	10,568
1980	962	1,324	181	1,577	182	158	516	1,349	3,275	9,525
1981	828	1,389	108	1,551	175	160	549	1,081	2,445	8,285
1982	829	1,313	141	1,598	159	138	541	1,047	2,029	7,795
1983	904	1,142	66	1,537	167	112	495	791	2,204	7,420
1984	992	1,203	58	1,691	178	160	538	889	2,317	8,025
1985	1,029	1,119	44	1,690	166	218	575	748	2,149	7,738
1986	1,086	1,160	32	1,603	162	206	581	736	2,313	7,880
1987	1,130	1,141	28	1,709	183	206	646	582	2,440	8,065
1988	1,136	1,130	30	1,772	177	193	675	546	2,681	8,339
1989	1,096	1,139	30	1,748	181	199	660	410	2,658	8,120
1990	1,170	1,150	12	1,608	186	185	714	411	2,840	8,278
1991	1,077	1,078	11	1,749	167	193	693	334	2,685	7,987
1992	1,102	1,107	10	1,860	170	194	798	387	2,953	8,581
1993	1,149	1,117	13	1,794	173	180	725	446	2,821	8,418
1994	1,173	1,111	17	1,997	181	192	723	419	2,988	8,801
1995	1,178	1,131	15	2,019	178	200	721	337	2,834	8,614
1996	1,176	1,187	18	2,089	173	200	757	335	3,119	9,053
1997	1,224	1,203	19	2,134	182	212	727	291	3,298	9,290
1998	1,263	1,211	22	2,048	191	199	858	230	3,093	9,116
1999	1,324	1,187	13	2,256	193	152	936	207	3,128	9,396
2000	1,276	1,200	16	2,271	190	150	796	241	2,981	9,120
2001	1,257	1,300	23	2,054	174	295	858	203	3,056	9,220
2002	1,240	1,204	14	2,200	172	309	842	190	3,041	9,213
2003	1,220	1,136	24	2,068	159	324	825	220	3,260	9,237
2004	1,304	1,214	28	2,181	161	372	934	249	3,429	9,872
2005	1,323	1,264	R39	R2,047	160	R356	889	281	3,320	R9,680
2006	R1,261	R1,263	R30	RP2,062	RP156	RP360	R934	R239	RP3,416	RP9,720
2007	P1,188	P1,276	P17	P2,083	P154	P361	P909	P251	P3,296	P9,535

[1] Finished motor gasoline. Through 1963, also includes special naphthas. Beginning in 1993, also includes ethanol blended into motor gasoline.

[2] Pentanes plus, petrochemical feedstocks, still gas (refinery gas), waxes, and miscellaneous products. Beginning in 1964, also includes special naphthas. Beginning in 1983, also includes crude oil burned as fuel.

R=Revised. P=Preliminary.

Notes: • Data are estimates. • For total heat content of petroleum consumption by all sectors, see data for heat content of petroleum products supplied in Table 5.12. Petroleum products supplied is an approximation of petroleum consumption and is synonymous with the term "petroleum consumption" in Tables 5.13a-d and 5.14a-c. • Totals may not equal sum of components due to independent rounding.

Web Pages: • For all data beginning in 1949, see http://www.eia.doe.gov/emeu/aer/petro.html. • For related information, see http://www.eia.doe.gov/emeu/sedr/contents.html.

Sources: Tables 5.13b, A1, and A3.

Table 5.14c Heat Content of Petroleum Consumption: Transportation and Electric Power Sectors, Selected Years, 1949-2007 (Trillion Btu)

Year	Transportation Sector									Electric Power Sector [1]			
	Aviation Gasoline	Distillate Fuel Oil	Jet Fuel: Kerosene Type	Jet Fuel: Total [2]	Liquefied Petroleum Gases	Lubricants	Motor Gasoline [3]	Residual Fuel Oil	Total	Distillate Fuel Oil [4]	Petroleum Coke	Residual Fuel Oil [5]	Total
1949	172	405	0	0	2	120	4,298	1,156	6,152	28	NA	387	415
1950	199	480	0	0	3	141	4,664	1,201	6,690	32	NA	440	472
1955	354	791	0	301	14	155	6,175	1,009	8,800	32	NA	439	471
1960	298	892	188	739	20	152	7,183	844	10,126	22	NA	530	553
1965	222	1,093	691	1,215	33	149	8,386	770	11,868	29	NA	693	722
1970	100	1,569	1,486	1,973	44	147	10,716	761	15,310	141	19	1,958	2,117
1971	90	1,701	1,554	2,061	50	147	11,173	701	15,923	200	18	2,277	2,495
1972	85	1,941	1,617	2,091	52	158	11,918	645	16,891	310	19	2,768	3,097
1973	83	2,222	1,707	2,131	48	163	12,455	727	17,831	273	15	3,226	3,515
1974	82	2,202	1,566	2,001	44	156	12,217	697	17,399	309	19	3,038	3,365
1975	71	2,121	1,619	2,029	42	155	12,485	711	17,614	226	2	2,937	3,166
1976	67	2,288	1,613	2,002	45	172	13,107	824	18,506	243	2	3,232	3,477
1977	70	2,489	1,684	2,090	48	172	13,464	908	19,241	283	3	3,614	3,901
1978	71	2,679	1,750	2,138	52	184	13,927	990	20,041	276	12	3,699	3,987
1979	70	2,905	1,795	2,186	21	193	13,221	1,228	19,825	178	8	3,097	3,283
1980	64	2,795	1,754	2,179	17	172	12,383	1,398	19,008	169	5	2,459	2,634
1981	56	2,901	1,671	2,058	32	165	12,379	1,219	18,811	124	4	2,073	2,202
1982	47	2,790	1,661	2,069	32	150	12,312	1,020	18,420	89	4	1,474	1,568
1983	48	2,905	1,736	2,141	38	157	12,482	821	18,593	96	8	1,440	1,544
1984	44	2,948	1,977	2,414	40	168	12,600	807	19,020	88	8	1,190	1,286
1985	50	3,170	2,079	2,497	28	156	12,784	786	19,471	85	7	998	1,090
1986	59	3,218	2,287	2,682	26	153	13,174	870	20,182	83	9	1,359	1,452
1987	46	3,335	2,444	2,843	21	173	13,499	900	20,816	90	10	1,157	1,257
1988	49	3,626	2,565	2,982	22	167	13,802	919	21,567	109	12	1,442	1,563
1989	48	3,687	2,658	3,059	22	171	13,749	971	21,706	152	16	1,535	1,703
1990	45	3,661	2,774	3,129	22	176	13,575	1,016	21,625	97	30	1,163	1,289
1991	42	3,601	2,681	3,025	20	157	13,503	1,026	21,373	84	29	1,085	1,198
1992	41	3,684	2,718	3,001	18	161	13,699	1,070	21,674	74	45	872	991
1993	38	3,796	2,809	3,028	19	163	14,126	901	22,072	86	79	959	1,124
1994	38	4,032	3,063	3,154	32	171	14,293	883	22,603	120	70	869	1,059
1995	40	4,195	3,099	3,132	17	168	14,607	911	23,069	108	81	566	755
1996	37	4,469	3,268	3,274	15	163	14,837	851	23,647	109	80	628	817
1997	40	4,672	3,307	3,308	13	172	14,999	712	23,917	111	102	715	927
1998	35	4,812	3,359	3,357	17	180	15,463	674	24,537	136	124	1,047	1,306
1999	39	5,001	3,466	3,462	13	182	15,855	665	25,218	140	112	959	1,211
2000	36	5,165	3,580	3,580	11	179	15,960	888	25,820	175	99	871	1,144
2001	35	5,292	3,427	3,426	13	164	16,041	586	25,556	171	103	1,003	1,277
2002	34	5,392	3,354	3,340	13	162	16,465	677	26,084	127	175	659	961
2003	30	5,666	3,266	3,265	16	150	16,597	571	26,296	161	175	869	1,205
2004	31	5,932	3,382	3,383	18	152	16,959	740	27,214	111	222	879	1,212
2005	35	6,076	3,475	3,475	R27	151	R17,043	837	R27,644	115	243	876	1,235
2006	R33	R6,414	R3,379	R3,379	RP27	RP147	RP17,216	R906	RP28,123	R74	R214	R361	R648
2007	P32	P6,480	P3,358	P3,358	P27	P145	P17,290	P951	P28,283	P92	P168	P399	P660

[1] Electricity-only and combined-heat-and-power (CHP) plants within the NAICS 22 category whose primary business is to sell electricity, or electricity and heat, to the public. Through 1988, data are for electric utilities only; beginning in 1989, data are for electric utilities and independent power producers.

[2] Through 1951, naphtha-type jet fuel is included in the products from which jet fuel was blended: in 1952, 71 percent gasoline, 17 percent kerosene, and 12 percent distillate fuel oil. Beginning in 1952, includes naphtha-type jet fuel. Beginning in 1957, also includes kerosene-type jet fuel. Beginning in 2005, includes kerosene-type jet fuel only.

[3] Finished motor gasoline. Through 1963, also includes special naphthas. Beginning in 1993, also includes ethanol blended into motor gasoline.

[4] Fuel oil nos. 1, 2, and 4. For 1949-1979, data are for gas turbine and internal combustion plant use of petroleum. For 1980-2000, electric utility data also include small amounts of kerosene and jet fuel.

[5] Fuel oil nos. 5 and 6. For 1949-1979, data are for steam plant use of petroleum. For 1980-2000, electric utility data also include a small amount of fuel oil no. 4.

R=Revised. P=Preliminary. NA=Not available.

Notes: • Data for "Transportation Sector" are estimates. • For total heat content of petroleum consumption by all sectors, see data for heat content of petroleum products supplied in Table 5.12. Petroleum products supplied is an approximation of petroleum consumption and is synonymous with the term "petroleum consumption" in Tables 5.13a-d and 5.14a-c. • Totals may not equal sum of components due to independent rounding.

Web Pages: • For all data beginning in 1949, see http://www.eia.doe.gov/emeu/aer/petro.html. • For related information, see http://www.eia.doe.gov/emeu/sedr/contents.html.

Sources: Tables 5.13c, 5.13d, A1, and A3.

Figure 5.15 Fuel Oil and Kerosene Sales, 1984-2006

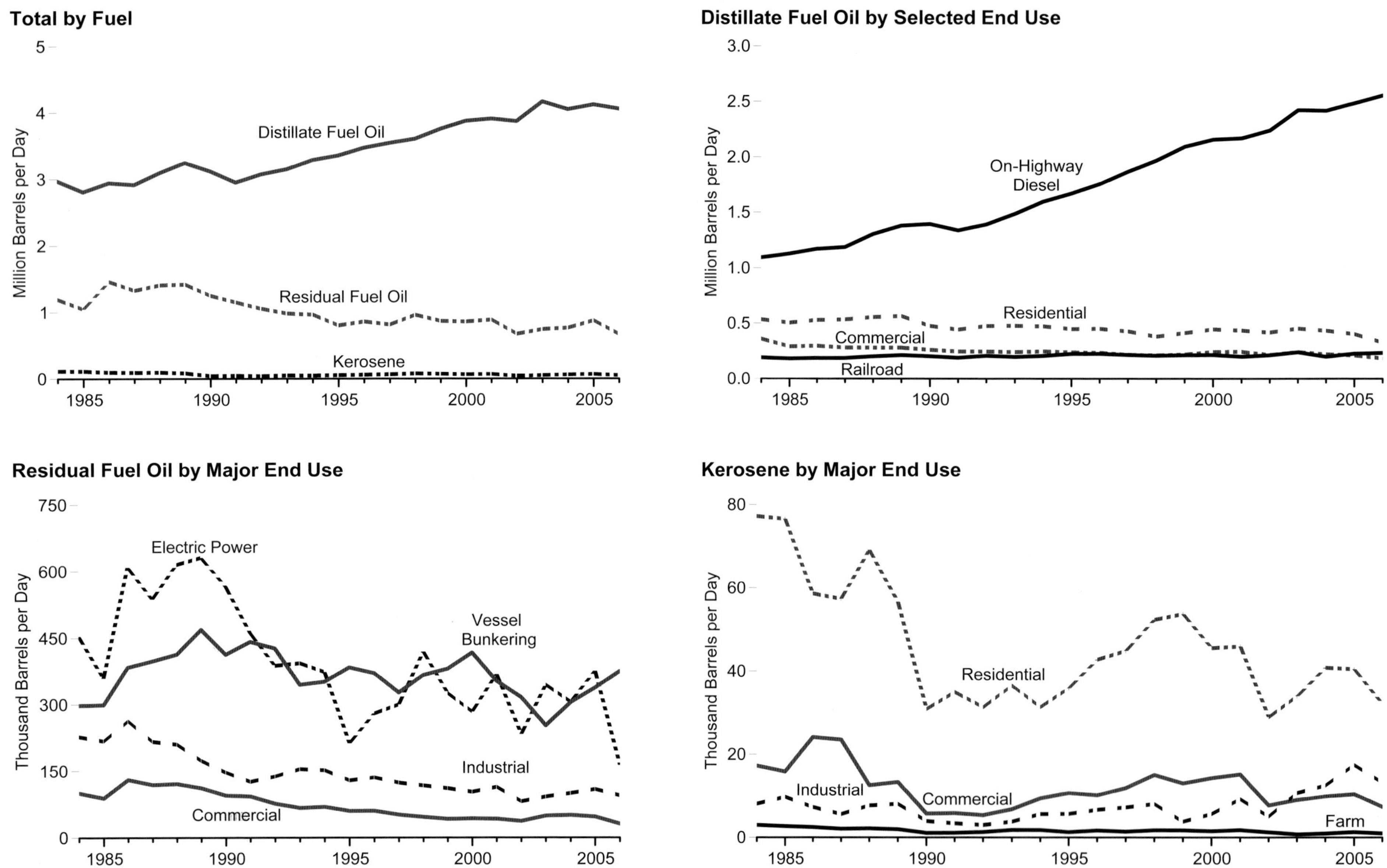

Note: Because vertical scales differ, graphs should not be compared.

Source: Table 5.15.

Table 5.15 Fuel Oil and Kerosene Sales, 1984-2006

(Thousand Barrels per Day)

Year	Distillate Fuel Oil												
	Residential	Commercial	Industrial	Oil Company	Farm	Electric Power [1]	Railroad	Vessel Bunkering	On-Highway Diesel	Military	Off-Highway Diesel	Other	Total
1984	534	360	166	55	208	42	192	115	1,093	46	114	46	2,971
1985	504	291	159	45	202	34	182	111	1,127	43	99	11	2,809
1986	528	296	175	41	218	38	186	127	1,169	47	108	10	2,944
1987	534	280	184	40	196	37	186	122	1,185	46	102	5	2,917
1988	554	279	167	41	206	47	201	130	1,304	54	109	4	3,095
1989	564	279	178	45	219	[1]58	211	147	1,378	56	110	2	3,248
1990	475	260	169	49	222	50	203	135	1,393	46	118	(s)	3,120
1991	442	246	151	48	206	39	188	133	1,336	53	107	(s)	2,949
1992	474	245	150	43	228	35	206	144	1,391	42	114	(s)	3,075
1993	475	241	139	46	222	36	196	141	1,485	32	137	(s)	3,150
1994	472	246	148	44	213	43	205	143	1,594	40	140	(s)	3,289
1995	447	237	146	45	227	39	224	153	1,668	30	142	– –	3,357
1996	450	234	149	48	234	43	224	162	1,754	30	146	– –	3,472
1997	426	216	151	56	231	41	214	168	1,867	28	149	– –	3,546
1998	380	211	161	51	222	55	207	169	1,967	23	162	– –	3,608
1999	411	218	162	43	223	53	211	158	2,091	23	162	– –	3,756
2000	444	241	152	45	225	66	214	147	2,155	20	168	– –	3,877
2001	433	243	161	49	234	88	198	133	2,167	26	177	– –	3,908
2002	416	215	156	50	223	49	212	136	2,238	23	154	– –	3,871
2003	452	240	156	33	209	75	239	145	2,420	27	169	– –	4,165
2004	432	220	151	31	207	54	198	139	2,415	23	179	– –	4,050
2005	401	210	160	31	210	59	225	131	2,482	18	193	– –	4,120
2006	325	183	161	42	213	43	232	124	2,552	21	162	– –	4,057

Year	Residual Fuel Oil								Kerosene					
	Commercial	Industrial	Oil Company	Electric Power [1]	Vessel Bunkering	Military	Other [2]	Total	Residential	Commercial	Industrial	Farm	Other	Total
1984	100	228	81	454	298	6	26	1,194	77	17	8	3	10	115
1985	89	218	62	359	299	8	13	1,048	77	16	10	3	9	114
1986	130	263	52	610	384	[E]7	15	1,462	59	24	7	2	6	98
1987	[3,R]119	217	44	537	398	10	3	[3,R]1,328	57	24	6	2	6	95
1988	[3,R]121	211	36	616	413	8	4	[3,R]1,409	69	13	8	2	5	96
1989	[3,R]112	174	24	[1]632	469	6	2	[3,R]1,419	57	13	8	2	4	84
1990	[3,R]95	147	21	566	413	7	2	[3,R]1,250	31	6	4	1	1	43
1991	93	126	20	461	442	8	1	1,150	35	6	3	1	1	46
1992	77	138	18	388	427	6	1	1,054	31	5	3	1	(s)	41
1993	67	155	17	394	345	5	(s)	983	37	7	4	2	1	50
1994	69	152	16	374	351	4	(s)	967	31	9	6	2	1	49
1995	60	129	14	213	384	3	(s)	804	36	11	6	1	(s)	54
1996	60	136	11	280	371	4	1	862	43	10	7	2	(s)	62
1997	52	124	10	300	327	3	(s)	816	45	12	7	1	(s)	66
1998	47	117	8	420	367	2	(s)	961	52	15	8	2	1	78
1999	42	111	8	326	381	2	(s)	869	54	13	4	2	1	73
2000	43	103	10	284	417	2	(s)	859	46	14	6	2	(s)	67
2001	42	114	9	368	353	1	(s)	888	46	15	9	2	(s)	72
2002	37	82	7	233	316	(s)	(s)	676	29	8	5	1	(s)	43
2003	49	92	5	344	253	1	(s)	744	34	9	11	1	(s)	55
2004	51	100	3	306	305	2	(s)	767	41	10	13	1	(s)	64
2005	47	109	5	376	338	2	(s)	877	40	10	17	1	(s)	70
2006	31	95	4	163	375	1	(s)	670	32	7	13	1	(s)	54

[1] Electricity-only and combined-heat-and-power (CHP) plants within the NAICS 22 category whose primary business is to sell electricity, or electricity and heat, to the public.
[2] Sales to railroads are included in "Other."
[3] Value has been revised since publication in the reports cited after "Sources."
E = Annual estimate based on eleven months of data. – – = Not applicable. (s)=Less than 0.5 thousand barrels per day.

Web Page: See http://www.eia.doe.gov/oil_gas/petroleum/info_glance/petroleum.html for related information.

Sources: • 1984—EIA, *Petroleum Marketing Annual 1988* (October 1989), Tables A1-A3, and unpublished revision. • 1985-2001—EIA, *Fuel Oil and Kerosene Sales*, annual reports, Tables 1-3, and unpublished revisions. • 2002 forward—EIA, *Fuel Oil and Kerosene Sales 2006* (December 2007), Tables 1-3.

Figure 5.16 Petroleum Primary Stocks by Type

Total, Crude Oil[1], and Products, 1949-2007

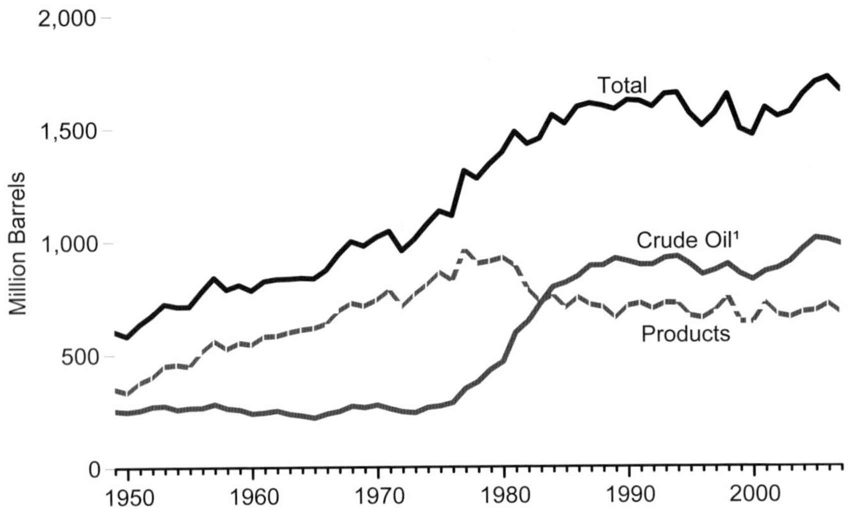

Total Stocks and Strategic Petroleum Reserve (SPR) Stocks, 1949-2007

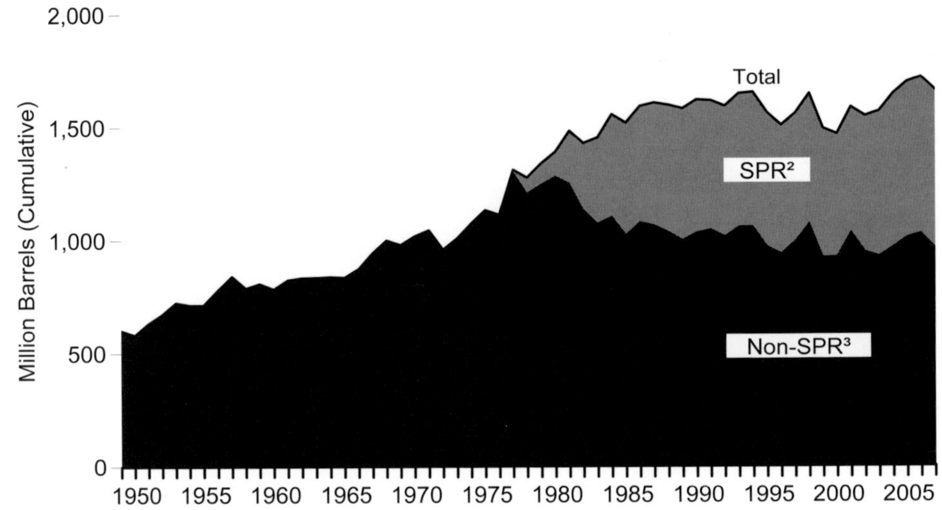

By Type, 2007

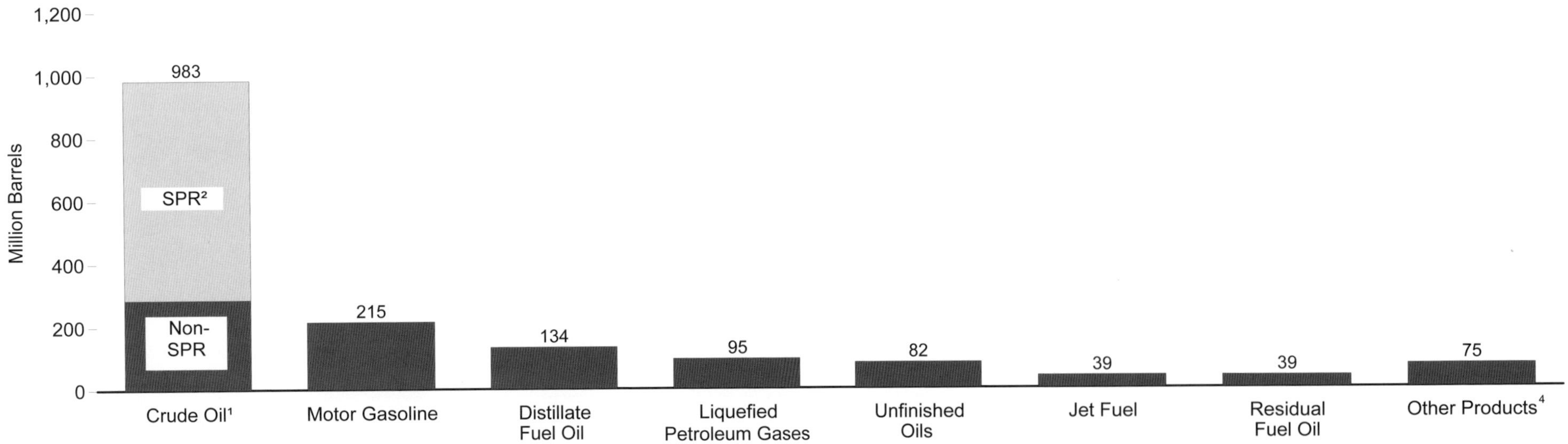

[1] Includes lease condensate and crude oil stored in the Strategic Petroleum Reserve (SPR).

[2] Crude oil stocks in the SPR include non-U.S. stocks held under foreign or commercial storage agreements. See Figure 5.17 for additional information about the SPR.

[3] Crude oil and products.

[4] Asphalt and road oil, aviation gasoline and blending components, kerosene, lubricants, naphtha-type jet fuel, pentanes plus, petrochemical feedstocks, petroleum coke, special naphthas, waxes, other hydrocarbons and oxygenates, and miscellaneous products.

Notes: • Stocks are at end of year. • Because vertical scales differ, graphs should not be compared.

Sources: Table 5.16.

Table 5.16 Petroleum Primary Stocks by Type, Selected Years, 1949-2007

(Million Barrels)

Year	Crude Oil and Lease Condensate			Petroleum Products									Total Petroleum
	SPR [1]	Non-SPR [2,3]	Total [3]	Distillate Fuel Oil [4]	Jet Fuel [5]	Liquefied Petroleum Gases: Propane [6]	Liquefied Petroleum Gases: Total	Motor Gasoline [7]	Residual Fuel Oil	Unfinished Oils	Other Products [8]	Total Products	
1949	0	253	253	75	(5)	(9)	1	110	60	66	37	350	603
1950	0	248	248	72	(5)	(9)	2	116	41	70	34	334	583
1955	0	266	266	111	3	(9)	7	165	39	68	55	449	715
1960	0	240	240	138	7	(9)	23	195	45	62	76	545	785
1965	0	220	220	155	19	(9)	30	175	56	89	92	616	836
1970	0	276	276	195	28	(9)	67	209	54	99	89	741	1,018
1971	0	260	260	191	28	(9)	95	219	60	101	92	784	1,044
1972	0	246	246	154	25	(9)	86	213	55	95	84	713	959
1973	0	242	242	196	29	65	99	209	53	99	80	766	1,008
1974	0	265	265	200	29	69	113	218	60	106	82	809	1,074
1975	0	271	271	209	30	82	125	235	74	106	82	862	1,133
1976	0	285	285	186	32	74	116	231	72	110	78	826	1,112
1977	7	340	348	250	35	81	136	258	90	113	82	964	1,312
1978	67	309	376	216	34	87	132	238	90	109	82	901	1,278
1979	91	339	430	229	39	64	111	237	96	118	82	911	1,341
1980	108	358	466	205	42	65	120	261	92	124	82	926	1,392
1981	230	363	594	192	41	76	135	253	78	111	80	890	1,484
1982	294	350	644	179	37	54	94	235	66	105	70	786	1,430
1983	379	344	723	140	39	48	101	222	49	108	72	731	1,454
1984	451	345	796	161	42	58	101	243	53	94	67	760	1,556
1985	493	321	814	144	40	39	74	223	50	107	67	705	1,519
1986	512	331	843	155	50	63	103	233	47	94	68	750	1,593
1987	541	349	890	134	50	48	97	226	47	93	70	718	1,607
1988	560	330	890	124	44	50	97	228	45	100	70	707	1,597
1989	580	341	921	106	41	32	80	213	44	106	70	660	1,581
1990	586	323	908	132	52	49	98	220	49	99	63	712	1,621
1991	569	325	893	144	49	48	92	219	50	98	72	724	1,617
1992	575	318	893	141	43	39	89	216	43	95	73	699	1,592
1993	587	335	922	141	40	51	106	226	44	88	78	725	1,647
1994	592	337	929	145	47	46	99	215	42	91	84	724	1,653
1995	592	303	895	130	40	43	93	202	37	86	79	668	1,563
1996	566	284	850	127	40	43	86	195	46	88	76	658	1,507
1997	563	305	868	138	44	44	89	210	40	89	81	692	1,560
1998	571	324	895	156	45	65	115	216	45	91	85	752	1,647
1999	567	284	852	125	41	43	89	193	36	86	70	641	1,493
2000	541	286	826	118	45	41	83	196	36	87	77	641	1,468
2001	550	312	862	145	42	66	121	210	41	88	78	724	1,586
2002	599	278	877	134	39	53	106	209	31	76	76	671	1,548
2003	638	269	907	137	39	50	94	207	38	76	71	661	1,568
2004	676	286	961	126	40	55	104	218	42	81	72	683	1,645
2005	685	324	1,008	136	42	57	109	208	37	86	71	689	1,698
2006	689	R312	R1,001	144	39	62	113	R212	42	84	85	R719	R1,720
2007P	697	286	983	134	39	52	95	215	39	82	75	679	1,662

[1] "SPR" is the Strategic Petroleum Reserve, which began in 1977. Crude oil stocks in the SPR include non-U.S. stocks held under foreign or commercial storage agreements.

[2] All crude oil and lease condensate stocks other than those in "SPR."

[3] Beginning in 1981, includes stocks of Alaskan crude oil in transit.

[4] Does not include stocks that are held in the Northeast Heating Oil Reserve.

[5] Through 1951, naphtha-type jet fuel is included in the products from which it was blended: in 1952, 71 percent gasoline, 17 percent kerosene, and 12 percent distillate fuel oil. Through 1964, kerosene-type jet fuel is included with kerosene in "Other Products." Beginning in 2005, naphtha-type jet fuel is included in "Other Products."

[6] Includes propylene.

[7] Finished motor gasoline, motor gasoline blending components, and gasohol. Through 1963, also includes aviation gasoline and special naphthas.

[8] Asphalt and road oil, aviation gasoline blending components, kerosene, lubricants, pentanes plus, petrochemical feedstocks, petroleum coke, waxes, other hydrocarbons and oxygenates, and miscellaneous products. Through 1964, also includes kerosene-type jet fuel. Beginning in 1964, also includes aviation gasoline and special naphthas. Beginning in 2005, also includes naphtha-type jet fuel.

[9] Included in "Liquefied Petroleum Gases Total."

R=Revised. P=Preliminary.

Notes: • Stocks are at end of year. • Totals may not equal sum of components due to independent rounding.

Web Pages: • For all data beginning in 1949, see http://www.eia.doe.gov/emeu/aer/petro.html. • For related information, see http://www.eia.doe.gov/oil_gas/petroleum/info_glance/petroleum.html

Sources: • 1949-1975—Bureau of Mines, Mineral Industry Surveys, *Petroleum Statement, Annual,* annual reports. • 1976-1980—Energy Information Administration (EIA), Energy Data Reports, *Petroleum Statement, Annual,* annual reports. • 1981-2006—EIA, *Petroleum Supply Annual,* annual reports. • 2007—EIA, *Petroleum Supply Monthly* (February 2008).

Figure 5.17 Strategic Petroleum Reserve, 1977-2007

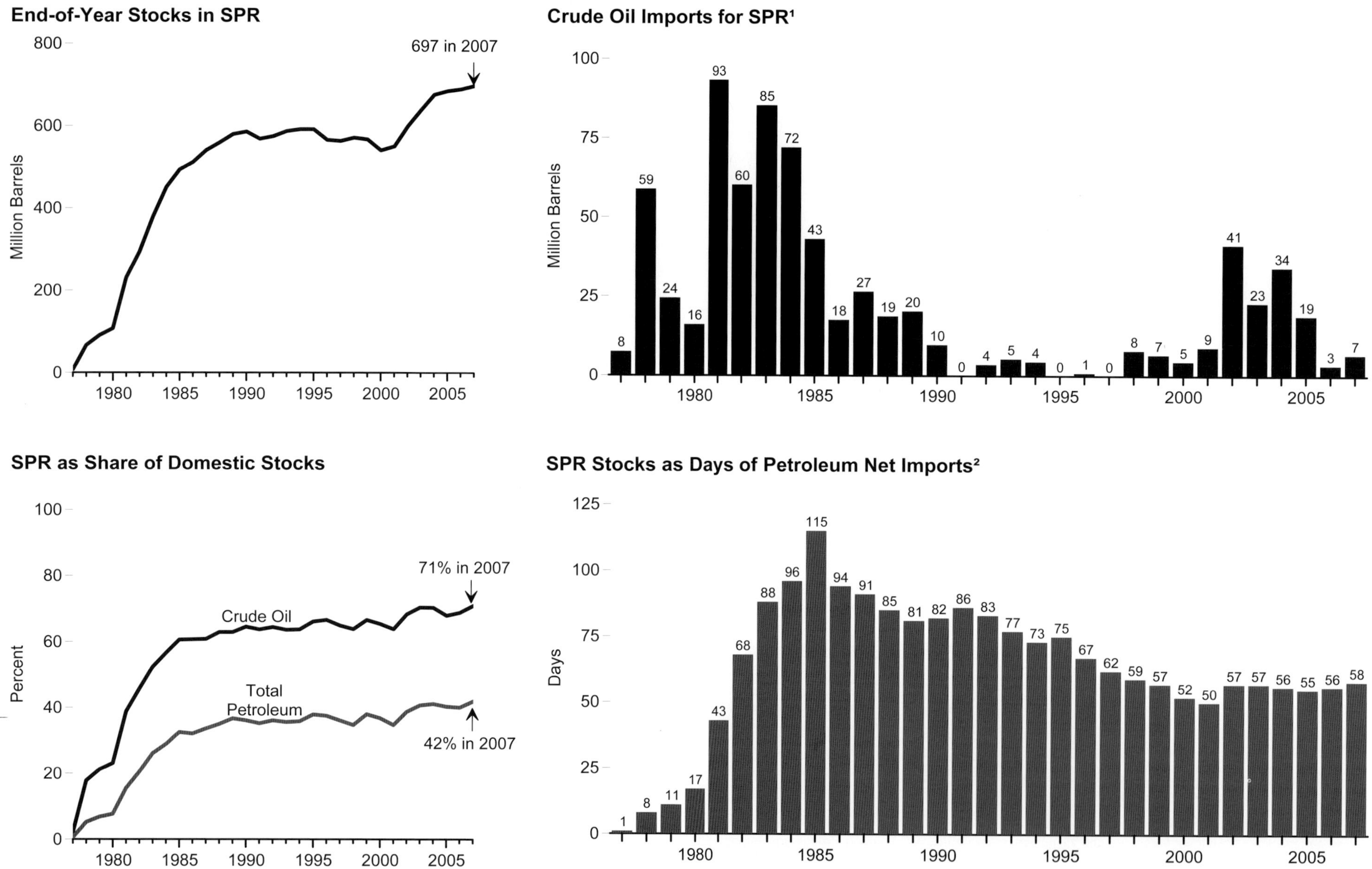

[1] Imported by SPR and imported by others for SPR.

[2] Derived by dividing end-of-year SPR stocks by annual average daily net imports of all petroleum.

Notes: • SPR=Strategic Petroleum Reserve. • Because vertical scales differ, graphs should not be compared.

Source: Table 5.17.

Table 5.17 Strategic Petroleum Reserve, 1977-2007

(Million Barrels, Except as Noted)

Year	Foreign Crude Oil Receipts		Domestic Crude Oil Receipts		Withdrawals		End-of-Year Stocks			Days of Petroleum Net Imports [5]
	Imported by SPR	Imported by Others [1,2]	Purchases	Exchanges [2]	Sales	Exchanges	Quantity [3]	Percent of Crude Oil [4] Stocks	Percent of Total Petroleum Stocks	
1977	7.54	0.00	[6]0.37	0.00	0.00	0.00	7.46	2.1	0.6	1
1978	58.80	.00	.00	.00	.00	.00	66.86	17.8	5.2	8
1979	24.43	.00	(s)	.00	.00	.00	91.19	21.2	6.8	11
1980	16.07	.00	1.30	.00	.00	.00	107.80	23.1	7.7	17
1981	93.30	.00	28.79	.00	.00	.00	230.34	38.8	15.5	43
1982	60.19	.00	3.79	.00	.00	.00	293.83	45.7	20.5	68
1983	85.29	.00	.42	.00	.00	.00	379.09	52.4	26.1	88
1984	72.04	.00	.05	.00	.00	.00	450.51	56.6	28.9	96
1985	43.12	.00	.17	.00	.00	.00	493.32	60.6	32.5	115
1986	17.56	.00	1.21	.00	.00	.00	511.57	60.7	32.1	94
1987	26.52	.00	2.69	.00	.00	.00	540.65	60.8	33.6	91
1988	18.76	.00	.01	.00	.00	.00	559.52	62.9	35.0	85
1989	20.35	.00	.00	.00	.00	.00	579.86	62.9	36.7	81
1990	9.77	.00	.00	.00	3.91	.00	585.69	64.5	36.1	82
1991	.00	.00	.00	.00	17.22	.00	568.51	63.7	35.2	86
1992	3.59	.00	2.60	.00	.00	.00	574.72	64.4	36.1	83
1993	5.37	.00	6.96	.00	.00	.00	587.08	63.6	35.6	77
1994	4.49	.00	.11	.00	.00	.00	591.67	63.7	35.8	73
1995	.00	.00	.00	.00	.00	.00	591.64	66.1	37.9	75
1996	.00	.90	.00	.00	25.82	.90	565.82	66.6	37.5	67
1997	.00	.00	.00	.00	2.33	.00	563.43	64.9	36.1	62
1998	.00	7.98	.00	.00	.00	.00	571.41	63.8	34.7	59
1999	3.04	3.60	.00	1.42	.00	10.75	567.24	66.6	38.0	57
2000	3.01	1.50	.00	2.29	.00	[7]33.35	540.68	65.4	36.8	52
2001	3.91	5.07	.58	.00	.00	.00	550.24	63.8	34.7	50
2002	5.77	35.59	.00	7.64	.00	.00	599.09	68.3	38.7	57
2003	.00	22.94	.00	16.40	.00	.00	638.39	70.4	40.7	57
2004	.00	34.24	.00	8.47	.00	5.44	675.60	70.3	41.1	56
2005	.00	18.88	.00	8.41	11.03	9.82	684.54	67.9	40.3	55
2006	.00	3.31	.00	2.44	.00	1.57	688.61	R68.8	40.0	56
2007	.00	6.67	.00	1.68	.00	.00	696.94	70.9	41.9	58

[1] Imported crude oil received represents volumes of imported crude oil received at SPR storage facilities for which the costs associated with the importation and delivery of crude oil are the responsibility of the commercial importer under contract to supply the SPR.

[2] The values shown for 1998 and 1999 represent an exchange agreement in which SPR received approximately 8.5 million barrels of high quality oil in exchange for approximately 11 million barrels of lower quality crude oil shipped from SPR during 1999 and 2000. Also, beginning in 1999, a portion of the crude oil in-kind royalties from Federal leases in the Gulf of Mexico was transferred to the Department of Energy and exchanged with commercial entities for crude oil to fill the SPR. Crude oil exchange barrels delivered to SPR could be either domestic or imported as long as the crude oil met the specification requirements of SPR. All exchange barrels of imported crude oil are included in "Foreign Crude Oil Receipts, Imported by Others," while exchange barrels of domestic crude oil are included in "Domestic Crude Oil Receipts, Exchanges."

[3] Stocks do not include imported quantities in transit to SPR terminals, pipeline fill, and above-ground storage.

[4] Includes lease condensate stocks.

[5] Derived by dividing end-of-year SPR stocks by annual average daily net imports of all petroleum. Calculated prior to rounding.

[6] The quantity of domestic fuel oil which was in storage prior to injection of foreign crude oil.

[7] Includes 30 million barrels released to increase heating oil stocks in exchange for a like quantity plus a bonus percentage to be returned in 2001 and 2002, as well as additional barrels to create a Northeast Home Heating Oil Reserve.

R=Revised. (s)=Less than 0.005 million barrels.

Note: "SPR" is the Strategic Petroleum Reserve—petroleum stocks maintained by the Federal Government for use during periods of major supply interruption.

Web Page: See http://www.eia.doe.gov/oil_gas/petroleum/info_glance/petroleum.html for related information.

Sources: **Imported by SPR** and **End-of-Year Stocks, Quantity:** • 1977-1980—Energy Information Administration (EIA), Energy Data Report, *Petroleum Statement, Annual,* annual reports. • 1981-2006—EIA, *Petroleum Supply Annual,* annual reports. • 2007—EIA, *Petroleum Supply Monthly* (February 2008). **Imported by Others, Domestic Crude Oil Receipts,** and **Withdrawals:** U.S. Department of Energy, Assistant Secretary for Fossil Energy, unpublished data. **All Other Data:** Calculated.

Figure 5.18 Crude Oil Domestic First Purchase Prices

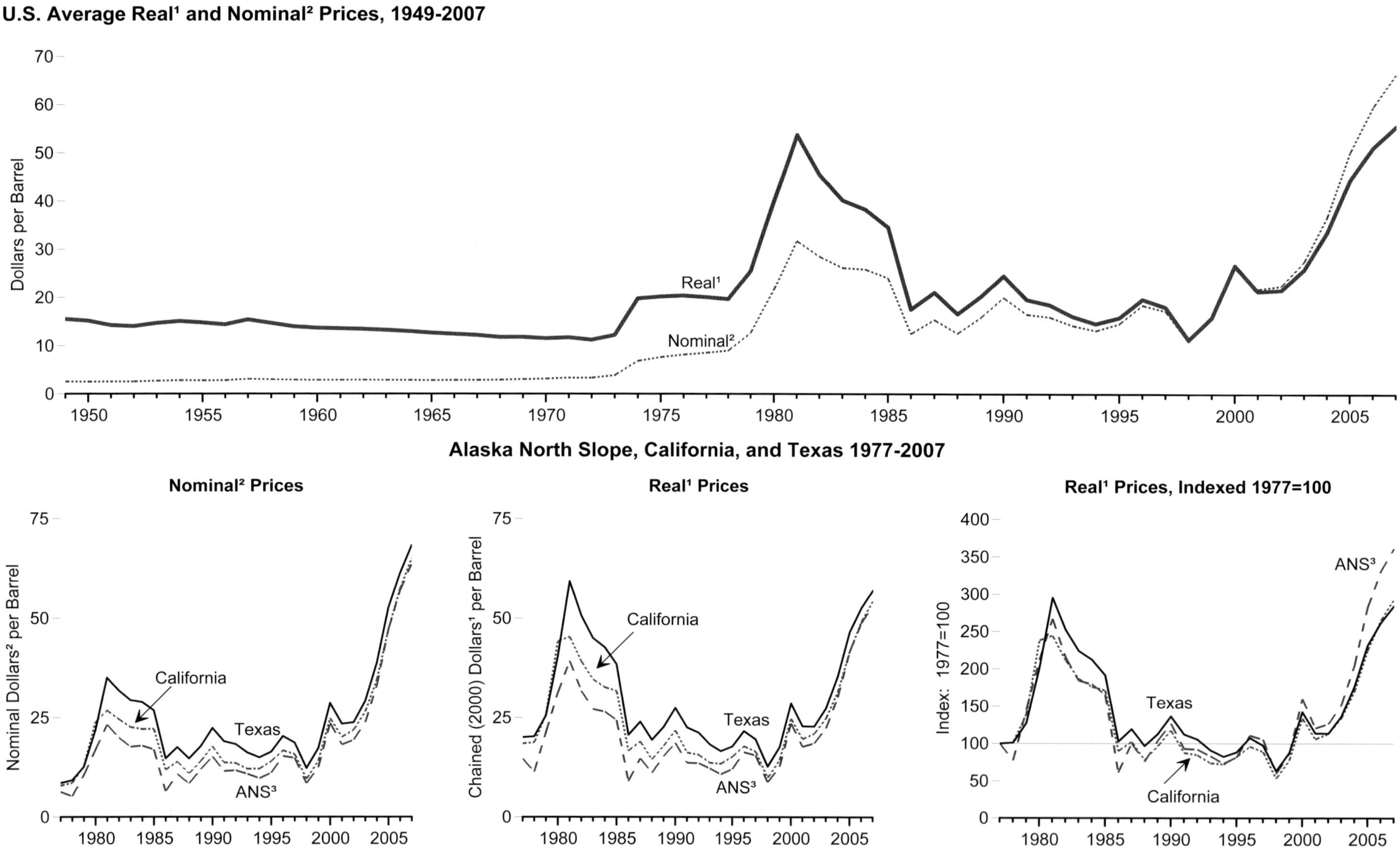

[1] In chained (2000) dollars, calculated by using gross domestic product implicit price deflators. See Table D1.

[2] See "Nominal Dollars" in Glossary.

[3] Alaska North Slope.

Source: Table 5.18.

Table 5.18 Crude Oil Domestic First Purchase Prices, Selected Years, 1949-2007
(Dollars per Barrel)

Year	Alaska North Slope		California		Texas		U.S. Average	
	Nominal [1]	Real [2]	Nominal [1]	Real [2]	Nominal [1]	Real [2]	Nominal [1]	Real [2]
1949	– –	– –	[R]NA	[R]NA	[R]NA	[R]NA	2.54	15.53
1950	– –	– –	[R]NA	[R]NA	[R]NA	[R]NA	2.51	15.18
1955	– –	– –	[R]NA	[R]NA	[R]NA	[R]NA	2.77	14.78
1960	[R]NA	[R]NA	[R]NA	[R]NA	[R]NA	[R]NA	2.88	13.69
1965	[R]NA	[R]NA	[R]NA	[R]NA	[R]NA	[R]NA	2.86	12.69
1970	[R]NA	[R]NA	[R]NA	[R]NA	[R]NA	[R]NA	3.18	11.55
1971	[R]NA	[R]NA	[R]NA	[R]NA	[R]NA	[R]NA	3.39	11.73
1972	[R]NA	[R]NA	[R]NA	[R]NA	[R]NA	[R]NA	3.39	11.24
1973	[R]NA	[R]NA	[R]NA	[R]NA	[R]NA	[R]NA	3.89	12.21
1974	[R]NA	[R]NA	[R]NA	[R]NA	[R]NA	[R]NA	6.87	19.78
1975	[R]NA	[R]NA	[R]NA	[R]NA	[R]NA	[R]NA	7.67	20.18
1976	[R]NA	[R]NA	[R]NA	[R]NA	[R]NA	[R]NA	8.19	20.38
1977	[3]6.29	[3]14.71	7.92	18.53	8.58	20.07	8.57	20.05
1978	5.21	11.39	8.58	18.75	9.29	20.30	9.00	19.67
1979	10.57	21.33	12.78	25.79	12.65	25.53	12.64	25.51
1980	16.87	31.22	23.87	44.17	21.84	40.41	21.59	39.95
1981	23.23	39.29	26.80	45.33	35.06	59.30	31.77	53.74
1982	19.92	31.76	24.58	39.19	31.77	50.65	28.52	45.47
1983	17.69	27.13	22.61	34.67	29.35	45.01	26.19	40.16
1984	17.91	26.47	22.09	32.65	28.87	42.67	25.88	38.25
1985	16.98	24.36	22.14	31.76	26.80	38.44	24.09	34.56
1986	6.45	9.05	11.90	16.70	14.73	20.67	12.51	17.56
1987	10.83	14.80	13.92	19.02	17.55	23.98	15.40	21.04
1988	8.43	11.14	10.97	14.49	14.71	19.43	12.58	16.62
1989	12.00	15.28	14.06	17.90	17.81	22.67	15.86	20.19
1990	15.23	18.67	17.81	21.83	22.37	27.42	20.03	24.55
1991	11.57	13.70	13.72	16.25	19.04	22.55	16.54	19.59
1992	11.73	13.58	13.55	15.69	18.32	21.21	15.99	18.51
1993	10.84	12.27	12.11	13.70	16.19	18.32	14.25	16.12
1994	9.77	10.82	12.12	13.43	14.98	16.60	13.19	14.61
1995	11.12	12.07	14.00	15.20	16.38	17.78	14.62	15.87
1996	15.32	16.32	16.72	17.82	20.31	21.64	18.46	19.67
1997	14.84	15.55	15.78	16.54	18.66	19.56	17.23	18.06
1998	8.47	8.78	9.55	9.90	12.28	12.73	10.87	11.27
1999	12.46	12.73	14.08	14.39	17.29	17.67	15.56	15.90
2000	23.62	23.62	24.82	24.82	28.60	28.60	26.72	26.72
2001	18.18	17.75	20.11	19.64	23.41	22.86	21.84	21.33
2002	19.37	18.59	21.87	20.99	23.77	22.81	22.51	21.61
2003	23.78	22.35	26.43	24.84	29.13	27.38	27.56	25.90
2004	33.03	[R]30.17	34.47	[R]31.49	38.79	[R]35.44	36.77	[R]33.59
2005	47.05	[R]41.64	47.08	[R]41.66	52.61	[R]46.56	50.28	[R]44.50
2006	[R]56.86	[R]48.78	57.34	[R]49.19	61.31	[R]52.60	[R]59.69	[R]51.21
2007[P]	63.69	53.22	65.07	54.38	68.31	57.08	66.52	55.59

[1] See "Nominal Dollars" in Glossary.
[2] In chained (2000) dollars, calculated by using gross domestic product implicit price deflators in Table D1. See "Chained Dollars" in Glossary.
[3] Average for July through December only.
R=Revised. P=Preliminary. NA=Not available. – – = Not applicable.
Note: Prices are for the marketed first sales price of domestic crude oil. See Note 5, "Crude Oil Domestic First Purchase Prices," at end of section.
Web Pages: • For all data beginning in 1949, see http://www.eia.doe.gov/emeu/aer/petro.html. • For related information, see http://www.eia.doe.gov/oil_gas/petroleum/info_glance/petroleum.html
Sources: • 1949-1973—Bureau of Mines, *Minerals Yearbook,* "Crude Petroleum and Petroleum Products" chapter. • 1974-January 1976—Federal Energy Administration (FEA), Form FEA-90, "Crude Petroleum Production Monthly Report." • February 1976-1977—FEA, Form FEA-P-124, "Domestic Crude Oil Purchaser's Monthly Report." • 1978-1982—Energy Information Administration (EIA), Form ERA-182, "Domestic Crude Oil First Purchaser's Report." • 1983 forward—EIA, *Petroleum Marketing Monthly* (April 2008), Table 18.

Figure 5.19 Landed Costs of Crude Oil Imports From Selected Countries

Total, 1973[1]-2007

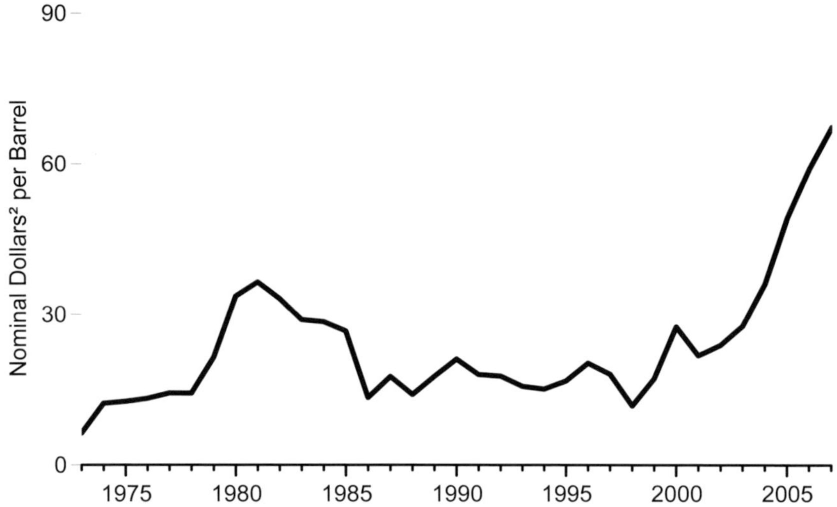

By Selected Country, 2007

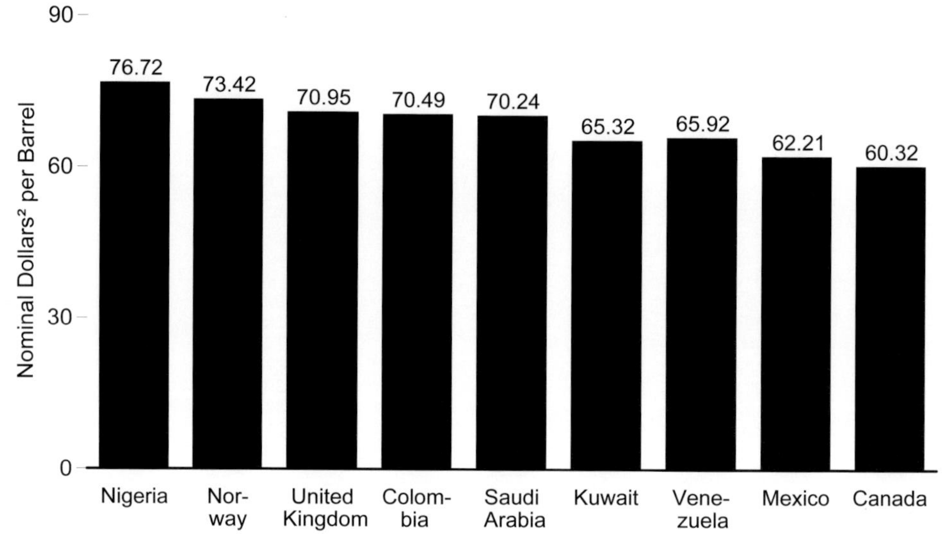

By Selected OPEC Country, 1973[1]-2007

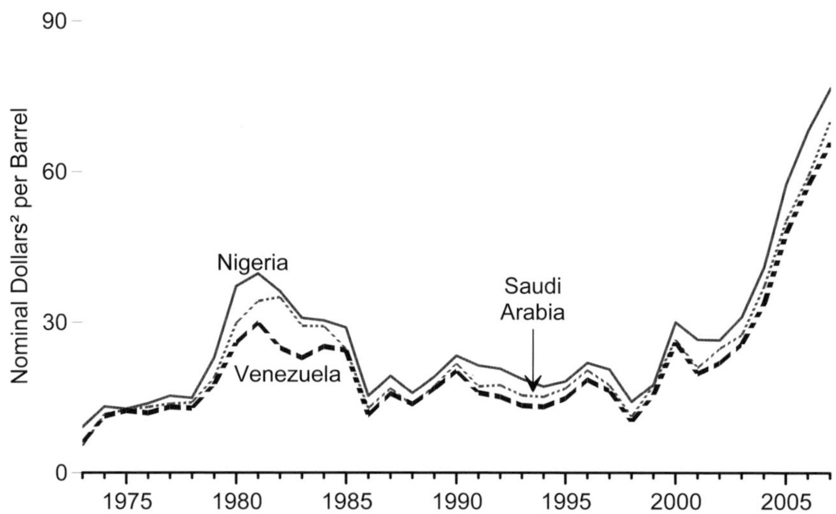

By Selected Non-OPEC Country, 1973[1]-2007

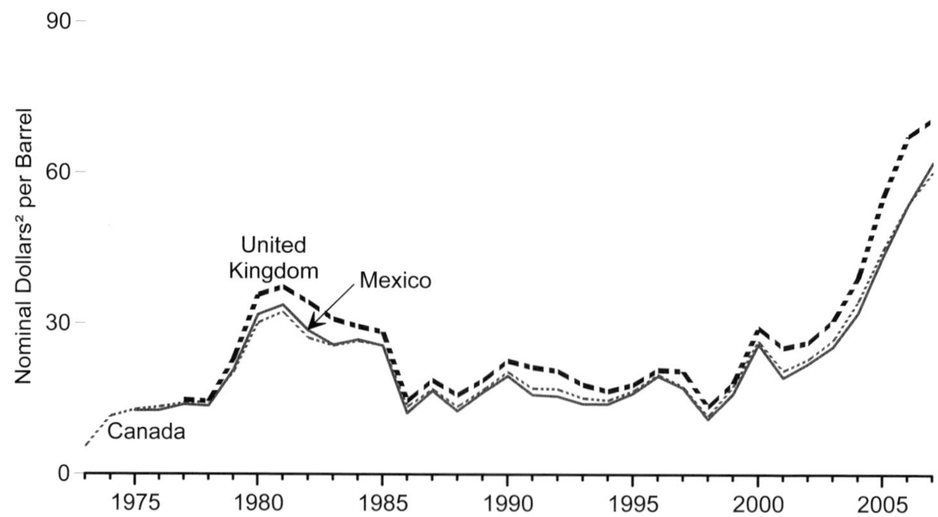

[1] 1973 cost is based on October, November, and December data only.
[2] See "Nominal Dollars" in Glossary.

Note: OPEC=Organization of the Petroleum Exporting Countries.
Source: Table 5.19

Table 5.19 Landed Costs of Crude Oil Imports From Selected Countries, 1973-2007

(Norminal Dollars [1] per Barrel)

Year	Persian Gulf [3]	Selected OPEC [2] Countries: Kuwait	Nigeria	Saudi Arabia	Venezuela	Total OPEC [4]	Selected Non-OPEC [2] Countries: Canada	Colombia	Mexico	Norway	United Kingdom	Total Non-OPEC [4]	Total
1973 [5]	5.91	W	9.08	5.37	5.99	6.85	5.33	W	–	–	–	5.64	6.41
1974	12.21	W	13.16	11.63	11.25	12.49	11.48	W	W	–	–	11.81	12.32
1975	12.64	W	12.70	12.50	12.36	12.70	12.84	–	12.61	12.80	–	12.70	12.70
1976	13.03	W	13.81	13.06	11.89	13.32	13.36	–	12.64	13.74	W	13.35	13.32
1977	13.85	W	15.29	13.69	13.11	14.35	14.13	–	13.82	14.93	14.83	14.42	14.36
1978	14.01	W	14.88	13.94	12.84	14.34	14.41	–	13.56	14.68	14.53	14.38	14.35
1979	20.42	W	22.97	18.95	17.65	21.29	20.22	–	20.77	22.55	22.97	22.10	21.45
1980	30.59	W	37.15	29.80	25.92	33.56	30.11	W	31.77	36.82	35.68	33.99	33.67
1981	34.61	–	39.66	34.20	29.91	36.60	32.32	–	33.70	38.70	37.29	36.14	36.47
1982	34.94	–	36.16	34.99	24.93	34.81	27.15	–	28.63	34.70	34.25	31.47	33.18
1983	29.37	–	30.85	29.27	22.94	29.84	25.63	–	25.78	30.72	30.87	28.08	28.93
1984	29.07	W	30.36	29.20	25.19	29.06	26.56	–	26.85	30.05	29.45	28.14	28.54
1985	25.50	–	28.96	24.72	24.43	26.86	25.71	–	25.63	28.32	28.36	26.53	26.67
1986	12.92	11.70	15.29	12.84	11.52	13.46	13.43	12.85	12.17	15.98	14.63	13.52	13.49
1987	17.47	18.14	19.32	16.81	15.76	17.64	17.04	18.43	16.69	19.10	18.78	17.66	17.65
1988	13.51	12.84	15.88	13.37	13.66	14.18	13.50	14.47	12.58	15.43	15.82	13.96	14.08
1989	17.37	16.90	19.19	17.34	16.78	17.78	16.81	18.10	16.35	19.06	18.74	17.54	17.68
1990	20.55	17.01	23.33	21.82	20.31	21.23	20.48	22.34	19.64	21.11	22.65	20.98	21.13
1991	17.34	18.48	21.39	17.22	15.92	18.08	17.16	19.55	15.89	21.44	21.37	17.93	18.02
1992	17.58	16.99	20.78	17.48	15.13	17.81	17.04	18.46	15.60	20.90	20.63	17.67	17.75
1993	15.26	14.23	18.73	15.40	13.39	15.68	15.27	16.54	14.11	18.99	17.92	15.78	15.72
1994	15.00	14.49	17.21	15.11	13.12	15.08	14.83	15.80	14.09	17.09	16.64	15.29	15.18
1995	16.78	16.47	18.25	16.84	14.81	16.61	16.65	17.45	16.19	18.06	17.91	16.95	16.78
1996	20.45	20.32	21.95	20.49	18.59	20.14	19.94	22.02	19.64	21.34	20.88	20.47	20.31
1997	17.44	17.03	20.64	17.52	16.35	17.73	17.63	19.71	17.30	20.26	20.64	18.45	18.11
1998	11.18	11.00	14.14	11.16	10.16	11.46	11.62	13.26	11.04	13.83	13.55	12.22	11.84
1999	17.37	16.77	17.63	17.48	15.58	16.94	17.54	18.09	16.12	19.06	18.26	17.51	17.23
2000	26.77	26.28	30.04	26.58	26.05	27.29	26.69	29.68	26.03	30.13	29.26	27.80	27.53
2001	20.73	19.66	26.55	20.98	19.81	21.52	20.72	25.88	19.37	25.77	25.32	22.17	21.82
2002	24.13	23.04	26.45	24.77	21.93	23.83	22.98	25.28	22.09	26.60	26.35	23.97	23.91
2003	27.54	26.82	31.07	27.50	25.70	27.70	26.76	30.55	25.48	30.51	30.62	27.68	27.69
2004	36.53	35.89	40.95	37.11	33.79	36.84	34.51	39.03	32.25	39.92	39.28	35.29	36.07
2005	49.68	48.36	57.55	50.31	47.87	51.36	44.73	53.42	43.47	56.23	55.28	47.31	49.29
2006	R58.92	R57.64	R68.26	R59.19	R57.37	R61.21	R53.90	R62.13	R53.76	64.39	67.44	57.14	R59.11
2007P	69.14	65.32	76.72	70.24	65.92	70.38	60.32	70.49	62.21	73.42	70.95	63.65	67.35

[1] See "Nominal Dollars" in Glossary.
[2] See "Organization of the Petroleum Exporting Countries (OPEC)" in Glossary.
[3] Bahrain, Iran, Iraq, Kuwait, Qatar, Saudi Arabia, United Arab Emirates, and the Neutral Zone (between Kuwait and Saudi Arabia).
[4] On this table, "Total OPEC" for all years includes Algeria, Indonesia, Iran, Iraq, Kuwait, Libya, Nigeria, Qatar, Saudi Arabia, United Arab Emirates, and Venezuela; for 1973-1992, also includes Ecuador (although Ecuador rejoined OPEC in November 2007, on this table Ecuador is included in "Total Non-OPEC" for 2007); for 1974-1995, also includes Gabon (although Gabon was a member of OPEC for only 1975-1994); and beginning in 2007, also includes Angola. Data for all countries not included in "Total OPEC" are included in "Total Non-OPEC."
[5] Based on October, November, and December data only.
R=Revised. P=Preliminary. – = No data reported. W=Value withheld to avoid disclosure of individual company data.

Notes: • Data are for landed costs of crude oil imports only; they do not account for refined petroleum products imported into the United States. • See "Crude Oil Landed Cost" in Glossary. • Totals may not equal sum of components due to independent rounding.

Web Page: See http://www.eia.doe.gov/oil_gas/petroleum/info_glance/petroleum.html for related information.

Sources: • 1973-September 1977—Federal Energy Administration, Form FEA-F701-M-0, "Transfer Pricing Report." • October 1977-December 1978—Energy Information Administration (EIA), Form FEA-F701-M-0, "Transfer Pricing Report." • January 1979-September 1982—EIA, Form ERA-51, "Transfer Pricing Report." • October 1982-June 1984—EIA, Form EP-51, "Monthly Foreign Crude Oil Transaction Report." • July 1984 forward—EIA, *Petroleum Marketing Monthly* (March 2008), Table 22; and EIA, Form EIA-856, "Monthly Foreign Crude Oil Acquisition Report."

Figure 5.20 Value of Crude Oil Imports

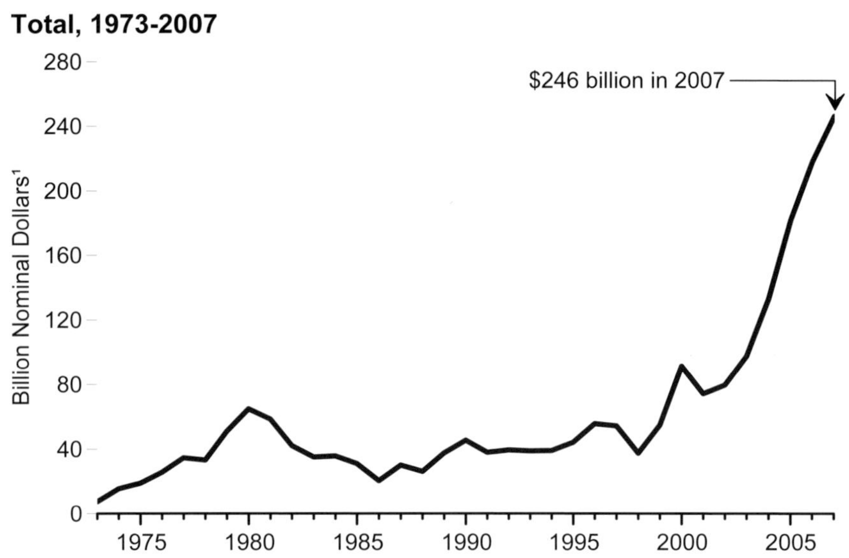

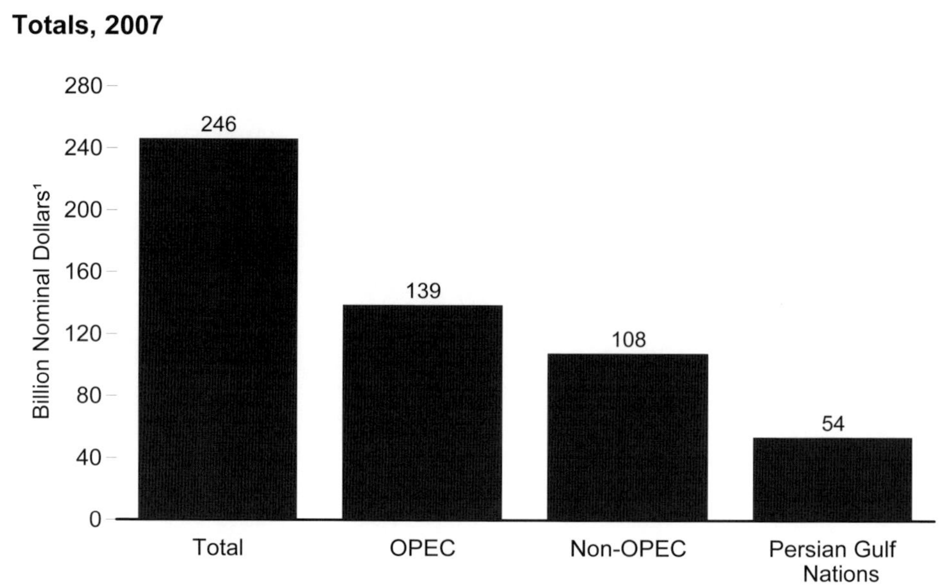

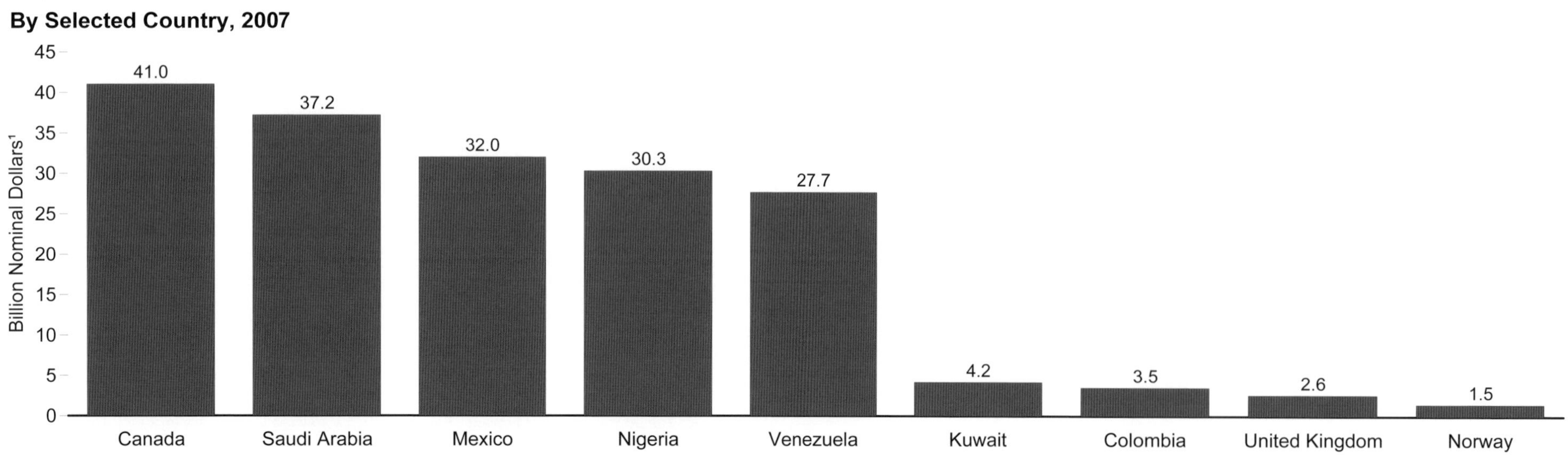

[1] See "Nominal Dollars" in Glossary.

Notes: • OPEC=Organization of the Petroleum Exporting Countries. • Because vertical scales differ, graphs should not be compared.

Source: Table 5.20.

Table 5.20 Value of Crude Oil Imports From Selected Countries, 1973-2007

(Billion Nominal Dollars [1])

Year	Persian Gulf [3]	Selected OPEC [2] Countries					Selected Non-OPEC [2] Countries						Total [5]
		Kuwait	Nigeria	Saudi Arabia	Venezuela	Total OPEC [4]	Canada	Colombia	Mexico	Norway	United Kingdom	Total Non-OPEC [4]	
1973	1.7	W	1.5	0.9	0.8	5.2	1.9	W	–	0.0	0.0	2.4	7.6
1974	4.4	W	3.3	1.9	1.3	11.6	3.3	.0	W	–	.0	4.1	15.6
1975	5.2	W	3.5	3.2	1.8	14.9	2.8	.0	.3	.1	–	4.1	19.0
1976	8.7	W	5.1	5.8	1.0	22.2	1.8	–	.4	.2	W	3.6	25.8
1977	12.2	W	6.3	6.9	1.2	29.6	1.4	.0	.9	.3	.5	5.1	34.7
1978	11.3	W	4.9	5.8	.8	27.1	1.3	.0	1.6	.6	.9	6.2	33.3
1979	15.3	W	9.0	9.3	1.9	39.7	2.0	.0	3.3	.6	1.7	11.3	51.0
1980	16.9	W	11.4	13.6	1.5	47.5	2.2	.0	5.9	1.9	2.3	17.4	64.9
1981	15.1	.0	8.8	13.9	1.6	39.0	1.9	.0	5.8	1.6	5.0	19.5	58.5
1982	8.4	–	6.7	6.8	1.4	22.0	2.1	.0	6.7	1.3	5.5	20.2	42.2
1983	4.3	–	3.4	3.4	1.4	16.1	2.6	.0	7.2	.7	4.1	19.1	35.2
1984	4.8	W	2.3	3.3	2.3	16.1	3.3	.0	6.5	1.2	4.1	19.7	35.8
1985	2.3	–	3.0	1.2	2.7	12.9	4.4	.0	6.7	.3	2.9	18.3	31.2
1986	3.8	.1	2.4	2.9	1.8	10.4	2.8	.3	2.8	.3	1.7	10.2	20.6
1987	6.0	.5	3.7	3.9	2.8	15.5	3.8	.8	3.7	.5	2.1	14.7	30.1
1988	6.7	.4	3.5	4.4	2.2	14.0	3.4	.6	3.1	.3	1.5	12.3	26.3
1989	11.0	1.0	5.6	7.1	3.0	21.9	3.9	.9	4.3	.9	1.1	15.8	37.7
1990	13.5	.5	6.7	9.5	4.9	27.2	4.8	1.1	4.9	.7	1.3	18.2	45.5
1991	11.0	(s)	5.3	10.7	3.9	22.3	4.7	.9	4.4	.6	.8	15.7	38.0
1992	10.5	.2	5.1	10.2	4.6	22.2	5.0	.7	4.5	.9	1.5	17.3	39.5
1993	9.1	1.8	4.9	7.2	4.9	20.7	5.0	.9	4.4	.9	2.0	18.3	38.9
1994	8.8	1.6	3.9	7.2	5.0	19.7	5.3	.8	4.8	1.2	2.4	19.4	39.1
1995	9.1	1.3	4.1	7.7	6.2	21.6	6.3	1.3	6.1	1.7	2.2	22.6	44.3
1996	11.1	1.8	4.8	9.4	8.9	25.3	7.8	1.8	8.7	2.3	1.6	30.5	55.8
1997	10.4	1.6	5.2	8.3	8.3	24.4	7.7	1.9	8.6	2.1	1.3	29.9	54.4
1998	8.3	1.2	3.6	5.7	5.1	17.4	5.4	1.7	5.3	1.1	.8	20.2	37.6
1999	15.0	1.5	4.0	8.8	6.5	26.1	7.5	3.0	7.4	1.8	1.9	28.8	54.9
2000	23.6	2.5	9.6	14.8	11.7	45.4	13.2	3.5	12.5	3.3	3.1	46.0	91.4
2001	20.2	1.7	8.2	12.3	9.3	38.1	10.3	2.5	9.9	2.6	2.3	36.2	74.3
2002	19.5	1.8	5.7	13.7	9.6	35.5	12.1	2.2	12.1	3.4	3.9	44.3	79.8
2003	24.4	2.0	9.4	17.3	11.1	46.3	15.1	1.8	14.6	2.0	4.0	51.4	97.7
2004	32.1	3.2	16.2	20.3	16.0	68.0	20.4	2.0	18.9	2.1	3.4	65.2	133.2
2005	40.0	4.0	22.6	26.5	21.7	90.3	26.7	3.0	24.7	2.4	4.5	91.9	182.2
2006	[R]46.5	3.8	[R]25.8	[R]30.7	23.9	[R]106.9	[R]35.5	3.2	30.9	2.3	3.2	[R]111.4	[R]218.3
2007[P]	53.6	4.2	30.3	37.2	27.7	138.6	41.0	3.5	32.0	1.5	2.6	107.7	246.2

[1] See "Nominal Dollars" in Glossary.
[2] See "Organization of the Petroleum Exporting Countries (OPEC)" in Glossary.
[3] Bahrain, Iran, Iraq, Kuwait, Qatar, Saudi Arabia, United Arab Emirates, and the Neutral Zone (between Kuwait and Saudi Arabia).
[4] On this table, "Total OPEC" for all years includes Algeria, Indonesia, Iran, Iraq, Kuwait, Libya, Nigeria, Qatar, Saudi Arabia, United Arab Emirates, and Venezuela; for 1973-1992, also includes Ecuador (although Ecuador rejoined OPEC in November 2007, on this table Ecuador is included in "Total Non-OPEC" for 2007); for 1974-1995, also includes Gabon (although Gabon was a member of OPEC for only 1975-1994); and beginning in 2007, also includes Angola. Data for all countries not included in "Total OPEC" are included in "Total Non-OPEC."
[5] Data shown here represent landed value; they differ from data in Table 3.7, which are data from U.S. Customs that represent crude oil value at the port of loading.

R=Revised. P=Preliminary. – = No data reported. (s)=Less than $0.05 billion. W=Value withheld to avoid disclosure of individual company data.

Notes: • Crude oil import volumes used to calculate values in this table are for the 50 States and the District of Columbia. • Totals may not equal sum of components due to independent rounding.

Web Page: See http://www.eia.doe.gov/oil_gas/petroleum/info_glance/petroleum.html for related information.

Sources: Calculated by using prices on Table 5.19 and volume data from the following sources: • 1973-1975—U.S. Department of the Interior, Bureau of Mines, *Petroleum Statement, Annual,* annual reports. • 1976-1980—Energy Information Administration (EIA), *Petroleum Statement, Annual,* annual reports. • 1981-2006—EIA, *Petroleum Supply Annual,* annual reports. • 2007—EIA, *Petroleum Supply Monthly* (February 2008).

Figure 5.21 Crude Oil Refiner Acquisition Costs, 1968-2007

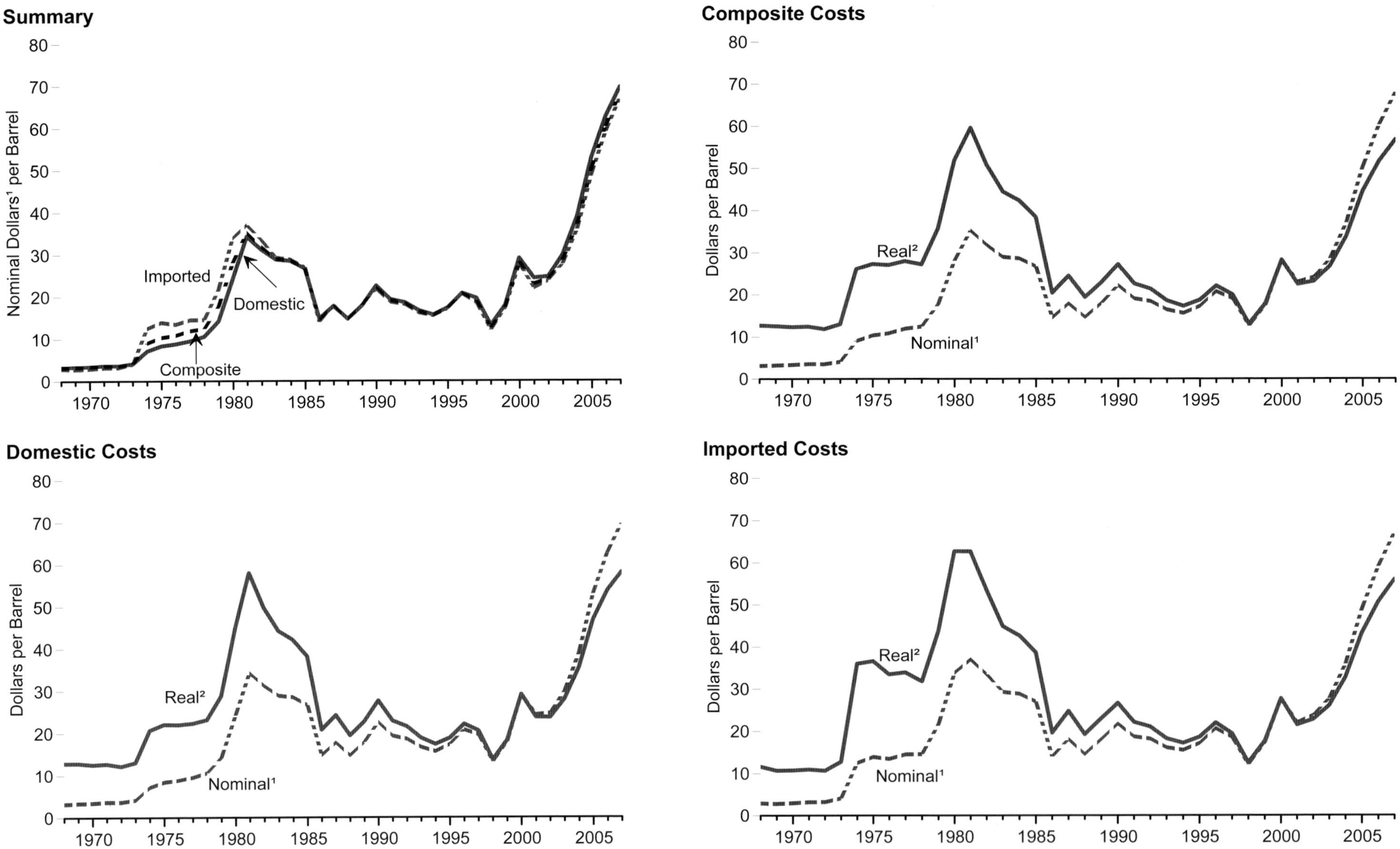

[1] See "Nominal Dollars" in Glossary.
[2] In chained (2000) dollars, calculated by using gross domestic product implicit price deflators. See Table D1.

Note: Because vertical scales differ, graphs should not be compared.
Source: Table 5.21.

Table 5.21 Crude Oil Refiner Acquisition Costs, 1968-2007

(Dollars per Barrel)

Year	Domestic		Imported		Composite	
	Nominal [1]	Real [2]	Nominal [1]	Real [2]	Nominal [1]	Real [2]
1968[E]	3.21	12.88	2.90	11.64	3.17	12.72
1969[E]	3.37	12.89	2.80	10.71	3.29	12.58
1970[E]	3.46	12.57	2.96	10.75	3.40	12.35
1971[E]	3.68	12.73	3.17	10.96	3.60	12.45
1972[E]	3.67	12.17	3.22	10.67	3.58	11.87
1973[E]	4.17	13.09	4.08	12.81	4.15	13.03
1974	7.18	20.68	12.52	36.05	9.07	26.12
1975	8.39	22.08	13.93	36.66	10.38	27.31
1976	8.84	21.99	13.48	33.54	10.89	27.09
1977	9.55	22.34	14.53	33.99	11.96	27.98
1978	10.61	23.19	14.57	31.84	12.46	27.23
1979	14.27	28.80	21.67	43.74	17.72	35.76
1980	24.23	44.83	33.89	62.71	28.07	51.94
1981	34.33	58.07	37.05	62.67	35.24	59.61
1982	31.22	49.77	33.55	53.49	31.87	50.81
1983	28.87	44.27	29.30	44.93	28.99	44.46
1984	28.53	42.17	28.88	42.69	28.63	42.32
1985	26.66	38.24	26.99	38.72	26.75	38.37
1986	14.82	20.80	14.00	19.65	14.55	20.42
1987	17.76	24.26	18.13	24.77	17.90	24.45
1988	14.74	19.47	14.56	19.24	14.67	19.38
1989	17.87	22.75	18.08	23.02	17.97	22.88
1990	22.59	27.69	21.76	26.67	22.22	27.23
1991	19.33	22.89	18.70	22.14	19.06	22.57
1992	18.63	21.57	18.20	21.07	18.43	21.33
1993	16.67	18.86	16.14	18.26	16.41	18.57
1994	15.67	17.36	15.51	17.18	15.59	17.27
1995	17.33	18.82	17.14	18.61	17.23	18.71
1996	20.77	22.13	20.64	21.99	20.71	22.07
1997	19.61	20.55	18.53	19.42	19.04	19.96
1998	13.18	13.66	12.04	12.48	12.52	12.98
1999	17.90	18.29	17.26	17.64	17.51	17.89
2000	29.11	29.11	27.70	27.70	28.26	28.26
2001	24.33	23.76	22.00	21.48	22.95	22.41
2002	24.65	23.66	23.71	22.76	24.10	23.13
2003	29.82	28.03	27.71	26.04	28.53	26.81
2004	38.97	R35.60	35.90	R32.80	36.98	R33.78
2005	52.94	R46.85	48.86	R43.24	50.24	R44.46
2006	R62.62	R53.72	R59.02	R50.63	R60.24	R51.68
2007[P]	69.63	58.19	67.02	56.01	67.93	56.77

[1] See "Nominal Dollars" in Glossary.
[2] In chained (2000) dollars, calculated by using gross domestic product implicit price deflators in Table D1. See "Chained Dollars" in Glossary.
R=Revised. P=Preliminary. E=Estimate.
Note: Costs are for crude oil to refiners, including transportation and other fees; they do not include crude oil purchased for the Strategic Petroleum Reserve. The cost for each category and for the composite is derived by dividing the sum of the total purchasing (acquisition) costs of all refiners by the total volume of all refiners' purchases.
Web Page: See http://www.eia.doe.gov/oil_gas/petroleum/info_glance/petroleum.html for related information.

Sources: • 1968-1973—Energy Information Administration (EIA) estimates. The cost of domestic crude oil was derived by adding estimated transportation costs to the reported average domestic first purchase value. The cost of imported crude oil was derived by adding an estimated ocean transport cost based on the published "Average Freight Rate Assessment" to the average "Free Alongside Ship" value published by the U.S. Bureau of the Census. The composite cost was derived by weighting domestic costs and imported costs on the basis of quantities produced and imported. • 1974-January 1976—Federal Energy Administration (FEA), Form FEA-96, "Monthly Cost Allocation Report." • February 1976-June 1978—FEA, Form FEA-P110-M-1, "Refiners' Monthly Cost Allocation Report." • July 1978-1982—EIA, Form ERA-49, "Domestic Crude Oil Entitlements Program Refiners' Monthly Report." • 1983 forward—EIA, *Petroleum Marketing Monthly* (April 2008), Table 1.

Figure 5.22 Refiner Sales Prices for Selected Petroleum Products, 1991-2007

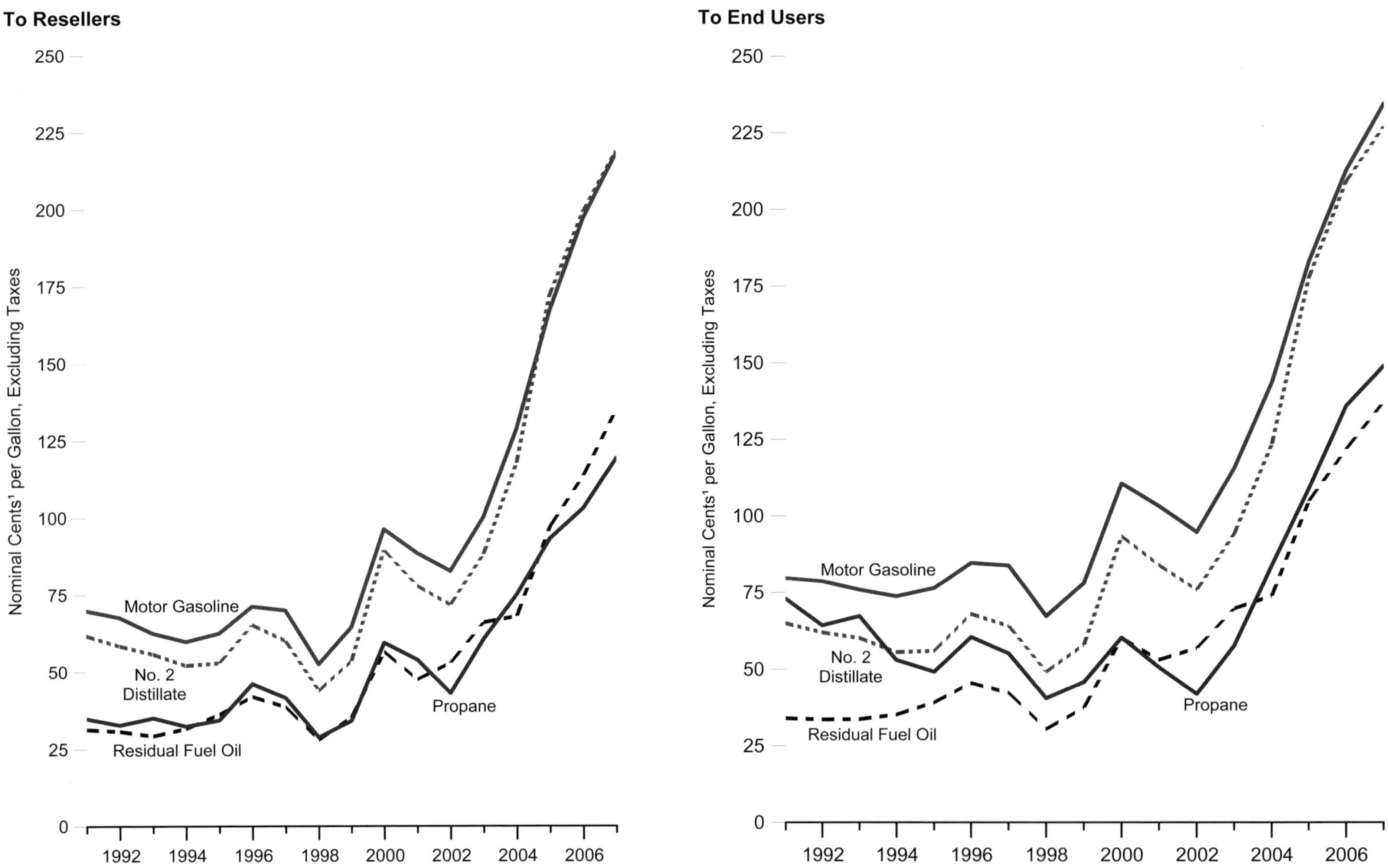

[1] See "Nominal Price" in Glossary.

Source: Table 5.22.

Table 5.22 Refiner Sales Prices and Refiner Margins for Selected Petroleum Products, 1991-2007

(Nominal Cents [1] per Gallon, Excluding Taxes)

Product	1991	1992	1993	1994	1995	1996	1997	1998	1999	2000	2001	2002	2003	2004	2005	2006	2007 [P]
Sales Prices to Resellers: [2]																	
Aviation Gasoline	100.1	99.1	96.5	93.3	97.5	105.5	106.5	91.2	100.7	133.0	125.6	114.6	128.8	162.7	207.6	249.0	274.6
Motor Gasoline	69.9	67.7	62.6	59.9	62.6	71.3	70.0	52.6	64.5	96.3	88.6	82.8	100.2	128.8	167.0	R196.9	218.2
Leaded Regular	65.7	69.3	NA	NA	NA	NA	NA	NA	NA	NA	NA	NA	NA	NA	NA	NA	NA
Unleaded Regular	67.2	64.5	59.3	56.6	59.3	68.5	67.3	49.9	62.0	94.2	86.5	80.6	98.1	126.9	165.4	195.0	216.0
Unleaded Midgrade	73.3	70.8	66.0	63.8	67.0	75.9	74.9	57.6	69.6	101.3	94.5	88.5	106.1	134.0	170.8	201.6	224.4
Premium	79.2	77.4	72.2	69.5	72.2	80.3	79.2	61.7	72.6	105.5	98.0	92.8	111.3	140.8	178.9	R211.7	235.7
Kerosene	72.2	63.2	60.4	61.8	58.0	71.4	65.3	46.5	55.0	96.9	82.1	75.2	95.5	127.1	175.7	R200.7	224.9
Jet Fuel, Kerosene-Type	65.0	60.5	57.7	53.4	53.9	64.6	61.3	45.0	53.3	88.0	76.3	71.6	87.1	120.8	172.3	R196.1	216.4
No. 1 Distillate	73.0	65.2	64.6	61.5	62.5	75.1	72.3	51.3	63.4	101.9	88.3	80.5	103.3	128.9	180.1	R204.4	243.1
No. 2 Distillate	61.8	58.5	55.9	52.2	53.0	65.3	60.2	43.9	53.6	89.6	77.9	71.8	88.2	117.8	172.0	R199.1	219.0
No. 2 Fuel Oil	62.2	57.9	54.4	50.6	51.1	63.9	59.0	42.2	49.3	88.6	75.6	69.4	88.1	112.5	162.3	R183.4	206.3
No. 2 Diesel Fuel	61.5	59.1	57.0	52.9	53.8	65.9	60.6	44.4	54.6	89.8	77.5	72.4	88.3	118.7	173.7	R201.2	220.4
No. 4 Fuel [3]	55.6	49.5	48.8	46.2	46.3	60.3	55.1	38.3	43.0	77.8	69.7	66.3	79.3	103.3	137.7	R139.5	155.1
Residual Fuel Oil	31.4	30.8	29.3	31.7	36.3	42.0	38.7	28.0	35.4	56.6	47.6	53.0	66.1	68.1	97.1	R113.6	135.0
Sulfur <= 1% [4]	36.4	35.1	33.7	34.5	38.3	45.6	41.5	29.9	38.2	62.7	52.3	54.6	72.8	76.4	111.5	120.2	140.6
Sulfur > 1% [4]	29.2	28.6	25.6	28.7	33.8	38.9	36.6	26.9	32.9	51.2	42.8	50.8	58.8	60.1	84.2	R108.5	131.4
Propane (Consumer Grade)	34.9	32.8	35.1	32.4	34.4	46.1	41.6	28.8	34.2	59.5	54.0	43.1	60.7	75.1	93.3	103.1	119.4
Sales Prices to End Users: [2]																	
Aviation Gasoline	104.7	102.7	99.0	95.7	100.5	111.6	112.8	97.5	105.9	130.6	132.3	128.8	149.3	181.9	223.1	268.2	284.9
Motor Gasoline	79.7	78.7	75.9	73.8	76.5	84.7	83.9	67.3	78.1	110.6	103.2	94.7	115.6	143.5	182.9	R212.8	234.5
Leaded Regular	71.5	78.5	NA	NA	NA	NA	NA	NA	NA	NA	NA	NA	NA	NA	NA	NA	NA
Unleaded Regular	76.1	74.3	71.2	68.9	71.7	80.7	79.8	63.0	74.2	107.3	99.7	91.2	112.3	140.4	180.2	R209.9	231.5
Unleaded Midgrade	84.3	82.7	80.5	78.5	80.8	89.6	89.5	72.8	83.5	116.8	110.0	101.0	121.8	149.9	189.3	R221.3	243.8
Premium	90.7	91.4	88.9	86.5	89.0	97.2	97.3	80.5	90.6	124.2	117.5	108.8	130.5	159.6	199.2	R232.0	255.1
Kerosene	83.8	78.8	75.4	66.0	58.9	74.0	74.5	50.1	60.5	112.3	104.5	99.0	122.4	116.0	195.7	R224.4	226.1
Jet Fuel, Kerosene-Type	65.2	61.0	58.0	53.4	54.0	65.1	61.3	45.2	54.3	89.9	77.5	72.1	87.2	120.7	173.5	199.8	216.7
No. 1 Distillate	74.0	66.6	66.6	64.0	62.0	72.6	68.9	55.1	62.1	98.8	90.2	82.8	101.7	126.2	183.2	R213.7	228.9
No. 2 Distillate	65.0	62.0	60.2	55.6	56.0	68.0	64.2	49.2	58.0	93.4	84.0	75.9	94.2	123.5	177.7	R209.1	227.1
No. 2 Fuel Oil	66.5	62.7	60.2	57.2	56.2	67.3	63.6	48.2	55.8	92.7	82.9	73.7	93.3	117.3	170.5	R198.2	224.1
No. 2 Diesel Fuel	64.8	61.9	60.2	55.4	56.0	68.1	64.2	49.4	58.4	93.5	84.2	76.2	94.4	124.3	178.6	R209.6	227.3
No. 4 Fuel [3]	58.0	52.6	50.1	50.1	50.5	60.3	56.5	42.8	47.4	76.9	67.9	65.7	85.6	101.7	W	W	W
Residual Fuel Oil	34.0	33.6	33.7	35.2	39.2	45.5	42.3	30.5	37.4	60.2	53.1	56.9	69.8	73.9	104.8	121.8	137.2
Sulfur <= 1% [4]	40.2	38.9	39.7	40.1	43.6	52.6	48.8	35.4	40.5	70.8	64.2	64.0	80.4	83.5	116.8	134.2	143.6
Sulfur > 1% [4]	30.6	31.2	30.3	33.0	37.7	43.3	40.3	28.7	36.2	56.6	49.2	54.4	65.1	69.2	97.4	R117.3	134.7
Propane (Consumer Grade)	73.0	64.3	67.3	53.0	49.2	60.5	55.2	40.5	45.8	60.3	50.6	41.9	57.7	83.9	108.9	R135.8	149.0
Refiner Margins [5]																	
Motor Gasoline	24.5	23.8	23.5	22.8	21.6	22.0	24.7	22.8	22.8	29.0	34.0	25.4	32.3	40.8	47.4	53.5	56.5
Jet Fuel, Kerosene-Type	19.6	16.5	18.6	16.3	12.9	15.3	16.0	15.2	11.6	20.7	21.7	14.2	19.2	32.8	52.7	R52.7	54.7
No. 2 Distillate	16.4	14.6	16.8	15.1	12.0	16.0	14.9	14.1	11.9	22.3	23.3	14.4	20.3	29.8	52.4	55.7	57.3
Residual Fuel Oil	-14.0	-13.2	-9.8	-5.4	-4.8	-7.2	-6.6	-1.8	-6.3	-10.7	-7.0	-4.4	-1.8	-19.9	-22.5	R-29.8	-26.7
Composite [6]	20.7	19.8	19.0	19.8	18.1	19.4	20.0	19.5	18.9	26.1	29.7	21.6	28.1	36.7	48.4	R53.0	55.4

[1] See "Nominal Price" in Glossary.

[2] Sales for resale (wholesale sales) are those made to purchasers who are other than ultimate consumers. Sales to end users are those made directly to the ultimate consumer, including bulk customers, such as agriculture, industry, and utilities, as well as residential and commercial customers.

[3] Includes No. 4 fuel oil and No. 4 diesel fuel.

[4] Sulfur content by weight.

[5] In this table, refiner margin is the difference between the composite refiner acquisition price of crude oil (see Table 5.21) and the price to resellers.

[6] A volume weighted average of the refiner prices to resellers for aviation gasoline, kerosene-type jet fuel, kerosene, motor gasoline, distillate fuel nos. 1, 2, and 4, and residual fuel oil.

R=Revised. P=Preliminary. NA=Not available. W=Value withheld to avoid disclosure of individual company data.

Web Page: See http://www.eia.doe.gov/oil_gas/petroleum/info_glance/petroleum.html for related information.

Source: Energy Information Administration, *Petroleum Marketing Monthly* (April 2008), Tables 1, 2, 4, 6, and 16.

Figure 5.23 All Sellers Sales Prices for Selected Petroleum Products, 2007

Motor Gasoline, Selected Grades

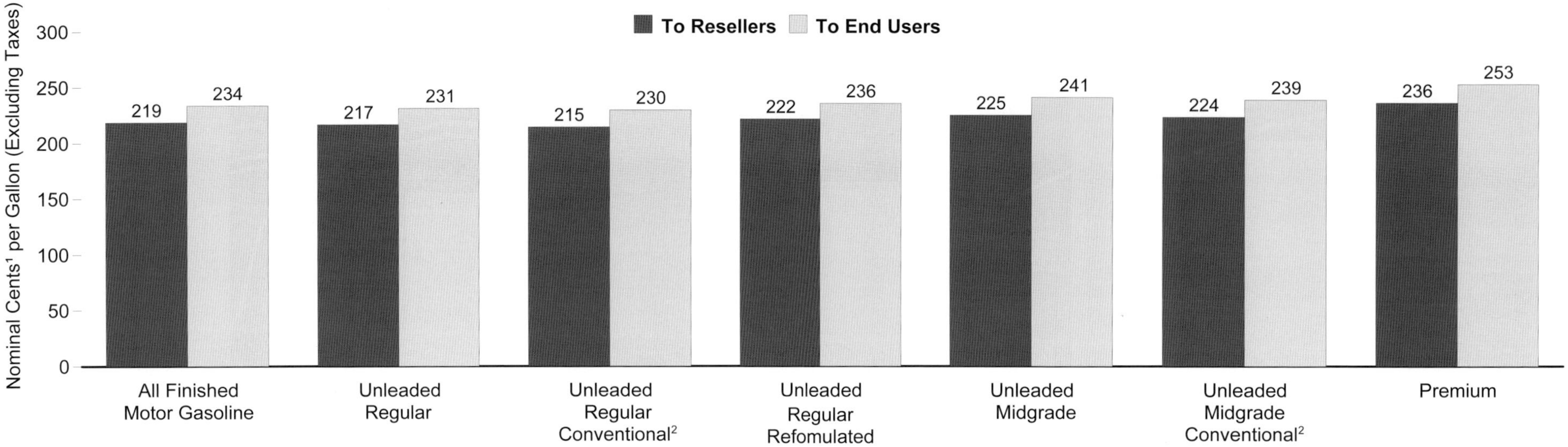

Distillate Fuel Oil, Residual Fuel Oil, and Propane

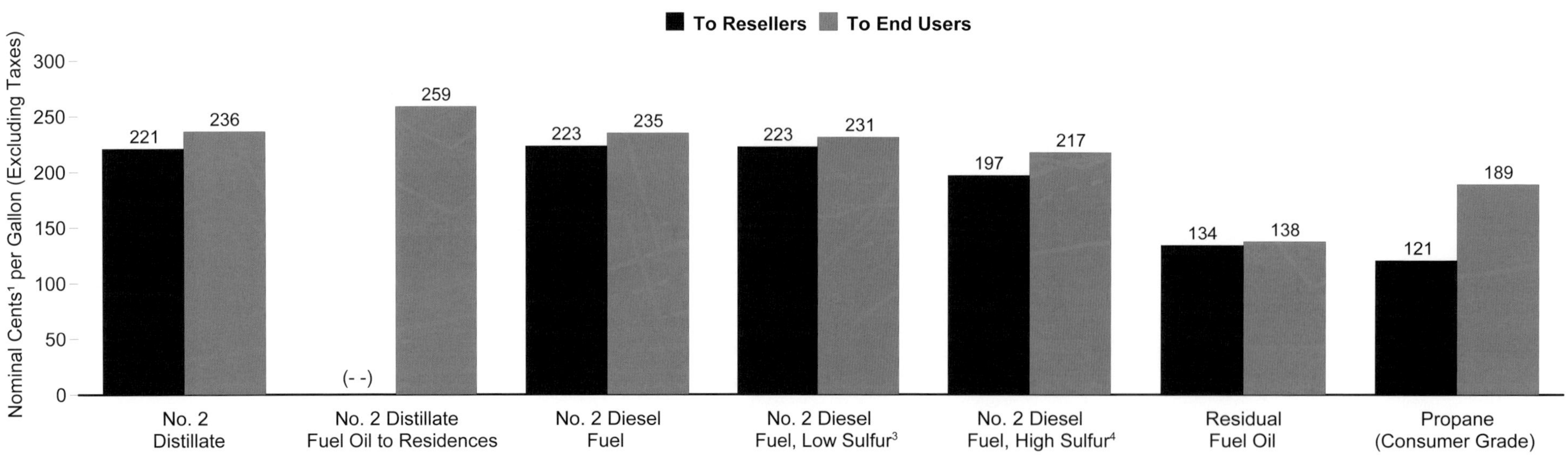

[1] See "Nominal Price" in Glossary.
[2] Includes oxygenated motor gasoline.
[3] > 15 and <= 500 parts per million by weight.
[4] > 500 parts per million by weight.

- - = Not applicable.
Notes: • Data are preliminary. • Because vertical scales differ, graphs should not be compared.
Source: Table 5.23.

Table 5.23 All Sellers Sales Prices for Selected Petroleum Products, 1991-2007

(Nominal Cents [1] per Gallon, Excluding Taxes)

Product	1991	1992	1993	1994	1995	1996	1997	1998	1999	2000	2001	2002	2003	2004	2005	2006	2007 [P]
Sales Prices to Resellers [2]																	
Motor Gasoline	70.8	68.0	62.8	60.2	63.0	71.5	70.3	53.0	64.5	96.6	88.8	83.2	100.1	128.8	167.5	[R]197.3	218.7
Unleaded Regular	68.2	64.9	59.7	57.1	59.9	68.9	67.7	50.4	62.1	94.6	86.8	81.3	98.2	127.1	165.9	195.6	216.7
Conventional [3]	NA	NA	NA	56.5	58.3	67.2	65.8	48.4	59.6	91.8	83.8	79.4	95.0	124.1	163.9	193.0	[3]214.6
Oxygenated [3]	NA	NA	NA	62.7	66.2	74.5	75.4	57.5	69.0	101.6	94.7	85.8	103.1	128.9	166.7	[R]195.1	([3])
Reformulated	NA	NA	NA	63.2	64.6	73.3	72.5	55.1	67.6	100.6	93.0	85.6	105.8	133.9	170.8	[R]202.2	221.7
Unleaded Midgrade	74.4	71.3	66.4	64.1	67.3	76.0	75.1	57.9	69.4	101.4	94.5	88.6	105.6	133.8	171.9	[R]202.4	225.2
Conventional [3]	NA	NA	NA	63.3	65.1	73.7	72.3	55.0	65.8	97.7	90.1	85.2	101.5	130.4	169.7	[R]199.2	[3]223.5
Oxygenated [3]	NA	NA	NA	68.9	71.1	78.9	79.1	59.9	69.5	102.1	96.5	88.5	104.3	130.5	168.2	199.9	([3])
Reformulated	NA	NA	NA	72.2	71.9	80.2	80.1	63.2	75.8	108.0	102.2	95.2	115.0	141.0	179.0	212.7	231.1
Premium	80.0	77.6	72.2	69.6	72.4	80.4	79.4	61.8	72.4	105.5	98.0	92.9	110.5	140.0	179.2	212.2	236.2
Conventional [3]	NA	NA	NA	68.6	69.5	77.7	76.4	58.7	68.8	101.3	93.3	89.7	105.5	135.1	175.6	[R]208.3	[3]232.9
Oxygenated [3]	NA	NA	NA	75.7	78.7	85.1	85.6	67.4	77.9	111.9	102.0	95.2	113.1	139.0	176.1	204.9	([3])
Reformulated	NA	NA	NA	76.9	77.9	85.1	84.5	67.1	78.7	111.7	105.4	98.6	118.9	146.7	184.4	[R]218.1	240.8
No. 2 Distillate	62.7	59.1	56.6	52.9	53.6	66.0	61.1	45.0	53.8	90.1	78.5	72.8	89.1	117.8	172.7	[R]199.9	221.0
No. 2 Diesel Fuel	NA	NA	NA	53.8	54.6	66.7	61.6	45.4	55.2	90.4	79.1	73.5	89.1	119.1	175.4	[R]202.9	223.4
Ultra-Low Sulfur (<= 15 ppm) [4]	NA	NA	NA	NA	NA	NA	NA	NA	NA	NA	NA	NA	NA	NA	NA	NA	225.9
Low Sulfur (> 15 and <= 500 ppm) [4]	NA	NA	NA	54.2	55.1	67.3	61.9	45.7	55.7	90.9	79.4	73.8	89.5	119.7	176.1	204.5	222.6
High Sulfur (> 500 ppm) [4]	NA	NA	NA	51.9	52.4	63.9	60.2	43.7	51.9	87.5	77.1	71.2	87.0	114.6	170.4	[R]191.5	197.0
Residual Fuel Oil	33.0	32.6	30.1	32.2	36.6	42.7	39.6	28.4	35.5	57.9	49.6	52.6	67.5	68.2	100.6	[R]114.1	134.2
Sulfur <= 1% [4]	37.9	36.8	34.1	35.0	38.3	46.1	42.4	30.5	38.2	63.8	54.2	54.8	73.2	74.0	110.7	120.5	137.7
Sulfur > 1% [4]	29.7	30.0	27.2	29.8	34.4	39.7	37.5	27.1	33.3	52.3	43.8	50.2	62.1	63.8	89.9	[R]108.7	131.4
Propane (Consumer Grade)	NA	NA	NA	33.6	35.4	47.1	42.6	29.7	35.4	60.3	55.6	44.0	61.5	76.1	93.9	[R]104.1	120.6
Sales Prices to End Users [2]																	
Motor Gasoline	81.1	78.7	75.3	72.9	76.1	84.3	83.1	66.0	76.2	109.1	102.2	94.3	113.5	142.3	182.8	212.1	233.7
Unleaded Regular	78.0	75.0	71.4	69.0	72.1	80.9	79.7	62.3	72.8	106.3	99.3	91.5	110.8	139.9	180.7	[R]209.8	231.3
Conventional [3]	NA	NA	NA	68.5	71.4	80.1	78.5	61.0	70.8	104.4	96.8	90.1	108.2	137.3	178.9	207.3	[3]229.7
Oxygenated [3]	NA	NA	NA	73.7	77.3	86.1	88.7	69.4	78.2	111.8	105.9	96.4	114.2	141.4	176.8	[R]207.6	([3])
Reformulated	NA	NA	NA	74.3	74.1	83.3	82.2	65.1	77.7	110.9	105.1	94.9	118.3	147.8	186.1	216.8	235.6
Unleaded Midgrade	NA	82.4	79.2	77.0	80.2	88.5	88.0	71.1	81.2	114.6	108.6	100.2	119.5	148.2	187.7	[R]218.3	241.2
Conventional [3]	NA	NA	NA	76.6	79.3	87.4	86.5	69.5	78.7	112.2	105.2	98.5	116.6	145.3	185.5	[R]215.1	[3]238.9
Oxygenated [3]	NA	NA	NA	82.1	83.8	92.9	96.4	76.3	85.3	118.5	112.0	103.1	119.3	145.9	182.4	[R]213.8	([3])
Reformulated	NA	NA	NA	85.1	82.9	91.6	91.5	74.8	86.9	119.7	115.6	104.2	127.7	156.9	194.5	[R]227.3	247.6
Premium	91.9	90.6	87.5	85.2	88.3	96.2	95.5	78.6	88.0	121.8	115.4	108.1	128.2	158.0	198.8	[R]230.3	252.8
Conventional [3]	NA	NA	NA	84.6	87.1	95.0	93.9	76.9	85.6	119.2	111.9	106.3	124.5	153.8	195.8	[R]226.3	[3]249.7
Oxygenated [3]	NA	NA	NA	90.8	93.8	101.9	105.4	84.5	94.0	127.9	121.8	112.8	130.7	156.8	191.5	[R]223.6	([3])
Reformulated	NA	NA	NA	93.7	91.4	99.1	98.8	82.2	93.1	126.7	121.7	111.6	135.5	166.0	204.3	[R]237.3	258.2
No. 2 Distillate	76.0	72.6	71.0	67.5	67.3	79.3	75.3	59.9	67.8	104.4	94.8	87.4	105.8	133.9	189.5	[R]217.6	236.3
No. 2 Distillate to Residences [5]	101.9	93.4	91.1	88.4	86.7	98.9	98.4	85.2	87.6	131.1	125.0	112.9	135.5	154.8	205.2	[R]236.5	259.0
No. 2 Diesel Fuel	NA	NA	NA	62.8	63.6	75.7	71.4	56.2	65.4	100.6	91.2	83.7	100.8	131.6	188.6	[R]216.5	234.8
Ultra-Low Sulfur (<= 15 ppm) [4]	NA	NA	NA	NA	NA	NA	NA	NA	NA	NA	NA	NA	NA	NA	NA	NA	239.4
Low Sulfur (> 15 and <= 500 ppm) [4]	NA	NA	NA	64.2	64.5	76.7	71.9	56.5	66.3	101.4	91.7	84.1	101.4	132.5	189.9	[R]218.3	231.0
High Sulfur (> 500 ppm) [4]	NA	NA	NA	59.8	61.4	73.2	69.8	55.5	62.0	98.1	89.2	82.2	98.6	128.1	183.3	[R]208.4	217.4
Residual Fuel Oil	34.7	34.6	34.1	35.8	39.7	46.4	42.9	31.1	37.8	60.9	53.3	56.1	69.6	72.5	104.1	121.2	137.5
Sulfur <= 1% [4]	40.0	39.4	39.3	40.3	43.3	52.9	47.2	35.6	40.6	68.3	62.0	61.2	78.5	79.4	114.8	132.2	144.4
Sulfur > 1% [4]	31.1	31.9	31.2	32.7	37.6	43.0	40.7	29.2	36.6	57.6	49.8	54.0	65.1	68.7	98.5	117.5	135.3
Propane (Consumer Grade)	NA	NA	NA	77.6	76.6	88.6	87.8	77.4	78.1	104.8	109.4	95.8	115.0	130.7	153.0	172.7	189.0

[1] See "Nominal Price" in Glossary.
[2] Sales for resale (wholesale sales) are those made to purchasers who are other than ultimate consumers. Sales to end users are those made directly to the ultimate consumer, including bulk customers, such as agriculture, industry, and utilities, as well as residential and commercial customers.
[3] Beginning in 2007, oxygenated motor gasoline is included with conventional motor gasoline.
[4] Sulfur content by weight ("ppm" is parts per million).
[5] See Note 6, "Historical Residential Heating Oil Prices," at end of section for historical data.

R=Revised. P=Preliminary. NA=Not available.

Web Page: See http://www.eia.doe.gov/oil_gas/petroleum/info_glance/petroleum.html for related information.

Sources: • 1991-2006—Energy Information Administration (EIA), *Petroleum Marketing Annual*, annual reports. • 2007—EIA, *Petroleum Marketing Monthly* (April 2008); EIA, Form EIA-782A, "Refiners'/Gas Plant Operators' Monthly Petroleum Product Sales Report"; and EIA, Form EIA-782B, "Resellers'/Retailers' Monthly Petroleum Product Sales Report."

Figure 5.24 Retail Motor Gasoline and On-Highway Diesel Fuel Prices

Motor Gasoline, All Grades, 1978-2007

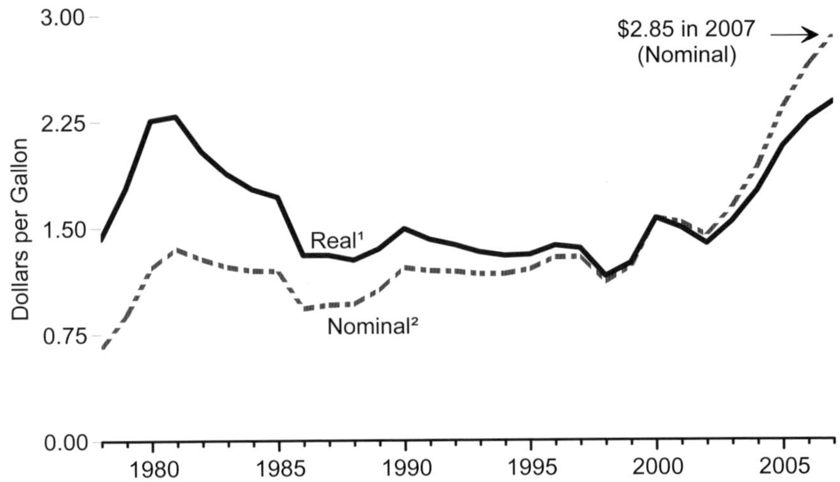

Regular Motor Gasoline by Area Type and On-Highway Diesel Fuel, 2007

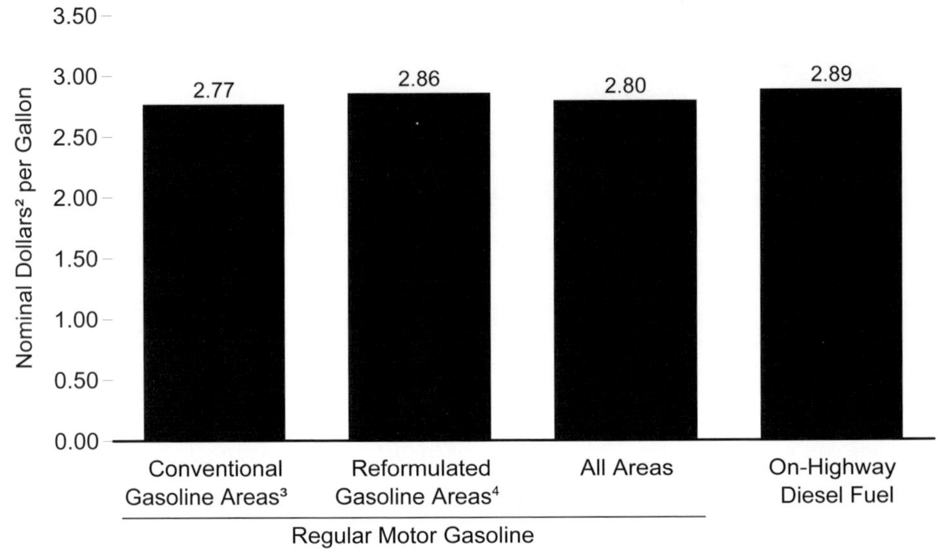

Motor Gasoline by Grade, 1949-2007

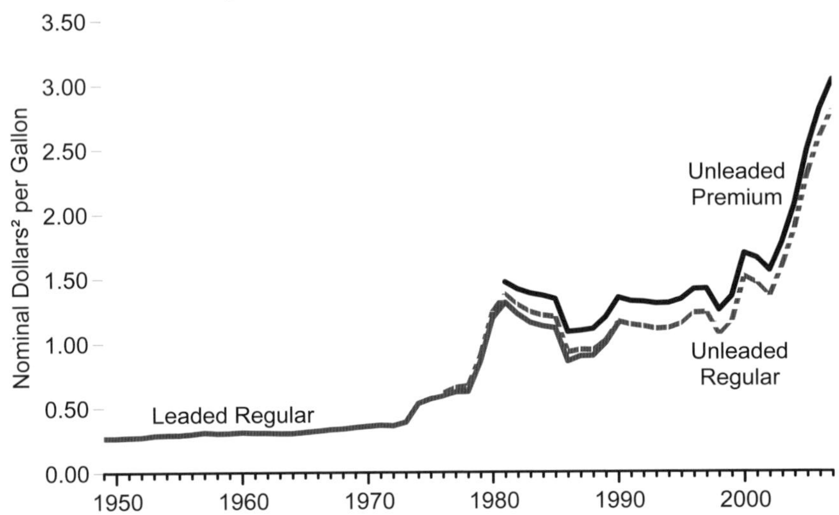

Motor Gasoline Price and Consumption, 1978-2007, Indexed to 1978

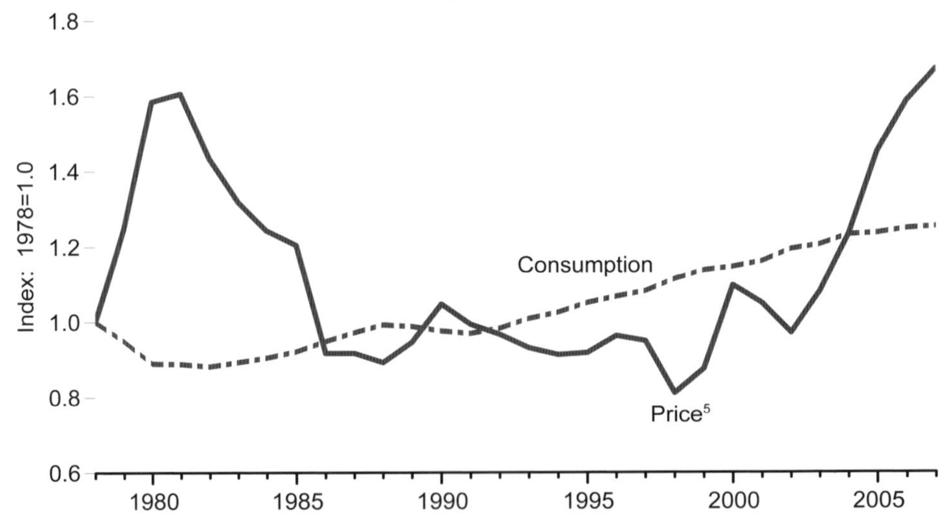

[1] In chained (2000) dollars, calculated by using gross domestic product implicit price deflators in Table D1. See "Chained Dollars" in Glossary.

[2] See "Nominal Dollars" in Glossary.

[3] Any area that does not require the sale of reformulated gasoline.

[4] "Reformulated Gasoline Areas" are ozone nonattainment areas designated by the Environmental Protection Agency that require the use of reformulated gasoline.

[5] All grades, in chained (2000) dollars.

Note: Because vertical scales differ, graphs should not be compared.

Sources: Tables 5.11 and 5.24.

Table 5.24 Retail Motor Gasoline and On-Highway Diesel Fuel Prices, Selected Years, 1949-2007

(Dollars per Gallon)

Year	Motor Gasoline by Grade								Regular Motor Gasoline by Area Type			On-Highway Diesel Fuel
	Leaded Regular		Unleaded Regular		Unleaded Premium		All Grades		Conventional Gasoline Areas [1,2]	Reformulated Gasoline Areas [3,4]	All Areas	
	Nominal [5]	Real [6]	Nominal [5]	Real [6]	Nominal [5]	Real [6]	Nominal [5]	Real [6]	Nominal [5]	Nominal [5]	Nominal [5]	Nominal [5]
1949	0.27	1.64	NA	NA	NA	NA	NA	NA	NA	NA	NA	NA
1950	.27	1.62	NA	NA	NA	NA	NA	NA	NA	NA	NA	NA
1955	.29	1.55	NA	NA	NA	NA	NA	NA	NA	NA	NA	NA
1960	.31	1.48	NA	NA	NA	NA	NA	NA	NA	NA	NA	NA
1965	.31	1.39	NA	NA	NA	NA	NA	NA	NA	NA	NA	NA
1970	.36	1.30	NA	NA	NA	NA	NA	NA	NA	NA	NA	NA
1971	.36	1.26	NA	NA	NA	NA	NA	NA	NA	NA	NA	NA
1972	.36	1.20	NA	NA	NA	NA	NA	NA	NA	NA	NA	NA
1973	.39	1.22	NA	NA	NA	NA	NA	NA	NA	NA	NA	NA
1974	.53	1.53	NA	NA	NA	NA	NA	NA	NA	NA	NA	NA
1975	.57	1.49	NA	NA	NA	NA	NA	NA	NA	NA	NA	NA
1976	.59	1.47	.61	1.53	NA	NA	NA	NA	NA	NA	NA	NA
1977	.62	1.46	.66	1.53	NA	NA	NA	NA	NA	NA	NA	NA
1978	.63	1.37	.67	1.46	NA	NA	.65	1.43	NA	NA	NA	NA
1979	.86	1.73	.90	1.82	NA	NA	.88	1.78	NA	NA	NA	NA
1980	1.19	2.20	1.25	2.30	NA	NA	1.22	2.26	NA	NA	NA	NA
1981	1.31	2.22	1.38	2.33	1.47	2.49	1.35	2.29	NA	NA	NA	NA
1982	1.22	1.95	1.30	2.07	1.42	2.26	1.28	2.04	NA	NA	NA	NA
1983	1.16	1.77	1.24	1.90	1.38	2.12	1.23	1.88	NA	NA	NA	NA
1984	1.13	1.67	1.21	1.79	1.37	2.02	1.20	1.77	NA	NA	NA	NA
1985	1.12	1.60	1.20	1.72	1.34	1.92	1.20	1.72	NA	NA	NA	NA
1986	.86	1.20	.93	1.30	1.09	1.52	.93	1.31	NA	NA	NA	NA
1987	.90	1.23	.95	1.30	1.09	1.49	.96	1.31	NA	NA	NA	NA
1988	.90	1.19	.95	1.25	1.11	1.46	.96	1.27	NA	NA	NA	NA
1989	1.00	1.27	1.02	1.30	1.20	1.52	1.06	1.35	NA	NA	NA	NA
1990	1.15	1.41	1.16	1.43	1.35	1.65	1.22	1.49	NA	NA	NA	NA
1991	NA	NA	1.14	1.35	1.32	1.56	1.20	1.42	1.10	NA	1.10	NA
1992	NA	NA	1.13	1.31	1.32	1.52	1.19	1.38	1.09	NA	1.09	NA
1993	NA	NA	1.11	1.25	1.30	1.47	1.17	1.33	[2]1.07	NA	1.07	NA
1994	NA	NA	1.11	1.23	1.31	1.45	1.17	1.30	[2]1.07	NA	1.08	NA
1995	NA	NA	1.15	1.25	1.34	1.45	1.21	1.31	[2]1.10	[4]1.16	1.11	1.11
1996	NA	NA	1.23	1.31	1.41	1.51	1.29	1.37	[2]1.19	[4]1.28	1.22	1.24
1997	NA	NA	1.23	1.29	1.42	1.48	1.29	1.35	[2]1.19	[4]1.25	1.20	1.20
1998	NA	NA	1.06	1.10	1.25	1.30	1.12	1.16	[2]1.02	[4]1.08	1.03	1.04
1999	NA	NA	1.17	1.19	1.36	1.39	1.22	1.25	[2]1.12	[4]1.20	1.14	1.12
2000	NA	NA	1.51	1.51	1.69	1.69	1.56	1.56	[2]1.46	[4]1.54	1.48	1.49
2001	NA	NA	1.46	1.43	1.66	1.62	1.53	1.50	1.38	1.50	1.42	1.40
2002	NA	NA	1.36	1.30	1.56	1.49	1.44	1.38	1.31	1.41	1.35	1.32
2003	NA	NA	1.59	1.50	1.78	1.67	1.64	1.54	1.52	1.66	1.56	1.51
2004	NA	NA	1.88	1.72	2.07	1.89	1.92	1.76	1.81	1.94	1.85	1.81
2005	NA	NA	2.30	R2.03	2.49	R2.20	2.34	2.07	2.24	2.34	2.27	2.40
2006	NA	NA	2.59	R2.22	2.81	R2.41	2.64	R2.26	2.53	2.65	2.57	2.71
2007	NA	NA	2.80	2.34	3.03	2.54	2.85	2.38	2.77	2.86	2.80	2.89

[1] Any area that does not require the sale of reformulated gasoline.

[2] For 1993-2000, data collected for oxygenated areas are included in "Conventional Gasoline Areas."

[3] "Reformulated Gasoline Areas" are ozone nonattainment areas designated by the Environmental Protection Agency that require the use of reformulated gasoline.

[4] For 1995-2000, data collected for combined oxygenated and reformulated areas are included in "Reformulated Gasoline Areas."

[5] See "Nominal Dollars" in Glossary.

[6] In chained (2000) dollars, calculated by using gross domestic product implicit price deflators in Table D1. See "Chained Dollars" in Glossary.

R=Revised. NA=Not available.

Note: See "Motor Gasoline Grades," "Motor Gasoline, Conventional," "Motor Gasoline, Oxygenated," and "Motor Gasoline, Reformulated" in Glossary.

Web Pages: • For all data beginning in 1949, see http://www.eia.doe.gov/emeu/aer/petro.html. • For related information, see http://www.eia.doe.gov/oil_gas/petroleum/info_glance/petroleum.html

Sources: **Motor Gasoline by Grade:** • 1949-1973—*Platt's Oil Price Handbook and Oilmanac, 1974,* 51st Edition. • 1974 forward—Energy Information Administration (EIA), annual averages of monthly data from the U.S. Department of Labor, Bureau of Labor Statistics, *U.S. City Average Gasoline Prices.* **Regular Motor Gasoline by Area Type:** EIA, weighted annual averages of data from "Weekly U.S. Retail Gasoline Prices, Regular Grade." **On-Highway Diesel Fuel:** EIA, weighted annual averages of data from "Weekly Retail On-Highway Diesel Prices."

Petroleum

Note 1. Petroleum Products Supplied and Petroleum Consumption. Total petroleum products supplied is the sum of the products supplied for each petroleum product, crude oil, unfinished oils, and gasoline blending components. For each of these, except crude oil, product supplied is calculated by adding refinery production, natural gas plant liquids production, new supply of other liquids, imports, and stock withdrawals, and subtracting stock additions, refinery inputs, and exports. Crude oil product supplied is the sum of crude oil burned on leases and at pipeline pump stations as reported on Form EIA-813, "Monthly Crude Oil Report." Prior to 1983, crude oil burned on leases and at pipeline pump stations was reported as either distillate or residual fuel oil and was included as product supplied for these products. Petroleum product supplied (see Table 5.11) is an approximation of petroleum consumption and is synonymous with the term "Petroleum Consumption" in Tables 5.13a-d. The sector allocation of product supplied in Tables 5.13a-d for products used in more than one sector is derived from sales to ultimate consumers by refiners, marketers, distributors, and dealers (see Energy Information Administration (EIA) report *Fuel Oil and Kerosene Sales*) and from EIA electric power sector petroleum consumption data (see Tables 8.7b and 8.7c).

Note 2. Adjustment to Total Petroleum Products Supplied. Accurate calculation of the quantity of petroleum products supplied to the domestic market is complicated by the recycling of products at the refinery, the renaming of products involved in a transfer, and the receipt of products from outside the primary supply system. Beginning in 1981, a single adjustment (always a negative quantity) is made to total product supplied to correct this accounting problem. The calculation of this adjustment, called "reclassified," involves only unfinished oils and gasoline blending components. It is the sum of their net changes in primary stocks (net withdrawals is a plus quantity; net additions is a minus quantity) plus imports minus net input to refineries.

Note 3. Changes Affecting Petroleum Production and Product Supplied Statistics. Beginning in January 1981, several Energy Information Administration survey forms and calculation methodologies were changed to reflect new developments in refinery and blending plant practices and to improve data integrity. Those changes affect production and product supplied statistics for motor gasoline, distillate fuel oil, and residual fuel oil, and stocks of motor gasoline. On the basis of those changes, motor gasoline production during the last half of 1980 would have averaged 289,000 barrels per day higher than that which was published on the old basis. Distillate and residual fuel oil production and product supplied for all of 1980 would have averaged, respectively, 105,000 and 54,000 barrels per day higher than the numbers that were published.

Note 4. Gross Input to Distillation Units. The methods of deriving Gross Input to Distillation Units (GIDU) in this report are as follows: for 1949-1966, GIDU is estimated by summing annual crude oil runs to stills, net unfinished oil reruns at refineries, and shipments of natural gasoline and plant condensate from natural gas processing plants to refineries; for 1967-1973, GIDU is estimated by summing annual crude oil runs to stills, net unfinished oil reruns, and refinery input of natural gasoline and plant condensate; for 1974-1980, GIDU is published annual data; and for 1981 forward, GIDU is the sum of reported monthly data.

Note 5. Crude Oil Domestic First Purchase Prices. Crude oil domestic first purchase prices were derived as follows: for 1949-1973, weighted average domestic first purchase values as reported by State agencies and calculated by the Bureau of Mines; for 1974 and 1975, weighted averages of a sample survey of major first purchasers' purchases; for 1976 forward, weighted averages of all first purchasers' purchases.

Note 6. Historical Residential Heating Oil Prices. Residential heating oil prices for 1956 through 1986 were formerly published in the *Annual Energy Review*. Those data, in cents per gallon, are: 1956—15.2; 1957—16.0; 1958—15.1; 1959—15.3; 1960—15.0; 1961—15.6; 1962—15.6; 1963—16.0; 1964—16.1; 1965—16.0; 1966—16.4; 1967—16.9; 1968—17.4; 1969—17.8; 1970—18.5; 1971—19.6; 1972—19.7; 1973—22.8; 1974—36.0; 1975—37.7; 1976—40.6; 1977—46.0; 1978—49.0; 1979—70.4; 1980—97.4; 1981—119.4; 1982—116.0; 1983—107.8; 1984—109.1; 1985—105.3; 1986—83.6; and 1987—80.3. The sources of these data are: 1956-1974—Bureau of Labor Statistics, "Retail Prices and Indexes of Fuels and Utilities for Residential Usage," monthly; January 1975–September 1977—Federal Energy Administration, Form FEA-P112-M-1, "No. 2 Heating Oil Supply/Price Monitoring Report"; October 1977–December 1977—Energy Information Administration (EIA), Form EIA-9, "No. 2 Heating Oil Supply/Price Monitoring Report"; 1978 forward—EIA, *Petroleum Marketing Annual*, Table 18.

6 Natural Gas

Natural gas pipeline, El Paso County, Texas. Source: U.S. Department of Energy.

Diagram 3. Natural Gas Flow, 2007
(Trillion Cubic Feet)

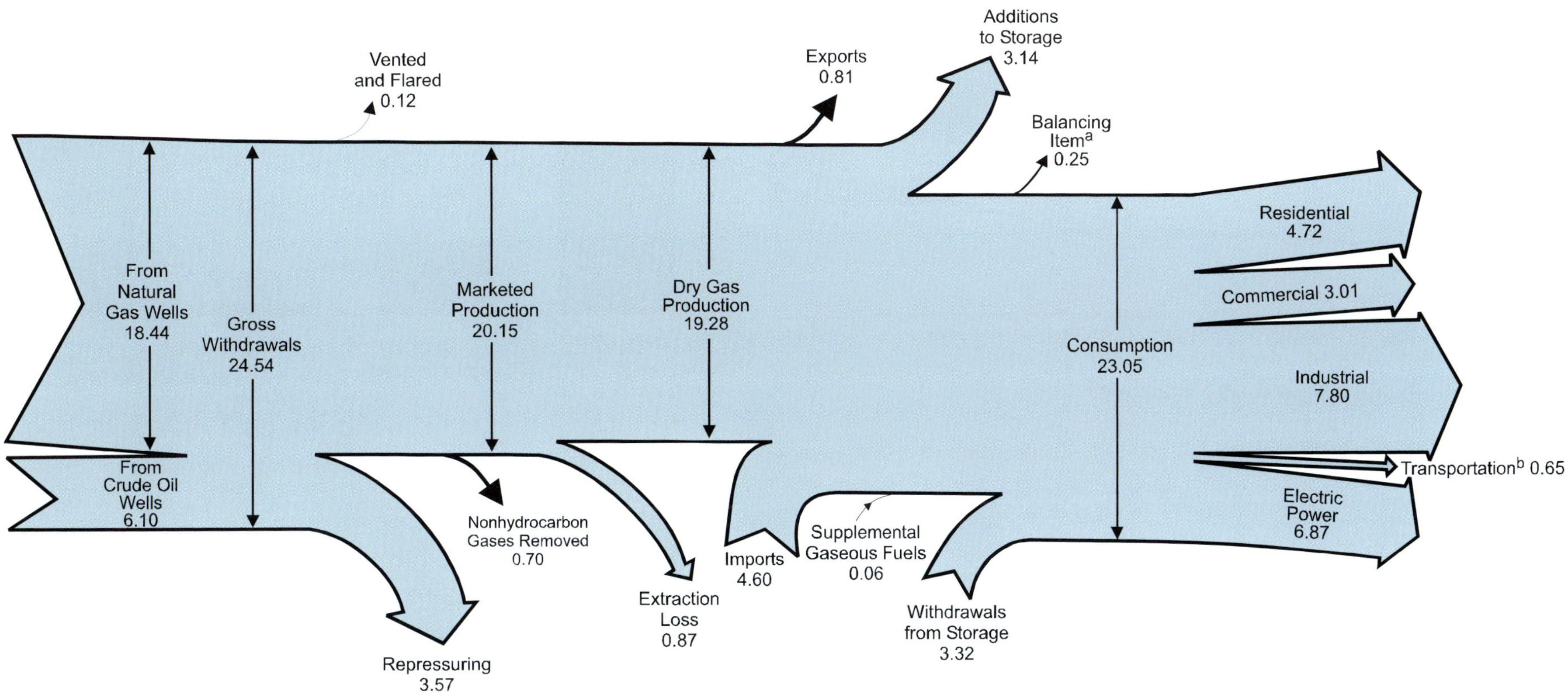

[a] Quantities lost and imbalances in data due to differences among data sources.

[b] Natural gas consumed in the operation of pipelines (primarily in compressors), and as fuel in the delivery of natural gas to consumers; plus a small quantity used as vehicle fuel.

Notes: • Data are preliminary. • Values are derived from source data prior to rounding for publication. • Totals may not equal sum of components due to independent rounding.

Sources: Tables 6.1, 6.2, and 6.5.

Figure 6.1 Natural Gas Overview

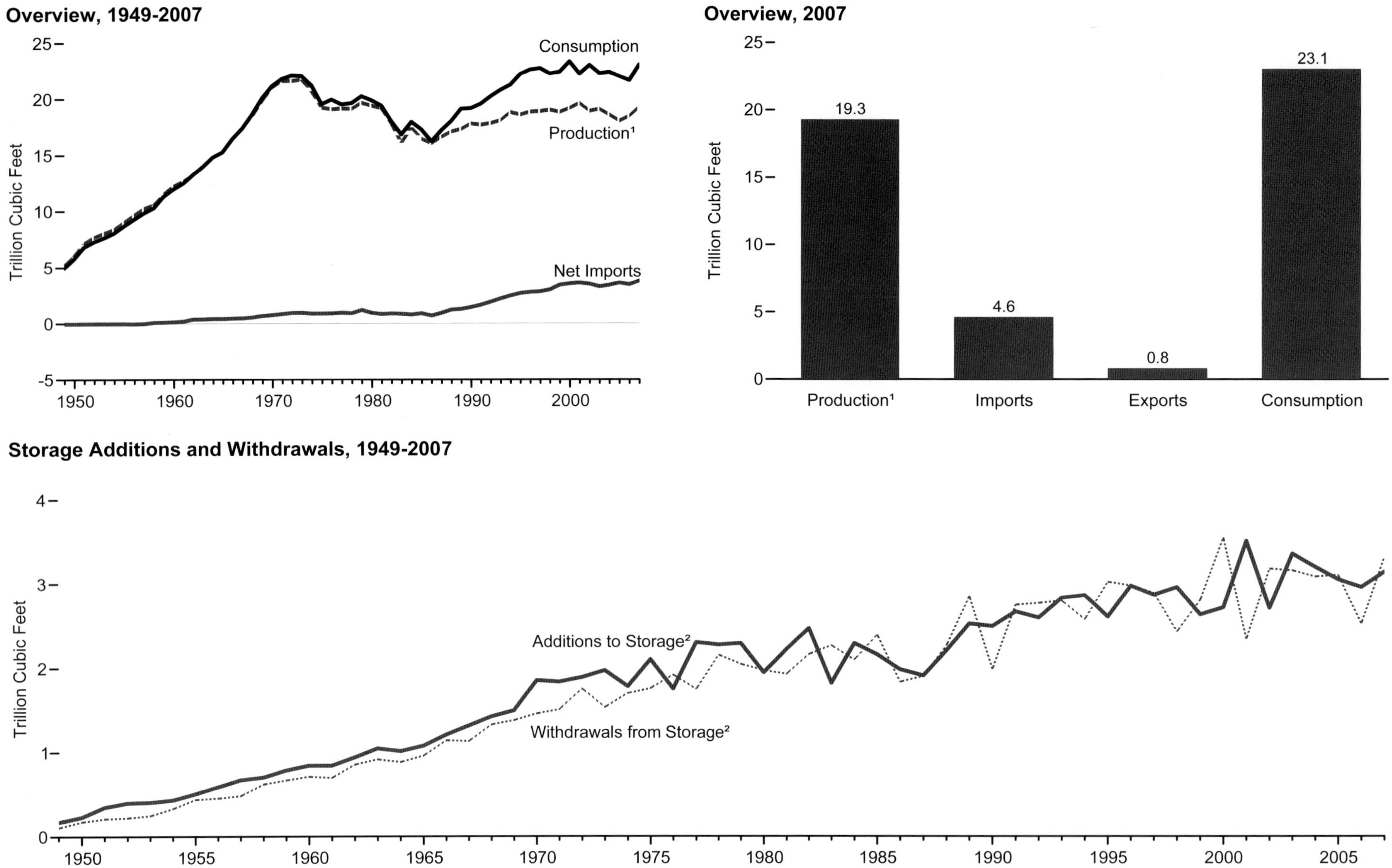

[1] Dry gas.
[2] Underground storage. For 1980-2006, also includes liquefied natural gas in above-ground tanks.

Note: Because vertical scales differ, graphs should not be compared.
Source: Table 6.1.

Table 6.1 Natural Gas Overview, Selected Years, 1949-2007

(Billion Cubic Feet)

Year	Dry Gas Production	Supplemental Gaseous Fuels [2]	Trade			Storage [1] Activity			Balancing Item [5]	Consumption [6]
			Imports	Exports	Net Imports [3]	Withdrawals	Additions	Net Withdrawals [4]		
1949	5,195	NA	0	20	-20	106	172	-66	-139	4,971
1950	6,022	NA	0	26	-26	175	230	-54	-175	5,767
1955	9,029	NA	11	31	-20	437	505	-68	-247	8,694
1960	12,228	NA	156	11	144	713	844	-132	-274	11,967
1965	15,286	NA	456	26	430	960	1,078	-118	-319	15,280
1970	21,014	NA	821	70	751	1,459	1,857	-398	-228	21,139
1971	21,610	NA	935	80	854	1,508	1,839	-332	-339	21,793
1972	21,624	NA	1,019	78	941	1,757	1,893	-136	-328	22,101
1973	21,731	NA	1,033	77	956	1,533	1,974	-442	-196	22,049
1974	20,713	NA	959	77	882	1,701	1,784	-84	-289	21,223
1975	19,236	NA	953	73	880	1,760	2,104	-344	-235	19,538
1976	19,098	NA	964	65	899	1,921	1,756	165	-216	19,946
1977	19,163	NA	1,011	56	955	1,750	2,307	-557	-41	19,521
1978	19,122	NA	966	53	913	2,158	2,278	-120	-287	19,627
1979	19,663	NA	1,253	56	1,198	2,047	2,295	-248	-372	20,241
1980	19,403	155	985	49	936	1,972	1,949	23	-640	19,877
1981	19,181	176	904	59	845	1,930	2,228	-297	-500	19,404
1982	17,820	145	933	52	882	2,164	2,472	-308	-537	18,001
1983	16,094	132	918	55	864	2,270	1,822	447	-703	16,835
1984	17,466	110	843	55	788	2,098	2,295	-197	-217	17,951
1985	16,454	126	950	55	894	2,397	2,163	235	-428	17,281
1986	16,059	113	750	61	689	1,837	1,984	-147	-493	16,221
1987	16,621	101	993	54	939	1,905	1,911	-6	-444	17,211
1988	17,103	101	1,294	74	1,220	2,270	2,211	59	-453	18,030
1989	17,311	107	1,382	107	1,275	2,854	2,528	326	101	[7]19,119
1990	17,810	123	1,532	86	1,447	1,986	2,499	-513	307	[7]19,174
1991	17,698	113	1,773	129	1,644	2,752	2,672	80	27	[7]19,562
1992	17,840	118	2,138	216	1,921	2,772	2,599	173	176	[7]20,228
1993	18,095	119	2,350	140	2,210	2,799	2,835	-36	401	20,790
1994	18,821	111	2,624	162	2,462	2,579	2,865	-286	139	21,247
1995	18,599	110	2,841	154	2,687	3,025	2,610	415	396	22,207
1996	18,854	109	2,937	153	2,784	2,981	2,979	2	860	22,609
1997	18,902	103	2,994	157	2,837	2,894	2,870	24	871	22,737
1998	19,024	102	3,152	159	2,993	2,432	2,961	-530	657	22,246
1999	18,832	98	3,586	163	3,422	2,808	2,636	172	-119	22,405
2000	19,182	90	3,782	244	3,538	3,550	2,721	829	-306	23,333
2001	19,616	86	3,977	373	3,604	2,344	3,510	-1,166	99	22,239
2002	18,928	68	4,015	516	3,499	3,180	2,713	467	45	23,007
2003	19,099	68	3,944	680	3,264	3,161	3,358	-197	44	22,277
2004	18,591	60	4,259	854	3,404	3,088	3,202	-114	448	22,389
2005	[R]18,051	64	4,341	729	3,612	[R]3,107	[R]3,055	[R]52	[R]232	[R]22,011
2006	[R]18,476	[R]66	[R]4,186	[R]724	[R]3,462	[R]2,527	[R]2,963	[R]-436	[R]85	[R]21,653
2007[P]	19,278	61	4,602	809	3,793	3,321	3,144	177	-253	23,055

[1] Underground storage. For 1980-2006, also includes liquefied natural gas in above-ground tanks.
[2] See Note 1, "Supplemental Gaseous Fuels," at end of section.
[3] Net imports equal imports minus exports. Minus sign indicates exports are greater than imports.
[4] Net withdrawals equal withdrawals minus additions. Minus sign indicates additions are greater than withdrawals.
[5] Quantities lost and imbalances in data due to differences among data sources. Since 1980, excludes intransit shipments that cross the U.S.-Canada border (i.e., natural gas delivered to its destination via the other country).
[6] See Note 2, "Natural Gas Consumption," at end of section.
[7] For 1989-1992, a small amount of consumption at independent power producers may be counted in both "Other Industrial" and "Electric Power Sector" on Table 6.5. See Note 3, "Natural Gas Consumption, 1989-1992," at end of section.

R=Revised. P=Preliminary. NA=Not available.

Notes: • Beginning with 1965, all volumes are shown on a pressure base of 14.73 p.s.i.a. at 60° F. For prior years, the pressure base was 14.65 p.s.i.a. at 60° F. • Totals may not equal sum of components due to independent rounding.

Web Pages: • For all data beginning in 1949, see http://www.eia.doe.gov/emeu/aer/natgas.html. • For related information, see http://www.eia.doe.gov/oil_gas/natural_gas/info_glance/natural_gas.html.

Sources: **Dry Gas Production:** Table 6.2. **Supplemental Gaseous Fuels:** • 1980-2002—Energy Information Administration (EIA), *Natural Gas Annual (NGA)*, annual reports. • 2003 forward—EIA, *Natural Gas Monthly (NGM)* (March 2008), Table 1. **Trade:** Table 6.3. **Storage Activity:** • 1949-2006—EIA, *NGA*, annual reports. • 2007—EIA, *NGM* (March 2008), Table 6. **Balancing Item:** Calculated as consumption minus dry gas production, supplemental gaseous fuels, net imports, and net withdrawals. **Consumption:** Table 6.5.

Figure 6.2 Natural Gas Production

Gross Withdrawals and Dry Gas Production, 1949-2007

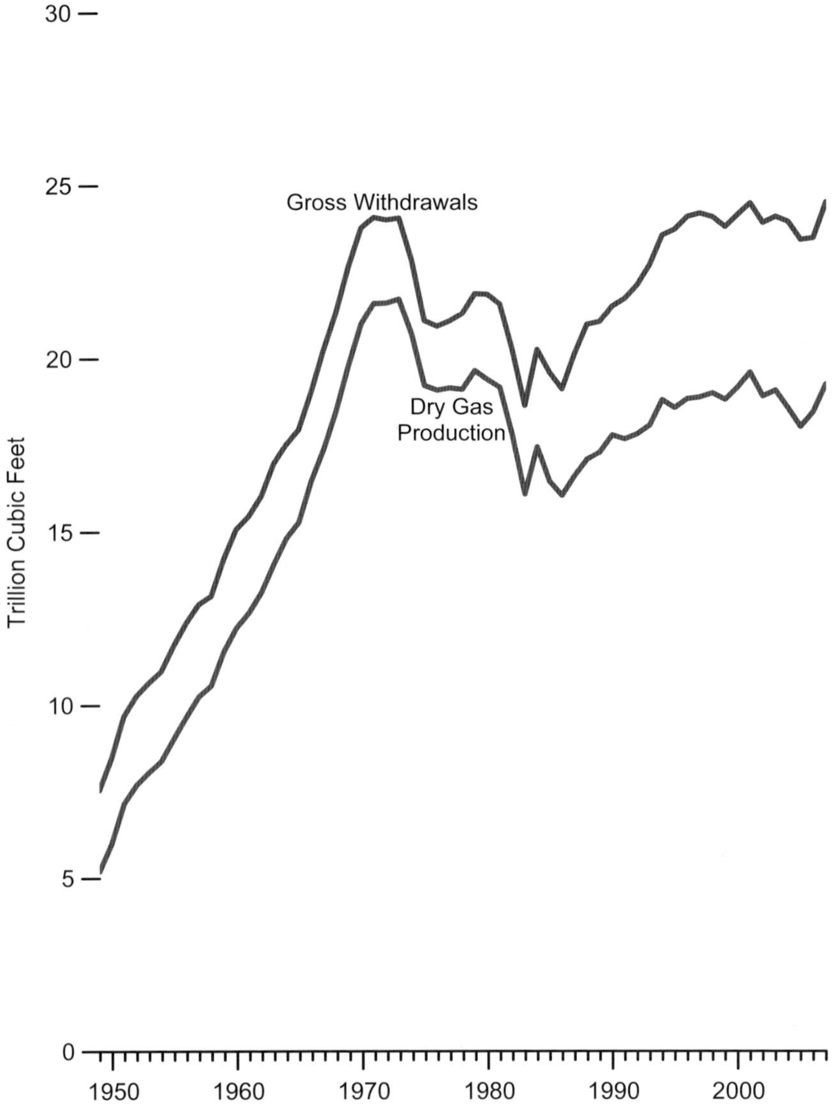

Production Flow, 2007
(Trillion Cubic Feet)

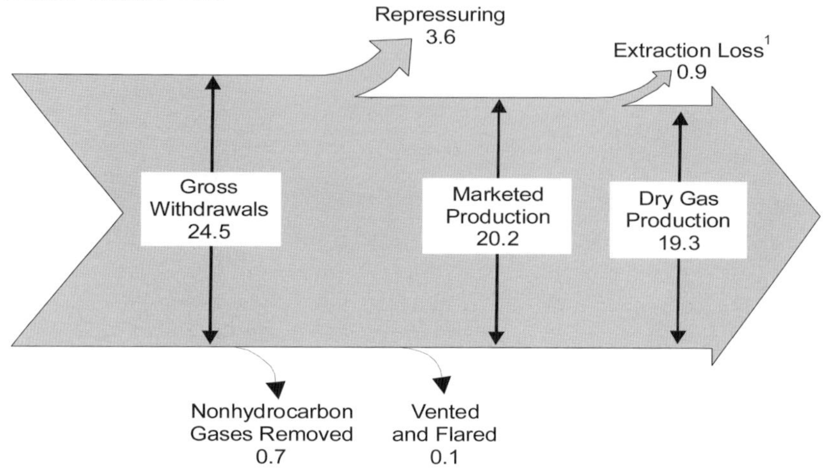

Gross Withdrawals by Well Type, 2007

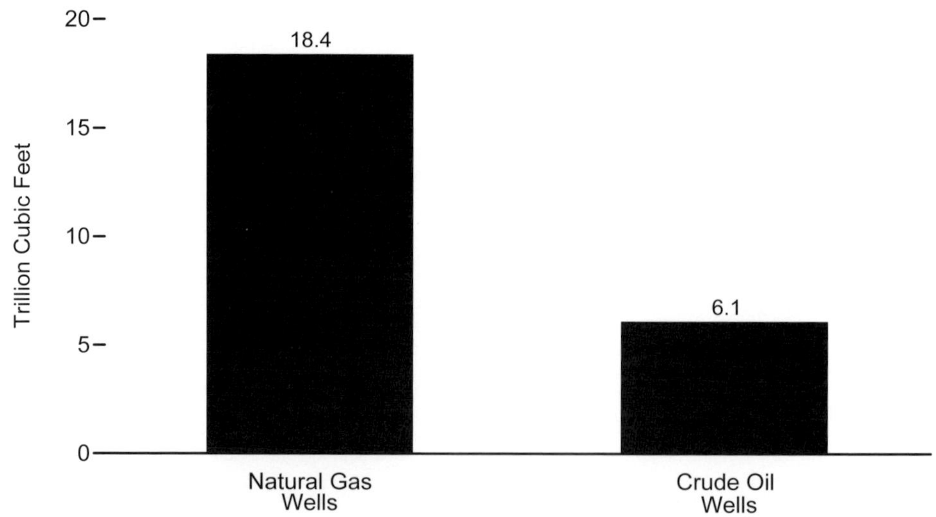

[1] Volume reduction resulting from the removal of natural gas plant liquids, which are transferred to petroleum supply.

Note: Because vertical scales differ, graphs should not be compared.
Source: Table 6.2.

Table 6.2 Natural Gas Production, Selected Years, 1949-2007

(Billion Cubic Feet)

Year	Natural Gas Gross Withdrawals: Natural Gas Wells	Natural Gas Gross Withdrawals: Crude Oil Wells	Natural Gas Gross Withdrawals: Total	Repressuring	Nonhydrocarbon Gases Removed	Vented and Flared	Marketed Production	Extraction Loss [1]	Dry Gas Production
1949	4,986	2,561	7,547	1,273	NA	854	5,420	224	5,195
1950	5,603	2,876	8,480	1,397	NA	801	6,282	260	6,022
1955	7,842	3,878	11,720	1,541	NA	774	9,405	377	9,029
1960	10,853	4,234	15,088	1,754	NA	563	12,771	543	12,228
1965	13,524	4,440	17,963	1,604	NA	319	16,040	753	15,286
1970	18,595	5,192	23,786	1,376	NA	489	21,921	906	21,014
1971	18,925	5,163	24,088	1,310	NA	285	22,493	883	21,610
1972	19,043	4,974	24,016	1,236	NA	248	22,532	908	21,624
1973	19,372	4,696	24,067	1,171	NA	248	22,648	917	21,731
1974	18,669	4,181	22,850	1,080	NA	169	21,601	887	20,713
1975	17,380	3,723	21,104	861	NA	134	20,109	872	19,236
1976	17,191	3,753	20,944	859	NA	132	19,952	854	19,098
1977	17,416	3,681	21,097	935	NA	137	20,025	863	19,163
1978	17,394	3,915	21,309	1,181	NA	153	19,974	852	19,122
1979	18,034	3,849	21,883	1,245	NA	167	20,471	808	19,663
1980	17,573	4,297	21,870	1,365	199	125	20,180	777	19,403
1981	17,337	4,251	21,587	1,312	222	98	19,956	775	19,181
1982	15,809	4,463	20,272	1,388	208	93	18,582	762	17,820
1983	14,153	4,506	18,659	1,458	222	95	16,884	790	16,094
1984	15,513	4,754	20,267	1,630	224	108	18,304	838	17,466
1985	14,535	5,071	19,607	1,915	326	95	17,270	816	16,454
1986	14,154	4,977	19,131	1,838	337	98	16,859	800	16,059
1987	14,807	5,333	20,140	2,208	376	124	17,433	812	16,621
1988	15,467	5,532	20,999	2,478	460	143	17,918	816	17,103
1989	15,709	5,366	21,074	2,475	362	142	18,095	785	17,311
1990	16,054	5,469	21,523	2,489	289	150	18,594	784	17,810
1991	16,018	5,732	21,750	2,772	276	170	18,532	835	17,698
1992	16,165	5,967	22,132	2,973	280	168	18,712	872	17,840
1993	16,691	6,035	22,726	3,103	414	227	18,982	886	18,095
1994	17,351	6,230	23,581	3,231	412	228	19,710	889	18,821
1995	17,282	6,462	23,744	3,565	388	284	19,506	908	18,599
1996	17,737	6,376	24,114	3,511	518	272	19,812	958	18,854
1997	17,844	6,369	24,213	3,492	599	256	19,866	964	18,902
1998	17,729	6,380	24,108	3,427	617	103	19,961	938	19,024
1999	17,590	6,233	23,823	3,293	615	110	19,805	973	18,832
2000	17,726	6,448	24,174	3,380	505	91	20,198	1,016	19,182
2001	18,129	6,371	24,501	3,371	463	97	20,570	954	19,616
2002	17,795	6,146	23,941	3,455	502	99	19,885	957	18,928
2003	17,882	6,237	24,119	3,548	499	98	19,974	876	19,099
2004	17,885	6,084	23,970	3,702	654	96	19,517	927	18,591
2005	R17,472	R5,985	R23,457	R3,700	711	119	R18,927	876	R18,051
2006	R17,942	R5,565	R23,507	R3,265	R731	R130	R19,382	R906	R18,476
2007	E18,437	E6,100	P24,536	E3,568	E701	E115	P20,151	P874	P19,278

[1] Volume reduction resulting from the removal of natural gas plant liquids, which are transferred to petroleum supply (see Tables 5.1 and 5.10).

R=Revised. P=Preliminary. E=Estimate. NA=Not available.

Notes: • Beginning with 1965 data, all volumes are shown on a pressure base of 14.73 p.s.i.a. at 60° F. For prior years, the pressure base was 14.65 p.s.i.a. at 60° F. • Totals may not equal sum of components due to independent rounding.

Web Pages: • For all data beginning in 1949, see http://www.eia.doe.gov/emeu/aer/natgas.html. • For related information, see http://www.eia.doe.gov/oil_gas/natural_gas/info_glance/natural_gas.html.

Sources: **Natural Gas Wells** and **Crude Oil Wells:** • 1949-1966—Bureau of Mines, *Minerals Yearbook*, "Natural Gas" chapter. • 1967-2006—Energy Information Administration (EIA), *Natural Gas Annual (NGA)*, annual reports. • 2007—EIA estimates. **Total Gross Withdrawals, Marketed Production, Extraction Loss,** and **Dry Gas Production:** • 1949-2002—EIA, *NGA*, annual reports. • 2003 forward—EIA, *Natural Gas Monthly* (March 2008), Table 1. **All Other Data:** • 1949-2006—EIA, *NGA*, annual reports. • 2007—EIA estimates.

Figure 6.3 Natural Gas Imports, Exports, and Net Imports

Trade Overview, 1949-2007

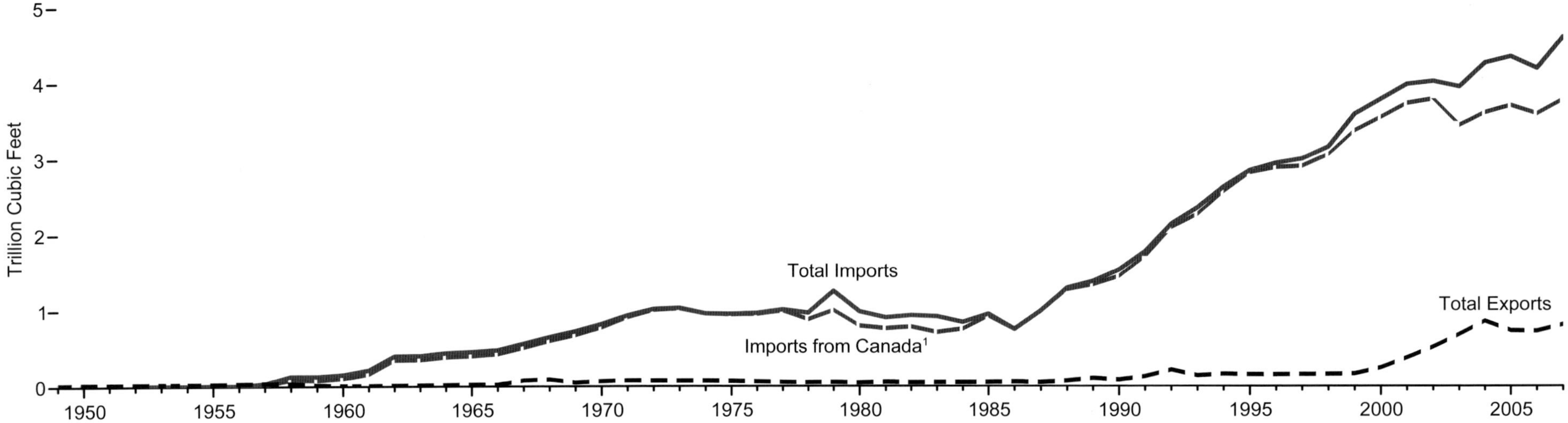

Trade, 2007

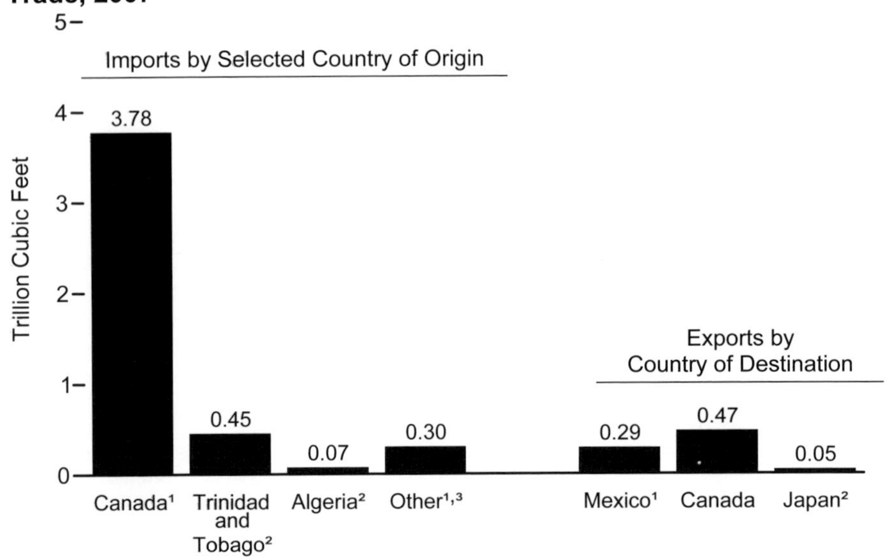

Net Imports as Share of Consumption, 1958-2007

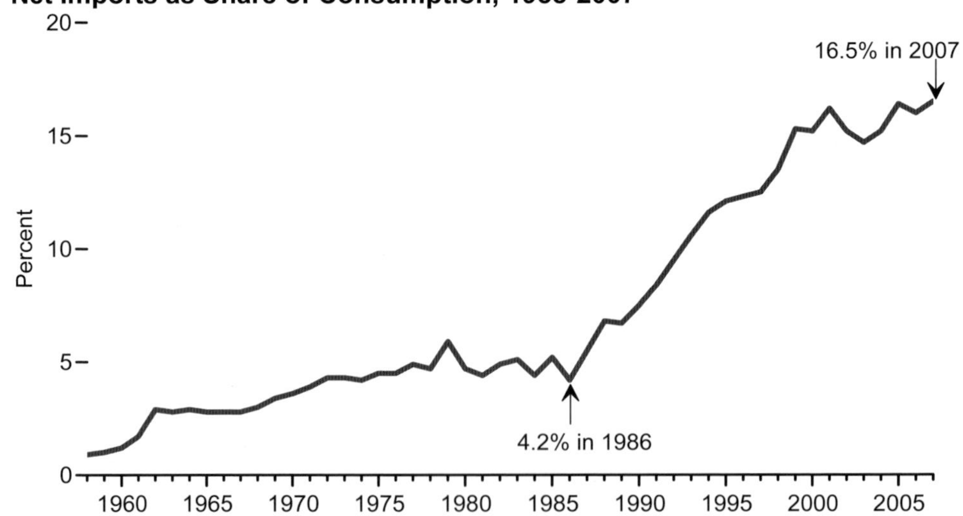

[1] Pipeline and liquefied natural gas.
[2] As liquefied natural gas.
[3] By pipeline and liquefied natural gas from Mexico, and as liquefied natural gas from Egypt, Nigeria, Mexico, Qatar, and Equatorial Guinea.
Source: Table 6.3.

Table 6.3 Natural Gas Imports, Exports, and Net Imports, Selected Years, 1949-2007

(Billion Cubic Feet, Except as Noted)

Year	Imports by Country of Origin										Exports by Country of Destination				Net Imports [1]	
	Algeria [2]	Canada [3]	Egypt [2]	Mexico [3]	Nigeria [2]	Oman [2]	Qatar [2]	Trinidad and Tobago [2]	Other [2,4]	Total	Canada [3]	Japan [2]	Mexico [3]	Total	Total	Percent of U.S. Consumption
1949	0	0	0	0	0	0	0	0	0	0	(s)	0	20	20	-20	([5])
1950	0	0	0	0	0	0	0	0	0	0	3	0	23	26	-26	([5])
1955	0	11	0	(s)	0	0	0	0	0	11	11	0	20	31	-20	([5])
1960	0	109	0	47	0	0	0	0	0	156	6	0	6	11	144	1.2
1965	0	405	0	52	0	0	0	0	0	456	18	0	8	26	430	2.8
1970	1	779	0	41	0	0	0	0	0	821	11	44	15	70	751	3.6
1971	1	912	0	21	0	0	0	0	0	935	14	50	16	80	854	3.9
1972	2	1,009	0	8	0	0	0	0	0	1,019	16	48	15	78	941	4.3
1973	3	1,028	0	2	0	0	0	0	0	1,033	15	48	14	77	956	4.3
1974	0	959	0	(s)	0	0	0	0	0	959	13	50	13	77	882	4.2
1975	5	948	0	0	0	0	0	0	0	953	10	53	9	73	880	4.5
1976	10	954	0	0	0	0	0	0	0	964	8	50	7	65	899	4.5
1977	11	997	0	2	0	0	0	0	0	1,011	(s)	52	4	56	955	4.9
1978	84	881	0	0	0	0	0	0	0	966	(s)	48	4	53	913	4.7
1979	253	1,001	0	0	0	0	0	0	0	1,253	(s)	51	4	56	1,198	5.9
1980	86	797	0	102	0	0	0	0	0	985	(s)	45	4	49	936	4.7
1981	37	762	0	105	0	0	0	0	0	904	(s)	56	3	59	845	4.4
1982	55	783	0	95	0	0	0	0	0	933	(s)	50	2	52	882	4.9
1983	131	712	0	75	0	0	0	0	0	918	(s)	53	2	55	864	5.1
1984	36	755	0	52	0	0	0	0	0	843	(s)	53	2	55	788	4.4
1985	24	926	0	0	0	0	0	0	0	950	(s)	53	2	55	894	5.2
1986	0	749	0	0	0	0	0	0	2	750	9	50	2	61	689	4.2
1987	0	993	0	0	0	0	0	0	0	993	3	49	2	54	939	5.5
1988	17	1,276	0	0	0	0	0	0	0	1,294	20	52	2	74	1,220	6.8
1989	42	1,339	0	0	0	0	0	0	0	1,382	38	51	17	107	1,275	6.7
1990	84	1,448	0	0	0	0	0	0	0	1,532	17	53	16	86	1,447	7.5
1991	64	1,710	0	0	0	0	0	0	0	1,773	15	54	60	129	1,644	8.4
1992	43	2,094	0	0	0	0	0	0	0	2,138	68	53	96	216	1,921	9.5
1993	82	2,267	0	2	0	0	0	0	0	2,350	45	56	40	140	2,210	10.6
1994	51	2,566	0	7	0	0	0	0	0	2,624	53	63	47	162	2,462	11.6
1995	18	2,816	0	7	0	0	0	0	0	2,841	28	65	61	154	2,687	12.1
1996	35	2,883	0	14	0	0	0	0	5	2,937	52	68	34	153	2,784	12.3
1997	66	2,899	0	17	0	0	0	0	12	2,994	56	62	38	157	2,837	12.5
1998	69	3,052	0	15	0	0	0	0	17	3,152	40	66	53	159	2,993	13.5
1999	76	3,368	0	55	0	0	20	51	17	3,586	39	64	61	163	3,422	15.3
2000	47	3,544	0	12	13	10	46	99	11	3,782	73	66	106	244	3,538	15.2
2001	65	3,729	0	10	38	12	23	98	2	3,977	167	66	141	373	3,604	16.2
2002	27	3,785	0	2	8	3	35	151	5	4,015	189	63	263	516	3,499	15.2
2003	53	3,437	0	0	50	9	14	378	3	3,944	271	66	343	680	3,264	14.7
2004	120	3,607	0	0	12	9	12	462	36	4,259	395	62	397	854	3,404	15.2
2005	97	3,700	73	9	8	2	3	439	9	4,341	358	65	305	729	3,612	R16.4
2006	17	R3,590	120	R13	57	0	0	389	0	R4,186	R341	61	R322	R724	R3,462	R16.0
2007P	74	3,777	114	54	95	0	18	451	18	4,602	472	47	288	[6]809	3,793	16.5

[1] Net imports equal imports minus exports.

[2] As liquefied natural gas.

[3] By pipeline, except for very small amounts of liquefied natural gas imported from Canada in 1973, 1977, and 1981, and exported to Mexico beginning in 1998.

[4] Australia in 1997-2001 and 2004; Brunei in 2002; Equatorial Guinea in 2007; Indonesia in 1986 and 2000; Malaysia in 1999 and 2002-2005; United Arab Emirates in 1996-2000; and Other (unassigned) in 2004.

[5] Not meaningful because there were net exports during this year.

[6] Includes 2 billion cubic feet to Russia.

R=Revised. P=Preliminary. (s)=Less than 0.5 billion cubic feet.

Note: Totals may not equal sum of components due to independent rounding.

Web Pages: • For all data beginning in 1949, see http://www.eia.doe.gov/emeu/aer/natgas.html. • For related information, see http://www.eia.doe.gov/oil_gas/natural_gas/info_glance/natural_gas.html.

Sources: **Percent of U.S. Consumption:** Calculated. **All Other Data:** • 1949-1954—Energy Information Administration (EIA), Office of Oil and Gas, Reserves and Natural Gas Division, unpublished data. • 1955-1971—EIA, Federal Power Commission, by telephone. • 1972-1987—EIA, Form FPC-14, "Annual Report for Importers and Exporters of Natural Gas." • 1988-2005—EIA, *Natural Gas Annual,* annual reports. • 2006 and 2007—EIA, *Natural Gas Monthly* (March 2008), Table 4.

Figure 6.4 Natural Gas Gross Withdrawals and Natural Gas Well Productivity, 1960-2007

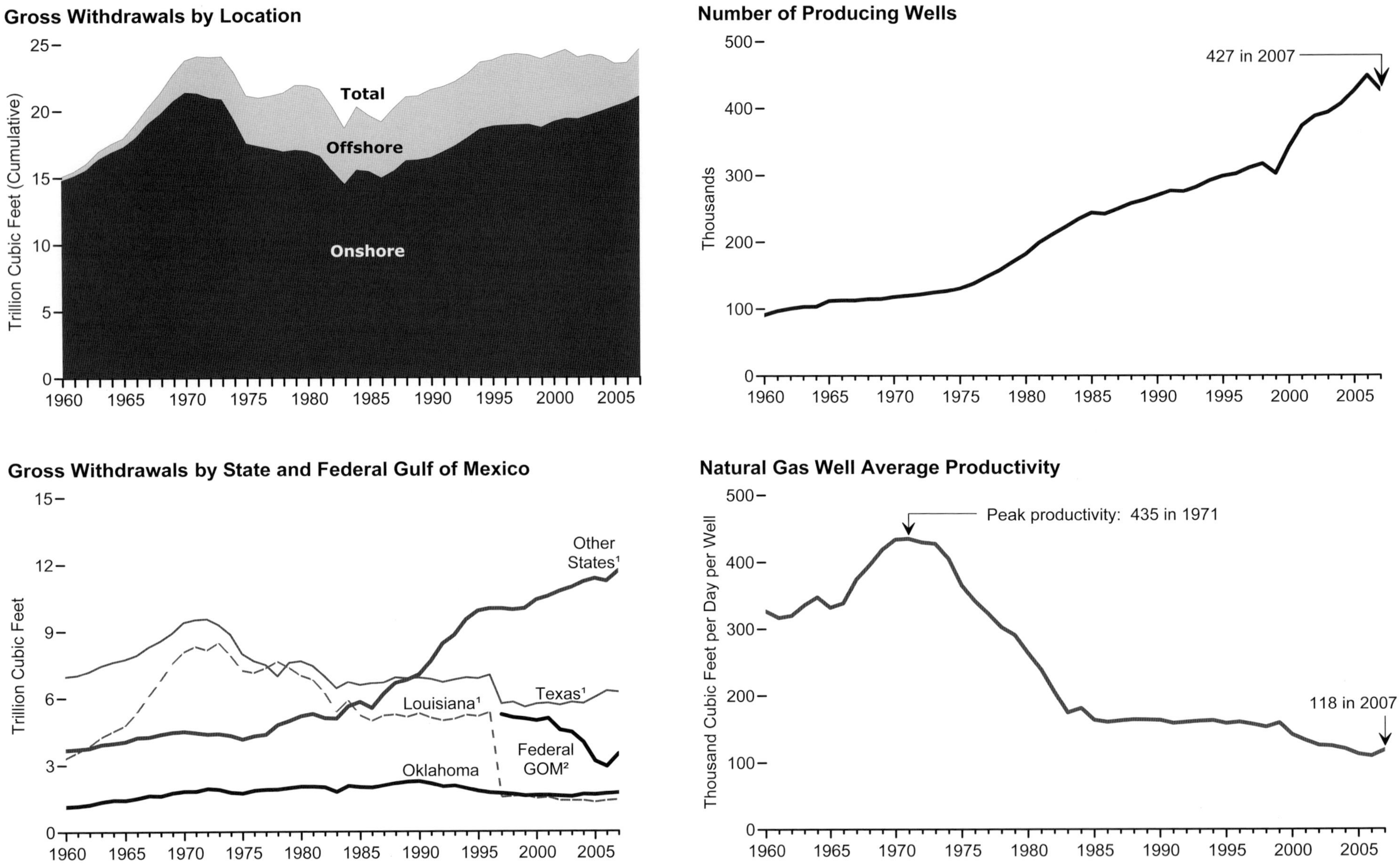

[1] Through 1996, includes gross withdrawals in Federal offshore areas of the Gulf of Mexico; beginning in 1997, these are included in "Federal Gulf of Mexico."
[2] Gulf of Mexico.

Note: Because vertical scales differ, graphs should not be compared.
Source: Table 6.4.

Table 6.4 Natural Gas Gross Withdrawals and Natural Gas Well Productivity, 1960-2007

(Billion Cubic Feet, Except as Noted)

Year	Natural Gas Gross Withdrawals From Crude Oil and Natural Gas Wells								Natural Gas Well Productivity		
	State					Location					
	Texas [1]	Louisiana [1]	Oklahoma	Other States [1]	Federal Gulf of Mexico [2]	Onshore [3]	Offshore [4]	Total	Gross Withdrawals From Natural Gas Wells	Producing Wells [5] (thousands)	Average Productivity (thousand cubic feet per day)
1960	6,965	3,313	1,133	3,677	([2])	14,815	273	15,088	10,853	91	326.7
1961	7,020	3,571	1,160	3,710	([2])	15,142	318	15,460	11,195	97	316.8
1962	7,199	3,854	1,222	3,764	([2])	15,587	452	16,039	11,702	100	319.8
1963	7,452	4,250	1,347	3,924	([2])	16,409	564	16,973	12,606	103	335.4
1964	7,622	4,515	1,423	3,975	([2])	16,914	622	17,536	13,106	103	347.4
1965	7,741	4,764	1,414	4,044	([2])	17,318	646	17,963	13,524	112	331.8
1966	7,935	5,365	1,502	4,232	([2])	18,026	1,007	19,034	13,894	112	338.4
1967	8,292	6,087	1,621	4,252	([2])	19,065	1,187	20,252	15,345	112	374.3
1968	8,566	6,778	1,607	4,375	([2])	19,801	1,524	21,325	16,540	114	395.1
1969	8,915	7,561	1,742	4,462	([2])	20,725	1,954	22,679	17,489	114	418.6
1970	9,399	8,076	1,811	4,501	([2])	21,368	2,419	23,786	18,595	117	433.6
1971	9,519	8,319	1,809	4,442	([2])	21,311	2,777	24,088	18,925	119	434.8
1972	9,550	8,160	1,928	4,378	([2])	20,978	3,039	24,016	19,043	121	429.4
1973	9,290	8,491	1,890	4,396	([2])	20,856	3,212	24,067	19,372	124	427.4
1974	8,859	7,920	1,757	4,314	([2])	19,335	3,515	22,850	18,669	126	404.9
1975	7,989	7,242	1,721	4,152	([2])	17,555	3,549	21,104	17,380	130	365.3
1976	7,666	7,143	1,842	4,293	([2])	17,348	3,596	20,944	17,191	138	341.5
1977	7,496	7,351	1,888	4,362	([2])	17,165	3,932	21,097	17,416	148	323.1
1978	6,988	7,639	1,892	4,790	([2])	16,953	4,356	21,309	17,394	157	302.7
1979	7,594	7,359	1,958	4,973	([2])	17,061	4,822	21,883	18,034	170	290.8
1980	7,656	7,008	2,019	5,187	([2])	16,967	4,902	21,870	17,573	182	263.8
1981	7,452	6,830	2,019	5,287	([2])	16,597	4,991	21,587	17,337	199	238.9
1982	6,976	6,217	1,985	5,094	([2])	15,499	4,773	20,272	15,809	211	205.5
1983	6,429	5,379	1,780	5,071	([2])	14,477	4,182	18,659	14,153	222	174.7
1984	6,712	5,888	2,046	5,620	([2])	15,560	4,707	20,267	15,513	234	181.2
1985	6,577	5,218	1,993	5,818	([2])	15,421	4,186	19,607	14,535	243	163.6
1986	6,656	4,965	1,972	5,538	([2])	14,945	4,186	19,131	14,154	242	160.6
1987	6,688	5,205	2,073	6,174	([2])	15,468	4,672	20,140	14,807	249	162.8
1988	6,919	5,248	2,167	6,665	([2])	16,253	4,747	20,999	15,467	257	164.3
1989	6,881	5,143	2,237	6,813	([2])	16,303	4,771	21,074	15,709	262	164.0
1990	6,907	5,303	2,258	7,054	([2])	16,476	5,047	21,523	16,054	269	163.4
1991	6,846	5,100	2,154	7,651	([2])	16,900	4,850	21,750	16,018	276	158.8
1992	6,708	4,977	2,017	8,429	([2])	17,361	4,772	22,132	16,165	275	160.4
1993	6,817	5,047	2,050	8,812	([2])	17,960	4,766	22,726	16,691	282	162.1
1994	6,912	5,226	1,935	9,508	([2])	18,585	4,996	23,581	17,351	292	162.9
1995	6,873	5,163	1,812	9,896	([2])	18,802	4,942	23,744	17,282	299	158.6
1996	7,028	5,351	1,735	9,999	([2])	18,867	5,246	24,114	17,737	302	160.6
1997	[1]5,730	[1]1,538	1,704	[1]9,999	5,242	18,897	5,316	24,213	17,844	311	157.2
1998	5,799	1,579	1,669	9,950	5,110	18,923	5,185	24,108	17,729	317	153.3
1999	5,575	1,599	1,594	10,002	5,053	18,692	5,131	23,823	17,590	302	159.4
2000	5,723	1,485	1,613	10,386	4,968	19,130	5,044	24,174	17,726	342	141.7
2001	5,752	1,525	1,615	10,542	5,066	19,364	5,137	24,501	18,129	373	133.1
2002	5,661	1,382	1,582	10,769	4,548	19,326	4,615	23,941	17,795	388	125.7
2003	5,791	1,378	1,558	10,944	4,447	19,614	4,505	24,119	17,882	393	124.6
2004	5,734	1,377	1,656	11,202	4,001	19,914	4,055	23,970	17,885	406	120.3
2005	[R]6,007	1,310	[R]1,639	[R]11,350	[R]3,151	[R]20,252	[R]3,205	[R]23,457	[R]17,472	[R]426	[R]112.4
2006	[R]6,292	[R]1,378	[R]1,689	[R]11,234	[R]2,914	[R]20,553	[R]2,955	[R]23,507	[R]17,942	[R]449	[R]109.6
2007	[E]6,238	[E]1,406	[E]1,724	[E]11,687	[E]3,482	[E]21,003	[E]3,533	[P]24,536	[E]18,437	[E]427	[E]118.3

[1] Through 1996, includes gross withdrawals in Federal offshore areas of the Gulf of Mexico; beginning in 1997, these are included in "Federal Gulf of Mexico."

[2] Gross withdrawals from Federal offshore areas of the Gulf of Mexico. Through 1996, these gross withdrawals are included in "Texas," "Louisiana," and "Other States."

[3] Includes State offshore gross withdrawals.

[4] Excludes State offshore gross withdrawals; includes Federal offshore (Outer Continental Shelf) gross withdrawals.

[5] As of December 31 each year.

R=Revised. P=Preliminary. E=Estimate.

Web Page: See http://www.eia.doe.gov/oil_gas/natural_gas/info_glance/natural_gas.html for related information.

Sources: **Offshore:** • 1960-1981—U.S. Geological Survey. • 1982-1985—U.S. Minerals Management Service, *Mineral Revenues—The 1989 Report on Receipts from Federal and Indian Leases,* and predecessor annual reports. • 1986-2006—Energy Information Administration (EIA), *Natural Gas Annual (NGA),* annual reports. • 2007—EIA estimate. **Total (Gross Withdrawals):** • 1960-2002—EIA, *NGA,* annual reports. • 2003 forward—EIA, *Natural Gas Monthly* (March 2008), Table 1. **Average Productivity:** Calculated as gross withdrawals from natural gas wells divided by the number of producing wells, and then divided by the number of days in the year. **All Other Data:** • 1960-1966—Bureau of Mines, *Natural Gas Production and Consumption.* • 1967-2006—EIA, *NGA,* annual reports and unpublished revisions. • 2007—EIA estimates.

Figure 6.5 Natural Gas Consumption by Sector

By Sector, 1949-2007

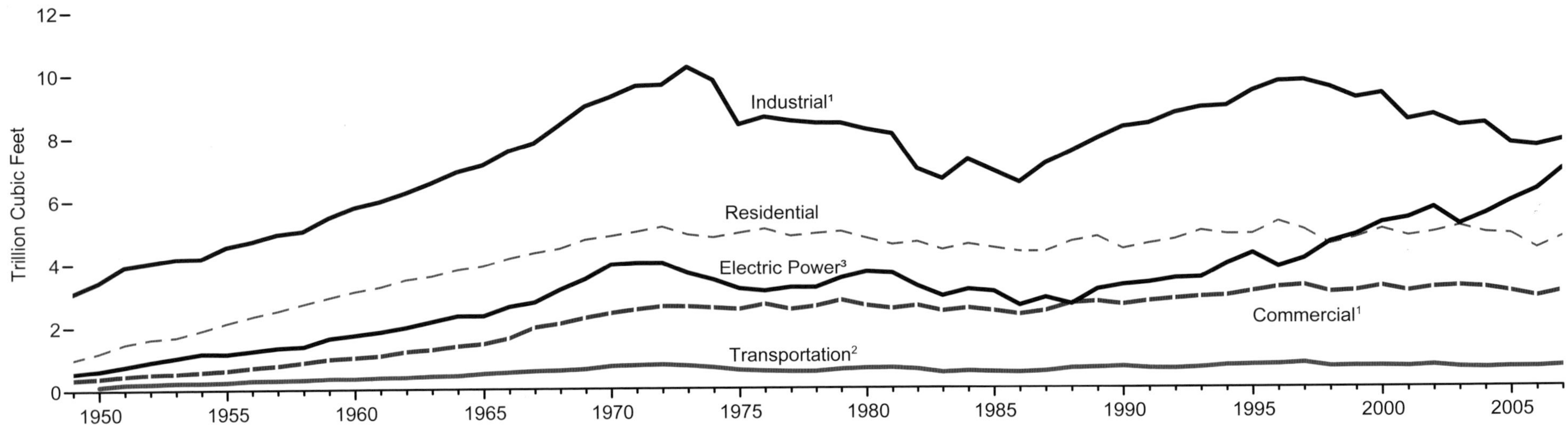

By Sector, 2007

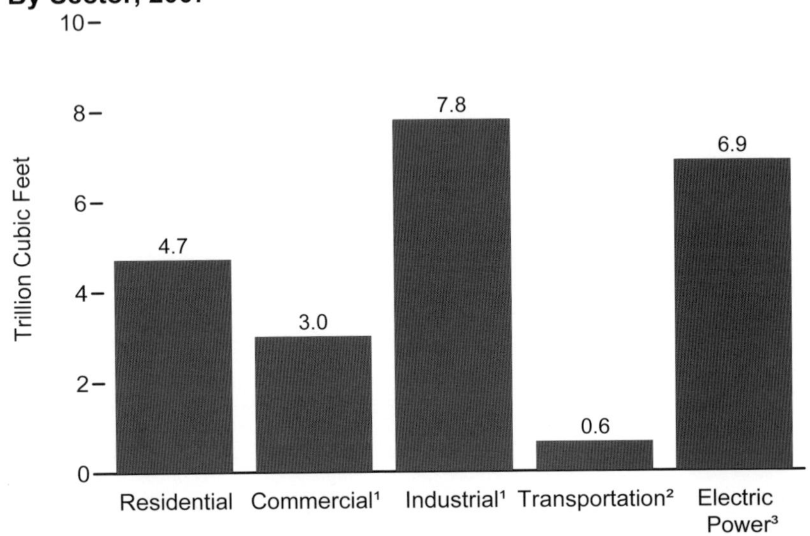

End Use and Electric Power Shares, 1950 and 2007

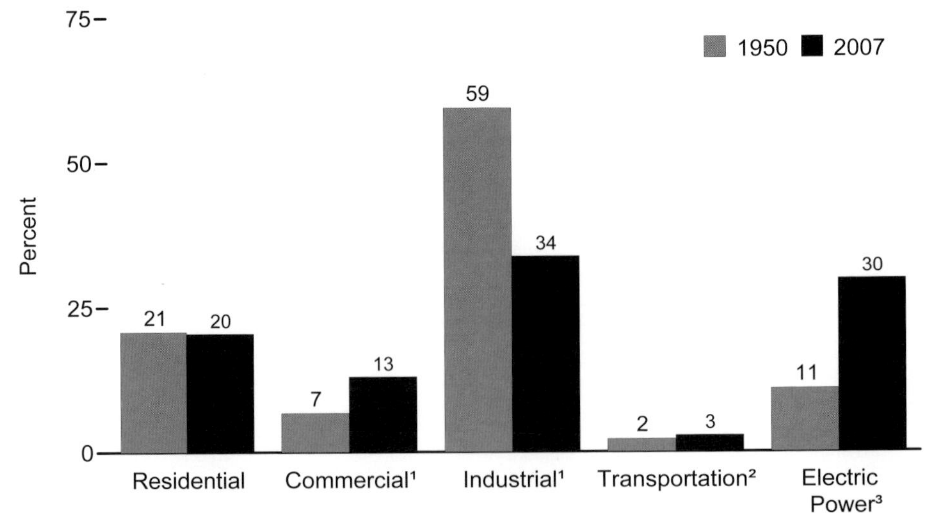

[1] Includes combined-heat-and-power plants and a small number of electricity-only plants.
[2] Natural gas consumed in the operation of pipelines (primarily in compressors), and as fuel in the delivery of natural gas to consumers; plus a small quantity used as vehicle fuel.
[3] Electricity-only and combined-heat-and-power plants whose primary business is to sell electricity, or electricity and heat, to the public.
Source: Table 6.5.

Table 6.5 Natural Gas Consumption by Sector, Selected Years, 1949-2007

(Billion Cubic Feet)

Year	Residential Sector	Commercial Sector			Industrial Sector					Transportation Sector			Electric Power Sector [1]			Total
		CHP [2]	Other [3]	Total	Lease and Plant Fuel	Other Industrial: CHP [4]	Other Industrial: Non-CHP [5]	Other Industrial: Total	Total	Pipelines [6] and Distribution [7]	Vehicle Fuel [8]	Total	Electricity Only	CHP	Total	
1949	993	(9)	348	348	835	(10)	2,245	2,245	3,081	NA	NA	NA	550	NA	550	4,971
1950	1,198	(9)	388	388	928	(10)	2,498	2,498	3,426	126	NA	126	629	NA	629	5,767
1955	2,124	(9)	629	629	1,131	(10)	3,411	3,411	4,542	245	NA	245	1,153	NA	1,153	8,694
1960	3,103	(9)	1,020	1,020	1,237	(10)	4,535	4,535	5,771	347	NA	347	1,725	NA	1,725	11,967
1965	3,903	(9)	1,444	1,444	1,156	(10)	5,955	5,955	7,112	501	NA	501	2,321	NA	2,321	15,280
1970	4,837	(9)	2,399	2,399	1,399	(10)	7,851	7,851	9,249	722	NA	722	3,932	NA	3,932	21,139
1971	4,972	(9)	2,509	2,509	1,414	(10)	8,181	8,181	9,594	743	NA	743	3,976	NA	3,976	21,793
1972	5,126	(9)	2,608	2,608	1,456	(10)	8,169	8,169	9,624	766	NA	766	3,977	NA	3,977	22,101
1973	4,879	(9)	2,597	2,597	1,496	(10)	8,689	8,689	10,185	728	NA	728	3,660	NA	3,660	22,049
1974	4,786	(9)	2,556	2,556	1,477	(10)	8,292	8,292	9,769	669	NA	669	3,443	NA	3,443	21,223
1975	4,924	(9)	2,508	2,508	1,396	(10)	6,968	6,968	8,365	583	NA	583	3,158	NA	3,158	19,538
1976	5,051	(9)	2,668	2,668	1,634	(10)	6,964	6,964	8,598	548	NA	548	3,081	NA	3,081	19,946
1977	4,821	(9)	2,501	2,501	1,659	(10)	6,815	6,815	8,474	533	NA	533	3,191	NA	3,191	19,521
1978	4,903	(9)	2,601	2,601	1,648	(10)	6,757	6,757	8,405	530	NA	530	3,188	NA	3,188	19,627
1979	4,965	(9)	2,786	2,786	1,499	(10)	6,899	6,899	8,398	601	NA	601	3,491	NA	3,491	20,241
1980	4,752	(9)	2,611	2,611	1,026	(10)	7,172	7,172	8,198	635	NA	635	3,682	NA	3,682	19,877
1981	4,546	(9)	2,520	2,520	928	(10)	7,128	7,128	8,055	642	NA	642	3,640	NA	3,640	19,404
1982	4,633	(9)	2,606	2,606	1,109	(10)	5,831	5,831	6,941	596	NA	596	3,226	NA	3,226	18,001
1983	4,381	(9)	2,433	2,433	978	(10)	5,643	5,643	6,621	490	NA	490	2,911	NA	2,911	16,835
1984	4,555	(9)	2,524	2,524	1,077	(10)	6,154	6,154	7,231	529	NA	529	3,111	NA	3,111	17,951
1985	4,433	(9)	2,432	2,432	966	(10)	5,901	5,901	6,867	504	NA	504	3,044	NA	3,044	17,281
1986	4,314	(9)	2,318	2,318	923	(10)	5,579	5,579	6,502	485	NA	485	2,602	NA	2,602	16,221
1987	4,315	(9)	2,430	2,430	1,149	(10)	5,953	5,953	7,103	519	NA	519	2,844	NA	2,844	17,211
1988	4,630	(9)	2,670	2,670	1,096	(10)	6,383	6,383	7,479	614	NA	614	2,636	NA	2,636	18,030
1989	4,781	30	2,688	2,718	1,070	914	[11]5,903	[11]6,816	7,886	629	NA	629	[11]2,791	[11]315	[11]3,105	[11]19,119
1990	4,391	46	2,576	2,623	1,236	1,055	[11]5,963	[11]7,018	8,255	660	(s)	660	[11]2,794	[11]451	[11]3,245	[11]19,174
1991	4,556	52	2,676	2,729	1,129	1,061	[11]6,170	[11]7,231	8,360	601	(s)	602	[11]2,822	[11]494	[11]3,316	[11]19,562
1992	4,690	62	2,740	2,803	1,171	1,107	[11]6,420	[11]7,527	8,698	588	2	590	[11]2,829	[11]619	[11]3,448	[11]20,228
1993	4,956	65	2,796	2,862	1,172	1,124	6,576	7,700	8,872	624	3	627	2,755	718	3,473	20,790
1994	4,848	72	2,823	2,895	1,124	1,176	6,613	7,790	8,913	685	3	689	3,065	838	3,903	21,247
1995	4,850	78	2,953	3,031	1,220	1,258	6,906	8,164	9,384	700	5	705	3,288	949	4,237	22,207
1996	5,241	82	3,076	3,158	1,250	1,289	7,146	8,435	9,685	711	6	718	2,824	983	3,807	22,609
1997	4,984	87	3,128	3,215	1,203	1,282	7,229	8,511	9,714	751	8	760	3,039	1,026	4,065	22,737
1998	4,520	87	2,912	2,999	1,173	1,355	6,965	8,320	9,493	635	9	645	3,544	1,044	4,588	22,246
1999	4,726	84	2,961	3,045	1,079	1,401	6,678	8,079	9,158	645	12	657	3,729	1,090	4,820	22,405
2000	4,996	85	3,098	3,182	1,151	1,386	6,757	8,142	9,293	642	13	655	4,093	1,114	5,206	23,333
2001	4,771	79	2,944	3,023	1,119	1,310	6,035	7,344	8,463	625	15	640	4,164	1,178	5,342	22,239
2002	4,889	74	3,070	3,144	1,113	1,240	6,267	7,507	8,620	667	15	682	4,258	1,413	5,672	23,007
2003	5,079	58	3,121	3,179	1,122	1,144	6,007	7,150	8,273	591	18	610	3,780	1,355	5,135	22,277
2004	4,869	72	3,057	3,129	1,098	1,191	6,052	7,243	8,341	566	21	587	4,142	1,322	5,464	22,389
2005	R4,827	75	R2,924	R2,999	1,112	1,084	R5,514	R6,597	R7,709	R584	R23	607	4,592	1,277	5,869	R22,011
2006	R4,368	R82	R2,753	R2,835	R1,124	R1,115	R5,380	R6,495	R7,618	R584	R25	R609	R5,091	R1,131	R6,222	R21,653
2007P	4,724	83	2,924	3,008	1,168	1,202	5,430	6,632	7,800	622	26	649	5,607	1,267	6,874	23,055

[1] Electricity-only and combined-heat-and-power (CHP) plants within the NAICS 22 category whose primary business is to sell electricity, or electricity and heat, to the public. Through 1988, data are for electric utilities only; beginning in 1989, data are for electric utilities and independent power producers. Electric utility CHP plants are included in "Electricity Only."

[2] Commercial combined-heat-and-power (CHP) and a small number of commercial electricity-only plants.

[3] All commercial sector fuel use other than that in "Commercial CHP."

[4] Industrial combined-heat-and-power (CHP) and a small number of industrial electricity-only plants.

[5] All industrial sector fuel use other than that in "Lease and Plant Fuel" and "Industrial CHP."

[6] Natural gas consumed in the operation of pipelines, primarily in compressors.

[7] Natural gas used as fuel in the delivery of natural gas to consumers.

[8] Vehicle fuel data do not reflect revised data shown in Table 10.4. See Note 4, "Natural Gas Vehicle Fuel," at end of section.

[9] Included in "Commercial Other."

[10] Included in "Industrial Non-CHP."

[11] For 1989-1992, a small amount of consumption at independent power producers may be counted in both "Other Industrial" and "Electric Power Sector." See Note 3, "Natural Gas Consumption, 1989-1992," at end of section.

R=Revised. P=Preliminary. NA=Not available. (s)=Less than 0.5 billion cubic feet.

Notes: • Data are for natural gas, plus a small amount of supplemental gaseous fuels. See Note 1, "Supplemental Gaseous Fuels," at end of section. • See Tables 8.5a-8.5d for the amount of natural gas used to produce electricity and Tables 8.6a-8.6c for the amount of natural gas used to produce useful thermal output. • See Note 2, "Natural Gas Consumption," at end of section. • Beginning with 1965, all volumes are shown on a pressure base of 14.73 p.s.i.a. at 60° F. For prior years, the pressure base was 14.65 p.s.i.a. at 60° F. • Totals may not equal sum of components due to independent rounding.

Web Pages: • For all data beginning in 1949, see http://www.eia.doe.gov/emeu/aer/natgas.html. • For related information, see http://www.eia.doe.gov/oil_gas/natural_gas/info_glance/natural_gas.html.

Sources: **Residential, Commercial Total, Lease and Plant Fuel, Other Industrial Total,** and **Pipelines and Distribution:** • 1949-2002—Energy Information Administration (EIA), *Natural Gas Annual (NGA)*, annual reports and unpublished revisions. • 2003 forward—EIA, *Natural Gas Monthly (NGM)* (March 2008), Table 2. **Commercial CHP** and **Industrial CHP:** Table 8.7c. **Vehicle Fuel:** • 1990 and 1991—EIA, *NGA 2000* (November 2001), Table 95. • 1992-1998—EIA, "Alternatives to Traditional Transportation Fuels 1999" (October 1999), Table 10, and "Alternatives to Traditional Transportation Fuels 2003" (February 2004), Table 10. Data for compressed natural gas and liquefied natural gas in gasoline-equivalent gallons were converted to cubic feet by multiplying by the motor gasoline conversion factor (see Table A3) and dividing by the natural gas end-use sectors conversion factor (see Table A4). • 1999-2002—EIA, *NGA*, annual reports. • 2003 forward—EIA, *NGM* (March 2008), Table 2. **Electric Power Sector:** Tables 8.5b, 8.5c, 8.6b, and 8.7b. **All Other Data:** Calculated.

Figure 6.6 Natural Gas Underground Storage

Underground Storage, 1954-2007

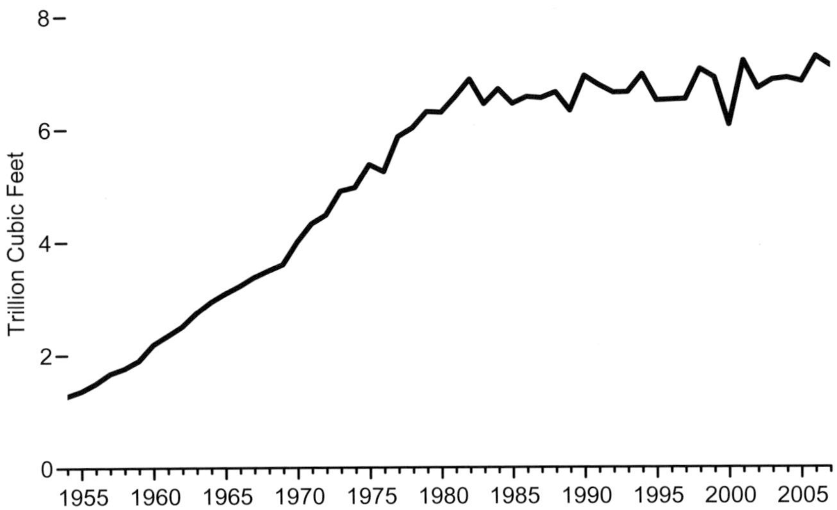

Underground Storage Capacity, 1975-2007

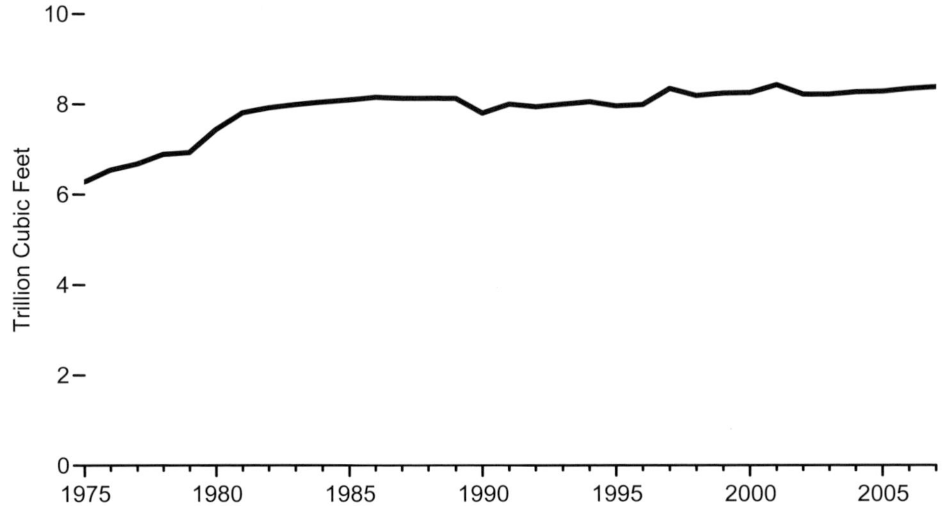

Base Gas and Working Gas in Underground Storage, 1954-2007

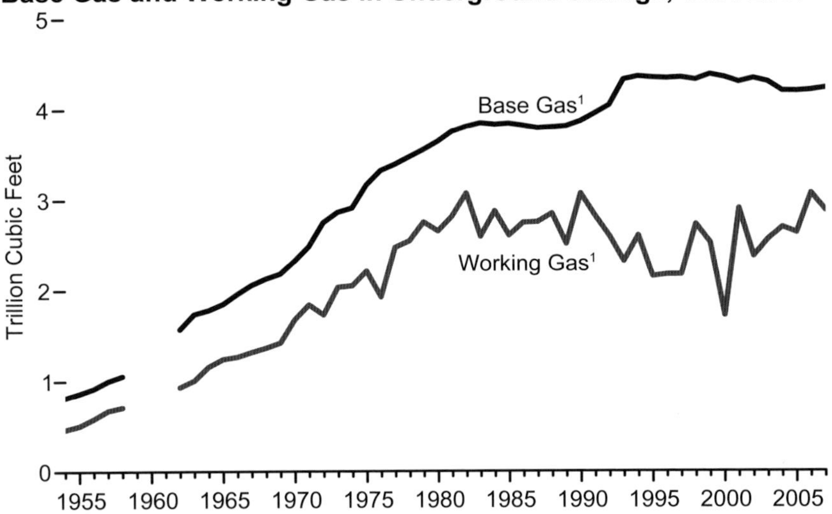

End-of-Year Storage as Share of Total Consumption, 1954-2007

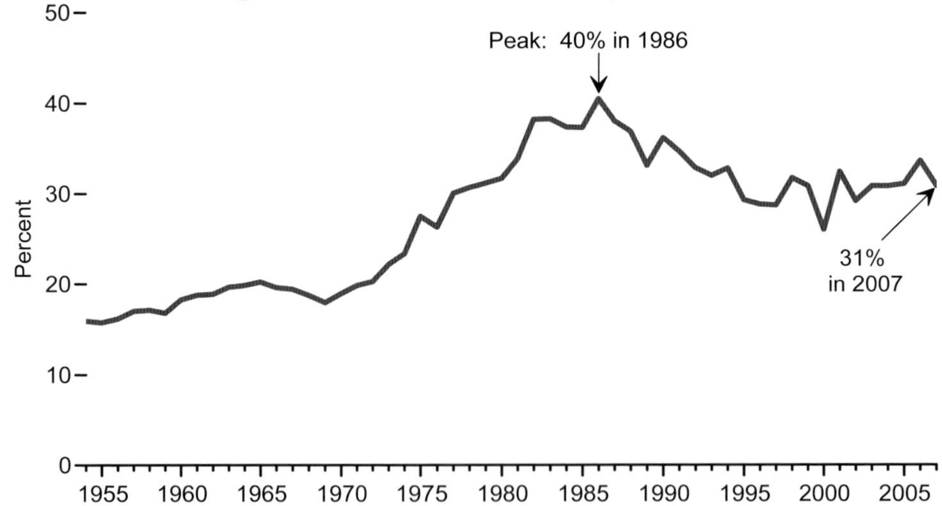

[1] Working-gas and base-gas data were not collected in 1959, 1960, and 1961.

Notes: • Storage is at end of year. • Because vertical scales differ, graphs should not be compared.

Sources: Tables 6.5 and 6.6.

Table 6.6 Natural Gas Underground Storage, 1954-2007

(Billion Cubic Feet)

Year	Natural Gas in Underground Storage									Natural Gas Underground Storage Capacity
	Base Gas [1]			Working Gas			Total			
	Traditional Storage	Salt Caverns	Total	Traditional Storage	Salt Caverns	Total	Traditional Storage	Salt Caverns	Total	
1954	NA	NA	817	NA	NA	465	NA	NA	1,281	NA
1955	NA	NA	863	NA	NA	505	NA	NA	1,368	NA
1956	NA	NA	919	NA	NA	583	NA	NA	1,502	NA
1957	NA	NA	1,001	NA	NA	673	NA	NA	1,674	NA
1958	NA	NA	1,056	NA	NA	708	NA	NA	1,764	NA
1959	NA	NA	NA	NA	NA	NA	NA	NA	1,901	NA
1960	NA	NA	NA	NA	NA	NA	NA	NA	2,184	NA
1961	NA	NA	NA	NA	NA	NA	NA	NA	2,344	NA
1962	NA	NA	1,571	NA	NA	933	NA	NA	2,504	NA
1963	NA	NA	1,738	NA	NA	1,007	NA	NA	2,745	NA
1964	NA	NA	1,781	NA	NA	1,159	NA	NA	2,940	NA
1965	NA	NA	1,848	NA	NA	1,242	NA	NA	3,090	NA
1966	NA	NA	1,958	NA	NA	1,267	NA	NA	3,225	NA
1967	NA	NA	2,058	NA	NA	1,318	NA	NA	3,376	NA
1968	NA	NA	2,128	NA	NA	1,366	NA	NA	3,495	NA
1969	NA	NA	2,181	NA	NA	1,421	NA	NA	3,602	NA
1970	NA	NA	2,326	NA	NA	1,678	NA	NA	4,004	NA
1971	NA	NA	2,485	NA	NA	1,840	NA	NA	4,325	NA
1972	NA	NA	2,751	NA	NA	1,729	NA	NA	4,480	NA
1973	NA	NA	2,864	NA	NA	2,034	NA	NA	4,898	NA
1974	NA	NA	2,912	NA	NA	2,050	NA	NA	4,962	NA
1975	NA	NA	3,162	NA	NA	2,212	NA	NA	5,374	6,280
1976	NA	NA	3,323	NA	NA	1,926	NA	NA	5,250	6,544
1977	NA	NA	3,391	NA	NA	2,475	NA	NA	5,866	6,678
1978	NA	NA	3,473	NA	NA	2,547	NA	NA	6,020	6,890
1979	NA	NA	3,553	NA	NA	2,753	NA	NA	6,306	6,929
1980	NA	NA	3,642	NA	NA	2,655	NA	NA	6,297	7,434
1981	NA	NA	3,752	NA	NA	2,817	NA	NA	6,569	7,805
1982	NA	NA	3,808	NA	NA	3,071	NA	NA	6,879	7,915
1983	NA	NA	3,847	NA	NA	2,595	NA	NA	6,442	7,985
1984	NA	NA	3,830	NA	NA	2,876	NA	NA	6,706	8,043
1985	NA	NA	3,842	NA	NA	2,607	NA	NA	6,448	8,087
1986	NA	NA	3,819	NA	NA	2,749	NA	NA	6,567	8,145
1987	NA	NA	3,792	NA	NA	2,756	NA	NA	6,548	8,124
1988	NA	NA	3,800	NA	NA	2,850	NA	NA	6,650	8,124
1989	NA	NA	3,812	NA	NA	2,513	NA	NA	6,325	8,120
1990	NA	NA	3,868	NA	NA	3,068	NA	NA	6,936	7,794
1991	NA	NA	3,954	NA	NA	2,824	NA	NA	6,778	7,993
1992	NA	NA	4,044	NA	NA	2,597	NA	NA	6,641	7,932
1993	NA	NA	4,327	NA	NA	2,322	NA	NA	6,649	7,989
1994	4,317	44	4,360	2,536	70	2,606	6,853	113	6,966	8,043
1995	4,290	60	4,349	2,082	72	2,153	6,371	131	6,503	7,953
1996	4,277	64	4,341	2,087	85	2,173	6,364	149	6,513	7,980
1997	4,283	67	4,350	2,092	83	2,175	6,375	150	6,525	8,332
1998	4,259	67	4,326	2,626	104	2,730	6,884	171	7,056	8,179
1999	4,314	69	4,383	2,423	100	2,523	6,738	169	6,906	8,229
2000	4,282	70	4,352	1,647	72	1,719	5,929	142	6,071	8,241
2001	4,224	77	4,301	2,789	115	2,904	7,013	191	7,204	8,415
2002	4,265	75	4,340	2,273	102	2,375	6,539	177	6,715	8,207
2003	4,227	76	4,303	2,438	125	2,563	6,665	201	6,866	8,206
2004	4,129	72	4,201	2,598	98	2,696	6,727	170	6,897	8,255
2005	4,122	78	4,200	2,513	123	2,635	6,635	201	6,835	8,268
2006	4,134	77	4,211	2,926	144	3,070	7,059	222	7,281	R8,330
2007P	4,154	80	4,234	2,756	123	2,879	6,910	203	7,113	8,368

[1] Includes native gas.

R=Revised. P=Preliminary. NA=Not available.

Notes: • Storage and capacity are at end of year. • Beginning with 1965, all volumes are shown on a pressure base of 14.73 p.s.i.a. at 60° F. For prior years, the pressure base was 14.65 p.s.i.a. at 60° F. • Totals may not equal sum of components due to independent rounding.

Web Page: See http://www.eia.doe.gov/oil_gas/natural_gas/info_glance/natural_gas.html for related information.

Sources: • 1954-1974—American Gas Association, *Gas Facts.* • 1975-1978—Federal Energy Administration, Form FEA-G318-M-O, "Underground Gas Storage Report," and Federal Power Commission, Form FPC-8, "Underground Gas Storage Report." • 1979-1984—Energy Information Administration (EIA), Form EIA-191, "Underground Gas Storage Report," and Federal Energy Regulatory Commission, Form FERC-8, "Underground Gas Storage Report." • 1985-2006—EIA, *Natural Gas Monthly (NGM),* monthly reports, and *Natural Gas Annual,* annual reports. • 2007—EIA, *NGM* (March 2008), Tables 6, 8, and 9, and Form EIA-191M, "Monthly Underground Gas Storage Report."

Figure 6.7 Natural Gas Wellhead, City Gate, and Imports Prices

Wellhead, City Gate, and Imports, 2007

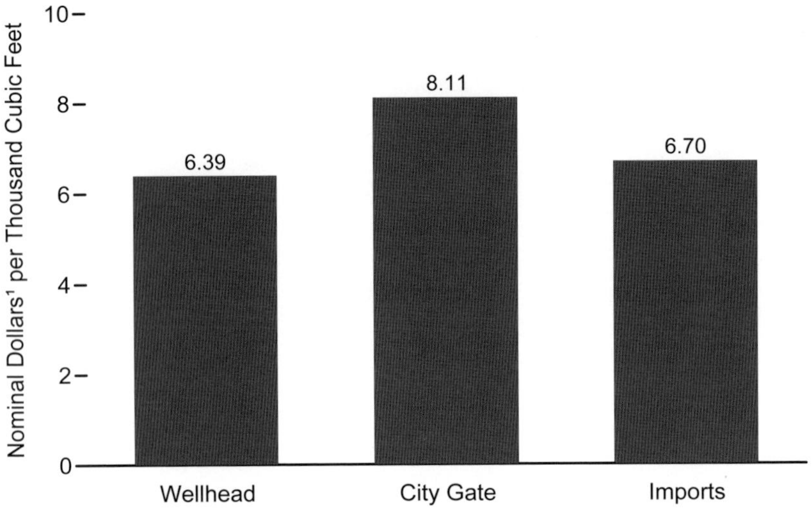

Wellhead, City Gate, and Imports, 1949-2007

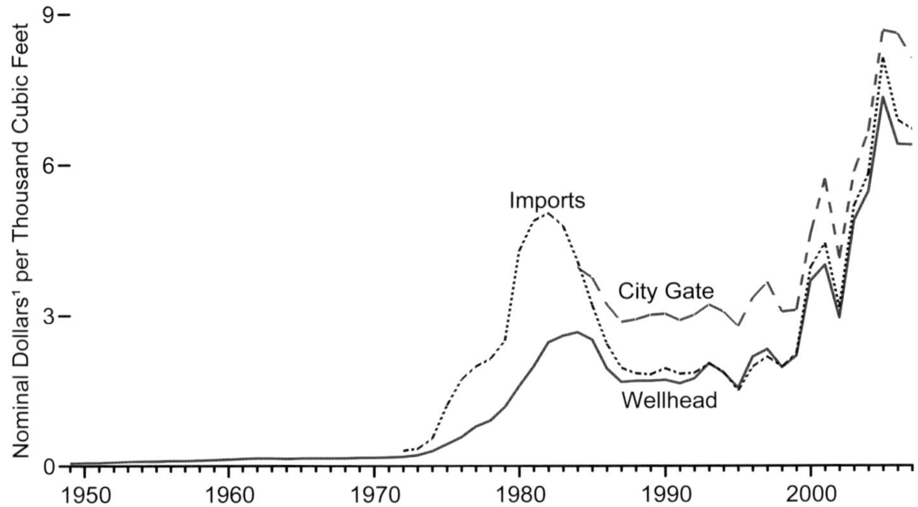

Wellhead, 1949-2007

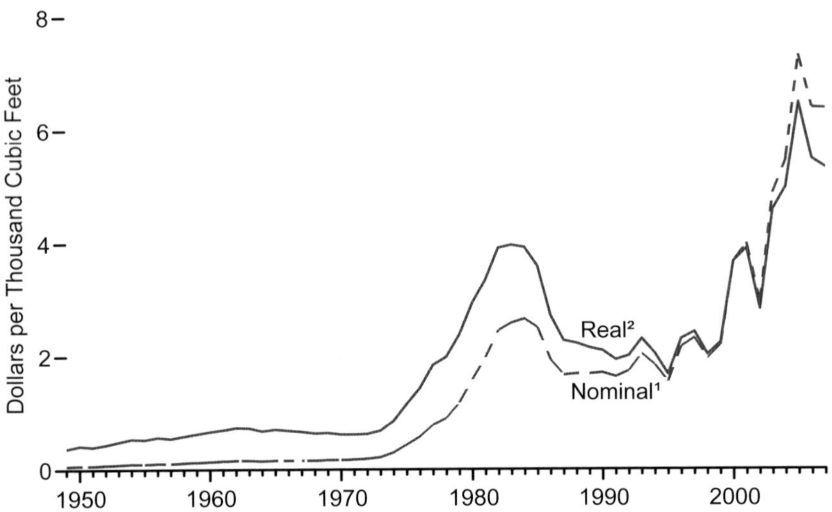

Imports, 1972-2007

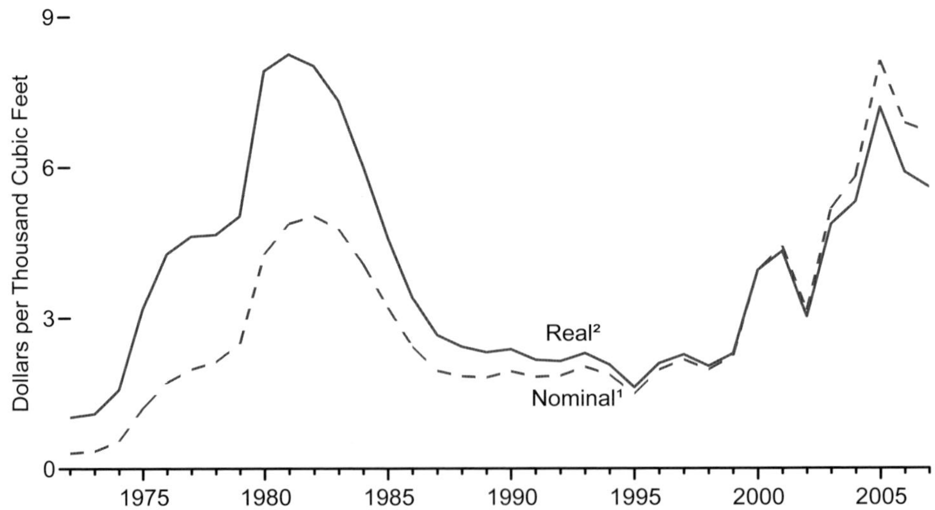

[1] See "Nominal Dollars" in Glossary.

[2] In chained (2000) dollars, calculated by using gross domestic product implicit price deflators. See Table D1.

Note: Because vertical scales differ, graphs should not be compared.

Source: Table 6.7.

Table 6.7 Natural Gas Wellhead, City Gate, and Imports Prices, Selected Years, 1949-2007

(Dollars per Thousand Cubic Feet)

Year	Wellhead [1]		City Gate [2]		Imports	
	Nominal [3]	Real [4]	Nominal [3]	Real [4]	Nominal [3]	Real [4]
1949	0.06	0.37	NA	NA	NA	NA
1950	.07	.42	NA	NA	NA	NA
1955	.10	.53	NA	NA	NA	NA
1960	.14	.67	NA	NA	NA	NA
1965	.16	.71	NA	NA	NA	NA
1970	.17	.62	NA	NA	NA	NA
1971	.18	.62	NA	NA	NA	NA
1972	.19	.63	NA	NA	.31	1.03
1973	.22	.69	NA	NA	.35	1.10
1974	.30	.86	NA	NA	.55	1.58
1975	.44	1.16	NA	NA	1.21	3.18
1976	.58	1.44	NA	NA	1.72	4.28
1977	.79	1.85	NA	NA	1.98	4.63
1978	.91	1.99	NA	NA	2.13	4.66
1979	1.18	2.38	NA	NA	2.49	5.03
1980	1.59	2.94	NA	NA	4.28	7.92
1981	1.98	3.35	NA	NA	4.88	8.25
1982	2.46	3.92	NA	NA	5.03	8.02
1983	2.59	3.97	NA	NA	4.78	7.33
1984	2.66	3.93	3.95	5.84	4.08	6.03
1985	2.51	3.60	3.75	5.38	3.21	4.60
1986	1.94	2.72	3.22	4.52	2.43	3.41
1987	1.67	2.28	2.87	3.92	1.95	2.66
1988	1.69	2.23	2.92	3.86	1.84	2.43
1989	1.69	2.15	3.01	3.83	1.82	2.32
1990	1.71	2.10	3.03	3.71	1.94	2.38
1991	1.64	1.94	2.90	3.43	1.83	2.17
1992	1.74	2.01	3.01	3.48	1.85	2.14
1993	2.04	2.31	3.21	3.63	2.03	2.30
1994	1.85	2.05	3.07	3.40	1.87	2.07
1995	1.55	1.68	2.78	3.02	1.49	1.62
1996	2.17	2.31	3.34	3.56	1.97	2.10
1997	2.32	2.43	3.66	3.84	2.17	2.27
1998	1.96	2.03	3.07	3.18	1.97	2.04
1999	2.19	2.24	3.10	3.17	2.24	2.29
2000	3.68	3.68	4.62	4.62	3.95	3.95
2001	4.00	3.91	5.72	5.59	4.43	4.33
2002	2.95	2.83	4.12	3.95	3.15	3.02
2003	4.88	4.59	5.85	5.50	5.17	4.86
2004	5.46	4.99	6.65	6.08	5.81	5.31
2005	7.33	[R]6.49	8.67	[R]7.67	8.12	[R]7.19
2006	[R]6.40	[R]5.49	[R]8.61	[R]7.39	[R]6.88	[R]5.90
2007	[P]6.39	[P]5.34	[P]8.11	[P]6.78	[E]6.70	[E]5.60

[1] See "Natural Gas Wellhead Price" in Glossary.
[2] See "City Gate" in Glossary.
[3] See "Nominal Dollars" in Glossary.
[4] In chained (2000) dollars, calculated by using gross domestic product implicit price deflators in Table D1. See "Chained Dollars" in Glossary.

R=Revised. P=Preliminary. E=Estimate. NA=Not available.

Web Pages: • For all data beginning in 1949, see http://www.eia.doe.gov/emeu/aer/natgas.html. • For related information, see http://www.eia.doe.gov/oil_gas/natural_gas/info_glance/natural_gas.html.

Sources: **Wellhead** and **City Gate:** • 1949-2002—Energy Information Administration (EIA), *Natural Gas Annual (NGA),* annual reports. • 2003 forward—EIA, *Natural Gas Monthly (NGM)* (March 2008), Table 3. **Imports:** • 1972 and 1973—Federal Power Commission (FPC), *Pipeline Imports and Exports of Natural Gas—Imports and Exports of LNG.* • 1974-1976—FPC, *United States Imports and Exports of Natural Gas,* annual reports. • 1977-2005—EIA, *NGA,* annual reports. • 2006—EIA, *NGM* (March 2008), Table 4. • 2007—EIA estimates.

Figure 6.8 Natural Gas Prices by Sector

Nominal[1] Prices, 2007

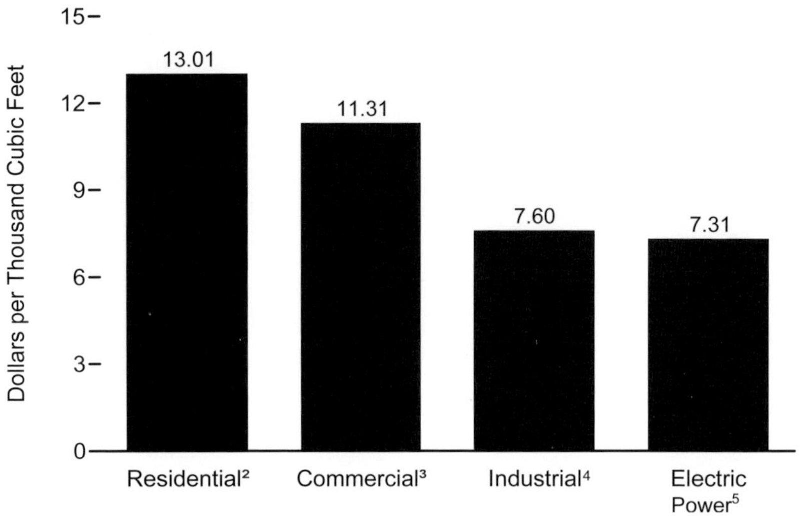

Real[6] Prices, Indexed, 1980-2007

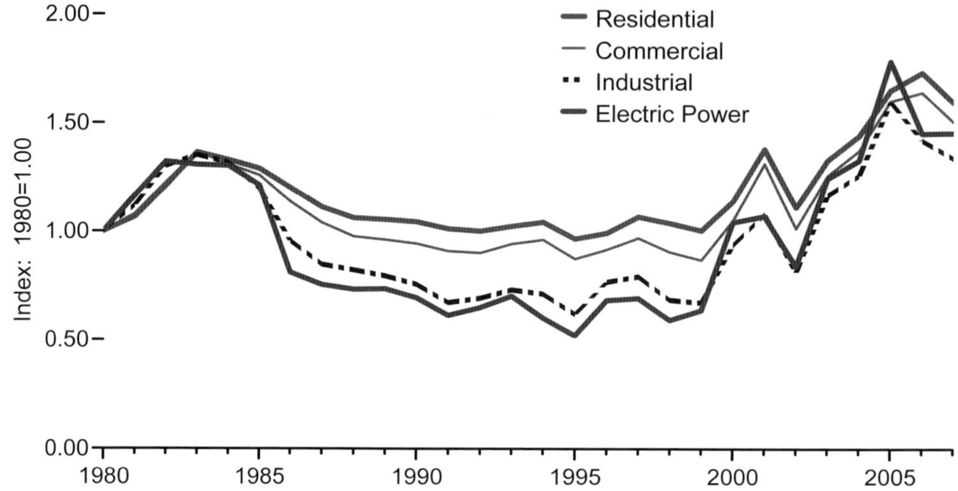

Real[6] Prices, 1967-2007

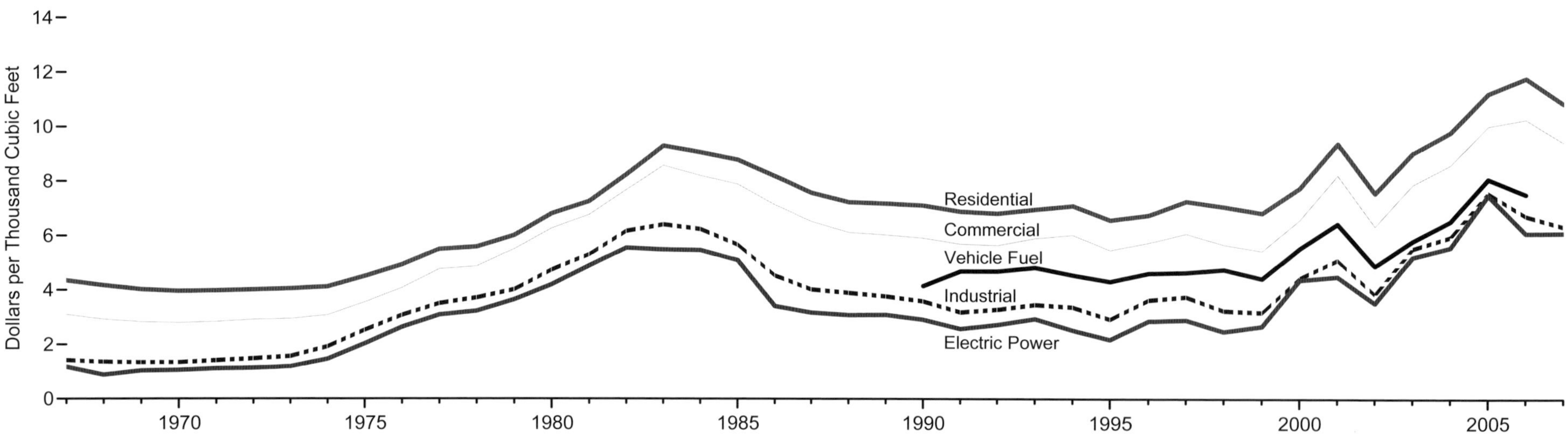

[1] See "Nominal Dollars" in Glossary.
[2] Based on 98.0 percent of volume delivered.
[3] Based on 79.1 percent of volume delivered.
[4] Based on 22.2 percent of volume delivered.
[5] Based on 93.2 percent of volume delivered.
[6] In chained (2000) dollars, calculated by using gross domestic product implicit price deflators. See Table D1.
Source: Table 6.8.

Table 6.8 Natural Gas Prices by Sector, 1967-2007

(Dollars per Thousand Cubic Feet)

Year	Residential Sector			Commercial Sector [1]			Industrial Sector [2]			Transportation Sector		Electric Power Sector [3]		
	Prices		Percentage of Sector [7]	Prices		Percentage of Sector [7]	Prices		Percentage of Sector [7]	Vehicle Fuel [4] Prices		Prices		Percentage of Sector [7]
	Nominal [5]	Real [6]		Nominal [5]	Real [6]		Nominal [5]	Real [6]		Nominal [5]	Real [6]	Nominal [5]	Real [6]	
1967	1.04	4.35	NA	0.74	3.10	NA	0.34	1.42	NA	NA	NA	0.28	1.17	NA
1968	1.04	4.17	NA	.73	2.93	NA	.34	1.36	NA	NA	NA	.22	.88	NA
1969	1.05	4.02	NA	.74	2.83	NA	.35	1.34	NA	NA	NA	.27	1.03	NA
1970	1.09	3.96	NA	.77	2.80	NA	.37	1.34	NA	NA	NA	.29	1.05	NA
1971	1.15	3.98	NA	.82	2.84	NA	.41	1.42	NA	NA	NA	.32	1.11	NA
1972	1.21	4.01	NA	.88	2.92	NA	.45	1.49	NA	NA	NA	.34	1.13	NA
1973	1.29	4.05	NA	.94	2.95	NA	.50	1.57	NA	NA	NA	.38	1.19	92.1
1974	1.43	4.12	NA	1.07	3.08	NA	.67	1.93	NA	NA	NA	.51	1.47	92.7
1975	1.71	4.50	NA	1.35	3.55	NA	.96	2.53	NA	NA	NA	.77	2.03	96.1
1976	1.98	4.93	NA	1.64	4.08	NA	1.24	3.08	NA	NA	NA	1.06	2.64	96.2
1977	2.35	5.50	NA	2.04	4.77	NA	1.50	3.51	NA	NA	NA	1.32	3.09	97.1
1978	2.56	5.59	NA	2.23	4.87	NA	1.70	3.72	NA	NA	NA	1.48	3.23	98.0
1979	2.98	6.01	NA	2.73	5.51	NA	1.99	4.02	NA	NA	NA	1.81	3.65	96.1
1980	3.68	6.81	NA	3.39	6.27	NA	2.56	4.74	NA	NA	NA	2.27	4.20	96.9
1981	4.29	7.26	NA	4.00	6.77	NA	3.14	5.31	NA	NA	NA	2.89	4.89	97.6
1982	5.17	8.24	NA	4.82	7.68	NA	3.87	6.17	85.1	NA	NA	3.48	5.55	92.6
1983	6.06	9.29	NA	5.59	8.57	NA	4.18	6.41	80.7	NA	NA	3.58	5.49	93.9
1984	6.12	9.05	NA	5.55	8.20	NA	4.22	6.24	74.7	NA	NA	3.70	5.47	94.4
1985	6.12	8.78	NA	5.50	7.89	NA	3.95	5.67	68.8	NA	NA	3.55	5.09	94.0
1986	5.83	8.18	NA	5.08	7.13	NA	3.23	4.53	59.8	NA	NA	2.43	3.41	91.7
1987	5.54	7.57	NA	4.77	6.52	93.1	2.94	4.02	47.4	NA	NA	2.32	3.17	91.6
1988	5.47	7.23	NA	4.63	6.12	90.7	2.95	3.90	42.6	NA	NA	2.33	3.08	89.6
1989	5.64	7.18	99.9	4.74	6.03	89.1	2.96	3.77	36.9	NA	NA	2.43	3.09	79.6
1990	5.80	7.11	99.2	4.83	5.92	86.6	2.93	3.59	35.2	3.39	4.15	2.38	2.92	76.8
1991	5.82	6.89	99.2	4.81	5.70	85.1	2.69	3.19	32.7	3.96	4.69	2.18	2.58	79.3
1992	5.89	6.82	99.1	4.88	5.65	83.2	2.84	3.29	30.3	4.05	4.69	2.36	2.73	76.5
1993	6.16	6.97	99.1	5.22	5.91	83.9	3.07	3.47	29.7	4.27	4.83	2.61	2.95	74.1
1994	6.41	7.10	99.1	5.44	6.03	79.3	3.05	3.38	25.5	4.11	4.55	2.28	2.53	73.4
1995	6.06	6.58	99.0	5.05	5.48	76.7	2.71	2.94	24.5	3.98	4.32	2.02	2.19	71.4
1996	6.34	6.76	99.0	5.40	5.75	77.6	3.42	3.64	19.4	4.34	4.62	2.69	2.87	68.4
1997	6.94	7.27	98.8	5.80	6.08	70.8	3.59	3.76	18.1	4.44	4.65	2.78	2.91	68.0
1998	6.82	7.07	97.7	5.48	5.68	67.0	3.14	3.25	16.1	4.59	4.76	2.40	2.49	63.7
1999	6.69	6.84	95.2	5.33	5.45	66.1	3.12	3.19	18.8	4.34	4.43	2.62	2.68	58.3
2000	7.76	7.76	92.6	6.59	6.59	63.9	4.45	4.45	19.8	5.54	5.54	4.38	4.38	50.5
2001	9.63	9.40	92.4	8.43	8.23	66.0	5.24	5.12	20.8	6.60	6.45	4.61	4.50	40.2
2002	7.89	7.57	97.9	6.63	6.36	77.4	4.02	3.86	22.7	5.10	4.90	[3]3.68	[3]3.53	[3]83.9
2003	9.63	9.05	97.5	8.40	7.89	78.2	5.89	5.54	22.1	6.19	5.82	5.57	5.23	91.2
2004	10.75	9.82	97.7	9.43	[R]8.61	78.0	6.53	5.97	23.7	7.16	6.54	6.11	5.58	89.8
2005	[R]12.70	[R]11.24	98.2	[R]11.34	[R]10.04	[R]82.1	8.56	[R]7.58	[R]24.1	[R]9.14	[R]8.09	[R]8.47	[R]7.50	[R]91.3
2006	[R]13.75	[R]11.80	[R]98.1	[R]11.99	[R]10.29	[R]80.7	[R]7.86	[R]6.74	[R]23.5	[R]8.78	[R]7.53	[R]7.11	[R]6.10	[R]93.4
2007	[P]13.01	[P]10.87	[E]98.0	[P]11.31	[P]9.45	[P]79.1	[P]7.60	[P]6.35	[P]22.2	NA	NA	[P]7.31	[P]6.11	[P]93.2

[1] Commercial sector, including commercial combined-heat-and-power (CHP) and commercial electricity-only plants.

[2] Industrial sector, including industrial combined-heat-and-power (CHP) and industrial electricity-only plants.

[3] Electricity-only and combined-heat-and-power (CHP) plants within the NAICS 22 category whose primary business is to sell electricity, or electricity and heat, to the public. Through 2001, data are for electric utilities only; beginning in 2002, data are for electric utilities and independent power producers. See Note 5, "Coverage of Electric Power Sector Natural Gas Prices," at end of section.

[4] Much of the natural gas delivered for vehicle fuel represents deliveries to fueling stations that are used primarily or exclusively by fleet vehicles. Thus, the prices are often those associated with the cost of gas in the operation of fleet vehicles.

[5] See "Nominal Dollars" in Glossary.

[6] In chained (2000) dollars, calculated by using gross domestic product implicit price deflators in Table D1. See "Chained Dollars" in Glossary.

[7] The percentage of the sector's consumption in Table 6.5 for which price data are available.

R=Revised. P=Preliminary. E=Estimate. NA=Not available.

Notes: • Prices are for natural gas, plus a small amount of supplemental gaseous fuels. • The average for each end-use sector is calculated by dividing the total value of the natural gas consumed by each sector by the total quantity consumed. • Prices are intended to include all taxes. • See Note 2, "Classification of Power Plants Into Energy-Use Sectors," at end of Section 8.

Web Page: See http://www.eia.doe.gov/oil_gas/natural_gas/info_glance/natural_gas.html for related information.

Sources: **Residential Percentage of Sector:** • 1989-2006—Energy Information Administration (EIA), Form EIA-176, "Annual Report of Natural and Supplemental Gas Supply and Disposition." • 2007—EIA estimate. **Vehicle Fuel:** EIA, *NGA*, annual reports. **Electric Power Price:** • 1967-2002—EIA, *NGA*, annual reports. • 2003-2006—EIA, *Natural Gas Monthly (NGM)* (March 2008), Table 3. • 2007—Federal Energy Regulatory Commission, Form FERC-423, "Monthly Report of Cost and Quality of Fuels for Electric Utility Plants," and EIA, Form EIA-423, "Monthly Cost and Quality of Fuels for Electric Plants Report." **Electric Power Percentage of Sector:** • 1973-2001—Calculated by EIA as the quantity of natural gas receipts by electric utilities reported on Form FERC-423, "Monthly Report of Cost and Quantity of Fuels for Electric Utility Plants" (and predecessor forms), divided by the quantity of natural gas consumed by the electric power sector (for 1973-1988, see Table 8.5b; for 1989-2001, see Table 8.7b). • 2002 forward—Calculated by EIA as the quantity of natural gas receipts by electric utilities and independent power producers reported on Forms FERC-423, "Monthly Report of Cost and Quantity of Fuels for Electric Utility Plants," and EIA-423, "Monthly Cost and Quality of Fuels for Electric Plants Report," divided by the quantity of natural gas consumed by the electric power sector (see Table 8.7b). **All Other Data:** • 1967-2002—EIA, *NGA*, annual reports. • 2003 forward—EIA, *NGM* (March 2008), Table 3.

Natural Gas

Note 1. Supplemental Gaseous Fuels. Supplemental gaseous fuels are any substances that, introduced into or commingled with natural gas, increase the volume available for disposition. Such substances include, but are not limited to, propane-air, refinery gas, coke oven gas, still gas, manufactured gas, biomass gas, or air or inert gases added for British thermal unit (Btu) stabilization.

Annual data beginning with 1980 are from the Energy Information Administration (EIA), *Natural Gas Annual*. Unknown quantities of supplemental gaseous fuels are included in consumption data for 1979 and earlier years.

Although the total amount of supplemental gaseous fuels consumed is known for 1980 forward, EIA estimates the amount consumed by each energy-use sector. It is assumed that supplemental gaseous fuels are commingled with natural gas consumed by the residential, commercial, other industrial, and electric power sectors, but are not commingled with natural gas used for lease and plant fuel, pipelines and distribution, or vehicle fuel. The estimated consumption of supplemental gaseous fuels by each sector (residential, commercial, other industrial, and electric power) is calculated as that sector's natural gas consumption (see Table 6.5) divided by the sum of natural gas consumption by the residential, commercial, other industrial, and electric power sectors (see Table 6.5). For estimated sectoral consumption of supplemental gaseous fuels in Btu, the residential, commercial, and other industrial values in cubic feet are multiplied by the "End-Use Sectors" conversion factors (see Table A4), and the electric power values in cubic feet are multiplied by the "Electric Power Sector" conversion factors (see Table A4). Total supplemental gaseous fuels consumption in Btu is calculated as the sum of the Btu values for the sectors.

Note 2. Natural Gas Consumption. Natural gas consumption statistics are compiled from surveys of natural gas production, transmission, and distribution companies and from surveys of electric power generation. Consumption by sector from these surveys is compiled on a national and individual State basis and then balanced with national and individual State supply data. Included in the data are the following: **Residential Sector**—Consumption by private households for space heating, cooking, and other household uses; **Commercial Sector**—Consumption by nonmanufacturing establishments; municipalities for institutional heating and lighting; and, through 1995, those engaged in agriculture, forestry, and fishing. The commercial sector includes generators that produce electricity and/or useful thermal output primarily to support the activities of the above-mentioned commercial establishments; **Industrial Sector**—Consumption by establishments engaged primarily in processing unfinished materials into another form of product (including mining; petroleum refining; manufacturing; and, beginning in 1996, agriculture, forestry, and fishing), and natural gas industry use for lease and plant fuel. The industrial sector includes generators that produce electricity and/or useful thermal output primarily to support the above-mentioned industrial activities; **Transportation Sector**—Natural gas transmission (pipeline) fuel, and natural gas delivered for use as vehicle fuel; and **Electric Power Sector (electric utilities and independent power producers)**—Consumption for electricity generation and useful thermal output at electricity-only and combined-heat-and-power (CHP) plants within the NAICS (North American Industry Classification System) 22 category whose primary business is to sell electricity, or electricity and heat, to the public.

Note 3. Natural Gas Consumption, 1989-1992. Prior to 1993, deliveries to nonutility generators were not separately collected from natural gas companies on Form EIA-176, "Annual Report of Natural and Supplemental Gas Supply and Disposition." As a result, for 1989-1992, those volumes are probably included in both the industrial and electric power sectors and double-counted in total consumption. In 1993, 0.28 trillion cubic feet was reported as delivered to nonutility generators.

Note 4. Natural Gas Vehicle Fuel. In Table 6.5, for 1992 forward, natural gas vehicle fuel data do not reflect revised data shown in Table 10.4. These revisions, in million cubic feet, are: 1992–2,112; 1993–2,860; 1994–3,222; 1995–4,619; 1996–6,111; 1997–8,393; 1998–9,416; 1999–10,398; 2000–11,461; 2001–13,788; 2002–15,810; 2003–17,417; 2004–21,466; 2005–22,556; and 2006–23,317.

Note 5. Coverage of Electric Power Sector Natural Gas Prices. For 1973-1982, data for electric power sector natural gas prices include all electric utility plants at which the generator nameplate capacity of all steam-electric units combined totaled 25 megawatts or greater. For 1974-1982, peaking units are also included and counted toward the 25-megawatt-or-greater total. For 1983-1990, data include all electric utility plants at which the generator nameplate capacity of all steam-electric units combined totaled 50 megawatts or greater. For 1991-2001, data include all electric utility plants at which the generator nameplate capacity of all steam-electric units and combined-cycle units together totaled 50 megawatts or greater. For 2002 forward, data include electric utility and independent power producer plants at which the total facility fossil-fueled nameplate generating capacity is 50 or more megawatts, regardless of unit type.

7 Coal

Coal yard, Curtis Bay, Maryland. Source: U.S. Department of Energy.

Diagram 4. Coal Flow, 2007
(Million Short Tons)

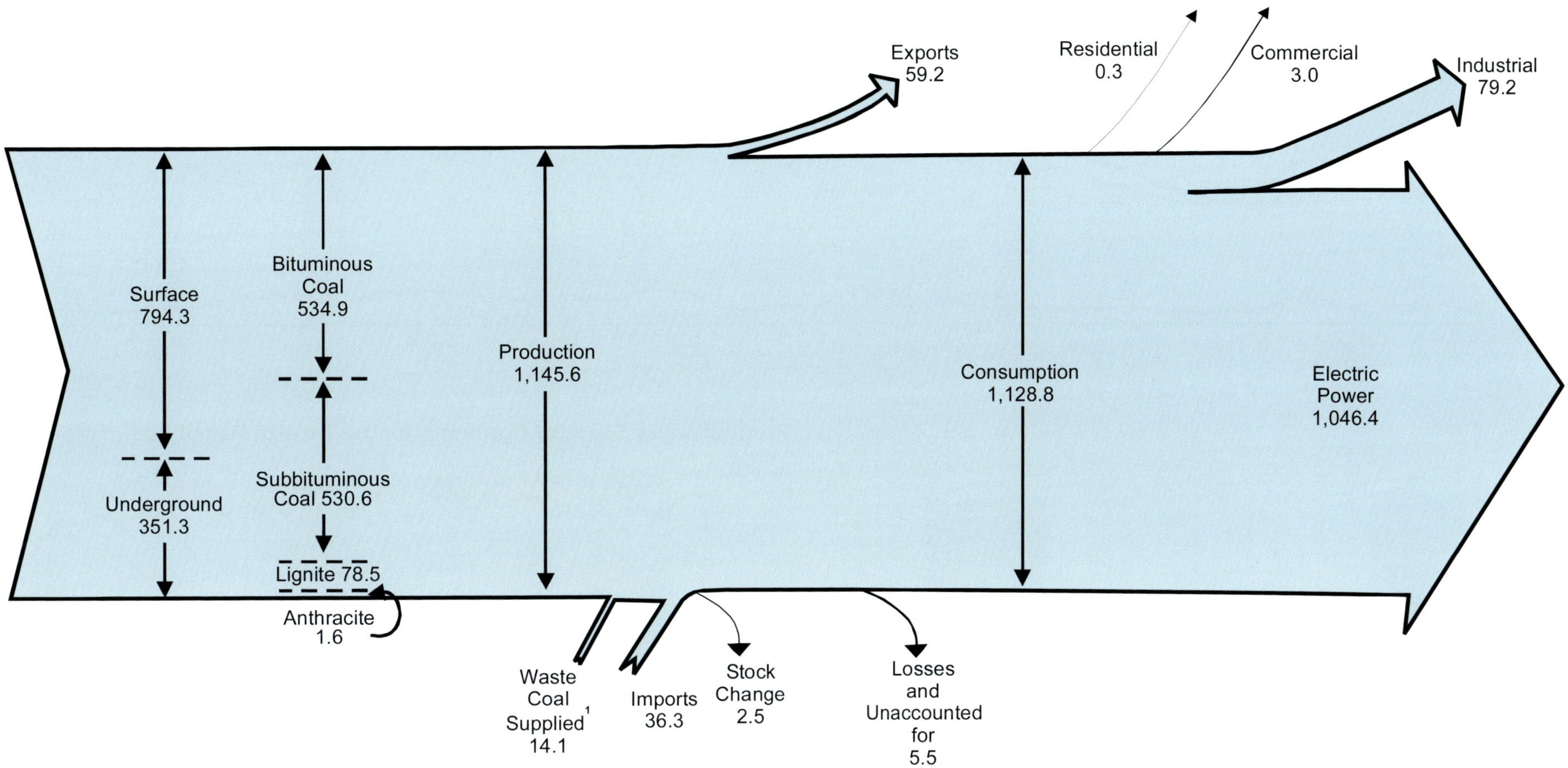

[1] Includes fine coal, coal obtained from a refuse bank or slurry dam, anthracite culm, bituminous gob, and lignite waste that are consumed by the electric power industrial sectors.

Notes: • Production categories are estimated; other data are preliminary. • Values are derived from source data prior to rounding for publication. • Totals may not equal sum of components due to independent rounding.

Sources: Tables 7.1, 7.2, and 7.3.

Figure 7.1 Coal Overview

Overview, 1949-2007

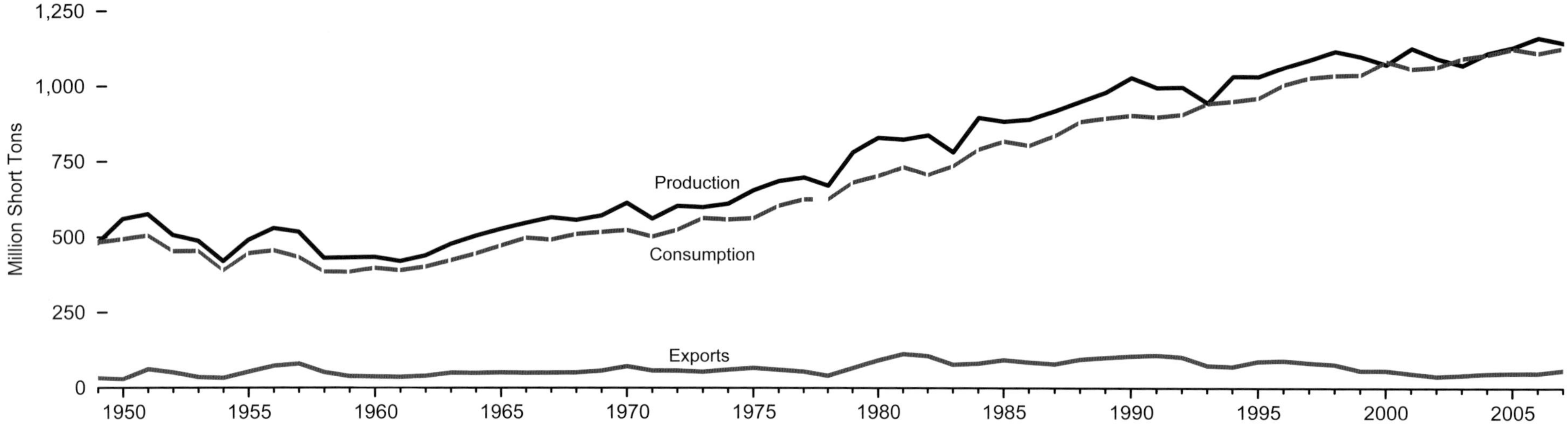

Overview, 2007

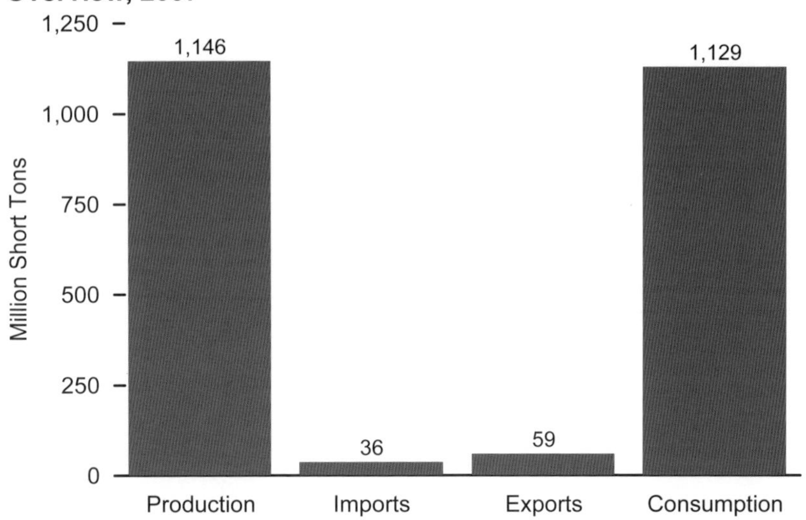

Production as Share of Consumption by Type of Fossil Fuel, 1949-2007

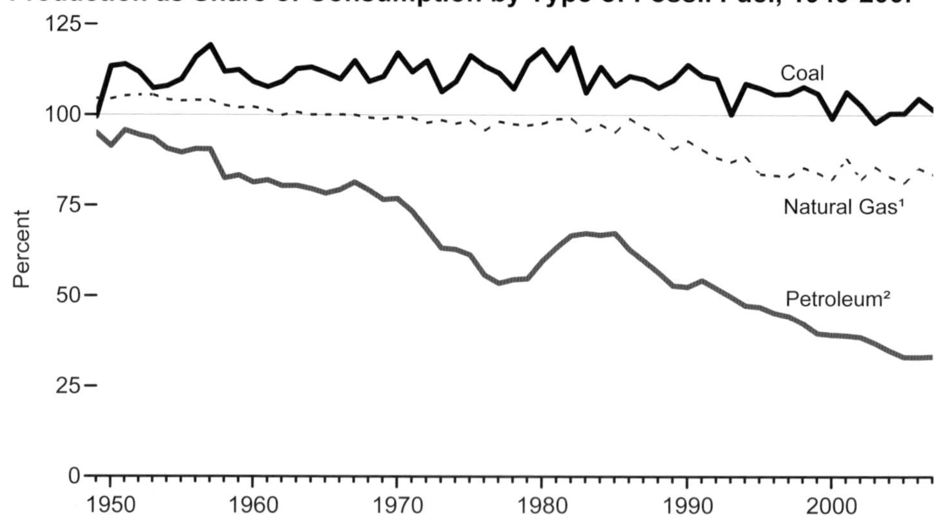

[1] Dry natural gas production as share of natural gas consumption.

[2] Crude oil and natural gas plant liquids production as share of petroleum products supplied.

Sources: Tables 5.1, 6.1, and 7.1.

Table 7.1 Coal Overview, Selected Years, 1949-2007

(Million Short Tons)

Year	Production [1]	Waste Coal Supplied [2]	Trade			Stock Change [4]	Losses and Unaccounted for [5]	Consumption
			Imports	Exports	Net Imports [3]			
1949	480.6	NA	0.3	32.8	-32.5	([6])	[6]-84.6	483.2
1950	560.4	NA	.4	29.4	-29.0	([6])	[6]9.5	494.1
1955	490.8	NA	.3	54.4	-54.1	([6])	[6]-6.3	447.0
1960	434.3	NA	.3	38.0	-37.7	([6])	[6]1.7	398.1
1965	527.0	NA	.2	51.0	-50.8	([6])	[6]2.2	472.0
1970	612.7	NA	(s)	71.7	-71.7	([6])	[6]6.6	523.2
1971	560.9	NA	.1	57.3	-57.2	([6])	[6]4.2	501.6
1972	602.5	NA	(s)	56.7	-56.7	([6])	[6]-4.3	524.3
1973	598.6	NA	.1	53.6	-53.5	([6])	[6]-17.9	562.6
1974	610.0	NA	2.1	60.7	-58.6	-8.9	2.0	558.4
1975	654.6	NA	.9	66.3	-65.4	32.2	-5.5	562.6
1976	684.9	NA	1.2	60.0	-58.8	8.5	13.8	603.8
1977	697.2	NA	1.6	54.3	-52.7	22.6	-3.4	625.3
1978	670.2	NA	3.0	40.7	-37.8	-4.9	12.1	625.2
1979	781.1	NA	2.1	66.0	-64.0	36.2	.4	680.5
1980	829.7	NA	1.2	91.7	-90.5	25.6	10.8	702.7
1981	823.8	NA	1.0	112.5	-111.5	-19.0	-1.4	732.6
1982	838.1	NA	.7	106.3	-105.5	22.6	3.1	706.9
1983	782.1	NA	1.3	77.8	-76.5	-29.5	-1.6	736.7
1984	895.9	NA	1.3	81.5	-80.2	28.7	-4.3	791.3
1985	883.6	NA	2.0	92.7	-90.7	-27.9	2.8	818.0
1986	890.3	NA	2.2	85.5	-83.3	4.0	-1.2	804.2
1987	918.8	NA	1.7	79.6	-77.9	6.5	-2.5	836.9
1988	950.3	NA	2.1	95.0	-92.9	-24.9	-1.3	883.6
1989	980.7	1.4	2.9	100.8	-98.0	-13.7	2.9	895.0
1990	1,029.1	3.3	2.7	105.8	-103.1	26.5	-1.7	904.5
1991	996.0	4.0	3.4	109.0	-105.6	-.9	-3.9	899.2
1992	997.5	6.3	3.8	102.5	-98.7	-3.0	.5	907.7
1993	945.4	8.1	8.2	74.5	-66.3	-51.9	-4.9	944.1
1994	1,033.5	8.2	8.9	71.4	-62.5	23.6	4.3	951.3
1995	1,033.0	8.6	9.5	88.5	-79.1	-.3	.6	962.1
1996	1,063.9	8.8	8.1	90.5	-82.4	-17.5	1.4	1,006.3
1997	1,089.9	8.1	7.5	83.5	-76.1	-11.3	3.7	1,029.5
1998	1,117.5	8.7	8.7	78.0	-69.3	24.2	-4.4	1,037.1
1999	1,100.4	8.7	9.1	58.5	-49.4	24.0	-2.9	1,038.6
2000	1,073.6	9.1	12.5	58.5	-46.0	-48.3	.9	1,084.1
2001	[1]1,127.7	10.1	19.8	48.7	-28.9	41.6	7.1	1,060.1
2002	1,094.3	9.1	16.9	39.6	-22.7	10.2	4.0	1,066.4
2003	1,071.8	10.0	25.0	43.0	-18.0	-26.7	-4.4	1,094.9
2004	1,112.1	11.3	27.3	48.0	-20.7	-11.5	6.9	1,107.3
2005	1,131.5	13.4	30.5	49.9	-19.5	-9.7	[R]9.1	[R]1,126.0
2006	[R]1,162.7	[R]14.4	36.2	49.6	-13.4	[R]42.6	[R]8.8	[R]1,112.3
2007[P]	1,145.6	14.1	36.3	59.2	-22.8	2.5	5.5	1,128.8

[1] Beginning in 2001, includes a small amount of refuse recovery (coal recaptured from a refuse mine, and cleaned to reduce the concentration of noncombustible materials).

[2] Waste coal (including fine coal, coal obtained from a refuse bank or slurry dam, anthracite culm, bituminous gob, and lignite waste) consumed by the electric power and industrial sectors. Beginning in 1989, waste coal supplied is counted as a supply-side item to balance the same amount of waste coal included in "Consumption."

[3] Net imports equal imports minus exports. Minus sign indicates exports are greater than imports.

[4] A negative value indicates a decrease in stocks; a positive value indicates an increase.

[5] "Losses and Unaccounted for" is calculated as the sum of production, imports, and waste coal supplied, minus exports, stock change, and consumption.

[6] Through 1973, stock change is included in "Losses and Unaccounted for."

R=Revised. P=Preliminary. NA=Not available. (s)=Less than 0.05 million short tons.

Notes: • See Note 1, "Coal Consumption," at end of section. • Totals may not equal sum of components due to independent rounding.

Web Pages: • For all data beginning in 1949, see http://www.eia.doe.gov/emeu/aer/coal.html. • For related information, see http://www.eia.doe.gov/fuelcoal.html.

Sources: **Production:** Table 7.2. **Waste Coal Supplied:** • 1989-1997—Energy Information Administration (EIA), Form EIA-867, "Annual Nonutility Power Producer Report." • 1998-2000—EIA, Form EIA-860B, "Annual Electric Generator Report—Nonutility." • 2001 forward—EIA, *Quarterly Coal Report October-December 2007* (March 2008), Table ES-1. **Imports:** • 1949-2000—U.S. Department of Commerce, Bureau of the Census, "Monthly Report IM145." • 2001 forward—EIA, *Quarterly Coal Report October-December 2007* (March 2008), Table ES-1. **Exports:** Table 7.4. **Stock Change:** Table 7.5. **Losses and Unaccounted for:** Calculated. **Consumption:** Table 7.3.

Figure 7.2 Coal Production, 1949-2007

Total

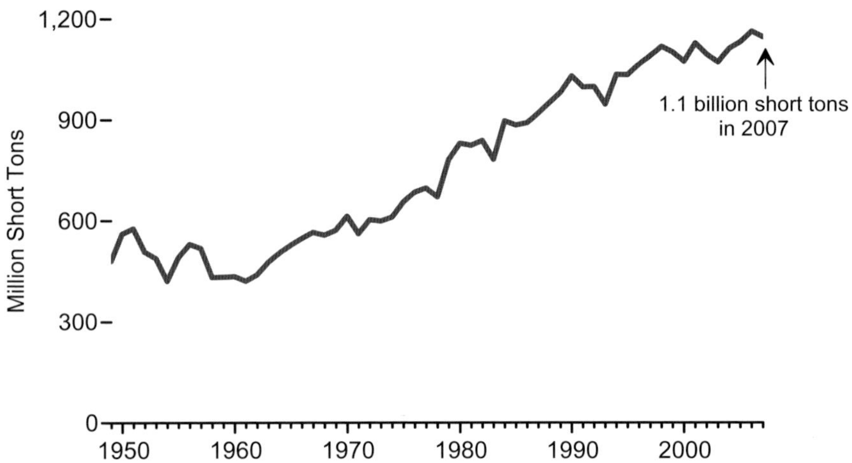

By Rank

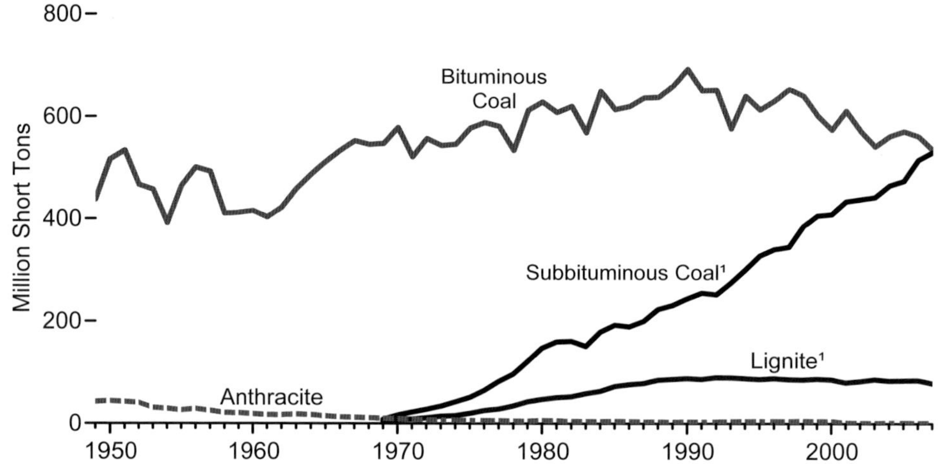

By Mining Method

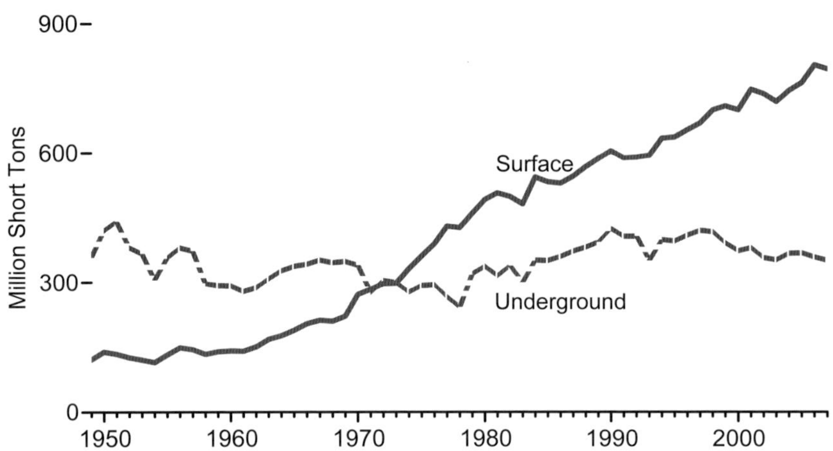

By Location

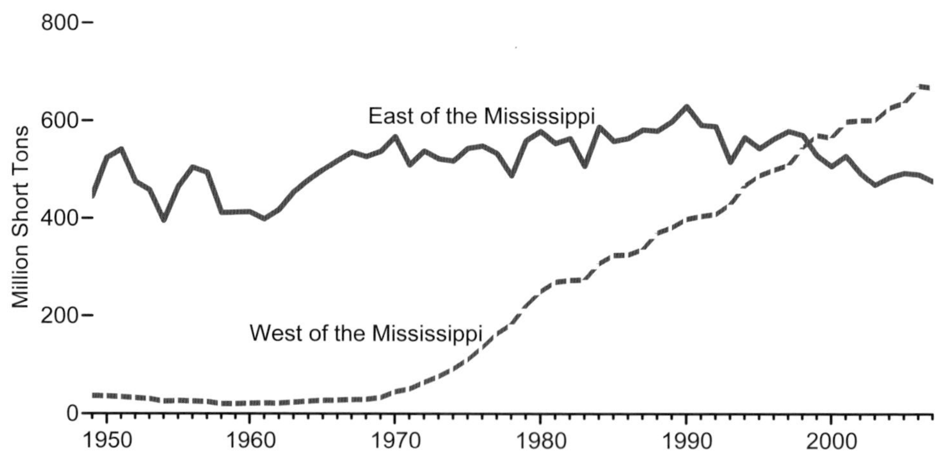

[1] Included in bituminous coal prior to 1969.

Note: Because vertical scales differ, graphs should not be compared.

Source: Table 7.2.

Table 7.2 Coal Production, Selected Years, 1949-2007

(Million Short Tons)

Year	Rank				Mining Method		Location		Total [1]
	Bituminous Coal [1]	Subbituminous Coal	Lignite	Anthracite [1]	Underground	Surface [1]	East of the Mississippi [1]	West of the Mississippi [1]	
1949	437.9	([2])	([2])	42.7	358.9	121.7	444.2	36.4	480.6
1950	516.3	([2])	([2])	44.1	421.0	139.4	524.4	36.0	560.4
1955	464.6	([2])	([2])	26.2	358.0	132.9	464.2	26.6	490.8
1960	415.5	([2])	([2])	18.8	292.6	141.7	413.0	21.3	434.3
1965	512.1	([2])	([2])	14.9	338.0	189.0	499.5	27.4	527.0
1970	578.5	16.4	8.0	9.7	340.5	272.1	567.8	44.9	612.7
1971	521.3	22.2	8.7	8.7	277.2	283.7	509.9	51.0	560.9
1972	556.8	27.5	11.0	7.1	305.0	297.4	538.2	64.3	602.5
1973	543.5	33.9	14.3	6.8	300.1	298.5	522.1	76.4	598.6
1974	545.7	42.2	15.5	6.6	278.0	332.1	518.1	91.9	610.0
1975	577.5	51.1	19.8	6.2	293.5	361.2	543.7	110.9	654.6
1976	588.4	64.8	25.5	6.2	295.5	389.4	548.8	136.1	684.9
1977	581.0	82.1	28.2	5.9	266.6	430.6	533.3	163.9	697.2
1978	534.0	96.8	34.4	5.0	242.8	427.4	487.2	183.0	670.2
1979	612.3	121.5	42.5	4.8	320.9	460.2	559.7	221.4	781.1
1980	628.8	147.7	47.2	6.1	337.5	492.2	578.7	251.0	829.7
1981	608.0	159.7	50.7	5.4	316.5	507.3	553.9	269.9	823.8
1982	620.2	160.9	52.4	4.6	339.2	499.0	564.3	273.9	838.1
1983	568.6	151.0	58.3	4.1	300.4	481.7	507.4	274.7	782.1
1984	649.5	179.2	63.1	4.2	352.1	543.9	587.6	308.3	895.9
1985	613.9	192.7	72.4	4.7	350.8	532.8	558.7	324.9	883.6
1986	620.1	189.6	76.4	4.3	360.4	529.9	564.4	325.9	890.3
1987	636.6	200.2	78.4	3.6	372.9	545.9	581.9	336.8	918.8
1988	638.1	223.5	85.1	3.6	382.2	568.1	579.6	370.7	950.3
1989	659.8	231.2	86.4	3.3	393.8	586.9	599.0	381.7	980.7
1990	693.2	244.3	88.1	3.5	424.5	604.5	630.2	398.9	1,029.1
1991	650.7	255.3	86.5	3.4	407.2	588.8	591.3	404.7	996.0
1992	651.8	252.2	90.1	3.5	407.2	590.3	588.6	409.0	997.5
1993	576.7	274.9	89.5	4.3	351.1	594.4	516.2	429.2	945.4
1994	640.3	300.5	88.1	4.6	399.1	634.4	566.3	467.2	1,033.5
1995	613.8	328.0	86.5	4.7	396.2	636.7	544.2	488.7	1,033.0
1996	630.7	340.3	88.1	4.8	409.8	654.0	563.7	500.2	1,063.9
1997	653.8	345.1	86.3	4.7	420.7	669.3	579.4	510.6	1,089.9
1998	640.6	385.9	85.8	5.3	417.7	699.8	570.6	547.0	1,117.5
1999	601.7	406.7	87.2	4.8	391.8	708.6	529.6	570.8	1,100.4
2000	574.3	409.2	85.6	4.6	373.7	700.0	507.5	566.1	1,073.6
2001	[1]611.3	434.4	80.0	[1]1.9	380.6	[1]747.1	[1]528.8	[1]598.9	[1]1,127.7
2002	572.1	438.4	82.5	1.4	357.4	736.9	492.9	601.4	1,094.3
2003	541.5	442.6	86.4	1.3	352.8	719.0	469.2	602.5	1,071.8
2004	561.5	465.4	83.5	1.7	367.6	744.5	484.8	627.3	1,112.1
2005	571.2	474.7	83.9	1.7	368.6	762.9	493.8	637.7	1,131.5
2006	[R]561.6	[R]515.3	84.2	1.5	[R]359.0	[R]803.7	[R]490.8	672.0	[R]1,162.7
2007	[E]534.9	[E]530.6	[E]78.5	[E]1.6	[E]351.3	[E]794.3	[E]477.2	[E]668.4	[P]1,145.6

[1] Beginning in 2001, includes a small amount of refuse recovery.
[2] Included in "Bituminous Coal."
R=Revised. P=Preliminary. E=Estimate.
Note: Totals may not equal sum of components due to independent rounding.
Web Pages: • For all data beginning in 1949, see http://www.eia.doe.gov/emeu/aer/coal.html. • For related information, see http://www.eia.doe.gov/fuelcoal.html.
Sources: • 1949-1975—Bureau of Mines, *Minerals Yearbook,* "Coal—Bituminous and Lignite" and "Coal—Pennsylvania Anthracite" chapters. • 1976—Energy Information Administration (EIA), Energy Data Reports, *Coal—Bituminous and Lignite in 1976* and *Coal—Pennsylvania Anthracite 1976.* • 1977 and 1978—EIA, Energy Data Reports, *Bituminous Coal and Lignite Production and Mine Operations—1977; 1978, Coal—Pennsylvania Anthracite 1977; 1978,* and *Coal Production,* annual reports. • 1979 and 1980—EIA, Energy Data Reports, *Weekly Coal Report* and *Coal Production,* annual reports. • 1981-1988—EIA, *Weekly Coal Production* and *Coal Production,* annual reports. • 1989-2000—EIA, *Coal Industry Annual,* annual reports. • 2001-2006—EIA, *Annual Coal Report,* annual reports. • 2007—EIA, *Quarterly Coal Report October-December 2007* (March 2008), Table 1; EIA, Form EIA-7A, "Coal Production Report"; and U.S. Department of Labor, Mine Safety and Health Administration, Form 7000-2, "Quarterly Mine Employment and Coal Production Report."

Figure 7.3 Coal Consumption by Sector

By Sector, 1949-2007

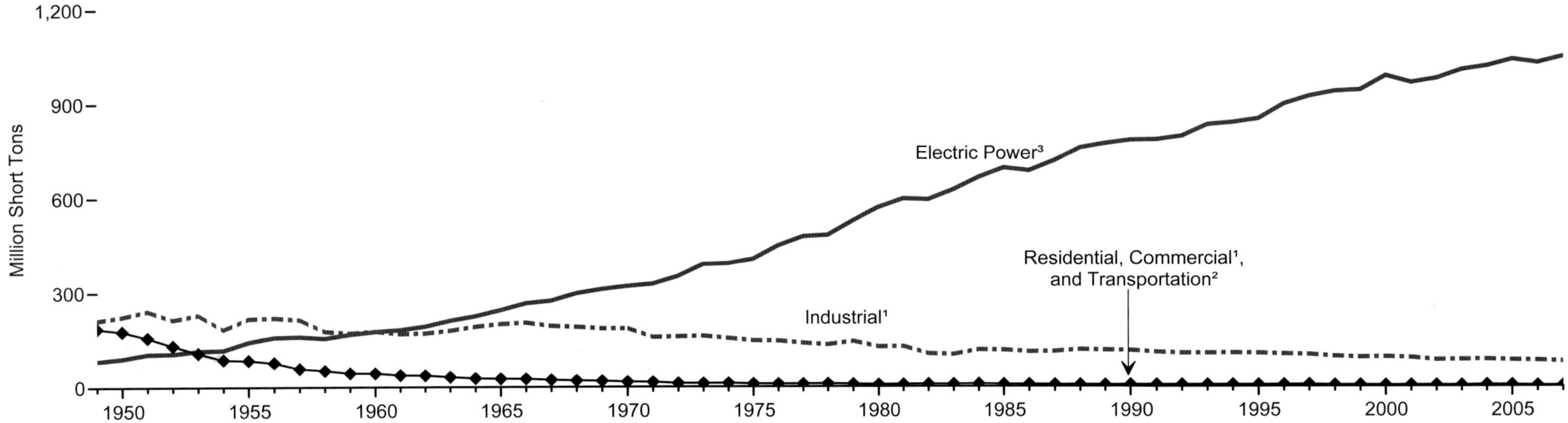

By Sector, 2007

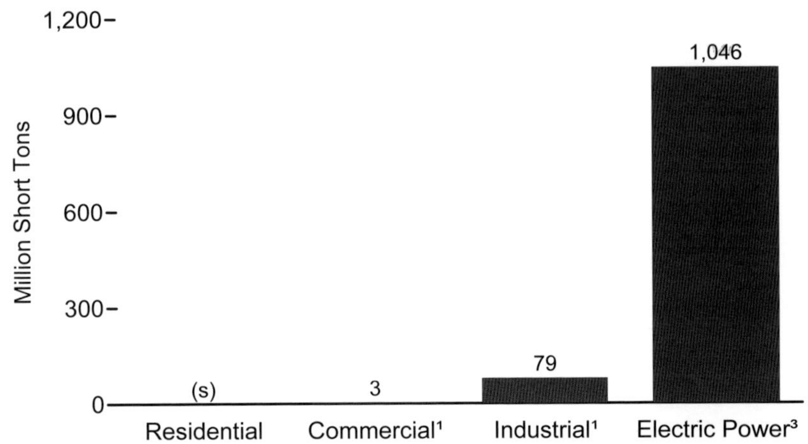

Sector Shares, 1949 and 2007

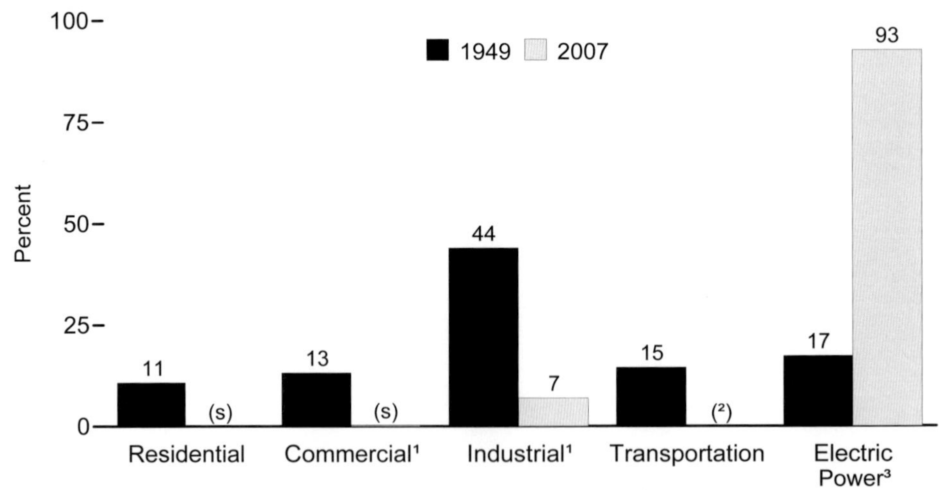

[1] Includes combined-heat-and-power (CHP) plants and a small number of electricity-only plants.

[2] For 1978 forward, small amounts of transportation sector use are included in "Industrial."

[3] Electricity-only and combined-heat-and-power (CHP) plants whose primary business is to sell electricity, or electricity and heat, to the public.

(s)=Less than 0.5.

Source: Table 7.3.

Table 7.3 Coal Consumption by Sector, Selected Years, 1949-2007

(Million Short Tons)

Year	Residential Sector [1]	Commercial Sector [1]			Industrial Sector					Transportation Sector	Electric Power Sector [2]			Total
						Other Industrial								
		CHP [3]	Other [4]	Total	Coke Plants	CHP [5]	Non-CHP [6]	Total	Total		Electricity Only	CHP	Total	
1949	52.4	(7)	64.1	64.1	91.4	(8)	121.2	121.2	212.6	70.2	84.0	NA	84.0	483.2
1950	51.6	(7)	63.0	63.0	104.0	(8)	120.6	120.6	224.6	63.0	91.9	NA	91.9	494.1
1955	35.6	(7)	32.9	32.9	107.7	(8)	110.1	110.1	217.8	17.0	143.8	NA	143.8	447.0
1960	24.2	(7)	16.8	16.8	81.4	(8)	96.0	96.0	177.4	3.0	176.7	NA	176.7	398.1
1965	14.6	(7)	11.0	11.0	95.3	(8)	105.6	105.6	200.8	.7	244.8	NA	244.8	472.0
1970	9.0	(7)	7.1	7.1	96.5	(8)	90.2	90.2	186.6	.3	320.2	NA	320.2	523.2
1971	7.4	(7)	7.8	7.8	83.2	(8)	75.6	75.6	158.9	.2	327.3	NA	327.3	501.6
1972	5.0	(7)	6.7	6.7	87.7	(8)	72.9	72.9	160.6	.2	351.8	NA	351.8	524.3
1973	4.1	(7)	7.0	7.0	94.1	(8)	68.0	68.0	162.1	.1	389.2	NA	389.2	562.6
1974	3.7	(7)	7.8	7.8	90.2	(8)	64.9	64.9	155.1	.1	391.8	NA	391.8	558.4
1975	2.8	(7)	6.6	6.6	83.6	(8)	63.6	63.6	147.2	(s)	406.0	NA	406.0	562.6
1976	2.6	(7)	6.3	6.3	84.7	(8)	61.8	61.8	146.5	(s)	448.4	NA	448.4	603.8
1977	2.5	(7)	6.4	6.4	77.7	(8)	61.5	61.5	139.2	(s)	477.1	NA	477.1	625.3
1978	2.2	(7)	7.3	7.3	71.4	(8)	63.1	63.1	134.5	(8)	481.2	NA	481.2	625.2
1979	1.7	(7)	6.7	6.7	77.4	(8)	67.7	67.7	145.1	(8)	527.1	NA	527.1	680.5
1980	1.4	(7)	5.1	5.1	66.7	(8)	60.3	60.3	127.0	(8)	569.3	NA	569.3	702.7
1981	1.3	(7)	6.1	6.1	61.0	(8)	67.4	67.4	128.4	(8)	596.8	NA	596.8	732.6
1982	1.4	(7)	6.8	6.8	40.9	(8)	64.1	64.1	105.0	(8)	593.7	NA	593.7	706.9
1983	1.4	(7)	7.1	7.1	37.0	(8)	66.0	66.0	103.0	(8)	625.2	NA	625.2	736.7
1984	1.7	(7)	7.4	7.4	44.0	(8)	73.7	73.7	117.8	(8)	664.4	NA	664.4	791.3
1985	1.7	(7)	6.1	6.1	41.1	(8)	75.4	75.4	116.4	(8)	693.8	NA	693.8	818.0
1986	1.8	(7)	5.9	5.9	35.9	(8)	75.6	75.6	111.5	(8)	685.1	NA	685.1	804.2
1987	1.6	(7)	5.3	5.3	37.0	(8)	75.2	75.2	112.1	(8)	717.9	NA	717.9	836.9
1988	1.6	(7)	5.6	5.6	41.9	(8)	76.3	76.3	118.1	(8)	758.4	NA	758.4	883.6
1989	1.3	1.1	3.7	4.9	40.5	24.9	51.3	76.1	116.6	(8)	767.4	4.8	772.2	895.0
1990	1.3	1.2	4.2	5.4	38.9	27.8	48.5	76.3	115.2	(8)	774.2	8.4	782.6	904.5
1991	1.1	1.2	3.8	5.0	33.9	27.0	48.4	75.4	109.3	(8)	773.2	10.7	783.9	899.2
1992	1.1	1.2	3.9	5.0	32.4	28.2	45.8	74.0	106.4	(8)	781.2	13.9	795.1	907.7
1993	1.1	1.4	3.7	5.1	31.3	28.9	46.0	74.9	106.2	(8)	816.6	15.1	831.6	944.1
1994	.9	1.3	3.8	5.1	31.7	29.7	45.5	75.2	106.9	(8)	821.2	17.1	838.4	951.3
1995	.8	1.4	3.6	5.1	33.0	29.4	43.7	73.1	106.1	(8)	832.9	17.3	850.2	962.1
1996	.7	1.7	3.6	5.3	31.7	29.4	42.3	71.7	103.4	(8)	878.8	18.1	896.9	1,006.3
1997	.7	1.7	4.0	5.8	30.2	29.9	41.7	71.5	101.7	(8)	904.2	17.1	921.4	1,029.5
1998	.5	1.4	2.9	4.3	28.2	28.6	38.9	67.4	95.6	(8)	920.4	16.3	936.6	1,037.1
1999	.6	1.5	2.8	4.3	28.1	27.8	37.0	64.7	92.8	(8)	924.7	16.2	940.9	1,038.6
2000	.5	1.5	2.1	3.7	28.9	28.0	37.2	65.2	94.1	(8)	967.1	18.7	985.8	1,084.1
2001	.5	1.4	2.4	3.9	26.1	25.8	39.5	65.3	91.3	(8)	946.1	18.4	964.4	1,060.1
2002	.5	1.4	2.5	3.9	23.7	26.2	34.5	60.7	84.4	(8)	960.1	17.4	977.5	1,066.4
2003	.6	1.8	1.9	3.7	24.2	24.8	36.4	61.3	85.5	(8)	983.5	21.6	1,005.1	1,094.9
2004	R.5	1.9	R2.7	4.6	23.7	26.6	35.6	62.2	85.9	(8)	994.8	21.5	1,016.3	1,107.3
2005	.4	1.9	R2.4	R4.3	23.4	25.9	34.5	60.3	83.8	(8)	1,015.6	21.8	1,037.5	R1,126.0
2006	R.3	1.9	R1.1	R3.0	23.0	R25.3	R34.2	R59.5	R82.4	(8)	R1,004.8	R21.9	R1,026.6	R1,112.3
2007P	.3	1.9	1.0	3.0	22.7	24.1	32.4	56.5	79.2	(8)	1,024.5	21.9	1,046.4	1,128.8

[1] See Note 2, "Residential and Commercial Coal Consumption Estimates," at end of section.
[2] Electricity-only and combined-heat-and-power (CHP) plants within the NAICS 22 category whose primary business is to sell electricity, or electricity and heat, to the public. Through 1988, data are for electric utilities only; beginning in 1989, data are for electric utilities and independent power producers. Electric utility CHP plants are included in "Electricity Only."
[3] Commercial combined-heat-and-power (CHP) and a small number of commercial electricity-only plants, such as those at hospitals and universities.
[4] All commercial sector fuel use other than that in "Commercial CHP."
[5] Industrial combined-heat-and-power (CHP) and a small number of industrial electricity-only plants.
[6] All industrial sector fuel use other than that in "Coke Plants" and "Industrial CHP."
[7] Included in "Commercial Other."
[8] Included in "Industrial Non-CHP."

R=Revised. P=Preliminary. NA=Not available. (s)=Less than 0.05 million short tons.

Notes: • See Tables 8.5a-8.5d for the amount of coal used to produce electricity and Tables 8.6a-8.6c for the amount of coal used to produce useful thermal output. • See Note 1, "Coal Consumption," at end of section. • See Note 2, "Classification of Power Plants Into Energy-Use Sectors," at end of section 8. • Totals may not equal sum of components due to independent rounding.

Web Pages: • For all data beginning in 1949, see http://www.eia.doe.gov/emeu/aer/coal.html. • For related information, see http://www.eia.doe.gov/fuelcoal.html.

Sources: **Commercial CHP** and **Industrial CHP:** Table 8.7c. **Electric Power Sector:** Tables 8.5b, 8.5c, 8.6b, and 8.7b. **All Other Data:** • 1949-1975—Bureau of Mines (BOM), *Minerals Yearbook*, "Coal—Bituminous and Lignite" and "Coal—Pennsylvania Anthracite" chapters. • 1976—Energy Information Administration (EIA), Energy Data Reports, *Coal—Bituminous and Lignite in 1976* and *Coal—Pennsylvania Anthracite 1976.* • 1977 and 1978—EIA, Energy Data Reports, *Coal—Pennsylvania Anthracite 1977; 1978,* and *Weekly Coal Report.* • 1979 and 1980—EIA, Energy Data Report, *Weekly Coal Report.* • 1981-2000—EIA, *Quarterly Coal Report (QCR) October-December,* quarterly reports. 2001 forward—EIA, *QCR October-December 2007* (March 2008), Table 25.

Figure 7.4 Coal Exports by Country of Destination

Total and Europe, 1960-2007

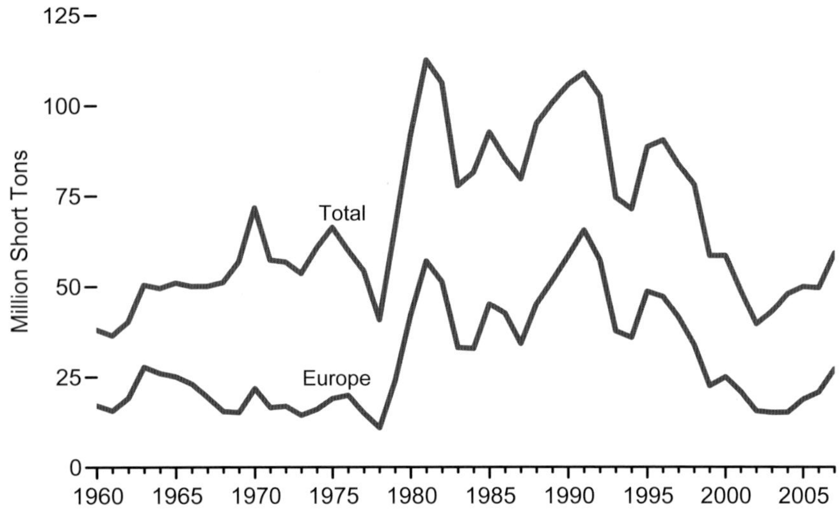

By Selected Country, 2007

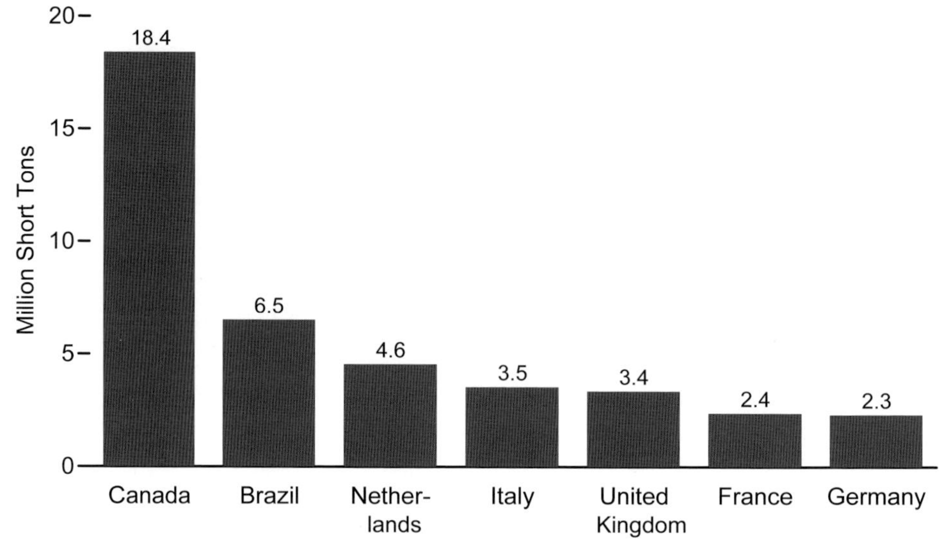

By Selected Country, 1960-2007

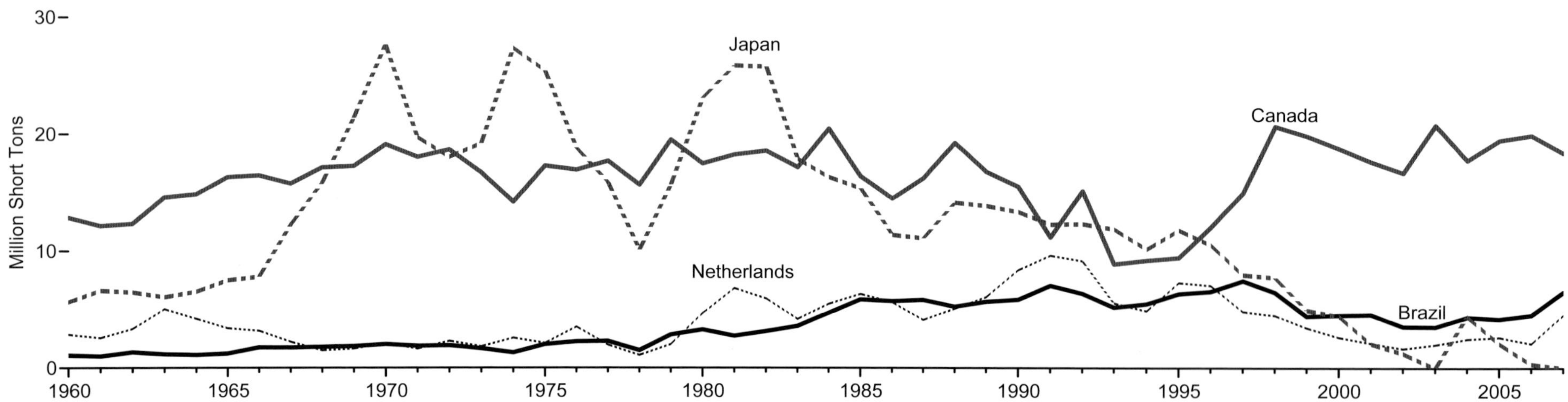

Note: Because vertical scales differ, graphs should not be compared.

Source: Table 7.4.

Table 7.4 Coal Exports by Country of Destination, 1960-2007

(Million Short Tons)

Year	Canada	Brazil	Europe											Japan	Other	Total
			Belgium [1]	Denmark	France	Germany [2]	Italy	Nether-lands	Spain	Turkey	United Kingdom	Other	Total			
1960	12.8	1.1	1.1	0.1	0.8	4.6	4.9	2.8	0.3	NA	–	2.4	17.1	5.6	1.3	38.0
1961	12.1	1.0	1.0	.1	.7	4.3	4.8	2.6	.2	NA	–	2.0	15.7	6.6	1.0	36.4
1962	12.3	1.3	1.3	(s)	.9	5.1	6.0	3.3	.8	NA	(s)	1.8	19.1	6.5	1.0	40.2
1963	14.6	1.2	2.7	(s)	2.7	5.6	7.9	5.0	1.5	NA	–	2.4	27.7	6.1	.9	50.4
1964	14.8	1.1	2.3	(s)	2.2	5.2	8.1	4.2	1.4	NA	–	2.6	26.0	6.5	1.1	49.5
1965	16.3	1.2	2.2	(s)	2.1	4.7	9.0	3.4	1.4	NA	(s)	2.3	25.1	7.5	.9	51.0
1966	16.5	1.7	1.8	(s)	1.6	4.9	7.8	3.2	1.2	NA	(s)	2.5	23.1	7.8	1.0	50.1
1967	15.8	1.7	1.4	–	2.1	4.7	5.9	2.2	1.0	NA	–	2.1	19.4	12.2	1.0	50.1
1968	17.1	1.8	1.1	–	1.5	3.8	4.3	1.5	1.5	NA	–	1.9	15.5	15.8	.9	51.2
1969	17.3	1.8	.9	–	2.3	3.5	3.7	1.6	1.8	NA	–	1.3	15.2	21.4	1.2	56.9
1970	19.1	2.0	1.9	–	3.6	5.0	4.3	2.1	3.2	NA	(s)	1.8	21.8	27.6	1.2	71.7
1971	18.0	1.9	.8	–	3.2	2.9	2.7	1.6	2.6	NA	1.7	1.1	16.6	19.7	1.1	57.3
1972	18.7	1.9	1.1	–	1.7	2.4	3.7	2.3	2.1	NA	2.4	1.1	16.9	18.0	1.2	56.7
1973	16.7	1.6	1.2	–	2.0	1.6	3.3	1.8	2.2	NA	.9	1.3	14.4	19.2	1.6	53.6
1974	14.2	1.3	1.1	–	2.7	1.5	3.9	2.6	2.0	NA	1.4	.9	16.1	27.3	1.8	60.7
1975	17.3	2.0	.6	–	3.6	2.0	4.5	2.1	2.7	NA	1.9	1.6	19.0	25.4	2.6	66.3
1976	16.9	2.2	2.2	(s)	3.5	1.0	4.2	3.5	2.5	NA	.8	2.1	19.9	18.8	2.1	60.0
1977	17.7	2.3	1.5	.1	2.1	.9	4.1	2.0	1.6	NA	.6	2.1	15.0	15.9	3.5	54.3
1978	15.7	1.5	1.1	–	1.7	.6	3.2	1.1	.8	NA	.4	2.2	11.0	10.1	2.5	40.7
1979	19.5	2.8	3.2	.2	3.9	2.6	5.0	2.0	1.4	NA	1.4	4.4	23.9	15.7	4.1	66.0
1980	17.5	3.3	4.6	1.7	7.8	2.5	7.1	4.7	3.4	NA	4.1	6.0	41.9	23.1	6.0	91.7
1981	18.2	2.7	4.3	3.9	9.7	4.3	10.5	6.8	6.4	.6	2.3	8.2	57.0	25.9	8.7	112.5
1982	18.6	3.1	4.8	2.8	9.0	2.3	11.3	5.9	5.6	1.6	2.0	6.0	51.3	25.8	7.5	106.3
1983	17.2	3.6	2.5	1.7	4.2	1.5	8.1	4.2	3.3	1.6	1.2	4.7	33.1	17.9	6.1	77.8
1984	20.4	4.7	3.9	.6	3.8	.9	7.6	5.5	2.3	1.5	2.9	3.9	32.8	16.3	7.2	81.5
1985	16.4	5.9	4.4	2.2	4.5	1.1	10.3	6.3	3.5	2.2	2.7	8.1	45.1	15.4	9.9	92.7
1986	14.5	5.7	4.4	2.1	5.4	.8	10.4	5.6	2.6	2.4	2.9	5.9	42.6	11.4	11.4	85.5
1987	16.2	5.8	4.6	.9	2.9	.5	9.5	4.1	2.5	.8	2.6	5.8	34.2	11.1	12.3	79.6
1988	19.2	5.3	6.5	2.8	4.3	.7	11.1	5.1	2.5	2.0	3.7	6.4	45.1	14.1	11.3	95.0
1989	16.8	5.7	7.1	3.2	6.5	.7	11.2	6.1	3.3	1.7	4.5	7.2	51.6	13.8	12.9	100.8
1990	15.5	5.8	8.5	3.2	6.9	1.1	11.9	8.4	3.8	2.1	5.2	7.4	58.4	13.3	12.7	105.8
1991	11.2	7.1	7.5	4.7	9.5	1.7	11.3	9.6	4.7	2.2	6.2	8.2	65.5	12.3	13.0	109.0
1992	15.1	6.4	7.2	3.8	8.1	1.0	9.3	9.1	4.5	2.0	5.6	6.6	57.3	12.3	11.4	102.5
1993	8.9	5.2	5.2	.3	4.0	.5	6.9	5.6	4.1	1.6	4.1	5.3	37.6	11.9	11.0	74.5
1994	9.2	5.5	4.9	.5	2.9	.3	7.5	4.9	4.1	1.3	3.4	6.0	35.8	10.2	10.7	71.4
1995	9.4	6.4	4.5	2.1	3.7	2.0	9.1	7.3	4.7	2.0	4.7	8.7	48.6	11.8	12.4	88.5
1996	12.0	6.5	4.6	1.3	3.9	1.1	9.2	7.1	4.1	2.2	6.2	7.7	47.2	10.5	14.2	90.5
1997	15.0	7.5	4.3	.4	3.4	.9	7.0	4.8	4.1	2.1	7.2	7.1	41.3	8.0	11.8	83.5
1998	20.7	6.5	3.2	.3	3.2	1.2	5.3	4.5	3.2	1.6	5.9	5.3	33.8	7.7	9.4	78.0
1999	19.8	4.4	2.1	–	2.5	.6	4.0	3.4	2.5	.8	3.2	3.5	22.5	5.0	6.7	58.5
2000	18.8	4.5	2.9	.1	3.0	1.0	3.7	2.6	2.7	1.8	3.3	3.9	25.0	4.4	5.8	58.5
2001	17.6	4.6	2.8	–	2.2	.9	5.4	2.1	1.6	.9	2.5	2.4	20.8	2.1	3.6	48.7
2002	16.7	3.5	2.4	–	1.3	1.0	3.1	1.7	1.9	.6	1.9	1.8	15.6	1.3	2.6	39.6
2003	20.8	3.5	1.8	.3	1.3	.5	2.8	2.0	1.8	1.1	1.5	2.1	15.1	(s)	3.6	43.0
2004	17.8	4.4	1.7	.1	1.1	.6	2.1	2.5	1.5	1.3	2.0	2.3	15.2	4.4	6.2	48.0
2005	19.5	4.2	2.1	.1	1.3	.7	2.5	2.6	1.9	1.9	1.8	4.1	18.8	2.1	5.4	49.9
2006	19.9	4.5	2.2	.4	1.6	1.7	3.3	2.1	1.6	1.2	2.6	4.2	20.8	.3	4.1	49.6
2007P	18.4	6.5	2.1	.1	2.4	2.3	3.5	4.6	1.5	1.4	3.4	5.8	27.1	(s)	7.1	59.2

[1] Through 1999, includes Luxembourg.

[2] Through 1990, data for Germany are for the former West Germany only. Beginning in 1991, data for Germany are for the unified Germany, i.e., the former East Germany and West Germany.

P=Preliminary. NA = Not Available. – = No data reported. (s)=Less than 0.05 million short tons.

Note: Totals may not equal sum of components due to independent rounding.

Sources: • 1960-1988—U.S. Department of Commerce, Bureau of the Census. *U.S. Exports by Schedule B Commodities, EM 522.* • 1989-2000—Energy Information Administration (EIA), *Coal Industry Annual,* annual reports. • 2001 forward—EIA, *Quarterly Coal Report October-December,* quarterly reports.

Figure 7.5 Coal Stocks

Total and Electric Power Sector Stocks, 1949-2007

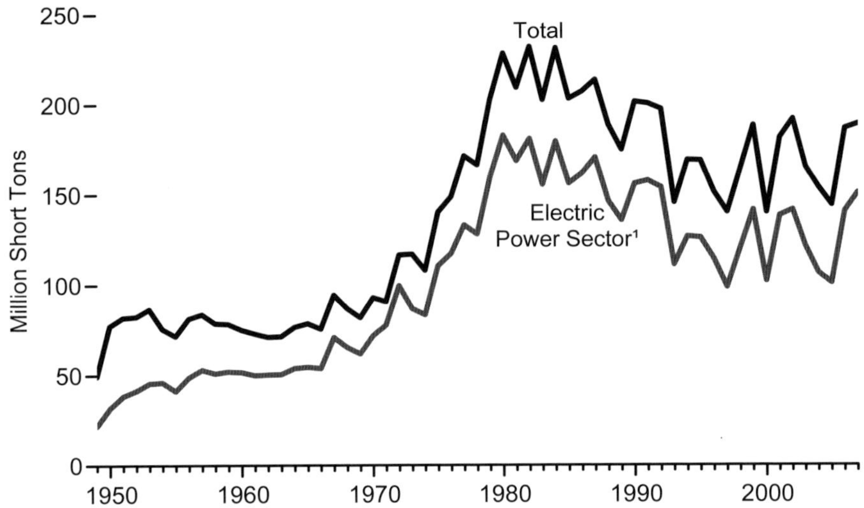

By Holding Entity, 2007

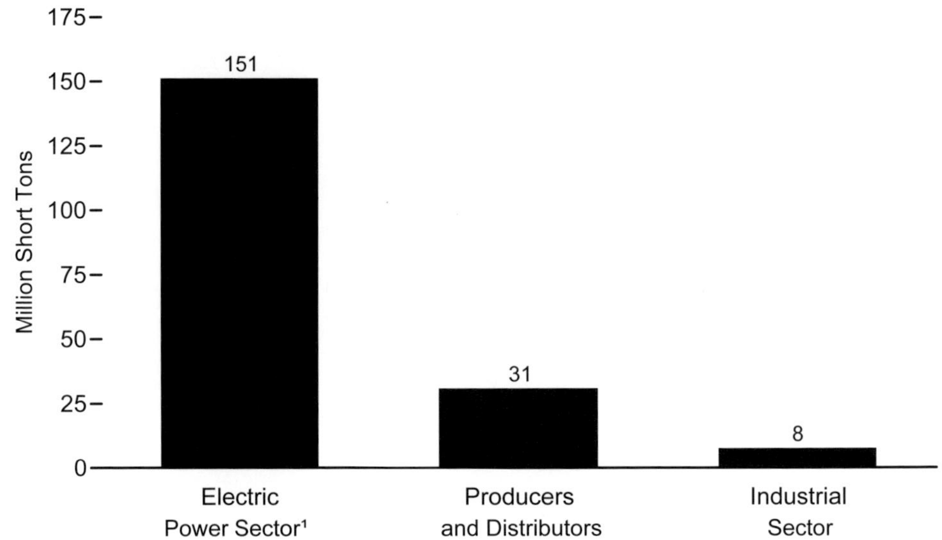

By Holding Entity, 1949-2007

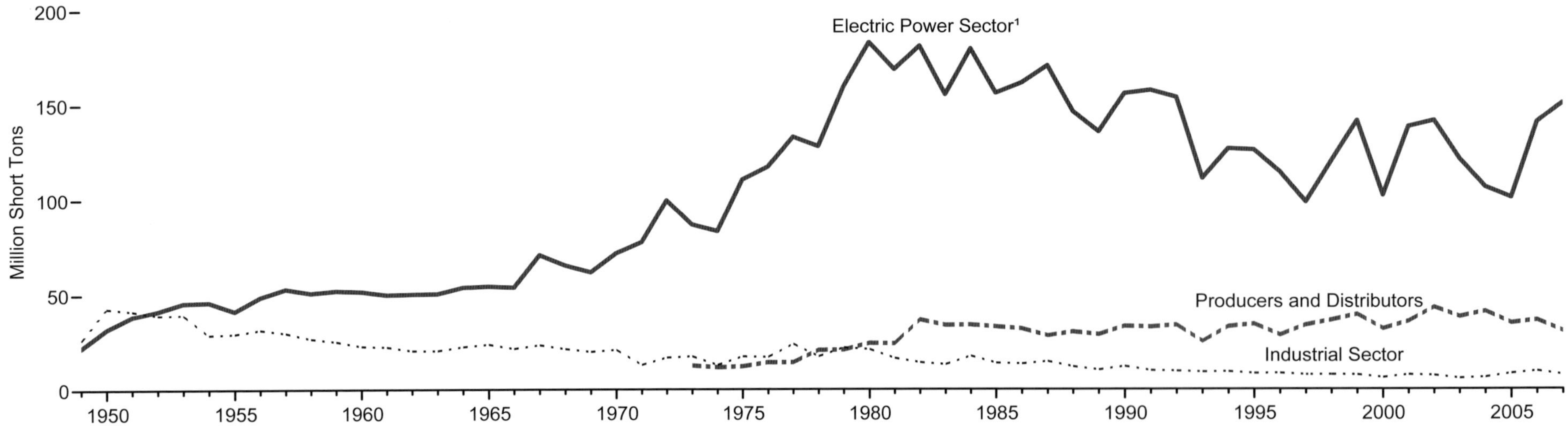

[1] Electricity-only and combined-heat-and-power (CHP) plants whose primary business is to sell electricity, or electricity and heat, to the public.

Notes: • Stocks are at end of year. • Because vertical scales differ, graphs should not be compared.
Source: Table 7.5.

Table 7.5 Coal Stocks by Sector, Selected Years, 1949-2007

(Million Short Tons)

Year	Producers and Distributors	Consumers							Total
		Residential and Commercial Sectors	Industrial Sector			Transportation Sector	Electric Power Sector [2]	Total	
			Coke Plants	Other [1]	Total				
1949	NA	1.4	10.0	16.1	26.0	([3])	22.1	49.5	49.5
1950	NA	2.5	16.8	26.2	43.0	([3])	31.8	77.3	77.3
1955	NA	1.0	13.4	15.9	29.3	([3])	41.4	71.7	71.7
1960	NA	.7	11.1	11.6	22.8	([3])	51.7	75.2	75.2
1965	NA	.4	10.6	13.1	23.8	([3])	54.5	78.6	78.6
1970	NA	.3	9.0	11.8	20.8	([3])	71.9	93.0	93.0
1971	NA	.3	7.3	5.6	12.9	([3])	77.8	91.0	91.0
1972	NA	.3	9.1	7.6	16.7	([3])	99.7	116.8	116.8
1973	12.5	.3	7.0	10.4	17.4	([3])	87.0	104.6	117.2
1974	11.6	.3	6.2	6.6	12.8	([3])	83.5	96.6	108.2
1975	12.1	.2	8.8	8.5	17.3	([3])	110.7	128.3	140.4
1976	14.2	.2	9.9	7.1	17.0	([3])	117.4	134.7	148.9
1977	14.2	.2	12.8	11.1	23.9	([3])	133.2	157.3	171.5
1978	20.7	.4	8.3	9.0	17.3	NA	128.2	145.9	166.6
1979	20.8	.3	10.2	11.8	21.9	NA	159.7	182.0	202.8
1980	24.4	NA	9.1	12.0	21.0	NA	183.0	204.0	228.4
1981	24.1	NA	6.5	9.9	16.4	NA	168.9	185.3	209.4
1982	36.8	NA	4.6	9.5	14.1	NA	181.1	195.3	232.0
1983	33.9	NA	4.3	8.7	13.1	NA	155.6	168.7	202.6
1984	34.1	NA	6.2	11.3	17.5	NA	179.7	197.2	231.3
1985	33.1	NA	3.4	10.4	13.9	NA	156.4	170.2	203.4
1986	32.1	NA	3.0	10.4	13.4	NA	161.8	175.2	207.3
1987	28.3	NA	3.9	10.8	14.7	NA	170.8	185.5	213.8
1988	30.4	NA	3.1	8.8	11.9	NA	146.5	158.4	188.8
1989	29.0	NA	2.9	7.4	10.2	NA	135.9	146.1	175.1
1990	33.4	NA	3.3	8.7	12.0	NA	156.2	168.2	201.6
1991	33.0	NA	2.8	7.1	9.8	NA	157.9	167.7	200.7
1992	34.0	NA	2.6	7.0	9.6	NA	154.1	163.7	197.7
1993	25.3	NA	2.4	6.7	9.1	NA	111.3	120.5	145.7
1994	33.2	NA	2.7	6.6	9.2	NA	126.9	136.1	169.4
1995	34.4	NA	2.6	5.7	8.3	NA	126.3	134.6	169.1
1996	28.6	NA	2.7	5.7	8.4	NA	114.6	123.0	151.6
1997	34.0	NA	2.0	5.6	7.6	NA	98.8	106.4	140.4
1998	36.5	NA	2.0	5.5	7.6	NA	120.5	128.1	164.6
1999	39.5	NA	1.9	5.6	7.5	NA	141.6	149.1	188.6
2000	31.9	NA	1.5	4.6	6.1	NA	102.3	108.4	140.3
2001	35.9	NA	1.5	6.0	7.5	NA	138.5	146.0	181.9
2002	43.3	NA	1.4	5.8	7.2	NA	141.7	148.9	192.1
2003	38.3	NA	.9	4.7	5.6	NA	121.6	127.2	165.5
2004	41.2	NA	1.3	4.8	6.2	NA	106.7	112.9	154.0
2005	35.0	NA	2.6	5.6	8.2	NA	101.1	109.3	144.3
2006	[R]36.5	NA	2.9	6.5	9.4	NA	[R]141.0	[R]150.4	[R]186.9
2007[P]	[E]30.8	NA	1.9	5.6	7.6	NA	151.1	158.7	189.4

[1] Through 1977, data are for stocks held by the manufacturing and transportation sectors. Beginning in 1978, data are for stocks held at manufacturing plants only.

[2] Electricity-only and combined-heat-and-power (CHP) plants within the NAICS 22 category whose primary business is to sell electricity, or electricity and heat, to the public. Through 1998, data are for electric utilities only; beginning in 1999, data are for electric utilities and independent power producers.

[3] Included in "Industrial Sector Other."

R=Revised. P=Preliminary. E=Estimate. NA=Not available.

Notes: • Stocks are at end of year. • Totals may not equal sum of components due to independent rounding.

Web Pages: • For all data beginning in 1949, see http://www.eia.doe.gov/emeu/aer/coal.html. • For related information, see http://www.eia.doe.gov/fuelcoal.html.

Sources: **Electric Power Sector:** Table 8.8. **All Other Data:** • 1949-1975—Bureau of Mines, *Minerals Yearbook,* "Coal—Bituminous and Lignite" and "Coal—Pennsylvania Anthracite" chapters. • 1976—Energy Information Administration (EIA), Energy Data Reports, *Coal—Bituminous and Lignite in 1976* and *Coal—Pennsylvania Anthracite 1976.* • 1977 and 1978—EIA, Energy Data Reports, *Coal—Pennsylvania Anthracite 1977; 1978,* and *Weekly Coal Report.* • 1979—EIA, Energy Data Report, *Weekly Coal Report.* • 1980-2000—EIA, *Quarterly Coal Report (QCR) October-December,* quarterly reports. • 2001 forward—EIA, *QCR October-December 2007* (March 2008), Table 29.

Figure 7.6 Coal Mining Productivity

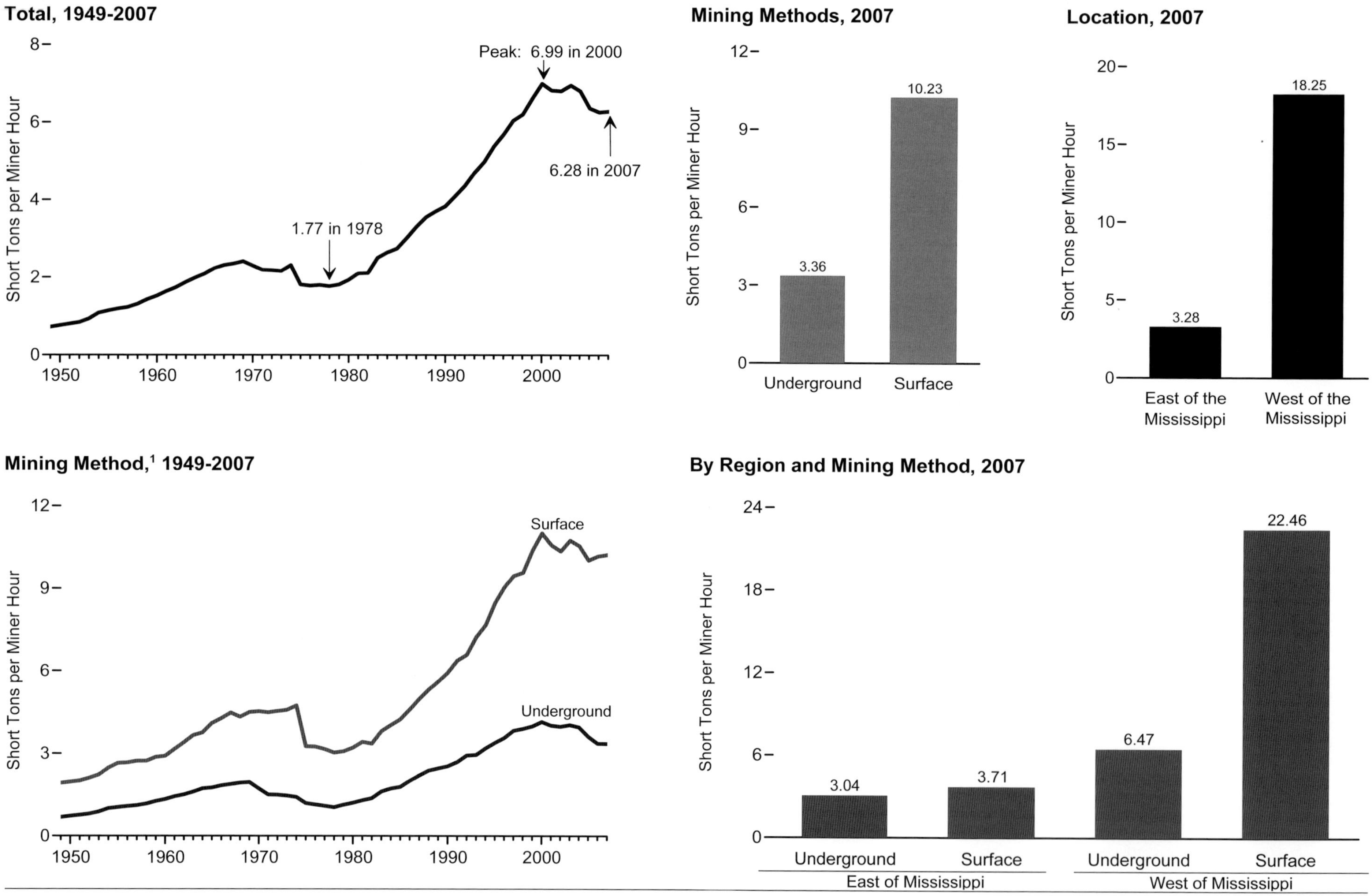

[1] For 1979 forward, includes all coal; prior to 1979, excludes anthracite.

Notes: • Beginning in 2001, surface mining includes a small amount of refuse recovery. • Because vertical scales differ, graphs should not be compared.

Source: Table 7.6.

Table 7.6 Coal Mining Productivity, Selected Years, 1949-2007

(Short Tons per Miner Hour [1])

Year	Mining Method		Location						
			East of the Mississippi			West of the Mississippi			
	Underground	Surface [2]	Underground	Surface [2]	Total [2]	Underground	Surface [2]	Total [2]	Total [2]
1949	[3]0.68	[3]1.92	NA	NA	NA	NA	NA	NA	0.72
1950	[3].72	[3]1.96	NA	NA	NA	NA	NA	NA	.76
1955	[3]1.04	[3]2.65	NA	NA	NA	NA	NA	NA	1.14
1960	[3]1.33	[3]2.91	NA	NA	NA	NA	NA	NA	1.52
1965	[3]1.75	[3]4.10	NA	NA	NA	NA	NA	NA	2.09
1970	[3]1.72	[3]4.53	NA	NA	NA	NA	NA	NA	2.30
1971	[3]1.50	[3]4.49	NA	NA	NA	NA	NA	NA	2.19
1972	[3]1.49	[3]4.54	NA	NA	NA	NA	NA	NA	2.18
1973	[3]1.46	[3]4.58	NA	NA	NA	NA	NA	NA	2.16
1974	[3]1.41	[3]4.74	NA	NA	NA	NA	NA	NA	2.31
1975	[3]1.19	[3]3.26	NA	NA	NA	NA	NA	NA	1.81
1976	[3]1.14	[3]3.25	NA	NA	NA	NA	NA	NA	1.78
1977	[3]1.09	[3]3.16	NA	NA	NA	NA	NA	NA	1.80
1978	[3]1.04	[3]3.03	NA	NA	NA	NA	NA	NA	1.77
1979	1.13	3.08	NA	NA	NA	NA	NA	NA	1.81
1980	1.20	3.21	NA	NA	NA	NA	NA	NA	1.93
1981	1.29	3.42	NA	NA	NA	NA	NA	NA	2.10
1982	1.37	3.36	NA	NA	NA	NA	NA	NA	2.11
1983	1.61	3.81	NA	NA	NA	NA	NA	NA	2.50
1984	1.72	4.03	1.69	2.56	1.98	2.49	8.15	7.07	2.64
1985	1.78	4.24	1.75	2.52	2.00	2.45	8.61	7.40	2.74
1986	2.00	4.60	1.96	2.75	2.21	2.80	9.02	7.90	3.01
1987	2.20	4.98	2.16	2.97	2.42	3.39	9.86	8.73	3.30
1988	2.38	5.32	2.32	2.99	2.54	3.55	10.73	9.38	3.55
1989	2.46	5.61	2.39	3.13	2.63	3.92	11.86	10.21	3.70
1990	2.54	5.94	2.46	3.32	2.73	4.01	12.26	10.41	3.83
1991	2.69	6.38	2.59	3.49	2.86	4.53	12.36	10.79	4.09
1992	2.93	6.59	2.82	3.61	3.07	4.85	12.49	11.03	4.36
1993	2.95	7.23	2.81	3.74	3.11	5.18	13.94	12.14	4.70
1994	3.19	7.67	3.02	3.85	3.28	5.93	15.19	13.22	4.98
1995	3.39	8.48	3.19	4.03	3.45	6.32	16.23	14.18	5.38
1996	3.57	9.05	3.36	4.25	3.63	7.03	17.89	15.66	5.69
1997	3.83	9.46	3.63	4.49	3.89	6.82	18.63	16.04	6.04
1998	3.90	9.58	3.69	4.31	3.89	6.76	18.82	16.27	6.20
1999	3.99	10.39	3.74	4.48	3.97	7.45	19.57	17.18	6.61
2000	4.15	11.01	3.89	4.82	4.18	7.66	20.04	17.62	6.99
2001	4.02	[2]10.58	3.71	[2]4.53	[2]3.98	8.39	[2]20.63	[2]18.32	[2]6.82
2002	3.98	10.36	3.67	4.22	3.86	7.80	20.67	18.06	6.80
2003	4.04	10.75	3.68	4.18	3.85	8.33	21.42	18.67	6.95
2004	3.96	10.55	3.59	3.95	3.72	8.22	22.04	19.00	6.80
2005	3.62	10.03	3.28	3.75	3.44	7.48	21.98	18.50	6.36
2006	[R]3.37	[R]10.18	[R]3.06	[R]3.74	3.29	[R]6.62	[R]22.26	18.33	[R]6.26
2007[P]	3.36	10.23	3.04	3.71	3.28	6.47	22.46	18.25	6.28

[1] Data through 1973 for bituminous coal, subbituminous coal, and lignite mines, and data through 1978 for anthracite mines, were originally reported in short tons per miner day. The data were converted to short tons per miner hour by assuming an eight-hour day. All remaining data were calculated by dividing total production by total labor hours worked by all mine employees except office workers.

[2] Beginning in 2001, includes a small amount of refuse recovery.

[3] Anthracite mining productivity is unavailable by underground and surface but is included in "Total."

R=Revised. P=Preliminary. NA=Not available.

Web Pages: • For all data beginning in 1949, see http://www.eia.doe.gov/emeu/aer/coal.html. • For related information, see http://www.eia.doe.gov/fuelcoal.html.

Sources: • 1949-1975—Bureau of Mines, *Minerals Yearbook,* "Coal—Bituminous and Lignite" and "Coal—Pennsylvania Anthracite" chapters. • 1976—Energy Information Administration (EIA), Energy Data Reports, *Coal—Bituminous and Lignite in 1976* and *Coal—Pennsylvania Anthracite 1976.* • 1977 and 1978—EIA, Energy Data Reports, *Bituminous Coal and Lignite Production and Mine Operations—1977; 1978* and *Coal—Pennsylvania Anthracite 1977; 1978.* • 1979—EIA, Energy Data Report, *Coal Production—1979.* • 1980-1988—EIA, *Coal Production,* annual reports. • 1989-2000—EIA, *Coal Industry Annual,* annual reports. • 2001-2006—EIA, *Annual Coal Report,* annual reports. • 2007—EIA, Form EIA-7A, "Coal Production Report," and U.S. Department of Labor, Mine Safety and Health Administration, Form 7000-2, "Quarterly Mine Employment and Coal Production Report."

Figure 7.7 Coke Overview

Production and Consumption, 1949-2007

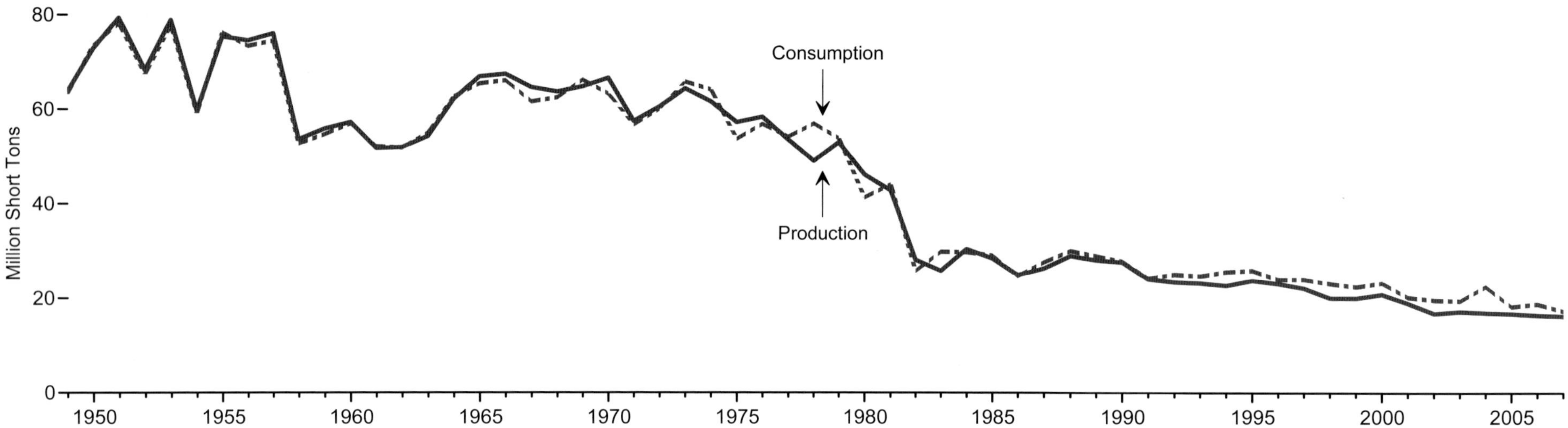

Overview, 2007

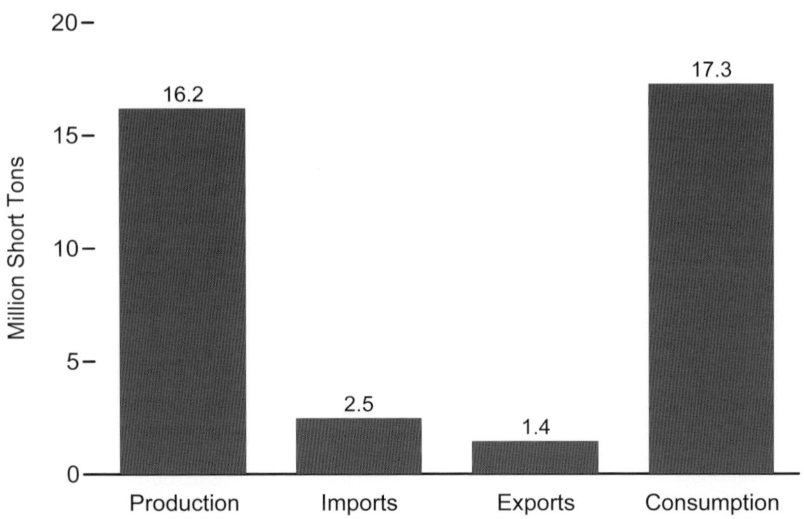

Trade, 1949-2007

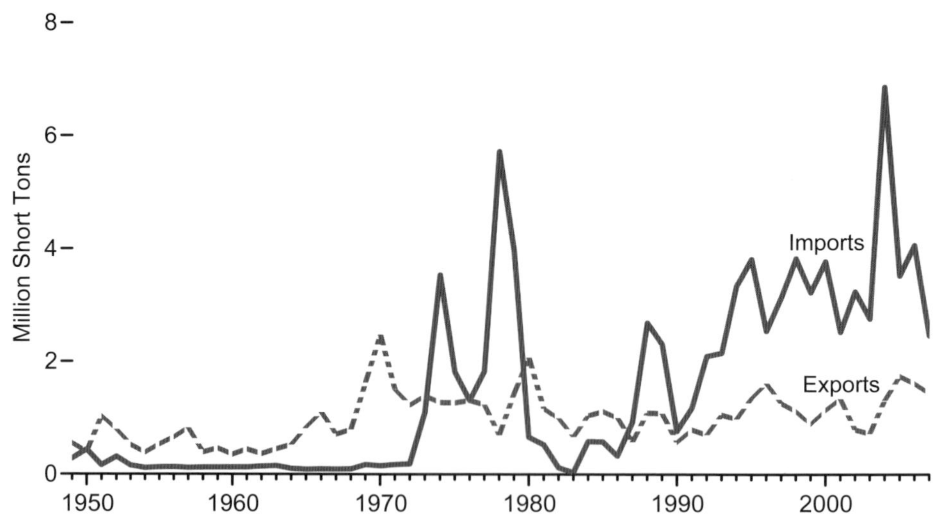

Note: Because vertical scales differ, graphs should not be compared.

Source: Table 7.7.

Table 7.7 Coke Overview, Selected Years, 1949-2007

(Million Short Tons)

Year	Production	Trade			Stock Change [2]	Consumption [3]
		Imports	Exports	Net Imports [1]		
1949	63.6	0.3	0.5	-0.3	0.2	63.2
1950	72.7	.4	.4	(s)	-.7	73.4
1955	75.3	.1	.5	-.4	-1.2	76.1
1960	57.2	.1	.4	-.2	.1	56.9
1965	66.9	.1	.8	-.7	.7	65.4
1970	66.5	.2	2.5	-2.3	1.0	63.2
1971	57.4	.2	1.5	-1.3	-.6	56.7
1972	60.5	.2	1.2	-1.0	-.6	60.0
1973	64.3	1.1	1.4	-.3	-1.7	65.8
1974	61.6	3.5	1.3	2.3	-.2	64.1
1975	57.2	1.8	1.3	.5	4.1	53.7
1976	58.3	1.3	1.3	(s)	1.5	56.8
1977	53.5	1.8	1.2	.6	(s)	54.1
1978	49.0	5.7	.7	5.0	-2.9	56.9
1979	52.9	4.0	1.4	2.5	1.7	53.8
1980	46.1	.7	2.1	-1.4	3.4	41.3
1981	42.8	.5	1.2	-.6	-1.9	44.0
1982	28.1	.1	1.0	-.9	1.5	25.8
1983	25.8	(s)	.7	-.6	-4.7	29.9
1984	30.4	.6	1.0	-.5	.2	29.7
1985	28.4	.6	1.1	-.5	-1.2	29.1
1986	24.9	.3	1.0	-.7	-.5	24.7
1987	26.3	.9	.6	.3	-1.0	27.7
1988	28.9	2.7	1.1	1.6	.5	30.0
1989	28.0	2.3	1.1	1.2	.3	28.9
1990	27.6	.8	.6	.2	(s)	27.8
1991	24.0	1.2	.8	.4	.2	24.2
1992	23.4	2.1	.7	1.4	-.2	25.0
1993	23.2	2.2	1.1	1.1	-.4	24.7
1994	22.7	3.3	1.0	2.4	-.5	25.6
1995	23.7	3.8	1.4	2.5	.4	25.8
1996	23.1	2.5	1.6	.9	(s)	24.0
1997	22.1	3.1	1.3	1.9	(s)	24.0
1998	20.0	3.8	1.1	2.7	-.4	23.1
1999	20.0	3.2	.9	2.3	-.1	22.4
2000	20.8	3.8	1.1	2.6	.2	23.2
2001	18.9	2.5	1.3	1.2	-.1	20.2
2002	16.8	3.2	.8	2.5	-.4	19.6
2003	17.2	2.8	.7	2.0	-.2	19.4
2004	16.9	6.9	1.3	5.6	(s)	22.5
2005	16.7	3.5	1.7	1.8	.3	18.2
2006	16.4	4.1	1.6	2.5	.1	18.8
2007P	16.2	2.5	1.4	1.0	-.1	17.3

[1] Net imports equal imports minus exports. Minus sign indicates exports are greater than imports.
[2] Producer and distributor stocks at end of year. A negative value indicates a decrease in stocks; a positive value indicates an increase.
[3] "Consumption" is calculated as the sum of production and imports minus exports and stock change.
P=Preliminary. (s)=Less than 0.05 million short tons.
Note: Totals may not equal sum of components due to independent rounding.

Web Pages: • For all data beginning in 1949, see http://www.eia.doe.gov/emeu/aer/coal.html. • For related information, see http://www.eia.doe.gov/fuelcoal.html.

Sources: • 1949-1975—Bureau of Mines, *Minerals Yearbook,* "Coke and Coal Chemicals" chapter. • 1976-1980—Energy Information Administration (EIA), Energy Data Report, *Coke and Coal Chemicals,* annual reports. • 1981-2000—EIA, *Quarterly Coal Report (QCR) October-December,* quarterly reports. • 2001 forward—EIA, *QCR October-December 2007* (March 2008), Table ES-2.

Figure 7.8 Coal Prices

Total, 1949-2007

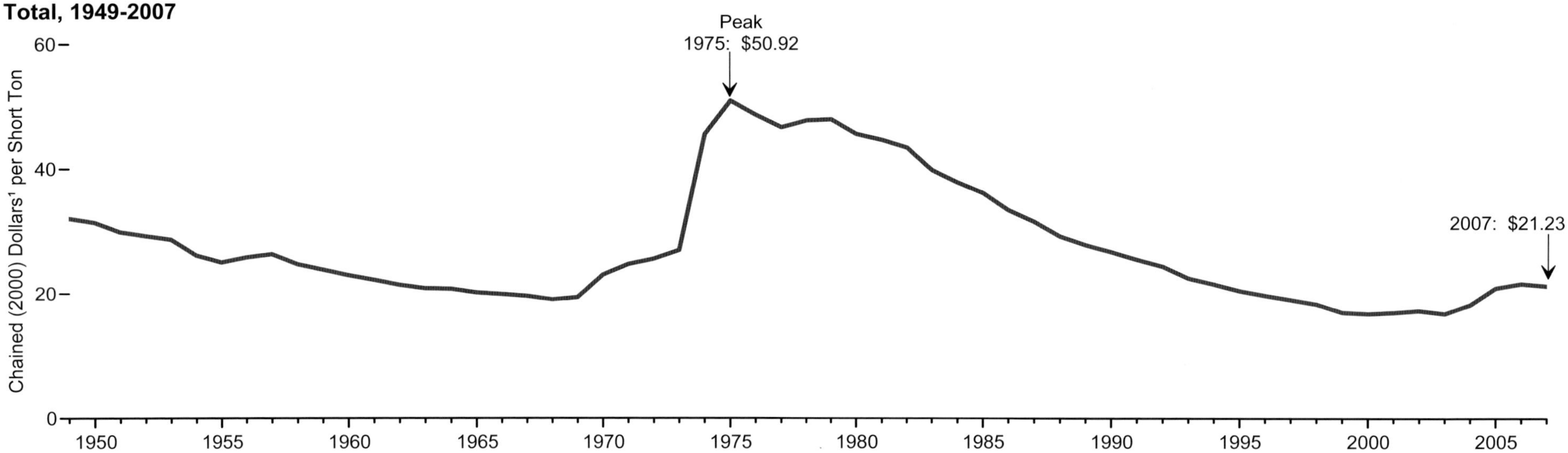

By Type, 1949-2007

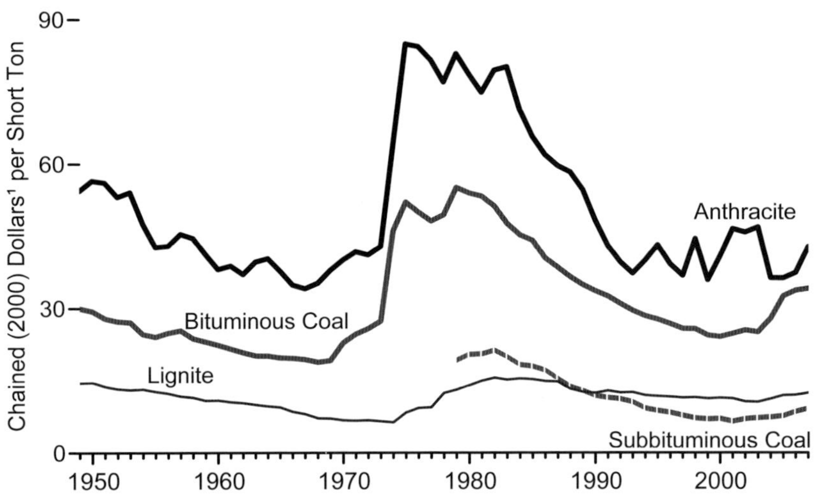

By Type, 2007

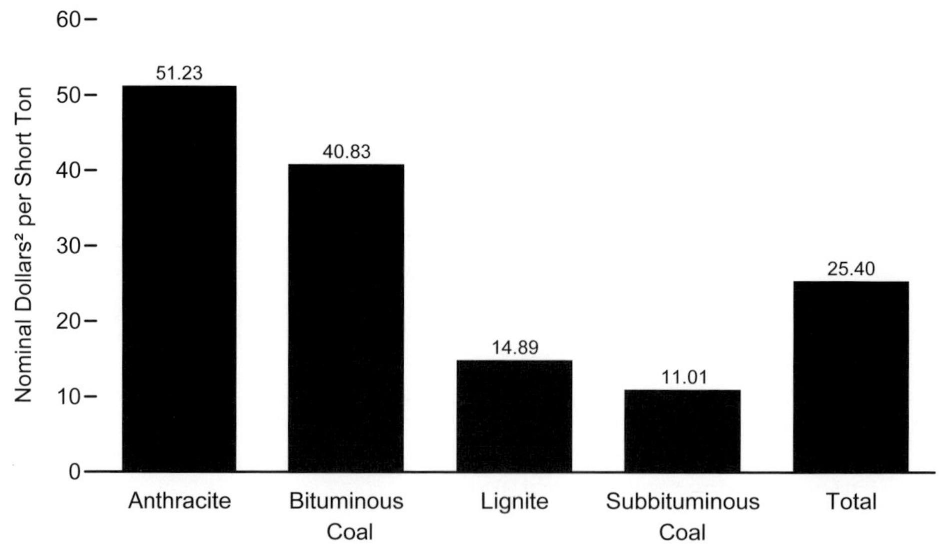

[1] Calculated by using gross domestic product implicit price deflators. See Table D1.
[2] See "Nominal Dollars" in Glossary.

Note: Because vertical scales differ, graphs should not be compared.
Source: Table 7.8.

Table 7.8 Coal Prices, Selected Years, 1949-2007

(Dollars per Short Ton)

Year	Bituminous Coal		Subbituminous Coal		Lignite [1]		Anthracite		Total	
	Nominal [2]	Real [3]	Nominal [2]	Real [3]	Nominal [2]	Real [3]	Nominal [2]	Real [3]	Nominal [2]	Real [3]
1949	[4]4.90	[4]29.97	([4])	([4])	2.37	14.49	8.90	54.43	5.24	32.05
1950	[4]4.86	[4]29.40	([4])	([4])	2.41	14.58	9.34	56.50	5.19	31.40
1955	[4]4.51	[4]24.06	([4])	([4])	2.38	12.70	8.00	42.68	4.69	25.02
1960	[4]4.71	[4]22.38	([4])	([4])	2.29	10.88	8.01	38.07	4.83	22.96
1965	[4]4.45	[4]19.75	([4])	([4])	2.13	9.45	8.51	37.76	4.55	20.19
1970	[4]6.30	[4]22.88	([4])	([4])	1.86	6.76	11.03	40.06	6.34	23.03
1971	[4]7.13	[4]24.66	([4])	([4])	1.93	6.68	12.08	41.78	7.15	24.73
1972	[4]7.78	[4]25.79	([4])	([4])	2.04	6.76	12.40	41.11	7.72	25.59
1973	[4]8.71	[4]27.35	([4])	([4])	2.09	6.56	13.65	42.86	8.59	26.97
1974	[4]16.01	[4]46.11	([4])	([4])	2.19	6.31	22.19	63.90	15.82	45.56
1975	[4]19.79	[4]52.08	([4])	([4])	3.17	8.34	32.26	84.89	19.35	50.92
1976	[4]20.11	[4]50.03	([4])	([4])	3.74	9.30	33.92	84.39	19.56	48.66
1977	[4]20.59	[4]48.16	([4])	([4])	4.03	9.43	34.86	81.54	19.95	46.66
1978	[4]22.64	[4]49.48	([4])	([4])	5.68	12.41	35.25	77.04	21.86	47.77
1979	27.31	55.12	9.55	19.27	6.48	13.08	41.06	82.87	23.75	47.93
1980	29.17	53.98	11.08	20.50	7.60	14.06	42.51	78.66	24.65	45.61
1981	31.51	53.30	12.18	20.60	8.85	14.97	44.28	74.90	26.40	44.66
1982	32.15	51.25	13.37	21.31	9.79	15.61	49.85	79.47	27.25	43.44
1983	31.11	47.71	13.03	19.98	9.91	15.20	52.29	80.19	25.98	39.84
1984	30.63	45.27	12.41	18.34	10.45	15.45	48.22	71.27	25.61	37.85
1985	30.78	44.15	12.57	18.03	10.68	15.32	45.80	65.70	25.20	36.15
1986	28.84	40.48	12.26	17.21	10.64	14.93	44.12	61.92	23.79	33.39
1987	28.19	38.51	11.32	15.47	10.85	14.82	43.65	59.63	23.07	31.52
1988	27.66	36.54	10.45	13.81	10.06	13.29	44.16	58.34	22.07	29.16
1989	27.40	34.88	10.16	12.93	9.91	12.62	42.93	54.65	21.82	27.78
1990	27.43	33.62	9.70	11.89	10.13	12.42	39.40	48.29	21.76	26.67
1991	27.49	32.55	9.68	11.46	10.89	12.90	36.34	43.03	21.49	25.45
1992	26.78	31.00	9.68	11.21	10.81	12.51	34.24	39.64	21.03	24.34
1993	26.15	29.59	9.33	10.56	11.11	12.57	32.94	37.27	19.85	22.46
1994	25.68	28.45	8.37	9.27	10.77	11.93	36.07	39.96	19.41	21.50
1995	25.56	27.75	8.10	8.79	10.83	11.76	39.78	43.19	18.83	20.44
1996	25.17	26.82	7.87	8.39	10.92	11.64	36.78	39.19	18.50	19.71
1997	24.64	25.82	7.42	7.78	10.91	11.43	35.12	36.81	18.14	19.01
1998	24.87	25.78	6.96	7.21	11.08	11.49	42.91	44.48	17.67	18.32
1999	23.92	24.44	6.87	7.02	11.04	11.28	35.13	35.90	16.63	16.99
2000	24.15	24.15	7.12	7.12	11.41	11.41	40.90	40.90	16.78	16.78
2001	25.36	24.77	6.67	6.51	11.52	11.25	47.67	46.55	17.38	16.97
2002	26.57	25.50	7.34	7.05	11.07	10.63	47.78	45.86	17.98	17.26
2003	26.73	25.12	7.73	7.26	11.20	10.53	49.87	46.87	17.85	16.78
2004	30.56	[R]27.92	8.12	7.42	12.27	11.21	39.77	[R]36.33	19.93	18.21
2005	36.80	[R]32.57	8.68	[R]7.68	13.49	[R]11.94	41.00	[R]36.28	23.59	[R]20.88
2006	[R]39.32	[R]33.73	[R]9.95	[R]8.54	[R]14.00	[R]12.01	[R]43.61	[R]37.41	[R]25.16	[R]21.58
2007[E]	40.83	34.12	11.01	9.20	14.89	12.44	51.23	42.81	25.40	21.23

[1] Because of withholding to protect company confidentiality, lignite prices exclude Texas for 1955-1977 and Montana for 1974-1978. As a result, lignite prices for 1974-1977 are for North Dakota only.

[2] See "Nominal Dollars" in Glossary.

[3] In chained (2000) dollars, calculated by using gross domestic product implicit price deflators in Table D1. See "Chained Dollars" in Glossary.

[4] Through 1978, subbituminous coal is included in "Bituminous Coal."

R=Revised. E=Estimate.

Note: Prices are free-on-board (F.O.B.) rail/barge prices, which are the F.O.B. prices of coal at the point of first sale, excluding freight or shipping and insurance costs. See "Free on Board (F.O.B.)" in Glossary.

Web Pages: • For all data beginning in 1949, see http://www.eia.doe.gov/emeu/aer/coal.html. • For related information, see http://www.eia.doe.gov/fuelcoal.html.

Sources: • 1949-1975—Bureau of Mines (BOM), *Minerals Yearbook*. • 1976—Energy Information Administration (EIA), Energy Data Report, *Coal—Bituminous and Lignite in 1976*, and BOM, *Minerals Yearbook*. • 1977 and 1978—EIA, Energy Data Reports, *Bituminous Coal and Lignite Production and Mine Operations*, and *Coal—Pennsylvania Anthracite*. • 1979—EIA, *Coal Production*, and Energy Data Report, *Coal—Pennsylvania Anthracite*. • 1980-1992—EIA, *Coal Production*, annual reports. • 1993-2000—EIA, *Coal Industry Annual*, annual reports and unpublished revisions. • 2001-2006—EIA, *Annual Coal Report*, annual reports. • 2007—EIA, Form EIA-7A, "Coal Production Report," and U.S. Department of Labor, Mine Safety and Health Administration, Form 7000-2, "Quarterly Mine Employment and Coal Production Report."

Coal

Note 1. Coal Consumption. Data in this report on the consumption of bituminous coal, subbituminous coal, lignite, anthracite, and waste coal are developed primarily from consumption data reported in surveys. Included are data reported by all electric power companies and coke plant companies. Data on coal consumption by all industrial and manufacturing establishments are based on consumption data obtained quarterly from coal users. Data on coal consumption by the residential and commercial sectors are based on distribution data obtained annually from coal distributors. Included in each sector's data are the following: Residential and Commercial Sectors—retail dealer sales to households and small commercial establishments; Industrial Sector—consumption at manufacturing plants, large commercial establishments, coking plants, and by agriculture, mining (other than coal mining), and construction industries; Transportation Sector—sales to railroads and for vessel bunkering; Electric Power Sector (electric utilities and independent power producers)—consumption for electricity generation and useful thermal output at electricity-only and CHP plants within the North American Industry Classification System (NAICS) 22 category whose primary business is to sell electricity, or electricity and heat, to the public.

Note 2. Residential and Commercial Coal Consumption Estimates. Coal consumption by the residential and commercial sectors is reported to the Energy Information Administration (EIA) for the two sectors combined; EIA estimates the amount consumed by the sectors individually. Previously, the breakdown was 40 percent residential and 60 percent commercial for each year. The current method results in variation over time. Beginning in 1949, a larger portion of the coal, 45 percent, is assigned to the residential sector; the share falls gradually over time and has been 11 percent since 2000. To create the estimate, it is first assumed that an occupied coal-heated housing unit consumes fuel at the same Btu rate as an oil-heated housing unit. Then, for the years in which data are available on the number of occupied housing units by heating source (1950, 1960, 1970, 1973–1981, and subsequent odd-numbered years (Table 2.7)), residential use of coal is estimated by the following steps: a ratio is created of the number of occupied housing units heated by coal to the number of housing units heated by oil; the ratio is multiplied by the Btu quantity of oil used by the residential sector to estimate the Btu quantity of coal used by the residential sector; and the residential sector's share of residential and commercial use is calculated. The 1950 share is applied to 1949, and the other missing years' shares are interpolated.

8 Electricity

High-tension power lines and towers. Source: U.S. Department of Energy.

Diagram 5. Electricity Flow, 2007
(Quadrillion Btu)

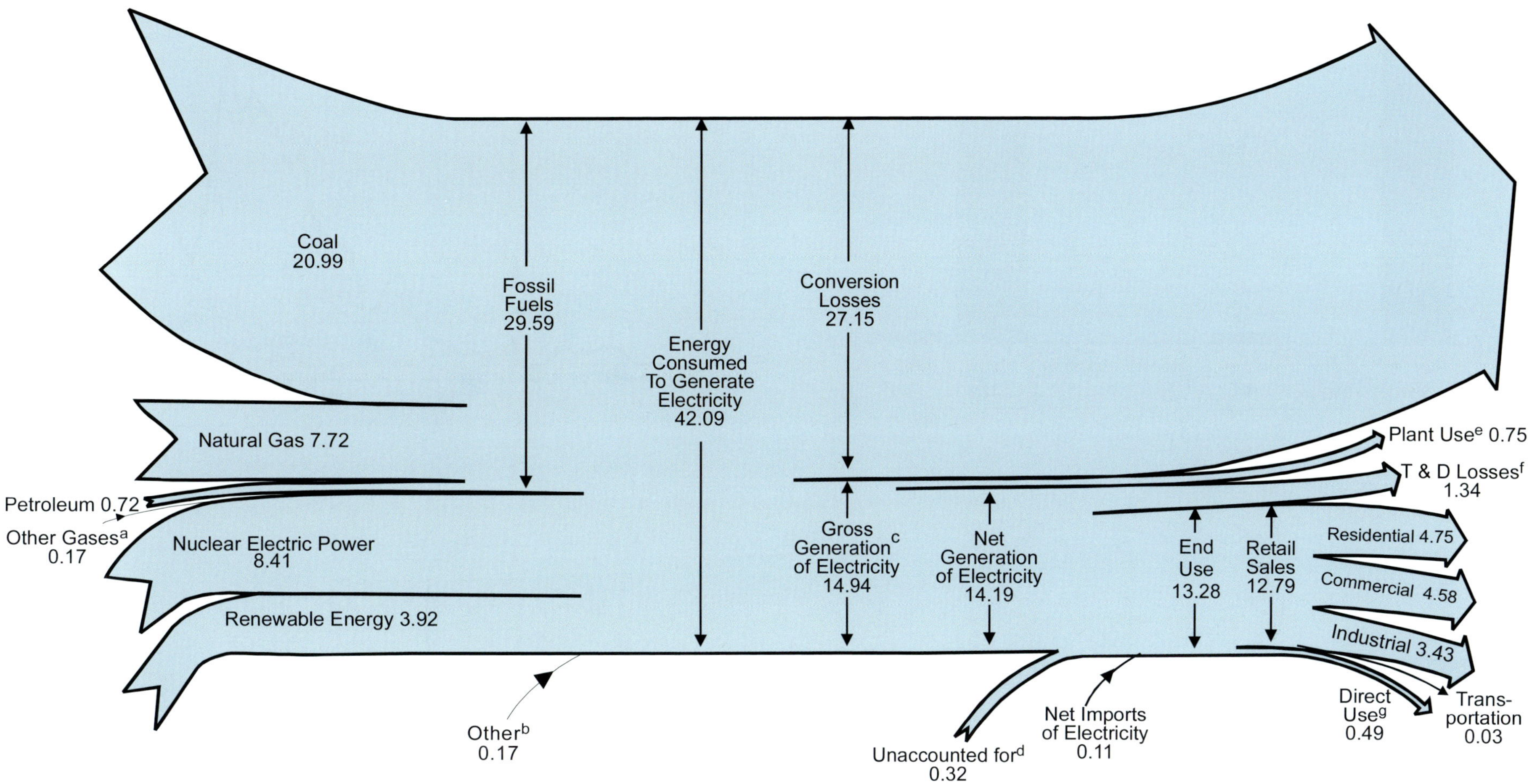

[a] Blast furnace gas, propane gas, and other manufactured and waste gases derived from fossil fuels.

[b] Batteries, chemicals, hydrogen, pitch, purchased steam, sulfur, miscellaneous technologies, and non-renewable waste (municipal solid waste from non-biogenic sources, and tire-derived fuels).

[c] Estimated as net generation divided by 0.95.

[d] Data collection frame differences and nonsampling error. Derived for the diagram by subtracting the "T & D Losses" estimate from "T & D Losses and Unaccounted for" derived from Table 8.1.

[e] Electric energy used in the operation of power plants, estimated as 5 percent of gross generation.

[f] Transmission and distribution losses (electricity losses that occur between the point of generation and delivery to the customer) are estimated as 9 percent of gross generation.

[g] Use of electricity that is 1) self-generated, 2) produced by either the same entity that consumes the power or an affiliate, and 3) used in direct support of a service or industrial process located within the same facility or group of facilities that house the generating equipment. Direct use is exclusive of station use.

Notes: • Data are preliminary. • See Note, "Electrical System Energy Losses," at the end of Section 2. • Values are derived from source data prior to rounding for publication. • Totals may not equal sum of components due to independent rounding.

Sources: Tables 8.1, 8.4a, 8.9, and A6 (column 4).

Figure 8.1 Electricity Overview

Overview, 2007

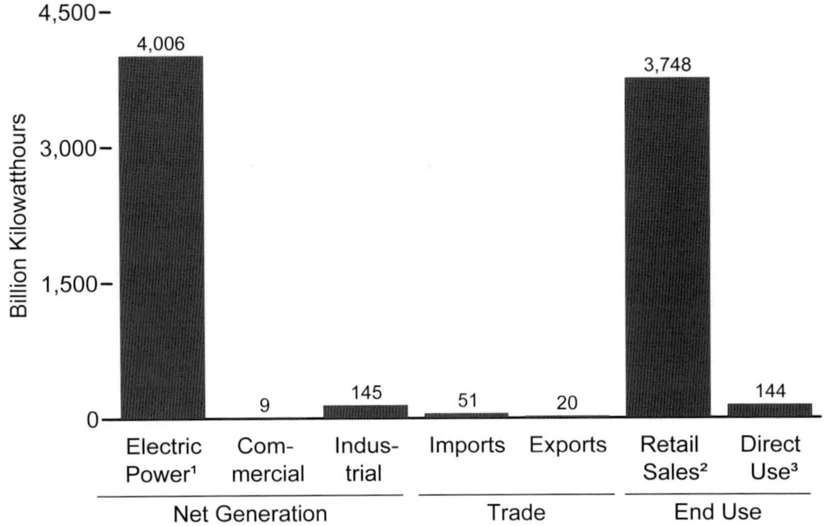

Electricity Trade, 1949-2007

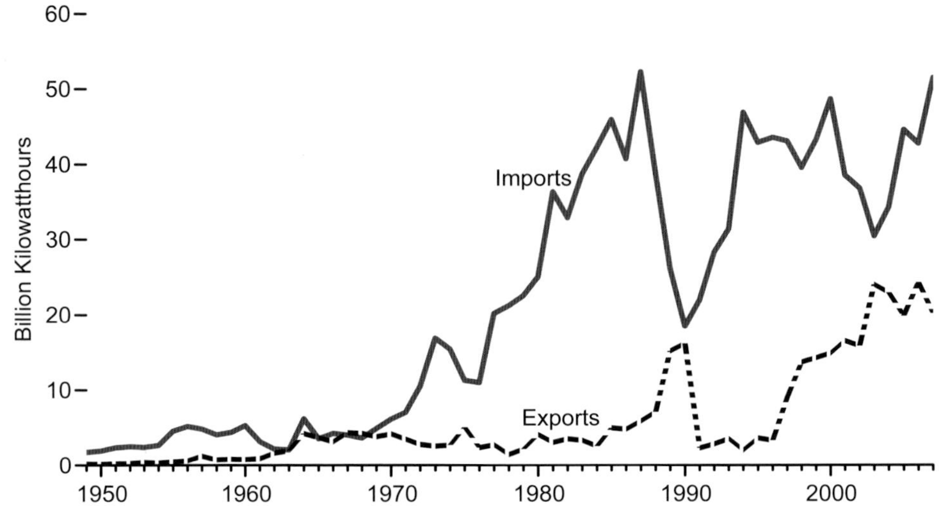

Net-Generation-to-End-Use Flow, 2007

(Billion Kilowatthours)

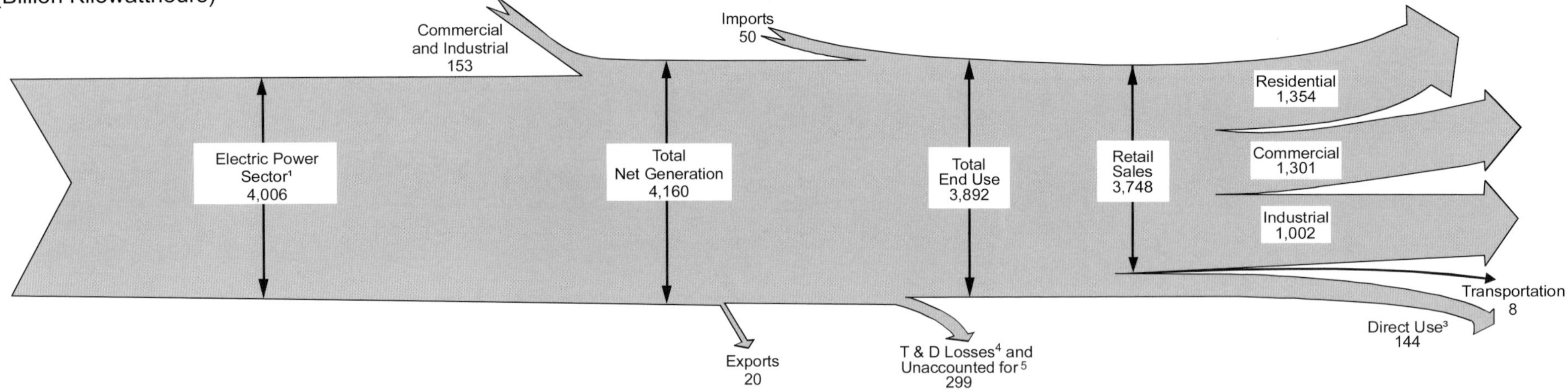

[1] Electricity-only and combined-heat-and-power plants within the NAICS 22 category whose primary business is to sell electricity, or electricity and heat, to the public.
[2] Electricity retail sales to ultimate customers reported by electric utilities and other energy service providers.
[3] See Table 8.1, footnote 8.
[4] Transmission and distribution losses (electricity losses that occur between the point of generation and delivery to the customer). See Note, "Electrical System Energy Losses," at the end of Section 2.
[5] Data collection frame differences and nonsampling error. Note: Because vertical scales differ, graphs should not be compared.
Sources: Tables 8.1 and 8.9.

Table 8.1 Electricity Overview, Selected Years, 1949-2007

(Billion Kilowatthours)

Year	Net Generation				Trade					T & D Losses [5] and Unaccounted for [6]	End Use		
	Electric Power Sector [2]	Commercial Sector [3]	Industrial Sector [4]	Total	Imports [1] From Canada	Imports [1] Total	Exports [1] To Canada	Exports [1] Total	Net Imports [1] Total		Retail Sales [7]	Direct Use [8]	Total
1949	291	NA	5	296	NA	2	NA	(s)	2	43	255	NA	255
1950	329	NA	5	334	NA	2	NA	(s)	2	44	291	NA	291
1955	547	NA	3	550	NA	5	NA	(s)	4	58	497	NA	497
1960	756	NA	4	759	NA	5	NA	1	5	76	688	NA	688
1965	1,055	NA	3	1,058	NA	4	NA	4	(s)	104	954	NA	954
1970	1,532	NA	3	1,535	NA	6	NA	4	2	145	1,392	NA	1,392
1971	1,613	NA	3	1,616	NA	7	NA	4	4	150	1,470	NA	1,470
1972	1,750	NA	3	1,753	NA	10	NA	3	8	166	1,595	NA	1,595
1973	1,861	NA	3	1,864	NA	17	NA	3	14	165	1,713	NA	1,713
1974	1,867	NA	3	1,870	NA	15	NA	3	13	177	1,706	NA	1,706
1975	1,918	NA	3	1,921	NA	11	NA	5	6	180	1,747	NA	1,747
1976	2,038	NA	3	2,041	NA	11	NA	2	9	194	1,855	NA	1,855
1977	2,124	NA	3	2,127	NA	20	NA	3	17	197	1,948	NA	1,948
1978	2,206	NA	3	2,209	NA	21	NA	1	20	211	2,018	NA	2,018
1979	2,247	NA	3	2,251	NA	23	NA	2	20	200	2,071	NA	2,071
1980	2,286	NA	3	2,290	NA	25	NA	4	21	216	2,094	NA	2,094
1981	2,295	NA	3	2,298	NA	36	NA	3	33	184	2,147	NA	2,147
1982	2,241	NA	3	2,244	NA	33	NA	4	29	187	2,086	NA	2,086
1983	2,310	NA	3	2,313	NA	39	NA	3	35	198	2,151	NA	2,151
1984	2,416	NA	3	2,419	NA	42	NA	3	40	173	2,286	NA	2,286
1985	2,470	NA	3	2,473	NA	46	NA	5	41	190	2,324	NA	2,324
1986	2,487	NA	3	2,490	NA	41	NA	5	36	158	2,369	NA	2,369
1987	2,572	NA	3	2,575	NA	52	NA	6	46	164	2,457	NA	2,457
1988	2,704	NA	3	2,707	NA	39	NA	7	32	161	2,578	NA	2,578
1989	[2]2,848	4	[4]115	2,967	NA	26	NA	15	11	223	2,647	109	2,756
1990	2,901	6	131	3,038	16	18	16	16	2	203	2,713	125	2,837
1991	2,936	6	133	3,074	20	22	2	2	20	207	2,762	124	2,886
1992	2,934	6	143	3,084	26	28	2	3	25	212	2,763	134	2,897
1993	3,044	7	146	3,197	29	31	3	4	28	224	2,861	139	3,001
1994	3,089	8	151	3,248	45	47	1	2	45	211	2,935	146	3,081
1995	3,194	8	151	3,353	41	43	2	4	39	229	3,013	151	3,164
1996	3,284	9	151	3,444	42	43	2	3	40	231	3,101	153	3,254
1997	3,329	9	154	3,492	43	43	7	9	34	224	3,146	156	3,302
1998	3,457	9	154	3,620	40	40	12	14	26	221	3,264	161	3,425
1999	3,530	9	156	3,695	43	43	13	14	29	240	3,312	172	3,484
2000	3,638	8	157	3,802	49	49	13	15	34	244	3,421	171	3,592
2001	3,580	7	149	3,737	38	39	16	16	22	202	3,394	163	3,557
2002	3,698	7	153	3,858	37	37	15	16	21	248	3,465	166	3,632
2003	3,721	7	155	3,883	29	30	24	24	6	228	3,494	168	3,662
2004	3,808	8	154	3,971	33	34	22	23	11	266	3,547	168	3,716
2005	3,902	8	145	4,055	43	45	19	20	25	[R]269	3,661	[R]150	[R]3,811
2006	[R]3,908	8	[R]148	[R]4,065	[R]42	[R]43	[R]23	[R]24	18	[R]266	[R]3,670	[R]147	[R]3,817
2007[P]	4,006	9	145	4,160	50	51	20	20	31	299	3,748	[E]144	3,892

[1] Electricity transmitted across U.S. borders. Net imports equal imports minus exports.

[2] Electricity-only and combined-heat-and-power (CHP) plants within the NAICS 22 category whose primary business is to sell electricity, or electricity and heat, to the public. Through 1988, data are for electric utilities only; beginning in 1989, data are for electric utilities and independent power producers.

[3] Commercial combined-heat-and-power (CHP) and commercial electricity-only plants.

[4] Industrial combined-heat-and-power (CHP) and industrial electricity-only plants. Through 1988, data are for industrial hydroelectric power only.

[5] Transmission and distribution losses (electricity losses that occur between the point of generation and delivery to the customer). See Note, "Electrical System Energy Losses," at end of Section 2.

[6] Data collection frame differences and nonsampling error.

[7] Electricity retail sales to ultimate customers by electric utilities and, beginning in 1996, other energy service providers.

[8] Use of electricity that is 1) self-generated, 2) produced by either the same entity that consumes the power or an affiliate, and 3) used in direct support of a service or industrial process located within the same facility or group of facilities that house the generating equipment. Direct use is exclusive of station use.

R=Revised. P=Preliminary. E=Estimate. NA=Not available. (s)=Less than 0.5 billion kilowatthours.

Notes: • See Note 1, "Coverage of Electricity Statistics," and Note 2, "Classification of Power Plants Into Energy-Use Sectors," at end of section. • Totals may not equal sum of components due to independent rounding.

Web Pages: • For all data beginning in 1949, see http://www.eia.doe.gov/emeu/aer/elect.html. • For related information, see http://www.eia.doe.gov/fuelelectric.html.

Sources: See end of section.

Figure 8.2a Electricity Net Generation, Total (All Sectors)

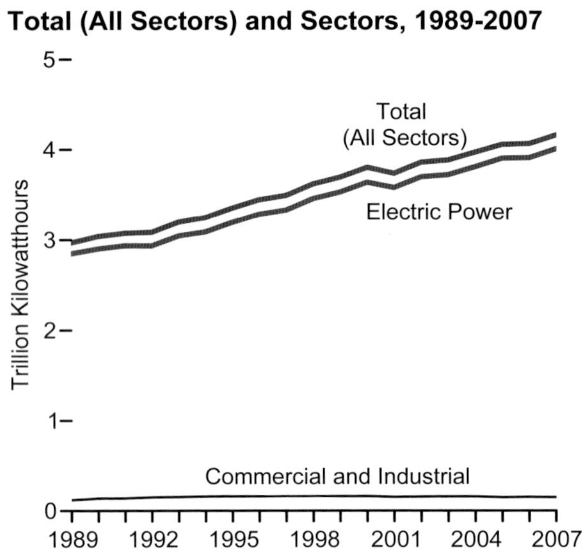

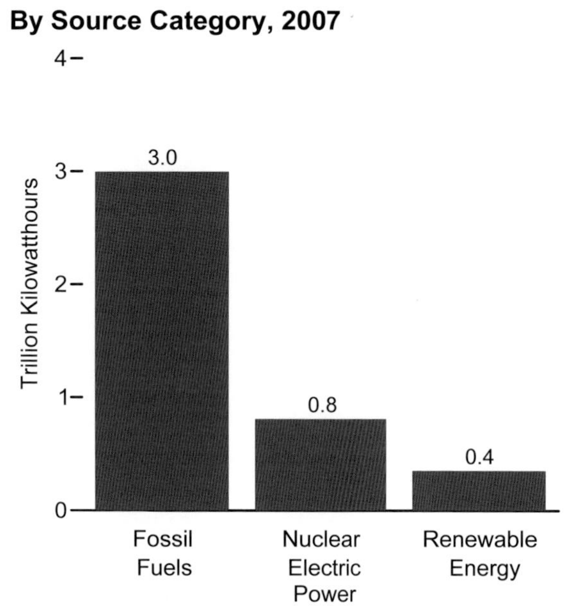

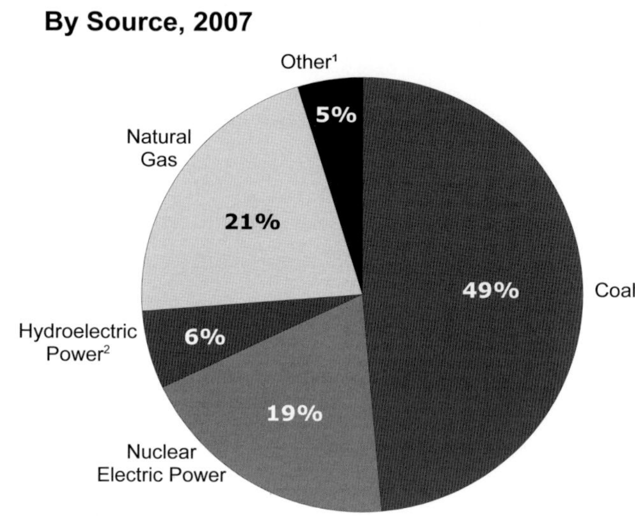

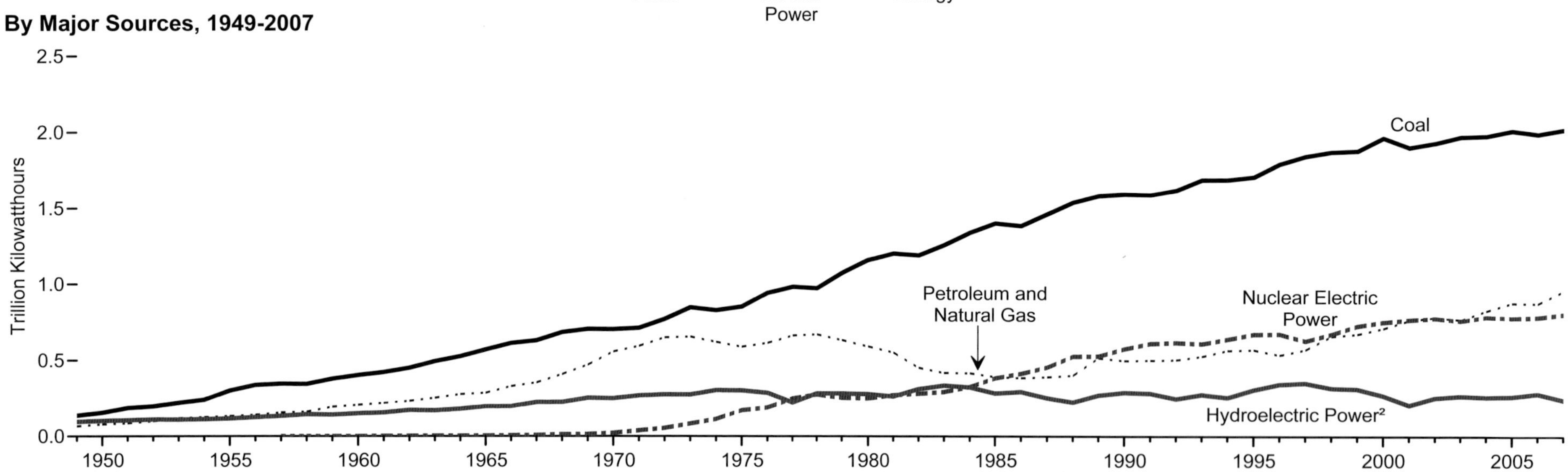

[1] Petroleum, wood, wind, waste, other gases, geothermal, solar, batteries, chemicals, hydrogen, pitch, puchased steam, sulfur, miscellaneous technologies, and non-renewable waste (municipal solid waste from non-biogenic sources, and tire-derived fuels).

[2] Conventional hydroelectric power and pumped storage.

Note: Because vertical scales differ, graphs should not be compared.

Sources: Tables 8.2a, 8.2b, and 8.2d.

Figure 8.2b Electricity Net Generation by Sector

By Sector, 2007

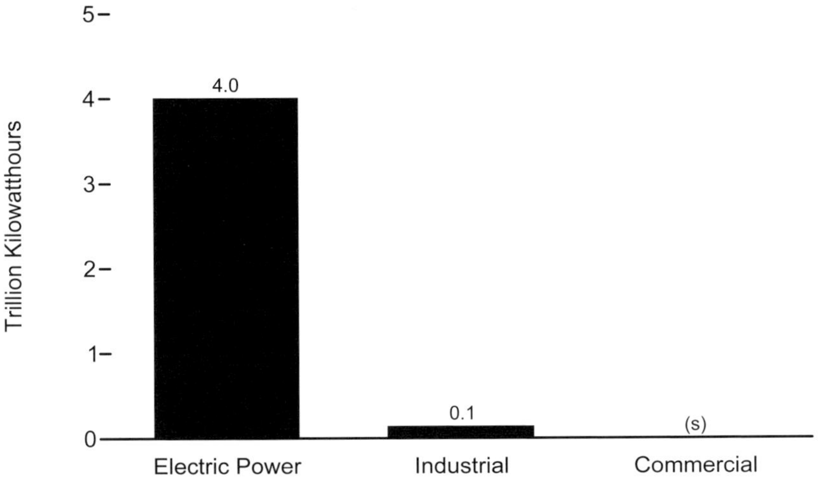

Electric Power Sector by Plant Type, 1989-2007

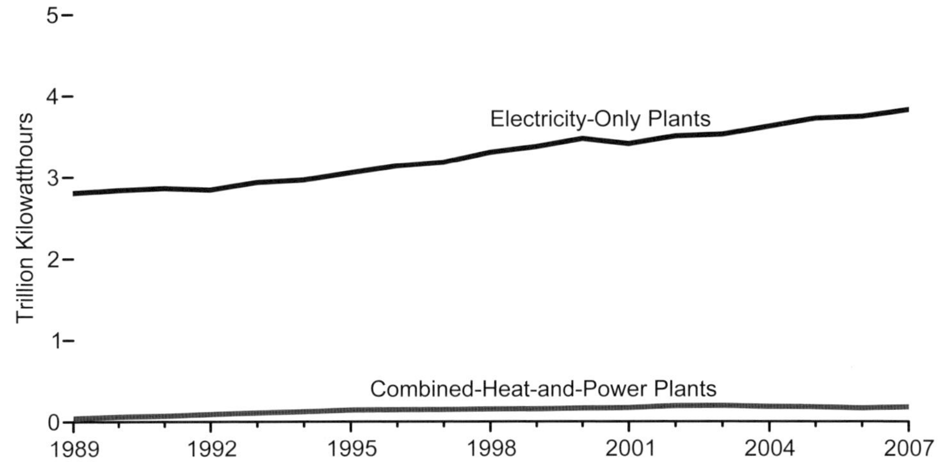

Industrial and Commercial Sectors, 2007

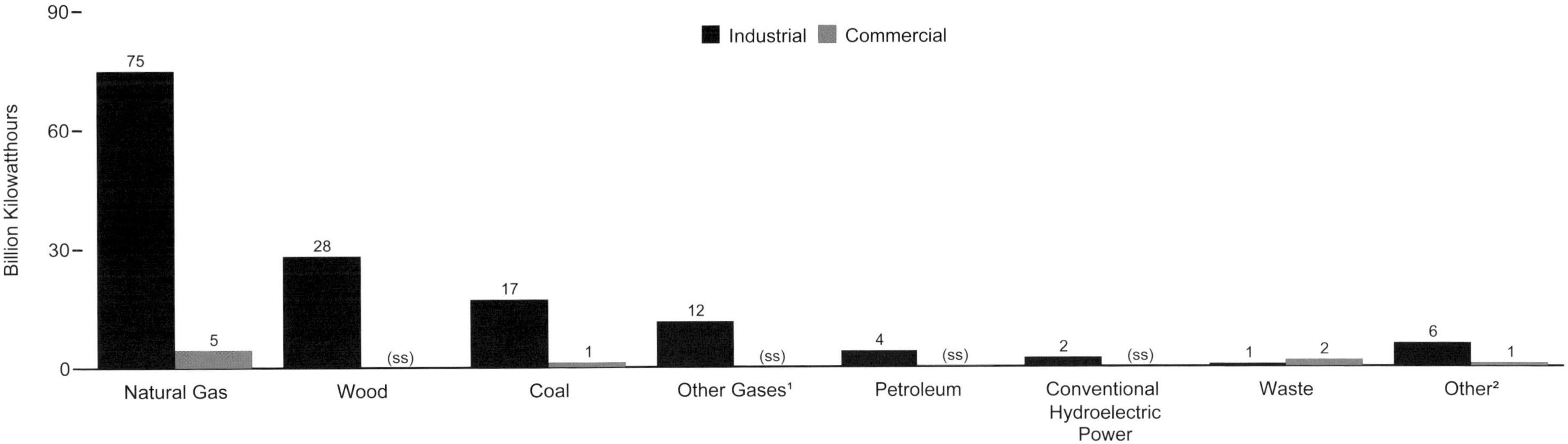

[1] Blast furnace gas, propane gas, and other manufactured and waste gases derived from fossil fuels.

[2] Batteries, chemicals, hydrogen, pitch, purchased steam, sulfur, miscellaneous technologies, and non-renewable waste (municipal solid waste from non-biogenic sources, and tire-derived fuels).

(s) = Less than 0.05 trillion kilowatthours.
(ss) = Less than 0.5 billion kilowatthours.
Note: Because vertical scales differ, graphs should not be compared.
Sources: Tables 8.2b-8.2d.

Table 8.2a Electricity Net Generation: Total (All Sectors), Selected Years, 1949-2007

(Sum of Tables 8.2b and 8.2d; Billion Kilowatthours)

Year	Fossil Fuels					Nuclear Electric Power	Hydro-electric Pumped Storage [5]	Renewable Energy							Other [9]	Total
	Coal [1]	Petroleum [2]	Natural Gas [3]	Other Gases [4]	Total			Conventional Hydroelectric Power	Biomass		Geo-thermal	Solar/PV [8]	Wind	Total		
									Wood [6]	Waste [7]						
1949	135.5	28.5	37.0	NA	201.0	0.0	([10])	94.8	0.4	NA	NA	NA	NA	95.2	NA	296.1
1950	154.5	33.7	44.6	NA	232.8	.0	([10])	100.9	.4	NA	NA	NA	NA	101.3	NA	334.1
1955	301.4	37.1	95.3	NA	433.8	.0	([10])	116.2	.3	NA	NA	NA	NA	116.5	NA	550.3
1960	403.1	48.0	158.0	NA	609.0	.5	([10])	149.4	.1	NA	(s)	NA	NA	149.6	NA	759.2
1965	570.9	64.8	221.6	NA	857.3	3.7	([10])	197.0	.3	NA	.2	NA	NA	197.4	NA	1,058.4
1970	704.4	184.2	372.9	NA	1,261.5	21.8	([10])	251.0	.1	.2	.5	NA	NA	251.8	NA	1,535.1
1971	713.1	220.2	374.0	NA	1,307.4	38.1	([10])	269.5	.1	.2	.5	NA	NA	270.4	NA	1,615.9
1972	771.1	274.3	375.7	NA	1,421.2	54.1	([10])	275.9	.1	.2	1.5	NA	NA	277.7	NA	1,753.0
1973	847.7	314.3	340.9	NA	1,502.9	83.5	([10])	275.4	.1	.2	2.0	NA	NA	277.7	NA	1,864.1
1974	828.4	300.9	320.1	NA	1,449.4	114.0	([10])	304.2	.1	.2	2.5	NA	NA	306.9	NA	1,870.3
1975	852.8	289.1	299.8	NA	1,441.7	172.5	([10])	303.2	(s)	.2	3.2	NA	NA	306.6	NA	1,920.8
1976	944.4	320.0	294.6	NA	1,559.0	191.1	([10])	286.9	.1	.2	3.6	NA	NA	290.8	NA	2,040.9
1977	985.2	358.2	305.5	NA	1,648.9	250.9	([10])	223.6	.3	.2	3.6	NA	NA	227.7	NA	2,127.4
1978	975.7	365.1	305.4	NA	1,646.2	276.4	([10])	283.5	.2	.1	3.0	NA	NA	286.8	NA	2,209.4
1979	1,075.0	303.5	329.5	NA	1,708.0	255.2	([10])	283.1	.3	.2	3.9	NA	NA	287.5	NA	2,250.7
1980	1,161.6	246.0	346.2	NA	1,753.8	251.1	([10])	279.2	.3	.2	5.1	NA	NA	284.7	NA	2,289.6
1981	1,203.2	206.4	345.8	NA	1,755.4	272.7	([10])	263.8	.2	.1	5.7	NA	NA	269.9	NA	2,298.0
1982	1,192.0	146.8	305.3	NA	1,644.1	282.8	([10])	312.4	.2	.1	4.8	NA	NA	317.5	NA	2,244.4
1983	1,259.4	144.5	274.1	NA	1,678.0	293.7	([10])	335.3	.2	.2	6.1	NA	(s)	341.7	NA	2,313.4
1984	1,341.7	119.8	297.4	NA	1,758.9	327.6	([10])	324.3	.5	.4	7.7	(s)	(s)	332.9	NA	2,419.5
1985	1,402.1	100.2	291.9	NA	1,794.3	383.7	([10])	284.3	.7	.6	9.3	(s)	(s)	295.0	NA	2,473.0
1986	1,385.8	136.6	248.5	NA	1,770.9	414.0	([10])	294.0	.5	.7	10.3	(s)	(s)	305.5	NA	2,490.5
1987	1,463.8	118.5	272.6	NA	1,854.9	455.3	([10])	252.9	.8	.7	10.8	(s)	(s)	265.1	NA	2,575.3
1988	1,540.7	148.9	252.8	NA	1,942.4	527.0	([10])	226.1	.9	.7	10.3	(s)	(s)	238.1	NA	2,707.4
1989[11]	1,583.8	164.5	352.6	7.9	2,108.8	529.4	([10])	272.0	27.2	9.2	14.6	.3	2.1	325.3	3.8	2,967.3
1990	1,594.0	126.6	372.8	10.4	2,103.8	576.9	-3.5	292.9	32.5	13.3	15.4	.4	2.8	357.2	3.6	3,038.0
1991	1,590.6	119.8	381.6	11.3	2,103.3	612.6	-4.5	289.0	33.7	15.7	16.0	.5	3.0	357.8	4.7	3,073.8
1992	1,621.2	100.2	404.1	13.3	2,138.7	618.8	-4.2	253.1	36.5	17.8	16.1	.4	2.9	326.9	3.7	3,083.9
1993	1,690.1	112.8	414.9	13.0	2,230.7	610.3	-4.0	280.5	37.6	18.3	16.8	.5	3.0	356.7	3.5	3,197.2
1994	1,690.7	105.9	460.2	13.3	2,270.1	640.4	-3.4	260.1	37.9	19.1	15.5	.5	3.4	336.7	3.7	3,247.5
1995	1,709.4	74.6	496.1	13.9	2,293.9	673.4	-2.7	310.8	36.5	20.4	13.4	.5	3.2	384.8	4.1	3,353.5
1996	1,795.2	81.4	455.1	14.4	2,346.0	674.7	-3.1	347.2	36.8	20.9	14.3	.5	3.2	423.0	3.6	3,444.2
1997	1,845.0	92.6	479.4	13.4	2,430.3	628.6	-4.0	356.5	36.9	21.7	14.7	.5	3.3	433.6	3.6	3,492.2
1998	1,873.5	128.8	531.3	13.5	2,547.1	673.7	-4.5	323.3	36.3	22.4	14.8	.5	3.0	400.4	3.6	3,620.3
1999	1,881.1	118.1	556.4	14.1	2,569.7	728.3	-6.1	319.5	37.0	22.6	14.8	.5	4.5	399.0	4.0	3,694.8
2000	1,966.3	111.2	601.0	14.0	2,692.5	753.9	-5.5	275.6	37.6	23.1	14.1	.5	5.6	356.5	4.8	3,802.1
2001	1,904.0	124.9	639.1	9.0	2,677.0	768.8	-8.8	217.0	35.2	14.5	13.7	.5	6.7	287.7	11.9	3,736.6
2002	1,933.1	94.6	691.0	11.5	2,730.2	780.1	-8.7	264.3	38.7	15.0	14.5	.6	10.4	343.4	13.5	3,858.5
2003	1,973.7	119.4	649.9	15.6	2,758.6	763.7	-8.5	275.8	37.5	15.8	14.4	.5	11.2	355.3	14.0	3,883.2
2004	1,978.6	R120.8	R708.9	16.8	2,825.0	788.5	-8.5	268.4	37.6	15.5	14.8	.6	14.1	351.0	14.5	3,970.6
2005	2,013.2	122.5	758.0	16.3	2,910.0	782.0	-6.6	270.3	38.7	15.5	14.7	.6	17.8	357.5	12.5	4,055.4
2006	R1,990.9	R64.4	R813.0	R16.1	R2,884.4	787.2	R-6.6	R289.2	R38.6	R16.1	R14.6	.5	R26.6	R385.7	R14.0	R4,064.7
2007P	2,020.6	65.7	893.2	15.4	2,994.9	806.5	-7.0	248.3	38.5	16.9	14.8	.6	32.1	351.3	13.8	4,159.5

[1] Anthracite, bituminous coal, subbituminous coal, lignite, waste coal, and coal synfuel.
[2] Distillate fuel oil, residual fuel oil, petroleum coke, jet fuel, kerosene, other petroleum, and waste oil.
[3] Natural gas, plus a small amount of supplemental gaseous fuels.
[4] Blast furnace gas, propane gas, and other manufactured and waste gases derived from fossil fuels.
[5] Pumped storage facility production minus energy used for pumping.
[6] Wood and wood-derived fuels.
[7] Municipal solid waste from biogenic sources, landfill gas, sludge waste, agricultural byproducts, and other biomass. Through 2000, also includes non-renewable waste (municipal solid waste from non-biogenic sources, and tire-derived fuels).
[8] Solar thermal and photovoltaic energy.
[9] Batteries, chemicals, hydrogen, pitch, purchased steam, sulfur, miscellaneous technologies, and, beginning in 2001, non-renewable waste (municipal solid waste from non-biogenic sources, and tire-derived fuels).
[10] Included in "Conventional Hydroelectric Power."
[11] Through 1988, all data except hydroelectric are for electric utilities only; hydroelectric data through 1988 include industrial plants as well as electric utilities. Beginning in 1989, data are for electric utilities, independent power producers, commercial plants, and industrial plants.

R=Revised. P=Preliminary. NA=Not available. (s)=Less than 0.05 billion kilowatthours.

Notes: • See Note 1, "Coverage of Electricity Statistics," at end of section. • Totals may not equal sum of components due to independent rounding.

Web Pages: • For all data beginning in 1949, see http://www.eia.doe.gov/emeu/aer/elect.html. • For related information, see http://www.eia.doe.gov/fuelelectric.html.

Sources: • 1949-1988—Table 8.2b for electric power sector, and Table 8.1 for industrial sector. • 1989 forward—Tables 8.2b and 8.2d.

Table 8.2b Electricity Net Generation: Electric Power Sector, Selected Years, 1949-2007

(Subset of Table 8.2a; Billion Kilowatthours)

	Fossil Fuels							Renewable Energy								
									Biomass							
Year	Coal [1]	Petroleum [2]	Natural Gas [3]	Other Gases [4]	Total	Nuclear Electric Power	Hydro-electric Pumped Storage [5]	Conventional Hydroelectric Power	Wood [6]	Waste [7]	Geo-thermal	Solar/PV [8]	Wind	Total	Other [9]	Total
1949	135.5	28.5	37.0	NA	201.0	0.0	([10])	89.7	0.4	NA	NA	NA	NA	90.1	NA	291.1
1950	154.5	33.7	44.6	NA	232.8	.0	([10])	95.9	.4	NA	NA	NA	NA	96.3	NA	329.1
1955	301.4	37.1	95.3	NA	433.8	.0	([10])	113.0	.3	NA	NA	NA	NA	113.3	NA	547.0
1960	403.1	48.0	158.0	NA	609.0	.5	([10])	145.8	.1	NA	(s)	NA	NA	146.0	NA	755.5
1965	570.9	64.8	221.6	NA	857.3	3.7	([10])	193.9	.3	NA	.2	NA	NA	194.3	NA	1,055.3
1970	704.4	184.2	372.9	NA	1,261.5	21.8	([10])	247.7	.1	.2	.5	NA	NA	248.6	NA	1,531.9
1971	713.1	220.2	374.0	NA	1,307.4	38.1	([10])	266.3	.1	.2	.5	NA	NA	267.2	NA	1,612.6
1972	771.1	274.3	375.7	NA	1,421.2	54.1	([10])	272.6	.1	.2	1.5	NA	NA	274.4	NA	1,749.7
1973	847.7	314.3	340.9	NA	1,502.9	83.5	([10])	272.1	.1	.2	2.0	NA	NA	274.4	NA	1,860.7
1974	828.4	300.9	320.1	NA	1,449.4	114.0	([10])	301.0	.1	.2	2.5	NA	NA	303.7	NA	1,867.1
1975	852.8	289.1	299.8	NA	1,441.7	172.5	([10])	300.0	(s)	.2	3.2	NA	NA	303.5	NA	1,917.6
1976	944.4	320.0	294.6	NA	1,559.0	191.1	([10])	283.7	.1	.2	3.6	NA	NA	287.6	NA	2,037.7
1977	985.2	358.2	305.5	NA	1,648.9	250.9	([10])	220.5	.3	.2	3.6	NA	NA	224.5	NA	2,124.3
1978	975.7	365.1	305.4	NA	1,646.2	276.4	([10])	280.4	.2	.1	3.0	NA	NA	283.7	NA	2,206.3
1979	1,075.0	303.5	329.5	NA	1,708.0	255.2	([10])	279.8	.3	.2	3.9	NA	NA	284.2	NA	2,247.4
1980	1,161.6	246.0	346.2	NA	1,753.8	251.1	([10])	276.0	.3	.2	5.1	NA	NA	281.5	NA	2,286.4
1981	1,203.2	206.4	345.8	NA	1,755.4	272.7	([10])	260.7	.2	.1	5.7	NA	NA	266.7	NA	2,294.8
1982	1,192.0	146.8	305.3	NA	1,644.1	282.8	([10])	309.2	.2	.1	4.8	NA	NA	314.4	NA	2,241.2
1983	1,259.4	144.5	274.1	NA	1,678.0	293.7	([10])	332.1	.2	.2	6.1	NA	(s)	338.6	NA	2,310.3
1984	1,341.7	119.8	297.4	NA	1,758.9	327.6	([10])	321.2	.5	.4	7.7	(s)	(s)	329.8	NA	2,416.3
1985	1,402.1	100.2	291.9	NA	1,794.3	383.7	([10])	281.1	.7	.6	9.3	(s)	(s)	291.9	NA	2,469.8
1986	1,385.8	136.6	248.5	NA	1,770.9	414.0	([10])	290.8	.5	.7	10.3	(s)	(s)	302.3	NA	2,487.3
1987	1,463.8	118.5	272.6	NA	1,854.9	455.3	([10])	249.7	.8	.7	10.8	(s)	(s)	262.0	NA	2,572.1
1988	1,540.7	148.9	252.8	NA	1,942.4	527.0	([10])	222.9	.9	.7	10.3	(s)	(s)	234.9	NA	2,704.3
1989[11]	1,562.4	159.0	297.3	.5	2,019.1	529.4	([10])	269.2	5.6	7.7	14.6	.3	2.1	299.5	.3	2,848.2
1990	1,572.1	118.9	309.5	.6	2,001.1	576.9	-3.5	289.8	7.0	11.5	15.4	.4	2.8	326.9	(s)	2,901.3
1991	1,568.8	112.8	317.8	.7	2,000.1	612.6	-4.5	286.0	7.7	13.9	16.0	.5	3.0	327.0	.4	2,935.6
1992	1,597.7	92.2	334.3	1.2	2,025.4	618.8	-4.2	250.0	8.5	15.9	16.1	.4	2.9	293.9	.5	2,934.4
1993	1,665.5	105.4	342.2	1.0	2,114.1	610.3	-4.0	277.5	9.2	16.2	16.8	.5	3.0	323.2	.4	3,043.9
1994	1,666.3	98.7	385.7	1.1	2,151.7	640.4	-3.4	254.0	9.2	17.0	15.5	.5	3.4	299.7	.2	3,088.7
1995	1,686.1	68.1	419.2	1.9	2,175.3	673.4	-2.7	305.4	7.6	18.0	13.4	.5	3.2	348.0	.2	3,194.2
1996	1,772.0	74.8	378.8	1.3	2,226.9	674.7	-3.1	341.2	8.4	17.8	14.3	.5	3.2	385.4	.2	3,284.1
1997	1,820.8	86.5	399.6	1.5	2,308.4	628.6	-4.0	350.6	8.7	18.5	14.7	.5	3.3	396.3	.1	3,329.4
1998	1,850.2	122.2	449.3	2.3	2,424.0	673.7	-4.5	317.9	8.6	19.2	14.8	.5	3.0	364.0	.2	3,457.4
1999	1,858.6	111.5	473.0	1.6	2,444.8	728.3	-6.1	314.7	9.0	19.5	14.8	.5	4.5	362.9	.1	3,530.0
2000	1,943.1	105.2	518.0	2.0	2,568.3	753.9	-5.5	271.3	8.9	20.3	14.1	.5	5.6	320.7	.1	3,637.5
2001	1,882.8	119.1	554.9	.6	2,557.5	768.8	-8.8	213.7	8.3	12.9	13.7	.5	6.7	256.0	6.5	3,580.1
2002	1,910.6	89.7	607.7	2.0	2,610.0	780.1	-8.7	260.5	9.0	13.1	14.5	.6	10.4	308.0	9.1	3,698.5
2003	1,952.7	113.7	567.3	2.6	2,636.4	763.7	-8.5	271.5	9.5	13.8	14.4	.5	11.2	321.0	8.6	3,721.2
2004	1,957.2	R114.7	R627.4	3.0	2,702.3	788.5	-8.5	265.1	9.7	13.1	14.8	.6	14.1	317.5	8.6	3,808.4
2005	1,992.1	116.8	683.3	4.0	2,796.1	782.0	-6.6	267.0	10.6	13.0	14.7	.6	17.8	323.7	7.0	3,902.2
2006	R1,969.8	R59.9	R734.4	R3.9	R2,768.1	787.2	R-6.6	R286.3	R10.3	R13.9	R14.6	.5	R26.6	R352.2	R7.1	R3,908.1
2007P	2,002.1	61.4	813.8	3.9	2,881.3	806.5	-7.0	246.0	10.4	14.6	14.8	.6	32.1	318.6	7.1	4,006.5

[1] Anthracite, bituminous coal, subbituminous coal, lignite, waste coal, and coal synfuel.

[2] Distillate fuel oil, residual fuel oil, petroleum coke, jet fuel, kerosene, other petroleum, and waste oil.

[3] Natural gas, plus a small amount of supplemental gaseous fuels.

[4] Blast furnace gas, propane gas, and other manufactured and waste gases derived from fossil fuels.

[5] Pumped storage facility production minus energy used for pumping.

[6] Wood and wood-derived fuels.

[7] Municipal solid waste from biogenic sources, landfill gas, sludge waste, agricultural byproducts, and other biomass. Through 2000, also includes non-renewable waste (municipal solid waste from non-biogenic sources, and tire-derived fuels).

[8] Solar thermal and photovoltaic energy.

[9] Batteries, chemicals, hydrogen, pitch, purchased steam, sulfur, miscellaneous technologies, and, beginning in 2001, non-renewable waste (municipal solid waste from non-biogenic sources, and tire-derived fuels).

[10] Included in "Conventional Hydroelectric Power."

[11] Through 1988, data are for electric utilities only. Beginning in 1989, data are for electric utilities and independent power producers.

R=Revised. P=Preliminary. NA=Not available. (s)=Less than 0.05 billion kilowatthours.

Notes: • The electric power sector comprises electricity-only and combined-heat-and-power (CHP) plants within the NAICS 22 category whose primary business is to sell electricity, or electricity and heat, to the public. • See Table 8.2d for commercial and industrial CHP and electricity-only data. • See Note 1, "Coverage of Electricity Statistics," and Note 2, "Classification of Power Plants Into Energy-Use Sectors," at end of section. • Totals may not equal sum of components due to independent rounding.

Web Pages: • For all data beginning in 1949, see http://www.eia.doe.gov/emeu/aer/elect.html. • For related information, see http://www.eia.doe.gov/fuelelectric.html.

Sources: • 1949-September 1977—Federal Power Commission, Form FPC-4, "Monthly Power Plant Report." • October 1977-1981—Federal Energy Regulatory Commission, Form FPC-4, "Monthly Power Plant Report." • 1982-1988—Energy Information Administration (EIA), Form EIA-759, "Monthly Power Plant Report." • 1989-1997—EIA, Form EIA-759, "Monthly Power Plant Report," and Form EIA-867, "Annual Nonutility Power Producer Report." • 1998-2000—EIA, Form EIA-759, "Monthly Power Plant Report," and Form EIA-860B, "Annual Electric Generator Report—Nonutility." • 2001-2003—EIA, Form EIA-906, "Power Plant Report." • 2004 forward—EIA, Form EIA-906, "Power Plant Report," and Form EIA-920, "Combined Heat and Power Plant Report."

Table 8.2c Electricity Net Generation: Electric Power Sector by Plant Type, 1989-2007

(Breakout of Table 8.2b; Billion Kilowatthours)

Year	Fossil Fuels					Nuclear Electric Power	Hydro-electric Pumped Storage[5]	Renewable Energy							Other[9]	Total
	Coal[1]	Petroleum[2]	Natural Gas[3]	Other Gases[4]	Total			Conventional Hydroelectric Power	Biomass: Wood[6]	Biomass: Waste[7]	Geo-thermal	Solar/PV[8]	Wind	Total		
Electricity-Only Plants[10]																
1989	1,554.0	158.3	266.9	–	1,979.3	529.4	([11])	269.2	4.2	6.9	14.6	0.3	2.1	297.3	–	2,805.9
1990	1,560.2	117.6	264.7	(s)	1,942.4	576.9	-3.5	289.8	5.6	10.4	15.4	.4	2.8	324.3	–	2,840.0
1991	1,551.9	112.2	267.8	(s)	1,931.9	612.6	-4.5	286.0	6.0	12.2	16.0	.5	3.0	323.7	–	2,863.6
1992	1,577.1	90.1	270.9	(s)	1,938.0	618.8	-4.2	250.0	6.6	14.4	16.1	.4	2.9	290.4	–	2,843.1
1993	1,642.1	100.6	267.2	(s)	2,009.9	610.3	-4.0	277.5	7.2	14.9	16.8	.5	3.0	319.8	–	2,935.9
1994	1,639.9	92.1	299.7	(s)	2,031.7	640.4	-3.4	254.0	7.6	15.4	15.5	.5	3.4	296.5	–	2,965.2
1995	1,658.0	62.0	317.4	(s)	2,037.4	673.4	-2.7	305.4	5.9	16.3	13.4	.5	3.2	344.7	–	3,052.8
1996	1,742.8	68.5	272.8	(s)	2,084.1	674.7	-3.1	341.2	6.5	16.1	14.3	.5	3.2	381.8	–	3,137.6
1997	1,793.2	80.3	291.1	(s)	2,164.6	628.6	-4.0	350.6	6.5	16.4	14.7	.5	3.3	392.0	–	3,181.3
1998	1,823.0	115.7	335.9	.1	2,274.6	673.7	-4.5	317.9	6.6	17.0	14.8	.5	3.0	359.8	–	3,303.6
1999	1,832.1	104.8	356.6	(s)	2,293.6	728.3	-6.1	314.7	7.3	17.1	14.8	.5	4.5	358.8	–	3,374.6
2000	1,910.6	98.0	399.4	.2	2,408.2	753.9	-5.5	271.3	7.3	17.6	14.1	.5	5.6	316.4	–	3,472.9
2001	1,851.8	113.2	427.0	(s)	2,392.0	768.8	-8.8	213.7	6.6	11.3	13.7	.5	6.7	252.6	5.9	3,410.5
2002	1,881.2	83.3	456.8	.2	2,421.5	780.1	-8.7	260.5	7.3	11.2	14.5	.6	10.4	304.3	7.6	3,504.8
2003	1,915.8	108.5	421.2	.3	2,445.7	763.7	-8.5	271.5	7.4	11.9	14.4	.5	11.2	317.0	7.6	3,525.5
2004	1,921.1	109.4	491.2	.4	2,522.0	788.5	-8.5	265.1	8.1	11.8	14.8	.6	14.1	314.5	7.6	3,624.1
2005	1,955.5	111.2	553.2	(s)	2,619.9	782.0	-6.6	267.0	8.5	11.7	14.7	.6	17.8	320.3	6.2	3,721.8
2006	[R]1,933.7	[R]55.2	[R]618.0	(s)	[R]2,607.0	787.2	[R]-6.6	[R]286.2	[R]8.3	[R]12.5	[R]14.6	.5	[R]26.6	[R]348.7	[R]6.3	[R]3,742.7
2007[P]	1,965.1	56.6	684.8	.1	2,706.6	806.5	-7.0	246.0	8.4	13.1	14.8	.6	32.1	315.1	6.3	3,827.5
Combined-Heat-and-Power Plants[12]																
1989	8.4	0.7	30.4	0.5	39.9	–	–	–	1.3	0.9	–	–	–	2.2	0.3	42.3
1990	11.9	1.3	44.8	.6	58.7	–	–	–	1.4	1.1	–	–	–	2.6	(s)	61.3
1991	16.9	.6	50.0	.7	68.2	–	–	–	1.7	1.6	–	–	–	3.3	.4	71.9
1992	20.7	2.2	63.4	1.2	87.4	–	–	–	1.9	1.5	–	–	–	3.4	.5	91.3
1993	23.4	4.8	75.0	1.0	104.2	–	–	–	2.0	1.4	–	–	–	3.4	.4	108.0
1994	26.4	6.6	86.0	1.1	120.1	–	–	–	1.6	1.6	–	–	–	3.2	.2	123.5
1995	28.1	6.1	101.7	1.9	137.9	–	–	–	1.7	1.7	–	–	–	3.4	.2	141.5
1996	29.2	6.3	105.9	1.3	142.7	–	–	–	1.9	1.7	–	–	–	3.6	.2	146.6
1997	27.6	6.2	108.5	1.5	143.7	–	–	–	2.2	2.1	–	–	–	4.3	.1	148.1
1998	27.2	6.6	113.4	2.3	149.4	–	–	–	2.0	2.3	–	–	–	4.2	.2	153.8
1999	26.6	6.7	116.4	1.6	151.2	–	–	–	1.7	2.4	–	–	–	4.1	.1	155.4
2000	32.5	7.2	118.6	1.8	160.2	–	–	–	1.6	2.7	–	–	–	4.3	.1	164.6
2001	31.0	6.0	128.0	.6	165.5	–	–	–	1.7	1.7	–	–	–	3.4	.6	169.5
2002	29.4	6.5	150.9	1.7	188.5	–	–	–	1.7	2.0	–	–	–	3.7	1.4	193.7
2003	36.9	5.2	146.1	2.4	190.6	–	–	–	2.1	1.9	–	–	–	4.0	1.1	195.7
2004	36.1	[R]5.3	[R]136.2	2.6	180.3	–	–	–	1.6	1.4	–	–	–	3.0	1.0	184.3
2005	36.5	5.6	130.1	3.9	176.2	–	–	(s)	2.1	1.3	–	–	–	3.4	.7	180.4
2006	[R]36.1	[R]4.7	[R]116.5	[R]3.9	[R]161.1	–	–	(s)	[R]2.0	1.4	–	–	–	3.5	.8	[R]165.4
2007[P]	37.1	4.8	129.0	3.8	174.7	–	–	(s)	2.0	1.5	–	–	–	3.5	.8	179.0

[1] Anthracite, bituminous coal, subbituminous coal, lignite, waste coal, and coal synfuel.
[2] Distillate fuel oil, residual fuel oil, petroleum coke, jet fuel, kerosene, other petroleum, and waste oil.
[3] Natural gas, plus a small amount of supplemental gaseous fuels.
[4] Blast furnace gas, propane gas, and other manufactured and waste gases derived from fossil fuels.
[5] Pumped storage facility production minus energy used for pumping.
[6] Wood and wood-derived fuels.
[7] Municipal solid waste from biogenic sources, landfill gas, sludge waste, agricultural byproducts, and other biomass. Through 2000, also includes non-renewable waste (municipal solid waste from non-biogenic sources, and tire-derived fuels).
[8] Solar thermal and photovoltaic energy.
[9] Batteries, chemicals, hydrogen, pitch, purchased steam, sulfur, miscellaneous technologies, and, beginning in 2001, non-renewable waste (municipal solid waste from non-biogenic sources, and tire-derived fuels).
[10] Electricity-only plants within the NAICS 22 category whose primary business is to sell electricity to the public. Data also include a small number of electric utility combined-heat-and-power (CHP) plants.
[11] Included in "Conventional Hydroelectric Power."
[12] Combined-heat-and-power (CHP) plants within the NAICS 22 category whose primary business is to sell electricity and heat to the public. Data do not include electric utility CHP plants—these are included under "Electricity-Only Plants."

R=Revised. P=Preliminary. – = No data reported. (s)=Less than 0.05 billion kilowatthours.

Notes: • See Table 8.2d for commercial and industrial CHP and electricity-only data. • See Note 1, "Coverage of Electricity Statistics," and Note 2, "Classification of Power Plants Into Energy-Use Sectors," at end of section. • Totals may not equal sum of components due to independent rounding.

Web Page: For related information, see http://www.eia.doe.gov/fuelelectric.html.

Sources: • 1989-1997—Energy Information Administration (EIA), Form EIA-759, "Monthly Power Plant Report," and Form EIA-867, "Annual Nonutility Power Producer Report." • 1998-2000—EIA, Form EIA-759, "Monthly Power Plant Report," and Form EIA-860B, "Annual Electric Generator Report—Nonutility." • 2001-2003—EIA, Form EIA-906, "Power Plant Report." • 2004 forward—EIA, Form EIA-906, "Power Plant Report," and Form EIA-920, "Combined Heat and Power Plant Report."

Table 8.2d Electricity Net Generation: Commercial and Industrial Sectors, 1989-2007

(Subset of Table 8.2a; Billion Kilowatthours)

Year	Fossil Fuels					Nuclear Electric Power	Hydro-electric Pumped Storage [5]	Renewable Energy							Other [9]	Total
	Coal [1]	Petroleum [2]	Natural Gas [3]	Other Gases [4]	Total			Conventional Hydroelectric Power	Biomass: Wood [6]	Biomass: Waste [7]	Geo-thermal	Solar/PV [8]	Wind	Total		
Commercial Sector [10]																
1989	0.7	0.6	2.2	0.1	3.6	–	–	0.1	0.1	0.5	–	–	–	0.7	–	4.3
1990	.8	.6	3.3	.1	4.8	–	–	.1	.1	.8	–	–	–	1.1	–	5.8
1991	.8	.4	3.2	.1	4.5	–	–	.1	.1	.9	–	–	–	1.1	(s)	5.7
1992	.7	.3	3.9	.1	5.0	–	–	.1	.1	1.0	–	–	–	1.2	(s)	6.2
1993	.9	.3	4.5	.1	5.8	–	–	.1	.1	1.0	–	–	–	1.2	(s)	7.0
1994	.8	.4	4.9	.1	6.3	–	–	.1	.1	1.2	–	–	–	1.3	–	7.6
1995	1.0	.4	5.2	–	6.5	–	–	.1	.1	1.5	–	–	–	1.7	(s)	8.2
1996	1.1	.4	5.2	(s)	6.7	–	–	.1	.1	2.2	–	–	–	2.4	(s)	9.0
1997	1.0	.4	4.7	(s)	6.2	–	–	.1	(s)	2.3	–	–	–	2.5	(s)	8.7
1998	1.0	.4	4.9	(s)	6.3	–	–	.1	(s)	2.3	–	–	–	2.5	–	8.7
1999	1.0	.4	4.6	(s)	6.0	–	–	.1	(s)	2.4	–	–	–	2.5	(s)	8.6
2000	1.1	.4	4.3	(s)	5.8	–	–	.1	(s)	2.0	–	–	–	2.1	(s)	7.9
2001	1.0	.4	4.4	(s)	5.9	–	–	.1	(s)	1.0	–	–	–	1.1	.5	7.4
2002	1.0	.4	4.3	(s)	5.7	–	–	(s)	(s)	1.1	–	–	–	1.1	.6	7.4
2003	1.2	.4	3.9	–	5.5	–	–	.1	(s)	1.3	–	–	–	1.4	.6	7.5
2004	1.3	.5	4.1	–	5.8	–	–	.1	(s)	1.5	–	–	–	1.6	.8	8.3
2005	1.3	.4	4.3	–	6.0	–	–	.1	(s)	1.7	–	–	–	1.8	.8	8.5
2006	1.3	.2	4.3	R(s)	R5.9	–	–	.1	(s)	R1.6	–	–	–	R1.7	.8	8.4
2007P	1.3	.2	4.5	(s)	6.0	–	–	.1	(s)	1.6	–	–	–	1.7	.8	8.5
Industrial Sector [11]																
1989	20.7	5.0	53.2	7.3	86.1	–	–	2.7	21.6	0.9	–	–	–	25.2	3.5	114.8
1990	21.1	7.2	60.0	9.6	97.9	–	–	3.0	25.4	.9	–	–	–	29.3	3.6	130.8
1991	21.0	6.5	60.6	10.5	98.6	–	–	2.8	25.9	.9	–	–	–	29.6	4.3	132.6
1992	22.7	7.6	65.9	12.0	108.2	–	–	2.9	27.9	.9	–	–	–	31.8	3.2	143.3
1993	23.7	7.0	68.2	11.9	110.9	–	–	2.9	28.4	1.1	–	–	–	32.3	3.1	146.3
1994	23.6	6.8	69.6	12.1	112.1	–	–	6.0	28.7	1.0	–	–	–	35.7	3.4	151.2
1995	22.4	6.0	71.7	11.9	112.1	–	–	5.3	28.9	.9	–	–	–	35.1	3.9	151.0
1996	22.2	6.3	71.0	13.0	112.5	–	–	5.9	28.4	.9	–	–	–	35.2	3.4	151.0
1997	23.2	5.6	75.1	11.8	115.8	–	–	5.7	28.2	.9	–	–	–	34.8	3.5	154.1
1998	22.3	6.2	77.1	11.2	116.8	–	–	5.3	27.7	.9	–	–	–	33.9	3.4	154.1
1999	21.5	6.1	78.8	12.5	118.9	–	–	4.8	28.1	.7	–	–	–	33.5	3.9	156.3
2000	22.1	5.6	78.8	11.9	118.4	–	–	4.1	28.7	.8	–	–	–	33.6	4.7	156.7
2001	20.1	5.3	79.8	8.5	113.6	–	–	3.1	26.9	.6	–	–	–	30.6	4.9	149.2
2002	21.5	4.4	79.0	9.5	114.4	–	–	3.8	29.6	.8	–	–	–	34.3	3.8	152.6
2003	19.8	5.3	78.7	13.0	116.8	–	–	4.2	28.0	.7	–	–	–	32.9	4.8	154.5
2004	20.1	5.6	77.4	13.7	116.9	–	–	3.2	27.8	.8	–	–	–	31.9	5.1	153.9
2005	19.8	5.4	70.4	12.4	107.9	–	–	3.2	28.1	.8	–	–	–	32.1	4.8	144.7
2006	19.9	R4.2	R74.3	R12.1	R110.4	–	–	R2.9	R28.3	R.6	–	–	–	R31.8	R6.0	R148.3
2007P	17.1	4.1	74.9	11.5	107.6	–	–	2.3	28.1	.6	–	–	–	31.0	5.9	144.5

[1] Anthracite, bituminous coal, subbituminous coal, lignite, waste coal, and coal synfuel.
[2] Distillate fuel oil, residual fuel oil, petroleum coke, jet fuel, kerosene, other petroleum, and waste oil.
[3] Natural gas, plus a small amount of supplemental gaseous fuels.
[4] Blast furnace gas, propane gas, and other manufactured and waste gases derived from fossil fuels.
[5] Pumped storage facility production minus energy used for pumping.
[6] Wood and wood-derived fuels.
[7] Municipal solid waste from biogenic sources, landfill gas, sludge waste, agricultural byproducts, and other biomass. Through 2000, also includes non-renewable waste (municipal solid waste from non-biogenic sources, and tire-derived fuels).
[8] Solar thermal and photovoltaic energy.
[9] Batteries, chemicals, hydrogen, pitch, purchased steam, sulfur, miscellaneous technologies, and, beginning in 2001, non-renewable waste (municipal solid waste from non-biogenic sources, and tire-derived fuels).
[10] Commercial combined-heat-and-power (CHP) and commercial electricity-only plants.
[11] Industrial combined-heat-and-power (CHP) and industrial electricity-only plants.

R=Revised. P=Preliminary. – = No data reported. (s)=Less than 0.05 billion kilowatthours.

Notes: • See Tables 8.2b and 8.2c for electric power sector electricity-only and CHP data. • See Note 1, "Coverage of Electricity Statistics," and Note 2, "Classification of Power Plants Into Energy-Use Sectors," at end of section. • Totals may not equal sum of components due to independent rounding.

Web Page: For related information, see http://www.eia.doe.gov/fuelelectric.html.

Sources: • 1989-1997—Energy Information Administration (EIA), Form EIA-867, "Annual Nonutility Power Producer Report." • 1998-2000—EIA, Form EIA-860B, "Annual Electric Generator Report—Nonutility." • 2001-2003—EIA, Form EIA-906, "Power Plant Report." • 2004 forward—EIA, Form EIA-906, "Power Plant Report," and Form EIA-920, "Combined Heat and Power Plant Report."

Figure 8.3 Useful Thermal Output at Combined-Heat-and-Power Plants

Total (All Sectors), 1989-2007

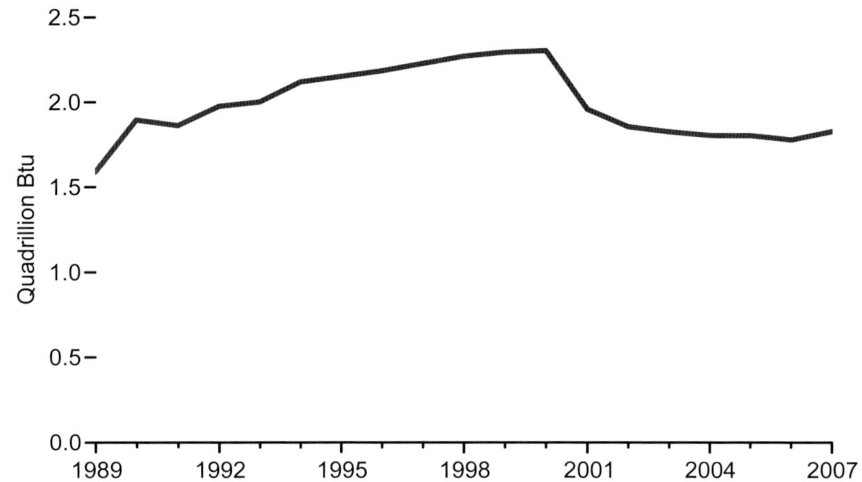

Total (All Sectors) by Source, 2007

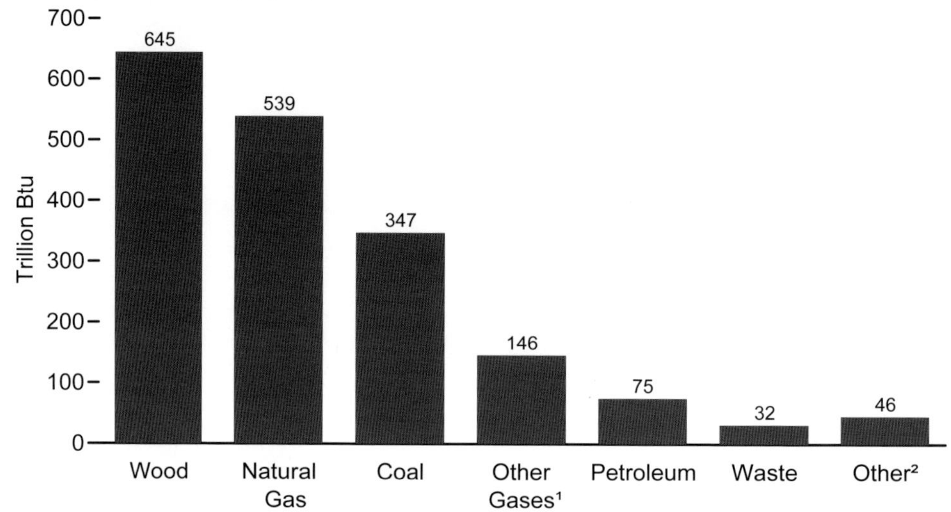

By Sector, 1989-2007

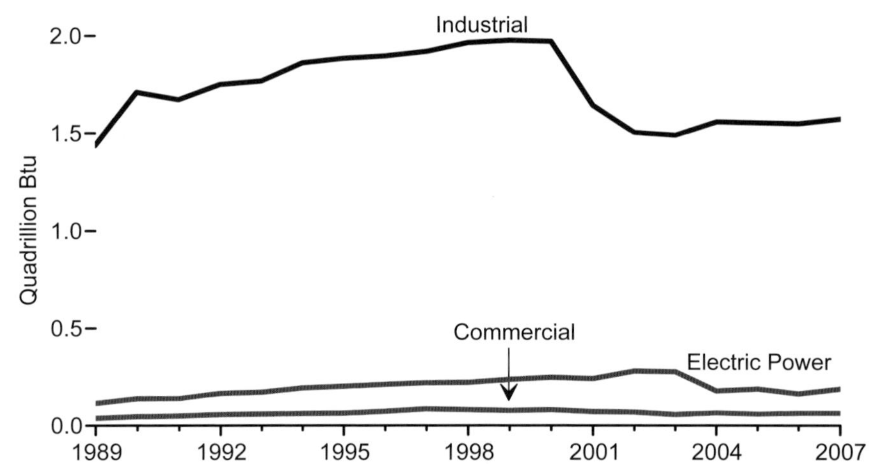

By Sector, 2007

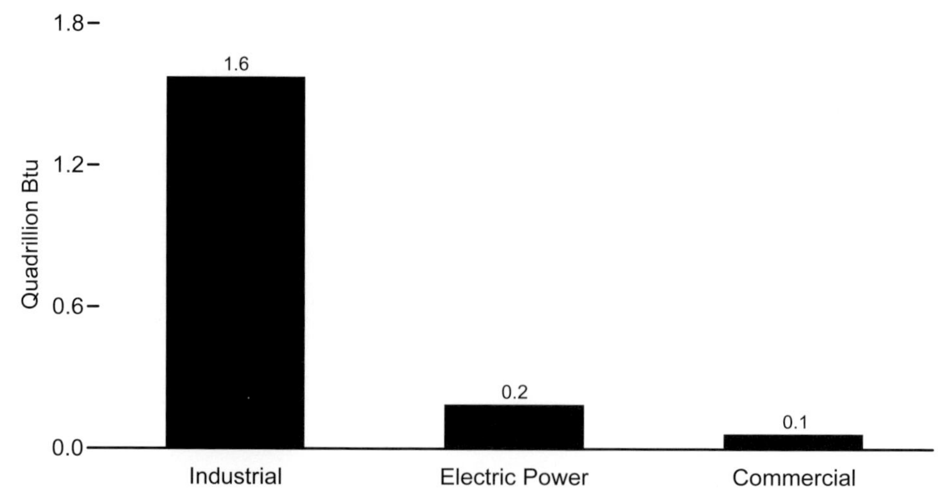

[1] Blast furnace gas, propane gas, and other manufactured and waste gases derived from fossil fuels.

[2] Batteries, chemicals, hydrogen, pitch, purchased steam, sulfur, miscellaneous technologies, and non-renewable waste (municipal solid waste from non-biogenic sources, and tire-derived fuels).

Note: Because vertical scales differ, graphs should not be compared.
Sources: Tables 8.3a–8.3c.

Table 8.3a Useful Thermal Output at Combined-Heat-and-Power Plants: Total (All Sectors), 1989-2007

(Sum of Tables 8.3b and 8.3c; Trillion Btu)

Year	Fossil Fuels					Renewable Energy			Other [7]	Total
						Biomass				
	Coal [1]	Petroleum [2]	Natural Gas [3]	Other Gases [4]	Total	Wood [5]	Waste [6]	Total		
1989	323	96	462	93	973	546	30	577	39	1,589
1990	363	127	538	141	1,169	651	36	687	40	1,896
1991	352	112	547	148	1,159	623	37	660	44	1,863
1992	367	117	592	160	1,236	658	40	698	42	1,976
1993	373	129	604	142	1,248	668	45	713	41	2,002
1994	388	133	646	144	1,309	722	45	767	42	2,119
1995	386	121	686	145	1,338	721	47	768	44	2,151
1996	392	133	711	150	1,385	701	55	756	43	2,184
1997	389	137	713	150	1,389	731	55	785	53	2,227
1998	382	136	782	167	1,466	700	57	757	46	2,269
1999	386	125	811	179	1,501	690	55	744	48	2,294
2000	384	108	812	184	1,488	707	56	764	50	2,302
2001	354	90	741	133	1,318	R557	28	585	55	1,958
2002	337	73	709	118	1,236	546	26	572	48	1,856
2003	333	85	610	110	1,139	597	35	632	55	1,826
2004	346	96	505	134	1,081	661	22	684	40	1,805
2005	357	97	445	137	1,036	697	30	727	41	1,804
2006	339	R78	R456	R128	R1,001	R701	R31	R732	R48	R1,780
2007P	347	75	539	146	1,107	645	32	676	46	1,829

[1] Anthracite, bituminous coal, subbituminous coal, lignite, waste coal, and coal synfuel.
[2] Distillate fuel oil, residual fuel oil, petroleum coke, jet fuel, kerosene, other petroleum, and waste oil.
[3] Natural gas, plus a small amount of supplemental gaseous fuels.
[4] Blast furnace gas, propane gas, and other manufactured and waste gases derived from fossil fuels.
[5] Wood and wood-derived fuels.
[6] Municipal solid waste from biogenic sources, landfill gas, sludge waste, agricultural byproducts, and other biomass. Through 2000, also includes non-renewable waste (municipal solid waste from non-biogenic sources, and tire-derived fuels).
[7] Batteries, chemicals, hydrogen, pitch, purchased steam, sulfur, miscellaneous technologies, and, beginning in 2001, non-renewable waste (municipal solid waste from non-biogenic sources, and tire-derived fuels).

R=Revised. P=Preliminary.

Notes: • Data do not include electric utility combined-heat-and-power (CHP) plants. • See Note 1, "Coverage of Electricity Statistics," at end of section. • See "Useful Thermal Output" in Glossary. • Totals may not equal sum of components due to independent rounding.

Web Page: For related information, see http://www.eia.doe.gov/fuelelectric.html.

Sources: Tables 8.3b and 8.3c.

Table 8.3b Useful Thermal Output at Combined-Heat-and-Power Plants: Electric Power Sector, 1989-2007

(Subset of Table 8.3a; Trillion Btu)

Year	Fossil Fuels					Renewable Energy			Other [7]	Total
						Biomass				
	Coal [1]	Petroleum [2]	Natural Gas [3]	Other Gases [4]	Total	Wood [5]	Waste [6]	Total		
1989	13	8	67	2	90	19	5	24	1	114
1990	21	9	80	4	114	18	6	25	(s)	138
1991	21	6	82	4	113	17	9	26	1	140
1992	28	6	102	5	140	17	8	25	2	167
1993	30	8	107	3	147	16	8	24	1	173
1994	37	9	119	5	170	15	10	24	1	195
1995	40	13	118	4	176	15	12	27	(s)	203
1996	43	12	121	4	180	16	16	33	(s)	213
1997	39	12	132	8	191	16	14	30	(s)	221
1998	43	6	142	5	196	10	16	26	(s)	222
1999	52	7	146	4	208	10	20	30	(s)	238
2000	53	7	158	5	223	6	19	26	(s)	249
2001	52	6	164	5	226	8	4	13	3	243
2002	40	4	214	6	264	8	5	13	5	281
2003	38	7	200	9	255	9	11	20	3	278
2004	22	1	130	16	169	6	2	8	1	179
2005	25	1	118	32	177	7	3	10	3	189
2006	R28	R1	R105	R17	R152	8	R2	R10	3	R165
2007P	27	1	125	22	174	9	2	12	3	189

[1] Anthracite, bituminous coal, subbituminous coal, lignite, waste coal, and coal synfuel.

[2] Distillate fuel oil, residual fuel oil, petroleum coke, jet fuel, kerosene, other petroleum, and waste oil.

[3] Natural gas, plus a small amount of supplemental gaseous fuels.

[4] Blast furnace gas, propane gas, and other manufactured and waste gases derived from fossil fuels.

[5] Wood and wood-derived fuels.

[6] Municipal solid waste from biogenic sources, landfill gas, sludge waste, agricultural byproducts, and other biomass. Through 2000, also includes non-renewable waste (municipal solid waste from non-biogenic sources, and tire-derived fuels).

[7] Batteries, chemicals, hydrogen, pitch, purchased steam, sulfur, miscellaneous technologies, and, beginning in 2001, non-renewable waste (municipal solid waste from non-biogenic sources, and tire-derived fuels).

R=Revised. P=Preliminary. (s)=Less than 0.5 trillion Btu.

Notes: • Data are for combined-heat-and-power (CHP) plants within the NAICS 22 category whose primary business is to sell electricity and heat to the public. Data do not include electric utility CHP plants. • See Table 8.3c for commercial and industrial CHP data. • See Note 1, "Coverage of Electricity Statistics," and Note 2, "Classification of Power Plants Into Energy-Use Sectors," at end of section. • See "Useful Thermal Output" in Glossary. • Totals may not equal sum of components due to independent rounding.

Web Page: For related information, see http://www.eia.doe.gov/fuelelectric.html.

Sources: • 1989-1997—Energy Information Administration (EIA), Form EIA-867, "Annual Nonutility Power Producer Report." • 1998-2000—EIA, Form EIA-860B, "Annual Electric Generator Report—Nonutility." • 2001-2003—EIA, Form EIA-906, "Power Plant Report." • 2004 forward—EIA, Form EIA-920, "Combined Heat and Power Plant Report."

Table 8.3c Useful Thermal Output at Combined-Heat-and-Power Plants: Commercial and Industrial Sectors, 1989-2007

(Subset of Table 8.3a; Trillion Btu)

Year	Fossil Fuels					Renewable Energy				
						Biomass				
	Coal [1]	Petroleum [2]	Natural Gas [3]	Other Gases [4]	Total	Wood [5]	Waste [6]	Total	Other [7]	Total
Commercial Sector [8]										
1989	14	4	10	(s)	27	(s)	10	10	–	38
1990	15	5	16	(s)	36	(s)	10	11	–	46
1991	16	4	21	(s)	41	(s)	9	9	(s)	50
1992	15	4	24	(s)	44	(s)	13	14	(s)	57
1993	18	4	23	(s)	45	(s)	14	14	(s)	59
1994	18	4	26	(s)	48	(s)	14	14	–	62
1995	17	3	29	–	48	(s)	15	15	(s)	63
1996	20	3	33	(s)	55	1	17	18	–	73
1997	22	4	40	(s)	66	1	19	20	–	86
1998	20	5	39	(s)	64	1	18	18	–	82
1999	20	3	37	(s)	61	1	17	17	–	78
2000	21	4	39	(s)	64	1	17	18	–	82
2001	18	4	35	–	58	1	8	8	6	72
2002	18	3	36	–	57	1	6	7	5	69
2003	23	3	17	–	42	1	8	8	6	57
2004	24	4	21	–	49	1	10	10	7	66
2005	21	3	22	–	47	(s)	8	8	5	60
2006	21	R2	R28	R (s)	R50	(s)	8	8	R6	R64
2007P	21	2	28	(s)	51	(s)	8	8	5	65
Industrial Sector [9]										
1989	297	84	385	90	856	527	15	542	38	1,437
1990	327	113	443	137	1,019	632	20	652	40	1,711
1991	315	103	444	144	1,005	606	19	625	44	1,674
1992	324	107	466	155	1,052	641	19	660	40	1,752
1993	325	117	475	139	1,055	652	23	675	39	1,769
1994	333	119	501	138	1,092	707	21	729	41	1,862
1995	329	105	540	140	1,114	706	20	726	44	1,884
1996	329	118	557	146	1,150	684	21	705	43	1,897
1997	328	121	541	142	1,132	713	22	735	53	1,920
1998	318	124	601	162	1,206	689	24	713	46	1,965
1999	313	115	629	175	1,233	679	18	697	48	1,978
2000	309	98	615	179	1,201	700	20	720	50	1,971
2001	284	80	542	128	1,034	548	16	564	46	1,644
2002	278	66	458	112	914	537	15	552	39	1,505
2003	272	75	393	101	842	588	16	604	46	1,491
2004	300	91	353	118	862	654	11	665	32	1,560
2005	311	92	305	105	813	689	20	709	33	1,555
2006	R290	R75	R323	R111	R799	R693	R21	R714	R39	R1,551
2007P	299	72	387	123	882	635	21	656	37	1,575

[1] Anthracite, bituminous coal, subbituminous coal, lignite, waste coal, and coal synfuel.
[2] Distillate fuel oil, residual fuel oil, petroleum coke, jet fuel, kerosene, other petroleum, and waste oil.
[3] Natural gas, plus a small amount of supplemental gaseous fuels.
[4] Blast furnace gas, propane gas, and other manufactured and waste gases derived from fossil fuels.
[5] Wood and wood-derived fuels.
[6] Municipal solid waste from biogenic sources, landfill gas, sludge waste, agricultural byproducts, and other biomass. Through 2000, also includes non-renewable waste (municipal solid waste from non-biogenic sources, and tire-derived fuels).
[7] Batteries, chemicals, hydrogen, pitch, purchased steam, sulfur, miscellaneous technologies, and, beginning in 2001, non-renewable waste (municipal solid waste from non-biogenic sources, and tire-derived fuels).
[8] Commercial combined-heat-and-power (CHP) plants.
[9] Industrial combined-heat-and-power (CHP) plants.

R=Revised. P=Preliminary. – = No data reported. (s)=Less than 0.5 trillion Btu.

Notes: • See Table 8.3b for electric power sector CHP data. • See Note 1, "Coverage of Electricity Statistics," and Note 2, "Classification of Power Plants Into Energy-Use Sectors," at end of section. • See "Useful Thermal Output" in Glossary. • Totals may not equal sum of components due to independent rounding.

Web Page: For related information, see http://www.eia.doe.gov/fuelelectric.html.

Sources: • 1989-1997—Energy Information Administration (EIA), Form EIA-867, "Annual Nonutility Power Producer Report." • 1998-2000—EIA, Form EIA-860B, "Annual Electric Generator Report—Nonutility." • 2001-2003—EIA, Form EIA-906, "Power Plant Report." • 2004 forward—EIA, Form EIA-920, "Combined Heat and Power Plant Report."

Figure 8.4 Consumption for Electricity Generation

Total and Major Fuel Categories, 1949-2007

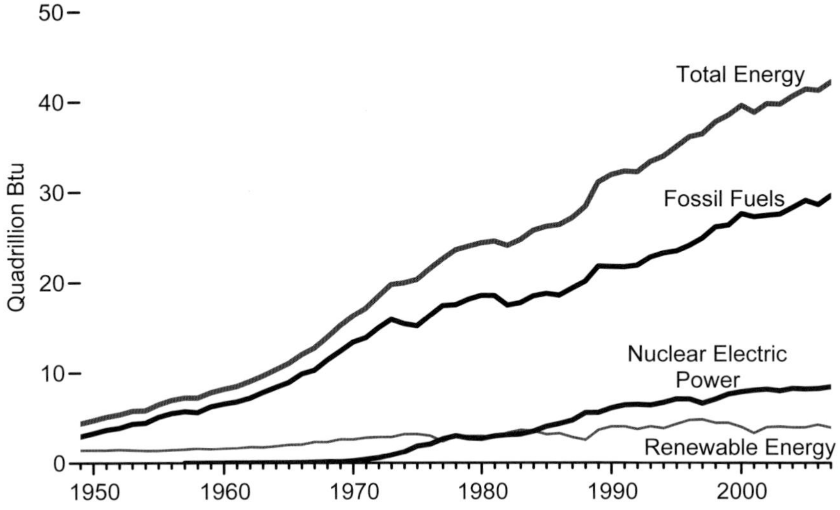

By Major Fuel, 2007

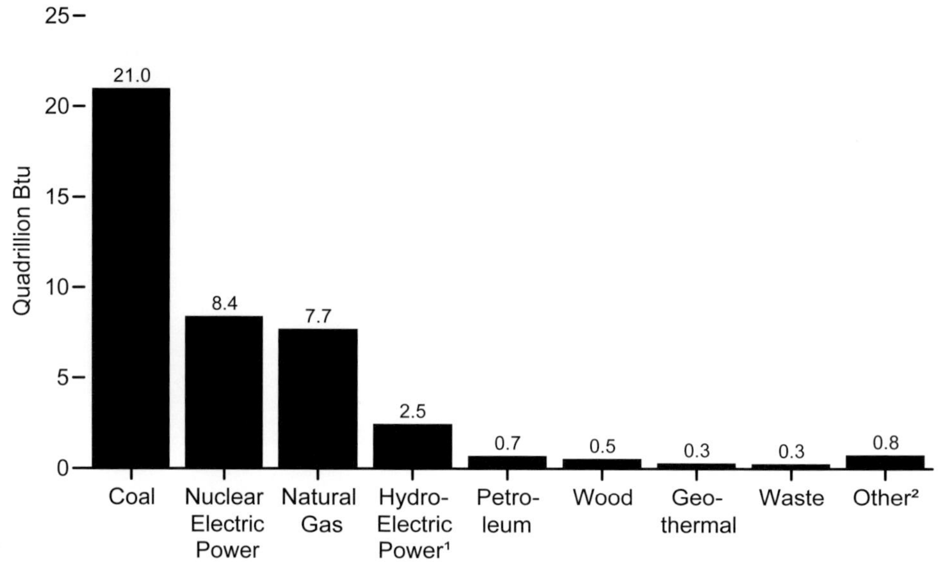

By Major Fuel, 1949-2007

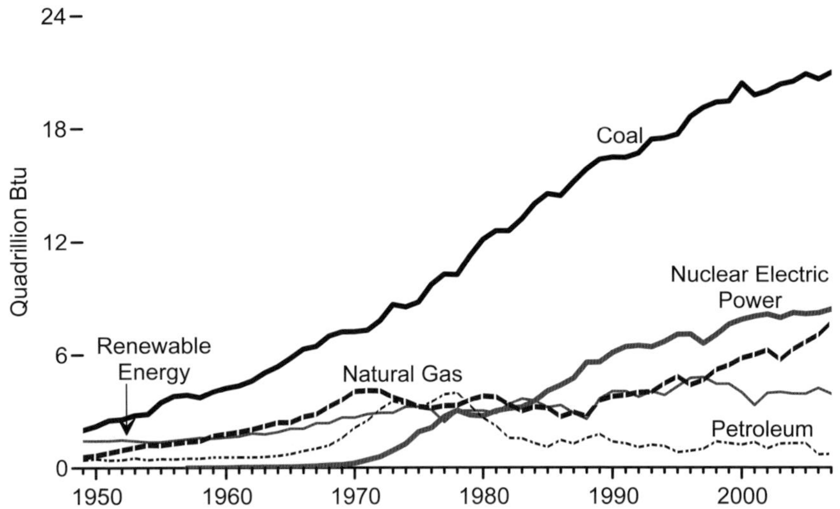

By Sector, 1989-2007

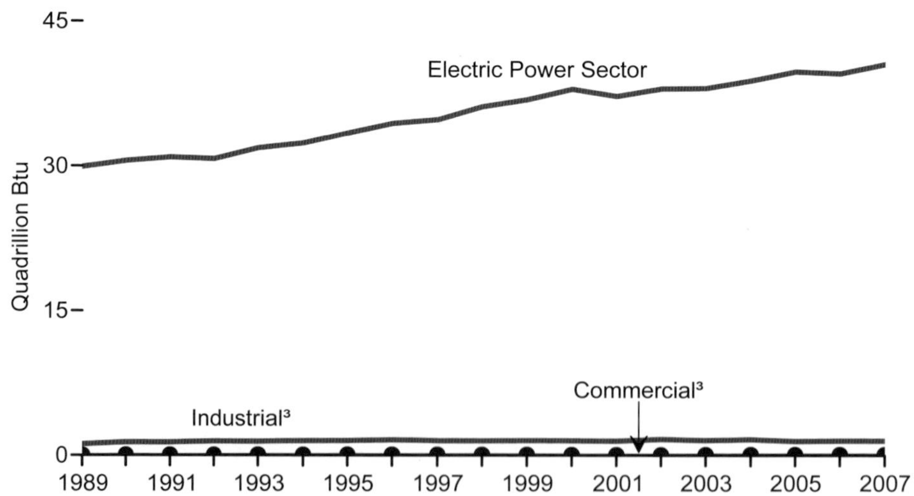

[1] Conventional hydroelectric power.
[2] Wind, other gases, electricity net imports, solar thermal and photovoltaic energy, batteries, chemicals, hydrogen, pitch, purchased steam, sulfur, miscellaneous technologies, and non-renewable waste (municipal solid waste from non-biogenic sources, and tire-derived fuels).
[3] Combined-heat-and-power plants and a small number of electricity-only plants.

Note: Because vertical scales differ, graphs should not be compared.
Sources: Tables 8.4a-8.4c.

Table 8.4a Consumption for Electricity Generation by Energy Source: Total (All Sectors), Selected Years, 1949-2007

(Sum of Tables 8.4b and 8.4c; Trillion Btu)

Year	Fossil Fuels					Nuclear Electric Power [5]	Renewable Energy							Other [9]	Electricity Net Imports [10]	Total
	Coal [1]	Petroleum [2]	Natural Gas [3]	Other Gases [4]	Total		Conventional Hydroelectric Power [5]	Biomass: Wood [6]	Biomass: Waste [7]	Geo-thermal [5]	Solar/PV [5,8]	Wind [5]	Total			
1949	1,995	415	569	NA	2,979	0	1,425	6	NA	NA	NA	NA	1,431	NA	5	4,415
1950	2,199	472	651	NA	3,322	0	1,415	5	NA	NA	NA	NA	1,421	NA	6	4,749
1955	3,458	471	1,194	NA	5,123	0	1,360	3	NA	NA	NA	NA	1,363	NA	14	6,500
1960	4,228	553	1,785	NA	6,565	6	1,608	2	NA	1	NA	NA	1,610	NA	15	8,197
1965	5,821	722	2,395	NA	8,938	43	2,059	3	NA	4	NA	NA	2,066	NA	(s)	11,047
1970	7,227	2,117	4,054	NA	13,399	239	2,634	1	2	11	NA	NA	2,649	NA	7	16,293
1971	7,299	2,495	4,099	NA	13,893	413	2,824	1	2	12	NA	NA	2,839	NA	12	17,158
1972	7,811	3,097	4,084	NA	14,992	584	2,864	1	2	31	NA	NA	2,899	NA	26	18,501
1973	8,658	3,515	3,748	NA	15,921	910	2,861	1	2	43	NA	NA	2,907	NA	49	19,788
1974	8,534	3,365	3,519	NA	15,418	1,272	3,177	1	2	53	NA	NA	3,232	NA	43	19,966
1975	8,786	3,166	3,240	NA	15,191	1,900	3,155	(s)	2	70	NA	NA	3,227	NA	21	20,339
1976	9,720	3,477	3,152	NA	16,349	2,111	2,976	1	2	78	NA	NA	3,057	NA	29	21,547
1977	10,262	3,901	3,284	NA	17,446	2,702	2,333	3	2	77	NA	NA	2,416	NA	59	22,623
1978	10,238	3,987	3,297	NA	17,522	3,024	2,937	2	1	64	NA	NA	3,005	NA	67	23,618
1979	11,260	3,283	3,613	NA	18,156	2,776	2,931	3	2	84	NA	NA	3,020	NA	69	24,021
1980	12,123	2,634	3,810	NA	18,567	2,739	2,900	3	2	110	NA	NA	3,014	NA	71	24,392
1981	12,583	2,202	3,768	NA	18,553	3,008	2,758	3	1	123	NA	NA	2,885	NA	113	24,559
1982	12,582	1,568	3,342	NA	17,491	3,131	3,266	2	1	105	NA	NA	3,374	NA	100	24,096
1983	13,213	1,544	2,998	NA	17,754	3,203	3,527	2	2	129	NA	(s)	3,661	NA	121	24,738
1984	14,019	1,286	3,220	NA	18,526	3,553	3,386	5	4	165	(s)	(s)	3,560	NA	135	25,774
1985	14,542	1,090	3,160	NA	18,792	4,076	2,970	8	7	198	(s)	(s)	3,183	NA	140	26,191
1986	14,444	1,452	2,691	NA	18,586	4,380	3,071	5	7	219	(s)	(s)	3,303	NA	122	26,392
1987	15,173	1,257	2,935	NA	19,365	4,754	2,635	8	7	229	(s)	(s)	2,879	NA	158	27,157
1988	15,850	1,563	2,709	NA	20,123	5,587	2,334	10	8	217	(s)	(s)	2,569	NA	108	28,387
1989	[11]16,359	[11]1,757	[11]3,582	90	[11]21,789	[11]5,602	[12]2,837	[11]345	[11]151	[11]308	[11]3	[11]22	[11]3,665	39	37	31,133
1990	16,477	1,367	3,791	112	21,747	6,104	3,046	442	211	326	4	29	4,058	36	8	31,954
1991	16,460	1,276	3,861	125	21,723	6,422	3,016	425	247	335	5	31	4,058	59	67	32,329
1992	16,686	1,076	3,999	141	21,903	6,479	2,617	481	283	338	4	30	3,752	40	87	32,261
1993	17,424	1,203	4,027	136	22,790	6,410	2,892	485	288	351	5	31	4,052	34	95	33,381
1994	17,485	1,135	4,476	136	23,233	6,694	2,683	498	301	325	5	36	3,848	40	153	33,968
1995	17,687	813	4,840	133	23,473	7,075	3,205	480	316	280	5	33	4,318	42	134	35,043
1996	18,650	888	4,400	159	24,097	7,087	3,590	513	324	300	5	33	4,765	37	137	36,123
1997	19,128	985	4,658	119	24,890	6,597	3,640	484	339	309	5	34	4,811	36	116	36,451
1998	19,417	1,378	5,205	125	26,124	7,068	3,297	475	332	311	5	31	4,450	36	88	37,767
1999	19,467	1,285	5,441	126	26,320	7,610	3,268	490	332	312	5	46	4,452	41	99	38,522
2000	20,411	1,212	5,818	126	27,567	7,862	2,811	496	330	296	5	57	3,995	46	115	39,586
2001	19,789	1,347	6,001	97	27,235	8,033	2,242	486	228	289	6	70	3,320	160	75	38,823
2002	19,997	1,014	6,250	131	27,392	8,143	2,689	605	257	305	6	105	3,967	191	72	39,764
2003	20,367	1,266	5,736	156	27,525	7,959	2,825	519	249	303	5	115	4,016	193	22	39,715
2004	20,508	1,282	6,281	187	28,257	8,222	2,690	534	254	311	6	142	3,936	176	39	40,629
2005	20,904	1,296	6,671	177	29,048	8,160	2,703	482	252	309	6	178	3,929	161	84	41,383
2006	R20,637	R696	R7,060	R181	R28,574	R8,214	R2,869	R523	R262	R306	5	R264	R4,229	R155	R63	R41,235
2007P	20,990	715	7,716	166	29,587	8,415	2,463	548	276	312	6	319	3,924	169	107	42,201

[1] Anthracite, bituminous coal, subbituminous coal, lignite, waste coal, and coal synfuel.
[2] Distillate fuel oil, residual fuel oil, petroleum coke, jet fuel, kerosene, other petroleum, and waste oil.
[3] Natural gas, plus a small amount of supplemental gaseous fuels.
[4] Blast furnace gas, propane gas, and other manufactured and waste gases derived from fossil fuels.
[5] Values are converted from kilowatthours to Btu using the approximate heat rates in Table A6.
[6] Wood and wood-derived fuels.
[7] Municipal solid waste from biogenic sources, landfill gas, sludge waste, agricultural byproducts, and other biomass. Through 2000, also includes non-renewable waste (municipal solid waste from non-biogenic sources, and tire-derived fuels).
[8] Solar thermal and photovoltaic energy.
[9] Batteries, chemicals, hydrogen, pitch, purchased steam, sulfur, miscellaneous technologies, and, beginning in 2001, non-renewable waste (municipal solid waste from non-biogenic sources, and tire-derived fuels).
[10] Net imports equal imports minus exports. See Note 3, "Electricity Imports and Exports," at end of section.
[11] Through 1988, data are for electric utilities only. Beginning in 1989, data are for electric utilities, independent power producers, commercial plants, and industrial plants.
[12] Through 1988, data are for electric utilities and industrial plants. Beginning in 1989, data are for electric utilities, independent power producers, commercial plants, and industrial plants.

R=Revised. P=Preliminary. NA=Not available. (s)=Less than 0.5 trillion Btu.

Notes: • Data are for energy consumed to produce electricity. Data also include energy consumed to produce useful thermal output at a small number of electric utility combined-heat-and-power (CHP) plants. • This table no longer shows energy consumption by hydroelectric pumped storage plants. The change was made because most of the electricity used to pump water into elevated storage reservoirs is generated by plants other than pumped-storage plants; thus, the associated energy is already accounted for in other data columns in this table (such as "Conventional Hydroelectric Power," "Coal," "Natural Gas," and so on). • See Note 1, "Coverage of Electricity Statistics," at end of section. • Totals may not equal sum of components due to independent rounding.

Web Pages: • For all data beginning in 1949, see http://www.eia.doe.gov/emeu/aer/elect.html. • For related information, see http://www.eia.doe.gov/fuelelectric.html.

Sources: • 1949-1988—Table 8.4b for electric power sector, and Tables 8.1 and A6 for industrial sector. • 1989 forward—Tables 8.4b and 8.4c.

Table 8.4b Consumption for Electricity Generation by Energy Source: Electric Power Sector, Selected Years, 1949-2007 (Subset of Table 8.4a; Trillion Btu)

Year	Fossil Fuels					Nuclear Electric Power [5]	Renewable Energy							Other [9]	Electricity Net Imports [10]	Total
	Coal [1]	Petroleum [2]	Natural Gas [3]	Other Gases [4]	Total		Conventional Hydroelectric Power [5]	Biomass		Geo-thermal [5]	Solar/PV [5,8]	Wind [5]	Total			
								Wood [6]	Waste [7]							
1949	1,995	415	569	NA	2,979	0	1,349	6	NA	NA	NA	NA	1,355	NA	5	4,339
1950	2,199	472	651	NA	3,322	0	1,346	5	NA	NA	NA	NA	1,351	NA	6	4,679
1955	3,458	471	1,194	NA	5,123	0	1,322	3	NA	NA	NA	NA	1,325	NA	14	6,461
1960	4,228	553	1,785	NA	6,565	6	1,569	2	NA	1	NA	NA	1,571	NA	15	8,158
1965	5,821	722	2,395	NA	8,938	43	2,026	3	NA	4	NA	NA	2,033	NA	(s)	11,014
1970	7,227	2,117	4,054	NA	13,399	239	2,600	1	2	11	NA	NA	2,615	NA	7	16,259
1971	7,299	2,495	4,099	NA	13,893	413	2,790	1	2	12	NA	NA	2,806	NA	12	17,124
1972	7,811	3,097	4,084	NA	14,992	584	2,829	1	2	31	NA	NA	2,864	NA	26	18,466
1973	8,658	3,515	3,748	NA	15,921	910	2,827	1	2	43	NA	NA	2,873	NA	49	19,753
1974	8,534	3,365	3,519	NA	15,418	1,272	3,143	1	2	53	NA	NA	3,199	NA	43	19,933
1975	8,786	3,166	3,240	NA	15,191	1,900	3,122	(s)	2	70	NA	NA	3,194	NA	21	20,307
1976	9,720	3,477	3,152	NA	16,349	2,111	2,943	1	2	78	NA	NA	3,024	NA	29	21,513
1977	10,262	3,901	3,284	NA	17,446	2,702	2,301	3	2	77	NA	NA	2,383	NA	59	22,591
1978	10,238	3,987	3,297	NA	17,522	3,024	2,905	2	1	64	NA	NA	2,973	NA	67	23,587
1979	11,260	3,283	3,613	NA	18,156	2,776	2,897	3	2	84	NA	NA	2,986	NA	69	23,987
1980	12,123	2,634	3,810	NA	18,567	2,739	2,867	3	2	110	NA	NA	2,982	NA	71	24,359
1981	12,583	2,202	3,768	NA	18,553	3,008	2,725	3	1	123	NA	NA	2,852	NA	113	24,525
1982	12,582	1,568	3,342	NA	17,491	3,131	3,233	2	1	105	NA	NA	3,341	NA	100	24,063
1983	13,213	1,544	2,998	NA	17,754	3,203	3,494	2	2	129	NA	(s)	3,627	NA	121	24,705
1984	14,019	1,286	3,220	NA	18,526	3,553	3,353	5	4	165	(s)	(s)	3,527	NA	135	25,741
1985	14,542	1,090	3,160	NA	18,792	4,076	2,937	8	7	198	(s)	(s)	3,150	NA	140	26,158
1986	14,444	1,452	2,691	NA	18,586	4,380	3,038	5	7	219	(s)	(s)	3,270	NA	122	26,359
1987	15,173	1,257	2,935	NA	19,365	4,754	2,602	8	7	229	(s)	(s)	2,846	NA	158	27,124
1988	15,850	1,563	2,709	NA	20,123	5,587	2,302	10	8	217	(s)	(s)	2,536	NA	108	28,354
1989	[11]16,121	[11]1,697	[11]3,107	7	[11]20,932	[11]5,602	[11]2,808	[11]75	[11]126	[11]308	[11]3	[11]22	[11]3,342	2	37	29,916
1990	16,235	1,281	3,233	6	20,755	6,104	3,014	106	180	326	4	29	3,658	(s)	8	30,526
1991	16,223	1,199	3,296	6	20,725	6,422	2,985	104	217	335	5	31	3,677	4	67	30,895
1992	16,431	990	3,407	12	20,840	6,479	2,586	120	252	338	4	30	3,329	3	87	30,738
1993	17,159	1,122	3,426	12	21,719	6,410	2,861	129	255	351	5	31	3,632	3	95	31,859
1994	17,215	1,056	3,851	12	22,134	6,694	2,620	134	269	325	5	36	3,389	2	153	32,372
1995	17,416	743	4,179	18	22,356	7,075	3,149	106	282	280	5	33	3,855	2	134	33,423
1996	18,375	810	3,730	16	22,930	7,087	3,528	117	280	300	5	33	4,264	2	137	34,420
1997	18,855	917	3,981	14	23,768	6,597	3,581	117	292	309	5	34	4,337	1	116	34,819
1998	19,162	1,306	4,520	23	25,011	7,068	3,241	125	287	311	5	31	4,000	2	88	36,168
1999	19,214	1,211	4,742	14	25,181	7,610	3,218	125	290	312	5	46	3,996	1	99	36,888
2000	20,153	1,145	5,120	19	26,438	7,862	2,768	126	294	296	5	57	3,547	1	115	37,963
2001	19,549	1,280	5,290	9	26,128	8,033	2,209	116	205	289	6	70	2,894	109	75	37,239
2002	19,733	955	5,522	25	26,235	8,143	2,650	141	224	305	6	105	3,430	137	72	38,016
2003	20,137	1,199	5,009	30	26,374	7,959	2,781	156	216	303	5	115	3,576	136	22	38,068
2004	20,277	1,216	5,443	38	26,975	8,222	2,656	157	220	311	6	142	3,493	136	39	38,865
2005	20,705	1,240	5,888	44	27,877	8,160	2,670	176	217	309	6	178	3,555	120	84	39,797
2006	[R]20,426	[R]647	[R]6,265	[R]43	[R]27,381	[R]8,214	[R]2,839	[R]172	[R]228	[R]306	5	[R]264	[R]3,814	[R]121	[R]63	[R]39,593
2007[P]	20,802	667	6,910	39	28,419	8,415	2,440	172	240	312	6	319	3,489	124	107	40,553

[1] Anthracite, bituminous coal, subbituminous coal, lignite, waste coal, and coal synfuel.
[2] Distillate fuel oil, residual fuel oil, petroleum coke, jet fuel, kerosene, other petroleum, and waste oil.
[3] Natural gas, plus a small amount of supplemental gaseous fuels.
[4] Blast furnace gas, propane gas, and other manufactured and waste gases derived from fossil fuels.
[5] Values are converted from kilowattthours to Btu using the approximate heat rates in Table A6.
[6] Wood and wood-derived fuels.
[7] Municipal solid waste from biogenic sources, landfill gas, sludge waste, agricultural byproducts, and other biomass. Through 2000, also includes non-renewable waste (municipal solid waste from non-biogenic sources, and tire-derived fuels).
[8] Solar thermal and photovoltaic energy.
[9] Batteries, chemicals, hydrogen, pitch, purchased steam, sulfur, miscellaneous technologies, and, beginning in 2001, non-renewable waste (municipal solid waste from non-biogenic sources, and tire-derived fuels).
[10] Net imports equal imports minus exports. See Note 3, "Electricity Imports and Exports," at end of section.
[11] Through 1988, data are for electric utilities only. Beginning in 1989, data are for electric utilities and independent power producers.

R=Revised. P=Preliminary. NA=Not available. (s)=Less than 0.5 trillion Btu.

Notes: • Data are for energy consumed to produce electricity. Data also include energy consumed to produce useful thermal output at a small number of electric utility combined-heat-and-power (CHP) plants. • The electric power sector comprises electricity-only and combined-heat-and-power (CHP) plants within the NAICS 22 category whose primary business is to sell electricity, or electricity and heat, to the public. • See Table 8.4c for commercial and industrial CHP and electricity-only data. • This table no longer shows energy consumption by hydroelectric pumped storage plants. The change was made because most of the electricity used to pump water into elevated storage reservoirs is generated by plants other than pumped-storage plants; thus, the associated energy is already accounted for in other data columns in this table (such as "Conventional Hydroelectric Power," "Coal," "Natural Gas," and so on). • See Note 1, "Coverage of Electricity Statistics," and Note 2, "Classification of Power Plants Into Energy-Use Sectors," at end of section. • Totals may not equal sum of components due to independent rounding.

Web Pages: • For all data beginning in 1949, see http://www.eia.doe.gov/emeu/aer/elect.html. • For related information, see http://www.eia.doe.gov/fuelelectric.html.

Sources: **Electricity Net Imports:** Tables 8.1 and A6. **All Other Data:** • 1949-1988—Tables 8.2b, 8.5b, A1, A4, A5, and A6. • 1989-1997—Energy Information Administration (EIA), Form EIA-759, "Monthly Power Plant Report," and Form EIA-867, "Annual Nonutility Power Producer Report." • 1998-2000—EIA, Form EIA-759, "Monthly Power Plant Report," and Form EIA-860B, "Annual Electric Generator Report—Nonutility." • 2001-2003—EIA, Form EIA-906, "Power Plant Report." • 2004 forward—EIA, Form EIA-906, "Power Plant Report," and Form EIA-920, "Combined Heat and Power Plant Report."

Table 8.4c Consumption for Electricity Generation by Energy Source: Commercial and Industrial Sectors, 1989-2007

(Subset of Table 8.4a; Trillion Btu)

Year	Fossil Fuels: Coal [1]	Fossil Fuels: Petroleum [2]	Fossil Fuels: Natural Gas [3]	Fossil Fuels: Other Gases [4]	Fossil Fuels: Total	Nuclear Electric Power	Renewable Energy: Conventional Hydroelectric Power [5]	Renewable Energy: Biomass: Wood [6]	Renewable Energy: Biomass: Waste [7]	Renewable Energy: Geo-thermal	Renewable Energy: Solar/PV [8]	Renewable Energy: Wind	Renewable Energy: Total	Other [9]	Electricity Net Imports	Total
Commercial Sector [10]																
1989	9	7	18	1	36	–	1	2	9	–	–	–	12	–	– –	47
1990	9	6	28	1	45	–	1	2	15	–	–	–	18	–	– –	63
1991	9	3	28	1	41	–	1	2	15	–	–	–	18	(s)	– –	59
1992	8	3	33	1	45	–	1	1	16	–	–	–	19	(s)	– –	64
1993	9	4	38	1	53	–	1	1	16	–	–	–	18	(s)	– –	71
1994	9	4	42	1	56	–	1	1	17	–	–	–	19	–	– –	75
1995	12	4	44	–	60	–	1	1	21	–	–	–	23	(s)	– –	83
1996	14	4	44	(s)	62	–	1	1	31	–	–	–	33	(s)	– –	95
1997	14	5	40	(s)	59	–	1	1	34	–	–	–	35	(s)	– –	94
1998	11	5	42	(s)	57	–	1	1	32	–	–	–	34	–	– –	91
1999	12	6	40	(s)	57	–	1	(s)	33	–	–	–	35	(s)	– –	92
2000	12	5	38	(s)	55	–	1	(s)	26	–	–	–	28	(s)	– –	82
2001	13	6	37	(s)	56	–	1	(s)	15	–	–	–	16	7	– –	79
2002	9	4	31	(s)	44	–	(s)	(s)	18	–	–	–	19	11	– –	73
2003	13	5	39	–	58	–	1	(s)	19	–	–	–	21	11	– –	89
2004	13	7	47	–	67	–	1	1	22	–	–	–	24	13	– –	104
2005	17	6	49	–	71	–	1	1	25	–	–	–	26	13	– –	111
2006	16	[R]3	[R]50	–	68	–	1	1	26	–	–	–	[R]28	14	– –	110
2007[P]	16	2	51	–	69	–	1	1	27	–	–	–	29	13	– –	111
Industrial Sector [11]																
1989	229	53	456	83	821	–	28	267	15	–	–	–	311	37	– –	1,169
1990	233	80	530	104	947	–	31	335	16	–	–	–	382	36	– –	1,365
1991	228	74	537	118	957	–	30	318	14	–	–	–	362	55	– –	1,374
1992	246	84	559	128	1,017	–	31	359	15	–	–	–	405	37	– –	1,459
1993	256	77	562	123	1,019	–	30	355	17	–	–	–	401	31	– –	1,451
1994	261	75	584	123	1,043	–	62	364	14	–	–	–	440	38	– –	1,521
1995	259	66	617	114	1,057	–	55	373	13	–	–	–	440	40	– –	1,537
1996	261	74	626	143	1,104	–	61	394	13	–	–	–	468	35	– –	1,607
1997	260	63	637	105	1,064	–	58	367	14	–	–	–	439	36	– –	1,538
1998	245	67	643	102	1,056	–	55	349	13	–	–	–	417	35	– –	1,508
1999	242	68	660	112	1,081	–	49	364	8	–	–	–	422	39	– –	1,542
2000	245	61	660	107	1,074	–	42	369	10	–	–	–	421	45	– –	1,540
2001	227	62	674	88	1,051	–	33	370	7	–	–	–	410	44	– –	1,504
2002	255	55	697	106	1,113	–	39	464	15	–	–	–	518	43	– –	1,675
2003	217	61	687	127	1,093	–	43	362	13	–	–	–	419	46	– –	1,558
2004	217	59	790	148	1,215	–	33	376	11	–	–	–	419	27	– –	1,660
2005	182	51	734	133	1,100	–	32	306	9	–	–	–	347	28	– –	1,475
2006	[R]194	[R]46	[R]746	[R]138	[R]1,124	–	[R]29	[R]350	8	–	–	–	[R]386	[R]21	– –	[R]1,531
2007[P]	172	46	755	127	1,099	–	23	376	8	–	–	–	406	31	– –	1,537

[1] Anthracite, bituminous coal, subbituminous coal, lignite, waste coal, and coal synfuel.
[2] Distillate fuel oil, residual fuel oil, petroleum coke, jet fuel, kerosene, other petroleum, and waste oil.
[3] Natural gas, plus a small amount of supplemental gaseous fuels.
[4] Blast furnace gas, propane gas, and other manufactured and waste gases derived from fossil fuels.
[5] Values are converted from kilowatthours to Btu using the approximate heat rates in Table A6.
[6] Wood and wood-derived fuels.
[7] Municipal solid waste from biogenic sources, landfill gas, sludge waste, agricultural byproducts, and other biomass. Through 2000, also includes non-renewable waste (municipal solid waste from non-biogenic sources, and tire-derived fuels).
[8] Solar thermal and photovoltaic energy.
[9] Batteries, chemicals, hydrogen, pitch, purchased steam, sulfur, miscellaneous technologies, and, beginning in 2001, non-renewable waste (municipal solid waste from non-biogenic sources, and tire-derived fuels).
[10] Commercial combined-heat-and-power (CHP) and commercial electricity-only plants.
[11] Industrial combined-heat-and-power (CHP) and industrial electricity-only plants.

R=Revised. P=Preliminary. – – = Not applicable. – = No data reported. (s)=Less than 0.5 trillion Btu.

Notes: • Data are for energy consumed to produce electricity. • See Table 8.4b for electric power sector electricity-only and CHP data. • See Note 1, "Coverage of Electricity Statistics," and Note 2, "Classification of Power Plants Into Energy-Use Sectors," at end of section. • Totals may not equal sum of components due to independent rounding.

Web Page: For related information, see http://www.eia.doe.gov/fuelelectric.html.

Sources: • 1989-1997—Energy Information Administration (EIA), Form EIA-867, "Annual Nonutility Power Producer Report." • 1998-2000—EIA, Form EIA-860B, "Annual Electric Generator Report—Nonutility." • 2001-2003—EIA, Form EIA-906, "Power Plant Report." • 2004 forward—EIA, Form EIA-906, "Power Plant Report," and Form EIA-920, "Combined Heat and Power Plant Report."

Figure 8.5a Consumption of Combustible Fuels for Electricity Generation, 1989-2007

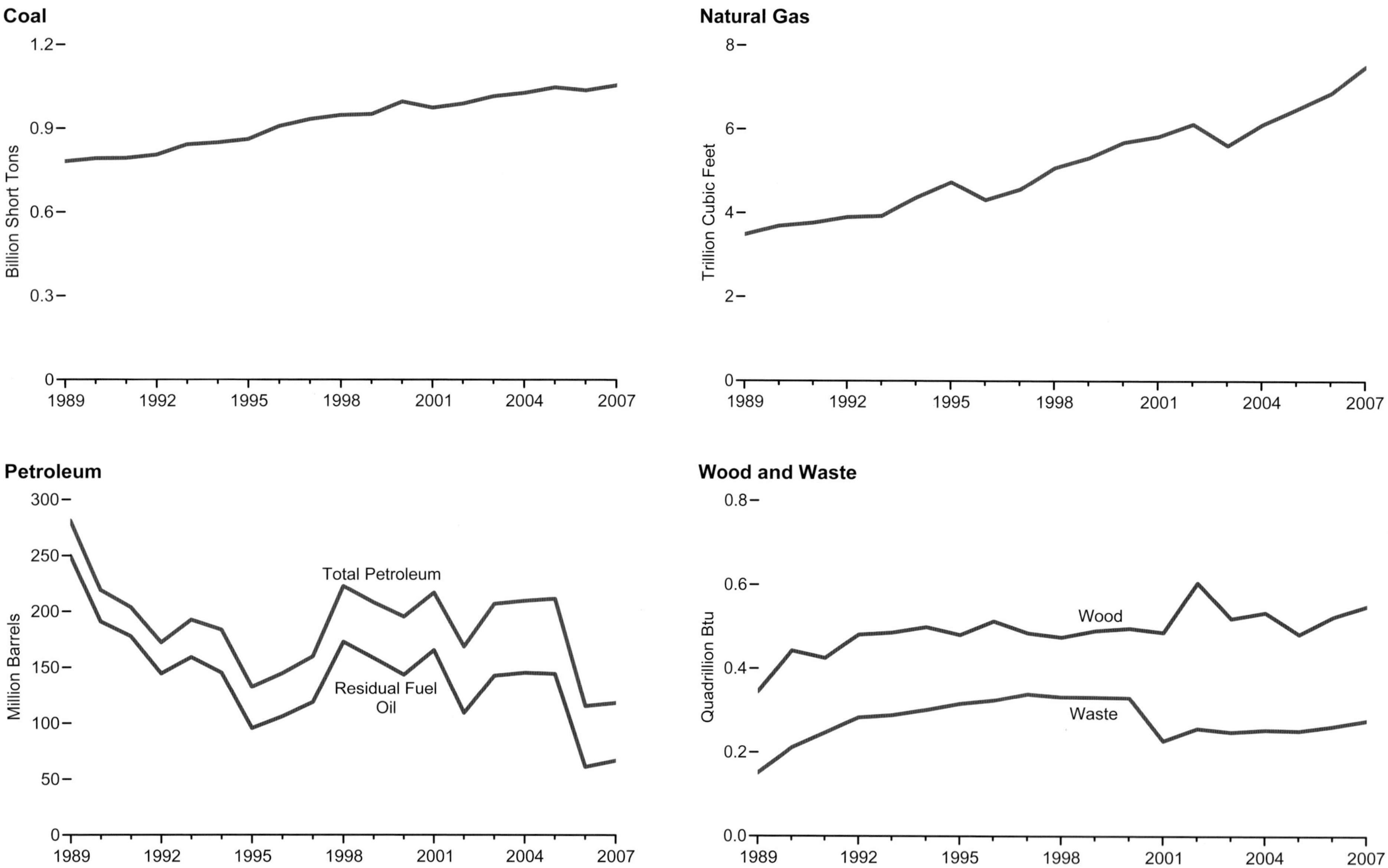

Source: Table 8.5a.

Figure 8.5b Consumption of Combustible Fuels for Electricity Generation by Sector, 2007

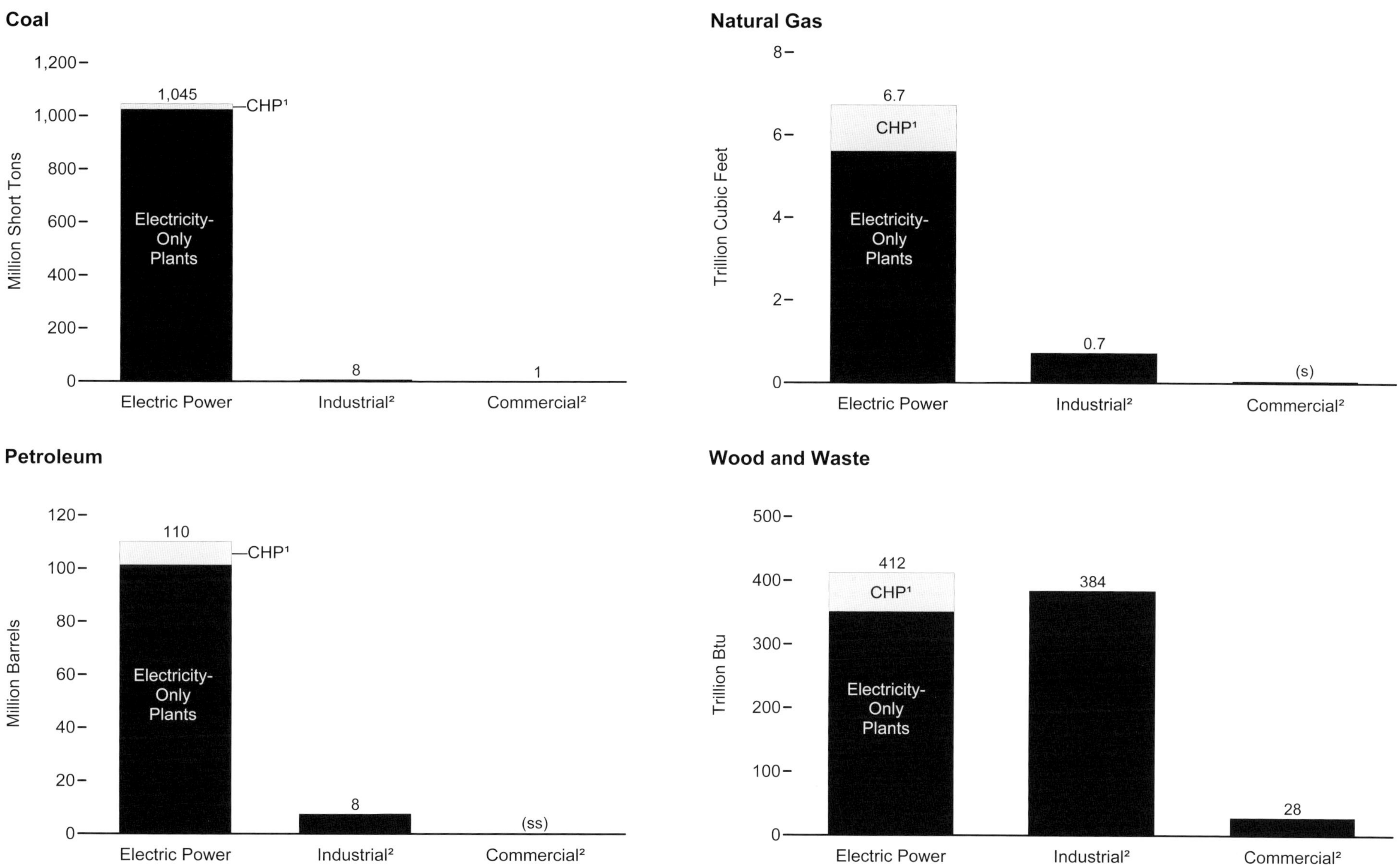

[1] Combined-heat-and-power plants.
[2] Combined-heat-and-power and electricity-only plants.

(s)=Less than 0.05 trillion cubic feet.
(ss)=Less than 0.5 million barrels.
Sources: Tables 8.5b-8.5d.

Table 8.5a Consumption of Combustible Fuels for Electricity Generation: Total (All Sectors), Selected Years, 1949-2007

(Sum of Tables 8.5b and 8.5d)

Year	Coal [1]	Petroleum: Distillate Fuel Oil [2]	Petroleum: Residual Fuel Oil [3]	Petroleum: Other Liquids [4]	Petroleum: Petroleum Coke [5]	Petroleum: Total [5]	Natural Gas [6]	Other Gases [7]	Biomass: Wood [8]	Biomass: Waste [9]	Other [10]
	Thousand Short Tons	Thousand Barrels	Thousand Barrels	Thousand Barrels	Thousand Short Tons	Thousand Barrels	Million Cubic Feet	Trillion Btu	Trillion Btu	Trillion Btu	Trillion Btu
1949	83,963	4,767	61,534	NA	NA	66,301	550,121	NA	6	NA	NA
1950	91,871	5,423	69,998	NA	NA	75,421	628,919	NA	5	NA	NA
1955	143,759	5,412	69,862	NA	NA	75,274	1,153,280	NA	3	NA	NA
1960	176,685	3,824	84,371	NA	NA	88,195	1,724,762	NA	2	NA	NA
1965	244,788	4,928	110,274	NA	NA	115,203	2,321,101	NA	3	NA	NA
1970	320,182	24,123	311,381	NA	636	338,686	3,931,860	NA	1	2	NA
1971	327,301	34,283	362,187	NA	605	399,496	3,976,018	NA	1	2	NA
1972	351,768	53,465	440,294	NA	627	496,895	3,976,913	NA	1	2	NA
1973	389,212	47,058	513,190	NA	507	562,781	3,660,172	NA	1	2	NA
1974	391,811	53,128	483,146	NA	625	539,399	3,443,428	NA	1	2	NA
1975	405,962	38,907	467,221	NA	70	506,479	3,157,669	NA	(s)	2	NA
1976	448,371	41,843	514,077	NA	68	556,261	3,080,868	NA	1	2	NA
1977	477,126	48,837	574,869	NA	98	624,193	3,191,200	NA	3	2	NA
1978	481,235	47,520	588,319	NA	398	637,830	3,188,363	NA	2	1	NA
1979	527,051	30,691	492,606	NA	268	524,636	3,490,523	NA	3	2	NA
1980	569,274	29,051	391,163	NA	179	421,110	3,681,595	NA	3	2	NA
1981	596,797	21,313	329,798	NA	139	351,806	3,640,154	NA	3	1	NA
1982	593,666	15,337	234,434	NA	149	250,517	3,225,518	NA	2	1	NA
1983	625,211	16,512	228,984	NA	261	246,804	2,910,767	NA	2	2	NA
1984	664,399	15,190	189,289	NA	252	205,736	3,111,342	NA	5	4	NA
1985	693,841	14,635	158,779	NA	231	174,571	3,044,083	NA	8	7	NA
1986	685,056	14,326	216,156	NA	313	232,046	2,602,370	NA	5	7	NA
1987	717,894	15,367	184,011	NA	348	201,116	2,844,051	NA	8	7	NA
1988	758,372	18,769	229,327	NA	409	250,141	2,635,613	NA	10	8	NA
1989[11]	781,672	27,733	249,820	303	667	281,192	3,485,429	90	345	151	39
1990	792,457	18,143	190,849	437	1,914	218,997	3,691,563	112	442	211	36
1991	793,666	16,564	177,780	380	1,789	203,669	3,764,778	125	425	247	59
1992	805,140	14,493	144,467	759	2,504	172,241	3,899,718	141	481	283	40
1993	842,153	16,845	159,059	715	3,169	192,462	3,928,653	136	485	288	34
1994	848,796	22,365	145,225	929	3,020	183,618	4,367,148	136	498	301	40
1995	860,594	19,615	95,507	680	3,355	132,578	4,737,871	133	480	316	42
1996	907,209	20,252	106,055	1,712	3,322	144,626	4,312,458	159	513	324	37
1997	931,949	20,309	118,741	237	4,086	159,715	4,564,770	119	484	339	36
1998	946,295	25,062	172,728	549	4,860	222,640	5,081,384	125	475	332	36
1999	949,802	25,951	158,187	974	4,552	207,871	5,321,984	126	490	332	41
2000	994,933	31,675	143,381	1,450	3,744	195,228	5,691,481	126	496	330	46
2001	972,691	31,150	165,312	855	3,871	216,672	5,832,305	97	486	228	160
2002	987,583	23,286	109,235	1,894	6,836	168,597	6,126,062	131	605	257	191
2003	1,014,058	29,672	142,518	2,947	6,303	206,653	5,616,135	156	519	249	193
2004	1,026,018	20,669	145,171	3,959	7,942	209,508	6,116,574	187	534	254	176
2005	1,045,878	21,163	144,234	3,303	8,511	211,256	6,486,761	177	482	252	161
2006	R1,035,346	R13,372	R61,019	R2,612	R7,673	R115,370	R6,869,624	R181	R523	R262	R155
2007P	1,053,346	16,605	66,701	3,699	6,222	118,115	7,507,446	166	548	276	169

[1] Anthracite, bituminous coal, subbituminous coal, lignite, waste coal, and coal synfuel.
[2] Fuel oil nos. 1, 2, and 4. For 1949-1979, data are for gas turbine and internal combustion plant use of petroleum. For 1980-2000, electric utility data also include small amounts of kerosene and jet fuel.
[3] Fuel oil nos. 5 and 6. For 1949-1979, data are for steam plant use of petroleum. For 1980-2000, electric utility data also include a small amount of fuel oil no. 4.
[4] Jet fuel, kerosene, other petroleum liquids, and waste oil.
[5] Petroleum coke is converted from short tons to barrels by multiplying by 5.
[6] Natural gas, plus a small amount of supplemental gaseous fuels.
[7] Blast furnace gas, propane gas, and other manufactured and waste gases derived from fossil fuels.
[8] Wood and wood-derived fuels.
[9] Municipal solid waste from biogenic sources, landfill gas, sludge waste, agricultural byproducts, and other biomass. Through 2000, also includes non-renewable waste (municipal solid waste from non-biogenic sources, and tire-derived fuels).
[10] Batteries, chemicals, hydrogen, pitch, purchased steam, sulfur, miscellaneous technologies, and, beginning in 2001, non-renewable waste (municipal solid waste from non-biogenic sources, and tire-derived fuels).
[11] Through 1988, data are for electric utilities only. Beginning in 1989, data are for electric utilities, independent power producers, commercial plants, and industrial plants.

R=Revised. P=Preliminary. NA=Not available. (s)=Less than 0.5 trillion Btu.

Notes: • Data are for fuels consumed to produce electricity. Data also include fuels consumed to produce useful thermal output at a small number of electric utility combined-heat-and-power (CHP) plants. • See Note 1, "Coverage of Electricity Statistics," at end of section. • Totals may not equal sum of components due to independent rounding.

Web Pages: • For all data beginning in 1949, see http://www.eia.doe.gov/emeu/aer/elect.html. • For related information, see http://www.eia.doe.gov/fuelelectric.html.

Sources: Tables 8.5b and 8.5d.

Table 8.5b Consumption of Combustible Fuels for Electricity Generation: Electric Power Sector, Selected Years, 1949-2007 (Subset of Table 8.5a)

Year	Coal [1]	Petroleum: Distillate Fuel Oil [2]	Petroleum: Residual Fuel Oil [3]	Petroleum: Other Liquids [4]	Petroleum: Petroleum Coke [5]	Petroleum: Total [5]	Natural Gas [6]	Other Gases [7]	Biomass: Wood [8]	Biomass: Waste [9]	Other [10]
	Thousand Short Tons	Thousand Barrels	Thousand Barrels	Thousand Barrels	Thousand Short Tons	Thousand Barrels	Million Cubic Feet	Trillion Btu	Trillion Btu	Trillion Btu	Trillion Btu
1949	83,963	4,767	61,534	NA	NA	66,301	550,121	NA	6	NA	NA
1950	91,871	5,423	69,998	NA	NA	75,421	628,919	NA	5	NA	NA
1955	143,759	5,412	69,862	NA	NA	75,274	1,153,280	NA	3	NA	NA
1960	176,685	3,824	84,371	NA	NA	88,195	1,724,762	NA	2	NA	NA
1965	244,788	4,928	110,274	NA	NA	115,203	2,321,101	NA	3	NA	NA
1970	320,182	24,123	311,381	NA	636	338,686	3,931,860	NA	1	2	NA
1971	327,301	34,283	362,187	NA	605	399,496	3,976,018	NA	1	2	NA
1972	351,768	53,465	440,294	NA	627	496,895	3,976,913	NA	1	2	NA
1973	389,212	47,058	513,190	NA	507	562,781	3,660,172	NA	1	2	NA
1974	391,811	53,128	483,146	NA	625	539,399	3,443,428	NA	1	2	NA
1975	405,962	38,907	467,221	NA	70	506,479	3,157,669	NA	(s)	2	NA
1976	448,371	41,843	514,077	NA	68	556,261	3,080,868	NA	1	2	NA
1977	477,126	48,837	574,869	NA	98	624,193	3,191,200	NA	3	2	NA
1978	481,235	47,520	588,319	NA	398	637,830	3,188,363	NA	2	1	NA
1979	527,051	30,691	492,606	NA	268	524,636	3,490,523	NA	3	2	NA
1980	569,274	29,051	391,163	NA	179	421,110	3,681,595	NA	3	2	NA
1981	596,797	21,313	329,798	NA	139	351,806	3,640,154	NA	3	1	NA
1982	593,666	15,337	234,434	NA	149	250,517	3,225,518	NA	2	1	NA
1983	625,211	16,512	228,984	NA	261	246,804	2,910,767	NA	2	2	NA
1984	664,399	15,190	189,289	NA	252	205,736	3,111,342	NA	5	4	NA
1985	693,841	14,635	158,779	NA	231	174,571	3,044,083	NA	8	7	NA
1986	685,056	14,326	216,156	NA	313	232,046	2,602,370	NA	5	7	NA
1987	717,894	15,367	184,011	NA	348	201,116	2,844,051	NA	8	7	NA
1988	758,372	18,769	229,327	NA	409	250,141	2,635,613	NA	10	8	NA
1989[11]	771,551	26,036	242,708	9	517	271,340	3,023,513	7	75	126	2
1990	781,301	16,394	183,285	25	1,008	204,745	3,147,289	6	106	180	(s)
1991	782,653	14,255	171,629	58	974	190,810	3,216,056	6	104	217	4
1992	793,390	12,469	137,681	118	1,490	157,719	3,324,963	12	120	252	3
1993	829,851	14,559	151,407	213	2,571	179,034	3,344,239	12	129	255	3
1994	836,113	20,241	137,198	667	2,256	169,387	3,758,484	12	134	269	2
1995	847,854	18,066	88,895	441	2,452	119,663	4,093,773	18	106	282	2
1996	894,400	18,472	98,795	567	2,467	130,168	3,659,810	16	117	280	2
1997	919,009	18,646	112,423	130	3,201	147,202	3,903,195	14	117	292	1
1998	934,126	23,166	165,875	411	3,999	209,447	4,415,813	23	125	287	2
1999	937,888	23,875	151,921	514	3,607	194,345	4,643,775	14	125	290	1
2000	982,713	29,722	138,047	403	3,155	183,946	5,014,071	19	126	294	1
2001	961,523	29,056	159,150	374	3,308	205,119	5,142,493	9	116	205	109
2002	975,251	21,810	104,577	1,243	5,705	156,154	5,408,279	25	141	224	137
2003	1,003,036	27,441	137,361	1,937	5,719	195,336	4,909,248	30	156	216	136
2004	1,015,079	18,927	139,806	2,702	7,357	198,220	5,305,863	38	157	220	136
2005	1,036,140	19,587	139,376	2,634	8,066	201,926	5,724,912	44	176	217	120
2006	R1,025,107	R12,613	R57,322	R1,844	R7,092	R107,238	R6,096,981	R43	R172	R228	R121
2007P	1,044,995	15,781	63,501	2,894	5,590	110,127	6,725,136	39	172	240	124

[1] Anthracite, bituminous coal, subbituminous coal, lignite, waste coal, and coal synfuel.
[2] Fuel oil nos. 1, 2, and 4. For 1949-1979, data are for gas turbine and internal combustion plant use of petroleum. For 1980-2000, electric utility data also include small amounts of kerosene and jet fuel.
[3] Fuel oil nos. 5 and 6. For 1949-1979, data are for steam plant use of petroleum. For 1980-2000, electric utility data also include a small amount of fuel oil no. 4.
[4] Jet fuel, kerosene, other petroleum liquids, and waste oil.
[5] Petroleum coke is converted from short tons to barrels by multiplying by 5.
[6] Natural gas, plus a small amount of supplemental gaseous fuels.
[7] Blast furnace gas, propane gas, and other manufactured and waste gases derived from fossil fuels.
[8] Wood and wood-derived fuels.
[9] Municipal solid waste from biogenic sources, landfill gas, sludge waste, agricultural byproducts, and other biomass. Through 2000, also includes non-renewable waste (municipal solid waste from non-biogenic sources, and tire-derived fuels).
[10] Batteries, chemicals, hydrogen, pitch, purchased steam, sulfur, miscellaneous technologies, and, beginning in 2001, non-renewable waste (municipal solid waste from non-biogenic sources, and tire-derived fuels).
[11] Through 1988, data are for electric utilities only. Beginning in 1989, data are for electric utilities and independent power producers.

R=Revised. P=Preliminary. NA=Not available. (s)=Less than 0.5 trillion Btu.

Notes: • Data are for fuels consumed to produce electricity. Data also include fuels consumed to produce useful thermal output at a small number of electric utility combined-heat-and-power (CHP) plants. • The electric power sector comprises electricity-only and combined-heat-and-power (CHP) plants within the NAICS 22 category whose primary business is to sell electricity, or electricity and heat, to the public. • See Table 8.5d for commercial and industrial CHP and electricity-only data. • See Note 1, "Coverage of Electricity Statistics," and Note 2, "Classification of Power Plants Into Energy-Use Sectors," at end of section. • Totals may not equal sum of components due to independent rounding.

Web Pages: • For all data beginning in 1949, see http://www.eia.doe.gov/emeu/aer/elect.html. • For related information, see http://www.eia.doe.gov/fuelelectric.html.

Sources: • 1949-September 1977—Federal Power Commission, Form FPC-4, "Monthly Power Plant Report." • October 1977-1981—Federal Energy Regulatory Commission, Form FPC-4, "Monthly Power Plant Report." • 1982-1988—Energy Information Administration (EIA), Form EIA-759, "Monthly Power Plant Report." • 1989-1997—EIA, Form EIA-759, "Monthly Power Plant Report," and Form EIA-867, "Annual Nonutility Power Producer Report." • 1998-2000—EIA, Form EIA-759, "Monthly Power Plant Report," and Form EIA-860B, "Annual Electric Generator Report—Nonutility." • 2001-2003—EIA, Form EIA-906, "Power Plant Report." • 2004 forward—EIA, Form EIA-906, "Power Plant Report," and Form EIA-920, "Combined Heat and Power Plant Report."

Table 8.5c Consumption of Combustible Fuels for Electricity Generation: Electric Power Sector by Plant Type, 1989-2007 (Breakout of Table 8.5b)

Year	Coal [1]	Petroleum: Distillate Fuel Oil [2]	Petroleum: Residual Fuel Oil [3]	Petroleum: Other Liquids [4]	Petroleum: Petroleum Coke [5]	Petroleum: Total [5]	Natural Gas [6]	Other Gases [7]	Biomass: Wood [8]	Biomass: Waste [9]	Other [10]
	Thousand Short Tons	Thousand Barrels	Thousand Barrels	Thousand Barrels	Thousand Short Tons	Thousand Barrels	Million Cubic Feet	Trillion Btu	Trillion Btu	Trillion Btu	Trillion Btu
Electricity-Only Plants [11]											
1989	767,378	25,574	241,960	3	517	270,125	2,790,567	–	59	111	–
1990	774,213	14,956	181,231	17	1,008	201,246	2,794,110	(s)	87	162	–
1991	773,183	13,822	171,157	51	974	189,898	2,822,159	(s)	85	195	–
1992	781,186	11,998	135,779	48	1,320	154,428	2,828,996	(s)	94	232	–
1993	816,558	13,460	149,287	11	1,553	170,521	2,755,093	(s)	101	237	–
1994	821,209	16,693	134,666	52	1,193	157,375	3,064,561	(s)	112	248	–
1995	832,928	16,169	86,584	133	1,082	108,297	3,287,571	(s)	84	262	–
1996	878,825	17,361	96,386	50	1,010	118,848	2,823,724	(s)	94	258	–
1997	904,245	17,702	109,989	30	1,687	136,156	3,039,227	1	91	266	–
1998	920,353	22,293	163,541	295	2,202	197,137	3,543,931	1	95	263	–
1999	924,692	22,877	149,193	380	1,891	181,905	3,729,175	1	105	264	–
2000	967,080	28,001	135,419	94	1,457	170,799	4,092,729	2	105	267	–
2001	946,068	27,695	157,090	26	1,827	193,945	4,163,930	(s)	96	179	98
2002	960,077	21,521	102,622	444	3,925	144,212	4,258,467	6	118	193	117
2003	983,538	25,951	136,050	936	4,794	186,904	3,780,314	6	127	185	120
2004	994,774	17,944	137,736	1,441	6,096	187,601	4,141,535	5	134	190	122
2005	1,015,640	18,689	137,082	1,676	6,876	191,827	4,592,271	(s)	143	189	108
2006	R1,004,769	R12,375	R55,192	R991	R5,988	R98,497	R5,091,049	(s)	R141	R198	R107
2007P	1,024,477	15,043	61,372	1,829	4,621	101,351	5,607,088	1	142	209	112
Combined-Heat-and-Power Plants [12]											
1989	4,173	462	747	6	–	1,215	232,946	7	16	16	2
1990	7,088	1,438	2,054	7	–	3,499	353,179	6	18	18	(s)
1991	9,470	433	473	7	–	912	393,898	6	20	22	4
1992	12,204	471	1,902	69	170	3,291	495,967	12	25	20	3
1993	13,293	1,098	2,120	202	1,018	8,513	589,147	12	28	18	3
1994	14,904	3,548	2,531	615	1,063	12,011	693,923	12	22	22	2
1995	14,926	1,898	2,311	307	1,370	11,366	806,202	18	22	20	2
1996	15,575	1,111	2,410	517	1,456	11,320	836,086	15	24	22	2
1997	14,764	944	2,434	100	1,514	11,046	863,968	14	26	26	1
1998	13,773	872	2,334	117	1,797	12,310	871,881	21	30	24	2
1999	13,197	998	2,728	134	1,716	12,440	914,600	14	20	26	1
2000	15,634	1,721	2,627	310	1,698	13,147	921,341	17	21	28	1
2001	15,455	1,360	2,059	347	1,482	11,175	978,563	9	20	26	11
2002	15,174	289	1,955	800	1,780	11,942	1,149,812	20	23	30	20
2003	19,498	1,491	1,311	1,002	926	8,431	1,128,935	23	29	31	16
2004	20,306	983	2,070	1,261	1,261	10,620	1,164,328	33	23	30	14
2005	20,500	898	2,293	958	1,190	10,099	1,132,641	44	32	28	12
2006	R20,337	R238	R2,130	R853	1,104	R8,740	R1,005,932	R42	31	R30	13
2007P	20,518	738	2,129	1,065	969	8,776	1,118,048	38	30	31	13

[1] Anthracite, bituminous coal, subbituminous coal, lignite, waste coal, and coal synfuel.
[2] Fuel oil nos. 1, 2, and 4. Through 2000, electric utility data also include small amounts of kerosene and jet fuel.
[3] Fuel oil nos. 5 and 6. Through 2000, electric utility data also include a small amount of fuel oil no. 4.
[4] Jet fuel, kerosene, other petroleum liquids, and waste oil.
[5] Petroleum coke is converted from short tons to barrels by multiplying by 5.
[6] Natural gas, plus a small amount of supplemental gaseous fuels.
[7] Blast furnace gas, propane gas, and other manufactured and waste gases derived from fossil fuels.
[8] Wood and wood-derived fuels.
[9] Municipal solid waste from biogenic sources, landfill gas, sludge waste, agricultural byproducts, and other biomass. Through 2000, also includes non-renewable waste (municipal solid waste from non-biogenic sources, and tire-derived fuels).
[10] Batteries, chemicals, hydrogen, pitch, purchased steam, sulfur, miscellaneous technologies, and, beginning in 2001, non-renewable waste (municipal solid waste from non-biogenic sources, and tire-derived fuels).
[11] Electricity-only plants within the NAICS 22 category whose primary business is to sell electricity to the public. Data also include a small number of electric utility combined-heat-and-power (CHP) plants.
[12] Combined-heat-and-power (CHP) plants within the NAICS 22 category whose primary business is to sell electricity and heat to the public. Data do not include electric utility CHP plants—these are included under "Electricity-Only Plants."

R=Revised. P=Preliminary. – = No data reported. (s)=Less than 0.5.

Notes: • Data are for fuels consumed to produce electricity. Data also include fuels consumed to produce useful thermal output at a small number of electric utility combined-heat-and-power (CHP) plants. • See Table 8.5d for commercial and industrial CHP and electricity-only data. • See Note 1, "Coverage of Electricity Statistics," and Note 2, "Classification of Power Plants Into Energy-Use Sectors," at end of section. • Totals may not equal sum of components due to independent rounding.

Web Page: For related information, see http://www.eia.doe.gov/fuelelectric.html.

Sources: • 1989-1997—Energy Information Administration (EIA), Form EIA-759, "Monthly Power Plant Report," and Form EIA-867, "Annual Nonutility Power Producer Report." • 1998-2000—EIA, Form EIA-759, "Monthly Power Plant Report," and Form EIA-860B, "Annual Electric Generator Report—Nonutility." • 2001-2003—EIA, Form EIA-906, "Power Plant Report." • 2004 forward—EIA, Form EIA-906, "Power Plant Report," and Form EIA-920, "Combined Heat and Power Plant Report."

Table 8.5d Consumption of Combustible Fuels for Electricity Generation: Commercial and Industrial Sectors, 1989-2007 (Subset of Table 8.5a)

Year	Coal [1]	Petroleum: Distillate Fuel Oil [2]	Petroleum: Residual Fuel Oil [3]	Petroleum: Other Liquids [4]	Petroleum: Petroleum Coke [5]	Petroleum: Total [5]	Natural Gas [6]	Other Gases [7]	Biomass: Wood [8]	Biomass: Waste [9]	Other [10]
	Thousand Short Tons	Thousand Barrels	Thousand Barrels	Thousand Barrels	Thousand Short Tons	Thousand Barrels	Million Cubic Feet	Trillion Btu	Trillion Btu	Trillion Btu	Trillion Btu
					Commercial Sector [11]						
1989	414	882	282	–	–	1,165	17,987	1	2	9	–
1990	417	580	372	(s)	–	953	27,544	1	2	15	–
1991	403	430	146	(s)	–	576	26,806	1	2	15	(s)
1992	371	289	137	(s)	1	429	32,674	1	1	16	(s)
1993	404	384	279	4	1	672	37,435	1	1	16	(s)
1994	404	481	209	–	1	694	40,828	1	1	17	–
1995	569	493	152	(s)	1	649	42,700	–	1	21	(s)
1996	656	422	218	(s)	1	645	42,380	(s)	1	31	(s)
1997	630	583	200	–	1	790	38,975	(s)	1	34	(s)
1998	440	436	359	–	1	802	40,693	(s)	1	32	–
1999	481	506	421	–	1	931	39,045	(s)	(s)	33	(s)
2000	514	505	310	1	1	823	37,029	(s)	(s)	26	(s)
2001	532	520	469	2	6	1,023	36,248	(s)	(s)	15	7
2002	477	524	292	10	2	834	32,545	(s)	(s)	18	11
2003	582	553	326	3	2	894	38,480	–	(s)	19	11
2004	602	821	350	1	3	1,188	45,883	–	1	22	13
2005	770	588	333	1	3	939	47,851	–	1	25	13
2006	R743	R287	R176	(s)	4	R481	R48,384	–	1	26	14
2007P	745	228	134	(s)	5	387	49,651	–	1	27	13
					Industrial Sector [12]						
1989	9,707	815	6,830	294	150	8,688	443,928	83	267	15	37
1990	10,740	1,169	7,192	412	905	13,299	516,729	104	335	16	36
1991	10,610	1,879	6,004	322	815	12,283	521,916	118	318	14	55
1992	11,379	1,735	6,650	642	1,013	14,093	542,081	128	359	15	37
1993	11,898	1,902	7,373	498	597	12,755	546,978	123	355	17	31
1994	12,279	1,644	7,818	263	762	13,537	567,836	123	364	14	38
1995	12,171	1,056	6,460	239	902	12,265	601,397	114	373	13	40
1996	12,153	1,359	7,042	1,145	853	13,813	610,268	143	394	13	35
1997	12,311	1,079	6,118	107	884	11,723	622,599	105	367	14	36
1998	11,728	1,461	6,494	137	860	12,392	624,878	102	349	13	35
1999	11,432	1,571	5,845	460	944	12,595	639,165	112	364	8	39
2000	11,706	1,448	5,024	1,046	588	10,459	640,381	107	369	10	45
2001	10,636	1,574	5,693	479	557	10,530	653,565	88	370	7	44
2002	11,855	952	4,366	640	1,130	11,608	685,239	106	464	15	43
2003	10,440	1,678	4,831	1,006	582	10,424	668,407	127	362	13	46
2004	10,337	921	5,015	1,256	581	10,100	764,828	148	376	11	27
2005	8,969	988	4,525	668	442	8,392	713,999	133	306	9	28
2006	R9,496	R472	R3,521	R768	R578	R7,651	R724,259	R138	R350	8	R21
2007P	7,606	595	3,066	806	627	7,601	732,658	127	376	8	31

[1] Anthracite, bituminous coal, subbituminous coal, lignite, waste coal, and coal synfuel.
[2] Fuel oil nos. 1, 2, and 4.
[3] Fuel oil nos. 5 and 6.
[4] Jet fuel, kerosene, other petroleum liquids, and waste oil.
[5] Petroleum coke is converted from short tons to barrels by multiplying by 5.
[6] Natural gas, plus a small amount of supplemental gaseous fuels.
[7] Blast furnace gas, propane gas, and other manufactured and waste gases derived from fossil fuels.
[8] Wood and wood-derived fuels.
[9] Municipal solid waste from biogenic sources, landfill gas, sludge waste, agricultural byproducts, and other biomass. Through 2000, also includes non-renewable waste (municipal solid waste from non-biogenic sources, and tire-derived fuels).
[10] Batteries, chemicals, hydrogen, pitch, purchased steam, sulfur, miscellaneous technologies, and, beginning in 2001, non-renewable waste (municipal solid waste from non-biogenic sources, and tire-derived fuels).
[11] Commercial combined-heat-and-power (CHP) and commercial electricity-only plants.
[12] Industrial combined-heat-and-power (CHP) and industrial electricity-only plants.

R=Revised. P=Preliminary. – = No data reported. (s)=Less than 0.5.

Notes: • Data are for fuels consumed to produce electricity. • See Tables 8.5b and 8.5c for electric power sector electricity-only and CHP data. • See Note 1, "Coverage of Electricity Statistics," and Note 2, "Classification of Power Plants Into Energy-Use Sectors," at end of section. • Totals may not equal sum of components due to independent rounding.

Web Page: For related information, see http://www.eia.doe.gov/fuelelectric.html.

Sources: • 1989-1997—Energy Information Administration (EIA), Form EIA-867, "Annual Nonutility Power Producer Report." • 1998-2000—EIA, Form EIA-860B, "Annual Electric Generator Report—Nonutility." • 2001-2003—EIA, Form EIA-906, "Power Plant Report." • 2004 forward—EIA, Form EIA-906, "Power Plant Report," and Form EIA-920, "Combined Heat and Power Plant Report."

Figure 8.6 Estimated Consumption of Combustible Fuels for Useful Thermal Output at Combined-Heat-and-Power Plants by Sector, 1989-2007

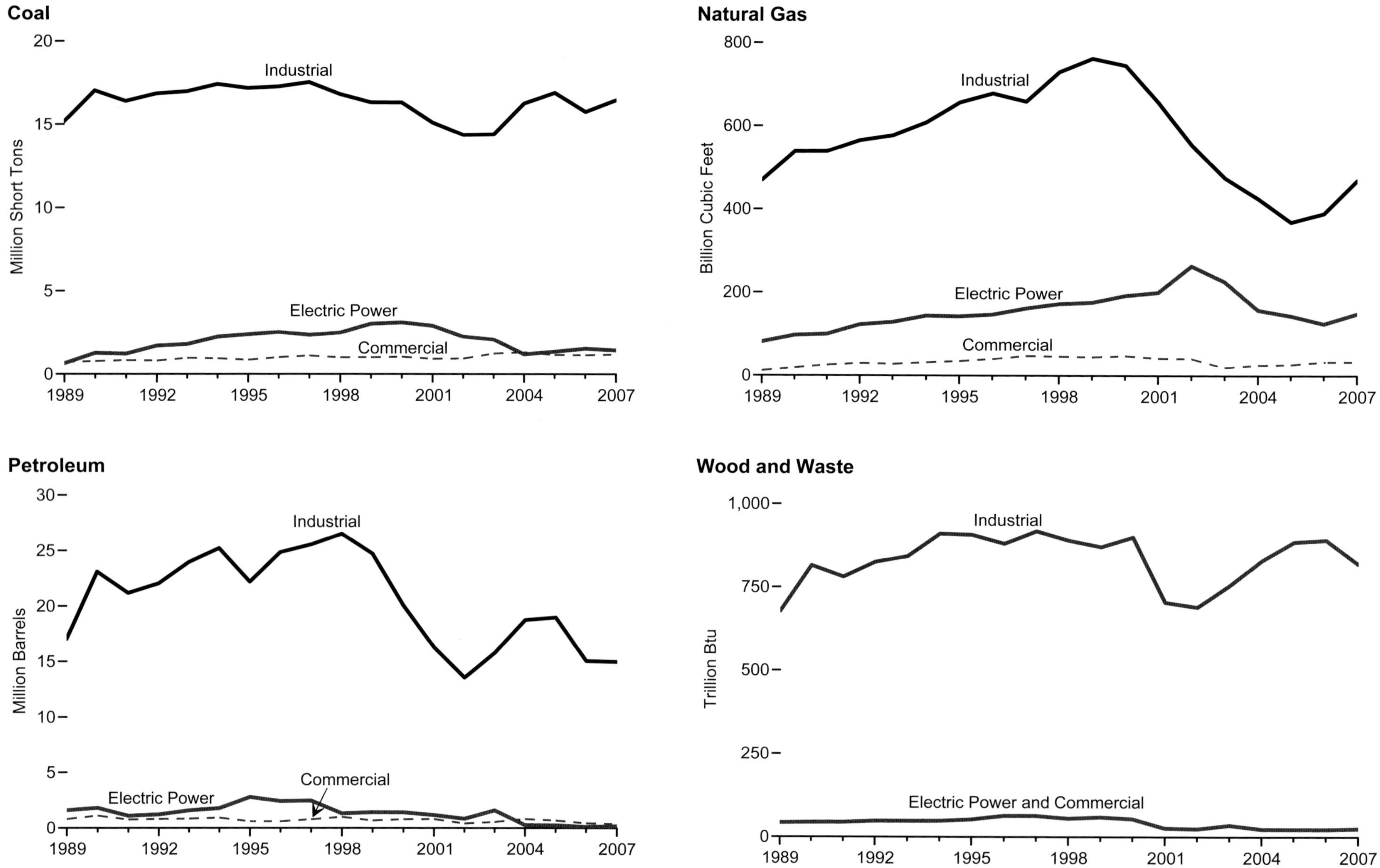

Sources: Tables 8.6b and 8.6c.

Table 8.6a Estimated Consumption of Combustible Fuels for Useful Thermal Output at Combined-Heat-and-Power Plants: Total (All Sectors), 1989-2007 (Sum of Tables 8.6b and 8.6c)

Year	Coal [1]	Petroleum					Natural Gas [6]	Other Gases [7]	Biomass		Other [10]
		Distillate Fuel Oil [2]	Residual Fuel Oil [3]	Other Liquids [4]	Petroleum Coke [5]	Total [5]			Wood [8]	Waste [9]	
	Thousand Short Tons	Thousand Barrels			Thousand Short Tons	Thousand Barrels	Million Cubic Feet	Trillion Btu	Trillion Btu		Trillion Btu
1989	16,510	1,410	16,391	353	247	19,391	563,307	116	683	38	49
1990	19,081	2,050	18,465	895	918	26,002	654,749	176	813	46	50
1991	18,458	3,027	15,293	835	777	23,039	663,963	185	779	46	55
1992	19,372	2,358	16,474	935	862	24,077	717,860	200	822	51	52
1993	19,750	2,449	17,933	857	1,031	26,394	733,584	178	836	56	51
1994	20,609	2,811	18,822	609	1,137	27,929	784,015	180	903	57	53
1995	20,418	2,082	16,661	642	1,235	25,562	834,382	181	902	59	55
1996	20,806	2,192	18,552	756	1,275	27,873	865,774	187	876	69	54
1997	21,005	2,584	15,882	289	2,009	28,802	868,569	188	913	68	67
1998	20,320	4,944	16,539	681	1,336	28,845	949,106	209	875	72	58
1999	20,373	4,665	14,133	838	1,437	26,822	982,958	224	862	68	60
2000	20,466	2,897	13,292	1,455	924	22,266	985,263	230	884	71	63
2001	18,944	2,574	11,826	563	661	18,268	898,286	166	696	35	69
2002	17,561	1,462	9,402	1,363	517	14,811	860,019	147	682	32	60
2003	17,720	2,153	10,341	1,629	763	17,939	721,267	138	746	44	69
2004	18,779	2,851	12,307	805	779	19,856	610,105	167	827	28	50
2005	19,402	3,283	12,681	966	601	19,937	541,206	171	871	38	51
2006	[R]18,437	[R]1,284	[R]8,827	[R]784	[R]948	[R]15,636	[R]549,335	[R]160	[R]876	[R]38	[R]59
2007[P]	19,084	1,796	7,564	878	1,063	15,554	652,073	182	806	39	57

[1] Anthracite, bituminous coal, subbituminous coal, lignite, waste coal, and coal synfuel.
[2] Fuel oil nos. 1, 2, and 4.
[3] Fuel oil nos. 5 and 6.
[4] Jet fuel, kerosene, other petroleum liquids, and waste oil.
[5] Petroleum coke is converted from short tons to barrels by multiplying by 5.
[6] Natural gas, plus a small amount of supplemental gaseous fuels.
[7] Blast furnace gas, propane gas, and other manufactured and waste gases derived from fossil fuels.
[8] Wood and wood-derived fuels.
[9] Municipal solid waste from biogenic sources, landfill gas, sludge waste, agricultural byproducts, and other biomass. Through 2000, also includes non-renewable waste (municipal solid waste from non-biogenic sources, and tire-derived fuels).
[10] Batteries, chemicals, hydrogen, pitch, purchased steam, sulfur, miscellaneous technologies, and, beginning in 2001, non-renewable waste (municipal solid waste from non-biogenic sources, and tire-derived fuels).

R=Revised. P=Preliminary.

Notes: • Estimates are for fuels consumed to produce useful thermal output; they exclude fuels consumed to produce electricity. • Data do not include electric utility combined-heat-and-power (CHP) plants. • See Note 1, "Coverage of Electricity Statistics," at end of section. • See "Useful Thermal Output" in Glossary. • Totals may not equal sum of components due to independent rounding.

Web Page: For related information, see http://www.eia.doe.gov/fuelelectric.html.

Sources: Tables 8.6b and 8.6c.

Table 8.6b Estimated Consumption of Combustible Fuels for Useful Thermal Output at Combined-Heat-and-Power Plants: Electric Power Sector, 1989-2007 (Subset of Table 8.6a)

Year	Coal [1]	Petroleum: Distillate Fuel Oil [2]	Petroleum: Residual Fuel Oil [3]	Petroleum: Other Liquids [4]	Petroleum: Petroleum Coke [5]	Petroleum: Total [5]	Natural Gas [6]	Other Gases [7]	Biomass: Wood [8]	Biomass: Waste [9]	Other [10]
	Thousand Short Tons	Thousand Barrels	Thousand Barrels	Thousand Barrels	Thousand Short Tons	Thousand Barrels	Million Cubic Feet	Trillion Btu	Trillion Btu	Trillion Btu	Trillion Btu
1989	639	120	1,471	1	–	1,591	81,670	3	24	6	1
1990	1,266	173	1,630	2	–	1,805	97,330	5	23	8	(s)
1991	1,221	104	995	1	–	1,101	99,868	5	21	11	1
1992	1,704	154	1,045	10	4	1,229	122,908	6	21	10	2
1993	1,794	290	1,074	27	40	1,591	128,743	4	21	10	2
1994	2,241	371	1,024	104	58	1,791	144,062	6	18	12	1
1995	2,376	486	1,127	58	222	2,784	142,753	5	19	15	(s)
1996	2,520	308	1,155	86	175	2,424	147,091	5	20	21	(s)
1997	2,355	343	1,246	23	171	2,466	161,608	10	20	17	(s)
1998	2,493	134	653	19	103	1,322	172,471	6	12	20	(s)
1999	3,033	183	572	30	128	1,423	175,757	4	13	25	(s)
2000	3,107	294	467	51	120	1,412	192,253	7	8	24	(s)
2001	2,910	219	355	3	119	1,171	199,808	6	10	5	4
2002	2,255	66	197	23	111	841	263,619	7	10	6	6
2003	2,080	190	919	88	80	1,596	225,967	12	11	14	4
2004	1,189	180	10	11	15	277	157,900	20	8	3	2
2005	1,345	88	33	51	17	258	144,233	40	9	3	3
2006	R1,529	R33	R23	R26	R9	R127	R125,119	R22	10	3	4
2007P	1,429	93	28	49	3	187	148,946	28	12	3	4

[1] Anthracite, bituminous coal, subbituminous coal, lignite, waste coal, and coal synfuel.
[2] Fuel oil nos. 1, 2, and 4.
[3] Fuel oil nos. 5 and 6.
[4] Jet fuel, kerosene, other petroleum liquids, and waste oil.
[5] Petroleum coke is converted from short tons to barrels by multiplying by 5.
[6] Natural gas, plus a small amount of supplemental gaseous fuels.
[7] Blast furnace gas, propane gas, and other manufactured and waste gases derived from fossil fuels.
[8] Wood and wood-derived fuels.
[9] Municipal solid waste from biogenic sources, landfill gas, sludge waste, agricultural byproducts, and other biomass. Through 2000, also includes non-renewable waste (municipal solid waste from non-biogenic sources, and tire-derived fuels).
[10] Batteries, chemicals, hydrogen, pitch, purchased steam, sulfur, miscellaneous technologies, and, beginning in 2001, non-renewable waste (municipal solid waste from non-biogenic sources, and tire-derived fuels).

R=Revised. P=Preliminary. – = No data reported. (s)=Less than 0.5.

Notes: • Estimates are for fuels consumed to produce useful thermal output; they exclude fuels consumed to produce electricity. • Data are for combined-heat-and-power (CHP) plants within the NAICS 22 category whose primary business is to sell electricity and heat to the public. Data do not include electric utility CHP plants. • See Table 8.6c for commercial and industrial CHP data. • See Note 1, "Coverage of Electricity Statistics," and Note 2, "Classification of Power Plants Into Energy-Use Sectors," at end of section. • See "Useful Thermal Output" in Glossary.• Totals may not equal sum of components due to independent rounding.

Web Page: For related information, see http://www.eia.doe.gov/fuelelectric.html.

Sources: • 1989-1997—Energy Information Administration (EIA), Form EIA-867, "Annual Nonutility Power Producer Report." • 1998-2000—EIA, Form EIA-860B, "Annual Electric Generator Report—Nonutility." • 2001-2003—EIA, Form EIA-906, "Power Plant Report." • 2004 forward—EIA, Form EIA-920, "Combined Heat and Power Plant Report."

Table 8.6c Estimated Consumption of Combustible Fuels for Useful Thermal Output at Combined-Heat-and-Power Plants: Commercial and Industrial Sectors, 1989-2007 (Subset of Table 8.6a)

Year	Coal [1]	Petroleum: Distillate Fuel Oil [2]	Petroleum: Residual Fuel Oil [3]	Petroleum: Other Liquids [4]	Petroleum: Petroleum Coke [5]	Petroleum: Total [5]	Natural Gas [6]	Other Gases [7]	Biomass: Wood [8]	Biomass: Waste [9]	Other [10]
	Thousand Short Tons	Thousand Barrels			Thousand Short Tons	Thousand Barrels	Million Cubic Feet	Trillion Btu	Trillion Btu		Trillion Btu
					Commercial Sector [11]						
1989	711	202	601	–	–	803	12,049	(s)	(s)	13	–
1990	773	389	715	(s)	–	1,104	18,913	(s)	(s)	13	–
1991	826	356	405	(s)	–	761	25,295	(s)	(s)	11	(s)
1992	804	259	538	(s)	2	807	29,672	(s)	1	16	(s)
1993	968	272	548	2	4	843	27,738	(s)	(s)	17	(s)
1994	940	534	379	–	4	931	31,457	(s)	(s)	17	–
1995	850	319	261	(s)	3	596	34,964	–	(s)	19	(s)
1996	1,005	260	328	(s)	3	601	40,075	(s)	1	22	(s)
1997	1,108	470	309	–	3	794	47,941	(s)	1	24	–
1998	1,002	418	573	–	3	1,006	46,527	(s)	1	22	–
1999	1,009	254	412	–	3	682	44,991	(s)	1	21	–
2000	1,034	403	366	2	4	792	47,844	(s)	1	21	–
2001	916	505	304	–	–	809	42,407	–	1	10	7
2002	929	248	108	28	6	416	41,430	–	1	8	6
2003	1,234	119	381	12	9	555	19,973	–	1	10	8
2004	1,315	294	477	20	6	821	26,189	–	1	12	8
2005	1,151	206	456	(s)	6	691	27,364	–	1	9	7
2006	R1,143	R79	R344	R (s)	6	R453	R33,877	R (s)	1	R10	7
2007P	1,179	65	285	(s)	7	387	33,708	(s)	(s)	10	7
					Industrial Sector [12]						
1989	15,160	1,088	14,320	352	247	16,997	469,588	113	659	19	48
1990	17,041	1,488	16,120	893	918	23,093	538,506	171	790	25	50
1991	16,412	2,567	13,893	834	777	21,177	538,800	180	758	23	55
1992	16,864	1,945	14,891	925	856	22,041	565,279	194	801	24	50
1993	16,988	1,887	16,311	829	987	23,960	577,103	174	815	29	49
1994	17,428	1,906	17,419	505	1,075	25,207	608,496	173	884	27	52
1995	17,192	1,277	15,272	584	1,010	22,182	656,665	175	882	25	55
1996	17,281	1,624	17,069	670	1,097	24,848	678,608	182	855	26	53
1997	17,542	1,772	14,328	267	1,835	25,541	659,021	178	892	27	67
1998	16,824	4,391	15,313	662	1,230	26,518	730,108	202	862	29	58
1999	16,330	4,228	13,148	808	1,307	24,718	762,210	219	849	23	60
2000	16,325	2,200	12,459	1,402	800	20,062	745,165	223	875	25	63
2001	15,119	1,850	11,167	560	542	16,287	656,071	160	685	20	58
2002	14,377	1,149	9,097	1,312	399	13,555	554,970	139	672	18	48
2003	14,406	1,844	9,041	1,529	675	15,788	475,327	126	735	21	57
2004	16,276	2,376	11,819	774	758	18,758	426,016	147	818	13	40
2005	16,906	2,989	12,192	915	578	18,987	369,609	131	861	25	41
2006	R15,765	R1,171	R8,460	R758	R933	R15,055	R390,338	R138	R866	R26	R49
2007P	16,477	1,638	7,251	828	1,053	14,979	469,420	154	794	27	46

[1] Anthracite, bituminous coal, subbituminous coal, lignite, waste coal, and coal synfuel.
[2] Fuel oil nos. 1, 2, and 4.
[3] Fuel oil nos. 5 and 6.
[4] Jet fuel, kerosene, other petroleum liquids, and waste oil.
[5] Petroleum coke is converted from short tons to barrels by multiplying by 5.
[6] Natural gas, plus a small amount of supplemental gaseous fuels.
[7] Blast furnace gas, propane gas, and other manufactured and waste gases derived from fossil fuels.
[8] Wood and wood-derived fuels.
[9] Municipal solid waste from biogenic sources, landfill gas, sludge waste, agricultural byproducts, and other biomass. Through 2000, also includes non-renewable waste (municipal solid waste from non-biogenic sources, and tire-derived fuels).
[10] Batteries, chemicals, hydrogen, pitch, purchased steam, sulfur, miscellaneous technologies, and, beginning in 2001, non-renewable waste (municipal solid waste from non-biogenic sources, and tire-derived fuels).
[11] Commercial combined-heat-and-power (CHP) plants.
[12] Industrial combined-heat-and-power (CHP) plants.

R=Revised. P=Preliminary. – = No data reported. (s)=Less than 0.5.

Notes: • Estimates are for fuels consumed to produce useful thermal output; they exclude fuels consumed to produce electricity. • See Table 8.6b for electric power sector CHP data. • See Note 1, "Coverage of Electricity Statistics," and Note 2, "Classification of Power Plants Into Energy-Use Sectors," at end of section. • See "Useful Thermal Output" in Glossary. • Totals may not equal sum of components due to independent rounding.

Web Page: For related information, see http://www.eia.doe.gov/fuelelectric.html.

Sources: • 1989-1997—Energy Information Administration (EIA), Form EIA-867, "Annual Nonutility Power Producer Report." • 1998-2000—EIA, Form EIA-860B, "Annual Electric Generator Report—Nonutility." • 2001-2003—EIA, Form EIA-906, "Power Plant Report." • 2004 forward—EIA, Form EIA-920, "Combined Heat and Power Plant Report."

Figure 8.7 Consumption of Combustible Fuels for Electricity Generation and Useful Thermal Output, 1989-2007

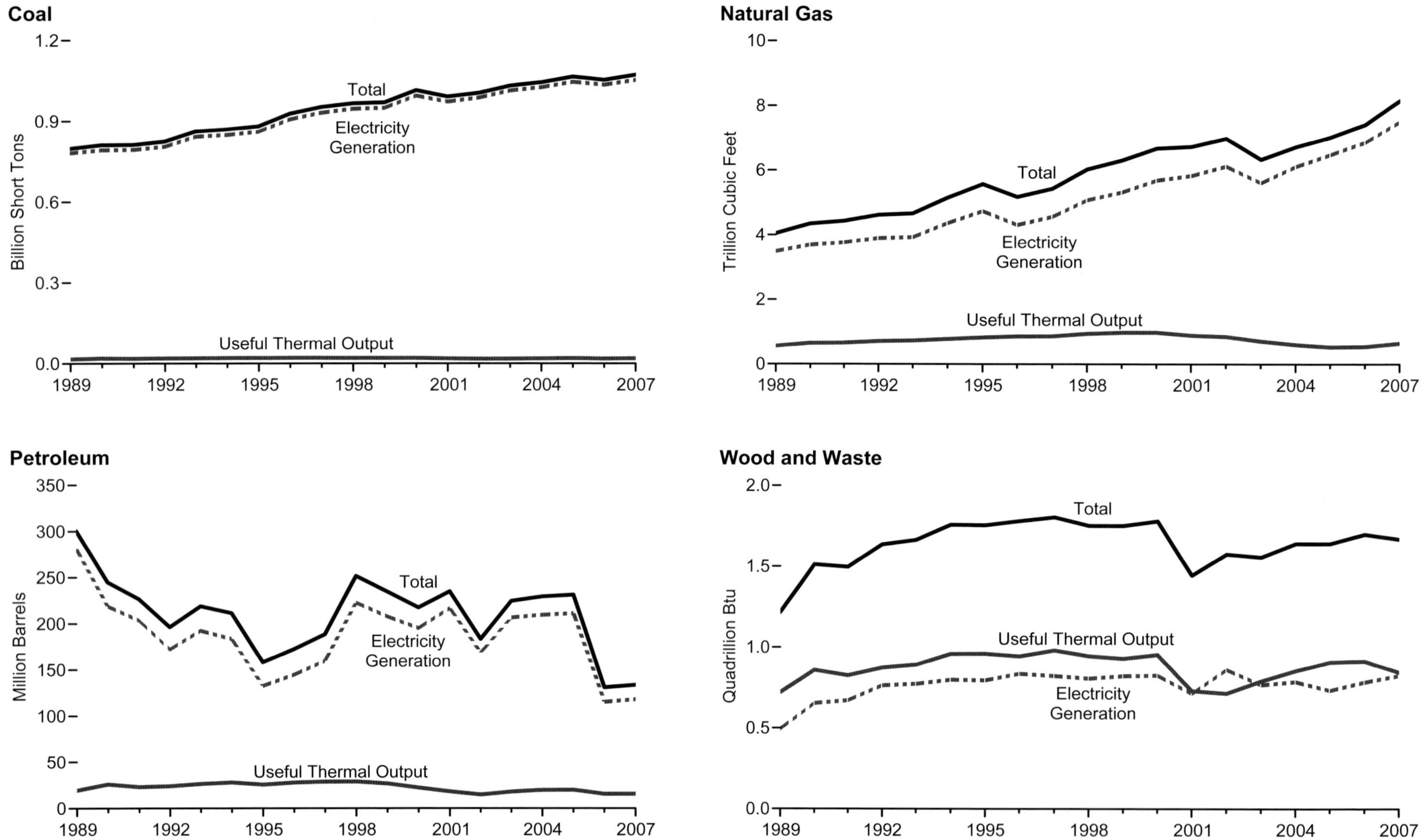

Sources: Tables 8.5a, 8.6a, and 8.7a.

Table 8.7a Consumption of Combustible Fuels for Electricity Generation and Useful Thermal Output: Total (All Sectors), 1989-2007 (Sum of Tables 8.7b and 8.7c)

Year	Coal[1]	Petroleum: Distillate Fuel Oil[2]	Petroleum: Residual Fuel Oil[3]	Petroleum: Other Liquids[4]	Petroleum: Petroleum Coke[5]	Petroleum: Total[5]	Natural Gas[6]	Other Gases[7]	Biomass: Wood[8]	Biomass: Waste[9]	Other[10]
	Thousand Short Tons	Thousand Barrels	Thousand Barrels	Thousand Barrels	Thousand Short Tons	Thousand Barrels	Million Cubic Feet	Trillion Btu	Trillion Btu	Trillion Btu	Trillion Btu
1989	798,181	29,143	266,211	656	915	300,583	4,048,736	206	1,028	189	88
1990	811,538	20,194	209,314	1,332	2,832	244,998	4,346,311	288	1,256	257	86
1991	812,124	19,590	193,073	1,215	2,566	226,708	4,428,742	311	1,204	292	114
1992	824,512	16,852	160,941	1,695	3,366	196,318	4,617,578	341	1,303	333	92
1993	861,904	19,293	176,992	1,571	4,200	218,855	4,662,236	314	1,321	344	85
1994	869,405	25,177	164,047	1,539	4,157	211,547	5,151,163	316	1,401	357	92
1995	881,012	21,697	112,168	1,322	4,590	158,140	5,572,253	313	1,382	374	97
1996	928,015	22,444	124,607	2,468	4,596	172,499	5,178,232	346	1,389	392	91
1997	952,955	22,893	134,623	526	6,095	188,517	5,433,338	307	1,397	407	103
1998	966,615	30,006	189,267	1,230	6,196	251,486	6,030,490	334	1,349	404	95
1999	970,175	30,616	172,319	1,812	5,989	234,694	6,304,942	350	1,352	400	101
2000	1,015,398	34,572	156,673	2,904	4,669	217,494	6,676,744	356	1,380	401	109
2001	991,635	33,724	177,137	1,418	4,532	234,940	6,730,591	263	1,182	263	229
2002	1,005,144	24,748	118,637	3,257	7,353	183,408	6,986,081	278	1,287	289	252
2003	1,031,778	31,825	152,859	4,576	7,067	224,593	6,337,402	294	1,266	293	262
2004	1,044,798	23,520	157,478	4,764	8,721	229,364	6,726,679	354	1,360	281	226
2005	1,065,281	24,446	156,915	4,270	9,113	231,193	7,027,967	348	1,353	289	213
2006	R1,053,783	R14,655	R69,846	R3,396	R8,622	R131,005	R7,418,959	R341	R1,399	R300	R215
2007P	1,072,430	18,401	74,265	4,577	7,285	133,668	8,159,519	348	1,354	315	226

[1] Anthracite, bituminous coal, subbituminous coal, lignite, waste coal, and coal synfuel.
[2] Fuel oil nos. 1, 2, and 4. Through 2000, electric utility data also include small amounts of kerosene and jet fuel.
[3] Fuel oil nos. 5 and 6. Through 2000, electric utility data also include a small amount of fuel oil no. 4.
[4] Jet fuel, kerosene, other petroleum liquids, and waste oil.
[5] Petroleum coke is converted from short tons to barrels by multiplying by 5.
[6] Natural gas, plus a small amount of supplemental gaseous fuels.
[7] Blast furnace gas, propane gas, and other manufactured and waste gases derived from fossil fuels.
[8] Wood and wood-derived fuels.
[9] Municipal solid waste from biogenic sources, landfill gas, sludge waste, agricultural byproducts, and other biomass. Through 2000, also includes non-renewable waste (municipal solid waste from non-biogenic sources, and tire-derived fuels).
[10] Batteries, chemicals, hydrogen, pitch, purchased steam, sulfur, miscellaneous technologies, and, beginning in 2001, non-renewable waste (municipal solid waste from non-biogenic sources, and tire-derived fuels).

R=Revised. P=Preliminary.

Notes: • Data are for fuels consumed to produce electricity and useful thermal output. • See Note 1, "Coverage of Electricity Statistics," at end of section. • See "Useful Thermal Output" in Glossary. • Totals may not equal sum of components due to independent rounding.

Web Page: For related information, see http://www.eia.doe.gov/fuelelectric.html.

Sources: Tables 8.7b and 8.7c.

Table 8.7b Consumption of Combustible Fuels for Electricity Generation and Useful Thermal Output: Electric Power Sector, 1989-2007 (Subset of Table 8.7a)

Year	Coal [1]	Petroleum					Natural Gas [6]	Other Gases [7]	Biomass		Other [10]
		Distillate Fuel Oil [2]	Residual Fuel Oil [3]	Other Liquids [4]	Petroleum Coke [5]	Total [5]			Wood [8]	Waste [9]	
	Thousand Short Tons	Thousand Barrels			Thousand Short Tons	Thousand Barrels	Million Cubic Feet	Trillion Btu	Trillion Btu		Trillion Btu
1989	772,190	26,156	244,179	10	517	272,931	3,105,183	9	100	132	3
1990	782,567	16,567	184,915	26	1,008	206,550	3,244,619	11	129	188	(s)
1991	783,874	14,359	172,625	59	974	191,911	3,315,925	11	126	229	4
1992	795,094	12,623	138,726	128	1,494	158,948	3,447,871	18	140	262	5
1993	831,645	14,849	152,481	239	2,611	180,625	3,472,982	16	150	265	5
1994	838,354	20,612	138,222	771	2,315	171,178	3,902,546	19	152	282	3
1995	850,230	18,553	90,023	499	2,674	122,447	4,236,526	24	125	296	2
1996	896,921	18,780	99,951	653	2,642	132,593	3,806,901	20	138	300	2
1997	921,364	18,989	113,669	152	3,372	149,668	4,064,803	24	137	309	1
1998	936,619	23,300	166,528	431	4,102	210,769	4,588,284	29	137	308	2
1999	940,922	24,058	152,493	544	3,735	195,769	4,819,531	19	138	315	1
2000	985,821	30,016	138,513	454	3,275	185,358	5,206,324	25	134	318	1
2001	964,433	29,274	159,504	377	3,427	206,291	5,342,301	15	126	211	113
2002	977,507	21,876	104,773	1,267	5,816	156,995	5,671,897	33	150	230	143
2003	1,005,116	27,632	138,279	2,026	5,799	196,932	5,135,215	41	167	230	140
2004	1,016,268	19,107	139,816	2,713	7,372	198,498	5,463,763	59	165	223	138
2005	1,037,485	19,675	139,409	2,685	8,083	202,184	5,869,145	84	185	221	123
2006	[R]1,026,636	[R]12,646	[R]57,345	[R]1,870	[R]7,101	[R]107,365	[R]6,222,100	[R]65	[R]182	[R]231	[R]125
2007[P]	1,046,424	15,874	63,529	2,943	5,594	110,314	6,874,082	67	184	243	128

[1] Anthracite, bituminous coal, subbituminous coal, lignite, waste coal, and coal synfuel.
[2] Fuel oil nos. 1, 2, and 4. Through 2000, electric utility data also include small amounts of kerosene and jet fuel.
[3] Fuel oil nos. 5 and 6. Through 2000, electric utility data also include a small amount of fuel oil no. 4.
[4] Jet fuel, kerosene, other petroleum liquids, and waste oil.
[5] Petroleum coke is converted from short tons to barrels by multiplying by 5.
[6] Natural gas, plus a small amount of supplemental gaseous fuels.
[7] Blast furnace gas, propane gas, and other manufactured and waste gases derived from fossil fuels.
[8] Wood and wood-derived fuels.
[9] Municipal solid waste from biogenic sources, landfill gas, sludge waste, agricultural byproducts, and other biomass. Through 2000, also includes non-renewable waste (municipal solid waste from non-biogenic sources, and tire-derived fuels).
[10] Batteries, chemicals, hydrogen, pitch, purchased steam, sulfur, miscellaneous technologies, and, beginning in 2001, non-renewable waste (municipal solid waste from non-biogenic sources, and tire-derived fuels).

R=Revised. P=Preliminary. (s)=Less than 0.5.

Notes: • Data are for fuels consumed to produce electricity and useful thermal output. • The electric power sector comprises electricity-only and combined-heat-and-power (CHP) plants within the NAICS 22 category whose primary business is to sell electricity, or electricity and heat, to the public. • See Table 8.7c for commercial and industrial CHP and electricity-only data. • See Note 1, "Coverage of Electricity Statistics," and Note 2, "Classification of Power Plants Into Energy-Use Sectors," at end of section. • See "Useful Thermal Output" in Glossary. • Totals may not equal sum of components due to independent rounding.

Web Page: For related information, see http://www.eia.doe.gov/fuelelectric.html.

Sources: • 1989-1997—Energy Information Administration (EIA), Form EIA-759, "Monthly Power Plant Report," and Form EIA-867, "Annual Nonutility Power Producer Report." • 1998-2000—EIA, Form EIA-759, "Monthly Power Plant Report," and Form EIA-860B, "Annual Electric Generator Report—Nonutility." • 2001-2003—EIA, Form EIA-906, "Power Plant Report." • 2004 forward—EIA, Form EIA-906, "Power Plant Report," and Form EIA-920, "Combined Heat and Power Plant Report."

Table 8.7c Consumption of Combustible Fuels for Electricity Generation and Useful Thermal Output: Commercial and Industrial Sectors, 1989-2007 (Subset of Table 8.7a)

Year	Coal [1]	Petroleum: Distillate Fuel Oil [2]	Petroleum: Residual Fuel Oil [3]	Petroleum: Other Liquids [4]	Petroleum: Petroleum Coke [5]	Petroleum: Total [5]	Natural Gas [6]	Other Gases [7]	Biomass: Wood [8]	Biomass: Waste [9]	Other [10]
	Thousand Short Tons	Thousand Barrels	Thousand Barrels	Thousand Barrels	Thousand Short Tons	Thousand Barrels	Million Cubic Feet	Trillion Btu	Trillion Btu	Trillion Btu	Trillion Btu
Commercial Sector [11]											
1989	1,125	1,085	883	–	–	1,967	30,037	1	2	22	–
1990	1,191	969	1,087	(s)	–	2,056	46,458	1	2	28	–
1991	1,228	786	551	(s)	–	1,337	52,101	1	2	26	(s)
1992	1,175	548	675	(s)	2	1,235	62,346	1	2	32	(s)
1993	1,373	656	828	6	5	1,515	65,173	1	2	33	(s)
1994	1,344	1,015	588	–	4	1,625	72,285	1	1	35	–
1995	1,419	812	413	(s)	4	1,245	77,664	–	1	40	(s)
1996	1,660	682	545	(s)	4	1,246	82,455	(s)	2	53	(s)
1997	1,738	1,053	509	–	4	1,584	86,915	(s)	2	58	(s)
1998	1,443	854	932	–	4	1,807	87,220	(s)	2	54	–
1999	1,490	759	834	–	4	1,613	84,037	(s)	1	54	(s)
2000	1,547	908	676	3	6	1,615	84,874	(s)	1	47	(s)
2001	1,448	1,026	773	2	6	1,832	78,655	(s)	1	25	15
2002	1,405	771	400	38	8	1,250	73,975	(s)	1	26	17
2003	1,816	671	708	16	11	1,449	58,453	–	1	29	18
2004	1,917	1,115	827	21	9	2,009	72,072	–	2	34	21
2005	1,922	794	789	1	9	1,630	75,215	–	1	34	20
2006	R1,886	R366	R520	(s)	10	R935	R82,261	R(s)	1	R36	R21
2007P	1,924	294	419	(s)	12	774	83,358	(s)	1	37	20
Industrial Sector [12]											
1989	24,867	1,903	21,150	646	397	25,685	913,516	195	926	35	85
1990	27,781	2,657	23,312	1,305	1,824	36,392	1,055,235	275	1,125	41	86
1991	27,021	4,446	19,897	1,156	1,592	33,460	1,060,716	298	1,076	37	110
1992	28,244	3,680	21,540	1,567	1,870	36,135	1,107,361	322	1,161	39	87
1993	28,886	3,788	23,684	1,326	1,583	36,715	1,124,081	297	1,169	46	80
1994	29,707	3,550	25,238	768	1,838	38,744	1,176,332	296	1,248	41	89
1995	29,363	2,333	21,732	823	1,912	34,448	1,258,063	290	1,255	38	95
1996	29,434	2,983	24,111	1,815	1,950	38,661	1,288,876	325	1,249	39	89
1997	29,853	2,851	20,445	374	2,719	37,265	1,281,620	283	1,259	41	102
1998	28,553	5,852	21,807	800	2,090	38,910	1,354,986	305	1,211	42	93
1999	27,763	5,799	18,993	1,268	2,251	37,312	1,401,374	331	1,213	31	99
2000	28,031	3,648	17,483	2,448	1,388	30,520	1,385,546	331	1,244	35	108
2001	25,755	3,424	16,860	1,039	1,099	26,817	1,309,636	248	1,054	27	101
2002	26,232	2,101	13,463	1,953	1,529	25,163	1,240,209	245	1,136	34	92
2003	24,846	3,522	13,872	2,535	1,257	26,212	1,143,734	253	1,097	34	103
2004	26,613	3,298	16,835	2,030	1,339	28,857	1,190,844	296	1,193	24	67
2005	25,875	3,977	16,718	1,583	1,020	27,380	1,083,607	264	1,166	34	70
2006	R25,262	R1,643	R11,981	R1,526	R1,511	R22,706	R1,114,597	R277	R1,216	R33	R69
2007P	24,082	2,233	10,317	1,634	1,679	22,580	1,202,079	281	1,169	35	78

[1] Anthracite, bituminous coal, subbituminous coal, lignite, waste coal, and coal synfuel.
[2] Fuel oil nos. 1, 2, and 4.
[3] Fuel oil nos. 5 and 6.
[4] Jet fuel, kerosene, other petroleum liquids, and waste oil.
[5] Petroleum coke is converted from short tons to barrels by multiplying by 5.
[6] Natural gas, plus a small amount of supplemental gaseous fuels.
[7] Blast furnace gas, propane gas, and other manufactured and waste gases derived from fossil fuels.
[8] Wood and wood-derived fuels.
[9] Municipal solid waste from biogenic sources, landfill gas, sludge waste, agricultural byproducts, and other biomass. Through 2000, also includes non-renewable waste (municipal solid waste from non-biogenic sources, and tire-derived fuels).
[10] Batteries, chemicals, hydrogen, pitch, purchased steam, sulfur, miscellaneous technologies, and, beginning in 2001, non-renewable waste (municipal solid waste from non-biogenic sources, and tire-derived fuels).
[11] Commercial combined-heat-and-power (CHP) and commercial electricity-only plants.
[12] Industrial combined-heat-and-power (CHP) and industrial electricity-only plants.

R=Revised. P=Preliminary. – = No data reported. (s)=Less than 0.5.

Notes: • Data are for fuels consumed to produce electricity and useful thermal output. • See Table 8.7b for electric power sector electricity-only and CHP data. • See Note 1, "Coverage of Electricity Statistics," and Note 2, "Classification of Power Plants Into Energy-Use Sectors," at end of section. • See "Useful Thermal Output" in Glossary. • Totals may not equal sum of components due to independent rounding.

Web Page: For related information, see http://www.eia.doe.gov/fuelelectric.html.

Sources: • 1989-1997—Energy Information Administration (EIA), Form EIA-867, "Annual Nonutility Power Producer Report." • 1998-2000—EIA, Form EIA-860B, "Annual Electric Generator Report—Nonutility." • 2001-2003—EIA, Form EIA-906, "Power Plant Report." • 2004 forward—EIA, Form EIA-906, "Power Plant Report," and Form EIA-920, "Combined Heat and Power Plant Report."

Figure 8.8 Stocks of Coal and Petroleum: Electric Power Sector

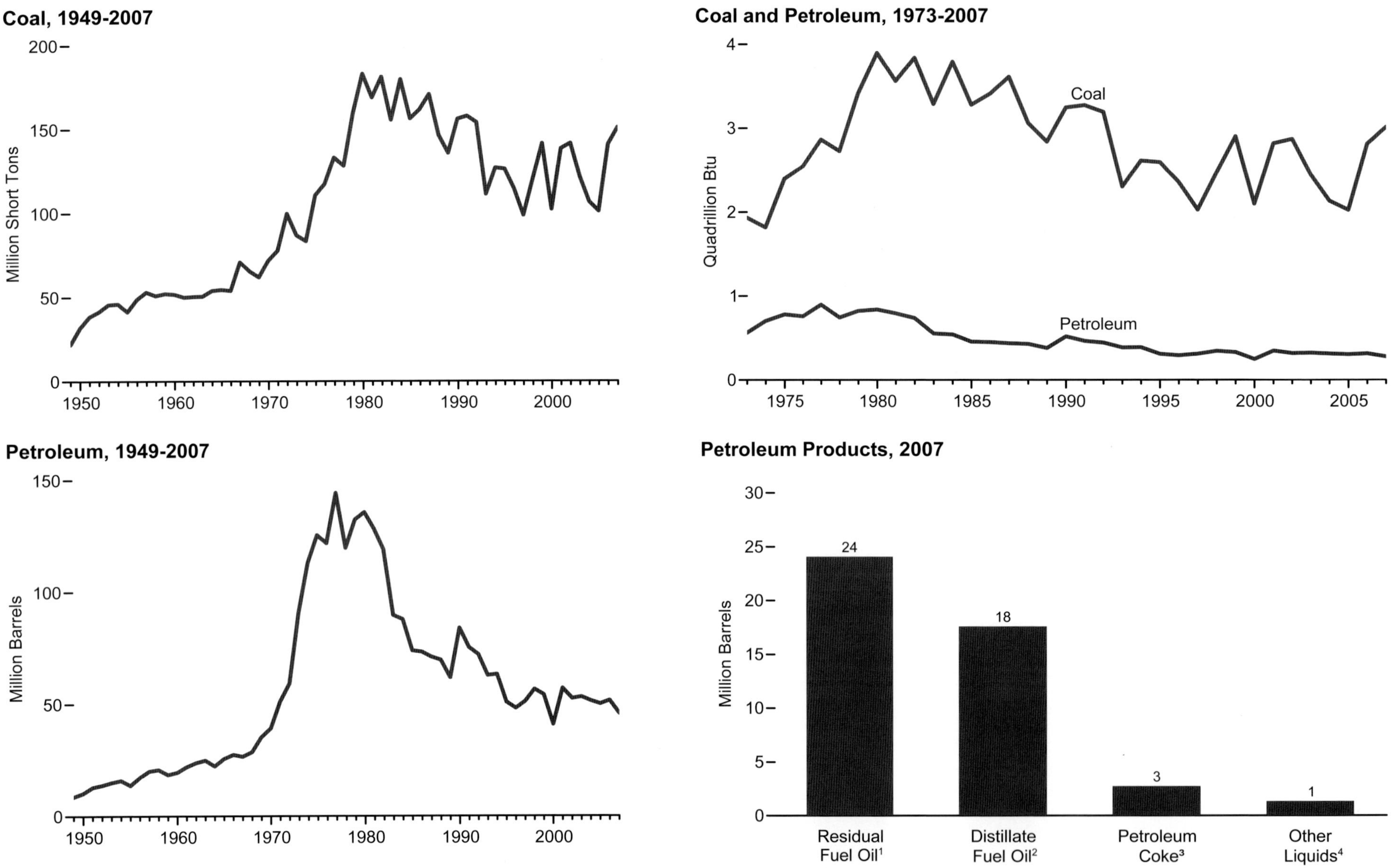

[1] Fuel oil nos. 5 and 6.
[2] Fuel oil nos. 1, 2, and 4.
[3] Petroleum coke, which is reported in short tons, is converted at a rate of 5 barrels per short ton.
[4] Jet fuel and kerosene.

Notes: • Stocks are at end of year. • Because vertical scales differ, graphs should not be compared.

Sources: Tables 8.8, A3, and A5.

Table 8.8 Stocks of Coal and Petroleum: Electric Power Sector, Selected Years, 1949-2007

		Petroleum				
	Coal [1]	Distillate Fuel Oil [2]	Residual Fuel Oil [3]	Other Liquids [4]	Petroleum Coke [5]	Total [5,6]
Year	Thousand Short Tons	Thousand Barrels			Thousand Short Tons	Thousand Barrels
1949	22,054	NA	NA	NA	NA	8,604
1950	31,842	NA	NA	NA	NA	10,201
1955	41,391	NA	NA	NA	NA	13,671
1960	51,735	NA	NA	NA	NA	19,572
1965	54,525	NA	NA	NA	NA	25,647
1970	71,908	NA	NA	NA	239	39,151
1971	77,778	NA	NA	NA	291	51,101
1972	99,722	NA	NA	NA	287	59,090
1973	86,967	10,095	79,121	NA	312	90,776
1974	83,509	15,199	97,718	NA	35	113,091
1975	110,724	16,432	108,825	NA	31	125,413
1976	117,436	14,703	106,993	NA	32	121,857
1977	133,219	19,281	124,750	NA	44	144,252
1978	128,225	16,386	102,402	NA	198	119,778
1979	159,714	20,301	111,121	NA	183	132,338
1980	183,010	30,023	105,351	NA	52	135,635
1981	168,893	26,094	102,042	NA	42	128,345
1982	181,132	23,369	95,515	NA	41	119,090
1983	155,598	18,801	70,573	NA	55	89,652
1984	179,727	19,116	68,503	NA	50	87,870
1985	156,376	16,386	57,304	NA	49	73,933
1986	161,806	16,269	56,841	NA	40	73,313
1987	170,797	15,759	55,069	NA	51	71,084
1988	146,507	15,099	54,187	NA	86	69,714
1989	135,860	13,824	47,446	NA	105	61,795
1990	156,166	16,471	67,030	NA	94	83,970
1991	157,876	16,357	58,636	NA	70	75,343
1992	154,130	15,714	56,135	NA	67	72,183
1993	111,341	15,674	46,770	NA	89	62,890
1994	126,897	16,644	46,344	NA	69	63,333
1995	126,304	15,392	35,102	NA	65	50,821
1996	114,623	15,216	32,473	NA	91	48,146
1997	98,826	15,456	33,336	NA	469	51,138
1998	120,501	16,343	37,451	NA	559	56,591
1999[7]	141,604	17,995	34,256	NA	372	54,109
2000	102,296	15,127	24,748	NA	211	40,932
2001	138,496	20,486	34,594	NA	390	57,031
2002	141,714	17,413	25,723	800	1,711	52,490
2003	121,567	19,153	25,820	779	1,484	53,170
2004	106,669	19,275	26,596	879	937	51,434
2005	101,137	18,778	27,624	1,012	530	50,062
2006	R140,964	R18,013	R28,823	R1,380	R674	R51,583
2007P	151,127	17,579	24,081	1,325	550	45,733

[1] Anthracite, bituminous coal, subbituminous coal, and lignite.

[2] Fuel oil nos. 1, 2, and 4. For 1973-1979, data are for gas turbine and internal combustion plant stocks of petroleum. For 1980-2000, electric utility data also include small amounts of kerosene and jet fuel.

[3] Fuel oil nos. 5 and 6. For 1973-1979, data are for steam plant stocks of petroleum. For 1980-2000, electric utility data also include a small amount of fuel oil no. 4.

[4] Jet fuel and kerosene. Through 2003, data also include a small amount of waste oil.

[5] Petroleum coke is converted from short tons to barrels by multiplying by 5.

[6] Distillate fuel oil and residual fuel oil; beginning in 1970, also includes petroleum coke; and beginning in 2002, also includes other liquids.

[7] Through 1998, data are for electric utilities only. Beginning in 1999, data are for electric utilities and independent power producers.

R=Revised. P=Preliminary. NA=Not available.

Notes: • Stocks are at end of year. • The electric power sector comprises electricity-only and combined-heat-and-power (CHP) plants within the NAICS 22 category whose primary business is to sell electricity, or electricity and heat, to the public. • See Note 1, "Coverage of Electricity Statistics," and Note 2, "Classification of Power Plants Into Energy-Use Sectors," at end of section. • Totals may not equal sum of components due to independent rounding.

Web Pages: • For all data beginning in 1949, see http://www.eia.doe.gov/emeu/aer/elect.html. • For related information, see http://www.eia.doe.gov/fuelelectric.html.

Sources: • 1949-September 1977—Federal Power Commission, Form FPC-4, "Monthly Power Plant Report." • October 1977-1981—Federal Energy Regulatory Commission, Form FPC-4, "Monthly Power Plant Report." • 1982-1988—Energy Information Administration (EIA), Form EIA-759, "Monthly Power Plant Report." • 1989-1997—EIA, Form EIA-759, "Monthly Power Plant Report," and Form EIA-867, "Annual Nonutility Power Producer Report." • 1998-2000—EIA, Form EIA-759, "Monthly Power Plant Report," and Form EIA-860B, "Annual Electric Generator Report—Nonutility." • 2001-2003—EIA, Form EIA-906, "Power Plant Report." • 2004 forward—EIA, Form EIA-906, "Power Plant Report," and Form EIA-920, "Combined Heat and Power Plant Report."

Figure 8.9 Electricity End Use

Overview, 1989-2007

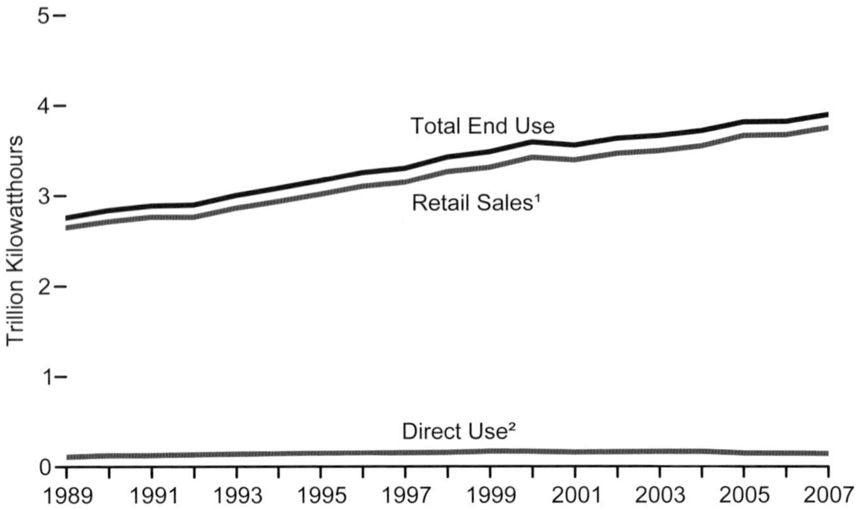

Retail Sales[1] by Sector, 2007

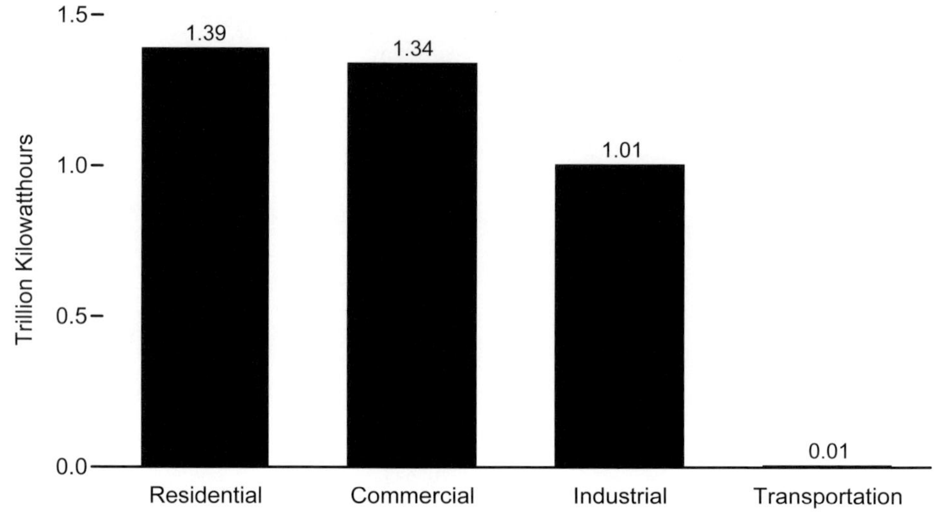

Retail Sales[1] by Sector, 1949-2007

[1] Electricity retail sales to ultimate customers reported by electric utilities and, beginning in 1996, other energy service providers.

[2] Use of electricity that is 1) self-generated, 2) produced by either the same entity that consumes the power or an affiliate, and 3) used in direct support of a service or industrial process located within the same facility or group of facilities that house the generating equipment. Direct use is exclusive of station use.

Note: Because vertical scales differ, graphs should not be compared.

Source: Table 8.9.

Table 8.9 Electricity End Use, Selected Years, 1949-2007

(Billion Kilowatthours)

Year	Retail Sales [1]					Direct Use [6]	Total End Use [7]	*Discontinued Retail Sales Series*	
	Residential	Commercial [2]	Industrial [3]	Transportation [4]	Total Retail Sales [5]			*Commercial (Old)* [8]	*Other (Old)* [9]
1949	67	[E]59	123	[E]6	255	NA	255	*45*	*20*
1950	72	[E]66	146	[E]7	291	NA	291	*51*	*22*
1955	128	[E]103	260	[E]6	497	NA	497	*79*	*29*
1960	201	[E]159	324	[E]3	688	NA	688	*131*	*32*
1965	291	[E]231	429	[E]3	954	NA	954	*200*	*34*
1970	466	[E]352	571	[E]3	1,392	NA	1,392	*307*	*48*
1971	500	[E]377	589	[E]3	1,470	NA	1,470	*329*	*51*
1972	539	[E]413	641	[E]3	1,595	NA	1,595	*359*	*56*
1973	579	[E]445	686	[E]3	1,713	NA	1,713	*388*	*59*
1974	578	[E]440	685	[E]3	1,706	NA	1,706	*385*	*58*
1975	588	[E]468	688	[E]3	1,747	NA	1,747	*403*	*68*
1976	606	[E]492	754	[E]3	1,855	NA	1,855	*425*	*70*
1977	645	[E]514	786	[E]3	1,948	NA	1,948	*447*	*71*
1978	674	[E]531	809	[E]3	2,018	NA	2,018	*461*	*73*
1979	683	543	842	3	2,071	NA	2,071	*473*	*73*
1980	717	559	815	3	2,094	NA	2,094	*488*	*74*
1981	722	596	826	3	2,147	NA	2,147	*514*	*85*
1982	730	609	745	3	2,086	NA	2,086	*526*	*86*
1983	751	620	776	4	2,151	NA	2,151	*544*	*80*
1984	780	664	838	4	2,286	NA	2,286	*583*	*85*
1985	794	689	837	4	2,324	NA	2,324	*606*	*87*
1986	819	715	831	4	2,369	NA	2,369	*631*	*89*
1987	850	744	858	5	2,457	NA	2,457	*660*	*88*
1988	893	784	896	5	2,578	NA	2,578	*699*	*90*
1989	906	811	926	5	2,647	109	2,756	*726*	*90*
1990	924	838	946	5	2,713	125	2,837	*751*	*92*
1991	955	855	947	5	2,762	124	2,886	*766*	*94*
1992	936	850	973	5	2,763	134	2,897	*761*	*93*
1993	995	885	977	5	2,861	139	3,001	*795*	*95*
1994	1,008	913	1,008	5	2,935	146	3,081	*820*	*98*
1995	1,043	953	1,013	5	3,013	151	3,164	*863*	*95*
1996	1,083	980	1,034	5	3,101	153	3,254	*887*	*98*
1997	1,076	1,027	1,038	5	3,146	156	3,302	*929*	*103*
1998	1,130	1,078	1,051	5	3,264	161	3,425	*979*	*104*
1999	1,145	1,104	1,058	5	3,312	172	3,484	*1,002*	*107*
2000	1,192	1,159	1,064	5	3,421	171	3,592	*1,055*	*109*
2001	1,202	1,191	997	6	3,394	163	3,557	*1,083*	*113*
2002	1,265	1,205	990	6	3,465	166	3,632	*1,104*	*106*
2003	1,276	1,199	1,012	7	3,494	168	3,662	– –	– –
2004	1,292	1,230	1,018	7	3,547	168	3,716	– –	– –
2005	1,359	1,275	1,019	8	3,661	[R]150	[R]3,811	– –	– –
2006	[R]1,352	[R]1,300	[R]1,011	[R]7	[R]3,670	[R]147	[R]3,817	– –	– –
2007[P]	1,392	1,343	1,006	8	3,748	[E]144	3,892	– –	– –

[1] Electricity retail sales to ultimate customers reported by electric utilities and, beginning in 1996, other energy service providers.
[2] Commercial sector, including public street and highway lighting, interdepartmental sales, and other sales to public authorities.
[3] Industrial sector. Through 2002, excludes agriculture and irrigation; beginning in 2003, includes agriculture and irrigation.
[4] Transportation sector, including sales to railroads and railways.
[5] The sum of "Residential," "Commercial," "Industrial," and "Transportation."
[6] Use of electricity that is 1) self-generated, 2) produced by either the same entity that consumes the power or an affiliate, and 3) used in direct support of a service or industrial process located within the same facility or group of facilities that house the generating equipment. Direct use is exclusive of station use.
[7] The sum of "Total Retail Sales" and "Direct Use."
[8] "Commercial (Old)" is a discontinued series—data are for the commercial sector, excluding public street and highway lighting, interdepartmental sales, and other sales to public authorities.
[9] "Other (Old)" is a discontinued series—data are for public street and highway lighting, interdepartmental sales, other sales to public authorities, agriculture and irrigation, and transportation including railroads and railways.
R=Revised. P=Preliminary. E=Estimate. NA=Not available. – – = Not applicable.
Note: Totals may not equal sum of components due to independent rounding.

Web Pages: • For all data beginning in 1949, see http://www.eia.doe.gov/emeu/aer/elect.html. • For related information, see http://www.eia.doe.gov/fuelelectric.html.

Sources: **Residential** and **Industrial:** • 1949-September 1977—Federal Power Commission, Form FPC-5, "Monthly Statement of Electric Operating Revenue and Income." • October 1977-February 1980—Federal Energy Regulatory Commission (FERC), Form FPC-5, "Monthly Statement of Electric Operating Revenue and Income." • March 1980-1982—FERC, Form FPC-5, "Electric Utility Company Monthly Statement." • 1983—Energy Information Administration (EIA), Form EIA-826, "Electric Utility Company Monthly Statement." • 1984-1992—EIA, Form EIA-861, "Annual Electric Utility Report." • 1993 forward—EIA, *Electric Power Monthly* (March 2008), Table 5.1. **Commercial:** • 1949-2002—Estimated by EIA as the sum of "Commercial (Old)" and the non-transportation portion of "Other (Old)." See estimation methodology at http://www.eia.doe.gov/emeu/states/sep_use/notes/use_elec.pdf. • 2003 forward—EIA, *Electric Power Monthly* (March 2008), Table 5.1. **Transportation:** • 1949-2002—Estimated by EIA as the transportation portion of "Other (Old)." See estimation methodology at http://www.eia.doe.gov/emeu/states/sep_use/notes/use_elec.pdf. • 2003 forward—EIA, *Electric Power Monthly* (March 2008), Table 5.1. **Direct Use:** • 1989-1994—EIA, Form EIA-867, "Annual Nonutility Power Producer Report." • 1995-2006—EIA, *Electric Power Annual 2006* (October 2007), Table 7.2. • 2007—Estimate based on the 2006 value adjusted by the percentage increase in commercial and industrial net generation on Table 8.1. **Commercial (Old)** and **Other (Old):** • 1949-2002—See sources for "Residential" and "Industrial."

Figure 8.10 Average Retail Prices of Electricity

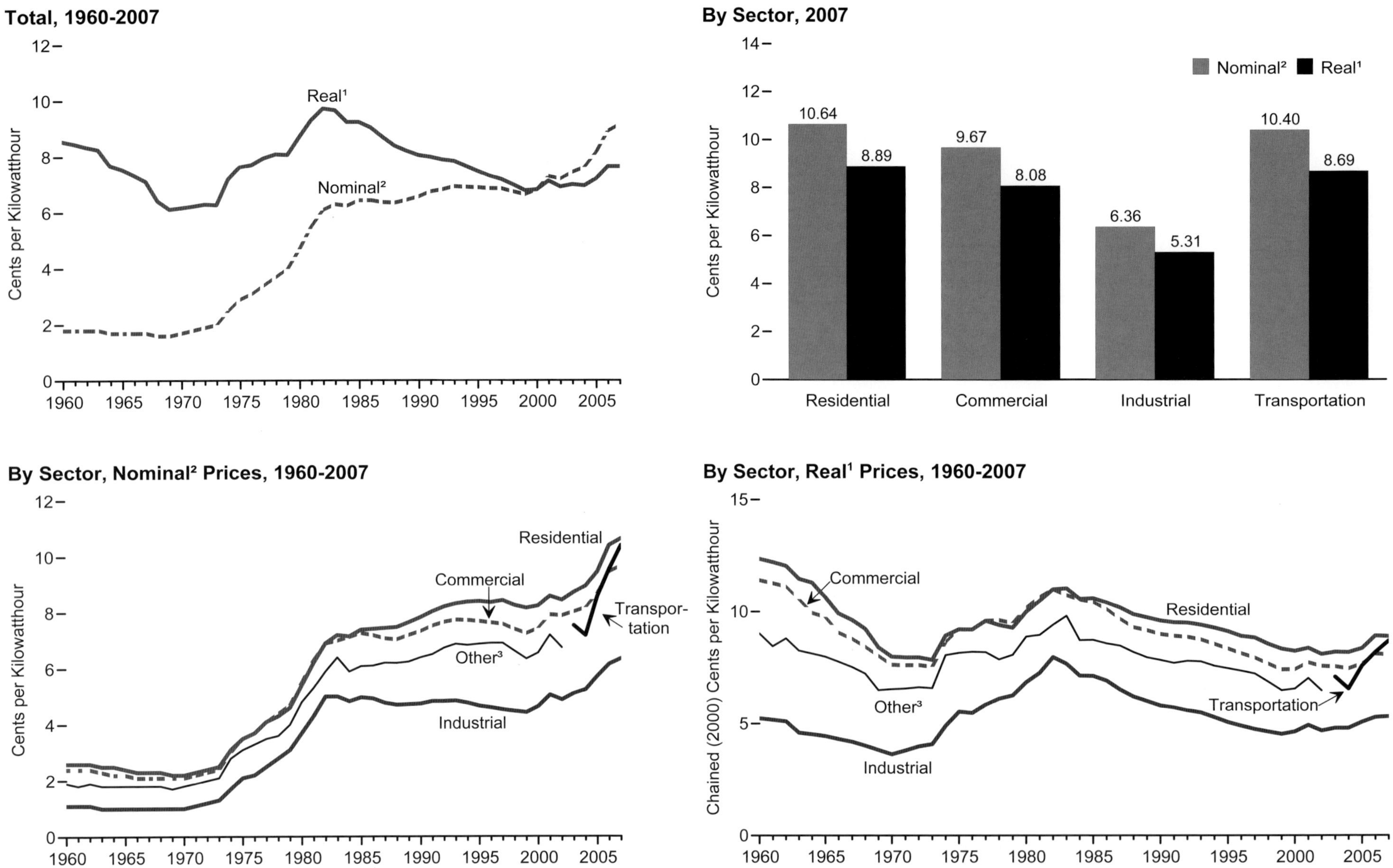

[1] In chained (2000) dollars, calculated by using gross domestic product implicit price deflators. See Table D1.
[2] See "Nominal Dollars" in Glossary.
[3] Public street and highway lighting, interdepartmental sales, other sales to public authorities, agriculture and irrigation, and transportation including railroads and railways.

Notes: • Taxes are included. • Because vertical scales differ, graphs should not be compared.

Source: Table 8.10.

Table 8.10 Average Retail Prices of Electricity, 1960-2007

(Cents per Kilowatthour, Including Taxes)

Year	Residential		Commercial [1]		Industrial [2]		Transportation [3]		Other [4]		Total	
	Nominal [5]	Real [6]	Nominal [5]	Real [6]	Nominal [5]	Real [6]	Nominal [5]	Real [6]	Nominal [5]	Real [6]	Nominal [5]	Real [6]
1960	2.6	12.4	2.4	11.4	1.1	5.2	NA	NA	1.9	9.0	1.8	8.6
1961	2.6	12.2	2.4	11.3	1.1	5.2	NA	NA	1.8	8.5	1.8	8.5
1962	2.6	12.1	2.4	11.1	1.1	5.1	NA	NA	1.9	8.8	1.8	8.4
1963	2.5	11.5	2.3	10.6	1.0	4.6	NA	NA	1.8	8.3	1.8	8.3
1964	2.5	11.3	2.2	9.9	1.0	4.5	NA	NA	1.8	8.1	1.7	7.7
1965	2.4	10.7	2.2	9.8	1.0	4.4	NA	NA	1.8	8.0	1.7	7.5
1966	2.3	9.9	2.1	9.1	1.0	4.3	NA	NA	1.8	7.8	1.7	7.3
1967	2.3	9.6	2.1	8.8	1.0	4.2	NA	NA	1.8	7.5	1.7	7.1
1968	2.3	9.2	2.1	8.4	1.0	4.0	NA	NA	1.8	7.2	1.6	6.4
1969	2.2	8.4	2.1	8.0	1.0	3.8	NA	NA	1.7	6.5	1.6	6.1
1970	2.2	8.0	2.1	7.6	1.0	3.6	NA	NA	1.8	6.5	1.7	6.2
1971	2.3	8.0	2.2	7.6	1.1	3.8	NA	NA	1.9	6.6	1.8	6.2
1972	2.4	8.0	2.3	7.6	1.2	4.0	NA	NA	2.0	6.6	1.9	6.3
1973	2.5	7.9	2.4	7.5	1.3	4.1	NA	NA	2.1	6.6	2.0	6.3
1974	3.1	8.9	3.0	8.6	1.7	4.9	NA	NA	2.8	8.1	2.5	7.2
1975	3.5	9.2	3.5	9.2	2.1	5.5	NA	NA	3.1	8.2	2.9	7.6
1976	3.7	9.2	3.7	9.2	2.2	5.5	NA	NA	3.3	8.2	3.1	7.7
1977	4.1	9.6	4.1	9.6	2.5	5.9	NA	NA	3.5	8.2	3.4	8.0
1978	4.3	9.4	4.4	9.6	2.8	6.1	NA	NA	3.6	7.9	3.7	8.1
1979	4.6	9.3	4.7	9.5	3.1	6.3	NA	NA	4.0	8.1	4.0	8.1
1980	5.4	10.0	5.5	10.2	3.7	6.9	NA	NA	4.8	8.9	4.7	8.7
1981	6.2	10.5	6.3	10.7	4.3	7.3	NA	NA	5.3	9.0	5.5	9.3
1982	6.9	11.0	6.9	11.0	5.0	8.0	NA	NA	5.9	9.4	6.1	9.7
1983	7.2	11.0	7.0	10.7	5.0	7.7	NA	NA	6.4	9.8	6.3	9.7
1984	7.15	10.57	7.13	10.54	4.83	7.14	NA	NA	5.90	8.72	6.25	9.24
1985	7.39	10.60	7.27	10.43	4.97	7.13	NA	NA	6.09	8.74	6.44	9.24
1986	7.42	10.41	7.20	10.11	4.93	6.92	NA	NA	6.11	8.58	6.44	9.04
1987	7.45	10.18	7.08	9.67	4.77	6.52	NA	NA	6.21	8.48	6.37	8.70
1988	7.48	9.88	7.04	9.30	4.70	6.21	NA	NA	6.20	8.19	6.35	8.39
1989	7.65	9.74	7.20	9.17	4.72	6.01	NA	NA	6.25	7.96	6.45	8.21
1990	7.83	9.60	7.34	9.00	4.74	5.81	NA	NA	6.40	7.84	6.57	8.05
1991	8.04	9.52	7.53	8.92	4.83	5.72	NA	NA	6.51	7.71	6.75	7.99
1992	8.21	9.50	7.66	8.87	4.83	5.59	NA	NA	6.74	7.80	6.82	7.89
1993	8.32	9.41	7.74	8.76	4.85	5.49	NA	NA	6.88	7.78	6.93	7.84
1994	8.38	9.28	7.73	8.56	4.77	5.28	NA	NA	6.84	7.58	6.91	7.66
1995	8.40	9.12	7.69	8.35	4.66	5.06	NA	NA	6.88	7.47	6.89	7.48
1996	8.36	8.91	7.64	8.14	4.60	4.90	NA	NA	6.91	7.36	6.86	7.31
1997	8.43	8.84	7.59	7.95	4.53	4.75	NA	NA	6.91	7.24	6.85	7.18
1998	8.26	8.56	7.41	7.68	4.48	4.64	NA	NA	6.63	6.87	6.74	6.99
1999	8.16	8.34	7.26	7.42	4.43	4.53	NA	NA	6.35	6.49	6.64	6.78
2000	8.24	8.24	7.43	7.43	4.64	4.64	NA	NA	6.56	6.56	6.81	6.81
2001	8.58	8.38	7.92	7.73	5.05	4.93	NA	NA	7.20	7.03	7.29	7.12
2002	8.44	8.10	7.89	7.57	4.88	4.68	NA	NA	6.75	6.48	7.20	6.91
2003	8.72	8.20	8.03	7.55	5.11	4.80	7.54	7.09	– –	– –	7.44	6.99
2004	8.95	8.18	8.17	R7.46	5.25	4.80	7.18	6.56	– –	– –	7.61	6.95
2005	9.45	R8.36	8.67	R7.67	5.73	R5.07	8.57	R7.58	– –	– –	8.14	R7.20
2006	10.40	R8.92	R9.46	R8.12	R6.16	R5.28	R9.54	R8.18	– –	– –	R8.90	R7.64
2007P	10.64	8.89	9.67	8.08	6.36	5.31	10.40	8.69	– –	– –	9.14	7.64

[1] Commercial sector. For 1960-2002, prices exclude public street and highway lighting, interdepartmental sales, and other sales to public authorities.

[2] Industrial sector. For 1960-2002, prices exclude agriculture and irrigation.

[3] Transportation sector, including railroads and railways.

[4] Public street and highway lighting, interdepartmental sales, other sales to public authorities, agriculture and irrigation, and transportation including railroads and railways.

[5] See "Nominal Dollars" in Glossary.

[6] In chained (2000) dollars, calculated by using gross domestic product implicit price deflators in Table D1. See "Chained Dollars" in Glossary.

R=Revised. P=Preliminary. NA=Not available. – – = Not applicable.

Notes: • Beginning in 2003, the category "Other" has been replaced by "Transportation," and the categories "Commercial" and "Industrial" have been redefined. • Data represent revenue from electricity retail sales divided by electricity retail sales. • Prices include State and local taxes, energy or demand charges, customer service charges, environmental surcharges, franchise fees, fuel adjustments, and other miscellaneous charges applied to end-use customers during normal billing operations. Prices do not include deferred charges, credits, or other adjustments, such as fuel or revenue from purchased power, from previous reporting periods. • Through 1979, data are for Classes A and B privately owned electric utilities only. For 1980-1982, data are for selected Class A utilities whose electric operating revenues were $100 million or more during the previous year. For 1983, data are for a selected sample of electric utilities. Beginning in 1984, data are for a census of electric utilities. Beginning in 1996, data also include energy service providers selling to retail customers.

Web Page: For related information, see http://www.eia.doe.gov/fuelelectric.html.

Sources: • 1960-September 1977—Federal Power Commission, Form FPC-5, "Monthly Statement of Electric Operating Revenues and Income." • October 1977-February 1980—Federal Energy Regulatory Commission (FERC), Form FPC-5, "Monthly Statement of Electric Operating Revenues and Income." • March 1980-1982—FERC, Form FERC-5, "Electric Utility Company Monthly Statement." • 1983—Energy Information Administration (EIA), Form EIA-826, "Electric Utility Company Monthly Statement." • 1984-1992—EIA, Form EIA-861, "Annual Electric Utility Report." • 1993 forward—EIA, *Electric Power Monthly* (March 2008), Table 5.3.

Figure 8.11a Electric Net Summer Capacity, Total (All Sectors)

Total, 1949-2007

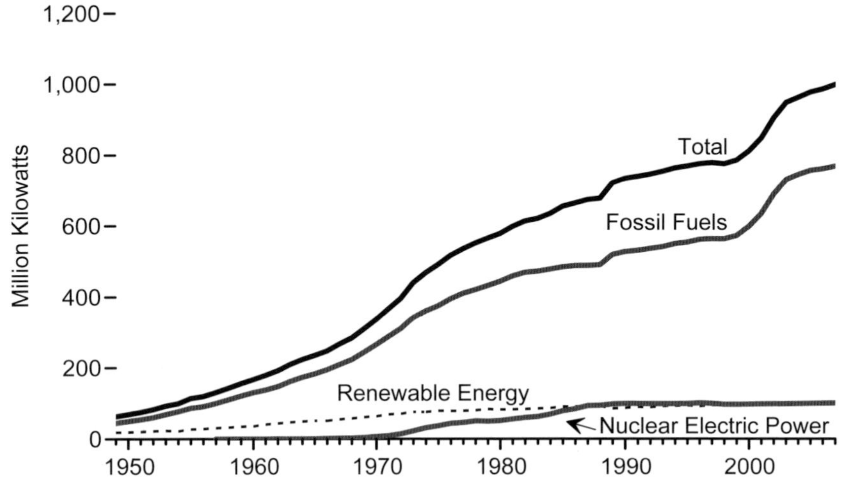

By Major Category, 2007

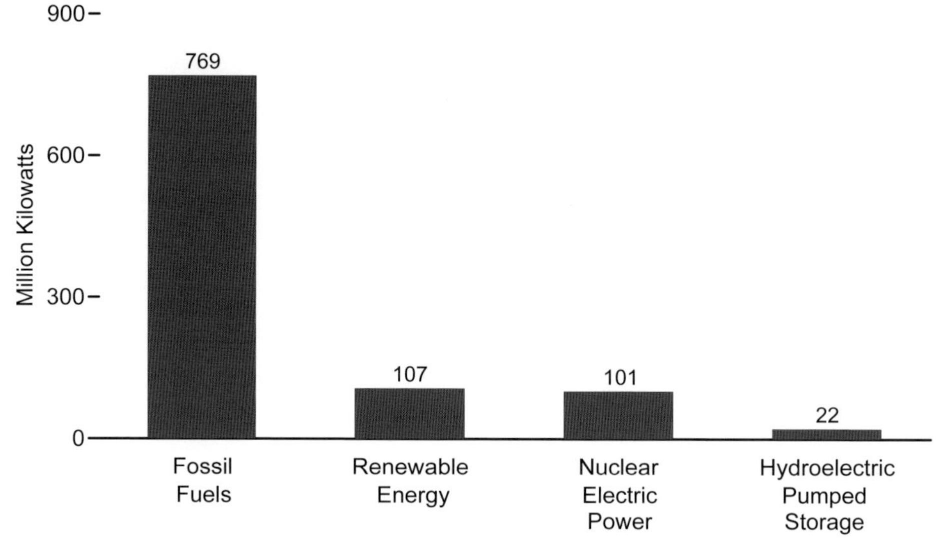

By Source, 2007

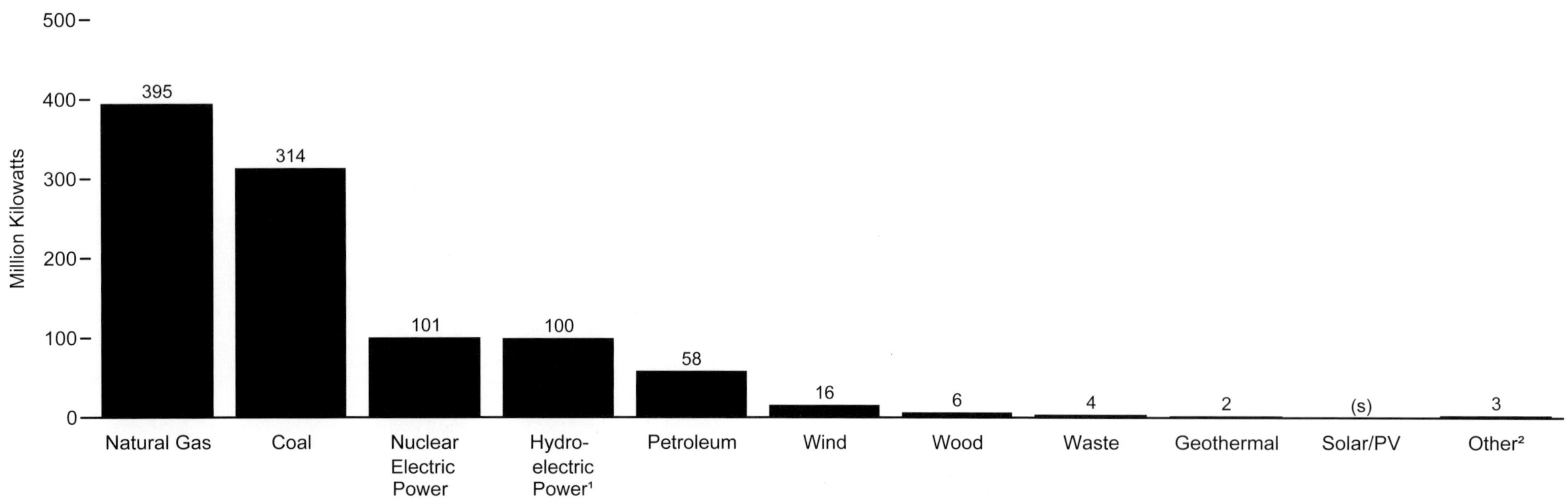

[1] Conventional and pumped storage.

[2] Blast furnace gas, propane gas, other manufactured and waste gases derived from fossil fuels, batteries, chemicals, hydrogen, pitch, purchased steam, sulfur, and miscellaneous technologies.

(s)=Less than 0.5 million kilowatts.

Note: Because vertical scales differ, graphs should not be compared.

Source: Table 8.11a.

Figure 8.11b Electric Net Summer Capacity by Sector

Total (All Sectors) and Sectors, 1989-2007

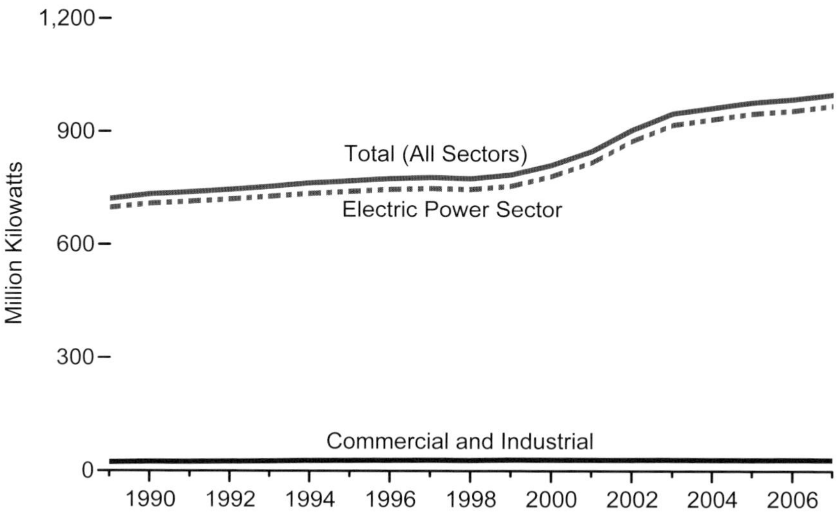

Electric Power Sector by Plant Type, 1989-2007

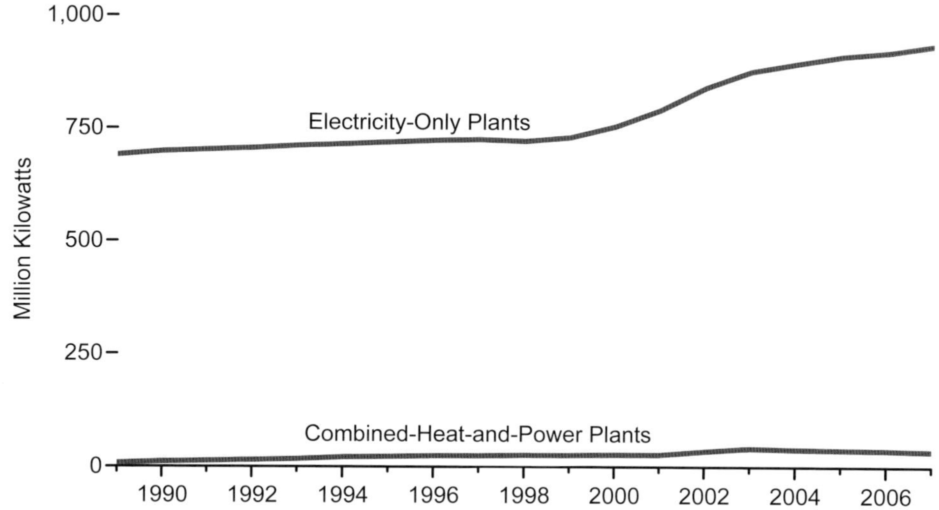

Commercial Sector, 2007

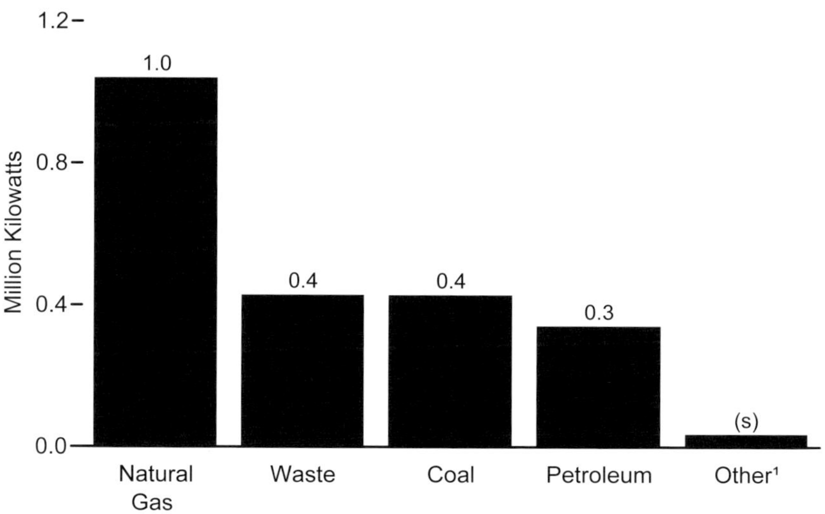

Industrial Sector, 2007

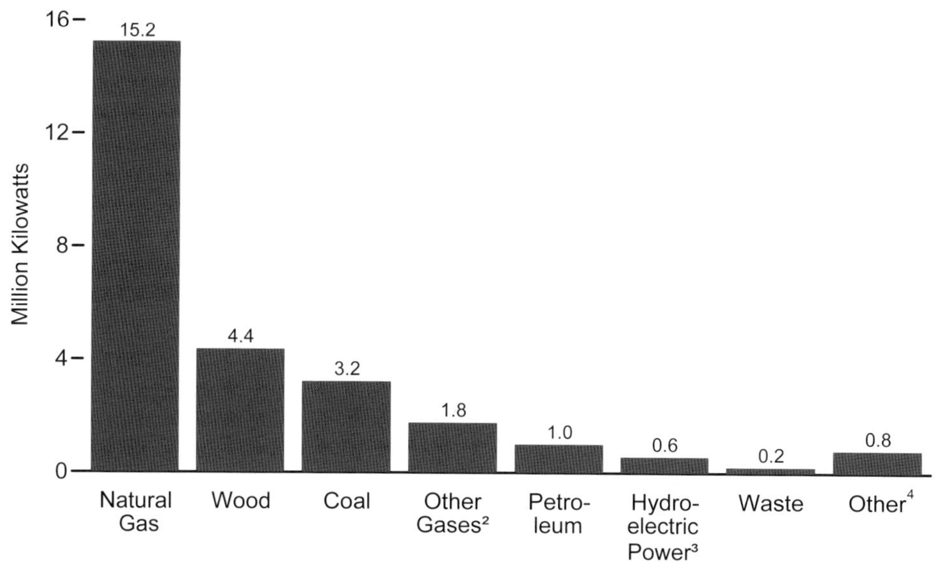

[1] Conventional hydroelectric power, wood, blast furnace gas, propane gas, and other manufactured and waste gases derived from fossil fuels.
[2] Blast furnace gas, propane gas, and other manufactured and waste gases derived from fossil fuels
[3] Conventional.
[4] Batteries, chemicals, hydrogen, pitch, purchased steam, sulfur, and miscellaneous technologies.
(s)=Less than 0.05 million kilowatts.
Note: Because vertical scales differ, graphs should not be compared.
Sources: Tables 8.11a-8.11d.

Table 8.11a Electric Net Summer Capacity: Total (All Sectors), Selected Years, 1949-2007

(Sum of Tables 8.11b and 8.11d; Million Kilowatts)

Year	Fossil Fuels					Nuclear Electric Power	Hydro-electric Pumped Storage	Renewable Energy							Other [8]	Total
	Coal [1]	Petroleum [2]	Natural Gas [3]	Other Gases [4]	Total			Conventional Hydroelectric Power	Biomass: Wood [5]	Biomass: Waste [6]	Geo-thermal	Solar/PV [7]	Wind	Total		
1949	NA	NA	NA	NA	44.9	0.0	(9)	18.5	(s)	(10)	NA	NA	NA	18.5	NA	63.4
1950	NA	NA	NA	NA	50.0	.0	(9)	19.2	(s)	(10)	NA	NA	NA	19.2	NA	69.2
1955	NA	NA	NA	NA	86.8	.0	(9)	27.4	(s)	(10)	NA	NA	NA	27.4	NA	114.2
1960	NA	NA	NA	NA	130.8	.4	(9)	35.8	.1	(10)	(s)	NA	NA	35.9	NA	167.1
1965	NA	NA	NA	NA	182.9	.8	(9)	51.0	.1	(10)	(s)	NA	NA	51.1	NA	234.8
1970	NA	NA	NA	NA	265.4	7.0	(9)	63.8	.1	(10)	.1	NA	NA	63.9	NA	336.4
1971	NA	NA	NA	NA	288.0	9.0	(9)	69.1	.1	(10)	.2	NA	NA	69.4	NA	366.4
1972	NA	NA	NA	NA	310.7	14.5	(9)	70.5	.1	(10)	.3	NA	NA	70.9	NA	396.0
1973	NA	NA	NA	NA	341.2	22.7	(9)	75.4	.1	(10)	.4	NA	NA	75.9	NA	439.8
1974	NA	NA	NA	NA	360.7	31.9	(9)	75.5	.1	(10)	.4	NA	NA	76.0	NA	468.5
1975	NA	NA	NA	NA	375.1	37.3	(9)	78.4	.1	(10)	.5	NA	NA	79.0	NA	491.3
1976	NA	NA	NA	NA	394.8	43.8	(9)	78.0	.1	(10)	.5	NA	NA	78.6	NA	517.2
1977	NA	NA	NA	NA	410.4	46.3	(9)	78.6	.1	(10)	.5	NA	NA	79.2	NA	535.9
1978	NA	NA	NA	NA	420.8	50.8	(9)	79.9	.1	(10)	.5	NA	NA	80.5	NA	552.1
1979	NA	NA	NA	NA	432.1	49.7	(9)	82.9	.1	(10)	.7	NA	NA	83.6	NA	565.5
1980	NA	NA	NA	NA	444.1	51.8	(9)	81.7	.1	(10)	.9	NA	NA	82.7	NA	578.6
1981	NA	NA	NA	NA	458.9	56.0	(9)	82.4	.1	(10)	.9	NA	(s)	83.4	NA	598.3
1982	NA	NA	NA	NA	469.6	60.0	(9)	83.0	.1	(10)	1.0	NA	(s)	84.1	NA	613.7
1983	NA	NA	NA	NA	472.8	63.0	(9)	83.9	.2	(10)	1.2	NA	(s)	85.3	NA	621.1
1984	NA	NA	NA	NA	478.6	69.7	(9)	85.3	.3	(10)	1.2	(11)	(s)	86.9	NA	635.1
1985	NA	NA	NA	NA	485.0	79.4	(9)	88.9	.2	.2	1.6	(11)	(s)	90.8	NA	655.2
1986	NA	NA	NA	NA	488.3	85.2	(9)	89.3	.2	.2	1.6	(11)	(s)	91.2	NA	664.8
1987	NA	NA	NA	NA	488.8	93.6	(9)	89.7	.2	.2	1.5	(11)	(s)	91.7	NA	674.1
1988	NA	NA	NA	NA	490.6	94.7	(9)	90.3	.2	.2	1.7	(11)	(s)	92.4	NA	677.7
1989 [12]	303.1	R79.1	R135.7	1.5	519.4	98.2	18.1	74.1	5.2	2.1	2.6	.2	1.5	85.7	.5	721.8
1990	307.4	R77.9	R140.8	1.6	527.8	99.6	19.5	73.9	5.5	2.5	2.7	.3	1.8	86.8	.5	734.1
1991	307.4	R74.2	R147.6	2.1	531.4	99.6	18.4	76.0	6.1	2.9	2.6	.3	1.9	89.9	.5	739.9
1992	309.4	R73.1	R152.2	2.1	536.7	99.0	21.2	74.8	6.2	3.0	2.9	.3	1.8	89.1	.5	746.5
1993	310.1	R71.1	R158.6	1.9	541.8	99.0	21.1	77.4	6.5	3.1	2.9	.3	1.8	92.1	.5	754.6
1994	311.4	R71.7	R164.8	2.1	550.0	99.1	21.2	78.0	6.7	3.3	3.0	.3	1.7	93.1	.5	764.0
1995	311.4	R66.6	R174.5	1.7	554.2	99.5	21.4	78.6	6.7	3.5	3.0	.3	1.7	93.9	.5	769.5
1996	313.4	R72.5	R174.1	1.7	561.7	100.8	21.1	76.4	6.8	3.6	2.9	.3	1.7	91.7	.5	775.9
1997	313.6	R72.5	R176.5	1.5	564.1	99.7	19.3	79.4	6.9	3.6	2.9	.3	1.6	94.8	.8	778.6
1998	315.8	R66.3	R180.3	1.5	563.9	97.1	19.5	79.2	6.8	3.7	2.9	.3	1.7	94.6	.8	775.9
1999	315.5	R60.1	R195.1	1.9	572.6	97.4	19.6	79.4	6.8	3.7	2.8	.4	2.3	95.3	1.0	785.9
2000	315.1	61.8	219.6	2.3	598.9	97.9	19.5	79.4	6.1	3.9	2.8	.4	2.4	94.9	.5	811.7
2001	314.2	66.2	252.8	1.7	634.9	98.2	19.7	78.9	5.9	R3.7	2.2	.4	3.9	R95.0	R.5	848.3
2002	315.4	59.7	312.5	2.0	689.5	98.7	20.4	79.4	5.8	3.8	2.3	.4	4.4	96.1	R.7	905.3
2003	313.0	60.7	355.4	2.0	731.2	99.2	20.5	78.7	5.9	3.8	2.1	.4	6.0	R96.8	R.7	948.4
2004	313.0	59.1	371.0	2.3	745.4	99.6	20.8	77.6	6.2	R3.5	2.2	.4	6.5	96.4	.7	962.9
2005	313.4	58.5	383.1	2.1	757.1	100.0	21.3	77.5	6.2	R3.6	2.3	.4	8.7	R98.7	R.9	978.0
2006	R313.0	R58.1	R388.3	R2.3	R761.6	R100.3	R21.5	R77.8	R6.4	R3.7	2.3	.4	R11.3	R101.9	R.9	R986.2
2007P	313.6	58.3	394.9	2.3	769.0	100.6	21.8	77.8	6.4	3.9	2.3	.5	15.6	106.6	.8	998.8

[1] Anthracite, bituminous coal, subbituminous coal, lignite, waste coal, and coal synfuel.
[2] Distillate fuel oil, residual fuel oil, petroleum coke, jet fuel, kerosene, other petroleum, and waste oil.
[3] Natural gas, plus a small amount of supplemental gaseous fuels.
[4] Blast furnace gas, propane gas, and other manufactured and waste gases derived from fossil fuels.
[5] Wood and wood-derived fuels.
[6] Municipal solid waste from biogenic sources, landfill gas, sludge waste, agricultural byproducts, and other biomass. For all years, also includes non-renewable waste (municipal solid waste from non-biogenic sources, and tire-derived fuels).
[7] Solar thermal and photovoltaic energy.
[8] Batteries, chemicals, hydrogen, pitch, purchased steam, sulfur, and miscellaneous technologies.
[9] Included in "Conventional Hydroelectric Power."
[10] Included in "Wood."
[11] Included in "Wind."
[12] Through 1988, data are for electric utilities only. Beginning in 1989, data are for electric utilities, independent power producers, commercial plants, and industrial plants.

R=Revised. P=Preliminary. NA=Not available. (s)=Less than 0.05 million kilowatts.

Notes: • Data are at end of year. • For plants that use multiple sources of energy, capacity is assigned to the predominant energy source. • See Note 1, "Coverage of Electricity Statistics," at end of section. • See "Generator Net Summer Capacity" in Glossary. • Totals may not equal sum of components due to independent rounding.

Web Pages: • For all data beginning in 1949, see http://www.eia.doe.gov/emeu/aer/elect.html. • For related information, see http://www.eia.doe.gov/fuelelectric.html.

Sources: Tables 8.11b and 8.11d.

Table 8.11b Electric Net Summer Capacity: Electric Power Sector, Selected Years, 1949-2007

(Subset of Table 8.11a; Million Kilowatts)

Year	Fossil Fuels					Nuclear Electric Power	Hydro-electric Pumped Storage	Renewable Energy							Other [8]	Total
	Coal [1]	Petroleum [2]	Natural Gas [3]	Other Gases [4]	Total			Conventional Hydroelectric Power	Biomass: Wood [5]	Biomass: Waste [6]	Geo-thermal	Solar/PV [7]	Wind	Total		
1949	NA	NA	NA	NA	44.9	0.0	([9])	18.5	(s)	([10])	NA	NA	NA	18.5	NA	63.4
1950	NA	NA	NA	NA	50.0	.0	([9])	19.2	(s)	([10])	NA	NA	NA	19.2	NA	69.2
1955	NA	NA	NA	NA	86.8	.0	([9])	27.4	(s)	([10])	NA	NA	NA	27.4	NA	114.2
1960	NA	NA	NA	NA	130.8	.4	([9])	35.8	.1	([10])	(s)	NA	NA	35.9	NA	167.1
1965	NA	NA	NA	NA	182.9	.8	([9])	51.0	.1	([10])	(s)	NA	NA	51.1	NA	234.8
1970	NA	NA	NA	NA	265.4	7.0	([9])	63.8	.1	([10])	.1	NA	NA	63.9	NA	336.4
1971	NA	NA	NA	NA	288.0	9.0	([9])	69.1	.1	([10])	.2	NA	NA	69.4	NA	366.4
1972	NA	NA	NA	NA	310.7	14.5	([9])	70.5	.1	([10])	.3	NA	NA	70.9	NA	396.0
1973	NA	NA	NA	NA	341.2	22.7	([9])	75.4	.1	([10])	.4	NA	NA	75.9	NA	439.8
1974	NA	NA	NA	NA	360.7	31.9	([9])	75.5	.1	([10])	.4	NA	NA	76.0	NA	468.5
1975	NA	NA	NA	NA	375.1	37.3	([9])	78.4	.1	([10])	.5	NA	NA	79.0	NA	491.3
1976	NA	NA	NA	NA	394.8	43.8	([9])	78.0	.1	([10])	.5	NA	NA	78.6	NA	517.2
1977	NA	NA	NA	NA	410.4	46.3	([9])	78.6	.1	([10])	.5	NA	NA	79.2	NA	535.9
1978	NA	NA	NA	NA	420.8	50.8	([9])	79.9	.1	([10])	.5	NA	NA	80.5	NA	552.1
1979	NA	NA	NA	NA	432.1	49.7	([9])	82.9	.1	([10])	.7	NA	NA	83.6	NA	565.5
1980	NA	NA	NA	NA	444.1	51.8	([9])	81.7	.1	([10])	.9	NA	NA	82.7	NA	578.6
1981	NA	NA	NA	NA	458.9	56.0	([9])	82.4	.1	([10])	.9	NA	(s)	83.4	NA	598.3
1982	NA	NA	NA	NA	469.6	60.0	([9])	83.0	.1	([10])	1.0	NA	(s)	84.1	NA	613.7
1983	NA	NA	NA	NA	472.8	63.0	([9])	83.9	.2	([10])	1.2	NA	(s)	85.3	NA	621.1
1984	NA	NA	NA	NA	478.6	69.7	([9])	85.3	.3	([10])	1.2	([11])	(s)	86.9	NA	635.1
1985	NA	NA	NA	NA	485.0	79.4	([9])	88.9	.2	.2	1.6	([11])	(s)	90.8	NA	655.2
1986	NA	NA	NA	NA	488.3	85.2	([9])	89.3	.2	.2	1.6	([11])	(s)	91.2	NA	664.8
1987	NA	NA	NA	NA	488.8	93.6	([9])	89.7	.2	.2	1.5	([11])	(s)	91.7	NA	674.1
1988	NA	NA	NA	NA	490.6	94.7	([9])	90.3	.2	.2	1.7	([11])	(s)	92.4	NA	677.7
1989[12]	298.0	R78.1	R125.4	.4	501.9	98.2	18.1	73.6	1.1	1.7	2.6	.2	1.5	80.7	–	698.8
1990	302.3	R76.8	R129.9	.4	509.3	99.6	19.5	73.3	1.2	2.1	2.7	.3	1.8	81.4	(s)	709.9
1991	302.5	R73.0	R137.1	.7	513.3	99.6	18.4	75.4	1.3	2.5	2.6	.3	1.9	84.0	–	715.3
1992	304.3	R71.8	R141.0	.7	517.9	99.0	21.2	74.2	1.4	2.5	2.9	.3	1.8	83.1	–	721.2
1993	305.0	R69.9	R146.9	.7	522.5	99.0	21.1	76.8	1.5	2.6	2.9	.3	1.8	85.9	–	728.6
1994	306.1	R70.5	R152.5	.7	529.8	99.1	21.2	76.9	1.7	2.7	3.0	.3	1.7	86.4	–	736.5
1995	306.0	R65.4	R161.9	.3	533.7	99.5	21.4	77.4	1.8	3.0	3.0	.3	1.7	87.3	–	741.8
1996	308.1	R71.3	R161.4	.1	540.9	100.8	21.1	75.3	1.7	2.9	2.9	.3	1.7	84.9	–	747.7
1997	308.5	R71.0	R163.4	.2	543.1	99.7	19.3	78.3	1.8	2.9	2.9	.3	1.6	87.8	.2	750.1
1998	310.9	R65.0	R167.1	.1	543.0	97.1	19.5	78.0	1.8	3.0	2.9	.3	1.7	87.8	.2	747.6
1999	310.7	R58.6	R181.1	.2	550.7	97.4	19.6	78.3	1.8	3.0	2.8	.4	2.3	88.6	.2	756.5
2000	310.2	60.7	204.7	.3	575.9	97.9	19.5	78.2	1.7	3.3	2.8	.4	2.4	88.8	(s)	782.1
2001	309.8	64.7	236.8	.3	611.6	98.2	19.7	77.9	1.6	3.3	2.2	.4	3.9	R89.2	R.1	818.8
2002	311.0	58.6	296.6	.3	666.5	98.7	20.4	78.3	1.6	R3.3	2.3	.4	4.4	R90.2	R.1	875.8
2003	308.5	59.6	339.1	.3	707.6	99.2	20.5	77.9	1.6	3.3	2.1	.4	6.0	91.3	R.1	918.6
2004	308.8	58.0	355.2	.4	722.4	99.6	20.8	77.0	1.6	R2.9	2.2	.4	6.5	90.6	R.1	933.4
2005	309.0	R57.4	367.5	.3	734.3	100.0	21.3	76.9	R1.6	3.0	2.3	.4	8.7	92.9	R.1	948.6
2006	R309.2	R56.8	R372.0	R.4	R738.4	R100.3	R21.5	R77.1	1.7	3.1	2.3	.4	R11.3	R95.9	R.1	R956.2
2007P	309.9	56.9	378.6	.5	745.9	100.6	21.8	77.2	2.1	3.3	2.3	.5	15.6	101.0	(s)	969.4

[1] Anthracite, bituminous coal, subbituminous coal, lignite, waste coal, and coal synfuel.
[2] Distillate fuel oil, residual fuel oil, petroleum coke, jet fuel, kerosene, other petroleum, and waste oil.
[3] Natural gas, plus a small amount of supplemental gaseous fuels.
[4] Blast furnace gas, propane gas, and other manufactured and waste gases derived from fossil fuels.
[5] Wood and wood-derived fuels.
[6] Municipal solid waste from biogenic sources, landfill gas, sludge waste, agricultural byproducts, and other biomass. For all years, also includes non-renewable waste (municipal solid waste from non-biogenic sources, and tire-derived fuels).
[7] Solar thermal and photovoltaic energy.
[8] Batteries, chemicals, hydrogen, pitch, purchased steam, sulfur, and miscellaneous technologies.
[9] Included in "Conventional Hydroelectric Power."
[10] Included in "Wood."
[11] Included in "Wind."
[12] Through 1988, data are for electric utilities only. Beginning in 1989, data are for electric utilities and independent power producers.

R=Revised. P=Preliminary. NA=Not available. – = No data reported. (s)=Less than 0.05 million kilowatts.

Notes: • Data are at end of year. • For plants that use multiple sources of energy, capacity is assigned to the predominant energy source. • The electric power sector comprises electricity-only and combined-heat-and-power (CHP) plants within the NAICS 22 category whose primary business is to sell electricity, or electricity and heat, to the public. • See Table 8.11d for commercial and industrial CHP and electricity-only data. • See Note 1, "Coverage of Electricity Statistics," and Note 2, "Classification of Power Plants Into Energy-Use Sectors," at end of section. • See "Generator Net Summer Capacity" in Glossary. • Totals may not equal sum of components due to independent rounding.

Web Pages: • For all data beginning in 1949, see http://www.eia.doe.gov/emeu/aer/elect.html. • For related information, see http://www.eia.doe.gov/fuelelectric.html.

Sources: • 1949-1984—Energy Information Administration (EIA) estimates. • 1985-1988—EIA, Form EIA-860, "Annual Electric Generator Report." • 1989-1997—EIA, Form EIA-860, "Annual Electric Generator Report," and Form EIA-867, "Annual Nonutility Power Producer Report." • 1998-2000—EIA, Form EIA-860A, "Annual Electric Generator Report—Utility," and Form EIA-860B, "Annual Electric Generator Report—Nonutility." • 2001 forward—EIA, Form EIA-860, "Annual Electric Generator Report."

Table 8.11c Electric Net Summer Capacity: Electric Power Sector by Plant Type, 1989-2007

(Breakout of Table 8.11b; Million Kilowatts)

Year	Fossil Fuels					Nuclear Electric Power	Hydro-electric Pumped Storage	Renewable Energy							Other [8]	Total
	Coal [1]	Petroleum [2]	Natural Gas [3]	Other Gases [4]	Total			Conventional Hydroelectric Power	Biomass: Wood [5]	Biomass: Waste [6]	Geo-thermal	Solar/PV [7]	Wind	Total		
Electricity-Only Plants [9]																
1989	296.5	R78.0	R119.3	0.4	494.2	98.2	18.1	73.6	0.9	1.5	2.6	0.2	1.5	80.3	–	690.7
1990	299.9	R76.6	R121.8	.4	498.6	99.6	19.5	73.3	1.0	1.9	2.7	.3	1.8	80.9	(s)	698.6
1991	299.6	R72.6	R127.9	.7	500.8	99.6	18.4	75.4	1.1	2.2	2.6	.3	1.9	83.6	–	702.4
1992	300.8	R71.5	R130.2	.7	503.1	99.0	21.2	74.2	1.2	2.3	2.9	.3	1.8	82.7	–	706.0
1993	301.2	R69.3	R134.5	.7	505.7	99.0	21.1	76.8	1.2	2.4	2.9	.3	1.8	85.5	–	711.3
1994	301.6	R69.8	R136.6	.7	508.7	99.1	21.2	76.9	1.5	2.5	3.0	.3	1.7	85.9	–	715.0
1995	301.3	R64.7	R145.3	.3	511.5	99.5	21.4	77.4	1.5	2.7	3.0	.3	1.7	86.6	–	719.1
1996	303.1	R70.6	R143.1	.1	516.9	100.8	21.1	75.3	1.4	2.6	2.9	.3	1.7	84.2	–	723.0
1997	303.6	R70.2	R144.7	.2	518.7	99.7	19.3	78.3	1.5	2.5	2.9	.3	1.6	87.1	.2	725.0
1998	305.9	R64.2	R147.5	.1	517.5	97.1	19.5	78.0	1.4	2.6	2.9	.3	1.7	87.0	.2	721.4
1999	305.5	R57.5	R161.7	.2	525.0	97.4	19.6	78.3	1.5	2.6	2.8	.4	2.3	87.8	.2	730.0
2000	305.2	59.8	184.0	.1	549.0	97.9	19.5	78.2	1.5	2.8	2.8	.4	2.4	88.1	(s)	754.5
2001	305.2	63.8	215.5	.1	584.5	98.2	19.7	77.9	1.5	R2.9	2.2	.4	3.6	R88.4	R.1	790.9
2002	305.8	57.5	268.1	.1	631.5	98.7	20.4	78.3	1.4	2.9	2.3	.4	4.4	89.7	R.1	840.3
2003	303.0	58.6	304.2	.1	665.9	99.2	20.5	77.9	1.4	2.8	2.1	.4	6.0	90.6	R.1	876.3
2004	303.2	57.3	322.6	.1	683.2	99.6	20.8	77.0	1.5	2.6	2.2	.4	6.5	90.0	R.1	893.7
2005	R303.4	R56.9	R335.8	R (s)	R696.2	100.0	21.3	76.9	1.4	2.6	2.3	.4	8.7	92.3	R.1	R909.8
2006	R303.4	R55.8	R341.9	.1	R701.2	R100.3	R21.5	R77.1	R1.5	2.7	2.3	.4	R11.3	R95.3	R.1	R918.4
2007[P]	304.7	55.7	350.2	.3	710.9	100.6	21.8	77.2	1.9	2.8	2.3	.5	15.6	100.4	(s)	933.8
Combined-Heat-and-Power Plants [10]																
1989	1.5	0.2	6.1	–	7.7	–	–	–	0.2	0.2	–	–	–	0.4	–	8.1
1990	2.4	.2	8.1	–	10.7	–	–	–	.2	.2	–	–	–	.5	–	11.2
1991	2.9	.4	9.2	–	12.5	–	–	–	.2	.2	–	–	–	.5	–	12.9
1992	3.5	.3	10.9	(s)	14.7	–	–	–	.2	.2	–	–	–	.5	–	15.2
1993	3.8	.7	12.3	–	16.8	–	–	–	.2	.2	–	–	–	.5	–	17.3
1994	4.5	.7	15.9	–	21.0	–	–	–	.3	.2	–	–	–	.5	–	21.5
1995	4.8	.8	16.6	–	22.1	–	–	–	.4	.2	–	–	–	.6	–	22.7
1996	5.0	.7	18.4	–	24.0	–	–	–	.3	.3	–	–	–	.6	–	24.6
1997	4.9	.8	18.7	(s)	24.4	–	–	–	.3	.4	–	–	–	.7	–	25.1
1998	5.0	.8	19.6	–	25.5	–	–	–	.4	.4	–	–	–	.7	–	26.2
1999	5.2	1.1	19.4	–	25.7	–	–	–	.4	.4	–	–	–	.7	–	26.5
2000	5.0	.9	20.7	.3	26.9	–	–	–	.2	.5	–	–	–	.7	–	27.7
2001	4.6	1.0	21.2	.3	27.1	–	–	(s)	.1	.4	–	–	.3	.8	(s)	27.9
2002	5.2	1.1	28.5	.2	34.9	–	–	–	.1	.4	–	–	–	.6	–	35.5
2003	5.5	1.1	34.9	.2	41.7	–	–	(s)	.2	.5	–	–	–	.7	–	42.3
2004	5.6	.7	32.6	.3	39.2	–	–	(s)	.2	.4	–	–	–	.6	–	39.7
2005	R5.6	R.5	R31.7	R.3	R38.1	–	–	(s)	.2	.4	–	–	–	.6	–	R38.7
2006	R5.8	R1.0	R30.0	R.3	R37.2	–	–	(s)	.2	.4	–	–	–	.6	–	R37.8
2007[P]	5.2	1.2	28.4	.2	35.0	–	–	(s)	.2	.4	–	–	–	.6	–	35.6

[1] Anthracite, bituminous coal, subbituminous coal, lignite, waste coal, and coal synfuel.
[2] Distillate fuel oil, residual fuel oil, petroleum coke, jet fuel, kerosene, other petroleum, and waste oil.
[3] Natural gas, plus a small amount of supplemental gaseous fuels.
[4] Blast furnace gas, propane gas, and other manufactured and waste gases derived from fossil fuels.
[5] Wood and wood-derived fuels.
[6] Municipal solid waste from biogenic sources, landfill gas, sludge waste, agricultural byproducts, and other biomass. For all years, also includes non-renewable waste (municipal solid waste from non-biogenic sources, and tire-derived fuels).
[7] Solar thermal and photovoltaic energy.
[8] Batteries, chemicals, hydrogen, pitch, purchased steam, sulfur, and miscellaneous technologies.
[9] Electricity-only plants within the NAICS 22 category whose primary business is to sell electricity to the public. Data also include a small number of electric utility combined-heat-and-power (CHP) plants.
[10] Combined-heat-and-power (CHP) plants within the NAICS 22 category whose primary business is to sell electricity and heat to the public. Data do not include electric utility CHP plants—these are included under "Electricity-Only Plants."

R=Revised. P=Preliminary. – = No data reported. (s)=Less than 0.05 million kilowatts.

Notes: • Data are at end of year. • For plants that use multiple sources of energy, capacity is assigned to the predominant energy source. • See Table 8.11d for commercial and industrial CHP and electricity-only data. • See Note 1, "Coverage of Electricity Statistics," and Note 2, "Classification of Power Plants Into Energy-Use Sectors," at end of section. • See "Generator Net Summer Capacity" in Glossary. • Totals may not equal sum of components due to independent rounding.

Web Page: For related information, see http://www.eia.doe.gov/fuelelectric.html.

Sources: • 1989-1997—Energy Information Administration (EIA), Form EIA-860, "Annual Electric Generator Report," and Form EIA-867, "Annual Nonutility Power Producer Report." • 1998-2000—EIA, Form EIA-860A, "Annual Electric Generator Report—Utility," and Form EIA-860B, "Annual Electric Generator Report—Nonutility." • 2001 forward—EIA, Form EIA-860, "Annual Electric Generator Report."

Table 8.11d Electric Net Summer Capacity: Commercial and Industrial Sectors, 1989-2007

(Subset of Table 8.11a; Million Kilowatts)

Year	Fossil Fuels					Nuclear Electric Power	Hydro electric Pumped Storage	Renewable Energy							Other [8]	Total
	Coal [1]	Petroleum [2]	Natural Gas [3]	Other Gases [4]	Total			Conventional Hydroelectric Power	Biomass		Geo-thermal	Solar/PV [7]	Wind	Total		
									Wood [5]	Waste [6]						
	Commercial Sector [9]															
1989	0.3	0.2	0.6	–	1.0	–	–	(s)	(s)	0.2	–	–	–	0.2	–	1.2
1990	.3	.2	.7	–	1.2	–	–	(s)	(s)	.2	–	–	–	.2	–	1.4
1991	.2	.2	.7	–	1.1	–	–	(s)	(s)	.2	–	–	–	.3	–	1.3
1992	.2	.2	.8	–	1.2	–	–	(s)	(s)	.2	–	–	–	.3	–	1.5
1993	.3	.2	.9	–	1.3	–	–	(s)	(s)	.3	–	–	–	.3	–	1.6
1994	.3	.2	1.2	–	1.7	–	–	(s)	(s)	.3	–	–	–	.3	–	2.1
1995	.3	.2	1.2	–	1.8	–	–	(s)	(s)	.3	–	–	–	.3	–	2.1
1996	.3	.3	1.2	–	1.8	–	–	(s)	(s)	.4	–	–	–	.5	–	2.3
1997	.3	.4	1.2	–	1.9	–	–	(s)	(s)	.4	–	–	–	.5	–	2.3
1998	.3	.3	1.2	–	1.8	–	–	(s)	(s)	.5	–	–	–	.5	–	2.3
1999	.3	.4	1.1	–	1.8	–	–	(s)	(s)	.5	–	–	–	.5	–	2.3
2000	.3	.3	1.2	–	1.8	–	–	(s)	(s)	.4	–	–	–	.4	–	2.2
2001	.3	.3	1.9	–	2.5	–	–	(s)	(s)	.3	–	–	–	.4	–	2.9
2002	.3	.3	1.2	–	1.8	–	–	(s)	(s)	.4	–	–	–	.4	–	2.2
2003	.3	.3	1.0	–	1.7	–	–	(s)	(s)	.4	–	–	–	.4	–	2.1
2004	.4	.3	1.1	(s)	1.8	–	–	(s)	(s)	.4	–	–	–	.4	–	2.2
2005	.4	.3	1.0	(s)	1.8	–	–	(s)	(s)	.4	–	–	–	.5	–	2.2
2006	.4	.3	1.0	(s)	1.8	–	–	(s)	(s)	.4	–	–	–	.5	–	2.3
2007P	.4	.3	1.0	(s)	1.8	–	–	(s)	(s)	.4	–	–	–	.5	–	2.3
	Industrial Sector [10]															
1989	4.8	0.7	9.7	1.2	16.5	–	–	0.5	4.1	0.2	–	–	–	4.8	0.5	21.8
1990	4.8	.9	10.3	1.3	17.3	–	–	.6	4.3	.2	–	–	–	5.1	.5	22.9
1991	4.7	1.1	9.8	1.4	17.1	–	–	.6	4.8	.2	–	–	–	5.6	.5	23.2
1992	4.8	1.1	10.3	1.4	17.6	–	–	.6	4.8	.3	–	–	–	5.6	.5	23.8
1993	4.9	1.0	10.9	1.2	18.0	–	–	.6	5.0	.3	–	–	–	5.8	.5	24.3
1994	5.0	1.0	11.0	1.4	18.5	–	–	1.1	5.0	.3	–	–	–	6.3	.5	25.4
1995	5.0	1.0	11.3	1.4	18.7	–	–	1.1	4.9	.2	–	–	–	6.3	.5	25.5
1996	5.0	.9	11.5	1.6	19.0	–	–	1.1	5.1	.2	–	–	–	6.4	.5	25.9
1997	4.8	1.1	11.9	1.3	19.2	–	–	1.1	5.1	.2	–	–	–	6.5	.6	26.2
1998	4.6	1.0	12.0	1.5	19.1	–	–	1.1	5.0	.2	–	–	–	6.3	.6	26.0
1999	4.4	1.1	12.9	1.7	20.1	–	–	1.1	5.0	.2	–	–	–	6.2	.8	27.1
2000	4.6	.8	13.7	2.0	21.2	–	–	1.1	4.4	.2	–	–	–	5.7	.5	27.3
2001	4.2	1.1	14.1	1.3	20.7	–	–	1.0	4.2	.1	–	–	–	5.4	.4	26.6
2002	4.0	.7	14.7	1.8	21.2	–	–	1.0	4.3	.1	–	–	–	5.5	.6	27.3
2003	4.1	.7	15.3	1.7	21.9	–	–	.8	4.3	.1	–	–	–	5.2	.6	27.7
2004	3.8	.8	14.8	1.9	21.3	–	–	.6	4.5	.2	–	–	–	5.4	.7	27.4
2005	4.0	.8	14.5	1.8	21.0	–	–	.7	4.5	.2	–	–	–	5.4	.8	27.2
2006	R3.3	R1.0	R15.3	1.8	R21.4	–	–	.7	R4.7	.2	–	–	–	R5.6	.8	R27.8
2007P	3.2	1.0	15.2	1.8	21.2	–	–	.6	4.4	.2	–	–	–	5.1	.8	27.2

[1] Anthracite, bituminous coal, subbituminous coal, lignite, waste coal, and coal synfuel.
[2] Distillate fuel oil, residual fuel oil, petroleum coke, jet fuel, kerosene, other petroleum, and waste oil.
[3] Natural gas, plus a small amount of supplemental gaseous fuels.
[4] Blast furnace gas, propane gas, and other manufactured and waste gases derived from fossil fuels.
[5] Wood and wood-derived fuels.
[6] Municipal solid waste from biogenic sources, landfill gas, sludge waste, agricultural byproducts, and other biomass. For all years, also includes non-renewable waste (municipal solid waste from non-biogenic sources, and tire-derived fuels).
[7] Solar thermal and photovoltaic energy.
[8] Batteries, chemicals, hydrogen, pitch, purchased steam, sulfur, and miscellaneous technologies.
[9] Commercial combined-heat-and-power (CHP) and commercial electricity-only plants.
[10] Industrial combined-heat-and-power (CHP) and industrial electricity-only plants.

R=Revised. P=Preliminary. – = No data reported. (s)=Less than 0.05 million kilowatts.

Notes: • Data are at end of year. • For plants that use multiple sources of energy, capacity is assigned to the predominant energy source. • See Tables 8.11b and 8.11c for electric power sector electricity-only and CHP data. • See Note 1, "Coverage of Electricity Statistics," and Note 2, "Classification of Power Plants Into Energy-Use Sectors," at end of section. • See "Generator Net Summer Capacity" in Glossary. • Totals may not equal sum of components due to independent rounding.

Web Page: For related information, see http://www.eia.doe.gov/fuelelectric.html.

Sources: • 1989-1997—Energy Information Administration (EIA), Form EIA-867, "Annual Nonutility Power Producer Report." • 1998-2000—EIA, Form EIA-860B, "Annual Electric Generator Report—Nonutility." • 2001 forward—EIA, Form EIA-860, "Annual Electric Generator Report."

Figure 8.12 Electric Noncoincident Peak Load and Capacity Margin

U.S. Peak Load, 1986-2007

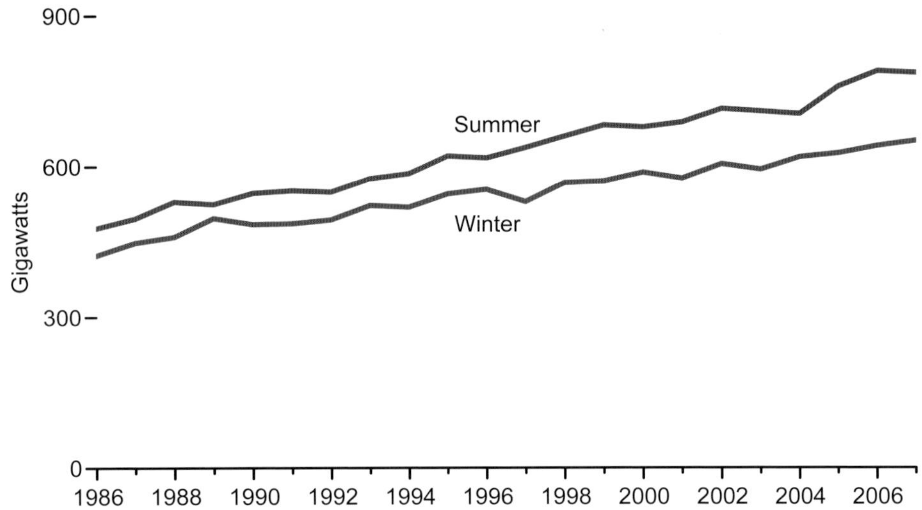

Capacity Margin, 1996-2007

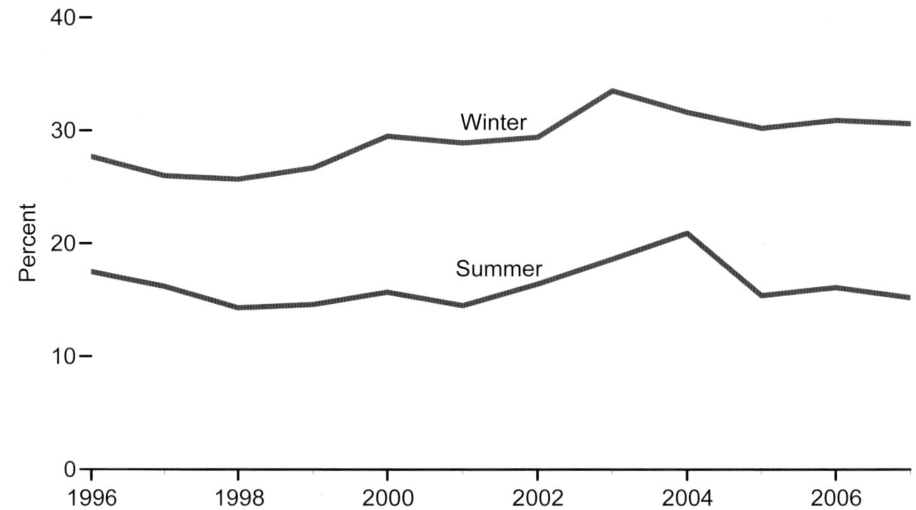

U.S. Peak Load by NERC Region, 2007

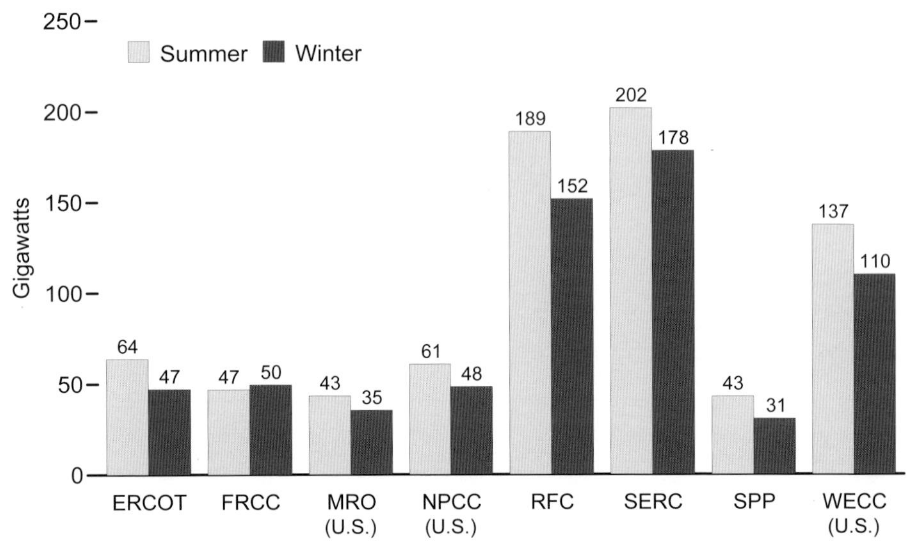

North American Electric Reliability Council Map for the United States

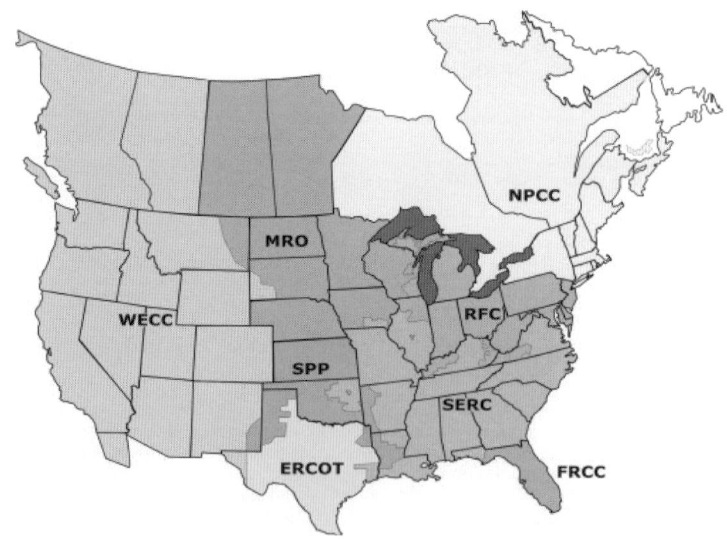

Notes: • Values for 2007 are forecast. • Noncoincident peak load is the sum of two or more peak loads on individual systems that do not occur at the same time interval. See Glossary for information on North American Electric Reliability Council (NERC). • Because vertical scales differ, graphs should not be compared.

Source: Table 8.12.

Table 8.12 Electric Noncoincident Peak Load and Capacity Margin, 1986-2007

(Megawatts, Except as Noted)

Year	Noncoincident Peak Load [1]															
	North American Electric Reliability Council Regions [2]											Contiguous United States	ASCC (Alaska)	Hawaii	U.S. Total	Capacity Margin [7] (percent)
	ECAR [3]	ERCOT	FRCC	MAAC [3]	MAIN [3]	MRO [4] (U.S.)	NPCC (U.S.)	RFC [5]	SERC	SPP	WECC [6] (U.S.)					
	Summer [8]															
1986	69,606	39,335	– –	37,564	35,943	21,029	39,026	– –	105,570	47,123	81,787	476,983	–	–	476,983	NA
1987	72,561	39,339	– –	40,526	37,446	23,162	42,651	– –	109,798	47,723	82,967	496,173	–	–	496,173	NA
1988	79,149	40,843	– –	43,110	41,139	24,899	45,245	– –	115,168	49,356	90,551	529,460	–	–	529,460	NA
1989	75,442	40,402	– –	41,614	39,460	24,336	45,031	– –	117,729	49,439	90,657	524,110	456	–	524,566	NA
1990	79,258	42,737	– –	42,613	40,740	24,994	44,116	– –	121,943	52,541	97,389	546,331	463	–	546,794	21.6
1991	81,224	41,870	– –	45,937	41,598	25,498	46,594	– –	124,716	51,885	92,096	551,418	471	–	551,889	20.9
1992	78,550	42,619	– –	43,658	38,819	22,638	43,658	– –	128,236	51,324	99,205	548,707	504	–	549,211	20.5
1993	80,930	44,255	– –	46,494	41,956	24,396	46,706	– –	135,704	57,106	97,809	575,356	511	–	575,867	19.9
1994	87,165	44,162	– –	46,019	42,562	27,000	47,581	– –	132,584	56,035	102,212	585,320	524	–	585,844	18.7
1995	92,619	46,618	– –	48,577	45,782	29,192	47,705	– –	146,569	59,595	103,592	620,249	622	–	620,871	18.9
1996	90,798	47,480	– –	44,302	46,402	28,253	45,094	– –	145,650	60,072	108,739	616,790	–	–	616,790	17.5
1997	93,492	50,541	35,375	49,464	45,887	29,787	49,269	– –	137,382	36,479	110,001	637,677	–	–	637,677	16.2
1998	93,784	54,666	38,730	48,445	47,509	30,722	49,566	– –	143,226	37,724	115,921	660,293	–	–	660,293	14.3
1999	99,239	55,529	37,493	51,645	51,535	31,903	52,855	– –	149,685	38,609	113,629	682,122	–	–	682,122	14.6
2000	92,033	57,606	37,194	49,477	52,552	28,605	50,057	– –	156,088	40,199	114,602	678,413	–	–	678,413	15.7
2001	100,235	55,201	39,062	54,015	56,344	28,321	55,949	– –	149,293	40,273	109,119	687,812	–	–	687,812	14.5
2002	102,996	56,248	40,696	55,569	56,396	29,119	56,012	– –	158,767	39,688	119,074	714,565	–	–	714,565	16.4
2003	98,487	59,996	40,475	53,566	56,988	28,831	55,018	– –	153,110	40,367	122,537	709,375	–	–	709,375	18.6
2004	95,300	58,531	42,383	52,049	53,439	29,351	52,549	– –	157,615	40,106	123,136	704,459	–	–	704,459	20.9
2005	– –	60,210	46,396	– –	– –	39,918	58,960	190,200	190,705	41,727	130,760	758,876	–	–	758,876	15.4
2006	– –	R62,339	R45,751	– –	– –	R42,194	R63,241	R191,920	R199,052	R42,882	R142,096	R789,475	–	–	R789,475	R16.1
2007F	– –	63,794	46,878	– –	– –	43,431	60,807	188,856	201,692	43,007	137,465	785,930	–	–	785,930	15.2
	Winter [9]															
1986	64,561	28,730	– –	32,807	28,036	18,850	37,976	– –	101,849	33,877	76,171	422,857	–	–	422,857	NA
1987	68,118	31,399	– –	35,775	30,606	19,335	41,902	– –	105,476	34,472	81,182	448,265	–	–	448,265	NA
1988	67,771	34,621	– –	36,363	30,631	20,162	42,951	– –	108,649	35,649	82,937	459,734	–	–	459,734	NA
1989	73,080	38,388	– –	38,161	33,770	21,360	42,588	– –	121,995	42,268	84,768	496,378	626	–	497,004	NA
1990	67,097	35,815	– –	36,551	32,461	21,113	40,545	– –	117,448	38,949	94,252	484,231	613	–	484,844	NA
1991	71,181	35,448	– –	37,983	33,420	21,432	41,866	– –	119,575	38,759	86,097	485,761	622	–	486,383	NA
1992	72,885	35,055	– –	37,915	31,289	21,866	41,125	– –	121,250	39,912	91,686	492,983	635	–	493,618	NA
1993	81,846	35,407	– –	41,406	34,966	21,955	42,063	– –	133,635	41,644	88,811	521,733	632	–	522,365	NA
1994	75,638	36,180	– –	40,653	33,999	23,033	42,547	– –	132,661	42,505	91,037	518,253	641	–	518,894	NA
1995	83,465	36,965	– –	40,790	35,734	23,429	42,755	– –	142,032	44,624	94,890	544,684	676	–	545,360	NA
1996	84,534	38,868	– –	40,468	37,162	24,251	41,208	– –	143,060	49,095	95,435	554,081	–	–	554,081	27.7
1997	75,670	37,966	33,076	37,217	34,973	25,390	41,338	– –	122,649	27,437	94,158	529,874	–	–	529,874	26.0
1998	84,401	41,876	39,975	36,532	37,410	26,080	44,199	– –	127,416	27,847	101,822	567,558	–	–	567,558	25.7
1999	86,239	39,164	40,178	40,220	39,081	25,200	45,227	– –	128,563	27,963	99,080	570,915	–	–	570,915	26.7
2000	84,546	44,641	38,606	43,256	41,943	24,536	43,852	– –	139,146	30,576	97,324	588,426	–	–	588,426	29.5
2001	85,485	44,015	40,922	39,458	40,529	21,815	42,670	– –	135,182	29,614	96,622	576,312	–	–	576,312	28.9
2002	87,300	45,414	45,635	46,551	42,412	23,645	46,009	– –	141,882	30,187	95,951	604,986	–	–	604,986	29.4
2003	86,332	42,702	36,841	45,625	41,719	24,134	48,079	– –	137,972	28,450	102,020	593,874	–	–	593,874	33.5
2004	91,800	44,010	44,839	45,905	42,929	24,526	48,176	– –	144,337	29,490	102,689	618,701	–	–	618,701	31.6
2005	– –	48,141	42,657	– –	– –	33,748	46,828	151,600	164,638	31,260	107,493	626,365	–	–	626,365	30.2
2006	– –	R50,402	R42,526	– –	– –	R34,677	R46,697	R149,631	R175,163	R30,792	R111,093	R640,981	–	–	R640,981	R30.9
2007F	– –	47,163	49,526	– –	– –	35,495	48,394	151,597	178,337	30,801	110,073	651,386	–	–	651,386	30.6

[1] Noncoincident peak load is the sum of two or more peak loads on individual systems that do not occur at the same time interval.

[2] See "North American Electric Reliablility Council (NERC)" in Glossary. Data include the U.S. portion of NERC only. See Figure 8.12 for an illustration of NERC regions.

[3] ECAR, MAAC, and MAIN dissolved at the end of 2005. Utility members joined other reliability regional councils.

[4] MRO was renamed from MAPP in 2004.

[5] ReliabilityFirst Corporation (RFC) came into existence on January 1, 2006. Many of the former utility members of ECAR, MAAC, and MAIN joined RFC.

[6] WECC was renamed from WSCC in 2002.

[7] The percent by which planned generating capacity resources are expected to be greater (or less) than estimated net internal demand at the time of expected peak summer (or winter) demand. Net internal demand does not include estimated demand for direct control load management and customers with interruptible service agreements. Data are for the contiguous United States only.

[8] The summer peak period is June through September.

[9] The winter peak period is December through February of the following year.

R=Revised. F=Forecast. NA=Not available. – – = Not applicable. – = No data reported.

Web Page: For related information, see http://www.eia.doe.gov/fuelelectric.html.

Sources: Energy Information Administration (EIA), *Electric Power Annual 2006* (October 2007), Tables 3.1-3.4; and EIA, Form EIA-411, "Coordinated Bulk Power Supply Program Report," and predecessor forms.

Figure 8.13 Electric Utility Demand-Side Management Programs

Actual Peakload Reductions Total, 1989-2006

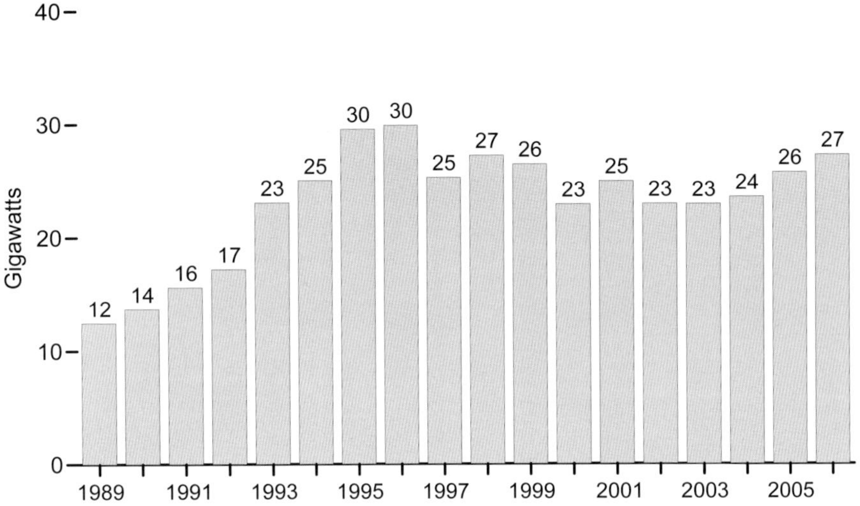

Actual Peakload Reductions, 2006

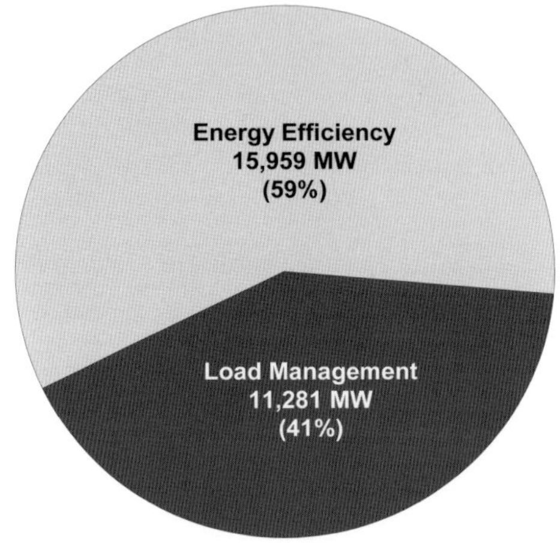

Total: 27,240 Megawatts (MW)

Energy Savings, 1989-2006

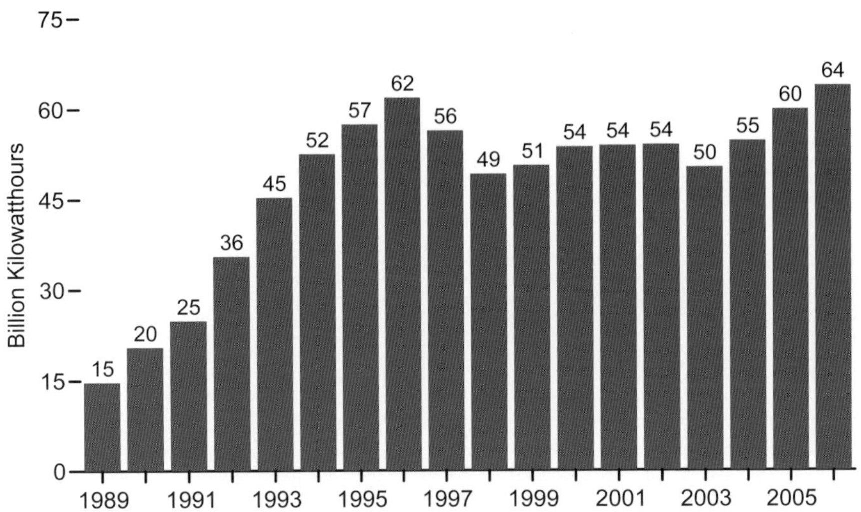

Costs, 1989-2006

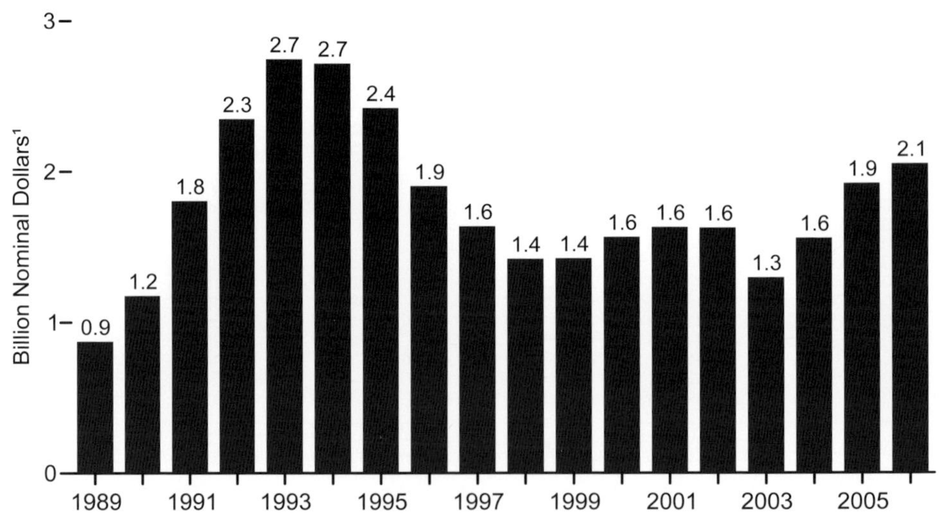

[1] See "Nominal Dollars" in Glossary.

Source: Table 8.13.

Table 8.13 Electric Utility Demand-Side Management Programs, 1989-2006

Year	Actual Peakload Reductions [1]			Energy Savings	Costs
	Energy Efficiency [2]	Load Management [3]	Total		
	Megawatts			Million Kilowatthours	Thousand Nominal Dollars [4]
1989	NA	NA	12,463	14,672	872,935
1990	NA	NA	13,704	20,458	1,177,457
1991	NA	NA	15,619	24,848	1,803,773
1992	7,890	9,314	17,204	35,563	2,348,094
1993	10,368	12,701	23,069	45,294	2,743,533
1994	11,662	13,340	25,001	52,483	2,715,657
1995	13,212	16,347	29,561	57,421	2,421,284
1996	14,243	15,650	29,893	61,842	1,902,197
1997	13,327	11,958	25,284	56,406	1,636,020
1998	13,591	13,640	27,231	49,167	1,420,920
1999	13,452	13,003	26,455	50,563	1,423,644
2000	12,873	10,027	22,901	53,701	1,564,901
2001	13,027	11,928	24,955	53,936	1,630,286
2002	13,420	9,516	22,936	54,075	1,625,537
2003	13,581	9,323	22,904	50,265	1,297,210
2004	14,272	9,260	23,532	54,710	1,557,466
2005	15,351	10,359	25,710	59,897	1,921,352
2006	15,959	11,281	27,240	63,817	2,051,394

[1] The actual reduction in peak load reflects the change in demand for electricity that results from a utility demand-side management (DSM) program that is in effect at the time that the utility experiences its actual peak load as opposed to the potential installed peakload reduction capacity. Differences between actual and potential peak reduction result from changes in weather, economic activity, and other variable conditions.

[2] "Energy Efficiency" refers to programs that are aimed at reducing the energy used by specific end-use devices and systems, typically without affecting the services provided. These programs reduce overall electricity consumption, often without explicit consideration for the timing of program-induced savings. Such savings are generally achieved by substituting technically more advanced equipment to produce the same level of end-use services (e.g., lighting, heating, motor drive) with less electricity. Examples include high-efficiency appliances, efficient lighting programs, high-efficiency heating, ventilating, and air conditioning systems or control modifications, efficient building design, advanced electric motor drives, and heat recovery systems.

[3] "Load Management" includes programs such as "Direct Load Control," "Interruptible Load Control," and, "Other Types" of DSM programs. "Direct Load Control" refers to program activities that can interrupt consumer load at the time of annual peak load by direct control of the utility system operator by interrupting power supply to individual appliances or equipment on consumer premises. This type of control usually involves residential consumers. "Interruptible Load Control" refers to program activities that, in accordance with contractual arrangements, can interrupt consumer load at times of seasonal peak load by direct control of the utility system operator or by action of the consumer at the direct request of the system operator. It usually involves commercial and industrial consumers. In some instances, the load reduction may be affected by direct action of the system operator (remote tripping) after notice to the consumer in accordance with contractual provisions. "Other Types" are programs that limit or shift peak loads from on-peak to off-peak time periods, such as space heating and water heating storage systems.

[4] See "Nominal Dollars" in Glossary.

NA=Not available.

Note: This table reports on the results of DSM programs operated by electric utilities. The decrease since 1998 in peakload reductions from DSM programs can be attributed in part to utilities cutting back or terminating these programs due to industry deregulation. Some State governments have created new programs to promote DSM. Examples include the "Energy $mart Loan Fund" administered by the New York Energy Research and Development Authority and the "Efficiency Vermont" program of the Vermont Public Service Board. Data on energy savings attributable to these non-utility programs are not collected by the Energy Information Administration.

Web Page: For related information, see http://www.eia.doe.gov/fuelelectric.html.

Sources: • 1989-1994—Energy Information Administration (EIA), Form EIA-861, "Annual Electric Utility Report." • 1995 forward—EIA, *Electric Power Annual 2006* (October 2007), Tables 9.1, 9.6, and 9.7.

Electricity

Note 1. Coverage of Electricity Statistics. Through 1984, data for electric utilities also include institutions (such as universities) and military facilities that generated electricity primarily for their own use; beginning in 1985, data for electric utilities exclude institutions and military facilities. Data for independent power producers, commercial plants, and industrial plants include plants with a generator nameplate capacity of 1 megawatt or greater; they exclude plants with a generator nameplate capacity less than 1 megawatt. Also excluded from the electricity statistics in Section 8 are data for residential and commercial self-generation from solar energy, except for the small amount sold to the grid and included in data for the electric power sector.

Note 2. Classification of Power Plants Into Energy-Use Sectors. The Energy Information Administration (EIA) classifies power plants (both electricity-only and combined-heat-and-power plants) into energy-use sectors based on the North American Industry Classification System (NAICS), which replaced the Standard Industrial Classification (SIC) system in 1997. Plants with a NAICS code of 22 are assigned to the Electric Power Sector. Those with NAICS codes beginning with 11 (agriculture, forestry, fishing, and hunting); 21 (mining, including oil and gas extraction); 23 (construction); 31-33 (manufacturing); 2212 (natural gas distribution); and 22131 (water supply and irrigation systems) are assigned to the Industrial Sector. Those with all other codes are assigned to the Commercial Sector. Form EIA-860, "Annual Electric Generator Report," asks respondents to indicate the primary purpose of the facility by assigning a NAICS code from the list at: http://www.eia.doe.gov/cneaf/electricity/forms/eia860/eia860.doc.

Note 3. Electricity Imports and Exports. Through the *Annual Energy Review* (*AER*) 2001, EIA estimated the proportions of traded electricity from fossil fuels and hydropower (and applied the fossil-fuel steam-electric-plant heat rate to convert from kilowatthours to Btu) and from geothermal (and applied the heat rate for geothermal energy plants). Beginning with the *AER* 2002, because of inadequate data, EIA is applying an overall rate of 3,412 Btu per kilowatthour to all traded electricity. In addition, electricity net imports derived from hydroelectric power and geothermal energy are no longer included in renewable energy consumption data. They continue to be included in total U.S. energy consumption as components of electricity net imports, with energy sources unspecified (see Tables 1.3 and 2.1f). This change between *AER* 2001 and *AER* 2002 resulted in a 0.0-to-0.5 quadrillion Btu drop in total renewable energy consumption from 1949 forward.

Table 8.1 Sources: Net Generation, Electric Power Sector: Table 8.2b. **Net Generation, Commercial Sector:** Table 8.2d. **Net Generation, Industrial Sector:** • 1949-September 1977—Federal Power Commission (FPC), Form FPC-4, "Monthly Power Plant Report," for plants with generating capacity exceeding 10 megawatts, and FPC, Form FPC-12C, "Industrial Electric Generating Capacity," for all other plants. • October 1977-1978—Federal Energy Regulatory Commission (FERC), Form FPC-4, "Monthly Power Plant Report," for plants with generating capacity exceeding 10 megawatts, and FERC, Form FPC-12C, "Industrial Electric Generating Capacity," for all other plants. • 1979—FERC, Form FPC-4, "Monthly Power Plant Report," for plants with generating capacity exceeding 10 megawatts, and EIA estimates for all other plants. • 1980-1988—Estimated by EIA as the average generation over the 6-year period of 1974-1979. • 1989 forward—Table 8.2d. **Net Generation, Total:** Table 8.2a. **Imports** and **Exports:** • 1949-September 1977—Unpublished FPC data. • October 1977-1980—Unpublished Economic Regulatory Administration (ERA) data. • 1981—Department of Energy (DOE), Office of Energy Emergency Operations, "Report on Electric Energy Exchanges with Canada and Mexico for Calendar Year 1981," April 1982 (revised June 1982). • 1982 and 1983—DOE, ERA, *Electricity Exchanges Across International Borders*. • 1984-1986—DOE, ERA, *Electricity Transactions Across International Borders*. • 1987 and 1988—DOE, ERA, Form ERA-781R, "Annual Report of International Electrical Export/Import Data." • 1989—DOE, Fossil Energy, Form FE-781R, "Annual Report of International Electrical Export/Import Data." • 1990 forward—National Energy Board of Canada, and DOE, Fossil Energy, Office of Fuels Programs, Form FE-781R, "Export/Import Data." For 2001 forward, data from the California Independent System Operator are used in combination with the Form FE-781R values to estimate electricity trade with Mexico. **T & D Losses and Unaccounted for:** Calculated as the sum of total net generation and imports minus total end use and exports. **End Use:** Table 8.9.

9 Nuclear Energy

Site of Shippingport atomic power station, the first commercial nuclear power plant in the United States (rectangular reactor building and foreground); background, Beaver Valley 1 and 2 nuclear power plants and Bruce Mansfield coal-fired power plant (southwestern Pennsylvania). Source: U.S. Department of Energy.

Figure 9.1 Nuclear Generating Units

Operable Units,[1] 1957-2007

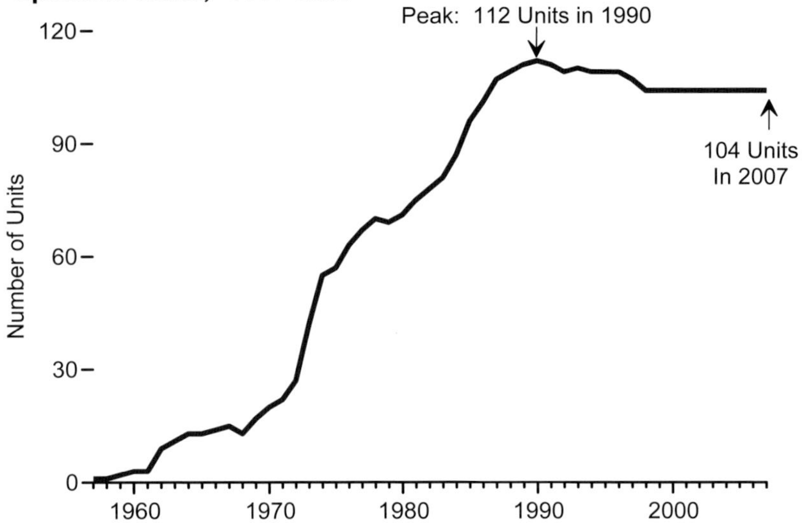

Full-Power Operating Licenses Issued,[2] 1957-2007

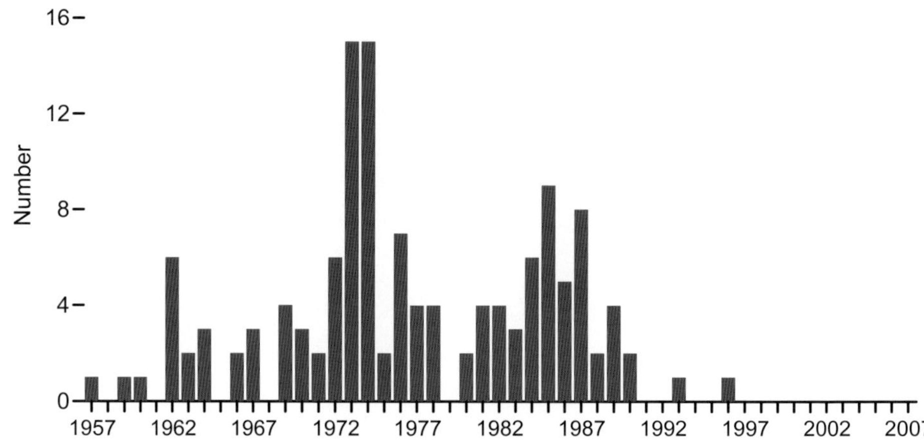

Status in 2007 of the 132 Full-Power Operating Licenses Issued from 1957-2007

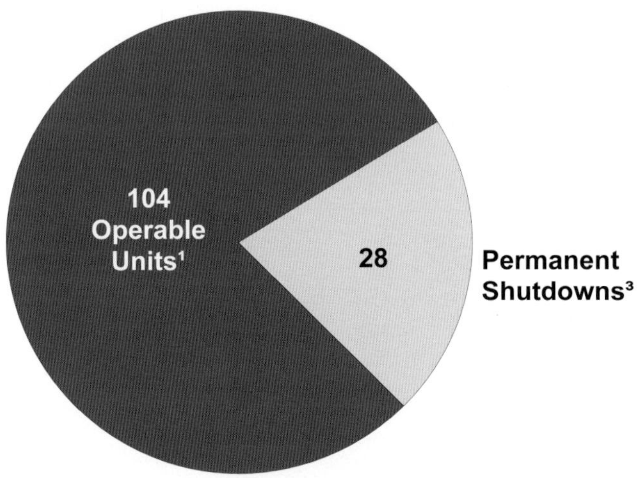

Permanent Shutdowns[3] by Year, 1957-2007

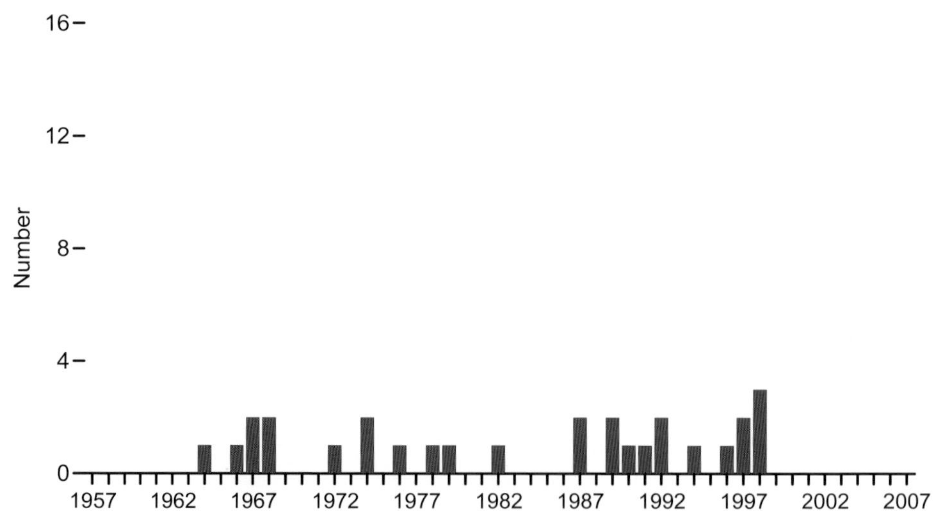

[1] Units holding full-power operating licenses, or equivalent permission to operate, at the end of the year.

[2] Issuance by regulatory authority of full-power operating license, or equivalent permission.

[3] Number of nuclear generating units ceasing operation permanently.

Notes: • Data are at end of year. • Because vertical scales differ, graphs should not be compared.

Source: Table 9.1.

Table 9.1 Nuclear Generating Units, 1955-2007

Year	Original Licensing Regulations (10 CFR Part 50) [1]			Current Licensing Regulations (10 CFR Part 52) [1]			Permanent Shutdowns [6]	Operable Units [7]
	Construction Permits Issued [2,3]	Low-Power Operating Licenses Issued [3,4]	Full-Power Operating Licenses Issued [3,5]	Early Site Permits Issued [3]	Combined License Applications Under Review	Combined Licenses Issued [3]		
1955	1	0	0	--	--	--	0	0
1956	3	0	0	--	--	--	0	0
1957	1	1	1	--	--	--	0	1
1958	0	0	0	--	--	--	0	1
1959	3	1	1	--	--	--	0	2
1960	7	1	1	--	--	--	0	3
1961	0	0	0	--	--	--	0	3
1962	1	7	6	--	--	--	0	9
1963	1	3	2	--	--	--	0	11
1964	3	2	3	--	--	--	1	13
1965	1	0	0	--	--	--	0	13
1966	5	1	2	--	--	--	1	14
1967	14	3	3	--	--	--	2	15
1968	23	0	0	--	--	--	2	13
1969	7	4	4	--	--	--	0	17
1970	10	4	3	--	--	--	0	20
1971	4	5	2	--	--	--	0	22
1972	8	6	6	--	--	--	1	27
1973	14	12	15	--	--	--	0	42
1974	23	14	15	--	--	--	2	55
1975	9	3	2	--	--	--	0	57
1976	9	7	7	--	--	--	1	63
1977	15	4	4	--	--	--	0	67
1978	13	3	4	--	--	--	1	70
1979	2	0	0	--	--	--	1	69
1980	0	5	2	--	--	--	0	71
1981	0	3	4	--	--	--	0	75
1982	0	6	4	--	--	--	1	78
1983	0	3	3	--	--	--	0	81
1984	0	7	6	--	--	--	0	87
1985	0	7	9	--	--	--	0	96
1986	0	7	5	--	--	--	0	101
1987	0	6	8	--	--	--	2	107
1988	0	1	2	--	--	--	0	109
1989	0	3	4	--	--	--	2	111
1990	0	1	2	--	--	--	1	112
1991	0	0	0	--	--	--	1	111
1992	0	0	0	--	--	--	2	109
1993	0	1	1	--	--	--	0	110
1994	0	0	0	--	--	--	1	109
1995	0	1	0	--	--	--	0	109
1996	0	0	1	--	--	--	1	109
1997	0	0	0	0	0	0	2	107
1998	0	0	0	0	0	0	3	104
1999	0	0	0	0	0	0	0	104
2000	0	0	0	0	0	0	0	104
2001	0	0	0	0	0	0	0	104
2002	0	0	0	0	0	0	0	104
2003	0	0	0	0	0	0	0	104
2004	0	0	0	0	0	0	0	104
2005	0	0	0	0	0	0	0	104
2006	0	0	0	0	0	0	0	104
2007	0	0	0	3	4	0	0	104
Total	177	132	132	3	4	0	28	--

[1] Data in columns 1-3 are based on the U.S. Nuclear Regulatory Commission (NRC) regulation 10 CFR Part 50. Data in columns 4-6 are based on the NRC regulation 10 CFR Part 52. See Note 1, "Pending Actions on Nuclear Generating Units," at end of section.
[2] Issuance by regulatory authority of a permit, or equivalent permission, to begin construction.
[3] Numbers reflect permits or licenses issued in a given year, not extant permits or licenses.
[4] Issuance by regulatory authority of license, or equivalent permission, to conduct testing but not to operate at full power.
[5] Issuance by regulatory authority of full-power operating license, or equivalent permission (note that some units receive full-power licenses the same year they receive low-power licenses). Units initially undergo low-power testing prior to commercial operation.
[6] Number of nuclear generating units, in a given year, ceasing operation permanently.
[7] Total of nuclear generating units holding full-power licenses, or equivalent permission to operate, at the end of the year (the number of operable units equals the cumulative number of units holding full-power licenses minus the cumulative number of permanent shutdowns).
– – = Not applicable.
Note: See Note 2, "Coverage of Nuclear Energy Statistics," at end of section.
Web Page: For related information, see http://www.eia.doe.gov/fuelnuclear.html.
Sources: See end of section.

Figure 9.2 Nuclear Power Plant Operations

Total Electricity and Nuclear Electricity Net Generation, 1957-2007

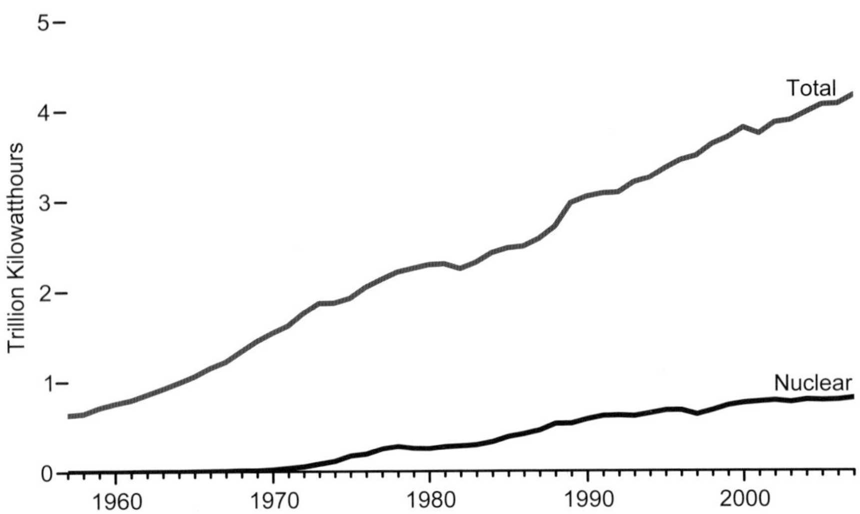

Nuclear Share of Total Electricity Net Generation, 1957-2007

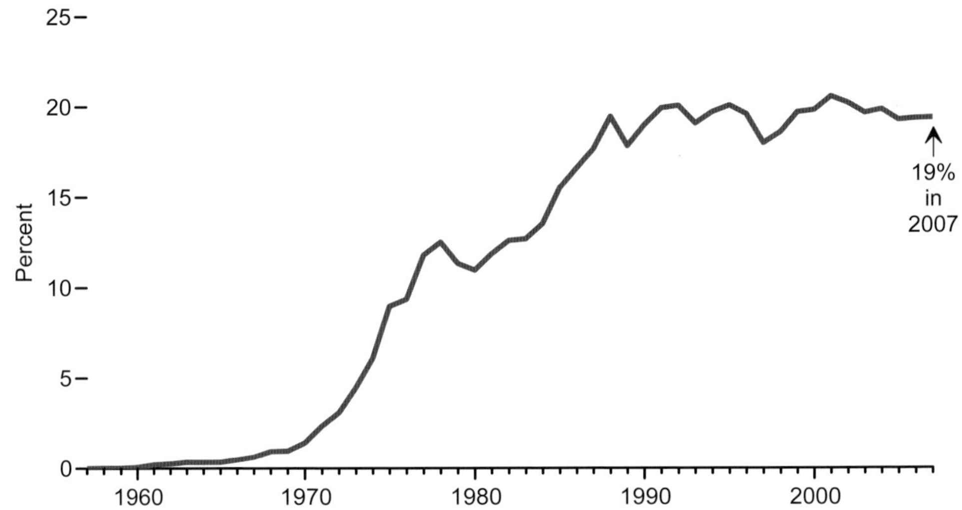

Net Summer Capacity of Operable Units, 1957-2007

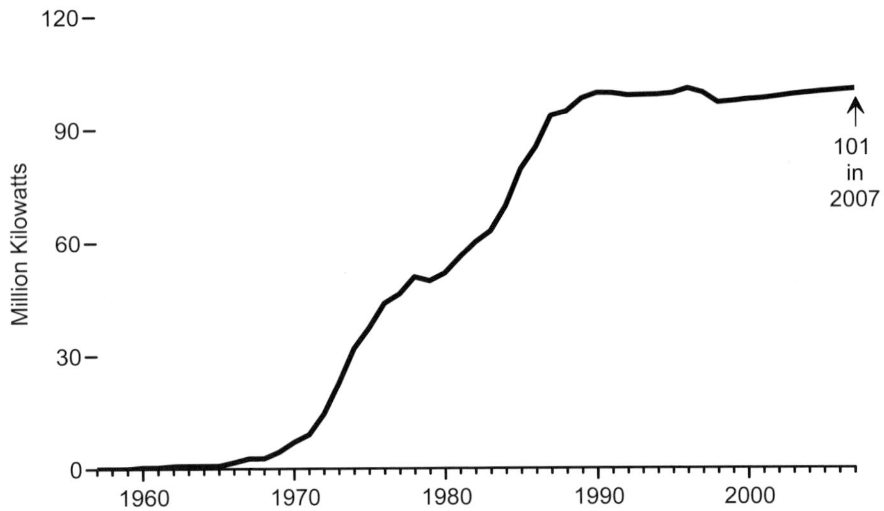

Capacity Factor, 1973-2007

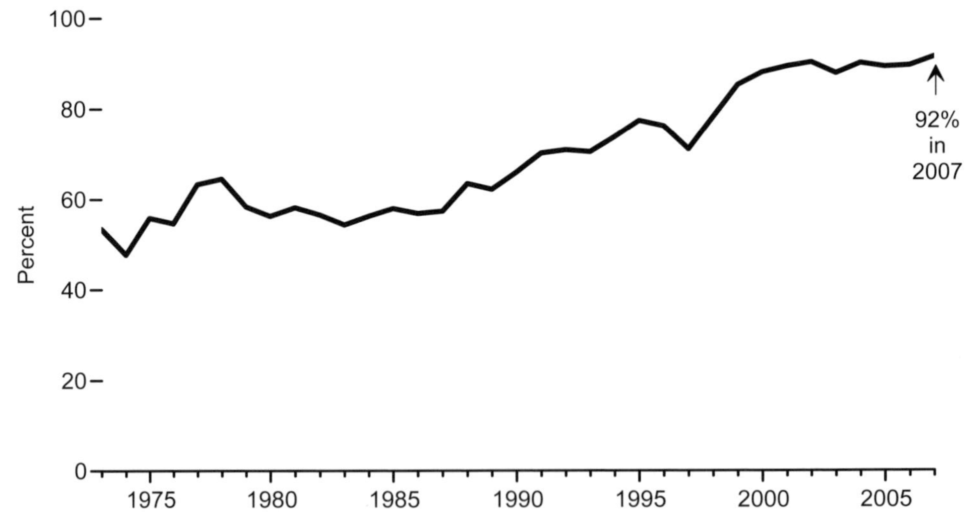

Note: Because vertical scales differ, graphs should not be compared.

Sources: Tables 8.1 and 9.2.

Table 9.2 Nuclear Power Plant Operations, 1957-2007

Year	Nuclear Electricity Net Generation	Nuclear Share of Total Electricity Net Generation	Net Summer Capacity of Operable Units [1]	Capacity Factor [2]
	Billion Kilowatthours	Percent	Million Kilowatts	Percent
1957	(s)	(s)	0.1	NA
1958	.2	(s)	.1	NA
1959	.2	(s)	.1	NA
1960	.5	.1	.4	NA
1961	1.7	.2	.4	NA
1962	2.3	.3	.7	NA
1963	3.2	.3	.8	NA
1964	3.3	.3	.8	NA
1965	3.7	.3	.8	NA
1966	5.5	.5	1.7	NA
1967	7.7	.6	2.7	NA
1968	12.5	.9	2.7	NA
1969	13.9	1.0	4.4	NA
1970	21.8	1.4	7.0	NA
1971	38.1	2.4	9.0	NA
1972	54.1	3.1	14.5	NA
1973	83.5	4.5	22.7	53.5
1974	114.0	6.1	31.9	47.8
1975	172.5	9.0	37.3	55.9
1976	191.1	9.4	43.8	54.7
1977	250.9	11.8	46.3	63.3
1978	276.4	12.5	50.8	64.5
1979	255.2	11.3	49.7	58.4
1980	251.1	11.0	51.8	56.3
1981	272.7	11.9	56.0	58.2
1982	282.8	12.6	60.0	56.6
1983	293.7	12.7	63.0	54.4
1984	327.6	13.5	69.7	56.3
1985	383.7	15.5	79.4	58.0
1986	414.0	16.6	85.2	56.9
1987	455.3	17.7	93.6	57.4
1988	527.0	19.5	94.7	63.5
1989	529.4	17.8	98.2	62.2
1990	576.9	19.0	99.6	66.0
1991	612.6	19.9	99.6	70.2
1992	618.8	20.1	99.0	70.9
1993	610.3	19.1	99.0	70.5
1994	640.4	19.7	99.1	73.8
1995	673.4	20.1	99.5	77.4
1996	674.7	19.6	100.8	76.2
1997	628.6	18.0	99.7	71.1
1998	673.7	18.6	97.1	78.2
1999	728.3	19.7	97.4	85.3
2000	753.9	19.8	97.9	88.1
2001	768.8	20.6	98.2	89.4
2002	780.1	20.2	98.7	90.3
2003	763.7	19.7	99.2	87.9
2004	788.5	19.9	99.6	90.1
2005	782.0	19.3	100.0	89.3
2006	787.2	19.4	R100.3	R89.6
2007	P806.5	P19.4	P100.6	E91.5

[1] At end of year. See "Generator Net Summer Capacity" in Glossary.
[2] See "Generator Capacity Factor" in Glossary.
R=Revised. P=Preliminary. E=Estimate. NA=Not available. (s)=Less than 0.05.
Note: See Note 2, "Coverage of Nuclear Energy Statistics," at end of section.
Web Page: For related information, see http://www.eia.doe.gov/fuelnuclear.html.

Sources: **Nuclear Electricity Net Generation** and **Nuclear Share of Electricity Net Generation:** Table 8.2a. **Net Summer Capacity of Operable Units:** Table 8.11a. **Capacity Factor:** • 1973-2006—Energy Information Administration, *Monthly Energy Review* (March 2008), Table 8.1. Annual capacity factors are weighted averages of monthly capacity factors. • 2007—Estimate based on annual generation and capacity values in this table.

Figure 9.3 Uranium Overview

Production and Trade, 1949-2007

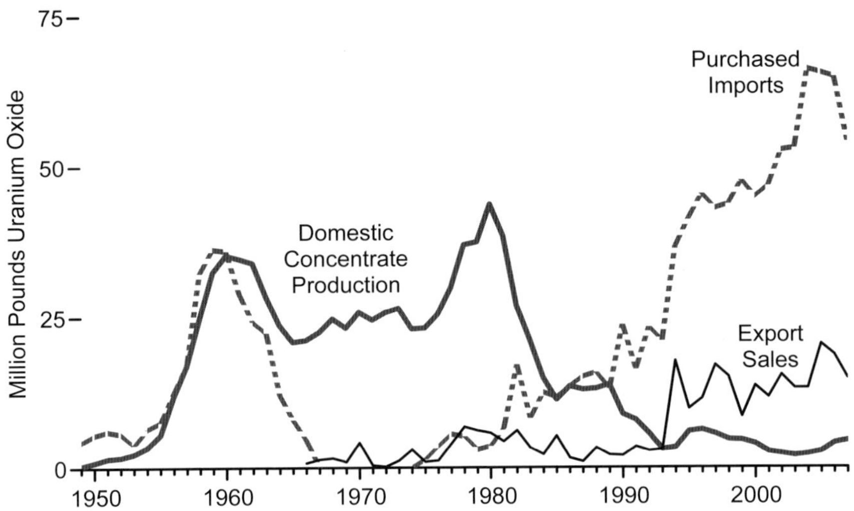

Production and Trade, 2007

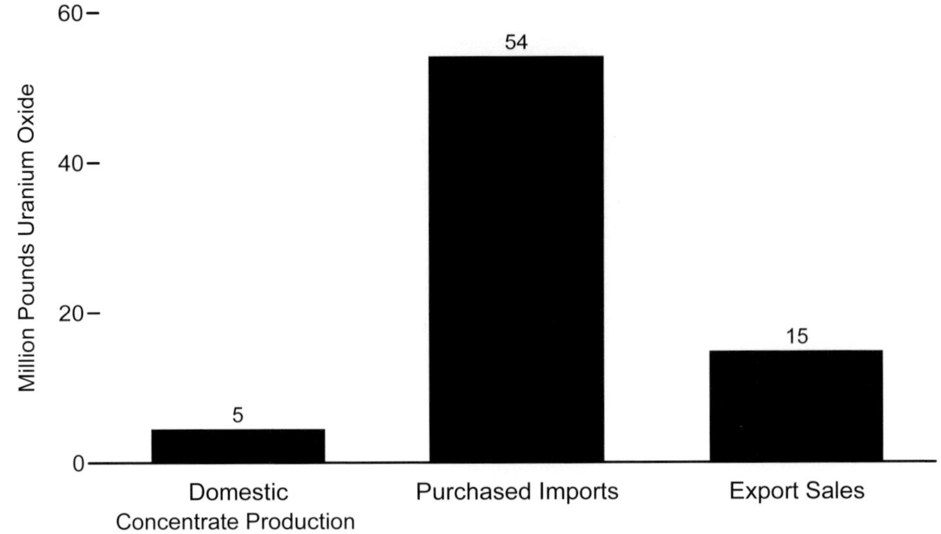

Inventories, End of Year 1981-2007

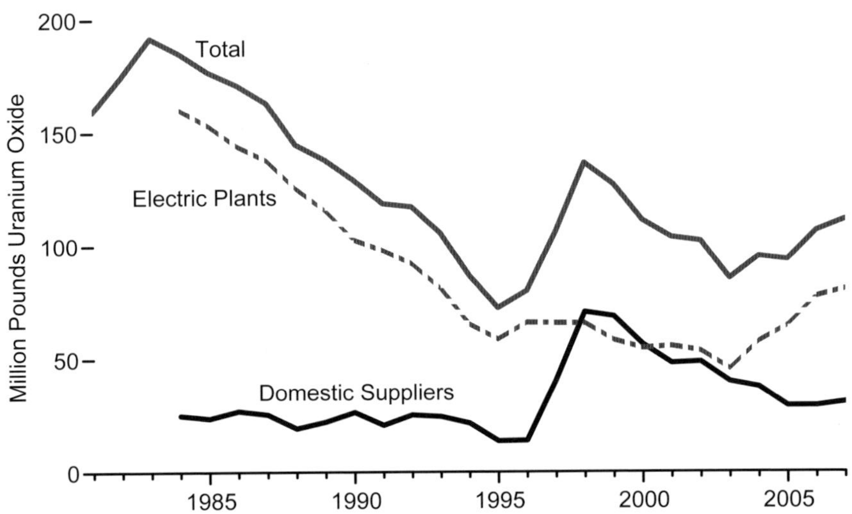

Average Prices, 1981-2007

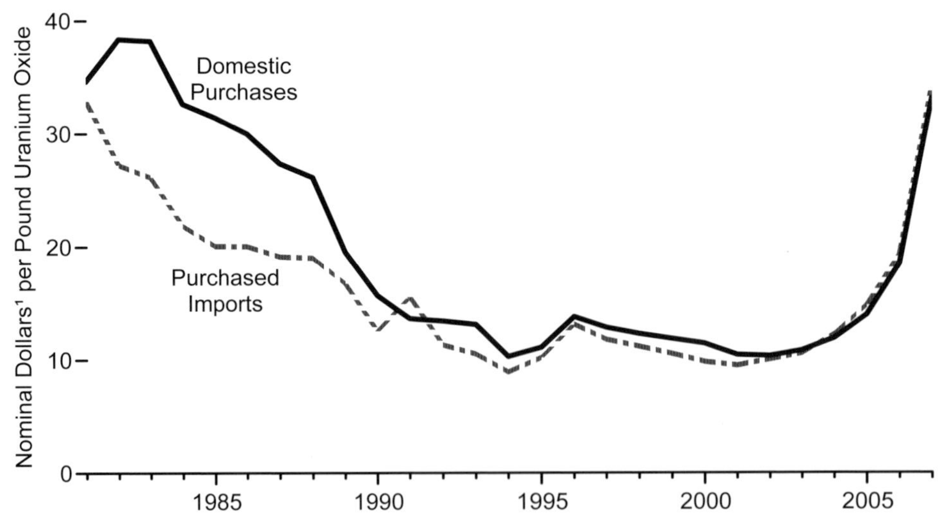

[1] See "Nominal Dollars" in Glossary.

Notes: • See "Uranium Oxide" in Glossary. • Because vertical scales differ, graphs should not be compared.

Source: Table 9.3.

Table 9.3 Uranium Overview, Selected Years, 1949-2007

Year	Domestic Concentrate Production [1]	Purchased Imports [2]	Export [2] Sales	Electric Plant Purchases From Domestic Suppliers	Loaded Into U.S. Nuclear Reactors [3]	Inventories			Average Price	
						Domestic Suppliers	Electric Plants	Total	Purchased Imports	Domestic Purchases
	Million Pounds Uranium Oxide								Nominal Dollars [4] per Pound Uranium Oxide	
1949	0.36	4.3	0.0	NA	NA	NA	NA	NA	NA	NA
1950	.92	5.5	.0	NA	NA	NA	NA	NA	NA	NA
1955	5.56	7.6	.0	NA	NA	NA	NA	NA	NA	NA
1960	35.28	36.0	.0	NA	NA	NA	NA	NA	NA	NA
1965	20.88	8.0	.0	NA	NA	NA	NA	NA	NA	NA
1970	25.81	.0	4.2	NA	NA	NA	NA	NA	– –	NA
1971	24.55	.0	.4	NA	NA	NA	NA	NA	– –	NA
1972	25.80	.0	.2	NA	NA	NA	NA	NA	– –	NA
1973	26.47	.0	1.2	NA	NA	NA	NA	NA	– –	NA
1974	23.06	.0	3.0	NA	NA	NA	NA	NA	– –	NA
1975	23.20	1.4	1.0	NA	NA	NA	NA	NA	NA	NA
1976	25.49	3.6	1.2	NA	NA	NA	NA	NA	NA	NA
1977	29.88	5.6	4.0	NA	NA	NA	NA	NA	NA	NA
1978	36.97	5.2	6.8	NA	NA	NA	NA	NA	NA	NA
1979	37.47	3.0	6.2	NA	NA	NA	NA	NA	NA	NA
1980	43.70	3.6	5.8	NA	NA	NA	NA	NA	NA	NA
1981	38.47	6.6	4.4	32.6	NA	NA	NA	159.2	32.90	34.65
1982	26.87	17.1	6.2	27.1	NA	NA	NA	174.8	27.23	38.37
1983	21.16	8.2	3.3	24.2	NA	NA	NA	191.8	26.16	38.21
1984	14.88	12.5	2.2	22.5	NA	25.0	160.2	185.2	21.86	32.65
1985	11.31	11.7	5.3	21.7	NA	23.7	153.2	176.9	20.08	31.43
1986	13.51	13.5	1.6	18.9	NA	27.0	144.1	171.1	20.07	30.01
1987	12.99	15.1	1.0	20.8	NA	25.4	137.8	163.2	19.14	27.37
1988	13.13	15.8	3.3	17.6	NA	19.3	125.5	144.8	19.03	26.15
1989	13.84	13.1	2.1	18.4	NA	22.2	115.8	138.1	16.75	19.56
1990	8.89	23.7	2.0	20.5	NA	26.4	102.7	129.1	12.55	15.70
1991	7.95	16.3	3.5	26.8	34.6	20.7	98.0	118.7	15.55	13.66
1992	5.65	23.3	2.8	23.4	43.0	25.2	92.1	117.3	11.34	13.45
1993	3.06	21.0	3.0	15.5	45.1	24.5	81.2	105.7	10.53	13.14
1994	3.35	36.6	17.7	22.7	40.4	21.5	65.4	86.9	8.95	10.30
1995	6.04	41.3	9.8	22.3	51.1	13.7	58.7	72.5	10.20	11.11
1996	6.32	45.4	11.5	23.7	46.2	13.9	66.1	80.0	13.15	13.81
1997	5.64	43.0	17.0	19.4	48.2	40.4	65.9	106.2	11.81	12.87
1998	4.71	43.7	15.1	21.6	38.2	70.7	65.8	136.5	11.19	12.31
1999	4.61	47.6	8.5	21.4	58.8	68.8	58.3	127.1	10.55	11.88
2000	3.96	44.9	13.6	24.3	51.5	56.5	54.8	111.3	9.84	11.45
2001	2.64	46.7	11.7	27.5	52.7	48.1	55.6	103.8	9.51	10.45
2002	2.34	52.7	15.4	22.7	57.2	48.7	53.5	102.1	10.05	10.35
2003	[5,E]2.00	53.0	13.2	21.7	62.3	39.9	45.6	85.5	10.59	10.84
2004	2.28	66.1	13.2	28.2	50.1	37.5	57.7	95.2	12.25	11.91
2005	2.69	65.5	20.5	27.3	58.3	29.1	64.7	93.8	14.83	13.98
2006	4.11	64.8	18.7	27.9	R51.7	29.1	R77.5	R106.6	19.31	18.54
2007	4.53	54.1	14.8	18.5	P47.2	P30.8	P80.8	P111.6	34.18	33.13

[1] See "Uranium Concentrate" in Glossary.
[2] Import quantities through 1970 are reported for fiscal years. Prior to 1968, the Atomic Energy Commission was the sole purchaser of all imported uranium oxide. Trade data prior to 1982 were for transactions conducted by uranium suppliers only. For 1982 forward, transactions by uranium buyers (consumers) have been included. Buyer imports and exports prior to 1982 are believed to be small.
[3] Does not include any fuel rods removed from reactors and later reloaded.
[4] See "Nominal Dollars" in Glossary.
[5] Value has been rounded to avoid disclosure of individual company data.
R=Revised. P=Preliminary. E=Estimate. NA=Not available. – – = Not applicable.

Note: See "Uranium Oxide" in Glossary.
Web Pages: • For all data beginning in 1949, see http://www.eia.doe.gov/emeu/aer/nuclear.html. • For related information, see http://www.eia.doe.gov/fuelnuclear.html.
Sources: • 1949-1966—U.S. Department of Energy, Grand Junction Office, *Statistical Data of the Uranium Industry,* Report No. GJO-100, annual reports. • 1967-2002—Energy Information Administration (EIA), *Uranium Industry Annual,* annual reports. • 2003 forward—EIA, "2007 Domestic Uranium Production Report" (May 2008), Table 3; EIA, "2007 Uranium Marketing Annual Report" (May 2008), Tables 5, 18, 19, 21, and 22; and EIA, Form EIA-858, "Uranium Marketing Annual Survey.

Nuclear Energy

Note 1. Pending Actions on Nuclear Generating Units. Much of Table 9.1 is based on the U.S Nuclear Regulatory Commission (NRC) regulation 10 CFR Part 50, which has in most instances been supplanted by 10 CFR Part 52 following the passage of the Energy Policy Act of 1992 and procedural reforms initiated in 1989 by the NRC. (This statement applies to permit and license procedures only.)

In 2007, the NRC issued three Early Site Permits (ESPs) under 10 CFR Part 52—for Clinton ESP Site in Illinois; Grand Gulf ESP Site in Mississippi; and North Anna ESP Site in Virginia. As of April 2008, the ESP application for the Vogtle ESP Site in Georgia was under review. No new ESP applications have been submitted since August 2006.

In 2007, the NRC had four Combined License (COL) applications under review—for Bellefonte 3 and 4 in Alabama; Calvert Cliffs 3 in Maryland; North Anna 3 in Virginia; and South Texas Project 3 and 4 in Texas. As of April 2008, an additional five COL applications were either under review or have been submitted to the NRC—for Grand Gulf 3 in Mississippi; Shearon Harris 2 and 3 in North Carolina; Virgil C. Summer 2 and 3 in South Carolina; Vogtle 3 and 4 in Georgia; and William States Lee III 1 and 2 in South Carolina. Of these nine COL applications, only States Lee III was for a completely new location.

As of April 2008, 11 applications for license extensions were under review by the NRC. The oldest application still pending, first submitted in July 2005, was for the oldest commercial reactor still in service, the Mark 1 Boiling Water Reactor at Oyster Creek. The most recent application, submitted in January 2008, was for Three Mile Island 1.

For more information on nuclear reactors, see http://www.nrc.gov/reactors.html.

Note 2. Coverage of Nuclear Energy Statistics. In 1997, the Energy Information Administration undertook a major revision of Table 9.1 to more fully describe the history of the U.S. commercial nuclear power industry. The time frame was extended back to the birth of the industry in 1953 and the data categories were revised for greater relevance to current industry conditions and trends. To acquire the data for the revised categories, it was necessary to develop a reactor unit database employing different sources than those used previously for Table 9.1 and still used for Table 9.2.

The data in Table 9.1 apply to commercial nuclear power units, which means that the units contributed power to the commercial electricity grid. A total of 259 units ever ordered was identified. Although most orders were placed by electric utilities, several units are or were ordered, owned, and operated wholly or in part by the Federal Government, including BONUS (Boiling Nuclear Superheater Power Station), Elk River, Experimental Breeder Reactor 2, Hallam, Hanford N, Piqua, and Shippingport.

A reactor is generally defined as operable in Table 9.1 while it possessed a full-power license from the Nuclear Regulatory Commission or its predecessor the Atomic Energy Commission, or equivalent permission to operate, at the end of the year. The definition is liberal in that it does not exclude units retaining full-power licenses during long, non-routine shutdowns. For example:

- In 1985, the five Tennessee Valley Authority units (Browns Ferry 1, 2, and 3 and Sequoyah 1 and 2) were shut down under a regulatory forced outage. Browns Ferry 1 was authorized by the NRC to restart in 2007, while the other units restarted in 1991, 1995, 1988, and 1988, respectively. All five units were counted as operable during the shutdowns.

- Shippingport was shut down from 1974 through 1976 for conversion to a light-water breeder reactor, but is counted as operable until its retirement in 1982.

- Calvert Cliffs 2 was shut down in 1989 and 1990 for replacement of pressurizer heater sleeves but is counted as operable during those years.

Exceptions to the rule are Shoreham and Three Mile Island 2. Shoreham was granted a full-power license in April 1989, but was shut down two months later and never restarted. In 1991, the license was changed to Possession Only. Although not operable at the end of the year, Shoreham is treated as operable during 1989 and shut down in 1990, because counting it as operable and shut down in the same year would introduce a statistical discrepancy in the tallies. A major accident closed Three Mile Island 2 in 1979, and although the unit retained its full-power license for several years, it is considered permanently shut down since that year.

Table 9.1 Sources: Operable Units: • 1955-1982–Compiled from various sources, primarily U.S. Department of Energy (DOE), Office of Nuclear Reactor Programs, "U.S. Central Station Nuclear Electric Generating Units: Significant Milestones." • 1983 forward–Energy Information Administration (EIA), Form EIA-860, "Annual Electric Generator Report," and predecessor forms. **All Other Data:** • 1955-1997–U.S. Atomic Energy Commission, *1973 Annual Report to Congress, Volume 2, Regulatory Activities*; Nuclear Energy Institute, *Historical Profile of U.S. Nuclear Power Development* (1988); EIA, *Commercial Nuclear Power 1991* (September 1991); DOE, *Nuclear Reactors Built, Being Built, and Planned: 1995*; U.S. Nuclear Regulatory Commission (NRC), *Information Digest* (1997 and 1998) and "Plant Status Report"; and various utility, Federal, and contractor officials. • 1998 forward–NRC, *Information Digest*, annual reports.

10 Renewable Energy

Grand Coulee Dam, Washington State. Source: U.S. Bureau of Reclamation.

Figure 10.1 Renewable Energy Consumption by Major Sources

Renewable Energy as Share of Total Primary Energy Consumption, 2007

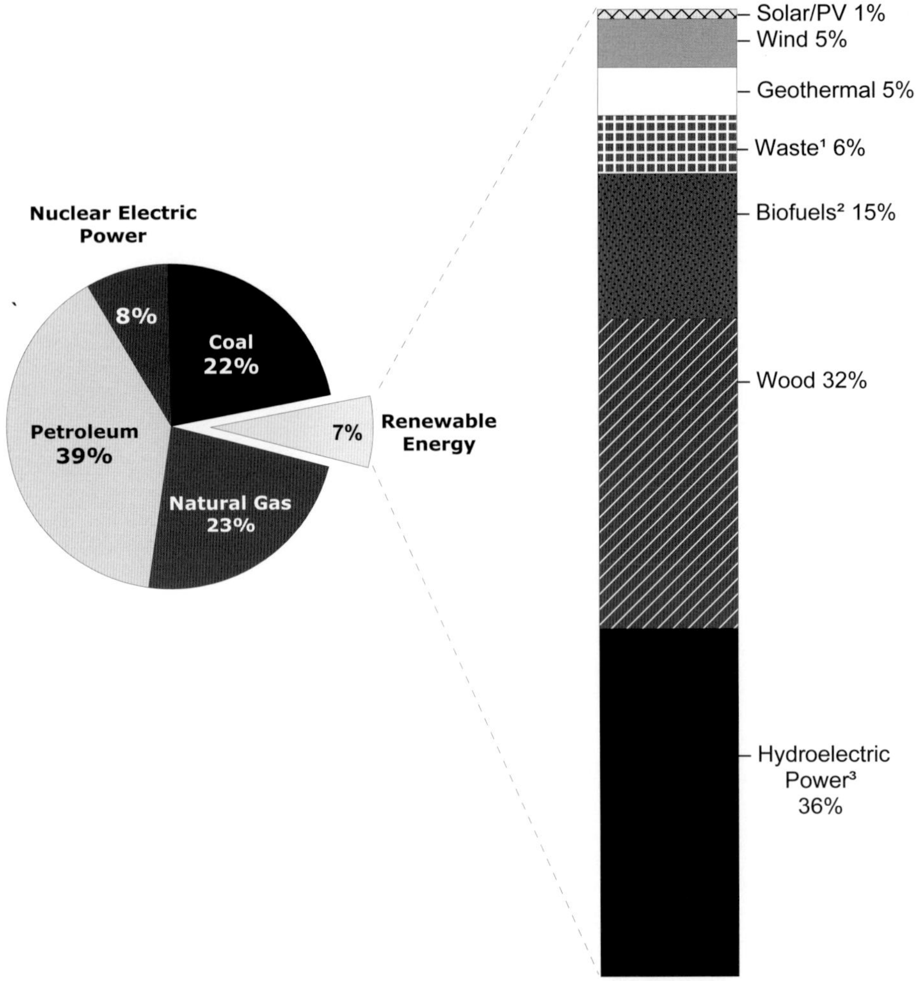

Renewable Energy Total Consumption and Major Sources, 1949-2007

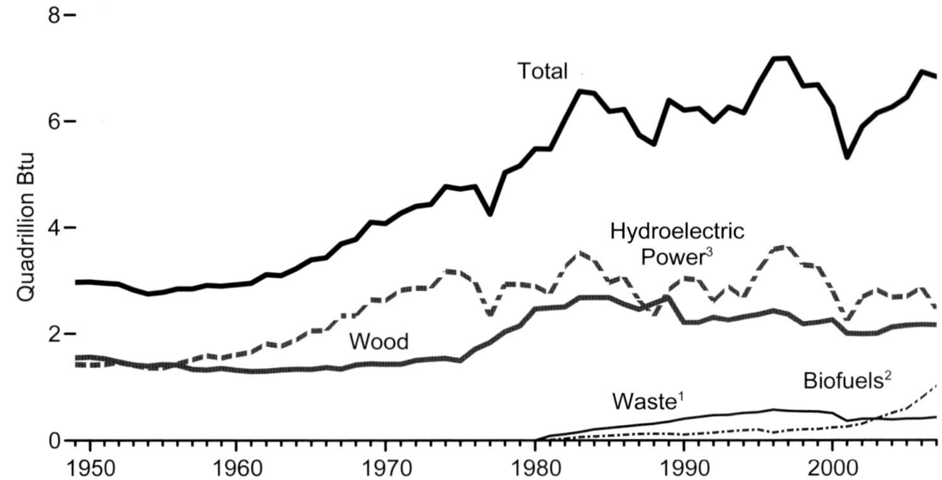

Renewable Energy Consumption by Source, 2007

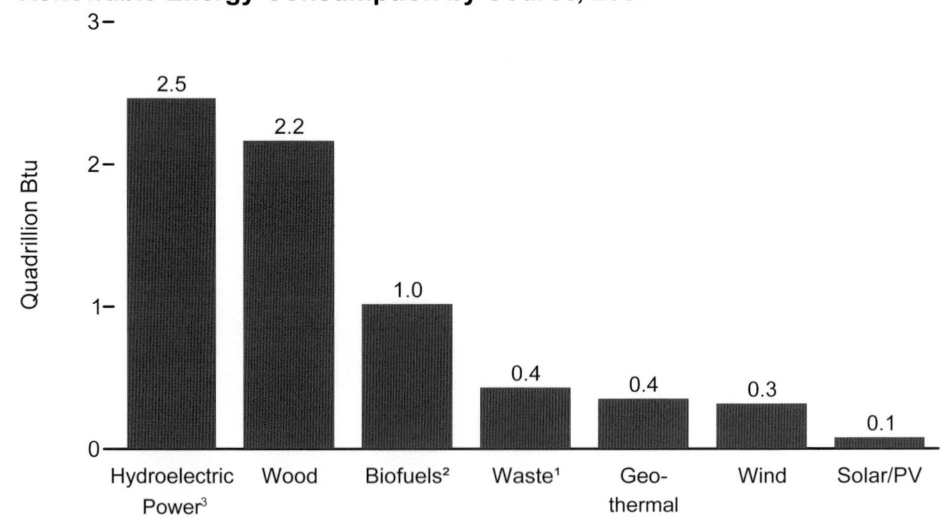

[1] Municipal solid waste from biogenic sources, landfill gas, sludge waste, agricultural byproducts, and other biomass. Through 2000, also includes non-renewable waste (municipal solid waste from non-biogenic sources, and tire-derived fuels).

[2] Fuel ethanol and biodiesel consumption, plus losses and co-products from the production of fuel ethanol and biodisel.

[3] Conventional hydroelectric power.

Note: Because vertical scales differ, graphs should not be compared.

Sources: Tables 1.3 and 10.1.

Table 10.1 Renewable Energy Production and Consumption by Primary Energy Source, Selected Years, 1949-2007

(Trillion Btu)

Year	Production [1]			Consumption								
	Biomass		Total Renewable Energy [4]	Hydro-electric Power [5]	Geo-thermal [6]	Solar/PV [7]	Wind [8]	Biomass				Total Renewable Energy
	Biofuels [2]	Total [3]						Wood [9]	Waste [10]	Biofuels [11]	Total	
1949	NA	1,549	2,974	1,425	NA	NA	NA	1,549	NA	NA	1,549	2,974
1950	NA	1,562	2,978	1,415	NA	NA	NA	1,562	NA	NA	1,562	2,978
1955	NA	1,424	2,784	1,360	NA	NA	NA	1,424	NA	NA	1,424	2,784
1960	NA	1,320	2,929	1,608	1	NA	NA	1,320	NA	NA	1,320	2,929
1965	NA	1,335	3,398	2,059	4	NA	NA	1,335	NA	NA	1,335	3,398
1970	NA	1,431	4,076	2,634	11	NA	NA	1,429	2	NA	1,431	4,076
1971	NA	1,432	4,268	2,824	12	NA	NA	1,430	2	NA	1,432	4,268
1972	NA	1,503	4,398	2,864	31	NA	NA	1,501	2	NA	1,503	4,398
1973	NA	1,529	4,433	2,861	43	NA	NA	1,527	2	NA	1,529	4,433
1974	NA	1,540	4,769	3,177	53	NA	NA	1,538	2	NA	1,540	4,769
1975	NA	1,499	4,723	3,155	70	NA	NA	1,497	2	NA	1,499	4,723
1976	NA	1,713	4,768	2,976	78	NA	NA	1,711	2	NA	1,713	4,768
1977	NA	1,838	4,249	2,333	77	NA	NA	1,837	2	NA	1,838	4,249
1978	NA	2,038	5,039	2,937	64	NA	NA	2,036	1	NA	2,038	5,039
1979	NA	2,152	5,166	2,931	84	NA	NA	2,150	2	NA	2,152	5,166
1980	NA	2,476	5,485	2,900	110	NA	NA	2,474	2	NA	2,476	5,485
1981	13	2,596	5,477	2,758	123	NA	NA	2,496	88	13	2,596	5,477
1982	35	2,664	6,034	3,266	105	NA	NA	2,510	119	35	2,664	6,034
1983	63	2,904	6,561	3,527	129	NA	(s)	2,684	157	63	2,904	6,561
1984	77	2,971	6,522	3,386	165	(s)	(s)	2,686	208	77	2,971	6,522
1985	93	3,016	6,185	2,970	198	(s)	(s)	2,687	236	93	3,016	6,185
1986	107	2,932	6,223	3,071	219	(s)	(s)	2,562	263	107	2,932	6,223
1987	123	2,875	5,739	2,635	229	(s)	(s)	2,463	289	123	2,875	5,739
1988	124	3,016	5,568	2,334	217	(s)	(s)	2,577	315	124	3,016	5,568
1989	126	3,160	6,391	2,837	317	55	22	2,680	354	126	3,160	6,391
1990	111	2,735	6,206	3,046	336	60	29	2,216	408	111	2,735	6,206
1991	129	2,782	6,238	3,016	346	63	31	2,214	440	129	2,782	6,238
1992	146	2,933	5,993	2,617	349	64	30	2,313	473	146	2,933	5,993
1993	171	2,910	6,263	2,892	364	66	31	2,260	479	171	2,910	6,262
1994	190	3,030	6,155	2,683	338	69	36	2,324	515	190	3,030	6,155
1995	200	3,102	6,703	3,205	294	70	33	2,370	531	202	3,104	6,705
1996	143	3,157	7,167	3,590	316	71	33	2,437	577	145	3,159	7,168
1997	190	3,111	7,180	3,640	325	70	34	2,371	551	187	3,108	7,178
1998	206	2,933	6,659	3,297	328	70	31	2,184	542	205	2,931	6,657
1999	215	2,969	6,683	3,268	331	69	46	2,214	540	213	2,967	6,681
2000	238	3,010	6,262	2,811	317	66	57	2,262	511	241	3,013	6,264
2001	260	2,629	5,318	2,242	311	65	70	2,006	364	258	2,627	R5,316
2002	R315	2,712	5,899	2,689	328	64	105	1,995	402	309	2,706	5,893
2003	412	R2,815	6,149	2,825	331	64	115	2,002	401	414	2,817	R6,150
2004	501	3,011	6,248	2,690	341	65	142	2,121	389	513	3,023	6,261
2005	R582	R3,141	R6,431	2,703	343	66	178	R2,156	403	R595	R3,154	R6,444
2006	R745	R3,324	R6,872	R2,869	R343	R72	R264	R2,172	R407	R795	R3,374	R6,922
2007P	988	3,584	6,800	2,463	353	80	319	2,165	431	1,018	3,615	6,830

[1] Production equals consumption for all renewable energy sources except biofuels.
[2] Total biomass inputs to the production of fuel ethanol and biodiesel.
[3] Wood and wood-derived fuels, biomass waste, fuel ethanol, and biodiesel.
[4] Hydroelectric power, geothermal, solar/PV, wind, and biomass.
[5] Conventional hydroelectricity net generation (converted to Btu using the fossil-fueled plants heat rate).
[6] Geothermal electricity net generation (converted to Btu using the geothermal energy plants heat rate), and geothermal heat pump and direct use energy.
[7] Solar thermal and photovoltaic electricity net generation (converted to Btu using the fossil-fueled plants heat rate), and solar thermal direct use energy.
[8] Wind electricity net generation (converted to Btu using the fossil-fueled plants heat rate).
[9] Wood and wood-derived fuels.
[10] Municipal solid waste from biogenic sources, landfill gas, sludge waste, agricultural byproducts, and other biomass. Through 2000, also includes non-renewable waste (municipal solid waste from non-biogenic sources, and tire-derived fuels).
[11] Fuel ethanol and biodiesel consumption, plus losses and co-products from the production of fuel ethanol and biodiesel.

R=Revised. P=Preliminary. NA=Not available. (s)=Less than 0.5 trillion Btu.

Notes: • See Note, "Renewable Energy Production and Consumption," at end of section. • See Table E1 for estimated renewable energy consumption for 1635-1945. • Totals may not equal sum of components due to independent rounding.

Web Pages: • For all data beginning in 1949, see http://www.eia.doe.gov/emeu/aer/renew.html. • For related information, see http://www.eia.doe.gov/fuelrenewable.html.

Sources: **Biofuels:** Table 10.3. **All Other Data:** Tables 10.2a-c.

Figure 10.2a Renewable Energy Consumption: End-Use Sectors, 1989-2007

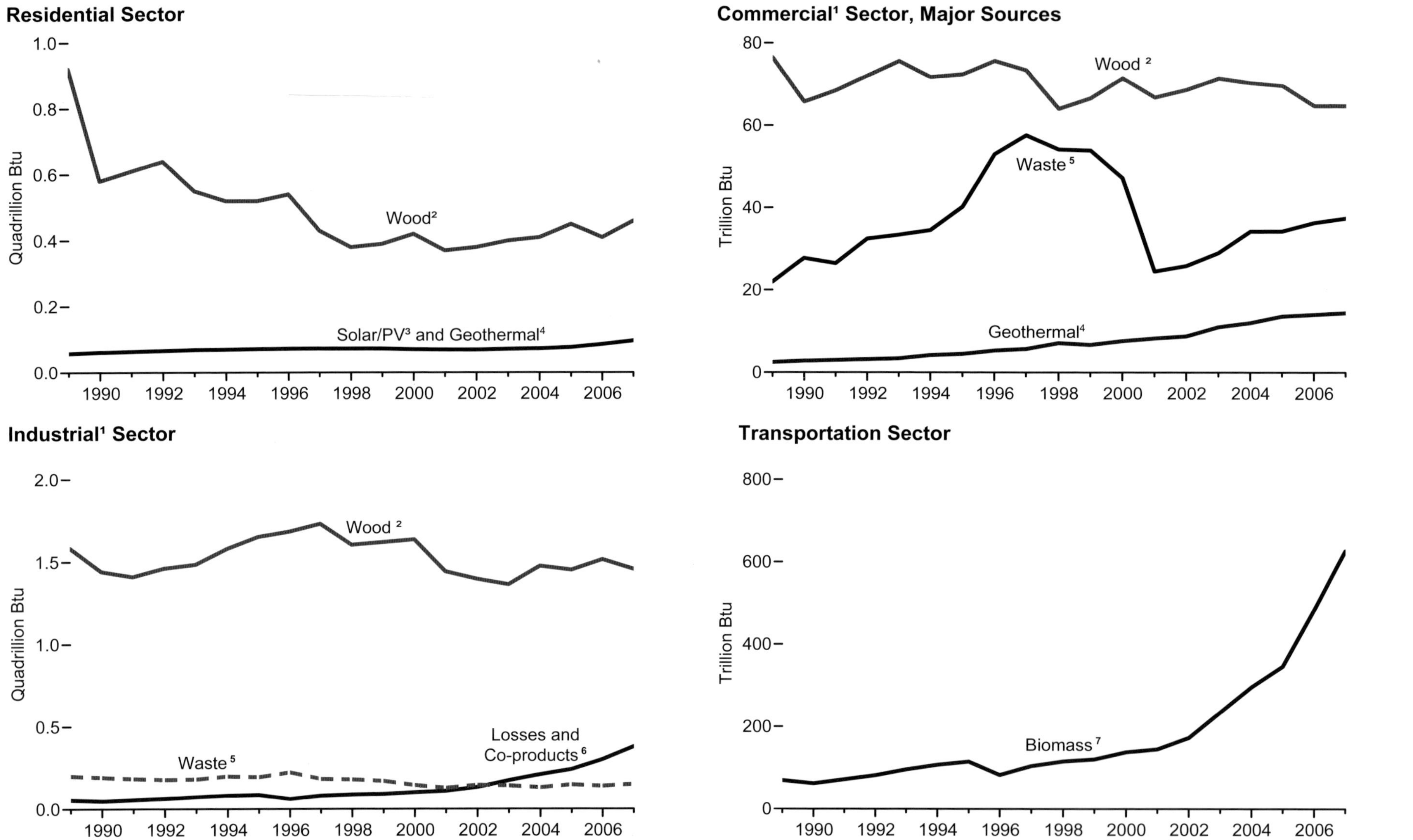

[1] Includes fuel used at combined-heat-and-power (CHP) plants and a small number of electricity-only plants.
[2] Wood and wood-derived fuels.
[3] Solar thermal direct use energy, and photovoltaic electricity generation. Includes small amounts of commercial sector use.
[4] Geothermal heat pump and direct use energy.
[5] Municipal solid waste from biogenic sources, landfill gas, sludge waste, agricultural byproducts, and other biomass. Through 2000, also includes non-renewable waste (municipal solid waste from non-biogenic sources, and tire-derived fuels).
[6] From the production of fuel ethanol and biodiesel.
[7] The ethanol portion of motor fuels (such as E10 and E85), and biofuels used as diesel fuel substitutes or additives.

Notes: • See related Figures 10.2b and 10.2c. • Because vertical scales differ, graphs should not be compared.

Sources: Tables 10.2a and 10.2b.

Figure 10.2b Renewable Energy Consumption: End-Use Sectors and Electric Power Sector

End-Use Sectors, 1949-2007

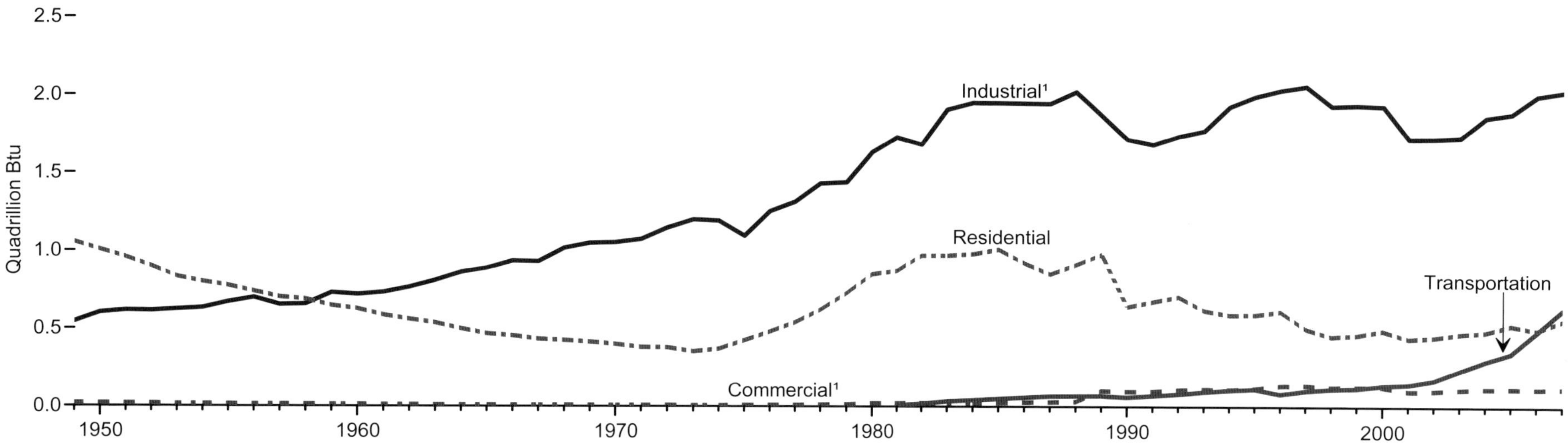

End-Use Sectors and Electric Power Sector, 2007

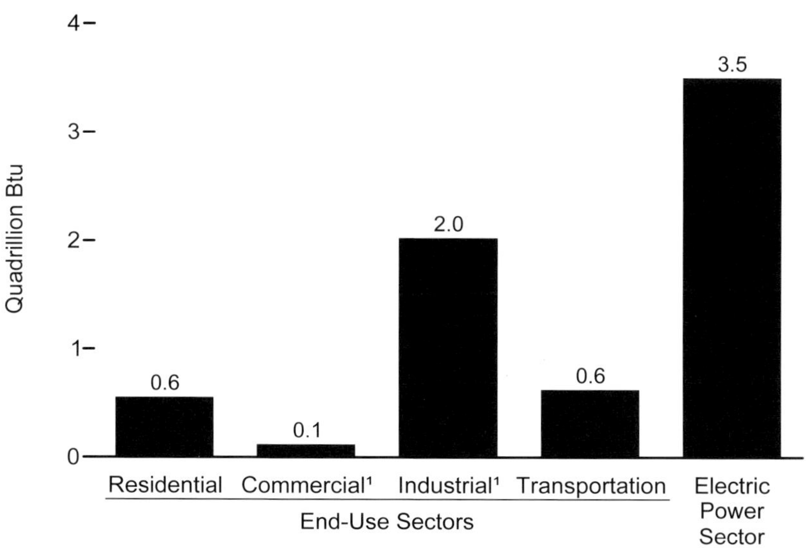

End-Use Sectors and Electric Power Sector Shares of Total Renewable Energy Consumption, 2007

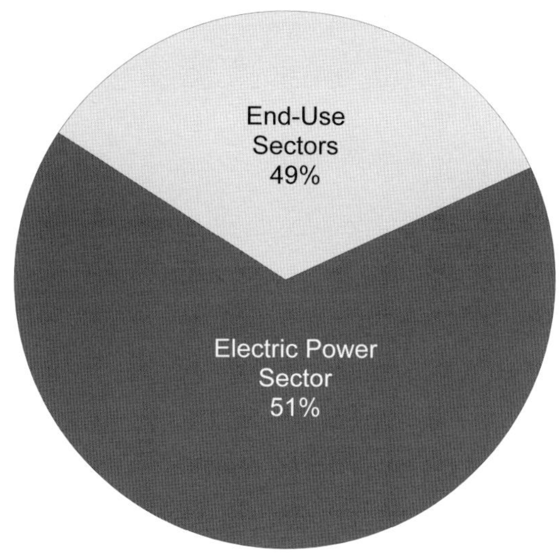

[1]Includes fuel use at combined-heat-and-power (CHP) plants and a small number of electricity-only plants.

Notes: • See related Figures 10.2a and 10.2c. • Because vertical scales differ, graphs should not be compared.

Sources: Tables 10.2a-10.2c.

Table 10.2a Renewable Energy Consumption: Residential and Commercial Sectors, Selected Years, 1949-2007

(Trillion Btu)

Year	Residential Sector				Commercial Sector [1]						
	Geo-thermal [2]	Solar/PV [3]	Biomass	Total	Hydro-electric Power [5]	Geo-thermal [2]	Biomass				Total
			Wood [4]				Wood [4]	Waste [6]	Fuel Ethanol [7]	Total	
1949	NA	NA	1,055	1,055	NA	NA	20	NA	NA	20	20
1950	NA	NA	1,006	1,006	NA	NA	19	NA	NA	19	19
1955	NA	NA	775	775	NA	NA	15	NA	NA	15	15
1960	NA	NA	627	627	NA	NA	12	NA	NA	12	12
1965	NA	NA	468	468	NA	NA	9	NA	NA	9	9
1970	NA	NA	401	401	NA	NA	8	NA	NA	8	8
1971	NA	NA	382	382	NA	NA	7	NA	NA	7	7
1972	NA	NA	380	380	NA	NA	7	NA	NA	7	7
1973	NA	NA	354	354	NA	NA	7	NA	NA	7	7
1974	NA	NA	371	371	NA	NA	7	NA	NA	7	7
1975	NA	NA	425	425	NA	NA	8	NA	NA	8	8
1976	NA	NA	482	482	NA	NA	9	NA	NA	9	9
1977	NA	NA	542	542	NA	NA	10	NA	NA	10	10
1978	NA	NA	622	622	NA	NA	12	NA	NA	12	12
1979	NA	NA	728	728	NA	NA	14	NA	NA	14	14
1980	NA	NA	850	850	NA	NA	21	NA	NA	21	21
1981	NA	NA	870	870	NA	NA	21	NA	(s)	21	21
1982	NA	NA	970	970	NA	NA	22	NA	(s)	22	22
1983	NA	NA	970	970	NA	NA	22	NA	(s)	22	22
1984	NA	NA	980	980	NA	NA	22	NA	(s)	22	22
1985	NA	NA	1,010	1,010	NA	NA	24	NA	(s)	24	24
1986	NA	NA	920	920	NA	NA	27	NA	(s)	27	27
1987	NA	NA	850	850	NA	NA	29	NA	1	30	30
1988	NA	NA	910	910	NA	NA	32	NA	1	33	33
1989	5	53	920	978	1	3	76	22	1	99	102
1990	6	56	580	641	1	3	66	28	1	94	98
1991	6	58	610	674	1	3	68	26	(s)	95	100
1992	6	60	640	706	1	3	72	32	(s)	105	109
1993	7	62	550	618	1	3	76	33	(s)	109	114
1994	6	64	520	590	1	4	72	35	(s)	106	112
1995	7	65	520	591	1	5	72	40	(s)	113	118
1996	7	65	540	612	1	5	76	53	(s)	129	135
1997	8	65	430	503	1	6	73	58	(s)	131	138
1998	8	65	380	452	1	7	64	54	(s)	118	127
1999	9	64	390	462	1	7	67	54	(s)	121	129
2000	9	61	420	490	1	8	71	47	(s)	119	128
2001	9	60	370	439	1	8	67	25	(s)	92	101
2002	10	59	380	449	(s)	9	69	26	(s)	95	104
2003	13	58	400	471	1	11	71	29	1	101	113
2004	14	59	410	483	1	12	70	34	1	105	118
2005	16	61	^{R}450	^{R}527	1	14	70	34	1	105	119
2006	18	^{R}67	^{R}410	^{R}495	1	14	65	^{R}36	1	^{R}102	^{R}117
2007^{P}	22	74	460	556	1	14	65	37	2	104	119

[1] Commercial sector, including commercial combined-heat-and-power (CHP) and commercial electricity-only plants. See Note 2, "Classification of Power Plants Into Energy-Use Sectors," at end of Section 8.

[2] Geothermal heat pump and direct use energy.

[3] Solar thermal direct use energy, and photovoltaic electricity net generation (converted to Btu using the fossil-fueled plants heat rate). Includes a small amount of commercial sector use.

[4] Wood and wood-derived fuels.

[5] Conventional hydroelectricity net generation (converted to Btu using the fossil-fueled plants heat rate).

[6] Municipal solid waste from biogenic sources, landfill gas, sludge waste, agricultural byproducts, and other biomass. Through 2000, also includes non-renewable waste (municipal solid waste from non-biogenic sources, and tire-derived fuels).

[7] The ethanol portion of motor fuels (such as E10) consumed by the commercial sector.

R=Revised. P=Preliminary. NA=Not available. (s)=Less than 0.5 trillion Btu.

Notes: • All values are estimated, except for commercial sector hydroelectric power and waste. • Totals may not equal sum of components due to independent rounding.

Web Pages: • For all data beginning in 1949, see http://www.eia.doe.gov/emeu/aer/renew.html. • For related information, see http://www.eia.doe.gov/fuelrenewable.html.

Sources: See end of section.

Table 10.2b Renewable Energy Consumption: Industrial and Transportation Sectors, Selected Years, 1949-2007

(Trillion Btu)

Year	Industrial Sector [1]								Transportation Sector		
			Biomass						Biomass		
	Hydro-electric Power [2]	Geo-thermal [3]	Wood [4]	Waste [5]	Fuel Ethanol [6]	Losses and Co-products [7]	Total	Total	Fuel Ethanol [8]	Biodiesel [9]	Total
1949	76	NA	468	NA	NA	NA	468	544	NA	NA	NA
1950	69	NA	532	NA	NA	NA	532	602	NA	NA	NA
1955	38	NA	631	NA	NA	NA	631	669	NA	NA	NA
1960	39	NA	680	NA	NA	NA	680	719	NA	NA	NA
1965	33	NA	855	NA	NA	NA	855	888	NA	NA	NA
1970	34	NA	1,019	NA	NA	NA	1,019	1,053	NA	NA	NA
1971	34	NA	1,040	NA	NA	NA	1,040	1,074	NA	NA	NA
1972	34	NA	1,113	NA	NA	NA	1,113	1,147	NA	NA	NA
1973	35	NA	1,165	NA	NA	NA	1,165	1,200	NA	NA	NA
1974	33	NA	1,159	NA	NA	NA	1,159	1,192	NA	NA	NA
1975	32	NA	1,063	NA	NA	NA	1,063	1,096	NA	NA	NA
1976	33	NA	1,220	NA	NA	NA	1,220	1,253	NA	NA	NA
1977	33	NA	1,281	NA	NA	NA	1,281	1,314	NA	NA	NA
1978	32	NA	1,400	NA	NA	NA	1,400	1,432	NA	NA	NA
1979	34	NA	1,405	NA	NA	NA	1,405	1,439	NA	NA	NA
1980	33	NA	1,600	NA	NA	NA	1,600	1,633	NA	NA	NA
1981	33	NA	1,602	87	(s)	6	1,695	1,728	7	NA	7
1982	33	NA	1,516	118	(s)	16	1,649	1,682	19	NA	19
1983	33	NA	1,690	155	(s)	28	1,874	1,907	34	NA	34
1984	33	NA	1,679	204	1	34	1,917	1,950	42	NA	42
1985	33	NA	1,645	230	1	41	1,917	1,950	51	NA	51
1986	33	NA	1,610	256	1	47	1,914	1,947	59	NA	59
1987	33	NA	1,576	282	1	54	1,912	1,945	67	NA	67
1988	33	NA	1,625	308	1	54	1,988	2,020	68	NA	68
1989	28	2	1,584	200	1	55	1,840	1,870	69	NA	69
1990	31	2	1,442	192	1	48	1,683	1,716	62	NA	62
1991	30	2	1,410	185	1	56	1,651	1,683	72	NA	72
1992	31	2	1,461	179	1	63	1,704	1,737	81	NA	81
1993	30	2	1,484	181	1	74	1,740	1,772	96	NA	96
1994	62	3	1,580	199	1	82	1,862	1,927	107	NA	107
1995	55	3	1,652	195	2	86	1,935	1,992	115	NA	115
1996	61	3	1,683	224	1	61	1,970	2,033	82	NA	82
1997	58	3	1,731	184	1	81	1,997	2,058	104	NA	104
1998	55	3	1,603	180	1	88	1,873	1,931	115	NA	115
1999	49	4	1,620	171	1	92	1,883	1,936	120	NA	120
2000	42	4	1,636	145	1	101	1,884	1,930	138	NA	138
2001	33	5	1,443	129	3	110	1,684	1,721	144	1	145
2002	39	5	1,396	146	3	133	1,679	1,723	171	1	172
2003	43	3	1,363	142	R5	174	1,684	1,731	233	2	235
2004	33	4	1,476	132	6	R210	1,824	1,861	292	R4	R296
2005	32	4	1,452	148	7	241	R1,848	R1,884	334	R12	R346
2006	R29	4	R1,515	R140	R9	R301	R1,966	R1,999	R451	32	R483
2007P	23	5	1,457	151	12	379	1,998	2,025	564	63	626

[1] Industrial sector, including industrial combined-heat-and-power (CHP) and industrial electricity-only plants. See Note 2, "Classification of Power Plants Into Energy-Use Sectors," at end of Section 8.

[2] Conventional hydroelectricity net generation (converted to Btu using the fossil-fueled plants heat rate).

[3] Geothermal heat pump and direct use energy.

[4] Wood and wood-derived fuels.

[5] Municipal solid waste from biogenic sources, landfill gas, sludge waste, agricultural byproducts, and other biomass. Through 2000, also includes non-renewable waste (municipal solid waste from non-biogenic sources, and tire-derived fuels).

[6] The ethanol portion of motor fuels (such as E10) consumed by the industrial sector.

[7] Losses and co-products from the production of fuel ethanol and biodiesel. Does not include natural gas, electricity, and other non-biomass energy used in the production of fuel ethanol and biodiesel—these are included in the industrial sector consumption statistics for the appropriate energy source.

[8] The ethanol portion of motor fuels (such as E10 and E85) consumed by the transportation sector.

[9] "Biodiesel is any liquid biofuel suitable as a diesel fuel substitute, additive, or extender. See "Biodiesel" in Glossary.

R=Revised. P=Preliminary. NA=Not available. (s)=Less than 0.5 trillion Btu.

Notes: • All values are estimated, except for industrial sector hydroelectric power in 1949-1978 and 1989 forward. • Totals may not equal sum of components due to independent rounding.

Web Pages: • For all data beginning in 1949, see http://www.eia.doe.gov/emeu/aer/renew.html. • For related information, see http://www.eia.doe.gov/fuelrenewable.html.

Sources: See end of section.

Figure 10.2c Renewable Energy Consumption: Electric Power Sector

Electric Power Sector Total and Hydroelectric Power, 1949-2007

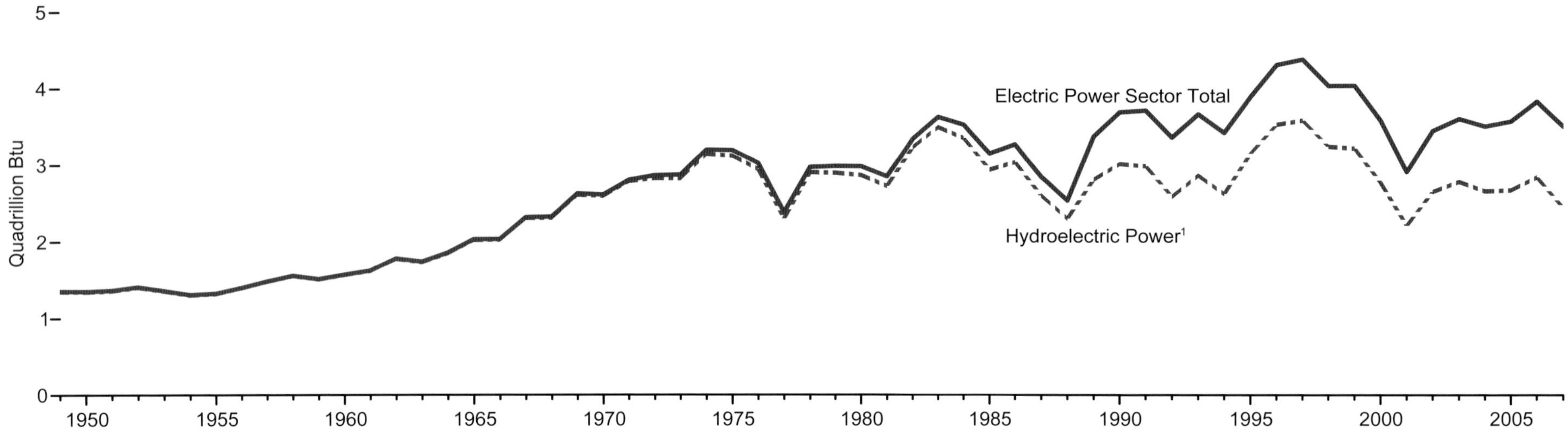

Non-Hydroelectric Power Sources, 1989-2007

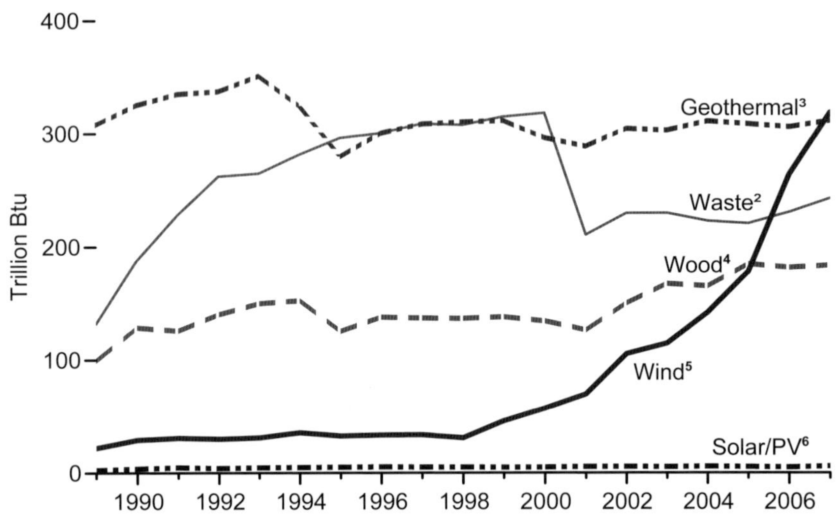

Non-Hydroelectric Power Sources, 2007

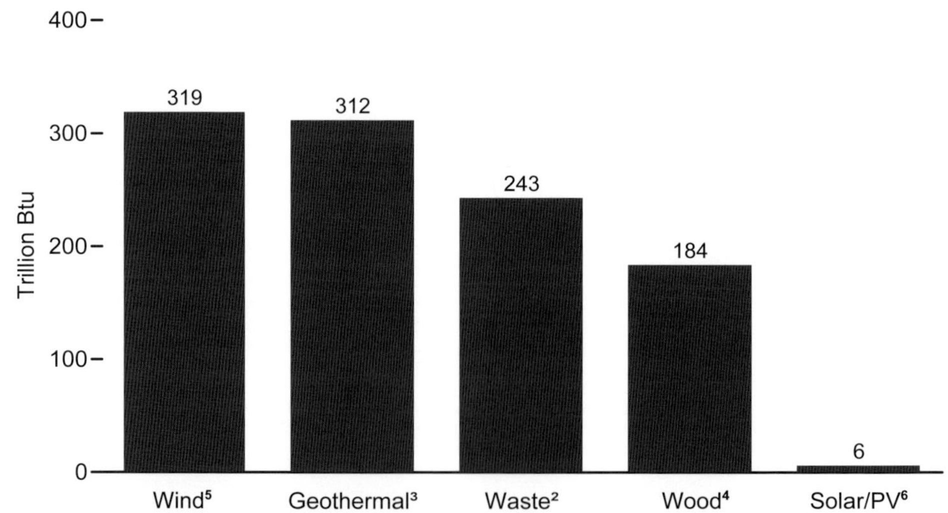

[1] Conventional hydroelectricity net generation.

[2] Municipal solid waste from biogenic sources, landfill gas, sludge waste, agricultural byproducts, and other biomass. Through 2000, also includes non-renewable waste (municipal solid waste from non-biogenic sources, and tire-derived fuels).

[3] Geothermal electricity net generation.

[4] Wood and wood-driven fuels.

[5] Wind electricity net generation.

[6] Solar thermal and photovoltaic electricity net generation.

Notes: • See related Figure 10.2a on the end-use sectors. • Because vertical scales differ, graphs should not be compared.

Source: Table 10.2c.

Table 10.2c Renewable Energy Consumption: Electric Power Sector, Selected Years, 1949-2007

(Trillion Btu)

Year	Hydroelectric Power [1]	Geothermal [2]	Solar/PV [3]	Wind [4]	Biomass			Total
					Wood [5]	Waste [6]	Total	
1949	1,349	NA	NA	NA	6	NA	6	1,355
1950	1,346	NA	NA	NA	5	NA	5	1,351
1955	1,322	NA	NA	NA	3	NA	3	1,325
1960	1,569	1	NA	NA	2	NA	2	1,571
1965	2,026	4	NA	NA	3	NA	3	2,033
1970	2,600	11	NA	NA	1	2	4	2,615
1971	2,790	12	NA	NA	1	2	3	2,806
1972	2,829	31	NA	NA	1	2	3	2,864
1973	2,827	43	NA	NA	1	2	3	2,873
1974	3,143	53	NA	NA	1	2	3	3,199
1975	3,122	70	NA	NA	(s)	2	2	3,194
1976	2,943	78	NA	NA	1	2	3	3,024
1977	2,301	77	NA	NA	3	2	5	2,383
1978	2,905	64	NA	NA	2	1	3	2,973
1979	2,897	84	NA	NA	3	2	5	2,986
1980	2,867	110	NA	NA	3	2	5	2,982
1981	2,725	123	NA	NA	3	1	4	2,852
1982	3,233	105	NA	NA	2	1	3	3,341
1983	3,494	129	NA	(s)	2	2	4	3,627
1984	3,353	165	(s)	(s)	5	4	9	3,527
1985	2,937	198	(s)	(s)	8	7	14	3,150
1986	3,038	219	(s)	(s)	5	7	12	3,270
1987	2,602	229	(s)	(s)	8	7	15	2,846
1988	2,302	217	(s)	(s)	10	8	17	2,536
1989[7]	2,808	308	3	22	100	132	232	3,372
1990	3,014	326	4	29	129	188	317	3,689
1991	2,985	335	5	31	126	229	354	3,710
1992	2,586	338	4	30	140	262	402	3,360
1993	2,861	351	5	31	150	265	415	3,662
1994	2,620	325	5	36	152	282	434	3,420
1995	3,149	280	5	33	125	296	422	3,889
1996	3,528	300	5	33	138	300	438	4,305
1997	3,581	309	5	34	137	309	446	4,375
1998	3,241	311	5	31	137	308	444	4,032
1999	3,218	312	5	46	138	315	453	4,034
2000	2,768	296	5	57	134	318	453	3,579
2001	2,209	289	6	70	126	211	337	2,910
2002	2,650	305	6	105	150	230	380	3,445
2003	2,781	303	5	115	167	230	397	3,601
2004	2,656	311	6	142	165	223	388	3,503
2005	2,670	309	6	178	185	221	406	3,568
2006	R2,839	R306	5	R264	R182	R231	R412	R3,827
2007P	2,440	312	6	319	184	243	427	3,503

[1] Conventional hydroelectricity net generation (converted to Btu using the fossil-fueled plants heat rate).

[2] Geothermal electricity net generation (converted to Btu using the geothermal energy plants heat rate).

[3] Solar thermal and photovoltaic electricity net generation (converted to Btu using the fossil-fueled plants heat rate).

[4] Wind electricity net generation (converted to Btu using the fossil-fueled plants heat rate).

[5] Wood and wood-derived fuels.

[6] Municipal solid waste from biogenic sources, landfill gas, sludge waste, agricultural byproducts, and other biomass. Through 2000, also includes non-renewable waste (municipal solid waste from non-biogenic sources, and tire-derived fuels).

[7] Through 1988, data are for electric utilities only. Beginning in 1989, data are for electric utilities and independent power producers.

R=Revised. P=Preliminary. NA=Not available. (s)=Less than 0.5 trillion Btu.

Notes: • The electric power sector comprises electricity-only and combined-heat-and-power (CHP) plants within the NAICS 22 category whose primary business is to sell electricity, or electricity and heat, to the public. • See Note 3, "Electricity Imports and Exports," at end of Section 8. • Totals may not equal sum of components due to independent rounding.

Web Pages: • For all data beginning in 1949, see http://www.eia.doe.gov/emeu/aer/renew.html. • For related information, see http://www.eia.doe.gov/fuelrenewable.html.

Sources: Tables 8.2b, 8.5b, 8.7b, and A6.

Figure 10.3 Fuel Ethanol and Biodiesel Overview

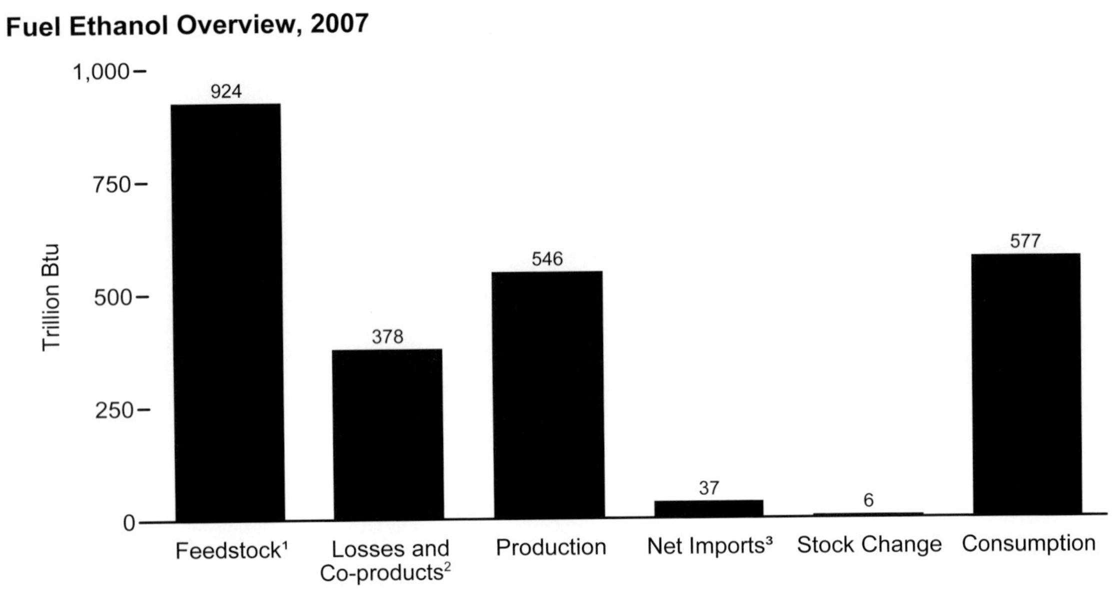

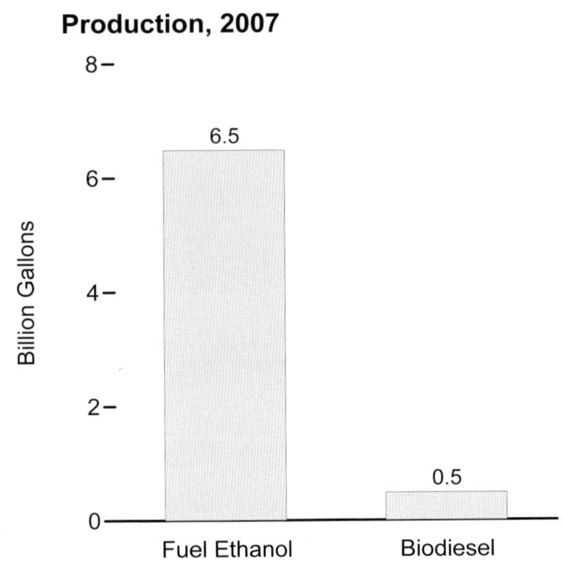

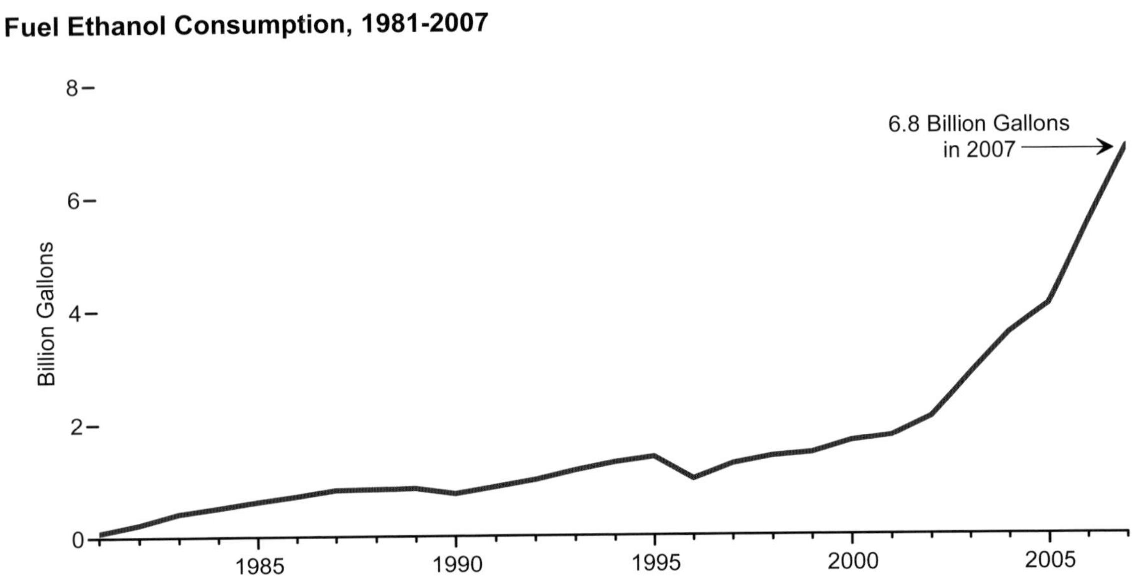

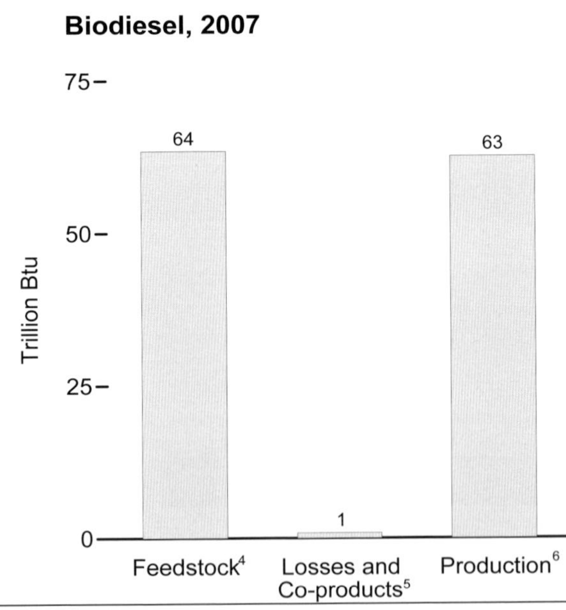

[1] Total corn and other biomass inputs to the production of fuel ethanol.
[2] Losses and co-products from the production of fuel ethanol.
[3] Fuel ethanol imports only. Data for fuel ethanol exports are not available.
[4] Total vegetable oil and other biomass inputs to the production of biodiesel.
[5] Losses and co-products from the production of biodiesel.
[6] Production of biodiesel for use as diesel fuel substitutes or additives.
Note: Because vertical scales differ, graphs should not be compared.
Source: Table 10.3.

Table 10.3 Fuel Ethanol and Biodiesel Overview, 1981-2007

	Fuel Ethanol													Biodiesel				
	Feed-stock [1]	Losses and Co-products [2]	Production			Net Imports [3]		Stocks [4]	Stock Change [5]		Consumption			Feed-stock [6]	Losses and Co-products [7]	Production [8]		
Year	TBtu	TBtu	Mbbl	MMgal	TBtu	Mbbl	TBtu	Mbbl	Mbbl	TBtu	Mbbl	MMgal	TBtu	TBtu	TBtu	Mbbl	MMgal	TBtu
1981	13	6	1,978	83	7	NA	NA	NA	NA	NA	1,978	83	7	NA	NA	NA	NA	NA
1982	35	16	5,369	225	19	NA	NA	NA	NA	NA	5,369	225	19	NA	NA	NA	NA	NA
1983	63	28	9,890	415	35	NA	NA	NA	NA	NA	9,890	415	35	NA	NA	NA	NA	NA
1984	77	34	12,150	510	43	NA	NA	NA	NA	NA	12,150	510	43	NA	NA	NA	NA	NA
1985	93	41	14,693	617	52	NA	NA	NA	NA	NA	14,693	617	52	NA	NA	NA	NA	NA
1986	107	47	16,954	712	60	NA	NA	NA	NA	NA	16,954	712	60	NA	NA	NA	NA	NA
1987	123	54	19,497	819	69	NA	NA	NA	NA	NA	19,497	819	69	NA	NA	NA	NA	NA
1988	124	54	19,780	831	70	NA	NA	NA	NA	NA	19,780	831	70	NA	NA	NA	NA	NA
1989	126	55	20,062	843	71	NA	NA	NA	NA	NA	20,062	843	71	NA	NA	NA	NA	NA
1990	111	48	17,802	748	63	NA	NA	NA	NA	NA	17,802	748	63	NA	NA	NA	NA	NA
1991	129	56	20,627	866	73	NA	NA	NA	NA	NA	20,627	866	73	NA	NA	NA	NA	NA
1992	146	63	23,453	985	83	NA	NA	1,791	NA	NA	23,453	985	83	NA	NA	NA	NA	NA
1993	171	74	27,484	1,154	97	244	1	2,114	323	1	27,405	1,151	97	NA	NA	NA	NA	NA
1994	190	82	30,689	1,289	109	279	1	2,393	279	1	30,689	1,289	109	NA	NA	NA	NA	NA
1995	200	86	32,325	1,358	114	387	1	2,186	-207	-1	32,919	1,383	117	NA	NA	NA	NA	NA
1996	143	61	23,178	973	82	313	1	2,065	-121	(s)	23,612	992	84	NA	NA	NA	NA	NA
1997	190	81	30,674	1,288	109	85	(s)	2,925	860	3	29,899	1,256	106	NA	NA	NA	NA	NA
1998	206	88	33,453	1,405	118	66	(s)	3,406	481	2	33,038	1,388	117	NA	NA	NA	NA	NA
1999	215	92	34,881	1,465	123	87	(s)	4,024	618	2	34,350	1,443	122	NA	NA	NA	NA	NA
2000	238	101	38,627	1,622	137	116	(s)	3,400	-624	-2	39,367	1,653	139	NA	NA	NA	NA	NA
2001	259	110	42,028	1,765	149	315	1	4,298	898	3	41,445	1,741	147	1	(s)	204	9	1
2002	313	133	50,956	2,140	180	306	1	6,200	1,902	7	49,360	2,073	175	1	(s)	250	10	1
2003	410	174	66,772	2,804	236	292	1	5,978	-222	-1	67,286	2,826	238	2	(s)	R338	14	2
2004	497	210	81,058	3,404	287	3,542	13	6,002	24	(s)	84,576	3,552	299	R4	(s)	666	28	R4
2005	570	241	92,961	3,904	329	3,234	11	5,563	-439	-2	96,634	4,059	342	R12	(s)	2,162	91	R12
2006	R712	R301	R116,294	4,884	R412	R17,408	R62	R8,760	R3,197	11	R130,505	5,481	R462	R32	R (s)	R5,963	250	R32
2007P	924	378	154,416	6,485	546	10,348	37	10,509	1,749	6	163,002	6,846	577	64	1	11,691	491	63

[1] Total corn and other biomass inputs to the production of fuel ethanol.

[2] Losses and co-products from the production of fuel ethanol. Does not include natural gas, electricity, and other non-biomass energy used in the production of fuel ethanol—these are included in the industrial sector consumption statistics for the appropriate energy source.

[3] Fuel ethanol imports only. Data for fuel ethanol exports are not available.

[4] Stocks are at end of year.

[5] A negative number indicates a decrease in stocks and a positive number indicates an increase.

[6] Total vegetable oil and other biomass inputs to the production of biodiesel.

[7] Losses and co-products from the production of biodiesel. Does not include natural gas, electricity, and other non-biomass energy used in the production of biodiesel—these are included in the industrial sector consumption statistics for the appropriate energy source.

[8] Production of biofuels for use as diesel fuel substitutes or additives. Biodiesel consumption equals biodiesel production.

R=Revised. P=Preliminary. NA=Not available. (s)=Less than 0.5 trillion Btu.

Notes: • Mbbl = thousand barrels. MMgal = million U.S. gallons. TBtu = trillion Btu. • Totals may not equal sum of components due to independent rounding.

Web Pages: For related information, see http://www.eia.doe.gov/oil_gas/petroleum/data_publications/monthly_oxygenate_telephone_report/motr.html, http://www.eia.doe.gov/oil_gas/petroleum/data_publications/petroleum_supply_monthly/psm.html, and http://www.census.gov/cir/www/311/m311k.html.

Sources: *(Note: For production, net imports, stocks, stock change, and consumption, data in thousand barrels are converted to million gallons by multiplying by 0.042; and are converted to trillion Btu by multiplying by the approximate heat content of fuel ethanol or biodiesel—see Table A3.)* **Fuel Ethanol Feedstock:** Calculated as fuel ethanol production in thousand barrels multiplied by the approximate heat content of fuel ethanol feedstock—see Table A3. **Fuel Ethanol Losses and Co-products:** Calculated as fuel ethanol feedstock minus fuel ethanol production. **Fuel Ethanol Production:** • 1981-1992—Fuel ethanol production is equal to fuel ethanol consumption—see sources for "Fuel Ethanol Consumption." • 1993-2004—Calculated as fuel ethanol consumption plus fuel ethanol stock change minus fuel ethanol net imports. These data differ slightly from the original production data from Energy Information Administration (EIA), Form EIA-819, "Monthly Oxygenate Report," and predecessor form, which were not reconciled and updated to be consistent with the final balance. • 2005 forward—EIA, Form EIA-819, "Monthly Oxygenate Report." **Fuel Ethanol Net Imports, Stocks, and Stock Change:** • 1992-2006—EIA, *Petroleum Supply Annual (PSA),* annual reports. • 2007—EIA, *Petroleum Supply Monthly (PSM),* monthly reports. **Fuel Ethanol Consumption:** • 1981-1989—EIA, *Estimates of U.S. Biofuels Consumption 1990,* Table 10; and EIA, Office of Coal, Nuclear, Electric and Alternate Fuels (CNEAF), estimates. • 1990-1992—EIA, *Estimates of U.S. Biomass Energy Consumption 1992,* Table D2; and EIA, CNEAF, estimates. • 1993-2004—EIA, *PSA,* annual reports, Tables 2 and 16. Calculated as ten percent of oxygenated finished motor gasoline field production (Table 2), plus fuel ethanol refinery input (Table 16). • 2005 and 2006—EIA, *PSA,* annual reports, Tables 1 and 15. Calculated as motor gasoline blending components adjustments (Table 1), plus finished motor gasoline adjustments (Table 1), plus fuel ethanol refinery and blender net inputs (Table 15). • 2007—EIA, *PSM,* monthly reports, Tables 1 and 27. Calculated as motor gasoline blending components adjustments (Table 1), plus finished motor gasoline adjustments (Table 1), plus fuel ethanol refinery and blender net inputs (Table 27). **Biodiesel Feedstock:** Calculated as biodiesel production in thousand barrels multiplied by the approximate heat content of biodiesel feedstock—see Table A3. **Biodiesel Losses and Co-products:** Calculated as biodiesel feedstock minus biodiesel production. **Biodiesel Production:** • 2001-2005—U.S. Department of Agriculture, Commodity Credit Corporation, Bioenergy Program records. Annual data are derived from quarterly data. • 2006—U.S. Department of Commerce, Bureau of the Census, "M311K - Fats and Oils: Production, Consumption, and Stocks," Table 3A, data for soybean oil consumed in methyl esters (biodiesel). In addition, EIA, Office of Integrated Analysis and Forecasting, estimates that 14.4 million gallons of yellow grease were consumed in methyl esters (biodiesel). EIA assumes that 7.65 pounds of vegetable oil are needed to make one gallon of biodiesel. • 2007—U.S. Department of Commerce, Bureau of the Census, "M311K - Fats and Oils: Production, Consumption, and Stocks," Table 3A, data for all fats and oils consumed in methyl esters (biodiesel). EIA assumes that 7.65 pounds of vegetable oil are needed to make one gallon of biodiesel.

Figure 10.4 Estimated Number of Alternative-Fueled Vehicles in Use and Alternative Fuel Consumption

Vehicles in Use, 1995-2006

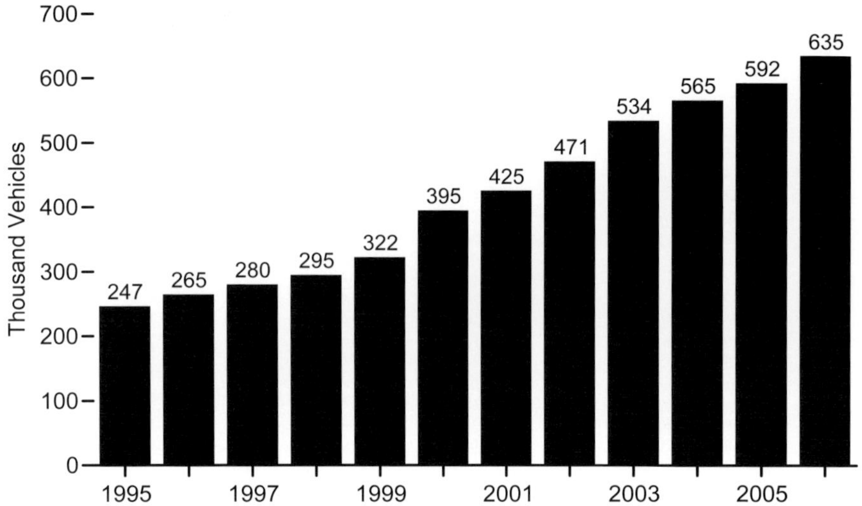

Vehicles in Use by Fuel Type, 2006

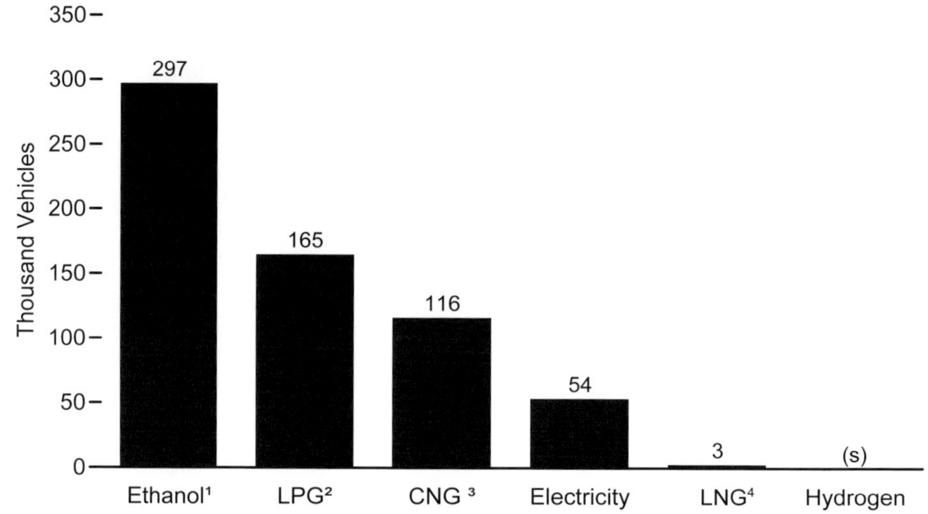

Fuel Consumption, 1995-2006

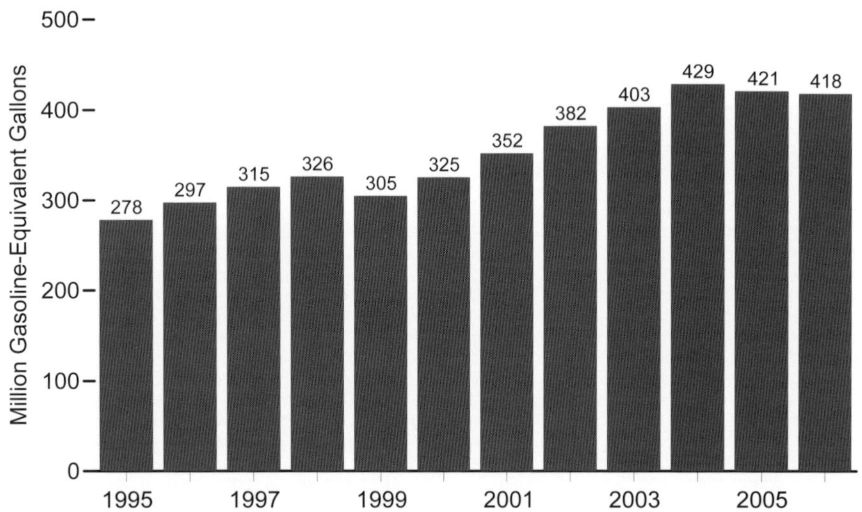

Fuel Consumption by Type, 2006

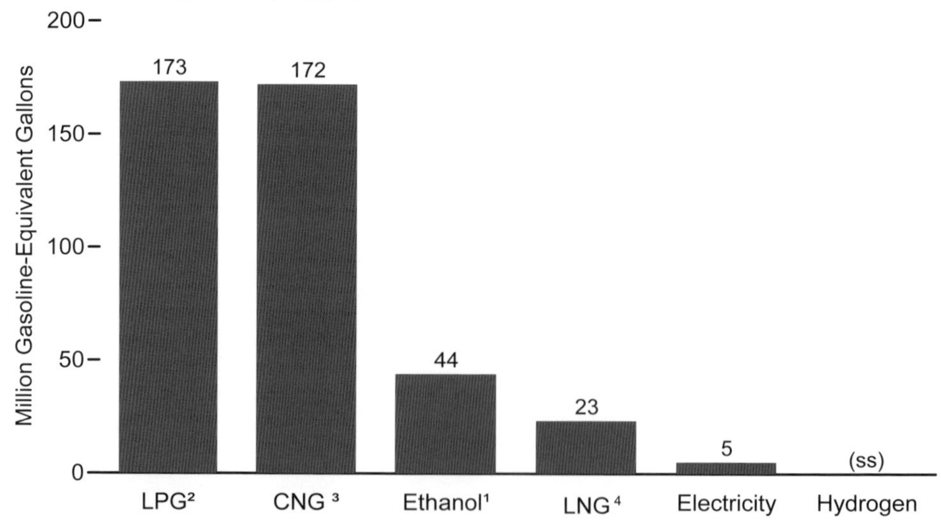

[1] Ethanol, 85 percent (E85). Includes only those E85 vehicles believed to be used as alternative-fueled vehicles, primarily fleet-operated vehicles; excludes other vehicles with E85-fueling capability.
[2] Liquefied petroleum gases.
[3] Compressed natural gas.
[4] Liquefied natural gas.
(s)=Fewer than 0.5 thousand vehicles.
(ss)=Less than 0.5 million gasoline-equivalent gallons.
Note: Because vertical scales differ, graphs should not be compared.
Source: Table 10.4.

Table 10.4 Estimated Number of Alternative-Fueled Vehicles in Use and Fuel Consumption, 1992-2006

Year	Alternative and Replacement Fuels [1]															
												Oxygenates [2]				
	Liquefied Petroleum Gases	Compressed Natural Gas	Liquefied Natural Gas	Methanol, 85 Percent (M85) [3]	Methanol, Neat (M100) [4]	Ethanol, 85 Percent (E85) [3,5]	Ethanol, 95 Percent (E95) [3]	Electricity [6]	Hydrogen	Other Fuels [7]	Total	Methyl Tertiary Butyl Ether [8]	Ethanol in Gasohol [9]	Total	Biodiesel [10]	Total
	Alternative-Fueled Vehicles in Use [11] (number)															
1992	NA	23,191	90	4,850	404	172	38	1,607	NA	NA	NA	NA	NA	NA	NA	NA
1993	NA	32,714	299	10,263	414	441	27	1,690	NA	NA	NA	NA	NA	NA	NA	NA
1994	NA	41,227	484	15,484	415	605	33	2,224	NA	NA	NA	NA	NA	NA	NA	NA
1995	172,806	50,218	603	18,319	386	1,527	136	2,860	0	0	246,855	NA	NA	NA	NA	NA
1996	175,585	60,144	663	20,265	172	4,536	361	3,280	0	0	265,006	NA	NA	NA	NA	NA
1997	175,679	68,571	813	21,040	172	9,130	347	4,453	0	0	280,205	NA	NA	NA	NA	NA
1998	177,183	78,782	1,172	19,648	200	12,788	14	5,243	0	0	295,030	NA	NA	NA	NA	NA
1999	178,610	91,267	1,681	18,964	198	24,604	14	6,964	0	0	322,302	NA	NA	NA	NA	NA
2000	181,994	100,750	2,090	10,426	0	87,570	4	11,830	0	0	394,664	NA	NA	NA	NA	NA
2001	185,053	111,851	2,576	7,827	0	100,303	0	17,847	0	0	425,457	NA	NA	NA	NA	NA
2002	187,680	120,839	2,708	5,873	0	120,951	0	33,047	0	0	471,098	NA	NA	NA	NA	NA
2003	190,369	114,406	2,640	0	0	179,090	0	47,485	9	0	533,999	NA	NA	NA	NA	NA
2004	182,864	118,532	2,717	0	0	211,800	0	49,536	43	0	565,492	NA	NA	NA	NA	NA
2005	173,795	117,699	2,748	0	0	246,363	0	51,398	119	3	592,125	NA	NA	NA	NA	NA
2006P	164,846	116,131	2,798	0	0	297,099	0	53,526	159	3	634,562	NA	NA	NA	NA	NA
	Fuel Consumption [12] (thousand gasoline-equivalent gallons)															
1992	NA	17,159	598	1,121	2,672	22	87	359	NA	NA	NA	1,175,964	719,408	1,895,372	NA	NA
1993	NA	22,035	1,944	1,671	3,321	49	82	288	NA	NA	NA	2,070,897	779,958	2,850,854	NA	NA
1994	NA	24,643	2,398	2,455	3,347	82	144	430	NA	NA	NA	2,020,455	868,113	2,888,569	NA	NA
1995	233,178	35,865	2,821	2,122	2,255	195	1,021	663	0	0	278,121	2,693,407	934,615	3,628,022	NA	3,906,142
1996	239,648	47,861	3,320	1,862	364	712	2,770	773	0	0	297,310	2,751,955	677,537	3,429,492	NA	3,726,802
1997	238,845	66,495	3,798	1,630	364	1,314	1,166	1,010	0	0	314,621	3,106,745	852,514	3,959,260	NA	4,273,880
1998	241,881	73,859	5,463	1,271	471	1,772	61	1,202	0	0	325,980	2,905,781	912,858	3,818,639	NA	4,144,620
1999	210,247	81,211	5,959	1,126	469	4,019	64	1,524	0	0	304,618	3,405,390	975,255	4,380,645	NA	4,685,263
2000	213,012	88,478	7,423	614	0	12,388	13	3,058	0	0	324,986	3,298,803	1,114,313	4,413,116	6,828	4,744,930
2001	216,319	106,584	9,122	461	0	15,007	0	4,066	0	0	351,558	3,354,949	1,173,323	4,528,272	7,089	4,886,919
2002	223,600	123,081	9,593	354	0	18,250	0	7,274	0	0	382,152	3,122,859	1,450,721	4,573,580	16,948	4,972,680
2003	224,697	133,222	13,503	0	0	26,376	0	5,141	2	0	402,941	2,368,400	1,919,572	4,287,972	R18,220	R4,709,133
2004	211,883	158,903	20,888	0	0	31,581	0	5,269	8	0	428,532	1,877,300	2,414,167	4,291,467	R28,244	R4,748,243
2005	188,171	166,878	22,409	0	0	38,074	0	5,219	25	2	420,778	1,654,500	2,756,663	4,411,163	R91,649	R4,923,590
2006P	173,130	172,011	23,474	0	0	44,041	0	5,104	41	2	417,803	435,000	3,729,168	4,164,168	260,606	4,842,577

[1] See "Alternative Fuel" and "Replacement Fuel" in Glossary.
[2] See "Oxygenates" in Glossary.
[3] Remaining portion is motor gasoline. Consumption data include the motor gasoline portion of the fuel.
[4] One hundred percent methanol.
[5] Includes only those E85 vehicles believed to be used as alternative-fuels vehicles (AFVs), primarily fleet-operated vechicles; excludes other vehicles with E85-fueling capability. In 1997, some vehicle manufacturers began including E85-fueling capability in certain model lines of vehicles. For 2006, the Energy Information Administration (EIA) estimates that the number of E85 vehicles that are capable of operating on E85, motor gasoline, or both, is about 6 million. Many of these AFVs are sold and used as traditional gasoline-powered vehicles.
[6] Excludes gasoline-electric hybrids.
[7] May include P-Series fuel or any other fuel designated by the Secretary of Energy as an alternative fuel in acordance with the Energy Policy Act of 1995.
[8] In addition to methyl tertiary butyl ether (MTBE), includes a very small amount of other ethers, primarily tertiary amyl methyl ether (TAME) and ethyl tertiary butyl ether (ETBE).
[9] Data do not include the motor gasoline portion of the fuel.
[10] "Biodiesel" may be used as a diesel fuel substitute or diesel fuel additive or extender. See "Biodiesel" in Glossary.
[11] "Vehicles in Use" data represent accumulated acquisitions, less retirements, as of the end of each calendar year; data do not include concept and demonstration vehicles that are not ready for delivery to end users. See "Alternative-Fueled Vehicle" in Glossary.
[12] Fuel consumption quantities are expressed in a common base unit of gasoline-equivalent gallons to allow comparisons of different fuel types. Gasoline-equivalent gallons do not represent gasoline displacement. Gasoline equivalent is computed by dividing the gross heat content of the replacement fuel by the gross heat content of gasoline (using an approximate heat content of 122,619 Btu per gallon) and multiplying the result by the replacement fuel consumption value. See "Heat Content" in Glossary.

R=Revised. P=Preliminary. NA=Not available.

Note: Totals may not equal sum of components due to independent rounding.

Web Page: For related information, see http://www.eia.doe.gov/fuelrenewable.html.

Sources: • 1992-1994—Science Applications International Corporation, "Alternative Transportation Fuels and Vehicles Data Development," unpublished final report prepared for the EIA, (McLean, VA, July 1996), and U.S. Department of Energy, Office of Energy Efficiency and Renewable Energy. Data were revised by using gross instead of net heat contents. For a table of gross and net heat contents, see EIA, *Alternatives to Traditional Transportation Fuels: An Overview* (June 1994), Table 22. • 1995-2002—EIA, "Alternatives to Traditional Transportation Fuels 2003 Estimated Data" (February 2004), Tables 1 and 10. Data were revised by using gross instead of net heat contents. • 2003 forward—EIA, "Alternatives to Traditional Transportation Fuels 2006" (May 2008), Tables V1 and C1.

Figure 10.5 Solar Thermal Collector Shipments by Type, Price, and Trade

Total Shipments, 1974-1984 and 1986-2006

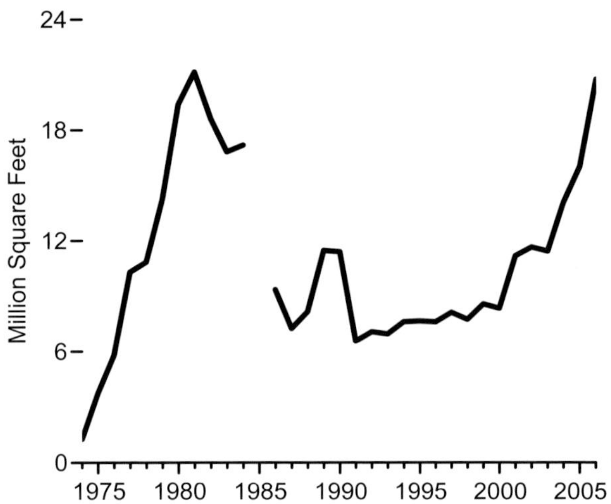

Trade, 1978-1984 and 1986-2006

Price of Total Shipments, 1986-2006

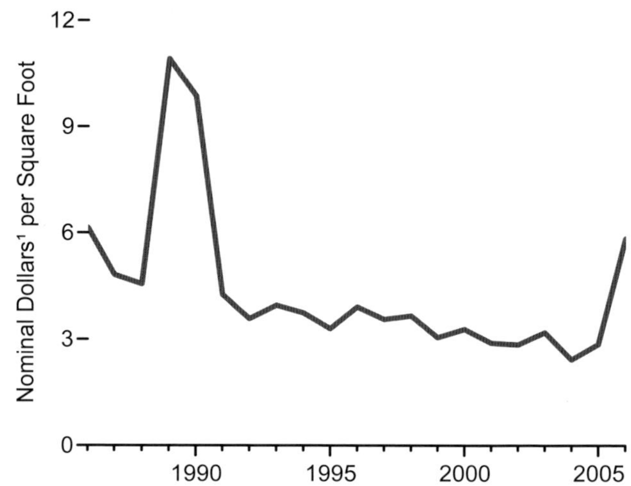

Number of U.S. Manufacturers by Type of Collector, 1974-1984 and 1986-2006

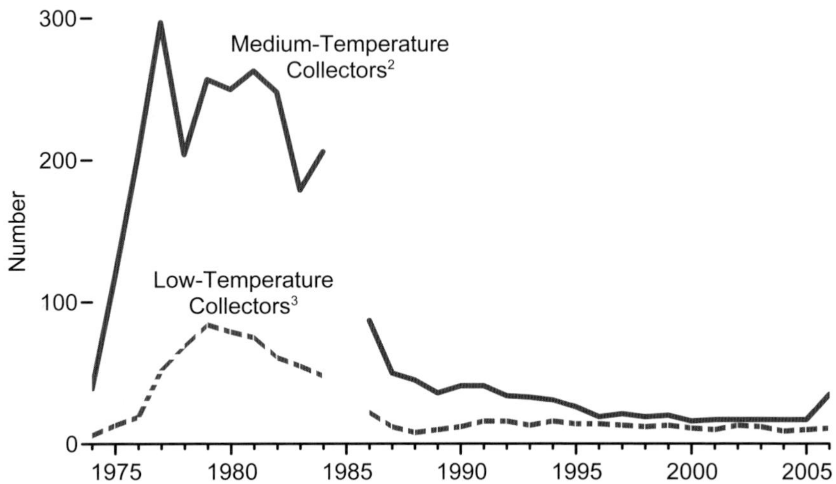

Average Annual Shipments per Manufacturer, 1974-1984 and 1986-2006

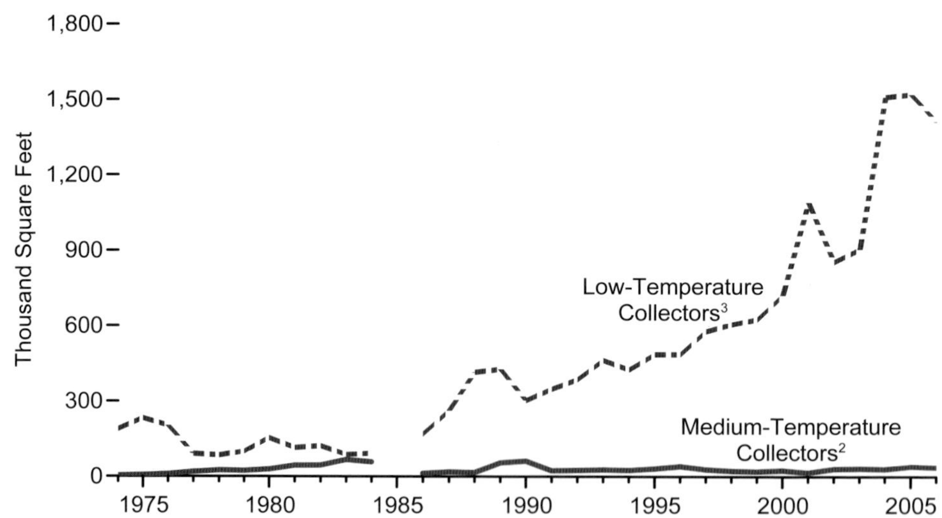

[1] See "Nominal Dollars" in Glossary.

[2] Collectors that generally operate in the temperature range of 140 degrees Fahrenheit to 180 degrees Fahrenheit but can also operate at temperatures as low as 110 degrees Fahrenheit. Special collectors–evacuated tube collectors or concentrating (focusing) collectors–are included in the medium-temperature category.

[3] Collectors that generally operate at temperatures below 110 degrees Fahrenheit.

Notes: • Shipments are for domestic and export shipments, and may include imports that subsequently were shipped to domestic or foreign customers. • Data were not collected for 1985. • Because vertical scales differ, graphs should not be compared.

Source: Table 10.5.

Table 10.5 Solar Thermal Collector Shipments by Type, Price, and Trade, 1974-2006

(Thousand Square Feet, Except as Noted)

Year	Low-Temperature Collectors [1]				Medium-Temperature Collectors [2]				High-Temperature Collectors [3]		Total Shipments		Trade	
	Number of U.S. Manu-facturers	Quantity Shipped	Shipments per Manu-facturer	Price [4] (nominal dollars [5] per square foot)	Number of U.S. Manu-facturers	Quantity Shipped	Shipments per Manu-facturer	Price [4] (nominal dollars [5] per square foot)	Quantity Shipped	Price [4] (nominal dollars [5] per square foot)	Quantity Shipped	Price [4] (nominal dollars [5] per square foot)	Imports	Exports
1974	6	1,137	190	NA	39	137	4	NA	NA	NA	1,274	NA	NA	NA
1975	13	3,026	233	NA	118	717	6	NA	NA	NA	3,743	NA	NA	NA
1976	19	3,876	204	NA	203	1,925	10	NA	NA	NA	5,801	NA	NA	NA
1977	52	4,743	91	NA	297	5,569	19	NA	NA	NA	10,312	NA	NA	NA
1978	69	5,872	85	NA	204	4,988	25	NA	NA	NA	10,860	NA	396	840
1979	84	8,394	100	NA	257	5,856	23	NA	NA	NA	14,251	NA	290	855
1980	79	12,233	155	NA	250	7,165	29	NA	NA	NA	19,398	NA	235	1,115
1981	75	8,677	116	NA	263	11,456	44	NA	NA	NA	21,133	NA	196	771
1982	61	7,476	123	NA	248	11,145	45	NA	NA	NA	18,621	NA	418	455
1983	55	4,853	88	NA	179	11,975	67	NA	NA	NA	16,828	NA	511	159
1984	48	4,479	93	NA	206	11,939	58	NA	773	NA	17,191	NA	621	348
1985[6]	NA	NA	NA	NA	NA	NA	NA	NA	NA	NA	NA	NA	NA	NA
1986	22	3,751	171	2.30	87	1,111	13	18.30	4,498	NA	9,360	6.14	473	224
1987	12	3,157	263	2.18	50	957	19	13.50	3,155	NA	7,269	4.82	691	182
1988	8	3,326	416	2.24	45	732	16	14.88	4,116	NA	8,174	4.56	814	158
1989	10	4,283	428	2.60	36	1,989	55	11.74	5,209	17.76	11,482	10.92	1,233	461
1990	12	3,645	304	2.90	41	2,527	62	7.68	5,237	15.74	11,409	9.86	1,562	245
1991	16	5,585	349	2.90	41	989	24	11.94	1	31.94	6,574	4.26	1,543	332
1992	16	6,187	387	2.50	34	897	26	10.96	2	75.66	7,086	3.58	1,650	316
1993	13	6,025	464	2.80	33	931	28	11.74	12	22.12	6,968	3.96	2,039	411
1994	16	6,823	426	2.54	31	803	26	13.54	2	177.00	7,627	3.74	1,815	405
1995	14	6,813	487	2.32	26	840	32	10.48	13	53.26	7,666	3.30	2,037	530
1996	14	6,821	487	2.67	19	785	41	14.48	10	18.75	7,616	3.91	1,930	454
1997	13	7,524	579	2.60	21	606	29	15.17	7	25.00	8,138	3.56	2,102	379
1998	12	7,292	607	2.83	19	443	23	15.17	21	53.21	7,756	3.66	2,206	360
1999	13	8,152	627	2.08	20	427	21	19.12	4	286.49	8,583	3.05	2,352	537
2000	11	7,948	723	2.09	16	400	25	W	5	W	8,354	3.28	2,201	496
2001	10	10,919	1,092	2.15	17	268	16	W	2	W	11,189	2.90	3,502	840
2002	13	11,126	856	1.97	17	535	31	W	2	W	11,663	2.85	3,068	659
2003	12	10,877	906	2.08	17	560	33	W	7	W	11,444	3.19	2,986	518
2004	9	13,608	1,512	1.80	17	506	30	19.30	0	– –	14,114	2.43	3,723	813
2005	10	15,224	1,522	2.00	17	702	41	W	115	W	16,041	2.86	4,546	1,361
2006	11	15,546	1,413	1.95	35	1,346	38	W	3,852	W	20,744	5.84	4,244	1,211

[1] Low-temperature collectors are solar thermal collectors that generally operate at temperatures below 110° F.

[2] Medium-temperature collectors are solar thermal collectors that generally operate in the temperature range of 140° F to 180° F but can also operate at temperatures as low as 110° F. Special collectors are included in this category. Special collectors are evacuated tube collectors or concentrating (focusing) collectors. They operate in the temperature range from just above ambient temperature (low concentration for pool heating) to several hundred degrees Fahrenheit (high concentration for air conditioning and specialized industrial processes).

[3] High-temperature collectors are solar thermal collectors that generally operate at temperatures above 180° F. High-temperature collector shipments are dominated by one manufacturer.

[4] Prices equal shipment value divided by quantity shipped. Value includes charges for advertising and warranties. Excluded are excise taxes and the cost of freight or transportation for the shipments.

[5] See "Nominal Dollars" in Glossary.

[6] No data are available for 1985.

NA=Not available. – – = Not applicable. W=Value withheld to avoid disclosure of proprietary company data.

Notes: • Shipments data are for domestic and export shipments, and may include imports that subsequently were shipped to domestic or foreign customers. • Manufacturers producing more than one type of collector are accounted for in both groups.

Web Page: For related information, see http://www.eia.doe.gov/fuelrenewable.html.

Sources: • 1974-1992—Energy Information Administration (EIA), *Solar Collector Manufacturing Activity,* annual reports, and Form CE-63A, "Annual Solar Thermal Collector Manufacturers Survey," and predecessor forms. • 1993-2001—EIA, *Renewable Energy Annual,* annual reports, and Form EIA-63A, "Annual Solar Thermal Collector Manufacturers Survey," and predecessor form. • 2002 forward—EIA, *Solar Thermal and Photovoltaic Collector Manufacturing Activities,* annual reports, and Form EIA-63A, "Annual Solar Thermal Collector Manufacturers Survey."

Figure 10.6 Solar Thermal Collector Shipments by End Use, Market Sector, and Type, 2006

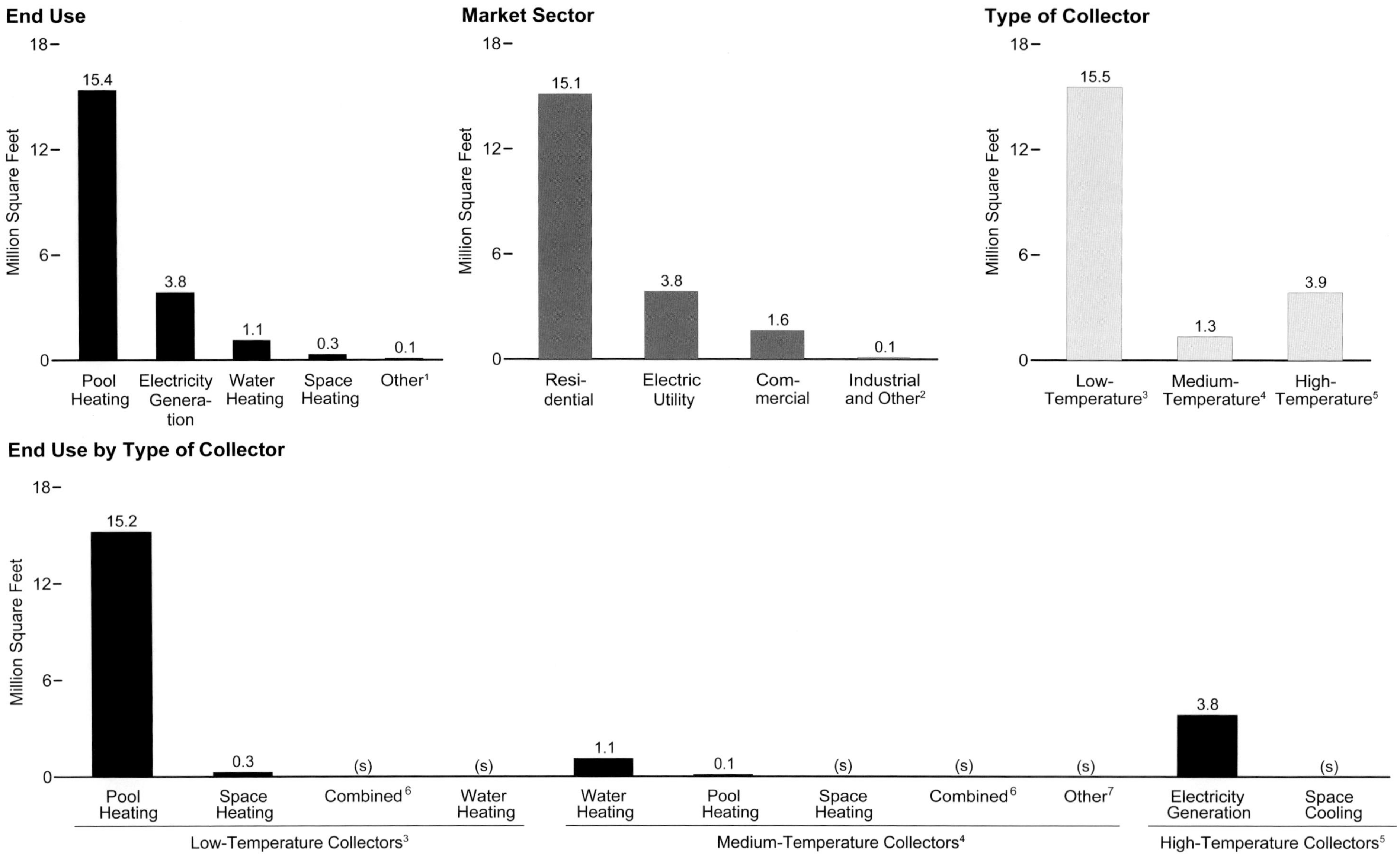

[1] Combined space and water heating, and space cooling.
[2] Other sectors, such as government, including the military but excluding space applications.
[3] Collectors that generally operate at temperatures below 110 degrees Fahrenheit.
[4] Collectors that generally operate in the temperature range of 140 degrees Fahrenheit to 180 degrees Fahrenheit but can also operate at temperatures as low as 110 degrees Fahrenheit.
[5] Collectors that generally operate at temperatures above 180 degrees Fahrenheit.
[6] Combined space and water heating.
[7] Space cooling and electricity generation.
(s)=Less than 0.05 million square feet.
Notes: • Data are for domestic and export shipments, and may include imports that subsequently were shipped to domestic or foreign customers.
Source: Table 10.6.

Table 10.6 Solar Thermal Collector Shipments by End Use, Market Sector, and Type, 2006
(Thousand Square Feet)

End Use	Low-Temperature Collectors [1]	Medium-Temperature Collectors [2]	High-Temperature Collectors [3]	Total
End-Use Total	**15,546**	**1,346**	**3,852**	**20,744**
Pool Heating	15,225	137	0	15,362
Water Heating	10	1,126	0	1,136
Space Heating	290	40	0	330
Space Cooling	0	3	0	3
Combined Space and Water Heating	21	38	7	66
Process Heating	0	0	0	0
Electricity Generation	0	2	3,845	3,847
Other [4]	0	0	0	0
Market Sector Total	**15,546**	**1,346**	**3,852**	**20,744**
Residential	13,906	1,217	0	15,123
Commercial	1,500	120	7	1,626
Industrial [5]	40	2	0	42
Electric Utility	0	0	3,845	3,845
Other [6]	100	7	0	107

[1] Low-temperature collectors are solar thermal collectors that generally operate at temperatures below 110° F.

[2] Medium-temperature collectors are solar thermal collectors that generally operate in the temperature range of 140° F to 180° F but can also operate at temperatures as low as 110° F. Special collectors are included in this category. Special collectors are evacuated tube collectors or concentrating (focusing) collectors. They operate in the temperature range from just above ambient temperature (low concentration for pool heating) to several hundred degrees Fahrenheit (high concentration for air conditioning and specialized industrial processes).

[3] High-temperature collectors are solar thermal collectors that generally operate at temperatures above 180° F. These are parabolic dish/trough collectors used primarily by independent power producers to generate electricity for the electric grid.

[4] Cooking, water pumping, water purification, desalinization, distillation, and other uses.

[5] Includes all independent power producers.

[6] Other sectors, such as government, including the military but excluding space applications.

Notes: • Data are for domestic and export shipments, and may include imports that subsequently were shipped to domestic or foreign customers. • Totals may not equal sum of components due to independent rounding.

Web Page: For related information, see http://www.eia.doe.gov/fuelrenewable.html.

Source: Energy Information Administration, *Solar Thermal and Photovoltaic Collector Manufacturing Activities 2006* (October 2007), Table 2.10.

Figure 10.7 Photovoltaic Cell and Module Shipments, Trade, and Prices

Number of U.S. Companies Reporting Shipments, 1982-2006

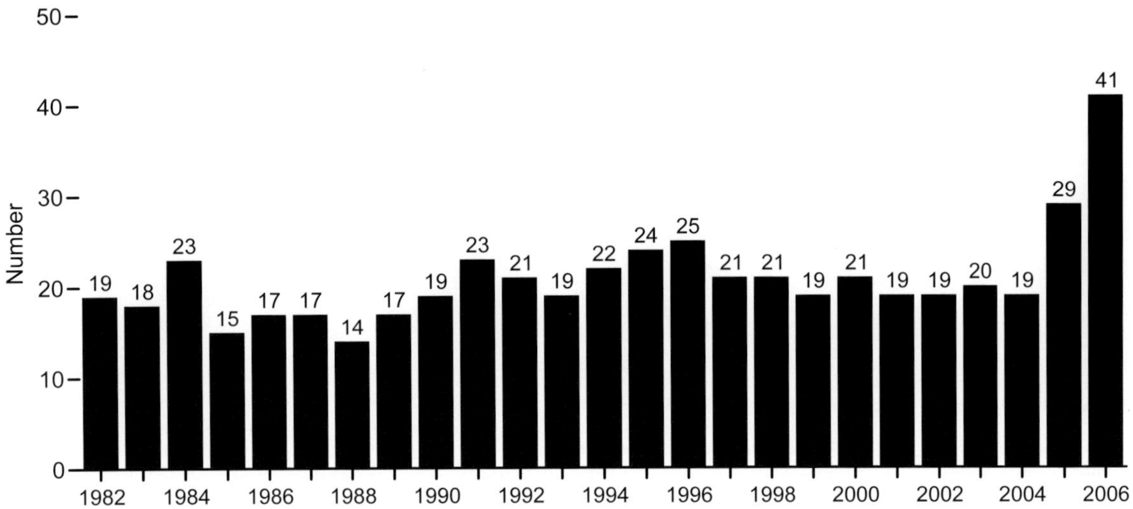

Total Shipments, 1982-2006

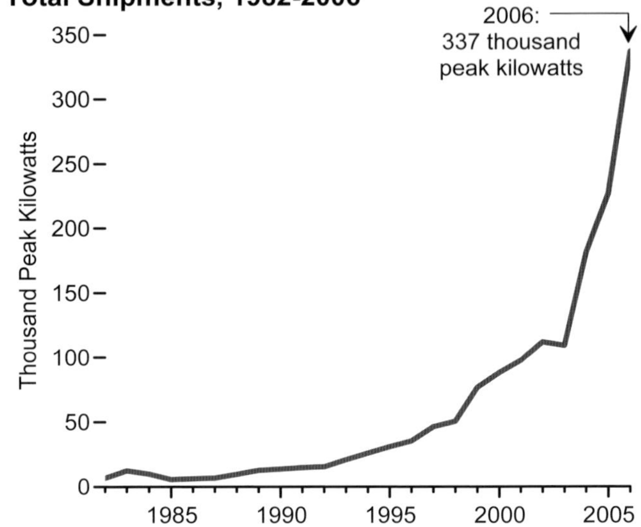

Trade, 1983-2006

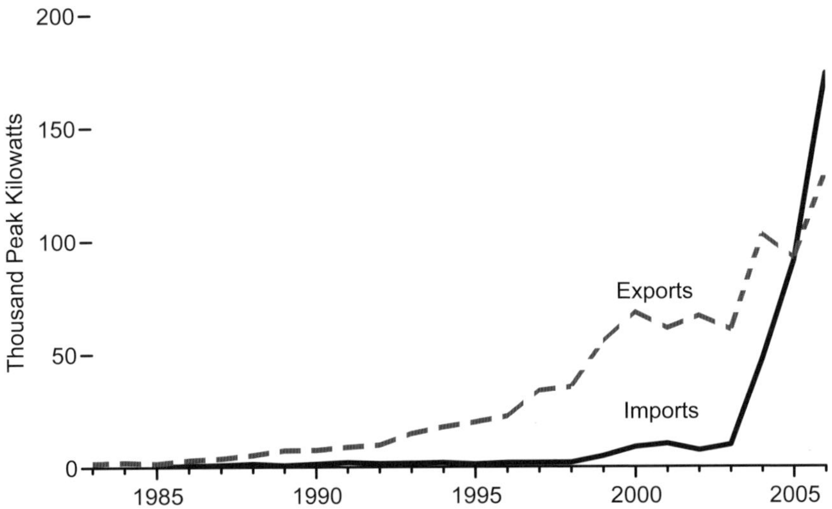

Prices, 1989-2006

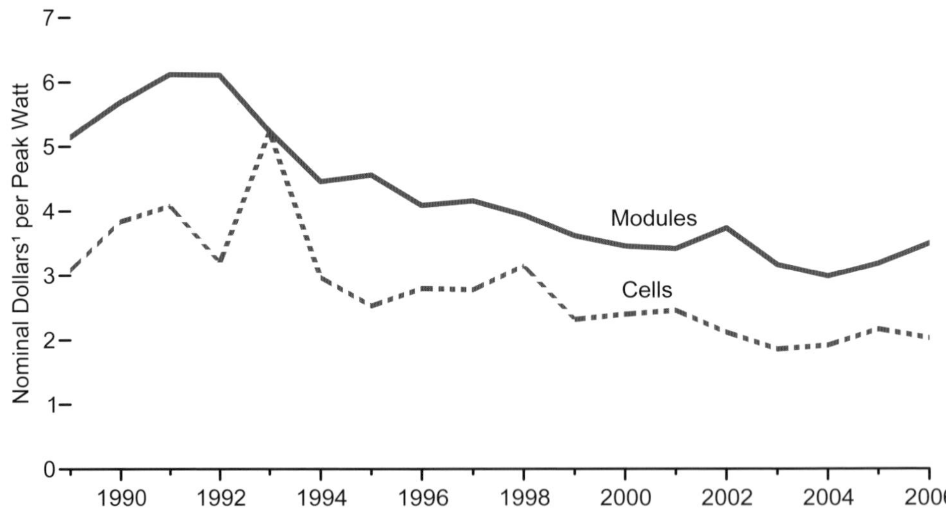

[1] See "Nominal Dollars" in Glossary.

Notes: • Shipments are for domestic and export shipments, and may include imports that subsequently were shipped to domestic and foreign customers. • Because vertical scales differ, graphs should not be compared.

Source: Table 10.7.

Table 10.7 Photovoltaic Cell and Module Shipments by Type, Trade, and Prices, 1982-2006

Year	U.S. Companies Reporting Shipments	Shipments			Trade		Prices [1]	
		Crystalline Silicon	Thin-Film Silicon	Total [2]	Imports	Exports	Modules	Cells
	Number	Peak Kilowatts [3]					Nominal Dollars [4] per Peak Watt [3]	
1982	19	NA	NA	6,897	NA	NA	NA	NA
1983	18	NA	NA	12,620	NA	1,903	NA	NA
1984	23	NA	NA	9,912	NA	2,153	NA	NA
1985	15	5,461	303	5,769	285	1,670	NA	NA
1986	17	5,806	516	6,333	678	3,109	NA	NA
1987	17	5,613	1,230	6,850	921	3,821	NA	NA
1988	14	7,364	1,895	9,676	1,453	5,358	NA	NA
1989	17	10,747	1,628	12,825	826	7,363	5.14	3.08
1990	[5]19	12,492	1,321	[5]13,837	1,398	7,544	5.69	3.84
1991	23	14,205	723	14,939	2,059	8,905	6.12	4.08
1992	21	14,457	1,075	15,583	1,602	9,823	6.11	3.21
1993	19	20,146	782	20,951	1,767	14,814	5.24	5.23
1994	22	24,785	1,061	26,077	1,960	17,714	4.46	2.97
1995	24	29,740	1,266	31,059	1,337	19,871	4.56	2.53
1996	25	33,996	1,445	35,464	1,864	22,448	4.09	2.80
1997	21	44,314	1,886	46,354	1,853	33,793	4.16	2.78
1998	21	47,186	3,318	50,562	1,931	35,493	3.94	3.15
1999	19	73,461	3,269	76,787	4,784	55,562	3.62	2.32
2000	21	85,155	2,736	88,221	8,821	68,382	3.46	2.40
2001	19	84,651	12,541	97,666	10,204	61,356	3.42	2.46
2002	19	104,123	7,396	112,090	7,297	66,778	3.74	2.12
2003	20	97,940	10,966	109,357	9,731	60,693	3.17	1.86
2004	19	159,138	21,978	181,116	47,703	102,770	2.99	1.92
2005	29	172,965	53,826	226,916	90,981	92,451	3.19	2.17
2006	41	233,518	101,766	337,268	173,977	130,757	3.50	2.03

[1] Prices equal shipment value divided by quantity shipped. Value includes charges for advertising and warranties. Excluded are excise taxes and the cost of freight or transportation for the shipments.
[2] Includes all types of photovoltaic cells and modules (single-crystal silicon, cast silicon, ribbon silicon, thin-film silicon, and concentrator silicon). Excludes cells and modules for space and satellite applications.
[3] See "Peak Kilowatt" and "Peak Watt" in Glossary.
[4] See "Nominal Dollars" in Glossary.
[5] Data were imputed for one nonrespondent who exited the industry during 1990.

NA=Not available.

Note: Shipments data are for domestic and export shipments, and may include imports that subsequently were shipped to domestic or foreign customers.

Web Page: For related information, see http://www.eia.doe.gov/fuelrenewable.html.

Sources: • 1982-1992—Energy Information Administration (EIA), *Solar Collector Manufacturing Activity,* annual reports. • 1993-2001—EIA, *Renewable Energy Annual,* annual reports. • 2002 forward—EIA, *Solar Thermal and Photovoltaic Collector Manufacturing Activities,* annual reports.

Figure 10.8 Photovoltaic Cell and Module Shipments by End Use and Market Sector, 2006

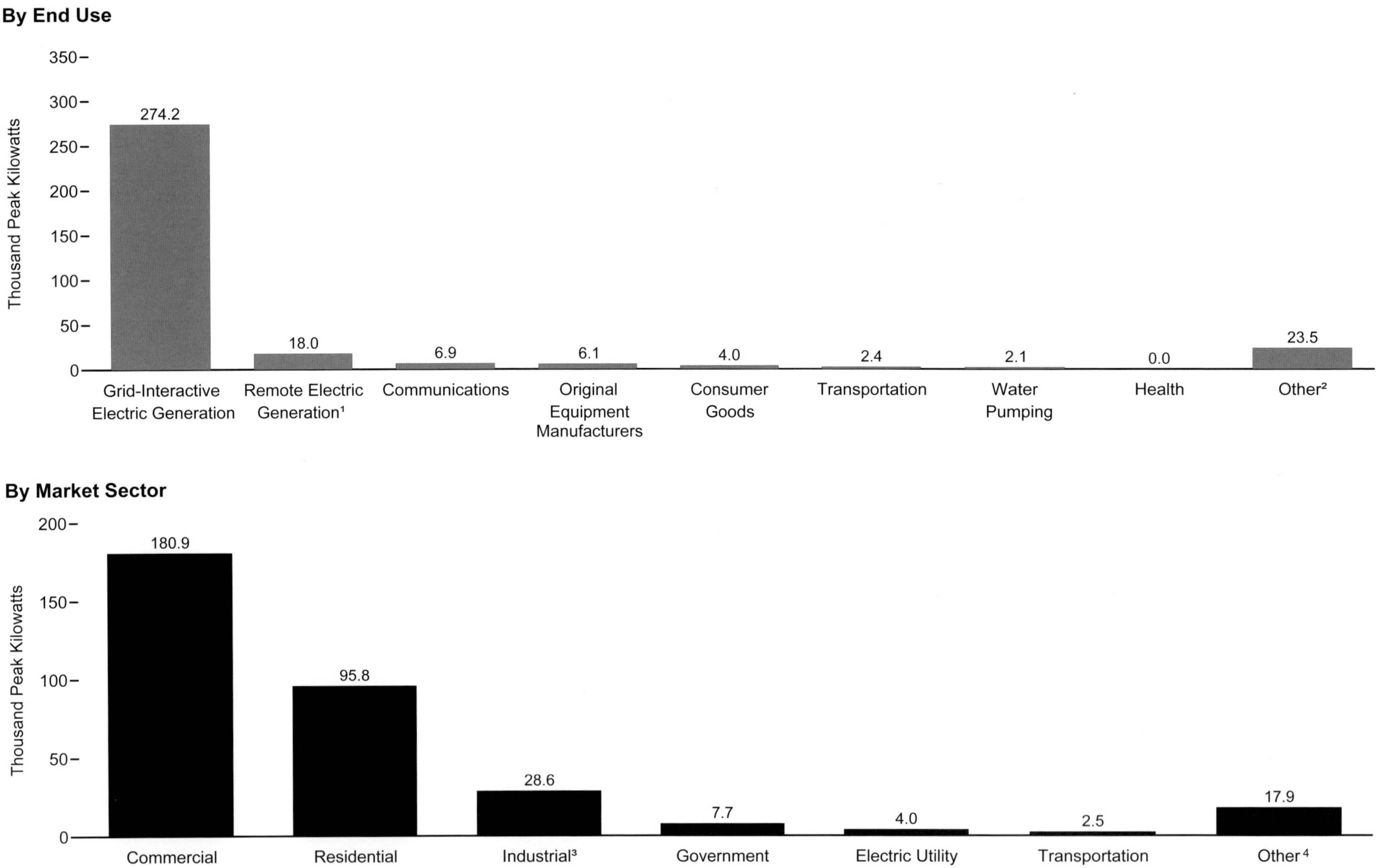

[1] Units designed for installations that do not interact with the electrical distribution system.
[2] Represents such applications as cooking food, desalinization, and distilling.
[3] Includes all independent power producers.
[4] Shipments for specialty purposes such as research.

Note: Data are for domestic and export shipments, and may include imports that subsequently were shipped to domestic or foreign customers.
Source: Table 10.8.

Table 10.8 Photovoltaic Cell and Module Shipments by End Use and Market Sector, 1989-2006

Year	End Use: Communications	End Use: Consumer Goods	End Use: Electric Generation[1]: Grid-Interactive	End Use: Electric Generation[1]: Remote	End Use: Health	End Use: Original Equipment Manufacturers[2]	End Use: Transportation	End Use: Water Pumping	End Use: Other[3]	Market Sector: Residential	Market Sector: Commercial	Market Sector: Government	Market Sector: Industrial[4]	Market Sector: Transportation	Market Sector: Electric Utility	Market Sector: Other[5]	Total
	Amount Shipped (peak kilowatts[6])																
1989	2,590	2,788	1,251	2,620	5	1,595	1,196	711	69	1,439	3,850	1,077	3,993	1,130	785	551	12,825
1990	4,340	2,484	469	3,097	5	1,119	1,069	1,014	240	1,701	6,086	1,002	2,817	974	826	432	13,837
1991	3,538	3,312	856	3,594	61	1,315	1,523	729	13	3,624	3,345	815	3,947	1,555	1,275	377	14,939
1992	3,717	2,566	1,227	4,238	67	828	1,602	809	530	4,154	2,386	1,063	4,279	1,673	1,553	477	15,583
1993	3,846	946	1,096	5,761	674	2,023	4,238	2,294	74	5,237	4,115	1,325	5,352	2,564	1,503	856	20,951
1994	5,570	3,239	2,296	9,253	79	1,849	2,128	1,410	254	6,632	5,429	2,114	6,855	2,174	2,364	510	26,077
1995	5,154	1,025	4,585	8,233	776	3,188	4,203	2,727	1,170	6,272	8,100	2,000	7,198	2,383	3,759	1,347	31,059
1996	6,041	1,063	4,844	10,884	977	2,410	5,196	3,261	789	8,475	5,176	3,126	8,300	3,995	4,753	1,639	35,464
1997	7,383	347	8,273	8,630	1,303	5,245	6,705	3,783	4,684	10,993	8,111	3,909	11,748	3,574	5,651	2,367	46,354
1998	8,280	1,198	14,193	8,634	1,061	5,044	6,356	4,306	1,491	15,936	8,460	2,808	13,232	3,440	3,965	2,720	50,562
1999	12,147	2,292	24,782	10,829	1,466	12,400	8,486	4,063	322	19,817	17,283	3,107	24,972	4,341	5,876	1,392	76,787
2000	12,269	2,870	21,713	14,997	2,742	12,153	12,804	5,644	3,028	24,814	13,692	4,417	28,808	5,502	6,298	4,690	88,221
2001	14,743	4,059	27,226	21,447	3,203	6,268	12,636	7,444	641	33,262	15,710	5,728	28,063	8,486	5,846	571	97,666
2002	17,290	3,400	33,983	21,693	4,202	7,869	16,028	7,532	93	29,315	20,578	8,565	32,218	12,932	7,640	841	112,090
2003	14,185	2,995	42,485	15,025	2,924	11,334	14,143	6,073	194	23,389	32,604	5,538	27,951	11,089	8,474	313	109,357
2004	11,348	6,444	129,265	18,371	341	6,452	1,380	1,322	6,193	53,928	74,509	3,257	30,493	1,380	3,233	14,316	181,116
2005	8,666	5,787	168,474	24,958	0	11,677	2,159	1,343	3,853	75,040	89,459	28,683	22,199	1,621	143	9,772	226,916
2006	6,888	4,030	274,197	18,003	0	6,132	2,438	2,093	23,487	95,815	180,852	7,688	28,618	2,458	3,981	17,857	337,268
	Percent of Total																
1989	20.2	21.7	9.8	20.4	(s)	12.4	9.3	5.5	0.5	11.2	30.0	8.4	31.1	8.8	6.1	4.3	100.0
1990	31.4	18.0	3.4	22.4	(s)	8.1	7.7	7.3	1.7	12.3	44.0	7.2	20.4	7.0	6.0	3.1	100.0
1991	23.7	22.2	5.7	24.1	.4	8.8	10.2	4.9	.1	24.3	22.4	5.5	26.4	10.4	8.5	2.5	100.0
1992	23.9	16.5	7.9	27.2	.4	5.3	10.3	5.2	3.4	26.7	15.3	6.8	27.5	10.7	10.0	3.1	100.0
1993	18.4	4.5	5.2	27.5	3.2	9.7	20.2	10.9	.4	25.0	19.6	6.3	25.5	12.2	7.2	4.1	100.0
1994	21.4	12.4	8.8	35.5	.3	7.1	8.2	5.4	1.0	25.4	20.8	8.1	26.3	8.3	9.1	2.0	100.0
1995	16.6	3.3	14.8	26.5	2.5	10.3	13.5	8.8	3.8	20.2	26.1	6.4	23.2	7.7	12.1	4.3	100.0
1996	17.0	3.0	13.7	30.7	2.8	6.8	14.7	9.2	2.2	23.9	14.6	8.8	23.4	11.3	13.4	4.6	100.0
1997	15.9	.7	17.8	18.6	2.8	11.3	14.5	8.2	10.1	23.7	17.5	8.4	25.3	7.7	12.2	5.1	100.0
1998	16.4	2.4	28.1	17.1	2.1	10.0	12.6	8.5	2.9	31.5	16.7	5.6	26.2	6.8	7.8	5.4	100.0
1999	15.8	3.0	32.3	14.1	1.9	16.1	11.1	5.3	.4	25.8	22.5	4.0	32.5	5.7	7.7	1.8	100.0
2000	13.9	3.3	24.6	17.0	3.1	13.8	14.5	6.4	3.4	28.1	15.5	5.0	32.7	6.2	7.1	5.3	100.0
2001	15.1	4.2	27.9	22.0	3.3	6.4	12.9	7.6	.7	34.1	16.1	5.9	28.7	8.7	6.0	.6	100.0
2002	15.4	3.0	30.3	19.4	3.7	7.0	14.3	6.7	.1	26.2	18.4	7.6	28.7	11.5	6.8	.8	100.0
2003	13.0	2.7	38.8	13.7	2.7	10.4	12.9	5.6	.2	21.4	29.8	5.1	25.6	10.1	7.7	.3	100.0
2004	6.3	3.6	71.4	10.1	.2	3.6	.8	.7	3.4	29.8	41.1	1.8	16.8	.8	1.8	7.9	100.0
2005	3.8	2.6	74.2	11.0	.0	5.1	1.0	.6	1.7	33.1	39.4	12.6	9.8	.7	.1	4.3	100.0
2006	2.0	1.2	81.3	5.3	.0	1.8	.7	.6	7.0	28.4	53.6	2.3	8.5	.7	1.2	5.3	100.0

[1] Grid-interactive means connection to the electrical distribution system; remote means electricity, for general use, that does not interact with the electrical distribution system, such as at an isolated residential site or mobile home. The other end uses in this table also include electricity generation but only for the specific use cited.

[2] "Original Equipment Manufacturers" are non-photovoltaic manufacturers that combine photovoltaic technology into existing or newly developed product lines.

[3] Represents such applications as cooking food, desalinization, and distilling.

[4] Includes all independent power producers.

[5] Shipments for specialty purposes such as research.

[6] See "Peak Kilowatt" in Glossary.

(s)=Less than 0.05 percent.

Notes: • Data are for domestic and export shipments, and may include imports that subsequently were shipped to domestic or foreign customers. • Totals may not equal sum of components due to independent rounding.

Web Page: For related information, see http://www.eia.doe.gov/fuelrenewable.html.

Sources: • 1989-1992—Energy Information Administration (EIA), *Solar Collector Manufacturing Activity*, annual reports. • 1993-2001—EIA, *Renewable Energy Annual*, annual reports. • 2002 forward—EIA, *Solar Thermal and Photovoltaic Collector Manufacturing Activities*, annual reports.

Renewable Energy

Note. Renewable Energy Production and Consumption. In Table 10.1, renewable energy consumption consists of: conventional hydroelectricity net generation (converted to Btu using the fossil-fueled plants heat rate); geothermal electricity net generation (converted to Btu using the geothermal plants heat rate), and geothermal heat pump and geothermal direct use energy; solar thermal and photovoltaic electricity net generation (converted to Btu using the fossil-fueled plants heat rate), and solar thermal direct use energy; wind electricity net generation (converted to Btu using the fossil-fueled plants heat rate); wood and wood-derived fuels consumption; biomass waste (municipal solid waste from biogenic sources, landfill gas, sludge waste, agricultural byproducts, and other biomass) consumption; fuel ethanol and biodiesel consumption; and losses and co-products from the production of fuel ethanol and biodiesel. Production is assumed to equal consumption for all renewable energy sources except biofuels (biofuels production comprises biomass inputs to the production of fuel ethanol and biodiesel).

Table 10.2a Sources

Residential Sector, Geothermal: Oregon Institute of Technology, Geo-Heat Center.
Residential Sector, Solar/PV: Energy Information Administration (EIA), Office of Coal, Nuclear, Electric and Alternate Fuels (CNEAF), estimates based on Form EIA-63A, "Annual Solar Thermal Collector Manufacturers Survey," and Form EIA-63B, "Annual Photovoltaic Module/Cell Manufacturers Survey."
Residential Sector, Wood: • 1949–1979: EIA, *Estimates of U.S. Wood Energy Consumption from 1949 to 1981*, Table A2. • 1980 forward: EIA, Form EIA-457, "Residential Energy Consumption Survey"; and EIA, CNEAF, estimates based on Form EIA-457 and regional heating degree-day data.
Commercial Sector, Hydroelectric Power: EIA, *Annual Energy Review (AER) 2007*, Tables 8.2d and A6.
Commercial Sector, Geothermal: Oregon Institute of Technology, Geo-Heat Center.
Commercial Sector, Wood: • 1949–1979: EIA, *Estimates of U.S. Wood Energy Consumption from 1949 to 1981*, Table A2. • 1980–1983: EIA, *Estimates of U.S. Wood Energy Consumption 1980-1983*, Table ES1. • 1984: EIA, CNEAF, estimate. • 1985–1988: Values interpolated. • 1989 forward: EIA, *AER 2007*, Table 8.7c; and EIA, CNEAF, estimates based on Form EIA-871, "Commercial Buildings Energy Consumption Survey."
Commercial Sector, Biomass Waste: EIA, *AER 2007*, Table 8.7c.
Commercial Sector, Ethanol: EIA, *AER 2007*, Tables 5.11, 5.13a, and 10.3. Calculated as commercial sector motor gasoline consumption (Table 5.13a) divided by total motor gasoline product supplied (Table 5.11), and then multiplied by fuel ethanol consumption (Table 10.3).

Table 10.2b Sources

Industrial Sector, Hydroelectric Power: • 1949–1988: EIA, *AER 2007*, Tables 8.1 and A6. • 1989 forward: EIA, *AER 2007*, Tables 8.2d and A6.
Industrial Sector, Geothermal: Oregon Institute of Technology, Geo-Heat Center.
Industrial Sector, Wood: • 1949–1979: EIA, *Estimates of U.S. Wood Energy Consumption from 1949 to 1981*, Table A2. • 1980–1983: EIA, *Estimates of U.S. Wood Energy Consumption 1980-1983*, Table ES1. • 1984: EIA, *Estimates of U.S. Biofuels Consumption 1990*, Table 1. • 1985 and 1986: Values interpolated. • 1987: EIA, *Estimates of Biofuels Consumption in the United States During 1987*, Table 2. • 1988: Value interpolated. • 1989 forward: EIA, *AER 2007*, Table 8.7c; and EIA, CNEAF, estimates based on Form EIA-846, "Manufacturing Energy Consumption Survey."
Industrial Sector, Biomass Waste: • 1981: EIA, *Estimates of U.S. Biofuels Consumption 1990*, Table 8, total waste consumption minus electric power sector waste consumption (see *AER 2007*, Table 10.2c). • 1982 and 1983: EIA, CNEAF, estimates for total waste consumption minus electric power sector waste consumption (see *AER 2007*, Table 10.2c). • 1984: EIA, *Estimates of U.S. Biofuels Consumption 1990*, Table 8, total waste consumption minus electric power sector waste consumption (see *AER 2007*, Table 10.2c). • 1985 and 1986: Values interpolated. • 1987: EIA, *Estimates of U.S. Biofuels Consumption 1990*, Table 8, total waste consumption minus electric power sector waste consumption (see *AER 2007*, Table 10.2c). • 1988: Value interpolated. • 1989 forward: EIA, *AER 2007*, Table 8.7c; and EIA, CNEAF, estimates based on information presented in Government Advisory Associates, *Resource Recovery Yearbook* and *Methane Recovery Yearbook*, and information provided by the U.S. Environmental Protection Agency, Landfill Methane Outreach Program.
Industrial Sector, Ethanol: EIA, *AER 2007*, Tables 5.11, 5.13b, and 10.3. Calculated as industrial sector motor gasoline consumption (Table 5.13b) divided by total motor gasoline product supplied (Table 5.11), and then multiplied by fuel ethanol consumption (Table 10.3).
Industrial Sector, Losses and Co-products: EIA, *AER 2007*, Table 10.3.
Transportation Sector, Ethanol: EIA, *AER 2007*, Tables 5.11, 5.13c, and 10.3. Calculated as transportation sector motor gasoline consumption (Table 5.13c) divided by total motor gasoline product supplied (Table 5.11), and then multiplied by fuel ethanol consumption (Table 10.3).
Transportation Sector, Biodiesel: EIA, *AER 2007*, Table 10.3.

11 International Energy

Drilling rig, Gansu Province, People's Republic of China. Source: U.S. Department of Energy.

Figure 11.1 World Primary Energy Production by Source

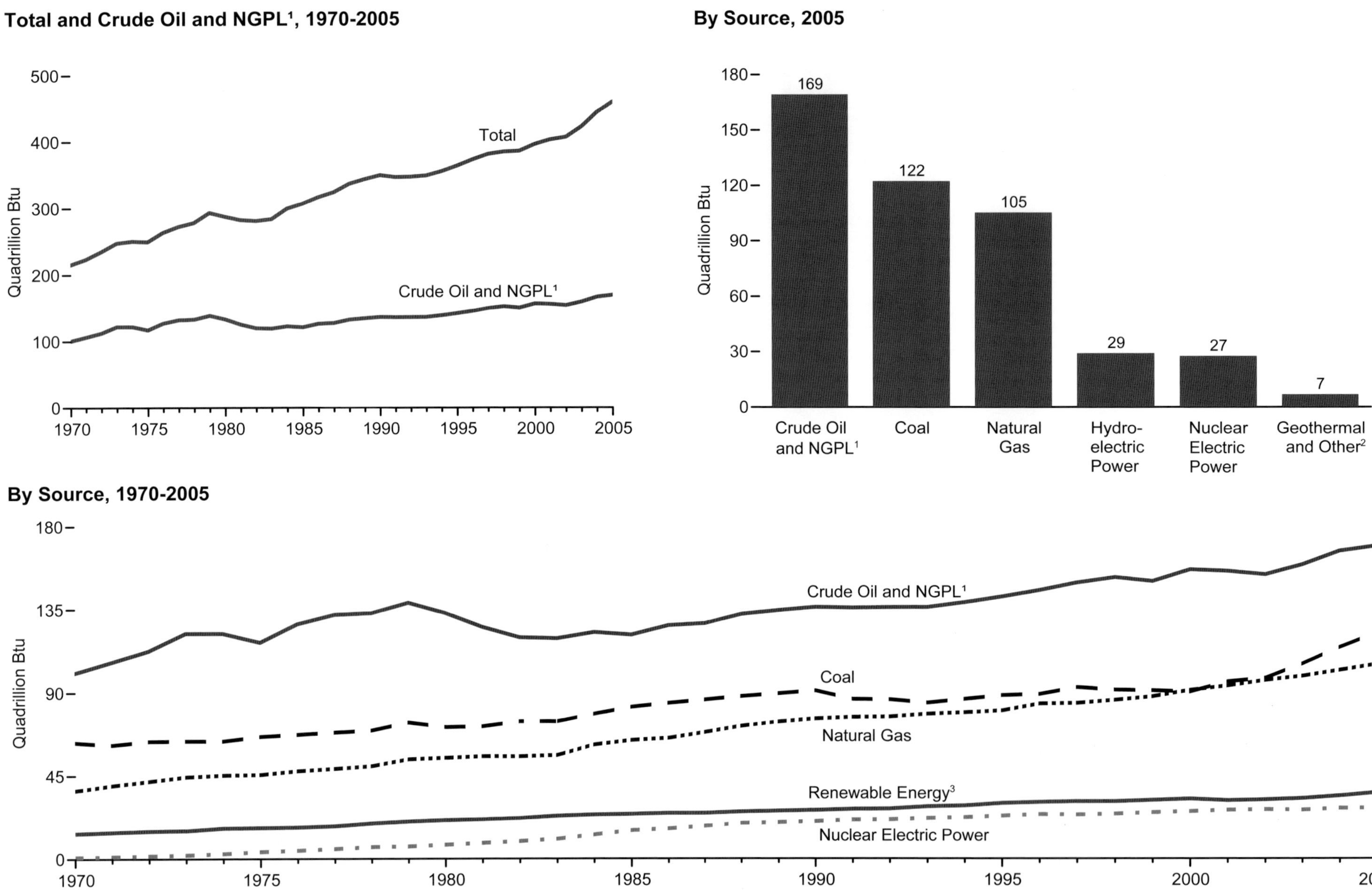

[1] Natural gas plant liquids.

[2] Net electricity generation from wood, waste, solar, and wind. Data for United States also include other renewable energy.

[3] Net electricity generation from hydroelectric power, geothermal, wood, waste, solar, and wind. Data for the United States also include other renewable energy.

Notes: • Crude oil includes lease condensate. • Because vertical scales differ, graphs should not be compared.

Source: Table 11.1.

Table 11.1 World Primary Energy Production by Source, 1970-2005

(Quadrillion Btu)

Year	Coal	Natural Gas [1]	Crude Oil [2]	Natural Gas Plant Liquids	Nuclear Electric Power [3]	Hydroelectric Power [3]	Geothermal [3] and Other [4]	Total
1970	62.96	37.09	97.09	3.61	0.90	12.15	1.59	215.39
1971	61.72	39.80	102.70	3.85	1.23	12.74	1.61	223.64
1972	63.65	42.08	108.52	4.09	1.66	13.31	1.68	234.99
1973	63.87	44.44	117.88	4.23	2.15	13.52	1.73	247.83
1974	63.79	45.35	117.82	4.22	2.86	14.84	1.76	250.64
1975	66.20	45.67	113.08	4.12	3.85	15.03	1.74	249.69
1976	67.32	47.62	122.92	4.24	4.52	15.08	1.97	263.67
1977	68.46	48.85	127.75	4.40	5.41	15.56	2.11	272.54
1978	69.56	50.26	128.51	4.55	6.42	16.80	2.32	278.41
1979	73.83	53.93	133.87	4.87	6.69	17.69	2.48	293.36
1980	R71.31	54.73	128.04	5.10	7.58	17.90	2.94	R287.59
1981	R71.72	55.56	120.11	5.37	8.53	18.26	3.10	R282.65
1982	R74.37	55.49	114.45	5.35	9.51	18.71	R3.29	R281.18
1983	R74.37	56.12	113.97	5.36	10.72	19.69	R3.59	R283.82
1984	R78.48	61.78	116.88	5.73	12.99	20.19	R3.73	R299.79
1985	R82.29	64.22	115.37	5.83	15.30	20.42	R3.82	R307.26
1986	R84.36	65.32	120.18	6.15	16.25	20.89	R3.83	R316.98
1987	R86.13	68.48	121.07	6.35	17.64	20.90	R3.85	R324.43
1988	R88.02	71.80	125.84	6.65	19.23	21.48	R4.02	R337.04
1989	R89.55	74.24	127.83	6.69	19.74	21.53	R4.39	R343.97
1990	R91.02	R75.90	129.35	6.87	20.36	22.35	R3.98	R349.83
1991	R86.38	R76.72	128.73	7.12	21.18	22.83	R4.08	R347.04
1992	R86.21	R76.92	128.93	R7.17	21.28	22.71	R4.35	R347.58
1993	R84.42	R78.43	128.72	R7.47	22.01	23.94	R4.38	R349.36
1994	R86.46	R79.19	R130.69	R8.10	22.41	24.15	R4.57	R355.58
1995	R88.52	R80.26	R133.43	R8.40	23.26	25.34	R4.72	R363.93
1996	R89.14	R84.01	R136.73	R8.59	24.11	25.79	R4.87	R373.24
1997	R92.82	R84.31	R140.63	R8.79	23.88	26.07	R4.98	R381.49
1998	R91.46	R85.97	R143.24	R9.02	24.32	R26.05	R4.98	R385.03
1999	R91.09	R87.87	R140.95	R9.26	25.09	R26.55	R5.17	R385.99
2000	R90.41	R91.32	R146.83	R9.63	25.66	R26.99	R5.43	R396.26
2001	R95.90	R93.66	R145.57	R10.10	26.39	R26.36	R5.21	R403.19
2002	R97.81	R96.67	R143.54	R10.28	26.68	R26.42	R5.54	R406.94
2003	R105.49	R98.87	R148.44	R10.74	26.45	R26.79	R5.92	R422.69
2004	R114.39	R102.03	R155.42	R11.10	R27.43	R27.65	R6.43	R444.45
2005P	122.25	105.33	157.80	11.47	27.47	29.00	6.81	460.14

[1] Dry production.
[2] Includes lease condensate.
[3] Net generation, i.e., gross generation less plant use.
[4] Includes net electricity generation from wood, waste, solar, and wind. Data for the United States also include other renewable energy.

R=Revised. P=Preliminary.

Notes: • Data in this table do not include recent updates for the United States (see Table 1.2) or for other countries (see the Energy Information Administration's "International Energy Annual 2006"). • See Note 1, "World Primary Energy Production," at end of section. • Totals may not equal sum of components due to independent rounding.

Web Page: For related information, see http://www.eia.doe.gov/international.

Sources: • 1970-1979—Energy Information Administration (EIA), International Energy Database. • 1980 forward—EIA, "International Energy Annual 2005" (June-October 2007), Tables F1-F9.

Figure 11.2 World Primary Energy Production by Region and Country

World and Top Producing Countries, 1996-2005

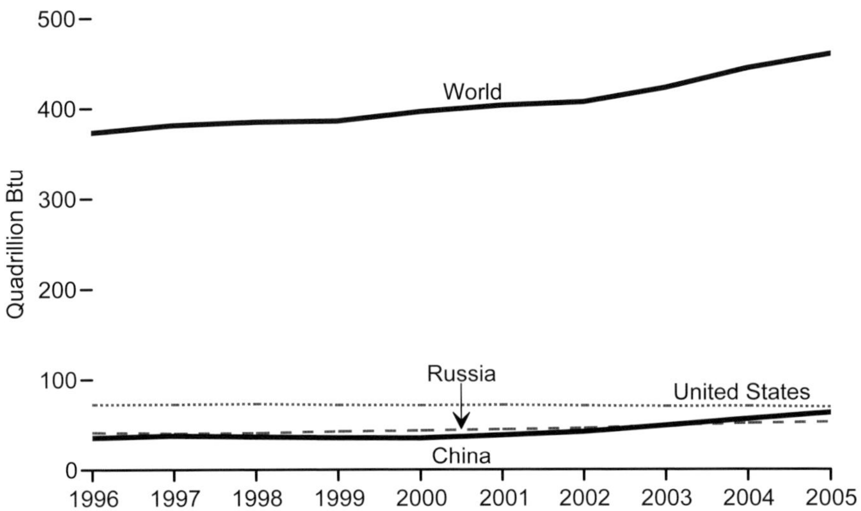

World Areas, 1996-2005

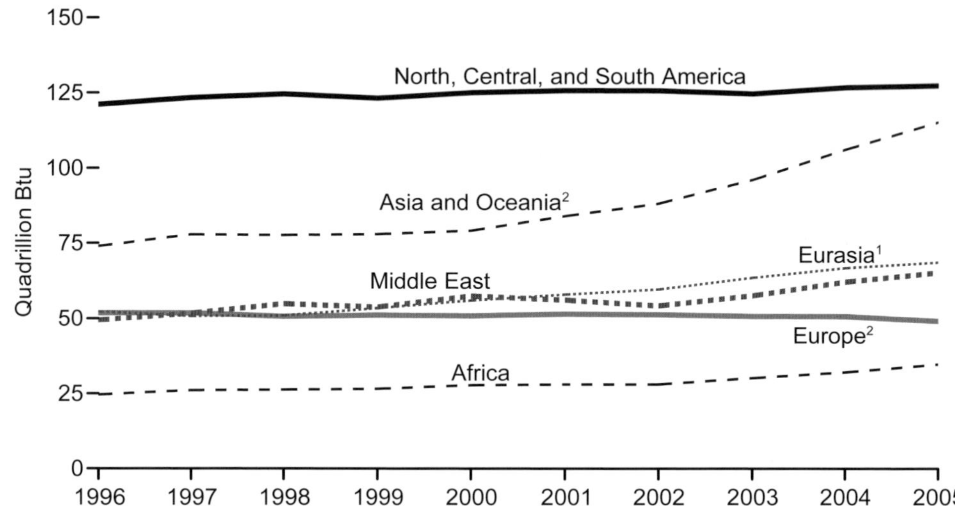

Top Producing Countries, 2005

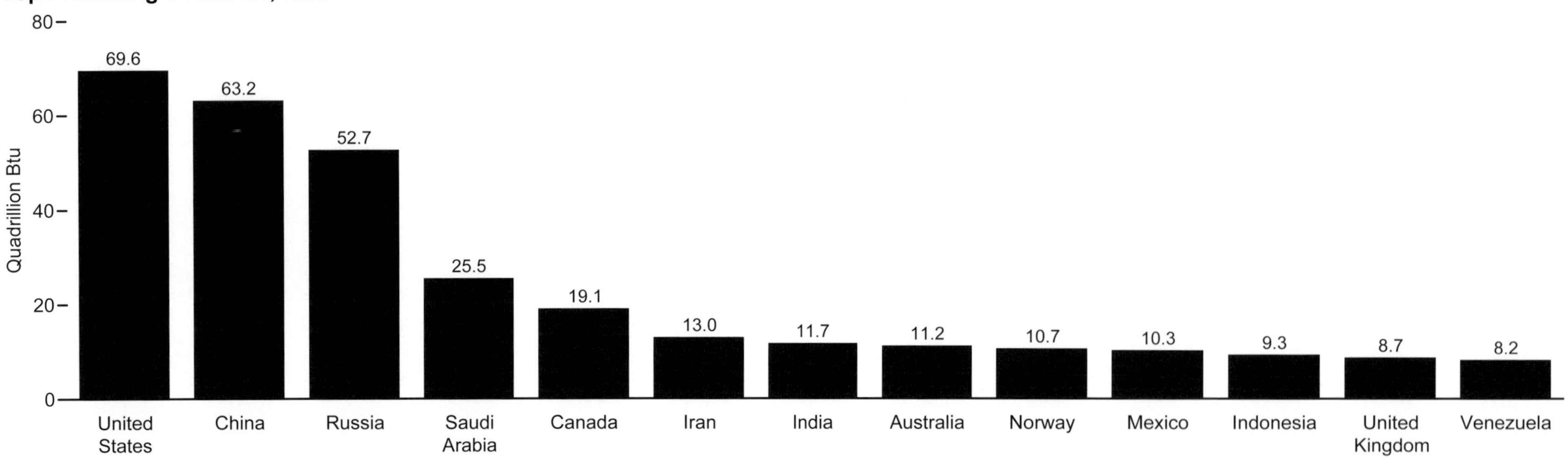

[1] Includes only countries that were part of the former U.S.S.R. See "U.S.S.R." in Glossary.
[2] Excludes countries that were part of the former U.S.S.R. See "U.S.S.R." in Glossary.

Note: Because vertical scales differ, graphs should not be compared.
Source: Table 11.2.

Table 11.2 World Primary Energy Production by Region, 1996-2005

(Quadrillion Btu)

Region and Country	1996	1997	1998	1999	2000	2001	2002	2003	2004	2005 P
North, Central, and South America	**R121.08**	**R123.27**	**R124.52**	**R123.12**	**R125.00**	**R125.67**	**R125.65**	**R124.62**	**R126.70**	**127.35**
Brazil	R4.83	R5.06	R5.48	R5.91	R6.40	R6.20	R6.75	R7.10	R7.19	7.71
Canada	17.22	17.48	17.43	17.68	18.12	R18.25	R18.57	R18.50	R18.87	19.09
Mexico	8.75	9.06	9.31	9.06	9.34	9.54	10.81	10.08	R10.28	10.26
United States	R72.64	R72.63	R73.04	R71.91	R71.49	R71.89	R70.94	R70.27	70.39	69.64
Venezuela	8.62	9.48	9.45	8.54	9.37	9.23	8.16	7.32	R8.06	8.23
Other	9.02	9.56	9.82	10.04	R10.27	R10.57	R10.43	R11.35	R11.90	12.42
Europe [1]	**R51.86**	**R51.77**	**R50.70**	**R51.09**	**R50.89**	**R51.49**	**R51.27**	**R50.74**	**R50.70**	**49.13**
France	R5.03	R4.92	R4.78	R4.92	R5.02	R5.14	R5.13	R5.15	R5.17	5.10
Germany	5.49	5.57	5.26	5.31	5.32	5.28	5.30	R5.29	R5.41	5.30
Netherlands	3.25	2.88	2.77	2.56	2.47	2.63	R2.62	2.53	2.94	2.71
Norway	9.26	9.59	9.33	9.53	10.27	10.28	10.69	10.64	R10.75	10.66
Poland	3.83	R3.83	3.36	R3.50	3.06	3.08	R3.08	R3.08	R3.06	2.98
United Kingdom	11.56	11.33	11.53	11.89	11.09	11.14	10.99	R10.62	R9.55	8.73
Other	R13.44	R13.65	R13.68	R13.38	R13.66	R13.94	R13.47	R13.43	R13.83	13.65
Eurasia [2]	**R52.27**	**R50.79**	**R51.02**	**R53.45**	**R56.00**	**R57.94**	**R59.67**	**R63.53**	**R66.76**	**68.60**
Kazakhstan	R2.41	R2.47	R2.43	R2.63	R3.32	R3.69	R3.98	R4.41	R5.08	5.48
Russia	R41.68	R40.63	R40.95	R42.64	R43.60	R44.77	R46.13	R49.11	R51.60	52.72
Ukraine	R3.03	3.01	R3.03	R3.08	R3.07	3.08	R3.07	R3.22	R3.22	3.21
Other	5.14	4.68	4.61	5.10	6.01	6.41	6.49	6.80	R6.86	7.20
Middle East	**49.43**	**51.72**	**54.88**	**53.80**	**57.48**	**56.16**	**R54.24**	**R57.58**	**R62.15**	**65.22**
Iran	9.65	9.84	9.90	10.00	10.40	10.67	10.45	11.36	R12.06	13.01
Iraq	1.39	2.60	4.71	5.47	5.62	5.22	4.42	2.84	4.38	4.11
Kuwait	4.94	4.85	5.02	4.60	5.04	4.81	4.58	5.14	5.71	6.12
Saudi Arabia	20.82	21.24	21.42	20.18	21.59	20.95	20.27	23.05	24.16	25.51
United Arab Emirates	6.34	6.50	6.61	6.25	6.77	6.59	6.50	7.13	7.42	7.59
Other	6.30	6.69	7.24	7.29	8.06	R7.92	R8.02	R8.05	R8.42	8.89
Africa	**R24.64**	**R26.09**	**R26.27**	**R26.57**	**R27.78**	**R28.01**	**R28.04**	**R30.15**	**R32.05**	**34.66**
Algeria	5.28	5.63	5.75	6.03	6.29	6.26	6.30	7.00	7.14	7.70
Libya	3.28	3.39	3.26	3.07	3.30	3.21	3.11	3.30	3.61	4.00
Nigeria	4.56	4.85	4.90	4.89	5.18	5.45	5.16	5.71	5.90	6.55
South Africa	4.86	5.44	5.52	5.43	5.58	5.62	5.52	5.91	6.06	6.05
Other	R6.66	R6.79	R6.83	R7.15	R7.44	R7.47	R7.94	R8.22	R9.33	10.37
Asia and Oceania [1]	**R73.97**	**R77.84**	**R77.64**	**R77.96**	**R79.11**	**R83.92**	**R88.07**	**R96.08**	**R106.10**	**115.16**
Australia	R7.59	R8.32	R8.67	R8.86	9.68	R10.26	10.51	10.35	10.56	11.23
China	R35.78	R37.97	R36.67	R35.71	R35.34	R38.48	R42.22	R49.02	R56.38	63.23
India	8.75	9.17	9.37	9.58	9.83	10.29	R10.04	R10.60	R11.24	11.73
Indonesia	7.42	7.41	7.56	8.02	7.87	8.09	8.32	8.55	R8.93	9.32
Japan	4.23	4.48	R4.59	R4.37	4.41	4.38	R4.07	R3.68	R4.16	4.10
Malaysia	2.84	3.01	3.14	3.16	3.21	3.31	3.44	3.84	R4.09	3.90
Other	R7.36	7.48	R7.65	R8.26	R8.78	R9.10	R9.49	R10.06	R10.74	11.66
World	**R373.24**	**R381.49**	**R385.03**	**R385.99**	**R396.26**	**R403.19**	**R406.94**	**R422.69**	**R444.45**	**460.14**

[1] Excludes countries that were part of the former U.S.S.R. See "U.S.S.R." in Glossary.

[2] Includes only countries that were part of the former U.S.S.R. See "U.S.S.R." in Glossary.

R=Revised. P=Preliminary.

Notes: • Data in this table do not include recent updates for the United States (see Table 1.2) or for other countries (see the Energy Information Administration's "International Energy Annual 2006"). • See Note 1, "World Primary Energy Production," at end of section. • World primary energy production includes production of crude oil (including lease condensate), natural gas plant liquids, dry natural gas, and coal; and net electricity generation from nuclear electric power, hydroelectric power, wood, waste, geothermal, solar, and wind. Data for the United States also include other renewable energy. • Totals may not equal sum of components due to independent rounding.

Web Page: For related information, see http://www.eia.doe.gov/international.

Source: Energy Information Administration, "International Energy Annual 2005" (June-October 2007), Table F1.

Figure 11.3 World Primary Energy Consumption

World and Top Consuming Countries, 1996-2005

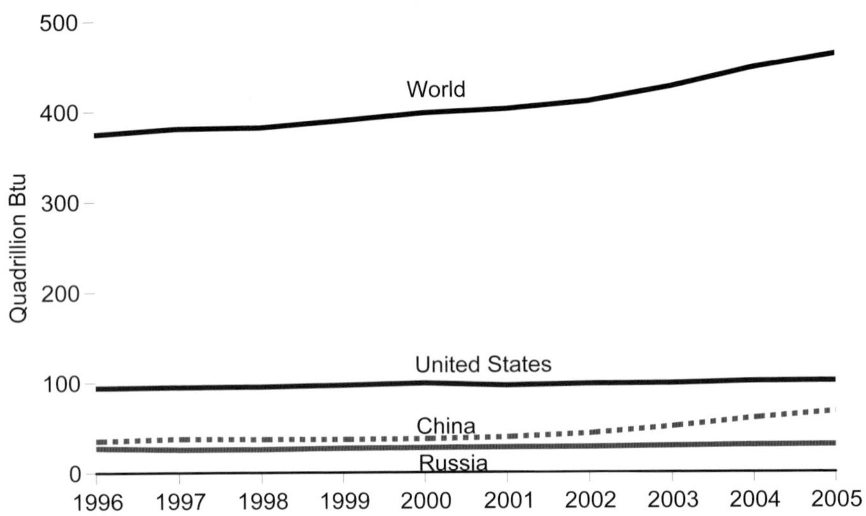

U.S. Share of World, 2005

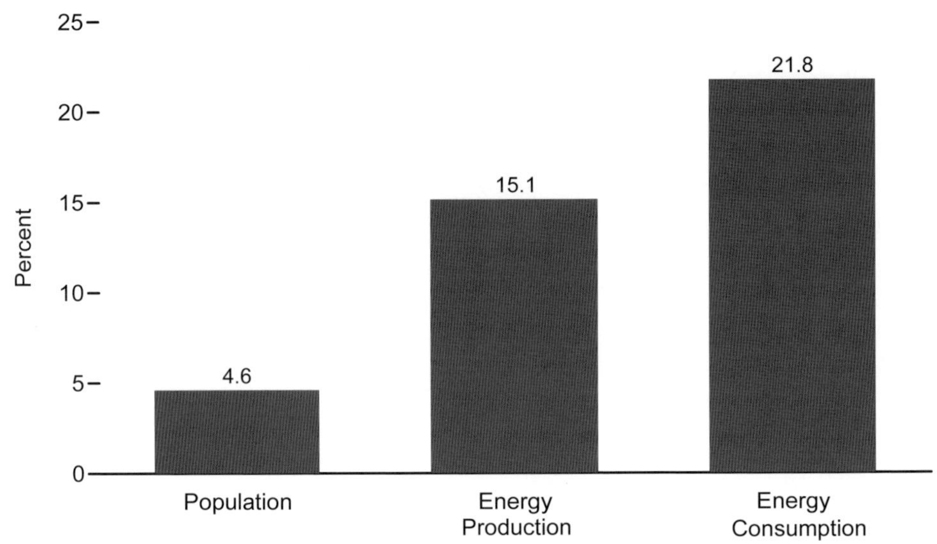

Regional Consumption Shares, 2005

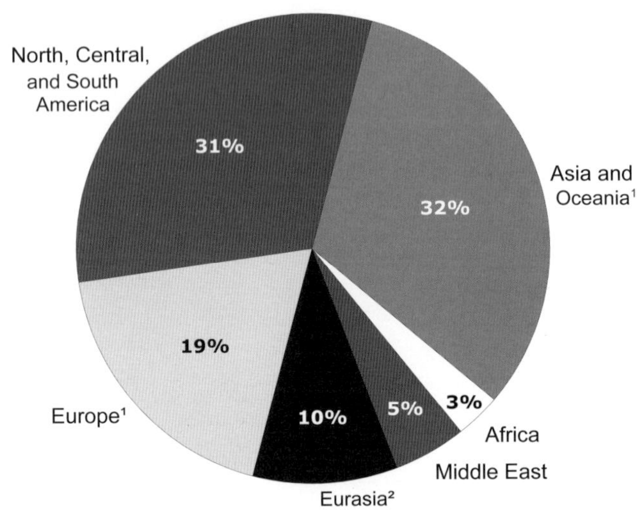

Production and Consumption by Region, 2005

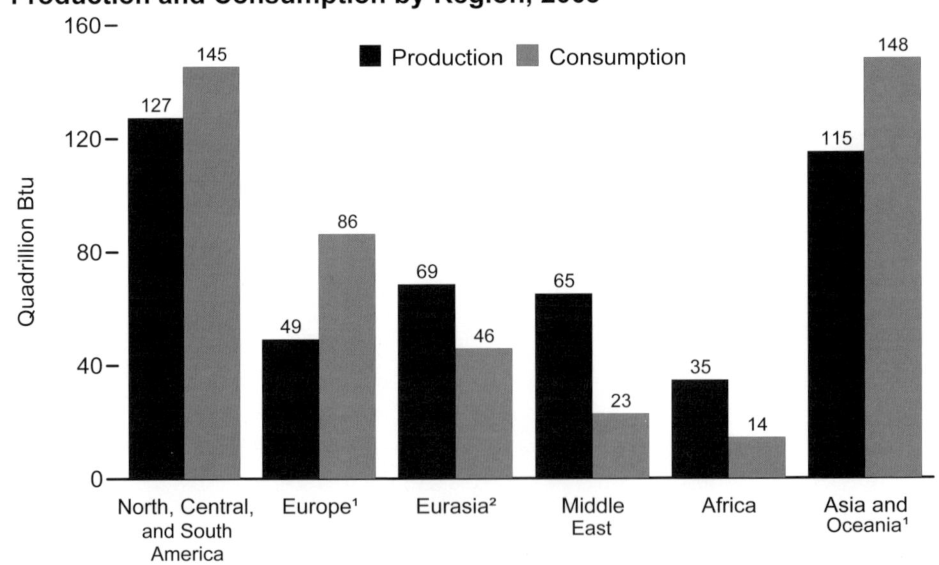

[1] Excludes countries that were part of the former U.S.S.R. See "U.S.S.R." in Glossary.
[2] Includes only countries that were part of the former U.S.S.R. See "U.S.S.R." in Glossary.

Note: Because vertical scales differ, graphs should not be compared.
Sources: Tables 11.2, 11.3, and D1.

Table 11.3 World Primary Energy Consumption by Region, 1996-2005

(Quadrillion Btu)

Region and Country	1996	1997	1998	1999	2000	2001	2002	2003	2004	2005 P
North, Central, and South America	**R130.77**	**R132.56**	**R133.63**	**R136.10**	**R139.10**	**R136.75**	**R138.66**	**R140.09**	**R143.42**	**145.30**
Argentina	2.39	R2.47	R2.58	R2.61	R2.67	2.61	2.47	R2.67	R2.78	2.93
Brazil	R7.45	R7.86	R8.11	8.29	R8.57	R8.49	8.60	8.71	R9.04	9.33
Canada	R12.54	R12.66	R12.36	R12.94	R12.93	R12.97	R13.33	R13.74	R14.03	14.31
Mexico	5.53	5.68	5.96	6.04	6.32	6.26	R6.30	R6.48	R6.53	6.88
United States	R94.17	R94.77	R95.18	R96.82	98.98	R96.33	R97.86	R98.21	R100.35	100.69
Venezuela	R2.58	2.66	2.85	2.73	2.77	3.03	2.93	2.72	R2.93	3.14
Other	R6.10	R6.47	6.59	6.68	R6.87	R7.06	R7.17	R7.57	R7.76	8.02
Europe [1]	**R79.48**	**R79.74**	**R80.32**	**R80.36**	**R81.42**	**R82.68**	**R82.41**	**R84.23**	**R85.78**	**86.29**
Belgium	2.53	2.60	2.65	2.60	2.69	2.67	2.64	2.72	R2.70	2.58
France	R10.42	R10.36	R10.58	R10.71	R10.85	R11.08	R11.00	11.12	R11.39	11.43
Germany	14.40	R14.36	R14.34	R14.13	14.26	14.62	R14.33	R14.60	R14.79	14.51
Italy	7.11	7.22	R7.43	7.56	7.63	7.68	7.70	7.98	R8.05	8.07
Netherlands	3.73	3.70	R3.70	R3.69	3.79	3.93	3.94	4.00	R4.11	4.24
Poland	R4.14	R4.08	R3.85	R3.98	R3.62	R3.45	R3.44	R3.60	R3.70	3.66
Spain	R4.48	R4.76	R4.99	R5.26	R5.62	R5.87	R5.95	R6.26	R6.47	6.59
Sweden	R2.20	2.25	R2.33	2.28	2.20	2.35	2.22	2.15	R2.28	2.34
Turkey	2.75	2.93	3.00	2.91	3.16	2.89	3.15	R3.32	R3.51	3.74
United Kingdom	10.05	R9.75	R9.75	R9.77	R9.71	9.82	R9.70	R9.90	R9.98	10.01
Other	R17.67	R17.74	R17.72	R17.48	R17.87	R18.30	R18.34	R18.58	R18.78	19.12
Eurasia [2]	**R41.53**	**R39.24**	**R38.97**	**R40.06**	**R40.79**	**R41.20**	**R41.81**	**R43.60**	**R44.77**	**45.82**
Russia	R27.71	R26.04	R26.16	R27.25	R27.71	R27.97	R28.18	R29.05	R29.94	30.29
Ukraine	6.34	6.07	R5.85	5.76	5.75	R5.64	R5.82	R6.28	R6.26	6.21
Uzbekistan	R1.88	R1.88	1.84	R1.86	1.94	2.03	R2.08	R2.10	R2.22	2.16
Other	5.60	5.26	R5.11	R5.19	R5.39	R5.56	R5.72	R6.17	R6.36	7.16
Middle East	**R14.57**	**R15.61**	**R16.27**	**R16.62**	**17.32**	**R17.95**	**R18.98**	**R19.76**	**R20.89**	**22.85**
Iran	R3.96	4.43	4.58	4.83	5.01	R5.39	R5.89	R6.18	R6.39	7.26
Saudi Arabia	4.10	R4.37	4.54	R4.60	R4.85	R5.14	R5.38	R5.76	R6.21	6.66
Other	R6.50	R6.81	R7.15	R7.18	R7.46	R7.42	R7.71	R7.82	R8.29	8.94
Africa	**R10.93**	**R11.40**	**R11.29**	**R11.61**	**R12.02**	**12.59**	**R12.72**	**R13.31**	**R13.96**	**14.43**
Egypt	R1.73	1.79	1.85	R1.92	R2.00	R2.22	2.27	R2.43	R2.59	2.75
South Africa	R4.17	R4.56	4.35	R4.46	R4.59	R4.66	R4.54	R4.88	R5.21	5.04
Other	R5.04	5.05	R5.09	R5.23	R5.43	R5.71	R5.90	R5.99	R6.16	6.64
Asia and Oceania [1]	**R97.66**	**R102.54**	**R101.65**	**R105.04**	**R107.48**	**R111.12**	**R116.14**	**R125.65**	**R138.78**	**148.10**
Australia	4.22	4.56	R4.59	4.82	R4.85	R5.02	R5.13	R5.13	R5.20	5.49
China	R35.66	R37.91	R37.32	R37.23	R37.49	R39.38	R43.28	R50.72	R59.91	67.09
India	R11.04	R11.64	R12.17	R12.99	R13.46	R13.94	R13.84	R14.29	R15.52	16.20
Indonesia	3.52	R3.66	3.56	R3.91	R4.06	4.46	R4.64	4.70	R5.12	5.36
Japan	R21.08	R21.60	R21.24	R21.76	R22.28	R22.10	R21.91	R22.04	R22.72	22.57
Malaysia	1.64	1.67	1.69	1.74	1.87	2.11	R2.18	R2.42	R2.66	2.55
South Korea	R6.80	R7.41	R6.83	7.55	R7.89	R8.10	8.42	R8.72	R8.98	9.28
Taiwan	3.06	R3.21	R3.40	R3.55	3.77	R3.86	R4.02	R4.21	R4.36	4.50
Thailand	2.44	R2.60	R2.44	2.50	2.58	2.70	2.94	R3.22	R3.46	3.63
Other	R8.18	R8.31	R8.43	R8.99	R9.22	R9.45	R9.76	R10.18	R10.84	11.43
World	**R374.93**	**R381.10**	**R382.13**	**R389.79**	**R398.13**	**R402.29**	**R410.71**	**R426.64**	**R447.60**	**462.80**

[1] Excludes countries that were part of the former U.S.S.R. See "U.S.S.R." in Glossary.

[2] Includes only countries that were part of the former U.S.S.R. See "U.S.S.R." in Glossary.

R=Revised. P=Preliminary.

Notes: • Data in this table do not include recent updates for the United States (see Table 1.3) or for other countries (see the Energy Information Administration's "International Energy Annual 2006"). • World primary energy consumption includes consumption of petroleum products (including natural gas plant liquids, and crude oil burned as fuel), dry natural gas, and coal (including net imports of coal coke); and the consumption of net electricity generated from nuclear electric power, hydroelectric power, wood, waste, geothermal, solar, and wind. It also includes, for the United States, the consumption of renewable energy by the end-use sectors. • Totals may not equal sum of components due to independent rounding.

Web Page: For related information, see http://www.eia.doe.gov/international.

Source: Energy Information Administration, "International Energy Annual 2005" (June-October 2007), Table E1.

Figure 11.4 World Crude Oil and Natural Gas Reserves, January 1, 2007

Crude Oil Reserves: *Oil and Gas Journal*

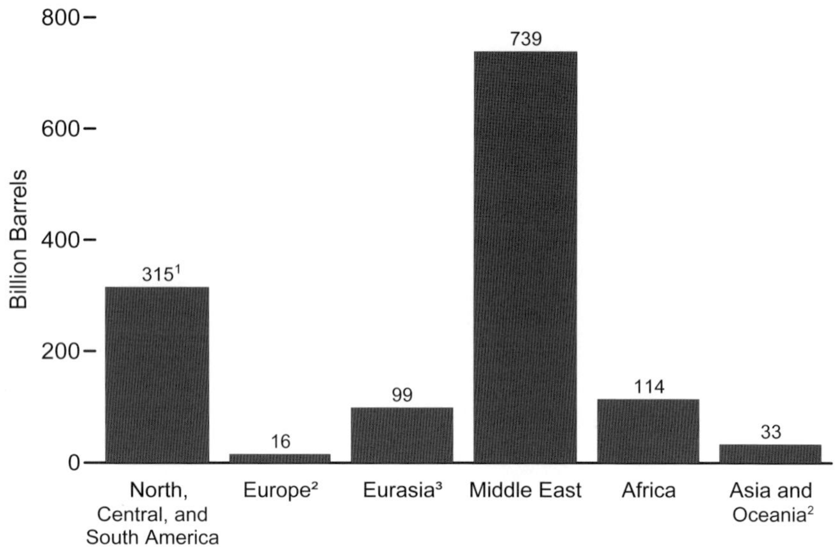

Crude Oil Reserves: *World Oil*

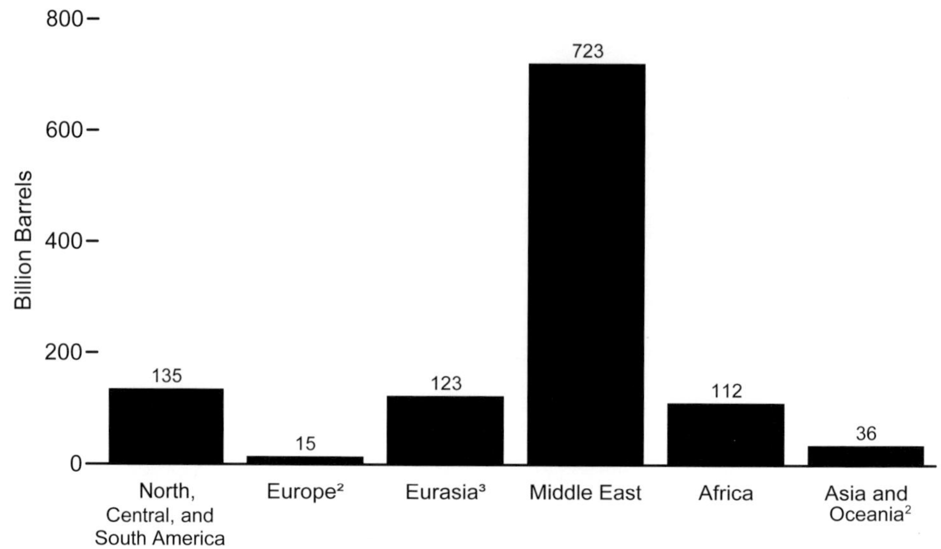

Natural Gas Reserves: *Oil and Gas Journal*

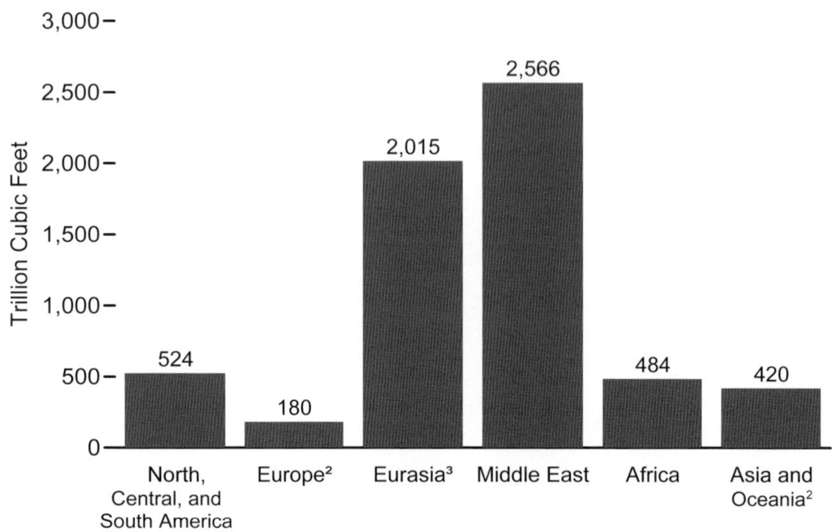

Natural Gas Reserves: *World Oil*

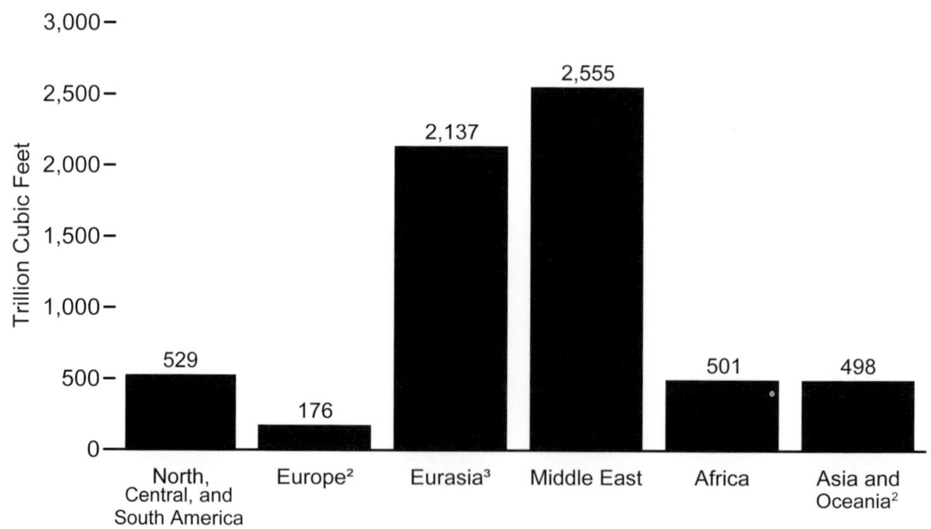

[1] Includes 174.0 billion barrels of bitumen in oil sands in Alberta, Canada.
[2] Excludes countries that were part of the former U.S.S.R. See "U.S.S.R." in Glossary.
[3] Includes only countries that were part of the former U.S.S.R. See "U.S.S.R." in Glossary.
Source: Table 11.4.

Table 11.4 World Crude Oil and Natural Gas Reserves, January 1, 2007

Region and Country	Crude Oil		Natural Gas	
	Oil & Gas Journal	*World Oil*	*Oil & Gas Journal*	*World Oil*
	Billion Barrels		Trillion Cubic Feet	
North America	**212.5**	**58.2**	**283.6**	**286.8**
Canada	[1]179.2	[2]25.6	57.9	56.8
Mexico	12.4	11.6	14.6	19.0
United States	21.0	21.0	211.1	211.1
Central and South America	**102.8**	**77.1**	**240.7**	**242.2**
Argentina	2.5	2.6	16.1	15.8
Bolivia	.4	.4	24.0	25.7
Brazil	11.8	12.3	10.8	12.3
Chile	.2	(s)	3.5	1.0
Colombia	1.5	1.4	4.0	6.7
Cuba	.1	.6	2.5	.6
Ecuador	4.5	4.9	NA	.3
Peru	.9	1.1	8.7	12.0
Trinidad and Tobago	.7	.6	18.8	16.7
Venezuela	80.0	52.9	152.4	151.1
Other [3]	.2	.2	(s)	(s)
Europe [4]	**15.8**	**14.5**	**180.3**	**175.7**
Austria	.1	.1	.6	.8
Croatia	.1	.1	1.1	1.0
Denmark	1.3	1.2	2.5	2.9
Germany	.4	.2	9.0	5.8
Hungary	(s)	.1	.3	2.0
Italy	.6	.4	5.8	3.7
Netherlands	.1	.2	50.0	50.8
Norway	7.8	7.1	82.3	81.3
Poland	.1	.3	5.8	4.9
Romania	.6	.5	2.2	4.5
Serbia	.1	NR	1.7	NR
United Kingdom	3.9	3.8	17.0	15.7
Other [3]	.8	.7	2.0	2.2
Eurasia [5]	**98.9**	**123.4**	**2,014.8**	**2,136.7**
Azerbaijan	7.0	NR	30.0	NR
Kazakhstan	30.0	NR	100.0	NR
Russia	60.0	74.4	1,680.0	1,688.8
Turkmenistan	.6	NR	100.0	NR
Ukraine	.4	NR	39.0	NR
Uzbekistan	.6	NR	65.0	NR
Other [3]	.3	48.9	.8	447.9
Middle East	**739.2**	**722.5**	**2,566.0**	**2,555.1**
Bahrain	.1	NR	3.3	NR
Iran	136.3	133.0	974.0	974.0
Iraq	115.0	125.1	112.0	90.0
Kuwait [6]	101.5	100.1	55.0	56.2
Oman	5.5	4.7	30.0	28.0
Qatar	15.2	20.4	910.5	905.5
Saudi Arabia [6]	262.3	262.3	240.0	252.5
Syria	2.5	2.9	8.5	12.8
United Arab Emirates	97.8	70.6	214.4	205.6
Yemen	3.0	2.8	16.9	17.0
Other [3]	(s)	.7	1.5	13.6
Africa	**114.1**	**111.7**	**484.4**	**500.7**
Algeria	12.3	11.9	161.7	161.2
Angola	8.0	9.3	2.0	4.1
Cameroon	.4	NR	3.9	NR
Congo (Brazzaville)	1.6	1.9	3.2	4.2
Egypt	3.7	3.6	58.5	67.4
Equatorial Guinea	1.1	1.8	1.3	3.4
Gabon	2.0	2.0	1.0	1.5
Libya	41.5	35.0	52.7	51.7
Mozambique	.0	.0	4.5	.0
Nigeria	36.2	37.2	181.9	184.5
Sudan	5.0	6.6	3.0	4.0
Tunisia	.4	.6	2.3	3.4
Other [3]	1.9	1.7	8.4	15.3
Asia and Oceania [4]	**33.4**	**36.0**	**419.5**	**497.9**
Australia	1.6	4.3	30.4	153.0
Bangladesh	(s)	NR	5.0	NR
Brunei	1.1	1.2	13.8	11.5
Burma	.1	.2	10.0	16.0
China	16.0	16.3	80.0	60.9
India	5.6	3.8	38.0	27.2
Indonesia	4.3	4.8	97.8	93.0
Japan	.1	NR	1.4	NR
Malaysia	3.0	2.8	75.0	60.0
New Zealand	.1	.1	.9	1.8
Pakistan	.3	.3	28.0	30.5
Papua New Guinea	.2	.2	12.2	13.7
Thailand	.3	.5	14.8	11.7
Vietnam	.6	1.3	6.8	8.3
Other [3]	.1	.2	5.5	10.4
World	**1,316.7**	**1,143.4**	**6,189.4**	**6,395.0**

[1] Comprises 5.2 billion barrels of conventional crude oil and condensate and 174.0 billion barrels of bitumen in Alberta's oil sands.
[2] Includes 20.7 billion barrels of oil sands and bitumen. Excludes another 153 billion barrels claimed by Canadian authorities.
[3] Includes data for those countries not separately reported.
[4] Excludes countries that were part of the former U.S.S.R. See "U.S.S.R." in Glossary.
[5] Includes only countries that were part of the former U.S.S.R. See "U.S.S.R." in Glossary.
[6] Data for Kuwait and Saudi Arabia include one-half of the reserves in the Neutral Zone between Kuwait and Saudi Arabia.

NA=Not available. NR=Not separately reported. (s)=Less than 0.05 billion barrels.
Notes: • All reserve figures except those for Eurasia and natural gas reserves in Canada are proved reserves recoverable with present technology and prices at the time of estimation. Eurasia and Canadian natural gas figures include proved and some probable reserves. • Totals may not equal sum of components due to independent rounding.
Web Page: For related information, see http://www.eia.doe.gov/international/oilreserves.html.
Source: Energy Information Administration, "International Energy Annual 2006" (June-July 2008), Table 8.1.

Figure 11.5 World Crude Oil Production

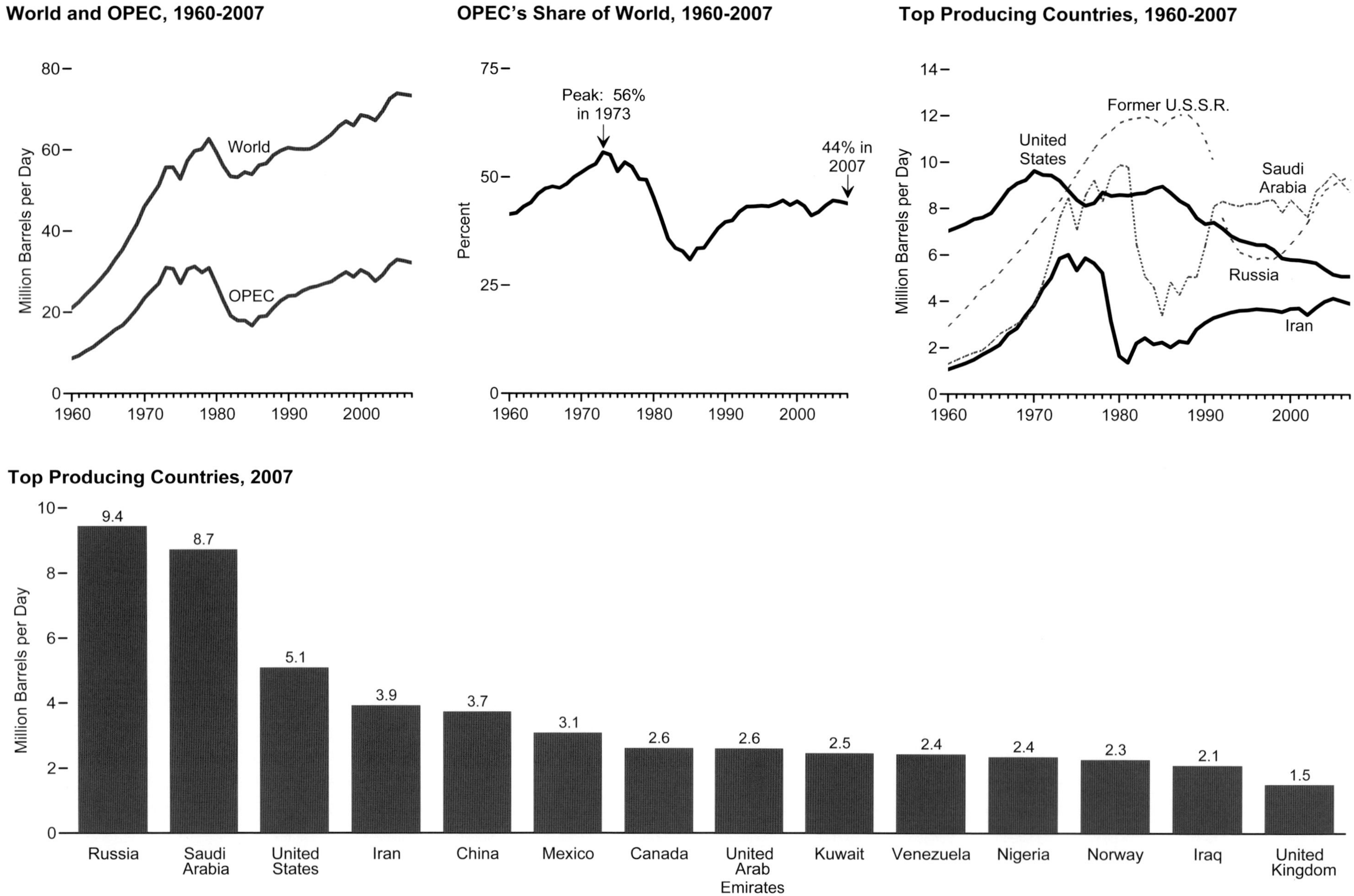

Notes: • OPEC=Organization of the Petroleum Exporting Countries. • Because vertical scales differ, graphs should not be compared.

Source: Table 11.5.

Table 11.5 World Crude Oil Production, 1960-2007

(Million Barrels per Day)

Year	Persian Gulf Nations [2]	Selected OPEC [1] Producers: Iran	Iraq	Kuwait [3]	Nigeria	Saudi Arabia [3]	United Arab Emirates	Venezuela	Total OPEC [4]	Selected Non-OPEC [1] Producers: Canada	China	Mexico	Norway	Former U.S.S.R.	Russia	United Kingdom	United States	Total Non-OPEC [4]	World
1960	5.27	1.07	0.97	1.69	0.02	1.31	0.00	2.85	R8.68	0.52	0.10	0.27	0.00	2.91	– –	(s)	7.04	R12.31	20.99
1961	5.65	1.20	1.01	1.74	.05	1.48	.00	2.92	R9.35	.61	.11	.29	.00	3.28	– –	(s)	7.18	R13.10	22.45
1962	6.19	1.33	1.01	1.96	.07	1.64	.01	3.20	R10.50	.67	.12	.31	.00	3.67	– –	(s)	7.33	R13.85	24.35
1963	6.82	1.49	1.16	2.10	.08	1.79	.05	3.25	11.51	.71	.13	.31	.00	4.07	– –	(s)	7.54	14.62	26.13
1964	7.61	1.71	1.26	2.30	.12	1.90	.19	3.39	12.98	.75	.18	.32	.00	4.60	– –	(s)	7.61	15.20	28.18
1965	8.37	1.91	1.32	2.36	.27	2.21	.28	3.47	R14.34	.81	.23	.32	.00	4.79	– –	(s)	7.80	R15.99	30.33
1966	9.32	2.13	1.39	2.48	.42	2.60	.36	3.37	R15.75	.88	.29	.33	.00	5.23	– –	(s)	8.30	R17.20	32.96
1967	9.91	2.60	1.23	2.50	.32	2.81	.38	3.54	R16.79	.96	.28	.36	.00	5.68	– –	(s)	8.81	R18.60	35.39
1968	10.91	2.84	1.50	2.61	.14	3.04	.50	3.60	R18.71	1.19	.30	.39	.00	6.08	– –	(s)	9.10	R19.93	38.63
1969	11.95	3.38	1.52	2.77	.54	3.22	.63	3.59	R20.85	1.13	.48	.46	.00	6.48	– –	(s)	9.24	R20.84	41.70
1970	13.39	3.83	1.55	2.99	1.08	3.80	.78	3.71	R23.42	1.26	.60	.49	.00	6.99	– –	(s)	9.64	R22.47	45.89
1971	15.77	4.54	1.69	3.20	1.53	4.77	1.06	3.55	R25.31	1.35	.78	.49	.01	7.48	– –	(s)	9.46	R23.21	48.52
1972	17.54	5.02	1.47	3.28	1.82	6.02	1.20	3.22	R27.11	1.53	.90	.51	.03	7.89	– –	(s)	9.44	R24.03	51.14
1973	20.67	5.86	2.02	3.02	2.05	7.60	1.53	3.37	R31.00	1.80	1.09	.47	.03	8.32	– –	(s)	9.21	R24.68	55.68
1974	21.28	6.02	1.97	2.55	2.26	8.48	1.68	2.98	R30.70	1.55	1.32	.57	.04	8.91	– –	(s)	8.77	R25.02	55.72
1975	18.93	5.35	2.26	2.08	1.78	7.08	1.66	2.35	R27.10	1.43	1.49	.71	.19	9.52	– –	.01	8.37	R25.73	52.83
1976	21.51	5.88	2.42	2.15	2.07	8.58	1.94	2.29	R30.62	1.31	1.67	.83	.28	10.06	– –	.25	8.13	R26.72	57.34
1977	21.73	5.66	2.35	1.97	2.09	9.25	2.00	2.24	R31.27	1.32	1.87	.98	.28	10.60	– –	.77	8.24	R28.44	59.71
1978	20.61	5.24	2.56	2.13	1.90	8.30	1.83	2.17	R29.80	1.32	2.08	1.21	.36	11.11	– –	1.08	8.71	R30.36	60.16
1979	21.07	3.17	3.48	2.50	2.30	9.53	1.83	2.36	R30.94	1.50	2.12	1.46	.40	11.38	– –	1.57	8.55	R31.73	62.67
1980	17.96	1.66	2.51	1.66	2.06	9.90	1.71	2.17	R26.96	1.44	2.11	1.94	.49	11.71	– –	1.62	8.60	R32.60	59.56
1981	15.25	1.38	1.00	1.13	1.43	9.82	1.47	2.10	R22.82	1.29	2.01	2.31	.47	11.85	– –	1.81	8.57	R33.23	56.05
1982	12.16	2.21	1.01	.82	1.30	6.48	1.25	1.90	R19.11	1.27	2.05	2.75	.49	11.91	– –	2.07	8.65	R34.34	53.45
1983	11.08	2.44	1.01	1.06	1.24	5.09	1.15	1.80	R17.91	1.36	2.12	2.69	.61	11.97	– –	2.29	8.69	R35.35	53.26
1984	10.78	2.17	1.21	1.16	1.39	4.66	1.15	1.80	R17.91	1.44	2.30	2.78	.71	11.86	– –	2.48	8.88	R36.59	54.50
1985	9.63	2.25	1.43	1.02	1.50	3.39	1.19	1.68	R16.69	1.47	2.51	2.75	.77	11.59	– –	2.53	8.97	R37.27	53.97
1986	11.70	2.04	1.69	1.42	1.47	4.87	1.33	1.79	R18.85	1.47	2.62	2.44	.84	11.90	– –	2.54	8.68	R37.35	56.20
1987	12.10	2.30	2.08	1.59	1.34	4.27	1.54	1.75	R19.05	1.54	2.69	2.55	.98	12.05	– –	2.41	8.35	R37.58	56.63
1988	13.46	2.24	2.69	1.49	1.45	5.09	1.57	1.90	R21.08	1.62	2.73	2.51	1.11	12.05	– –	2.23	8.14	R37.61	58.69
1989	14.84	2.81	2.90	1.78	1.72	5.06	1.86	1.91	R22.81	1.56	2.76	2.52	1.48	11.72	– –	1.80	7.61	R36.99	59.79
1990	15.28	3.09	2.04	1.18	1.81	6.41	2.12	2.14	R23.96	1.55	2.77	2.55	1.63	10.98	– –	1.82	7.36	R36.54	60.49
1991	14.74	3.31	.31	.19	1.89	8.12	2.39	2.38	R24.07	1.55	2.84	2.68	1.87	9.99	– –	1.80	7.42	R36.11	60.19
1992	15.97	3.43	.43	1.06	1.94	8.33	2.27	2.37	R25.25	1.61	2.85	2.67	2.13	– –	7.63	1.83	7.17	R34.87	60.12
1993	16.71	3.54	.51	1.85	1.96	8.20	2.16	2.45	R25.97	1.68	2.89	2.67	2.28	– –	6.73	1.92	6.85	R34.20	60.17
1994	16.96	3.62	.55	2.03	1.93	8.12	2.19	2.59	R26.41	1.75	2.94	2.69	2.57	– –	6.14	2.37	6.66	R34.69	R61.10
1995	17.21	3.64	.56	2.06	1.99	8.23	2.23	2.75	R27.04	1.81	2.99	2.62	2.77	– –	6.00	2.49	6.56	R35.34	R62.38
1996	17.37	3.69	.58	2.06	2.00	8.22	2.28	2.94	R27.57	1.84	3.13	2.86	3.09	– –	5.85	2.57	6.46	R36.19	R63.75
1997	18.10	3.66	1.16	2.01	2.13	8.36	2.32	3.28	R28.81	1.92	3.20	3.02	3.14	– –	5.92	2.52	6.45	R36.93	R65.74
1998	19.34	3.63	2.15	2.09	2.15	8.39	2.35	3.17	R29.88	1.98	3.20	3.07	3.01	– –	5.85	2.62	6.25	R37.08	R66.97
1999	18.67	3.56	2.51	1.90	2.13	7.83	2.17	2.83	R28.70	1.91	3.20	2.91	3.02	– –	6.08	2.68	5.88	R37.23	R65.92
2000	19.89	3.70	2.57	2.08	2.17	8.40	2.37	3.16	R30.41	1.98	3.25	3.01	3.22	– –	6.48	2.28	5.82	R38.09	R68.49
2001	19.10	3.72	2.39	2.00	2.26	8.03	2.21	3.01	R29.50	2.03	3.30	3.13	3.23	– –	6.92	2.28	5.80	R38.60	R68.10
2002	17.79	3.44	2.02	1.89	2.12	7.63	2.08	2.60	R27.64	2.17	3.39	3.18	3.13	– –	7.41	2.29	5.75	R39.53	R67.17
2003	19.06	3.74	1.31	2.14	2.28	8.78	2.35	2.34	R29.14	2.31	3.41	3.37	3.04	– –	8.13	2.09	5.68	R40.31	R69.45
2004	20.79	4.00	2.01	2.38	2.33	9.10	2.48	2.56	R31.50	2.40	3.49	3.38	2.95	– –	8.80	1.85	5.42	R41.01	R72.51
2005	21.50	4.14	1.88	2.53	2.63	9.55	2.54	2.56	R32.94	2.37	3.61	3.33	2.70	– –	9.04	1.65	5.18	R40.87	R73.81
2006	21.23	4.03	2.00	2.54	2.44	9.15	2.64	2.51	R32.61	2.53	R3.67	3.26	2.49	– –	9.25	1.49	R5.10	R40.93	R73.54
2007P	20.68	3.92	2.09	2.46	2.35	8.72	2.60	2.43	32.18	2.61	3.73	3.08	2.27	– –	9.44	1.50	5.10	41.09	73.27

[1] See "Organization of the Petroleum Exporting Countries (OPEC)" in Glossary. On this table, countries are classified as "OPEC" or "Non-OPEC" in all years based on their membership status in the most current year.

[2] Bahrain, Iran, Iraq, Kuwait, Qatar, Saudi Arabia, United Arab Emirates, and the Neutral Zone (between Kuwait and Saudi Arabia).

[3] Includes about one-half of the production in the Neutral Zone between Kuwait and Saudi Arabia.

[4] On this table, "Total OPEC" for all years includes Algeria, Angola, Ecuador, Indonesia, Iran, Iraq, Kuwait, Libya, Nigeria, Qatar, Saudi Arabia, United Arab Emirates, Venezuela, and the Neutral Zone (between Kuwait and Saudi Arabia). Data for all countries not included in "Total OPEC" are included in "Total Non-OPEC."

R=Revised. P=Preliminary. – – = Not applicable. (s)=Less than 0.005 million barrels per day.

Notes: • Data are for crude oil, including extra heavy crude oil, lease condensate, and liquids processed from Canadian oil sands; they exclude natural gas plant liquids. • Totals may not equal sum of components due to independent rounding.

Web Page: For related information, see http://www.eia.doe.gov/international.

Sources: **Selected OPEC Producers:** • 1960-1972—OPEC, *Annual Statistical Bulletin 1979.* • 1973-1979—Energy Information Administration (EIA), *International Energy Annual (IEA),* annual reports, and the International Energy Database. • 1980-2006—EIA, International Energy Database, April 24, 2008. • 2007—EIA, *Monthly Energy Review (MER)* (April 2008), Table 11.1a. **China:** • 1960-1972—Central Intelligence Agency, unpublished data. • 1973-1979—EIA, *IEA,* annual reports, and the International Energy Database. • 1980-2006—EIA, International Energy Database, April 24, 2008. • 2007—EIA, *MER* (April 2008), Table 11.1b. **Former U.S.S.R.:** • 1960-1969—U.S.S.R. Central Statistical Office, *Narodnoye Khozyaystvo SSSR* (National Economy USSR). • 1970-1979—EIA, *International Petroleum Monthly,* February 2001, Table 4.1c. • 1980-1991—EIA, International Energy Database, April 24, 2008. **Russia:** • 1992-2006—EIA, International Energy Database, April 24, 2008. • 2007—EIA, *MER* (April 2008), Table 11.1b. **United States:** Table 5.1. **All Other Data:** • 1960-1969—Bureau of Mines, *International Petroleum Annual, 1969.* • 1970-1972—EIA, *International Petroleum Annual, 1978.* • 1973-1979—EIA, *IEA,* annual reports, and the International Energy Database. • 1980-2006—EIA, International Energy Database, April 24 2008. • 2007—EIA, *MER* (April 2008), Tables 11.1a and 11.1b.

Figure 11.6 World Natural Gas Plant Liquids Production

Crude Oil and NGPL[1] Production, 1973-2007

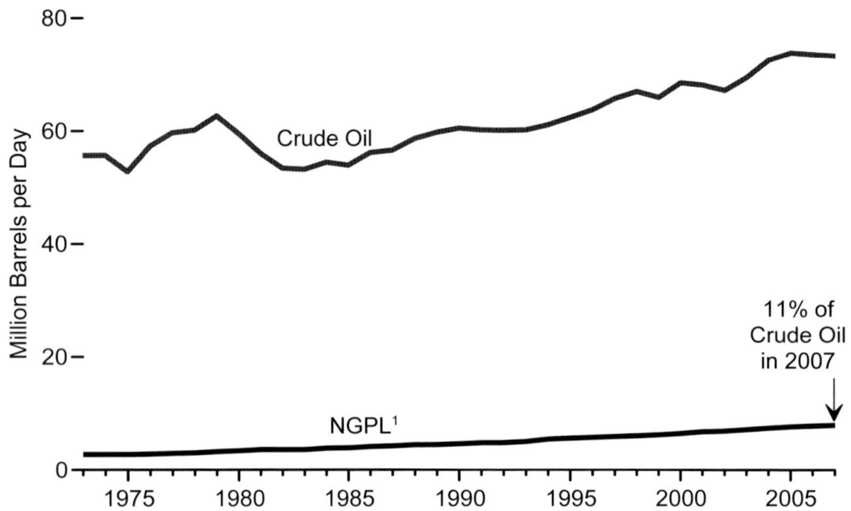

World and OPEC NGPL[1] Production, 1973-2007

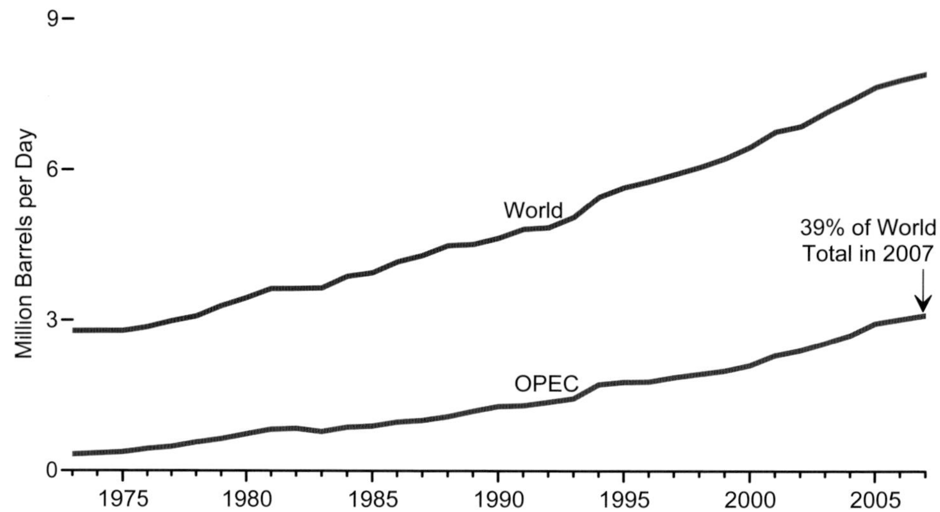

Top NGPL[1] Producing Countries, 2007

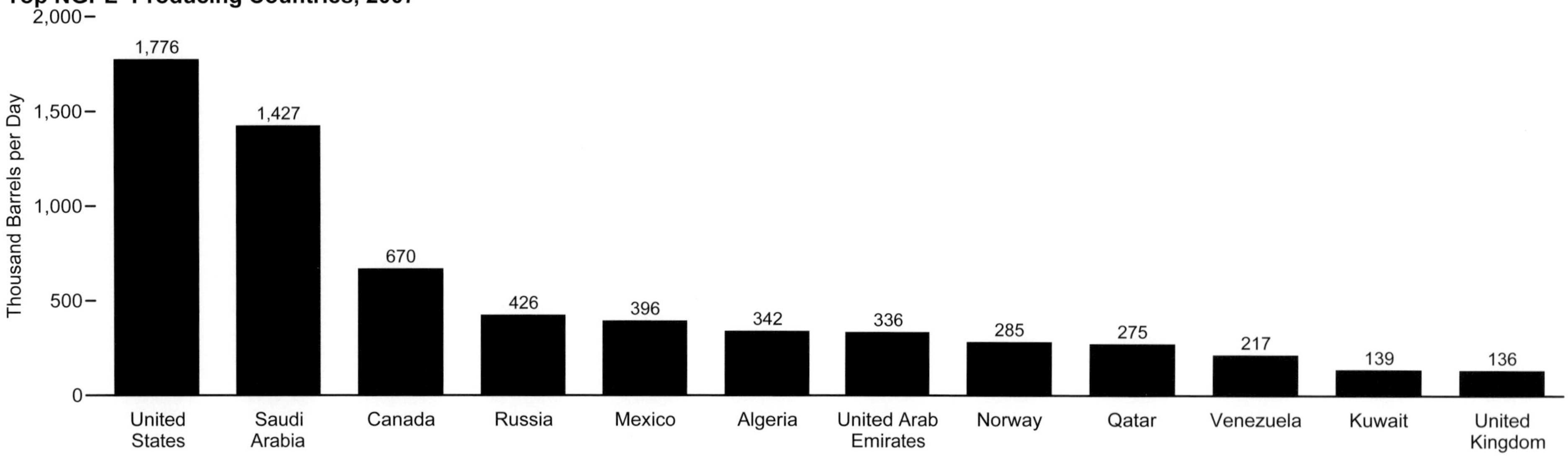

[1] Natural gas plant liquids.

Notes: • Crude oil includes lease condensate. • OPEC=Organization of the Petroleum Exporting Countries. • Because vertical scales differ, graphs should not be compared.

Sources: Tables 11.5 and 11.6.

Table 11.6 World Natural Gas Plant Liquids Production, 1973-2007

(Thousand Barrels per Day)

Year	Selected OPEC [1] Producers								
	Algeria	Indonesia	Iran	Kuwait [2]	Qatar	Saudi Arabia [2]	United Arab Emirates	Venezuela	Total OPEC [3]
1973	9	(s)	40	60	(s)	90	(s)	89	R325
1974	12	(s)	45	50	5	130	(s)	84	347
1975	20	(s)	45	50	10	140	(s)	76	R373
1976	24	10	45	50	10	185	(s)	77	442
1977	19	10	40	55	5	215	15	78	482
1978	25	30	45	75	5	250	30	61	R568
1979	30	40	10	95	10	303	30	69	R639
1980	36	70	9	95	10	369	35	60	R734
1981	49	95	9	60	24	433	60	55	R827
1982	58	80	9	40	30	430	90	60	R844
1983	56	94	8	55	25	330	120	57	R782
1984	105	75	10	67	28	355	130	57	R871
1985	120	44	10	54	30	375	160	63	R894
1986	120	30	15	75	22	385	185	97	R971
1987	140	30	20	95	24	418	145	94	R1,008
1988	120	30	20	100	30	499	130	98	R1,079
1989	130	72	27	105	24	503	130	108	R1,190
1990	130	77	35	65	40	620	135	114	R1,283
1991	140	76	50	0	50	680	146	117	R1,301
1992	140	75	50	34	55	713	144	113	R1,367
1993	145	78	55	53	55	704	146	143	R1,439
1994	140	80	55	85	50	951	150	146	R1,721
1995	145	76	60	95	55	961	160	149	R1,769
1996	150	80	60	85	50	968	160	150	R1,776
1997	160	85	70	109	70	982	160	143	R1,862
1998	155	87	75	115	85	1,020	170	145	R1,931
1999	190	87	75	115	111	1,010	160	170	R1,997
2000	230	90	75	115	133	1,008	200	175	R2,104
2001	250	82	80	120	150	1,051	290	200	R2,307
2002	270	80	86	125	160	1,095	300	202	R2,406
2003	280	78	95	124	201	1,220	310	163	R2,551
2004	292	73	105	130	250	1,310	279	180	R2,707
2005	295	68	102	130	265	1,460	300	206	2,943
2006	310	68	123	130	280	1,427	300	215	3,024
2007P	342	65	124	139	275	1,427	336	217	3,108

Year	Selected Non-OPEC [1] Producers										World
	Australia	Canada	Malaysia	Mexico	Norway	Former U.S.S.R.	Russia	United Kingdom	United States	Total Non-OPEC [3]	
1973	50	314	0	75	(s)	170	– –	5	1,738	R2,461	2,786
1974	50	314	0	80	(s)	190	– –	5	1,688	R2,442	2,790
1975	50	309	0	80	5	205	– –	15	1,633	2,419	2,791
1976	50	289	0	95	20	220	– –	15	1,604	2,425	2,867
1977	55	290	0	105	20	235	– –	30	1,618	R2,501	2,984
1978	60	281	0	115	35	255	– –	40	1,567	R2,512	3,080
1979	60	331	0	150	40	270	– –	45	1,584	R2,648	3,287
1980	60	331	0	193	42	285	– –	45	1,573	R2,712	3,446
1981	60	330	0	241	37	300	– –	50	1,609	R2,804	3,631
1982	52	318	0	255	39	315	– –	78	1,550	R2,788	3,632
1983	52	309	0	265	46	330	– –	111	1,559	R2,861	3,643
1984	54	336	10	257	45	340	– –	136	1,630	R3,007	3,878
1985	65	337	10	271	51	350	– –	145	1,609	R3,054	3,948
1986	60	328	9	352	66	440	– –	152	1,551	R3,192	4,163
1987	65	367	11	338	71	430	– –	162	1,595	R3,287	4,295
1988	67	381	11	370	83	450	– –	159	1,625	R3,410	4,489
1989	65	410	11	384	84	425	– –	140	1,546	R3,323	4,513
1990	63	426	12	428	86	425	– –	108	1,559	R3,357	4,640
1991	61	431	12	457	84	420	– –	141	1,659	R3,517	4,818
1992	56	460	13	454	85	– –	230	160	1,697	R3,480	R4,848
1993	55	506	17	459	95	– –	220	169	1,736	R3,618	R5,057
1994	56	529	17	461	123	– –	200	218	1,727	R3,741	R5,462
1995	52	581	20	447	137	– –	180	267	1,762	R3,885	R5,654
1996	62	596	20	423	141	– –	185	259	1,830	R3,998	R5,774
1997	71	636	50	388	139	– –	195	233	1,817	R4,052	R5,914
1998	70	651	90	424	127	– –	220	241	1,759	R4,131	R6,062
1999	72	653	85	439	120	– –	231	238	1,850	R4,241	R6,237
2000	70	699	65	438	124	– –	232	233	1,911	R4,361	R6,466
2001	74	709	70	433	188	– –	237	258	1,868	R4,460	R6,766
2002	80	698	75	408	203	– –	246	211	1,880	R4,471	R6,877
2003	79	724	80	418	222	– –	390	241	1,719	R4,602	R7,152
2004	83	658	80	442	234	– –	456	172	1,809	R4,686	R7,393
2005	92	645	94	426	271	– –	457	157	1,717	4,711	7,654
2006	88	685	90	427	286	– –	417	142	1,739	4,771	7,795
2007P	92	670	88	396	285	– –	426	136	1,776	4,808	7,916

[1] See "Organization of the Petroleum Exporting Countries (OPEC)" in Glossary. On this table, countries are classified as "OPEC" or "Non-OPEC" in all years based on their membership status in the most current year.

[2] Includes about one-half of the production in the Neutral Zone between Kuwait and Saudi Arabia.

[3] On this table, "Total OPEC" for all years includes Algeria, Angola, Ecuador, Indonesia, Iran, Iraq, Kuwait, Libya, Nigeria, Qatar, Saudi Arabia, United Arab Emirates, Venezuela, and the Neutral Zone (between Kuwait and Saudi Arabia). Data for all countries not included in "Total OPEC" are included in "Total Non-OPEC."

R=Revised. P=Preliminary. – – = Not applicable. (s)=Less than 500 barrels per day.

Note: Totals may not equal sum of components due to independent rounding.

Web Page: For related information, see http://www.eia.doe.gov/international.

Sources: • 1973-1979—Energy Information Administration (EIA), *International Energy Annual*, annual reports, and the International Energy Database. • 1980 forward—EIA, International Energy Database, April 11, 2008.

Figure 11.7 Crude Oil Prices by Selected Type

Selected Types, 1970-2008

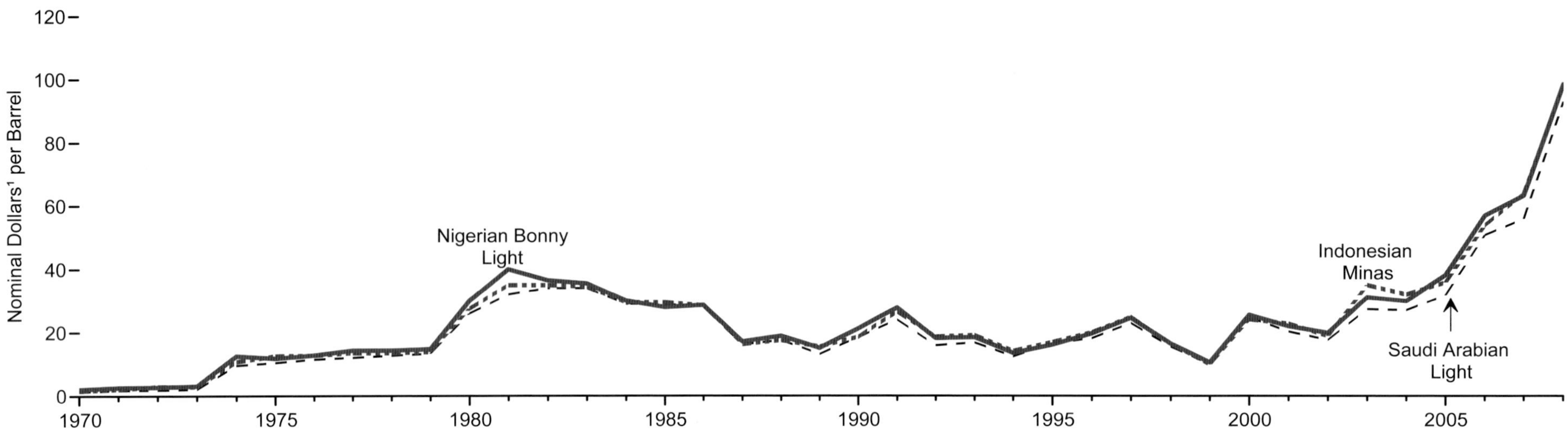

Selected Types, 2008

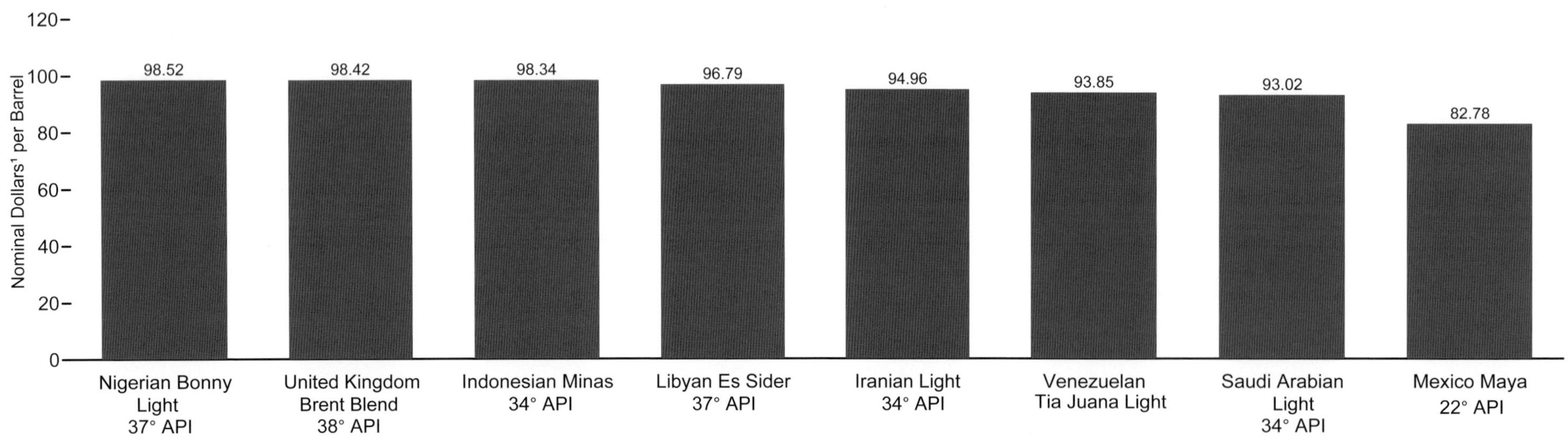

[1] See "Nominal Dollars" in Glossary.

Notes: • Prices are as of the Friday that is closest to January 1, except in 1987, when prices are as of the first Friday in February. See "API" and "API Gravity" in Glossary.

Source: Table 11.7.

Table 11.7 Crude Oil Prices by Selected Type, 1970-2008

(Nominal Dollars [1] per Barrel)

Year	Saudi Arabian Light-34° API	Iranian Light-34° API	Libyan [2] Es Sider-37° API	Nigerian [3] Bonny Light-37° API	Indonesian Minas-34° API	Venezuelan Tia Juana Light [4]	Mexico Maya-22° API	United Kingdom Brent Blend-38° API
1970	1.35	1.36	2.09	2.10	1.67	2.05	NA	NA
1971	1.75	1.76	2.80	2.65	2.18	2.45	NA	NA
1972	1.90	1.91	2.80	2.80	2.96	2.45	NA	NA
1973	2.10	2.11	3.10	3.10	2.96	2.60	NA	NA
1974	9.60	10.63	14.30	12.60	10.80	9.30	NA	NA
1975	10.46	10.67	11.98	11.80	12.60	11.00	NA	NA
1976	11.51	11.62	12.21	12.84	12.80	11.12	NA	NA
1977	12.09	12.81	13.74	14.33	13.55	12.72	NA	NA
1978	12.70	12.81	13.80	14.33	13.55	12.82	NA	NA
1979	13.34	13.45	14.52	14.80	13.90	13.36	15.45	15.70
1980	26.00	[5]30.37	34.50	29.97	27.50	25.20	28.00	26.02
1981	32.00	37.00	40.78	40.00	35.00	32.88	34.50	39.25
1982	34.00	34.20	36.50	36.50	35.00	32.88	26.50	36.60
1983	34.00	31.20	35.10	35.50	34.53	32.88	25.50	33.50
1984	29.00	28.00	30.15	30.00	29.53	27.88	25.00	30.00
1985	29.00	28.00	30.15	28.00	29.53	27.88	25.50	28.65
1986	28.00	28.05	30.15	28.65	28.53	28.05	21.93	26.00
1987	16.15	16.14	16.95	17.13	16.28	15.10	14.00	18.25
1988	17.52	15.55	18.52	18.92	17.56	17.62	11.10	18.00
1989	13.15	12.75	15.40	15.05	15.50	12.27	10.63	15.80
1990	18.40	18.20	20.40	21.20	18.55	24.69	17.05	21.00
1991	24.00	23.65	26.90	27.80	26.50	28.62	20.00	27.20
1992	15.90	15.50	17.20	18.20	18.65	19.67	10.75	17.75
1993	16.80	16.70	17.55	18.50	19.10	17.97	12.50	17.90
1994	12.40	12.40	12.55	13.50	14.15	12.97	9.01	13.15
1995	16.63	16.18	16.05	16.15	16.95	16.57	13.77	16.15
1996	18.20	17.73	19.20	19.70	20.05	18.52	15.79	19.37
1997	22.98	22.63	24.10	24.65	24.95	26.62	19.33	24.05
1998	15.50	14.93	16.72	16.50	16.50	15.93	10.81	15.89
1999	10.03	9.83	10.65	10.60	9.95	9.45	6.38	10.44
2000	24.78	24.63	25.85	25.55	24.15	24.85	20.20	25.10
2001	20.30	20.20	22.40	22.00	22.80	22.13	15.82	22.50
2002	17.68	18.90	19.63	19.88	18.89	17.78	14.30	21.20
2003	27.39	27.85	30.40	31.16	35.03	30.25	26.29	31.36
2004	27.08	28.67	29.47	29.97	32.10	30.10	24.37	29.73
2005	31.86	33.84	38.00	38.21	35.86	35.98	26.16	39.43
2006	50.86	52.56	55.89	56.97	53.95	52.52	42.93	57.25
2007	55.94	56.28	59.22	63.28	63.87	55.57	46.98	60.50
2008	93.02	94.96	96.79	98.52	98.34	93.85	82.78	98.42

[1] See "Nominal Dollars" in Glossary.
[2] For 1974 and 1975, prices are for crude oil with 40° API. For 1980, prices include $4.72 in retroactive charges and market premiums.
[3] Beginning in 1977, prices include 2 cents per barrel harbor dues.
[4] For 1970-1985, prices are for crude oil with 26° API. Beginning in 1986, prices are for crude oil with 31° API.
[5] For 1980, price includes $1.87 market premiums and credit charges.
NA=Not available.

Notes: • Prices are as of the Friday that is closest to January 1, except in 1987, when prices are as of the first Friday in February. • Prices are based on official government-selling prices, netback values, or spot market quotations. • Prices are usually free on board (f.o.b.) at the foreign port of lading. • See Tables 5.18, 5.19, and 5.21 for other types of crude oil prices for the United States, such as domestic first purchase prices, landed costs of crude oil imports, and refiner acquisition costs. • See "API" and "API Gravity" in Glossary.

Web Page: For related information, see http://www.eia.doe.gov/international.

Sources: • 1970-1978—Petroleum and Energy Intelligence Weekly, Inc., *Petroleum Intelligence Weekly*. • 1979 forward—Energy Information Administration, *Weekly Petroleum Status Report*.

Figure 11.8 Retail Motor Gasoline Prices in Selected Countries, 2007

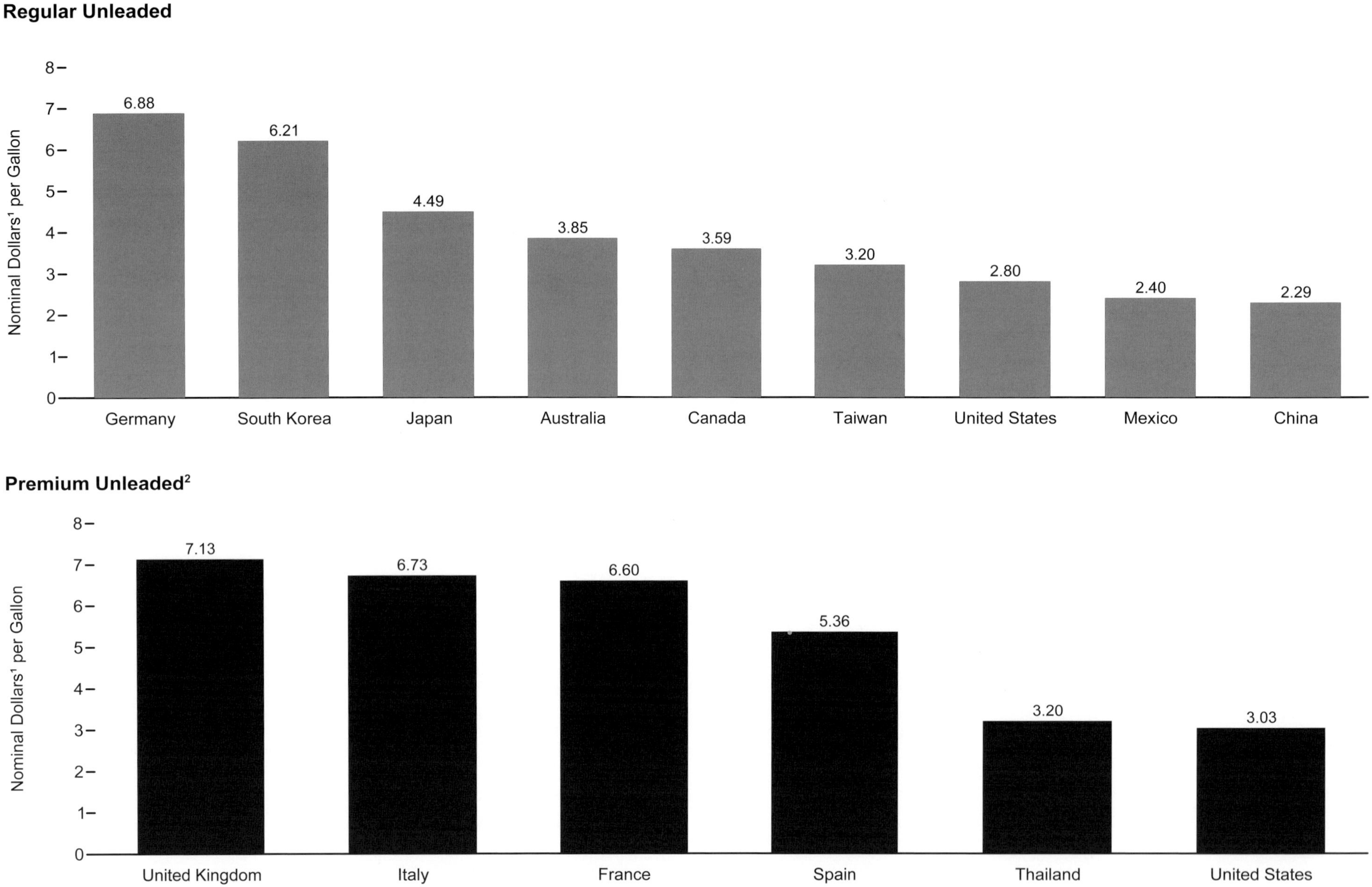

[1] See "Nominal Dollars" in Glossary.
[2] Research Octane Number (RON) of 95, except RON of 98 for the United States.

Source: Table 11.8.

Table 11.8 Retail Motor Gasoline Prices in Selected Countries, 1990-2007

(Nominal Dollars [1] per Gallon)

Year	Regular Unleaded									Premium Unleaded [2]						
	Australia	Canada	China	Germany	Japan	Mexico	South Korea	Taiwan	United States	France	Italy	South Africa	Spain	Thailand	United Kingdom	United States
1990	NA	1.87	NA	2.65	3.16	1.00	2.05	2.49	1.16	3.63	4.59	NA	NA	NA	2.82	1.35
1991	1.96	1.92	NA	2.90	3.46	1.30	2.49	2.39	1.14	3.45	4.50	NA	NA	NA	3.01	1.32
1992	1.89	1.73	NA	3.27	3.59	1.50	2.65	2.42	1.13	3.57	4.53	NA	3.50	1.35	3.06	1.32
1993	1.73	1.57	NA	3.07	4.02	1.56	2.88	2.27	1.11	3.41	3.68	NA	3.01	1.26	2.84	1.30
1994	1.84	1.45	NA	3.52	4.39	1.48	2.87	2.14	1.11	3.59	3.70	NA	2.99	1.21	2.99	1.31
1995	1.95	1.53	NA	3.96	4.43	1.11	2.94	2.23	1.15	4.26	4.00	NA	3.24	1.25	3.21	1.34
1996	2.12	1.61	NA	3.94	3.65	1.25	3.18	2.15	1.23	4.41	4.39	NA	3.32	1.49	3.34	1.41
1997	2.05	1.62	NA	3.53	3.27	1.47	3.34	2.23	1.23	4.00	4.07	1.72	3.01	1.27	3.83	1.42
1998	1.63	1.38	NA	3.34	2.83	1.49	3.04	1.86	1.06	3.87	3.84	1.51	2.80	1.09	4.06	1.25
1999	1.72	1.52	NA	3.42	3.27	1.79	3.80	1.86	1.17	3.85	3.87	1.55	2.82	1.22	4.29	1.36
2000	1.94	1.86	NA	3.45	3.65	2.01	4.18	2.15	1.51	3.80	3.77	1.78	2.86	1.38	4.58	1.69
2001	1.71	1.72	1.22	3.40	3.27	2.20	3.76	2.02	1.46	3.51	3.57	1.59	2.73	1.33	4.13	1.66
2002	1.76	1.70	1.21	3.67	3.15	2.24	3.84	1.93	1.36	3.62	3.74	1.41	2.90	1.35	4.16	1.56
2003	2.19	1.99	1.33	4.59	3.47	2.04	4.12	2.16	1.59	4.35	4.53	1.91	3.49	1.52	4.70	1.78
2004	2.72	2.37	1.48	5.24	3.93	2.03	4.51	2.46	1.88	4.99	5.30	2.58	4.09	1.76	5.56	2.07
2005	3.23	R2.89	1.70	5.66	4.28	2.22	5.28	R2.76	2.30	5.46	5.74	3.05	4.49	2.25	5.97	2.49
2006	3.54	3.26	2.11	6.03	4.47	2.31	R5.92	R3.05	2.59	5.88	6.10	R3.42	R4.84	R2.76	6.36	2.81
2007	3.85	3.59	2.29	6.88	4.49	2.40	6.21	3.20	2.80	6.60	6.73	NA	5.36	3.20	7.13	3.03

[1] See "Nominal Dollars" in Glossary.
[2] Research Octane Number (RON) of 95, except RON of 98 for the United States.
R=Revised. NA=Not available.

Notes: • Prices are those actually paid, i.e., net of rebates, and include transport costs and taxes which are not refundable. Prices in national currencies are converted to U.S. dollars using exchange rates published by the International Monetary Fund. • Prices for all countries, except the United States, have been converted from dollars per liter to dollars per gallon at 3.785412 liters per gallon. Comparisons between prices and price trends in different countries require care. They are of limited validity because of fluctuations in exchange rates, differences in product quality, marketing practices, market structures, and the extent to which the standard categories of sales are representative of total national sales for a given period.

Web Page: For related information, see http://www.eia.doe.gov/international.

Sources: • **United States:** Table 5.24. • **All Other Data:** International Energy Agency, Organization for Economic Cooperation and Development, *Energy Prices & Taxes, Quarterly Statistics, First Quarter 2008, Part II, Section D, and Part III, Section B.*

Figure 11.9 World Crude Oil Refining Capacity

World, OECD, and Non-OECD, 1970-2007

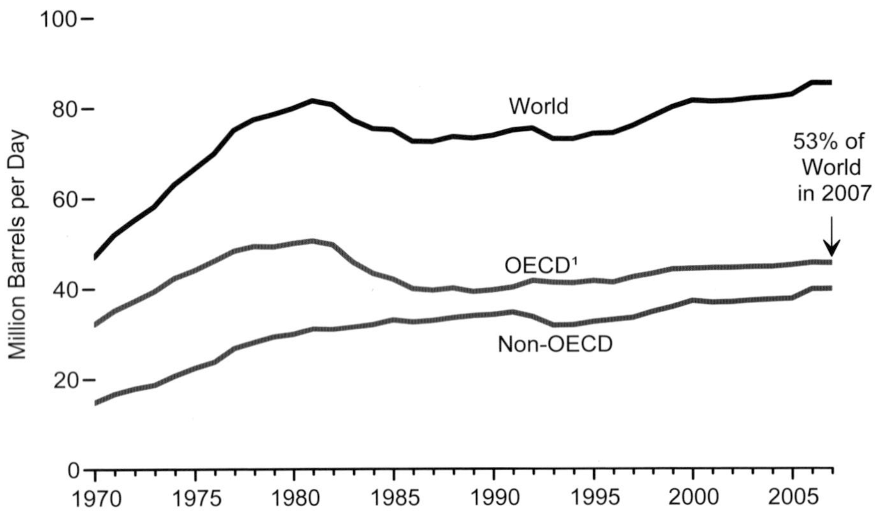

Top Refining Countries, 1970-2007

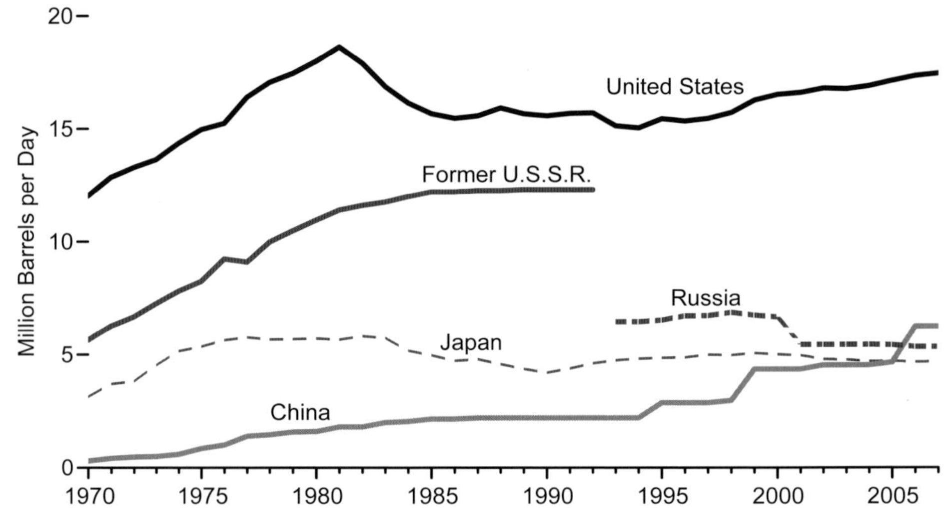

Top Refining Countries, 2007

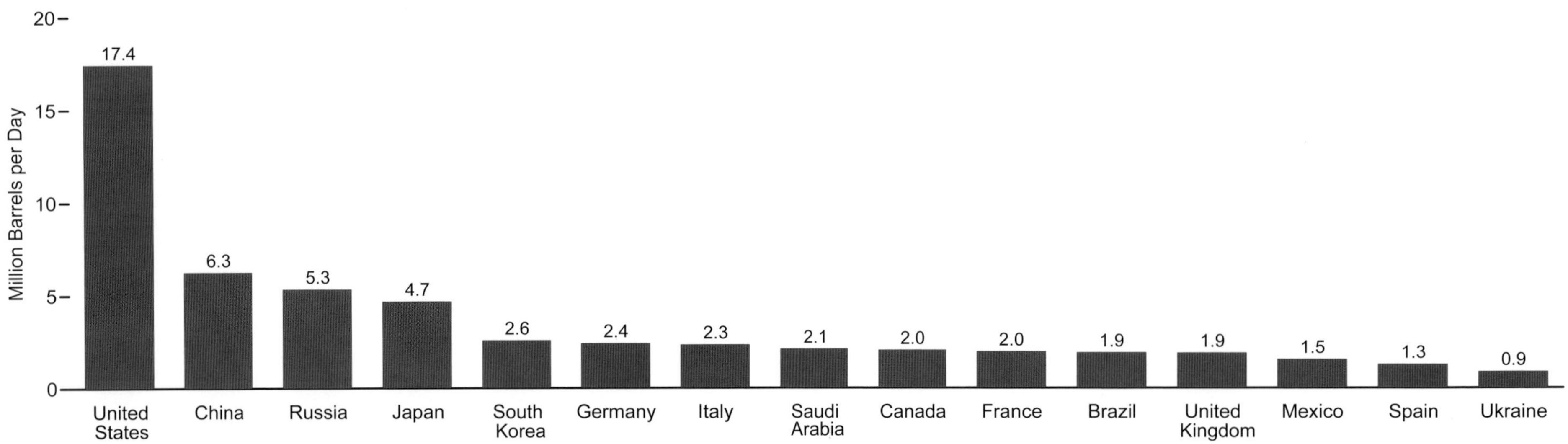

[1] Organization for Economic Cooperation and Development. See Glossary for membership.

Notes: • Capacity as of January 1. • Because vertical scales differ, graphs should not be compared.

Source: Table 11.9.

Table 11.9 World Crude Oil Refining Capacity, 1970-2007

(Million Barrels per Day)

	Selected OECD Countries											Selected Non-OECD Countries							
Year	Canada	France	Germany [1]	Italy	Japan	Mexico [2]	South Korea [2]	Spain	United Kingdom	United States	Total OECD [3]	Brazil	China	Former U.S.S.R.	Russia	Saudi Arabia	Ukraine	Total Non-OECD	World
1970	1.40	2.32	2.36	2.96	3.14	0.50	0.18	0.69	2.30	12.02	32.18	0.50	0.30	5.64	– –	0.38	– –	14.87	47.05
1971	1.45	2.53	2.54	3.24	3.70	.57	.25	.85	2.39	12.86	35.19	.51	.42	6.27	– –	.91	– –	16.73	51.92
1972	1.45	2.69	2.56	3.68	3.82	.59	.22	.87	2.59	13.29	37.29	.56	.48	6.68	– –	.51	– –	17.90	55.19
1973	1.73	2.95	2.70	3.59	4.53	.63	.43	1.03	2.47	13.64	39.45	.72	.50	7.26	– –	.43	– –	18.72	58.17
1974	1.79	3.14	2.83	3.88	5.15	.63	.42	1.16	2.76	14.36	42.39	.79	.60	7.81	– –	.43	– –	20.74	63.13
1975	1.88	3.34	2.99	3.95	5.35	.76	.43	1.17	2.78	14.96	44.07	.96	.85	8.24	– –	.61	– –	22.45	66.52
1976	2.02	3.31	3.10	4.08	5.63	.76	.44	1.32	2.89	15.24	46.16	.99	1.01	9.23	– –	.54	– –	23.77	69.93
1977	2.10	3.52	3.08	4.26	5.76	.94	.42	1.28	3.01	16.40	48.34	1.12	1.40	9.10	– –	.60	– –	26.77	75.11
1978	2.17	3.46	3.08	4.23	5.67	1.38	.48	1.27	2.91	17.05	49.37	1.16	1.46	9.98	– –	.59	– –	28.09	77.46
1979	2.23	3.47	3.10	4.20	5.68	1.24	.54	1.43	2.53	17.44	49.26	1.21	1.58	10.48	– –	.49	– –	29.34	78.60
1980	2.22	3.40	2.99	4.13	5.71	1.39	.60	1.46	2.53	17.99	50.03	1.21	1.60	10.95	– –	.49	– –	29.85	79.88
1981	2.17	3.34	3.02	4.09	5.66	1.39	.61	1.46	2.63	18.62	50.54	1.40	1.81	11.40	– –	.49	– –	31.06	81.60
1982	2.20	3.29	2.94	4.00	5.81	1.47	.76	1.52	2.48	17.89	49.74	1.41	1.81	11.60	– –	.49	– –	31.00	80.73
1983	2.02	2.87	2.47	3.28	5.73	1.29	.76	1.52	2.26	16.86	45.83	1.22	2.00	11.75	– –	.71	– –	31.48	77.30
1984	1.81	2.67	2.39	3.05	5.17	1.27	.78	1.49	2.09	16.14	43.33	1.30	2.05	12.00	– –	.86	– –	32.06	75.39
1985	1.87	2.39	2.17	3.10	4.97	1.27	.78	1.49	2.01	15.66	42.10	1.31	2.15	12.20	– –	.84	– –	33.08	75.18
1986	1.86	1.95	1.93	2.74	4.72	1.27	.78	1.37	1.79	15.46	39.99	1.31	2.15	12.20	– –	1.12	– –	32.62	72.61
1987	1.76	1.83	1.72	2.68	4.79	1.35	.86	1.31	1.78	15.57	39.63	1.32	2.20	12.26	– –	1.13	– –	32.93	72.56
1988	1.87	1.94	1.65	2.56	4.57	1.35	.82	1.31	1.80	15.92	40.07	1.41	2.20	12.26	– –	1.38	– –	33.54	73.61
1989	1.86	1.88	1.52	2.45	4.36	1.35	.88	1.29	1.80	15.66	39.31	1.41	2.20	12.30	– –	1.38	– –	33.99	73.30
1990	1.85	1.82	1.51	2.80	4.20	1.51	.87	1.29	1.83	15.57	39.66	1.40	2.20	12.30	– –	1.48	– –	34.21	73.86
1991	1.88	1.82	2.07	2.39	4.38	1.68	.87	1.32	1.87	15.68	40.25	1.41	2.20	12.30	– –	1.86	– –	34.75	75.00
1992	1.91	1.82	2.06	2.39	4.61	1.57	1.16	1.32	1.86	15.70	41.72	1.41	2.20	12.30	– –	1.86	– –	33.72	75.43
1993	1.87	1.85	2.23	2.42	4.74	1.52	1.15	1.30	1.84	15.12	41.28	1.40	2.20	– –	6.46	1.86	1.24	31.83	73.11
1994	1.88	1.86	2.27	2.26	4.81	1.52	1.15	1.28	1.87	15.03	41.18	1.25	2.20	– –	6.46	1.61	1.24	31.90	73.07
1995	1.91	1.77	2.32	2.26	4.85	1.52	1.17	1.28	1.87	15.43	41.62	1.25	2.87	– –	6.53	1.66	1.26	32.63	74.25
1996	1.85	1.78	2.13	2.28	4.87	1.52	1.24	1.33	1.89	15.33	41.31	1.26	2.87	– –	6.72	1.66	1.26	33.09	74.39
1997	1.85	1.79	2.11	2.26	4.99	1.52	2.21	1.30	1.94	15.45	42.47	1.26	2.87	– –	6.73	1.66	1.25	33.51	75.99
1998	1.85	1.87	2.18	2.45	4.97	1.52	2.54	1.29	1.83	15.71	43.23	1.66	2.97	– –	6.87	1.65	1.25	34.80	78.03
1999	1.87	1.95	2.25	2.45	5.06	1.53	2.54	1.32	1.85	16.26	44.19	1.77	4.35	– –	6.75	1.69	1.09	35.89	80.08
2000	1.91	1.90	2.28	2.34	5.00	1.53	2.54	1.32	1.79	16.51	44.33	1.78	4.35	– –	6.67	1.71	1.15	37.20	81.53
2001	1.91	1.90	2.26	2.36	4.96	1.53	2.56	1.29	1.77	16.60	44.48	1.92	4.35	– –	5.44	1.75	1.03	36.84	81.32
2002	1.94	1.90	2.26	2.28	4.79	1.53	2.56	1.29	1.78	16.79	44.50	1.79	4.53	– –	5.44	1.75	1.03	36.95	81.44
2003	1.98	1.90	2.27	2.30	4.77	1.68	2.56	1.32	1.79	16.76	44.69	1.87	4.53	– –	5.44	1.75	1.03	37.30	82.00
2004	1.99	1.95	2.29	2.31	4.70	1.68	2.54	1.27	1.82	16.89	44.75	1.91	4.53	– –	5.44	1.75	1.03	37.51	82.26
2005	2.02	1.95	2.32	2.32	4.71	1.68	2.58	1.27	1.83	17.13	45.12	1.92	4.65	– –	5.43	1.75	.88	37.67	82.80
2006	2.02	1.98	2.43	2.32	4.67	1.68	2.58	1.27	1.88	17.34	45.58	1.91	6.25	– –	5.34	2.10	.88	39.76	85.34
2007	2.04	1.96	2.42	2.34	4.68	1.54	2.58	1.27	1.89	17.44	45.52	1.91	6.25	– –	5.34	2.10	.88	39.83	85.36

[1] Through 1990, this is East and West Germany. Beginning in 1991, this is unified Germany.

[2] Mexico, which joined the OECD on May 18, 1994, and South Korea, which joined the OECD on December 12, 1996, are included in the OECD for all years shown in this table.

[3] Hungary and Poland, which joined the OECD on May 7, 1996, and November 22, 1996, respectively, are included in Total OECD beginning in 1992, the first year that data for these countries were available. The Czech Republic and Slovakia (or Slovak Republic), which joined the OECD on December 21, 1995 and December 14, 2000, respectively, are included in Total OECD beginning in 1992, the first year that data for these countries were available.

– – = Not applicable.

Notes: • OECD = Organization for Economic Cooperation and Development. See Glossary for membership. • Capacity data represent distillation capacity. • Capacity for all years is as of January 1. • Totals may not equal sum of components due to independent rounding.

Web Page: For related information, see http://www.eia.doe.gov/international.

Sources: **United States:** Table 5.9. **China** and **Former U.S.S.R.:** • 1970-1976—Ballinger Publishing Company, *The Energy Decade, 1970-1980, A Statistical and Graphic Chronicle.* • 1977 forward—PennWell Publishing Company, *Oil & Gas Journal.* **All Other Data:** PennWell Publishing Company, *Oil & Gas Journal.*

Figure 11.10 World Petroleum Consumption

World, 1960-2006

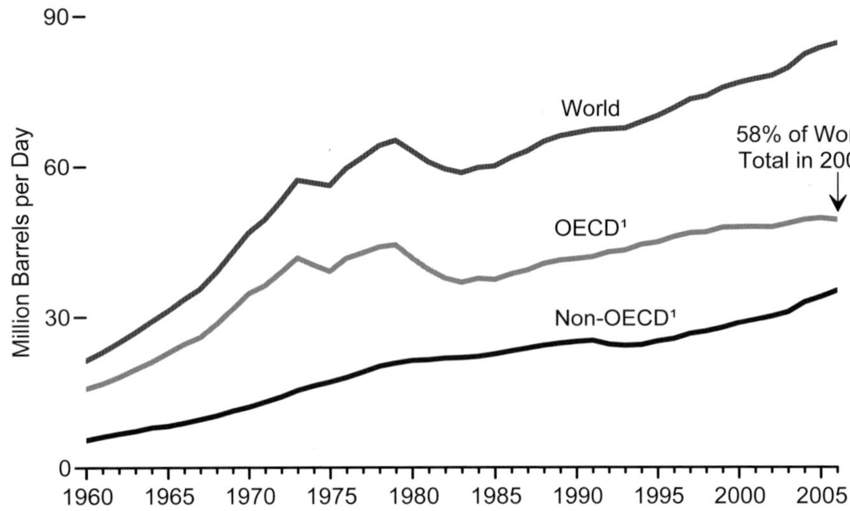

Top Consuming Countries, 1960-2006

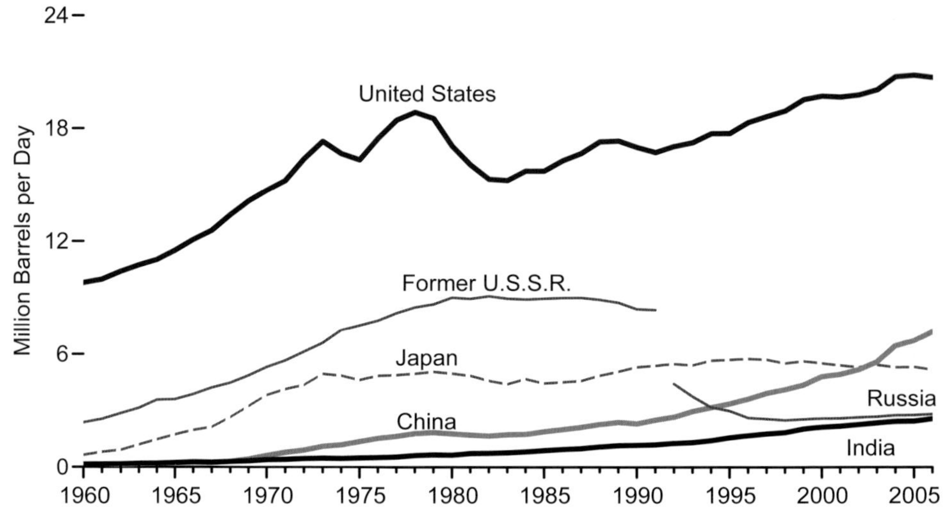

Selected Consumers, 2006

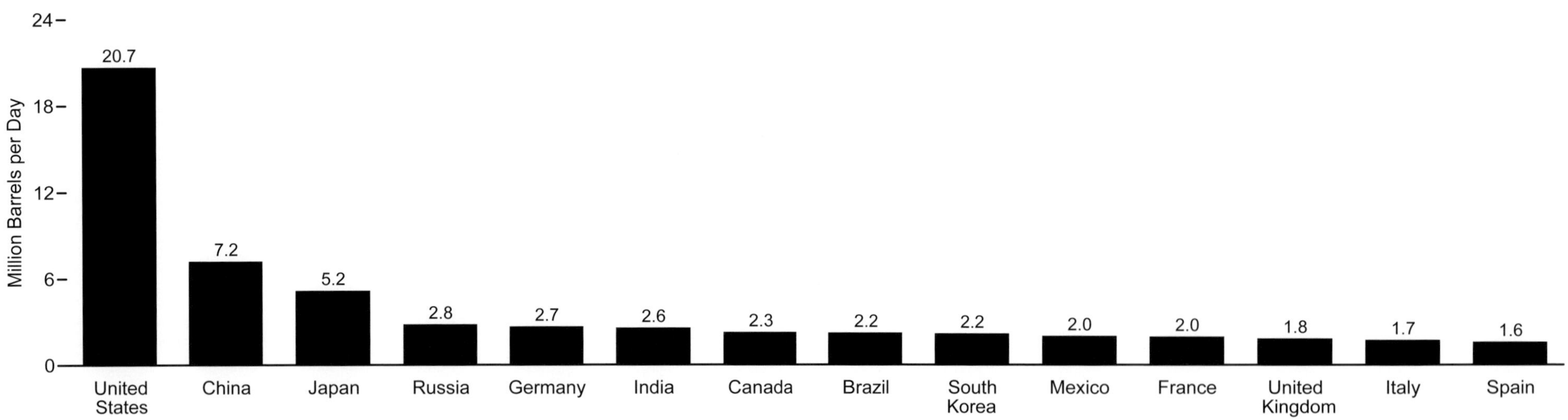

[1] Organization for Economic Cooperation and Development. See Glossary for membership.
Note: Because vertical scales differ, graphs should not be compared.

Source: Table 11.10.

Table 11.10 World Petroleum Consumption, 1960-2006

(Million Barrels per Day)

Year	Selected OECD Consumers											Selected Non-OECD Consumers						World
	Canada	France	Germany [1]	Italy	Japan	Mexico [2]	South Korea [2]	Spain	United Kingdom	United States	Total OECD [3]	Brazil	China	India	Former U.S.S.R.	Russia	Total Non-OECD	
1960	0.84	0.56	0.63	0.44	0.66	0.30	0.01	0.10	0.94	9.80	15.78	0.27	0.17	0.16	2.38	– –	5.56	21.34
1961	.87	.63	.79	.54	.82	.29	.02	.12	1.04	9.98	16.77	.28	.17	.17	2.57	– –	6.23	23.00
1962	.92	.73	1.00	.67	.93	.30	.02	.12	1.12	10.40	18.06	.31	.14	.18	2.87	– –	6.83	24.89
1963	.99	.86	1.17	.77	1.21	.31	.03	.12	1.27	10.74	19.60	.34	.17	.21	3.15	– –	7.32	26.92
1964	1.05	.98	1.36	.90	1.48	.33	.02	.20	1.36	11.02	21.05	.35	.20	.22	3.58	– –	8.03	29.08
1965	1.14	1.09	1.61	.98	1.74	.34	.03	.23	1.49	11.51	22.81	.33	.23	.25	3.61	– –	8.33	31.14
1966	1.21	1.19	1.80	1.08	1.98	.36	.04	.31	1.58	12.08	24.60	.38	.30	.28	3.87	– –	8.96	33.56
1967	1.25	1.34	1.86	1.19	2.14	.39	.07	.36	1.64	12.56	25.94	.38	.28	.26	4.22	– –	9.65	35.59
1968	1.34	1.46	1.99	1.40	2.66	.41	.10	.46	1.82	13.39	28.56	.46	.31	.31	4.48	– –	10.40	38.96
1969	1.42	1.66	2.33	1.69	3.25	.45	.15	.49	1.98	14.14	31.54	.48	.44	.34	4.87	– –	11.35	42.89
1970	1.52	1.94	2.83	1.71	3.82	.50	.20	.58	2.10	14.70	34.69	.53	.62	.40	5.31	– –	12.12	46.81
1971	1.56	2.12	2.94	1.84	4.14	.52	.23	.64	2.14	15.21	36.30	.58	.79	.42	5.66	– –	13.12	49.42
1972	1.66	2.32	3.10	1.95	4.36	.59	.24	.68	2.28	16.37	38.95	.66	.91	.46	6.12	– –	14.15	53.09
1973	1.73	2.60	3.32	2.07	4.95	.67	.28	.78	2.34	17.31	41.80	.78	1.12	.49	6.60	– –	15.43	57.24
1974	1.78	2.45	3.03	2.00	4.86	.71	.29	.86	2.21	16.65	40.38	.86	1.19	.47	7.28	– –	16.30	56.68
1975	1.78	2.25	2.96	1.86	4.62	.75	.31	.87	1.91	16.32	39.14	.92	1.36	.50	7.52	– –	17.06	56.20
1976	1.82	2.42	3.21	1.97	4.84	.83	.36	.97	1.89	17.46	41.72	1.00	1.53	.51	7.78	– –	17.96	59.67
1977	1.85	2.29	3.21	1.90	4.88	.88	.42	.94	1.91	18.43	42.78	1.02	1.64	.55	8.18	– –	19.05	61.83
1978	1.90	2.41	3.29	1.95	4.95	.99	.48	.98	1.94	18.85	43.98	1.11	1.79	.62	8.48	– –	20.18	64.16
1979	1.97	2.46	3.37	2.04	5.05	1.10	.53	1.02	1.97	18.51	44.39	1.18	1.84	.66	8.64	– –	20.84	65.22
1980	1.87	2.26	3.08	1.93	4.96	1.27	.54	.99	1.73	17.06	41.76	1.15	1.77	.64	9.00	– –	21.35	63.11
1981	1.77	2.02	2.80	1.87	4.85	1.40	.54	.94	1.59	16.06	39.49	1.09	1.71	.73	8.94	– –	21.45	60.94
1982	1.58	1.88	2.74	1.78	4.58	1.48	.53	1.00	1.59	15.30	37.77	1.06	1.66	.74	9.08	– –	21.78	59.54
1983	1.45	1.84	2.66	1.75	4.40	1.35	.56	1.01	1.53	15.23	36.91	.98	1.73	.77	8.95	– –	21.87	58.78
1984	1.52	1.77	2.56	1.72	4.67	[R]1.40	.55	.85	1.83	15.73	37.69	1.03	1.74	.82	8.91	– –	22.12	59.82
1985	1.53	1.75	2.65	[R]1.71	4.44	1.48	.55	.86	1.62	15.73	37.48	1.08	1.89	.89	8.95	– –	22.60	[R]60.09
1986	1.53	1.76	2.79	1.73	4.50	1.52	.59	.87	1.64	16.28	38.60	1.24	2.00	.95	8.98	– –	23.21	61.81
1987	1.61	1.79	2.72	1.82	4.57	1.58	.63	.90	1.61	16.67	39.34	1.26	2.12	.99	[R]8.99	– –	23.75	63.10
1988	1.68	1.80	2.72	1.83	4.85	1.60	.75	.98	1.69	17.28	40.65	1.30	2.27	1.08	8.89	– –	24.32	64.97
1989	1.75	1.84	2.58	1.90	5.06	[R]1.72	.86	.98	1.73	17.33	41.33	1.32	2.38	1.15	8.74	– –	24.75	66.08
1990	1.73	1.83	2.68	1.87	[R]5.30	1.75	1.05	1.01	1.78	16.99	[R]41.61	1.47	2.30	1.17	8.39	– –	25.07	[R]66.68
1991	1.67	1.94	2.83	1.86	[R]5.38	1.83	1.26	1.07	1.80	16.71	[R]42.00	1.48	2.50	1.19	8.35	– –	25.28	[R]67.28
1992	1.72	1.93	2.84	1.89	[R]5.47	1.86	1.53	1.10	[R]1.82	17.03	[R]42.95	1.52	2.66	1.27	– –	4.42	24.51	[R]67.46
1993	1.75	1.88	2.91	1.89	[R]5.40	1.84	1.68	1.05	1.83	17.24	[R]43.30	1.58	2.96	1.31	– –	3.75	24.30	[R]67.60
1994	1.77	[R]1.87	2.88	1.87	[R]5.65	1.93	1.84	1.12	1.83	17.72	[R]44.44	1.67	3.16	1.41	– –	3.18	24.43	[R]68.86
1995	1.81	1.92	2.88	1.94	[R]5.69	1.82	2.01	1.19	1.82	17.72	[R]44.90	1.79	3.36	1.57	– –	2.98	25.16	[R]70.07
1996	1.86	1.95	2.92	1.92	[R]5.74	1.79	2.10	1.20	1.85	18.31	[R]45.98	1.90	3.61	1.68	– –	2.62	25.65	[R]71.63
1997	1.95	1.97	2.92	1.93	[R]5.70	1.85	2.25	1.27	1.80	18.62	[R]46.72	2.03	3.92	1.77	– –	2.56	[R]26.65	[R]73.37
1998	1.94	2.04	2.92	1.94	[R]5.50	1.95	1.92	1.36	1.79	18.92	[R]46.89	2.10	4.11	1.84	– –	2.49	[R]27.12	[R]74.00
1999	2.03	2.03	2.84	1.89	[R]5.61	1.96	2.08	1.40	1.80	19.52	[R]47.81	2.13	4.36	2.03	– –	2.54	[R]27.86	[R]75.66
2000	2.03	2.00	2.77	1.85	[R]5.50	2.04	2.14	1.43	1.76	19.70	[R]47.88	2.17	4.80	2.13	– –	2.58	[R]28.79	[R]76.66
2001	2.06	2.05	2.81	1.84	[R]5.39	2.01	2.13	1.49	[R]1.74	19.65	[R]47.94	2.21	4.92	2.18	– –	2.59	[R]29.46	[R]77.40
2002	2.08	1.98	2.72	1.87	5.30	1.95	2.15	1.50	[R]1.73	19.76	[R]47.89	2.13	5.16	2.26	– –	2.64	[R]30.15	[R]78.04
2003	2.21	2.00	2.68	1.87	5.42	1.95	2.18	1.54	[R]1.76	20.03	[R]48.60	2.06	5.58	2.35	– –	2.68	[R]31.01	79.61
2004	2.30	2.01	2.67	1.79	5.29	2.00	2.16	1.57	[R]1.80	20.73	[R]49.36	[R]2.12	6.44	2.43	– –	2.75	[R]32.97	[R]82.33
2005	[R]2.30	[R]1.99	[R]2.65	[R]1.75	[R]5.31	[R]2.05	[R]2.19	[R]1.61	[R]1.83	20.80	[R]49.66	[R]2.17	[R]6.72	2.44	– –	[R]2.76	[R]33.99	[R]83.65
2006[P]	2.26	1.96	2.66	1.73	5.16	2.00	2.17	1.59	1.83	20.69	49.34	2.22	7.20	2.57	– –	2.81	35.29	84.62

[1] Through 1969, the data for Germany are for the former West Germany only. For 1970 through 1990, this is East and West Germany. Beginning in 1991, this is unified Germany.

[2] Mexico, which joined the OECD on May 18, 1994, and South Korea, which joined the OECD on December 12, 1996, are included in the OECD for all years shown in this table.

[3] Hungary and Poland, which joined the OECD on May 7, 1996, and November 22, 1996, respectively, are included in Total OECD beginning in 1970, the first year that data for these countries were available. Total OECD includes Czechoslovakia from 1970-1992, and Czech Republic and Slovakia from 1993 forward.

R=Revised. P=Preliminary. – – = Not applicable.

Notes: • OECD = Organization for Economic Cooperation and Development. See Glossary for membership. • Totals may not equal sum of components due to independent rounding.

Web Page: For related information, see http://www.eia.doe.gov/international.

Sources: **United States:** Table 5.1, "Petroleum Products Supplied." **All Other Data:** • 1960 forward—Energy Information Administration, International Energy Database, May 12, 2008.

Figure 11.11 World Dry Natural Gas Production

World and Top Producing Countries, 1997-2006

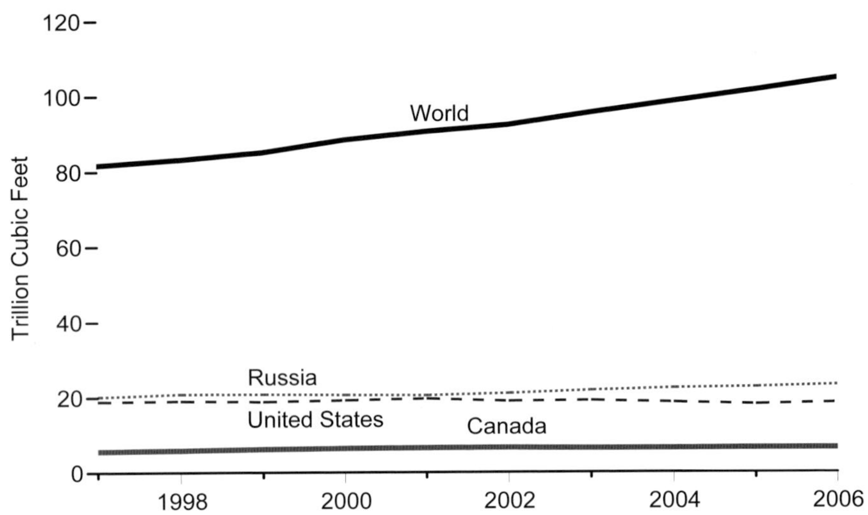

World Areas, 2006

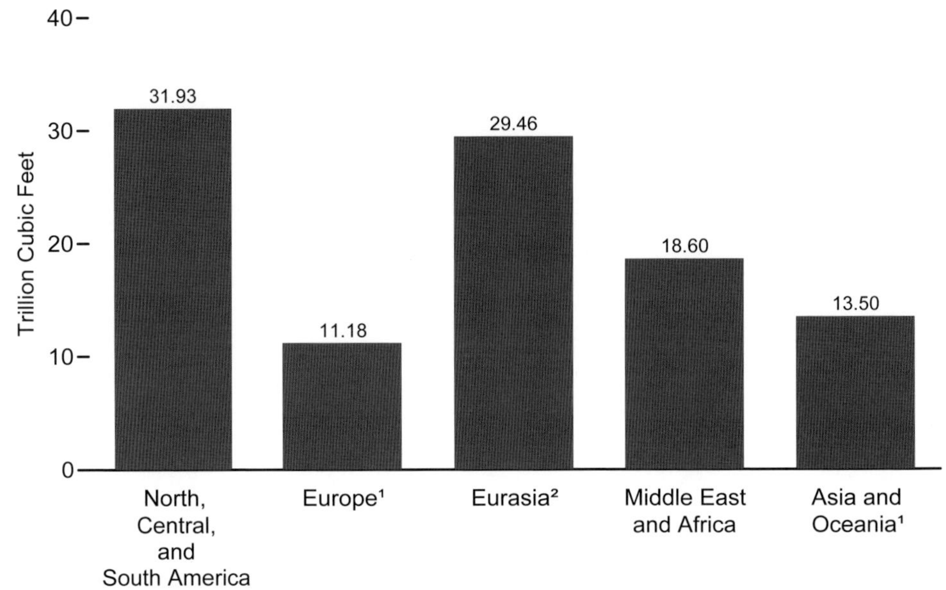

World Areas, 1997-2006

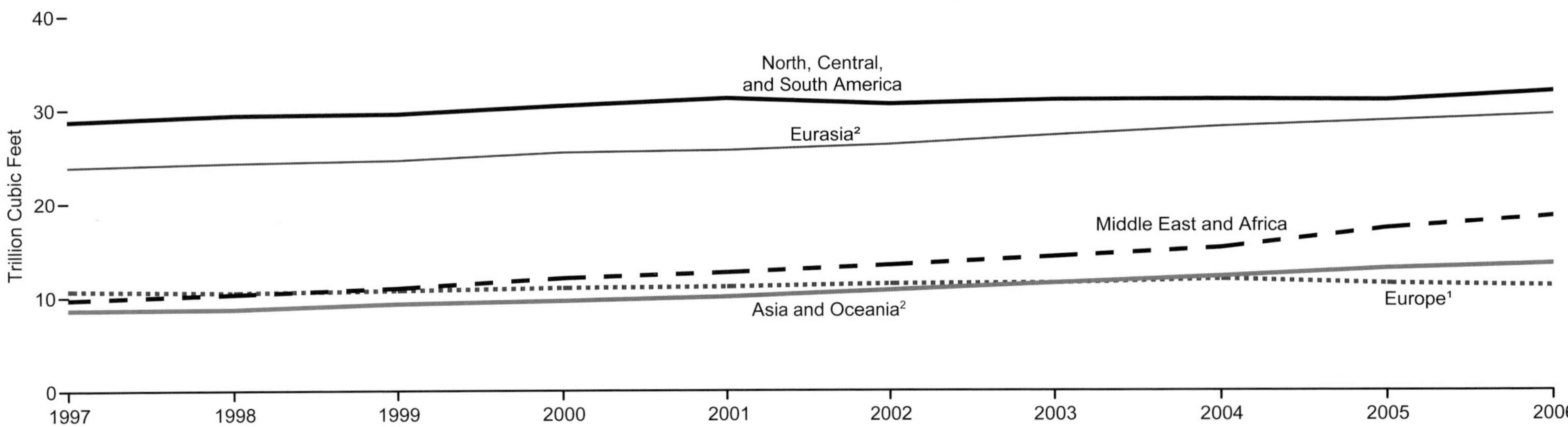

[1] Excludes countries that were part of the former U.S.S.R. See "U.S.S.R." in Glossary.
[2] Includes only countries that were part of the former U.S.S.R. See "U.S.S.R." in Glossary.

Note: Because vertical scales differ, graphs should not be compared.
Source: Table 11.11.

Table 11.11 World Dry Natural Gas Production, 1997-2006

(Trillion Cubic Feet)

Region and Country	1997	1998	1999	2000	2001	2002	2003	2004	2005	2006 [P]
North, Central, and South America	**28.75**	**29.39**	**29.53**	**30.40**	**31.17**	**30.56**	**31.02**	**R31.08**	**31.00**	**31.93**
Argentina	.97	1.04	1.22	1.32	1.31	1.28	1.45	1.58	1.61	1.63
Canada	5.76	5.98	6.27	6.47	6.60	6.63	6.45	6.48	6.56	6.55
Mexico	1.17	1.27	1.29	1.31	1.30	1.33	1.40	1.46	1.52	1.71
United States	18.90	19.02	18.83	19.18	19.62	18.93	19.10	18.59	18.05	18.48
Venezuela	.99	1.11	.95	.96	1.12	1.00	.86	.96	1.01	1.01
Other	.96	.96	.98	1.15	1.22	1.39	1.76	2.00	2.24	2.56
Europe [1]	**10.68**	**10.49**	**10.72**	**10.98**	**11.10**	**11.41**	**11.48**	**11.89**	**11.42**	**11.18**
Germany	.79	.77	.82	.78	.79	.79	.78	.73	.70	.69
Italy	.68	.67	.62	.59	.54	.52	.49	.46	.43	.39
Netherlands	2.99	2.84	2.65	2.56	2.75	2.68	2.57	3.04	2.78	2.73
Norway	1.62	1.63	1.76	1.87	1.95	2.41	2.70	2.95	3.07	3.20
Romania	.61	.52	.50	.48	.51	.47	.43	.42	.41	.42
United Kingdom	3.03	3.14	3.49	3.83	3.69	3.66	3.63	3.39	3.10	2.83
Other	.95	.92	.88	.88	.89	.88	.86	.92	.92	.92
Eurasia [2]	**23.88**	**24.31**	**24.59**	**25.43**	**25.65**	**26.26**	**27.25**	**28.16**	**28.79**	**29.46**
Russia	20.17	20.87	20.83	20.63	20.51	21.03	21.77	22.39	22.62	23.17
Turkmenistan	.90	.47	.79	1.64	1.70	1.89	2.09	2.07	2.22	2.23
Ukraine	.64	.64	.63	.64	.64	.65	.69	.68	.69	.69
Uzbekistan	1.74	1.94	1.96	1.99	2.23	2.04	2.03	2.11	2.11	2.22
Other	.44	.40	.39	.53	.57	.65	.68	.91	1.15	1.16
Middle East and Africa	**9.74**	**10.30**	**10.95**	**12.01**	**12.61**	**13.39**	**R14.29**	**15.23**	**17.31**	**18.60**
Algeria	2.43	2.60	2.88	2.94	2.79	2.80	2.85	2.83	3.11	3.07
Egypt	.48	.49	.52	.65	.87	.88	1.06	1.15	1.50	1.86
Iran	1.66	1.77	2.04	2.13	2.33	2.65	2.86	2.96	3.56	3.71
Qatar	.61	.69	.78	1.03	.95	1.04	1.11	1.38	1.62	1.75
Saudi Arabia	1.60	1.65	1.63	1.76	1.90	2.00	2.12	2.32	2.52	2.59
United Arab Emirates	1.28	1.31	1.34	1.36	1.39	1.53	1.58	1.63	1.66	1.67
Other	1.67	1.79	1.76	2.15	2.39	2.48	R2.71	2.95	3.35	3.94
Asia and Oceania [1]	**R8.64**	**R8.70**	**R9.29**	**R9.60**	**R10.02**	**R10.72**	**R11.47**	**R12.17**	**13.00**	**13.50**
Australia	1.06	1.10	1.12	1.16	1.19	1.23	1.27	1.31	1.44	1.51
China	.80	.82	.89	.96	1.07	1.15	1.21	1.44	1.76	2.07
India	.72	.76	.75	.79	.85	.93	.96	1.00	1.06	1.07
Indonesia	2.37	2.27	2.51	2.36	2.34	2.48	2.61	2.66	2.61	2.61
Malaysia	1.36	1.37	1.42	1.50	1.66	1.71	2.01	2.20	2.24	2.22
Pakistan	.70	.71	.78	.86	.77	.81	.89	.97	1.09	1.11
Other	R1.64	R1.68	R1.82	R1.97	R2.14	R2.40	R2.52	R2.59	2.81	2.91
World	**R81.70**	**R83.19**	**R85.08**	**R88.42**	**R90.56**	**R92.33**	**R95.52**	**R98.53**	**101.52**	**104.67**

[1] Excludes countries that were part of the former U.S.S.R. See "U.S.S.R." in Glossary.
[2] Includes only countries that were part of the former U.S.S.R. See "U.S.S.R." in Glossary.
R=Revised. P=Preliminary.
Note: Totals may not equal sum of components due to independent rounding.

Web Page: For related information, see http://www.eia.doe.gov/international.
Sources: **United States:** Table 6.1. **All Other Data:** Energy Information Administration, International Energy Database, May 22, 2008.

Figure 11.12 World Dry Natural Gas Consumption

World and Top Consuming Countries, 1980-2006

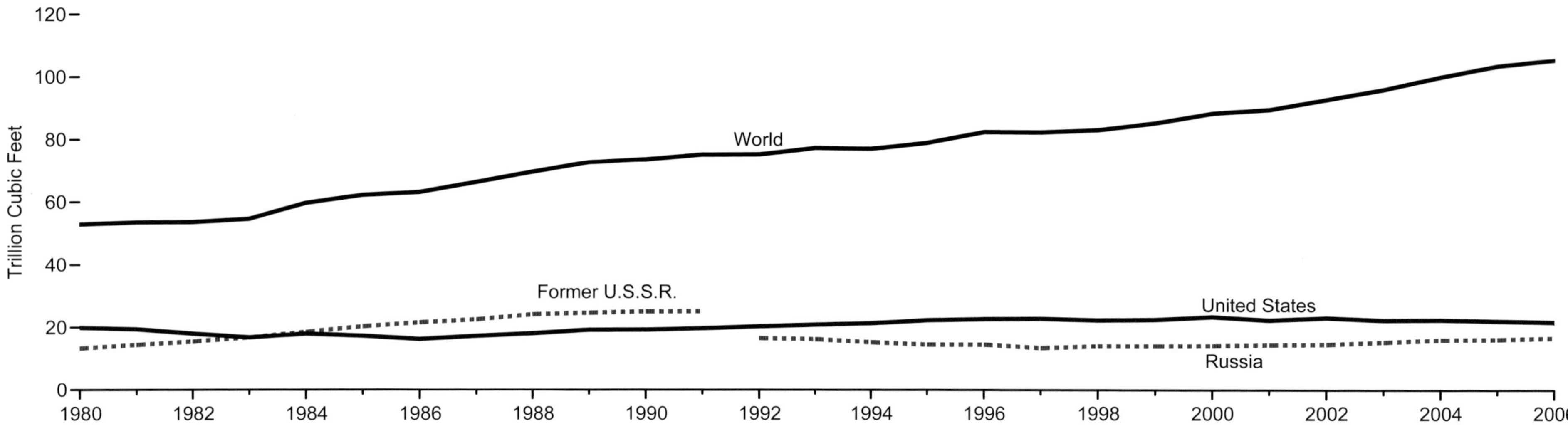

Top Consuming Countries, 2006

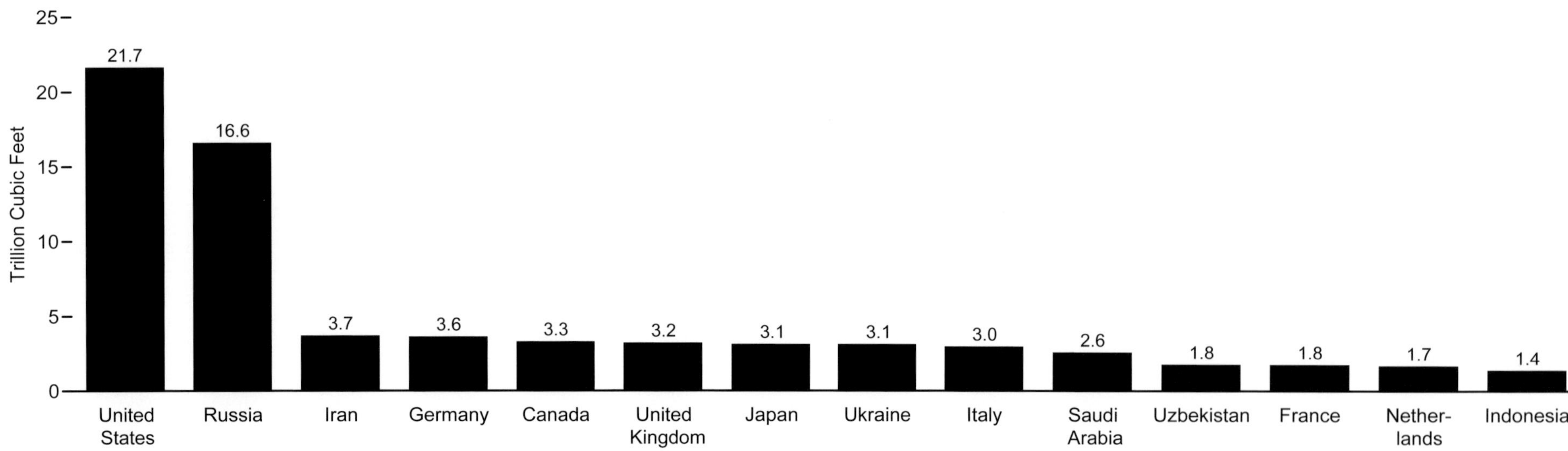

Note: Because vertical scales differ, graphs should not be compared.

Source: Table 11.12.

Table 11.12 World Dry Natural Gas Consumption, 1980-2006

(Billion Cubic Feet)

Year	Canada	France	Germany [1]	Indonesia	Iran	Italy	Japan	Nether-lands	Former U.S.S.R.	Russia	Saudi Arabia	Ukraine	United Kingdom	United States	Uzbek-istan	Other	World
1980	1,883	981	2,621	195	232	972	903	1,493	13,328	– –	334	– –	1,702	19,877	– –	8,369	52,890
1981	1,842	1,003	2,513	232	155	942	925	1,421	14,440	– –	564	– –	1,740	19,404	– –	8,333	53,513
1982	1,859	979	2,334	218	200	944	956	1,511	15,522	– –	430	– –	1,743	18,001	– –	8,931	53,628
1983	1,863	999	2,397	302	310	967	1,020	1,451	16,822	– –	418	– –	1,815	16,835	– –	9,427	54,626
1984	2,017	1,079	2,584	365	476	1,135	1,372	1,540	18,512	– –	620	– –	1,851	17,951	– –	10,189	59,692
1985	2,165	1,110	2,546	513	600	1,151	1,468	1,624	20,302	– –	716	– –	1,991	17,281	– –	10,777	62,244
1986	2,130	1,129	2,595	441	536	1,217	1,494	1,620	21,522	– –	890	– –	2,020	16,221	– –	11,303	63,118
1987	2,112	1,038	2,733	542	565	1,346	1,543	1,672	22,462	– –	946	– –	2,079	17,211	– –	12,062	66,312
1988	2,331	963	2,716	492	706	1,460	1,618	1,513	24,092	– –	1,028	– –	1,972	18,030	– –	12,628	69,548
1989	2,427	984	2,835	546	784	1,581	1,731	1,550	24,529	– –	1,052	– –	1,951	19,119	– –	13,549	72,638
1990	2,378	997	2,669	547	837	1,674	[R]2,028	1,535	24,961	– –	1,077	– –	2,059	19,174	– –	13,611	[R]73,547
1991	2,400	1,131	2,776	557	811	1,775	[R]2,175	1,715	25,014	– –	1,130	– –	2,218	19,562	– –	13,841	[R]75,106
1992	2,596	1,146	2,739	673	883	1,760	[R]2,177	1,669	– –	16,482	1,201	3,503	2,170	20,228	1,095	16,868	[R]75,190
1993	2,691	1,158	2,830	850	938	1,801	[R]2,217	1,714	– –	16,185	1,268	3,871	2,412	20,790	1,541	17,005	[R]77,269
1994	2,738	1,157	2,965	965	1,123	1,748	[R]2,334	1,654	– –	15,214	1,331	3,327	2,542	21,247	1,229	17,400	[R]76,974
1995	2,872	1,183	3,172	1,061	1,243	1,921	[R]2,409	1,701	– –	14,507	1,343	2,970	2,690	22,207	1,349	18,297	[R]78,925
1996	2,917	1,314	3,163	1,108	1,416	1,984	[R]2,524	1,874	– –	14,504	1,460	2,935	3,182	22,609	1,434	19,939	[R]82,364
1997	2,887	1,300	3,012	1,125	1,663	2,048	[R]2,590	1,763	– –	13,434	1,601	2,832	3,013	22,737	1,455	20,825	[R]82,286
1998	2,798	1,313	3,130	983	1,828	2,205	[R]2,676	1,752	– –	14,045	1,653	2,606	3,072	22,246	1,409	21,357	[R]83,071
1999	3,108	1,383	3,151	1,124	2,112	2,396	[R]2,818	1,705	– –	14,013	1,632	2,755	3,259	22,405	1,423	21,995	[R]85,278
2000	2,991	1,403	3,098	1,081	2,221	2,498	[R]2,914	1,725	– –	14,130	1,759	2,779	3,373	23,333	1,511	23,538	[R]88,356
2001	3,121	1,471	3,239	1,182	2,478	2,505	[R]2,902	1,769	– –	14,412	1,896	2,617	3,338	22,239	1,596	24,804	[R]89,571
2002	3,173	1,528	3,204	1,218	2,798	2,488	[R]3,100	1,767	– –	14,567	2,002	2,779	3,379	23,007	1,642	[R]26,119	[R]92,774
2003	3,373	1,511	3,566	1,222	2,910	[R]2,743	[R]3,100	1,775	– –	15,291	2,121	3,023	3,358	22,277	1,670	[R]28,072	[R]96,012
2004	[R]3,341	[R]1,690	3,576	1,309	3,021	2,847	[R]3,074	1,812	– –	16,022	2,319	3,051	3,477	[R]22,389	1,773	[R]30,345	[R]100,047
2005	3,404	1,740	3,566	1,325	3,616	3,046	3,081	1,747	– –	16,153	2,516	3,079	3,357	22,011	1,702	33,156	103,500
2006[P]	3,307	1,759	3,637	1,398	3,712	2,984	3,137	1,688	– –	16,598	2,594	3,136	3,225	21,653	1,769	34,836	105,434

[1] Through 1990, this is East and West Germany. Beginning in 1991, this is unified Germany.
R=Revised. P=Preliminary. – – = Not applicable.
Note: Totals may not equal sum of components due to independent rounding.

Web Page: For related information, see http://www.eia.doe.gov/international.
Sources: **United States:** Table 6.1. **All Other Data:** Energy Information Administration, International Energy Database, May 22, 2008.

Figure 11.13 World Recoverable Reserves of Coal, 2005

By Type

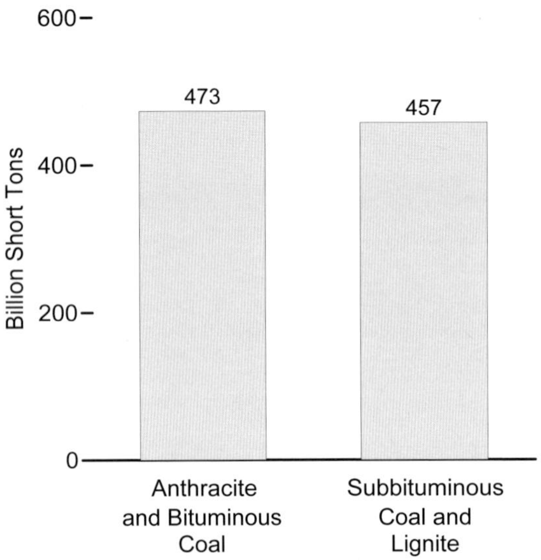

By Region

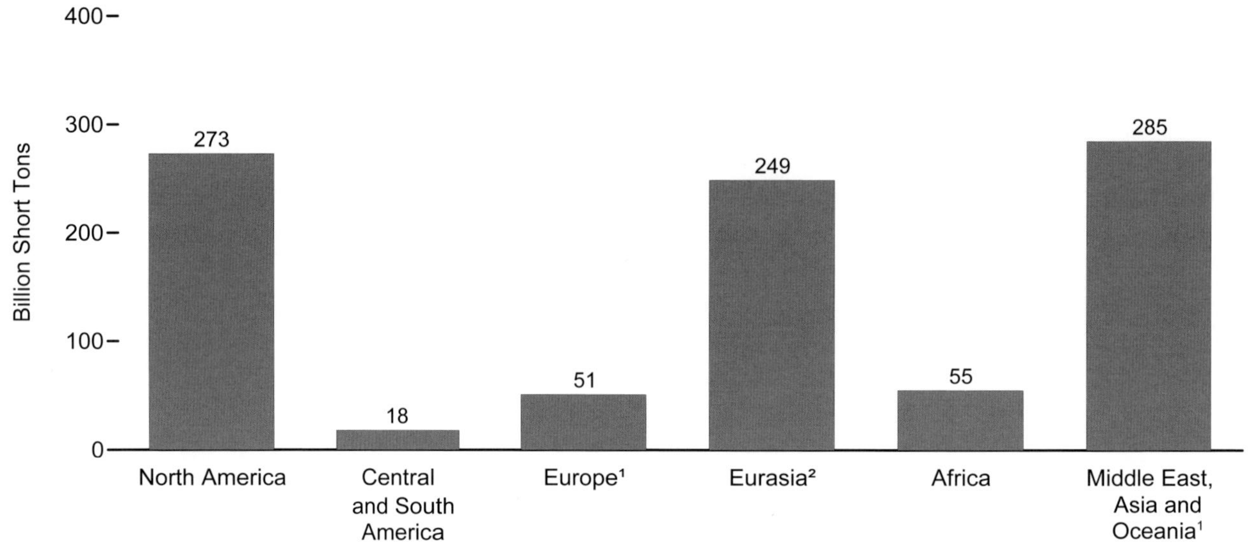

Top Reserves Countries

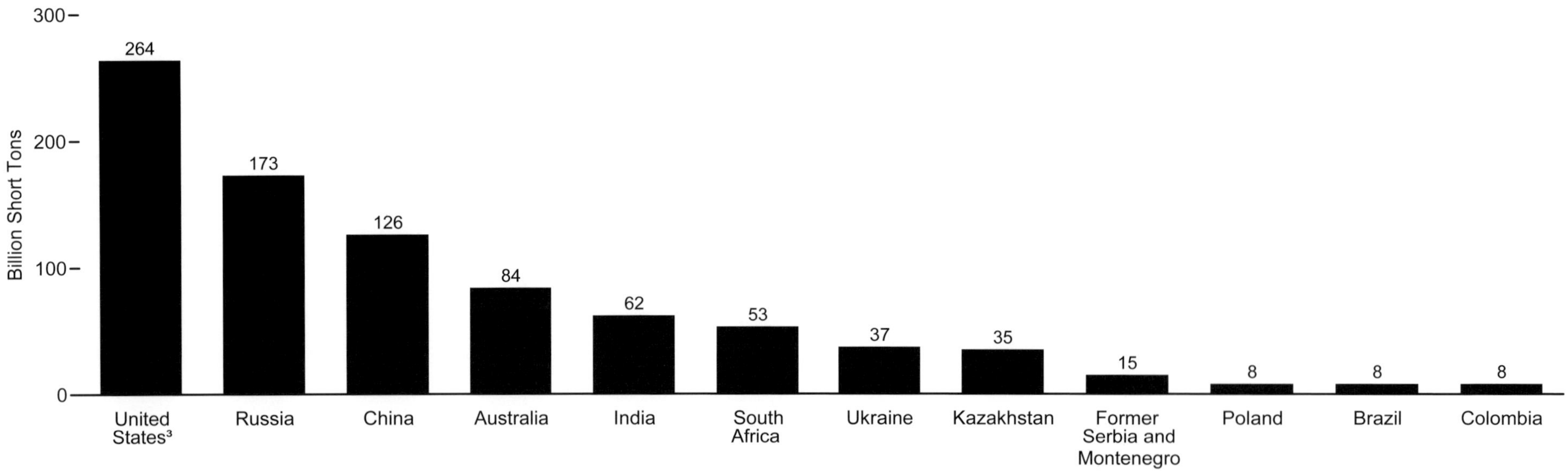

[1] Excludes countries that were part of the former U.S.S.R. See "U.S.S.R." in Glossary.
[2] Includes only countries that were part of the former U.S.S.R. See "U.S.S.R." in Glossary.
[3] U.S. reserves are at end of 2006, 1 year later than other data in this figure.

Notes: • Data are at end of year. • Because vertical scales differ, graphs should not be compared.
Source: Table 11.13.

Table 11.13 World Recoverable Reserves of Coal, 2005

(Million Short Tons)

Region and Country	Anthracite and Bituminous Coal	Subbituminous Coal and Lignite	Total
North America	**126,776**	**145,793**	**272,569**
Canada	3,826	3,425	7,251
Greenland	0	202	202
Mexico	948	387	1,335
United States [1]	122,001	141,780	263,781
Central and South America	**7,969**	**9,973**	**17,941**
Brazil	0	7,791	7,791
Chile	34	1,268	1,302
Colombia	7,251	420	7,671
Peru	154	0	154
Other	529	494	1,023
Europe [2]	**9,296**	**41,485**	**50,781**
Bulgaria	6	2,195	2,200
Czech Republic	1,844	3,117	4,962
Former Serbia and Montenegro	7	15,299	15,306
Germany	168	7,227	7,394
Greece	0	4,299	4,299
Hungary	219	3,420	3,640
Poland	6,627	1,642	8,270
Romania	13	452	465
Turkey	0	2,000	2,000
United Kingdom	171	0	171
Other	241	1,834	2,076
Eurasia [3]	**103,186**	**145,931**	**249,117**
Kazakhstan	31,052	3,450	34,502
Russia	54,110	118,964	173,074
Ukraine	16,922	20,417	37,339
Uzbekistan	1,102	2,205	3,307
Other	0	895	895
Africa	**54,488**	**192**	**54,680**
Botswana	44	0	44
South Africa	52,911	0	52,911
Zimbabwe	553	0	553
Other	980	192	1,172
Middle East, Asia, and Oceania [2]	**171,522**	**113,813**	**285,334**
Australia	40,896	43,541	84,437
China	68,564	57,651	126,215
India	57,585	4,694	62,278
Indonesia	1,897	2,874	4,771
North Korea	331	331	661
Pakistan	1	2,184	2,185
Thailand	0	1,493	1,493
Other	2,249	1,046	3,295
World	**473,236**	**457,186**	**930,423**

[1] U.S. data are as of the end of 2006, 1 year later than the other data on this table.
[2] Excludes countries that were part of the former U.S.S.R. See "U.S.S.R." in Glossary.
[3] Includes only countries that were part of the former U.S.S.R. See "U.S.S.R." in Glossary.

Notes: • Data are at end of year. • World Energy Council data represent "proved recoverable reserves," which are the tonnage within the "proved amount in place" that can be recovered (extracted from the earth in raw form) under present and expected local economic conditions with existing, available technology. • The Energy Information Administration (EIA) does not certify the international reserves data, but reproduces the information as a matter of convenience for the reader. • U.S. reserves represent estimated recoverable reserves from the Demonstrated Reserve Base, which includes both measured and indicated tonnage. The U.S. term "measured" approximates the term "proved," used by the World Energy Council. The U.S. "measured and indicated" data have been combined and cannot be recaptured as "measured alone." • Totals may not equal sum of components due to independent rounding.

Web Page: For related information, see http://www.eia.doe.gov/international.

Sources: **United States:** Based on EIA, *Annual Coal Report 2006,* Table 15 and Unpublished File Data of the Coal Reserves Data Base (April 2007). **All Other Data:** World Energy Council, *2007 Survey of Energy Resources.*

Figure 11.14 World Coal Production

World and Top Producing Countries, 1997-2006

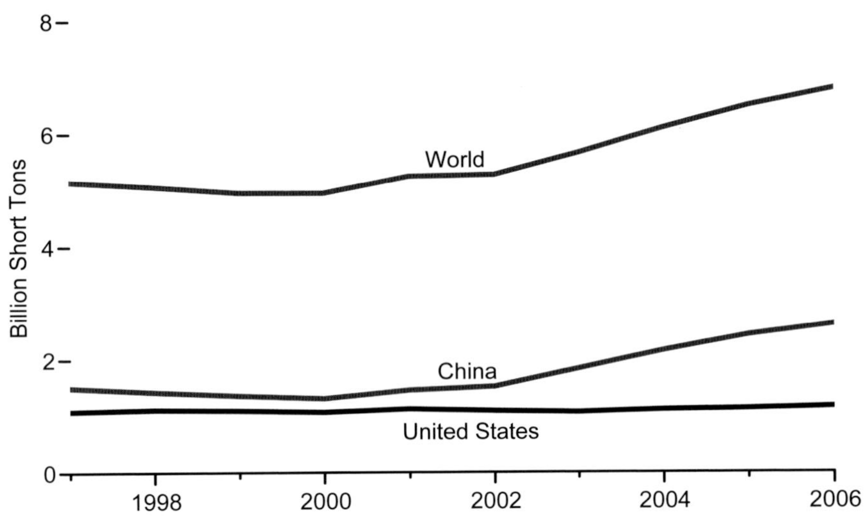

World Areas, 1997-2006

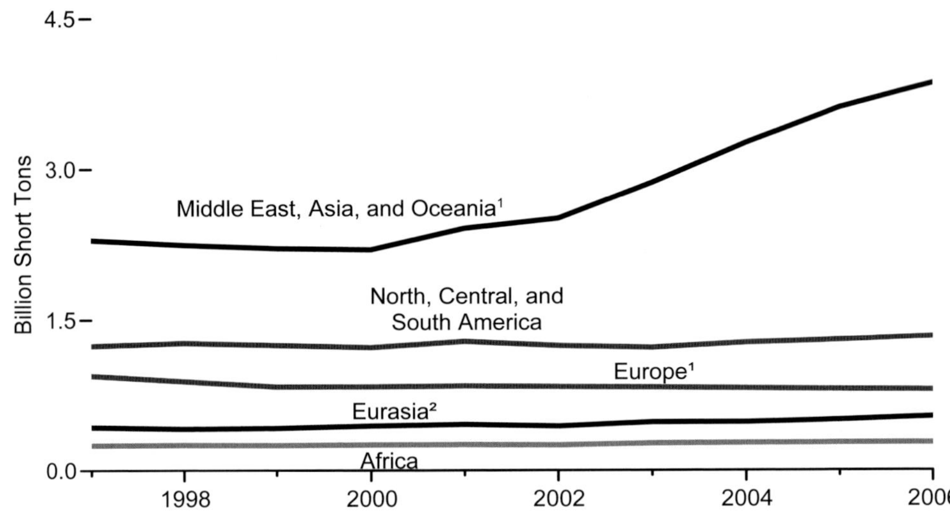

Top Producing Countries, 2006

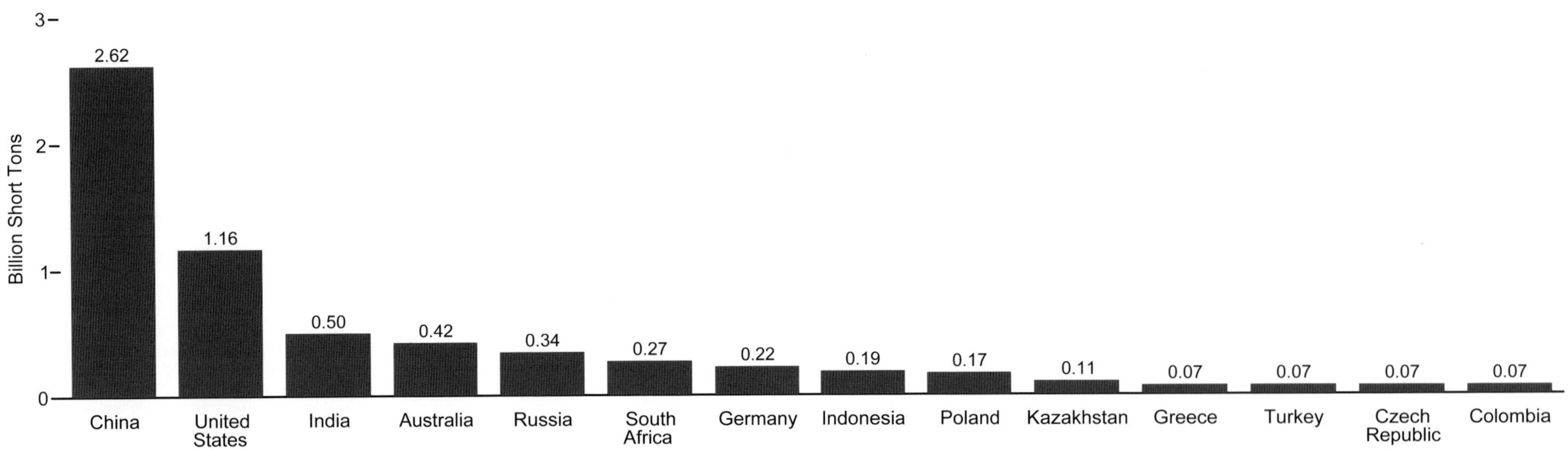

[1] Excludes countries that were part of the former U.S.S.R. See "U.S.S.R." in Glossary.
[2] Includes only countries that were part of the former U.S.S.R. See "U.S.S.R." in Glossary.

Note: Because vertical scales differ, graphs should not be compared.
Source: Table 11.14.

Table 11.14 World Coal Production, 1997-2006

(Million Short Tons)

Region and Country	1997	1998	1999	2000	2001	2002	2003	2004	2005	2006 P
North, Central, and South America	**1,237**	**1,266**	**1,243**	**1,221**	**1,281**	**1,238**	**1,219**	**R1,269**	**1,296**	**1,331**
Canada	87	83	80	76	78	73	R69	73	72	69
Colombia	36	37	36	42	48	44	55	59	65	70
Mexico	11	12	11	13	13	12	11	R11	12	13
United States	1,090	1,118	1,100	1,074	1,128	1,094	1,072	1,112	1,131	1,163
Other	13	16	15	17	15	14	13	R14	16	16
Europe [1]	**R941**	**R886**	**R832**	**R832**	**R839**	**R832**	**R825**	**R816**	**806**	**799**
Bosnia and Herzegovina	6	8	7	9	8	10	9	R9	10	10
Bulgaria	33	33	28	29	29	29	30	29	27	28
Czech Republic	82	74	65	72	73	70	R71	R69	69	70
Former Serbia and Montenegro	47	49	36	41	40	42	44	45	45	43
Germany	252	233	226	226	227	233	229	R235	229	223
Greece	65	67	68	70	73	78	75	R77	76	72
Hungary	17	16	16	15	15	14	15	12	11	11
Macedonia	8	9	8	8	9	8	8	R8	8	6
Poland	221	197	R188	179	180	178	R180	178	175	171
Romania	37	29	25	32	37	34	36	35	34	39
Slovenia	6	5	5	5	5	5	5	5	5	5
Spain	R35	R34	R32	R31	R30	R29	R27	R27	26	24
Turkey	66	74	74	70	68	R60	R54	51	64	72
United Kingdom	52	44	40	34	35	33	31	27	22	20
Other	R14	13	R12	R10	10	10	R10	R7	5	5
Eurasia [2]	**R431**	**R414**	**R421**	**R443**	**R457**	**R440**	**R479**	**R479**	**502**	**534**
Estonia	15	14	12	13	13	14	16	15	16	16
Kazakhstan	80	78	66	82	R87	R81	R94	96	95	106
Russia	R268	R252	R270	R276	R285	R273	R296	R299	321	341
Ukraine	65	66	R69	69	68	68	71	R66	67	68
Other	4	4	4	3	4	4	3	R4	4	4
Africa	**251**	**254**	**250**	**256**	**257**	**252**	**270**	**R273**	**276**	**276**
South Africa	244	247	243	249	251	246	264	268	270	269
Zimbabwe	4	5	5	5	R5	4	4	4	4	4
Other	2	2	2	2	2	2	2	2	2	2
Middle East, Asia, and Oceania [1]	**2,291**	**2,243**	**R2,210**	**R2,197**	**2,408**	**R2,507**	**R2,864**	**R3,260**	**3,609**	**3,852**
Australia	292	317	321	338	363	377	377	391	414	420
China	1,507	1,429	1,365	1,314	1,459	1,521	1,838	2,156	2,430	2,620
India	338	343	356	370	389	R401	R426	R455	473	497
Indonesia	R61	68	81	84	102	114	127	R146	168	186
Mongolia	5	6	R5	6	R6	R6	R6	R8	8	8
North Korea	30	27	31	33	34	32	32	R33	35	36
South Korea	5	5	5	5	4	4	4	4	3	3
Thailand	26	22	20	20	22	22	21	22	24	22
Vietnam	13	12	11	13	14	18	18	R28	36	42
Other	15	15	15	15	16	13	14	R17	19	18
World	**R5,151**	**R5,062**	**R4,955**	**R4,949**	**R5,243**	**R5,268**	**R5,657**	**R6,097**	**6,490**	**6,793**

[1] Excludes countries that were part of the former U.S.S.R. See "U.S.S.R." in Glossary.
[2] Includes only countries that were part of the former U.S.S.R. See "U.S.S.R." in Glossary.
R=Revised. P=Preliminary.
Notes: • Coal includes anthracite, subanthracite, bituminous coal, subbituminous coal, lignite, and brown coal. • Production from Estonia is oil shale. • Totals may not equal sum of components due to independent rounding.
Web Page: For related information, see http://www.eia.doe.gov/international.
Sources: **United States:** Table 7.1. **All Other Data:** Energy Information Administration, International Energy Database, May 1, 2008.

Figure 11.15 World Coal Consumption

World Total, 1980-2006

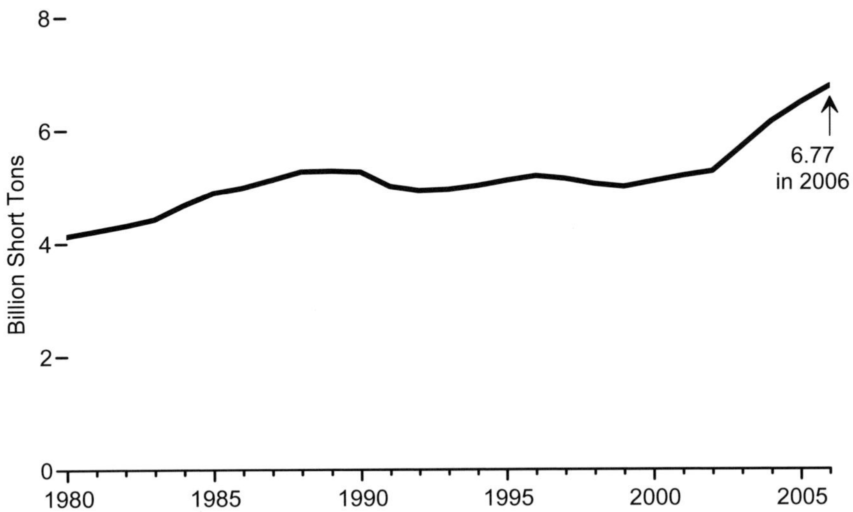

Selected Countries, 1980-2006

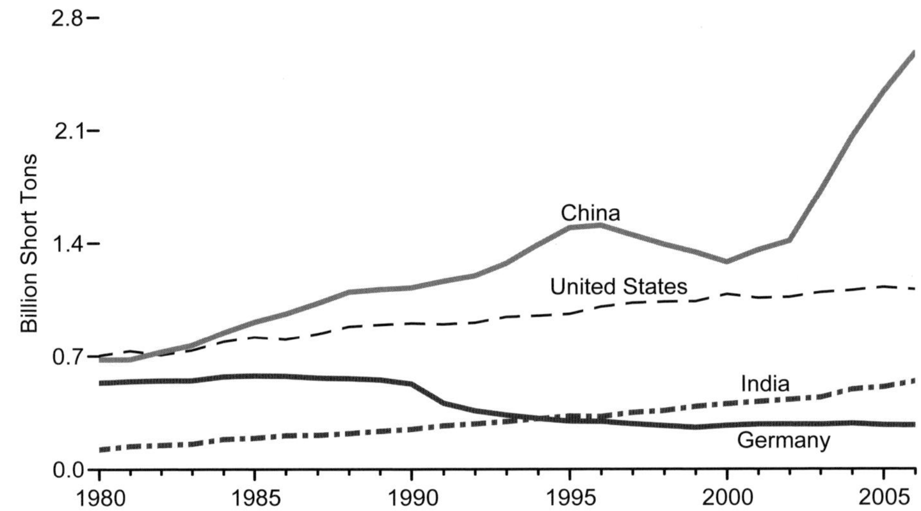

Top Consuming Countries, 2006

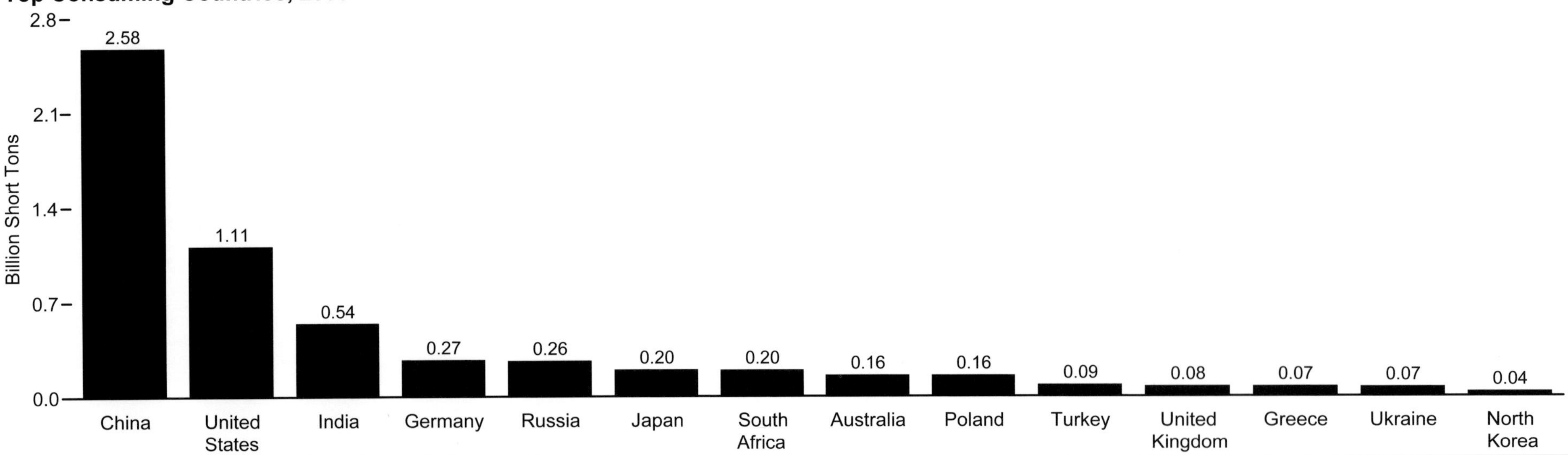

Note: Because vertical scales differ, graphs should not be compared.

Source: Table 11.15.

Table 11.15 World Coal Consumption, 1980-2006

(Million Short Tons)

Year	Australia	China	Germany [1]	Greece	India	Japan	North Korea	Poland	Former U.S.S.R.	Russia	South Africa	Turkey	Ukraine	United Kingdom	United States	Other	World
1980	74	679	535	26	R123	98	49	221	751	– –	105	R23	– –	134	703	R611	R4,131
1981	R74	680	544	30	R144	106	51	200	748	– –	116	R24	– –	130	733	R640	R4,221
1982	R80	726	548	31	R150	R104	54	208	771	– –	124	26	– –	122	707	R666	R4,316
1983	R77	768	549	36	R158	R99	56	213	764	– –	127	R30	– –	123	737	R691	R4,428
1984	R80	845	573	36	R187	R111	57	227	770	– –	137	35	– –	88	791	R740	R4,677
1985	86	R911	579	42	R194	R117	60	238	779	– –	142	46	– –	116	818	R767	R4,896
1986	R85	962	576	44	209	R107	59	247	803	– –	145	R55	– –	123	804	R765	R4,985
1987	93	1,027	565	49	R211	R110	57	258	807	– –	148	54	– –	129	837	R782	R5,127
1988	96	1,098	561	56	R222	R122	58	253	821	– –	151	R46	– –	123	884	R780	R5,271
1989	104	1,113	553	59	R236	123	57	242	777	– –	140	R61	– –	126	895	R796	R5,283
1990	104	1,124	528	59	R248	R127	54	202	848	– –	139	60	– –	119	904	R749	R5,266
1991	R102	1,165	408	59	R270	129	52	202	672	– –	144	64	– –	118	899	R725	R5,008
1992	R108	1,199	362	62	R282	125	46	192	– –	R347	147	66	R132	111	908	R847	R4,935
1993	R107	1,276	335	62	R296	127	42	R193	– –	R332	146	60	R115	96	944	R824	R4,956
1994	110	1,390	314	66	R313	R136	39	184	– –	R305	161	66	R92	91	951	802	R5,019
1995	112	1,495	298	64	R329	R141	36	185	– –	R290	162	67	R95	79	962	R798	R5,113
1996	120	1,509	296	66	R326	142	31	189	– –	R297	164	73	R74	77	1,006	R821	R5,192
1997	127	1,450	R281	66	R351	R148	30	182	– –	R268	172	80	72	69	1,030	R818	R5,144
1998	138	1,392	269	68	R362	R141	27	168	– –	R250	161	86	73	68	1,037	R810	R5,050
1999	141	1,343	258	68	R388	R153	31	162	– –	R258	R169	84	72	61	1,039	R774	R4,999
2000	141	1,282	270	72	R403	169	33	R158	– –	R264	R175	89	R72	R65	1,084	R819	R5,098
2001	141	1,357	278	75	417	R173	34	152	– –	R254	R179	80	R70	70	1,060	R851	R5,191
2002	145	1,413	R278	77	R430	R179	32	149	– –	R252	170	73	R71	64	1,066	R871	R5,272
2003	143	1,720	277	78	R443	185	33	155	– –	R256	R186	71	R77	R70	1,095	R917	R5,706
2004	R146	R2,057	R283	R79	R494	R203	33	R154	– –	R249	R199	R71	R69	67	1,107	R939	R6,150
2005	158	2,333	273	78	507	196	35	150	– –	258	193	85	68	68	1,126	953	6,483
2006P	156	2,578	272	74	543	198	35	155	– –	264	195	86	70	75	1,112	959	6,771

[1] Through 1990, this is East and West Germany. Beginning in 1991, this is unified Germany.
R=Revised. P=Preliminary. – – = Not applicable.
Note: Totals may not equal sum of components due to independent rounding.

Web Page: For related information, see http://www.eia.doe.gov/international.
Sources: **United States:** Table 7.1. **All Other Data:** Energy Information Administration, International Energy Database, May 2, 2008.

Figure 11.16 World Net Generation of Electricity

Net Generation by Type, 1980, 1990, and 2005

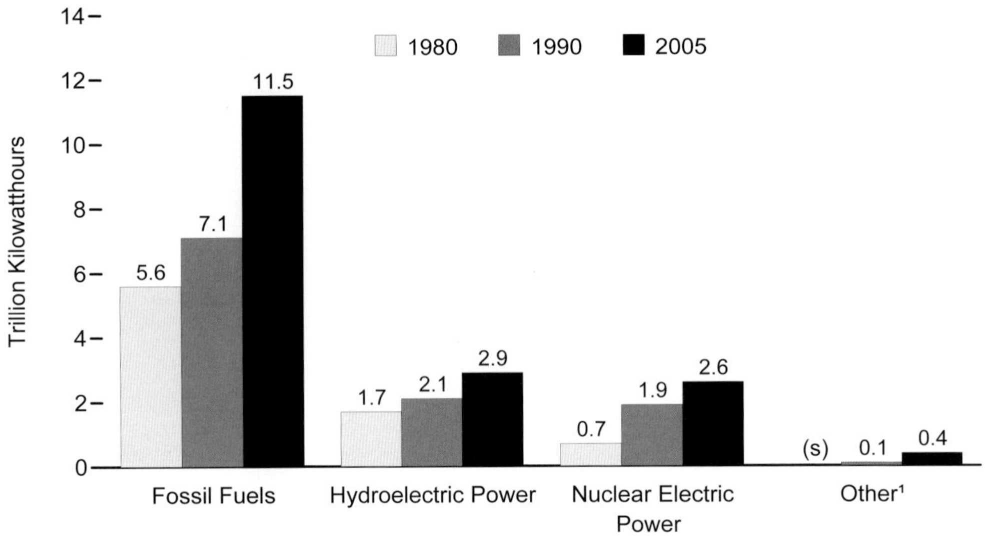

Top Net Generating Countries, 2005

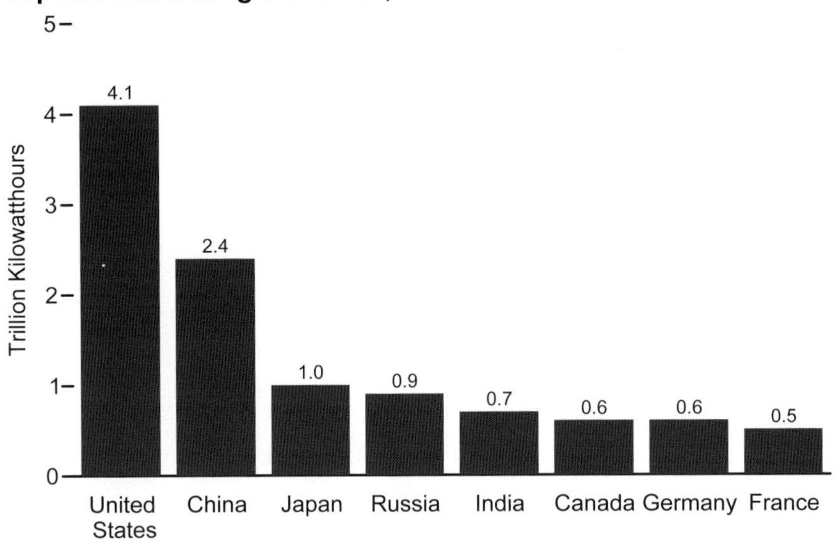

Net Generation by Type by Region, 2005

(Percent of Regional Total)

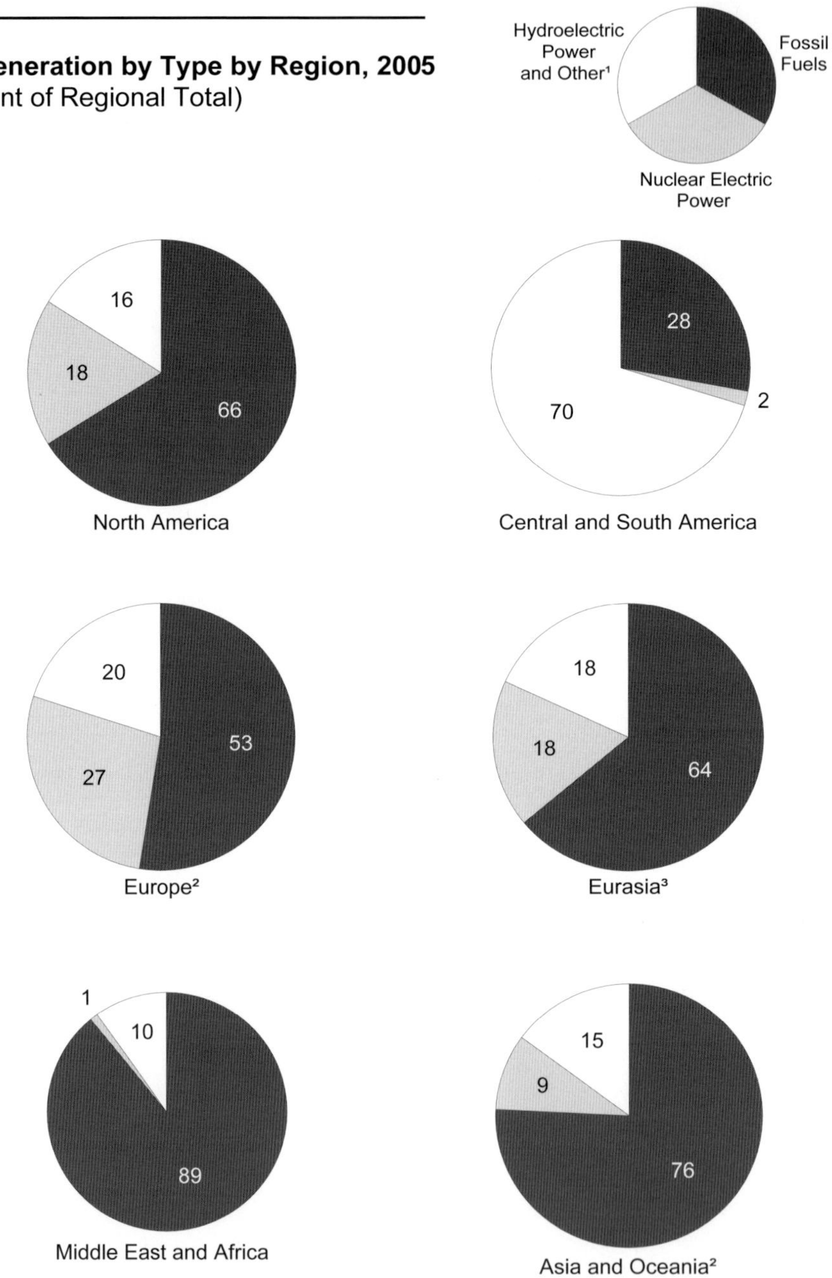

[1] Wood, waste, geothermal, solar, wind, batteries, chemicals, hydrogen, pitch, purchased steam, sulfur, and miscellaneous technologies.

[2] Excludes countries that were part of the former U.S.S.R. See "U.S.S.R." in Glossary.

[3] Includes countries that were part of the former U.S.S.R. See "U.S.S.R." in Glossary.

(s)=Less than 0.05 trillion kilowatthours.

Note: Because vertical scales differ, graphs should not be compared.

Source: Table 11.16.

Table 11.16 World Net Generation of Electricity by Type, 1980, 1990, and 2005

(Billion Kilowatthours)

Region and Country	Fossil Fuels			Nuclear Electric Power			Hydroelectric Power [1]			Total [2]		
	1980	1990	2005 P	1980	1990	2005 P	1980	1990	2005 P	1980	1990	2005 P
North America	**1,880.1**	**2,292.0**	**3,238.4**	**287.0**	**648.9**	**879.7**	**546.9**	**606.5**	**651.1**	**2,721.6**	**3,624.0**	**4,888.4**
Canada	79.8	101.9	152.2	35.9	69.2	87.4	251.0	293.9	359.9	367.9	468.7	609.6
Mexico	46.0	85.7	175.2	.0	2.8	10.3	16.7	23.2	27.5	63.6	116.6	222.4
United States	1,753.8	2,103.8	2,910.0	251.1	576.9	782.0	279.2	289.4	263.8	2,289.6	3,038.0	4,055.4
Other	.5	.7	1.0	.0	.0	.0	.0	.0	.0	.5	.7	1.0
Central and South America	**99.8**	**114.8**	**253.3**	**2.2**	**9.0**	**16.3**	**201.5**	**365.1**	**613.2**	**308.2**	**497.2**	**908.7**
Argentina	22.2	20.9	59.5	2.2	7.0	6.4	17.3	20.2	33.9	41.8	48.3	101.1
Brazil	7.5	8.1	34.1	.0	1.9	9.9	128.4	204.6	334.1	138.3	219.6	396.4
Paraguay	(s)	(s)	(s)	.0	.0	.0	.7	27.2	50.7	.8	27.2	50.7
Venezuela	17.6	21.0	24.9	.0	.0	.0	14.4	36.6	74.3	32.0	57.6	99.2
Other	52.4	64.8	134.9	.0	.0	.0	40.6	76.4	120.2	95.3	144.4	261.4
Europe [3]	**R 1,453.3**	**R 1,441.0**	**1,837.7**	**229.6**	**761.3**	**957.3**	**458.0**	**474.5**	**539.6**	**R 2,154.7**	**R 2,696.4**	**3,494.7**
Belgium	38.3	25.0	33.0	11.9	40.6	45.2	.3	.3	.3	50.8	66.5	80.8
Czech Republic	– –	– –	50.8	– –	– –	23.5	– –	– –	2.4	– –	– –	77.4
Finland	22.0	22.8	22.0	6.6	18.3	22.1	10.1	10.8	13.6	38.7	51.8	67.1
France	118.0	44.3	57.2	63.4	298.4	429.0	68.3	52.8	51.2	250.8	397.6	543.6
Germany	390.3	358.9	362.3	55.6	145.1	154.9	18.8	17.2	19.4	469.9	526.0	579.4
Italy	125.5	167.5	231.1	2.1	.0	.0	45.0	31.3	33.3	176.4	202.1	278.5
Netherlands	58.0	63.2	81.8	3.9	3.3	3.8	.0	.1	.1	62.9	67.7	94.3
Norway	.1	.2	.5	.0	.0	.0	82.7	119.9	134.4	82.9	120.4	135.8
Poland	111.1	125.0	142.1	.0	.0	.0	2.3	1.4	2.2	113.8	126.7	146.2
Romania	51.4	49.7	31.6	.0	.0	5.3	12.5	10.9	20.0	63.9	60.6	56.9
Spain	74.5	66.5	173.1	5.2	51.6	54.7	29.2	25.2	19.4	109.2	143.9	270.3
Sweden	10.1	3.2	3.7	25.3	64.8	68.6	58.1	71.8	72.1	94.3	141.5	153.2
Switzerland	.9	.6	1.0	12.9	22.4	22.2	32.5	29.5	30.9	46.4	53.0	56.1
Turkey	12.0	32.3	114.8	.0	.0	.0	11.2	22.9	39.2	23.3	55.2	154.2
United Kingdom	228.9	230.0	277.5	32.3	62.5	75.2	3.9	5.1	4.9	265.1	299.0	372.6
Other	R 212.1	R 251.6	255.3	10.3	54.4	52.9	83.2	75.4	96.3	R 306.3	R 384.3	428.2
Eurasia [4]	**1,037.1**	**1,204.1**	**843.6**	**72.9**	**201.3**	**235.8**	**184.0**	**230.7**	**244.7**	**1,294.0**	**1,636.1**	**1,327.3**
Kazakhstan	– –	– –	56.5	– –	– –	.0	– –	– –	7.8	– –	– –	64.2
Russia	– –	– –	588.4	– –	– –	140.2	– –	– –	172.9	– –	– –	904.4
Ukraine	– –	– –	79.7	– –	– –	83.3	– –	– –	12.4	– –	– –	175.4
Other	1,037.1	1,204.1	119.1	72.9	201.3	12.3	184.0	230.7	51.7	1,294.0	1,636.1	183.3
Middle East	**R 81.8**	**R 215.5**	**581.7**	**.0**	**.0**	**.0**	**9.6**	**9.6**	**21.0**	**R 91.4**	**R 225.1**	**602.7**
Iran	15.7	49.8	154.4	.0	.0	.0	5.6	6.0	15.9	21.3	55.9	170.4
Saudi Arabia	20.5	64.9	165.6	.0	.0	.0	.0	.0	.0	20.5	64.9	165.6
Other	R 45.7	R 100.7	261.7	.0	.0	.0	4.1	3.6	5.0	R 49.7	R 104.3	266.7
Africa	**128.8**	**243.6**	**430.3**	**.0**	**8.4**	**12.2**	**60.1**	**54.9**	**88.7**	**188.9**	**307.3**	**533.2**
Egypt	8.6	31.5	89.8	.0	.0	.0	9.7	9.9	12.1	18.3	41.4	102.5
South Africa	92.1	146.6	214.9	.0	8.4	12.2	1.0	1.0	.9	93.1	156.0	228.3
Other	28.2	65.5	125.7	.0	.0	.0	49.4	44.0	75.6	77.6	109.9	202.4
Asia and Oceania [3]	**907.7**	**1,626.8**	**4,270.2**	**92.7**	**279.9**	**524.3**	**262.7**	**404.1**	**735.3**	**1,268.0**	**2,333.0**	**5,589.1**
Australia	74.5	131.8	218.4	.0	.0	.0	12.8	14.0	15.5	87.7	146.4	236.7
China	227.9	465.2	1,922.1	.0	.0	50.3	57.6	125.1	397.0	285.5	590.3	2,371.8
India	69.7	198.9	539.2	3.0	5.6	15.7	46.5	70.9	99.0	119.3	275.5	661.6
Indonesia	10.6	35.3	103.4	.0	.0	.0	2.2	6.7	10.7	12.8	43.0	120.3
Japan	381.6	524.0	645.5	78.6	192.2	278.4	87.8	88.4	77.4	549.1	817.3	1,024.6
South Korea	29.8	45.5	222.7	3.3	50.2	139.4	1.5	4.6	3.6	34.6	100.4	366.2
Taiwan	31.3	43.6	164.5	7.8	31.6	38.0	2.9	8.2	7.8	42.0	83.3	210.3
Thailand	12.3	38.7	115.7	.0	.0	.0	1.3	4.9	5.7	13.6	43.7	124.6
Other	70.1	143.8	338.6	(s)	.4	2.4	50.0	81.2	118.5	123.5	233.2	472.9
World	**5,588.5**	**7,137.9**	**11,455.3**	**684.4**	**1,908.8**	**2,625.6**	**1,722.9**	**2,145.4**	**2,893.5**	**8,026.9**	**R 11,319.2**	**17,344.0**

[1] Excludes pumped storage, except for the United States.
[2] Wood, waste, geothermal, solar, wind, batteries, chemicals, hydrogen, pitch, purchased steam, sulfur, and miscellaneous technologies are included in total.
[3] Excludes countries that were part of the former U.S.S.R. See "U.S.S.R." in Glossary.
[4] Includes only countries that were part of the former U.S.S.R. See "U.S.S.R." in Glossary.

P=Preliminary. – – = Not applicable. (s)=Less than 0.05 billion kilowatthours.
Note: Totals may not equal sum of components due to independent rounding.
Web Page: For related information, see http://www.eia.doe.gov/international.
Sources: **United States:** Table 8.2a. **All Other Data:** Energy Information Administration, International Energy Database, April 24, 2008.

Figure 11.17 World Electrical Installed Capacity by Type

Installed Capacity by Type, 1980, 1990, and 2005

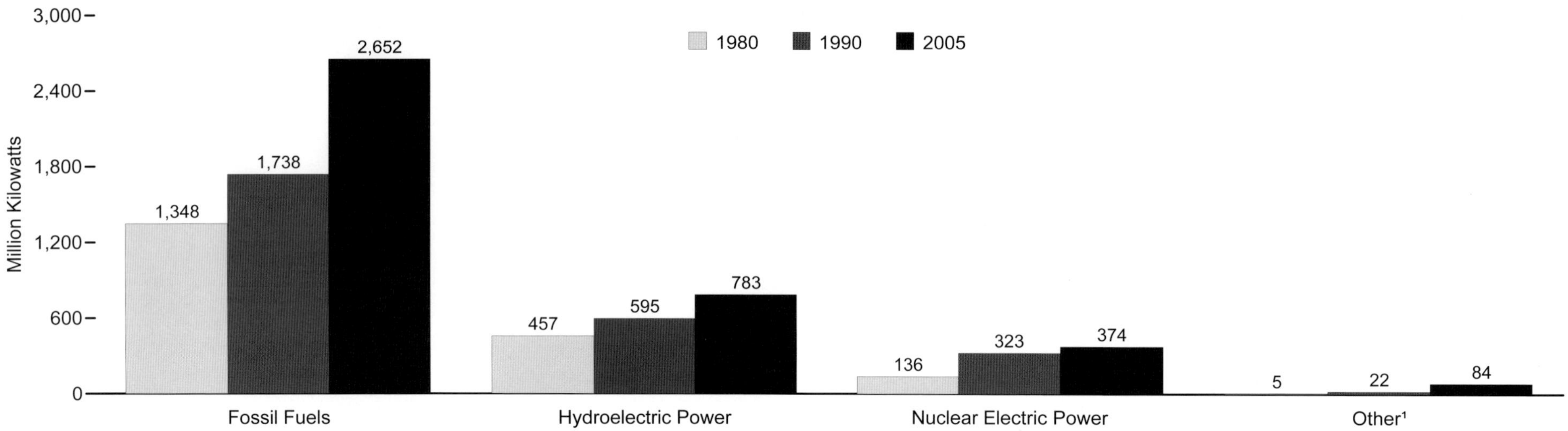

Installed Capacity and Net Generation Shares of Totals by Type, 2005

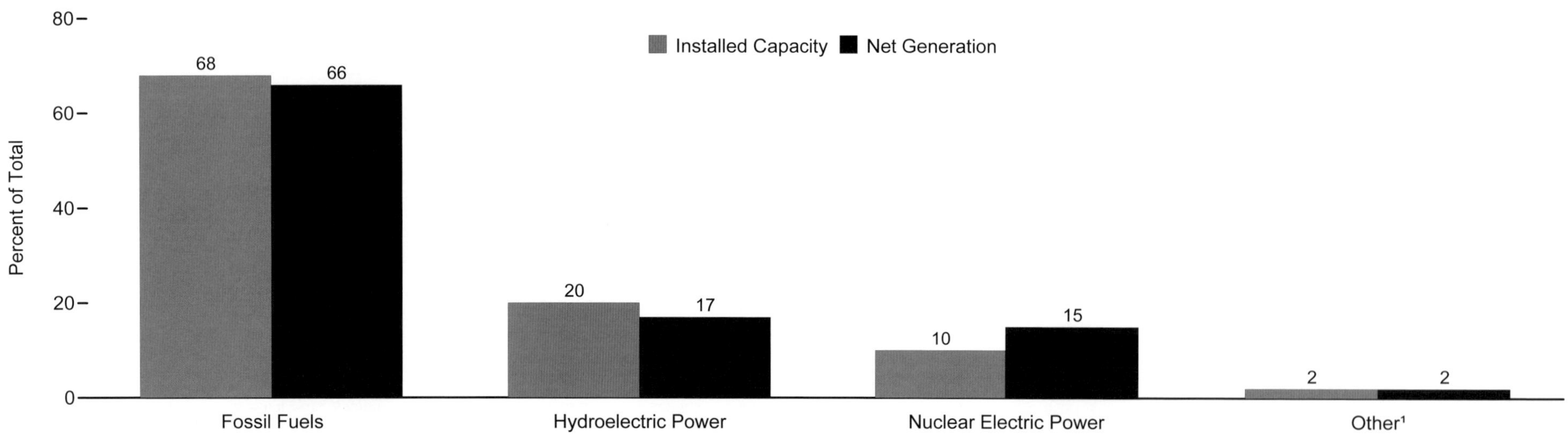

[1] Wood, waste, geothermal, solar, wind, batteries, chemicals, hydrogen, pitch, purchased steam, sulfur, and miscellaneous technologies.

Note: Shares are based on data prior to rounding for publication and may not sum exactly to 100 percent.

Sources: Tables 11.16 and 11.17.

Table 11.17 World Electrical Installed Capacity by Type, 1980, 1990, and 2005

(Million Kilowatts)

Region and Country	Fossil Fuels			Nuclear Electric Power			Hydroelectric Power [1]			Total [2]		
	1980	1990	2005 P	1980	1990	2005 P	1980	1990	2005 P	1980	1990	2005 P
North America	**481.7**	**574.9**	**831.4**	**57.7**	**112.2**	**114.2**	**135.7**	**159.1**	**180.1**	**676.6**	**861.3**	**1,150.5**
Canada	26.6	28.0	35.1	5.9	11.9	12.8	47.9	57.9	70.7	80.8	98.9	120.3
Mexico	10.8	18.9	39.0	.0	.7	1.4	6.1	7.8	10.6	17.0	28.0	51.9
United States [3]	444.1	527.8	757.1	51.8	99.6	100.0	81.7	93.4	98.9	578.6	734.1	978.0
Other	.2	.3	.3	.0	.0	.0	.0	.0	.0	.2	.3	.3
Central and South America	**36.0**	**44.9**	**80.2**	**.4**	**1.7**	**3.0**	**43.0**	**84.1**	**125.0**	**81.2**	**132.8**	**214.9**
Argentina	8.0	9.5	17.3	.4	1.0	1.0	3.6	6.6	9.9	12.0	17.2	28.2
Brazil	4.1	4.7	13.8	.0	.7	2.0	27.5	44.8	69.0	33.4	52.1	90.7
Paraguay	.1	(s)	(s)	.0	.0	.0	.2	5.8	7.4	.2	5.8	7.4
Venezuela	5.8	8.5	8.3	.0	.0	.0	2.7	10.0	13.9	8.5	18.5	22.1
Other	18.0	22.1	40.8	.0	.0	.0	9.0	16.9	24.8	27.1	39.2	66.4
Europe [4]	**R 351.2**	**R 386.6**	**455.3**	**46.5**	**125.7**	**137.5**	**135.3**	**153.9**	**169.5**	**R 533.8**	**R 669.3**	**803.4**
Belgium	8.2	7.2	8.0	1.7	5.5	5.8	.7	.1	.1	10.6	12.8	14.3
Czech Republic	– –	– –	11.5	– –	– –	3.8	– –	– –	1.0	– –	– –	16.3
Finland	6.3	7.8	10.8	2.2	2.4	2.7	2.4	2.6	3.0	11.0	12.7	16.6
France	30.1	22.8	27.8	14.4	52.5	63.4	16.4	20.3	21.0	61.1	95.9	112.7
Germany	84.1	87.5	78.4	10.4	24.5	20.6	7.9	8.7	4.1	102.5	121.6	120.4
Italy	27.6	37.5	57.8	1.4	.0	.0	15.8	12.6	16.7	45.3	50.6	77.3
Netherlands	16.9	16.8	20.2	.5	.5	.4	.0	(s)	(s)	17.4	17.3	21.8
Norway	.2	.2	.1	.0	.0	.0	19.8	25.7	26.1	20.0	26.0	26.5
Poland	23.0	26.1	29.4	.0	.0	.0	.6	.6	.9	23.6	26.8	30.3
Romania	12.7	17.3	13.3	.0	.0	.7	3.5	5.6	6.3	16.1	22.9	20.3
Spain	13.9	19.9	35.5	1.1	7.5	7.6	13.5	11.6	15.6	28.5	39.1	66.9
Sweden	7.9	7.2	7.6	4.6	9.9	9.5	14.9	15.8	16.3	27.4	33.5	33.9
Switzerland	.6	.8	.5	1.9	3.0	3.2	11.5	11.6	13.3	14.0	15.3	17.5
Turkey	3.0	9.2	24.1	.0	.0	.0	2.1	6.6	12.6	5.1	15.8	36.8
United Kingdom	65.6	59.1	62.0	6.5	11.4	11.9	2.5	1.4	1.5	74.5	72.0	78.1
Other	R 51.2	R 67.2	68.2	1.7	8.6	8.1	23.8	30.7	30.9	R 76.8	R 106.9	113.6
Eurasia [5]	**201.9**	**240.8**	**238.1**	**12.5**	**37.9**	**38.6**	**52.3**	**65.0**	**68.5**	**266.7**	**343.7**	**345.4**
Kazakhstan	– –	– –	14.9	– –	– –	.0	– –	– –	2.2	– –	– –	17.2
Russia	– –	– –	148.3	– –	– –	23.2	– –	– –	45.5	– –	– –	217.2
Ukraine	– –	– –	36.2	– –	– –	13.8	– –	– –	4.8	– –	– –	54.9
Other	201.9	240.8	38.7	12.5	37.9	1.6	52.3	65.0	15.9	266.7	343.7	56.1
Middle East	**R 27.6**	**R 67.6**	**115.6**	**.0**	**.0**	**.0**	**2.6**	**4.8**	**7.5**	**R 30.1**	**R 72.4**	**123.0**
Iran	9.4	15.5	37.9	.0	.0	.0	1.8	2.0	4.8	11.2	17.4	42.7
Saudi Arabia	5.9	19.1	30.5	.0	.0	.0	.0	.0	.0	5.9	19.1	30.5
Other	R 12.3	R 33.1	47.2	.0	.0	.0	.8	2.8	2.6	R 13.0	R 35.9	49.8
Africa	**30.5**	**57.4**	**82.8**	**.0**	**1.8**	**1.8**	**13.9**	**R 18.7**	**21.6**	**44.5**	**R 78.1**	**106.5**
Egypt	2.4	8.7	15.7	.0	.0	.0	2.4	2.7	2.7	4.9	11.5	18.5
South Africa	17.8	28.6	38.0	.0	1.8	1.8	.5	.6	.7	18.4	31.0	40.5
Other	10.3	20.1	29.1	.0	.0	.0	10.9	R 15.4	18.2	21.2	R 35.6	47.6
Asia and Oceania [4]	**218.9**	**365.4**	**849.0**	**18.5**	**43.9**	**79.1**	**74.4**	**109.3**	**211.1**	**312.5**	**520.2**	**1,149.7**
Australia	17.6	27.8	40.8	.0	.0	.0	6.2	7.3	7.8	23.8	35.1	49.3
China	45.6	92.1	329.5	.0	.0	6.8	20.3	34.6	105.2	65.9	126.6	442.4
India	20.7	51.9	99.8	.9	1.6	2.8	11.8	18.3	31.0	33.3	71.8	137.6
Indonesia	3.9	9.6	17.9	.0	.0	.0	1.0	3.0	4.6	4.9	12.7	23.2
Japan	94.3	119.1	177.5	15.7	29.4	47.1	19.6	20.4	22.0	129.8	169.1	247.9
South Korea	6.5	11.0	43.8	.6	7.6	16.7	.8	1.3	1.6	7.9	20.0	62.2
Taiwan	6.9	10.2	27.6	1.3	5.1	5.1	1.4	2.6	4.5	9.6	17.9	37.4
Thailand	2.6	6.0	22.4	.0	.0	.0	1.3	2.3	3.5	3.8	8.3	25.9
Other	20.8	37.7	89.7	.1	.1	.5	12.1	19.6	30.9	33.6	58.7	123.7
World	**1,347.8**	**1,737.6**	**2,652.3**	**135.5**	**323.1**	**374.2**	**457.2**	**R 594.9**	**783.2**	**1,945.5**	**R 2,677.8**	**3,893.3**

[1] Excludes pumped storage, except for the United States.
[2] Wood, waste, geothermal, solar, wind, batteries, chemicals, hydrogen, pitch, purchased steam, sulfur, and miscellaneous technologies are included in total.
[3] Net summer capability.
[4] Excludes countries that were part of the former U.S.S.R. See "U.S.S.R." in Glossary.
[5] Includes only countries that were part of the former U.S.S.R. See "U.S.S.R." in Glossary.

P=Preliminary. – – = Not applicable. (s)=Less than 0.05 million kilowatts.
Note: Totals may not equal sum of components due to independent rounding.
Web Page: For related information, see http://www.eia.doe.gov/international.
Source: **United States:** Table 8.11a **All Other Data:** Energy Information Administration, International Energy Database, April 24, 2008.

Figure 11.18 World Nuclear Electricity Net Generation

By Region, 2006

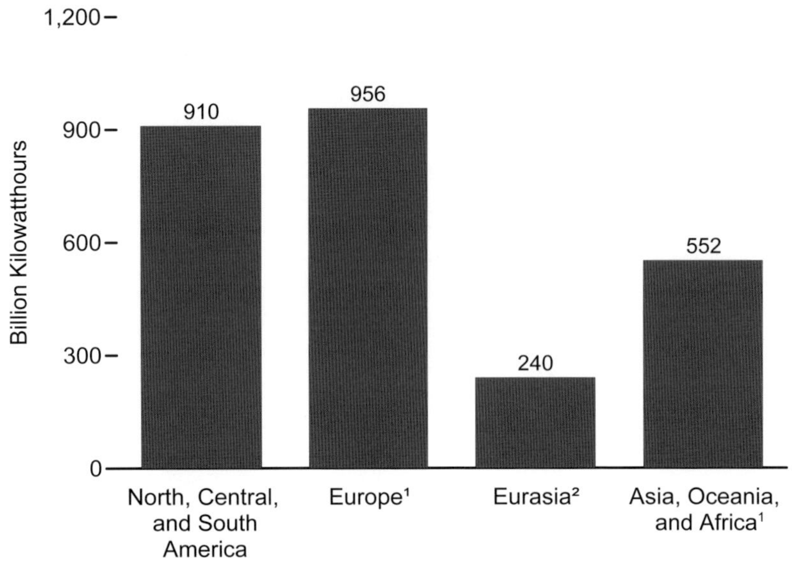

Top Net Generating Countries, 1997-2006

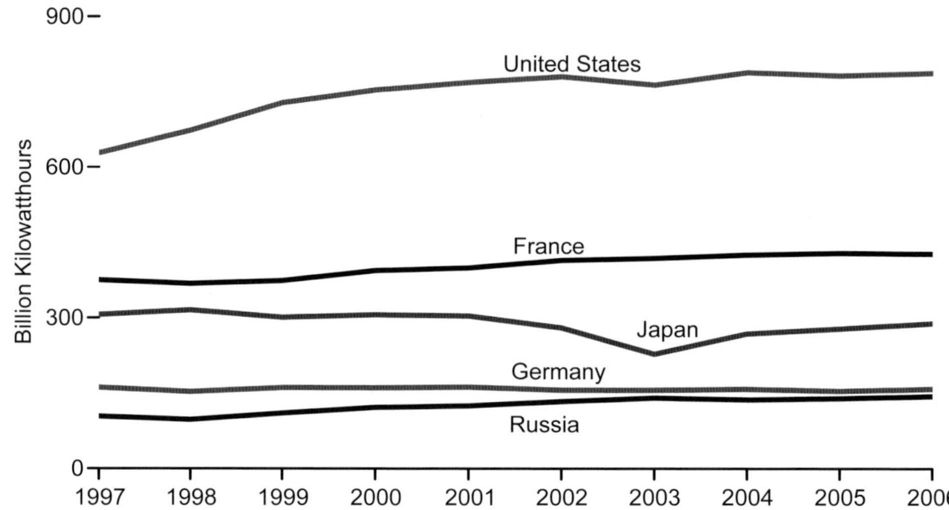

Top Net Generating Countries, 2006

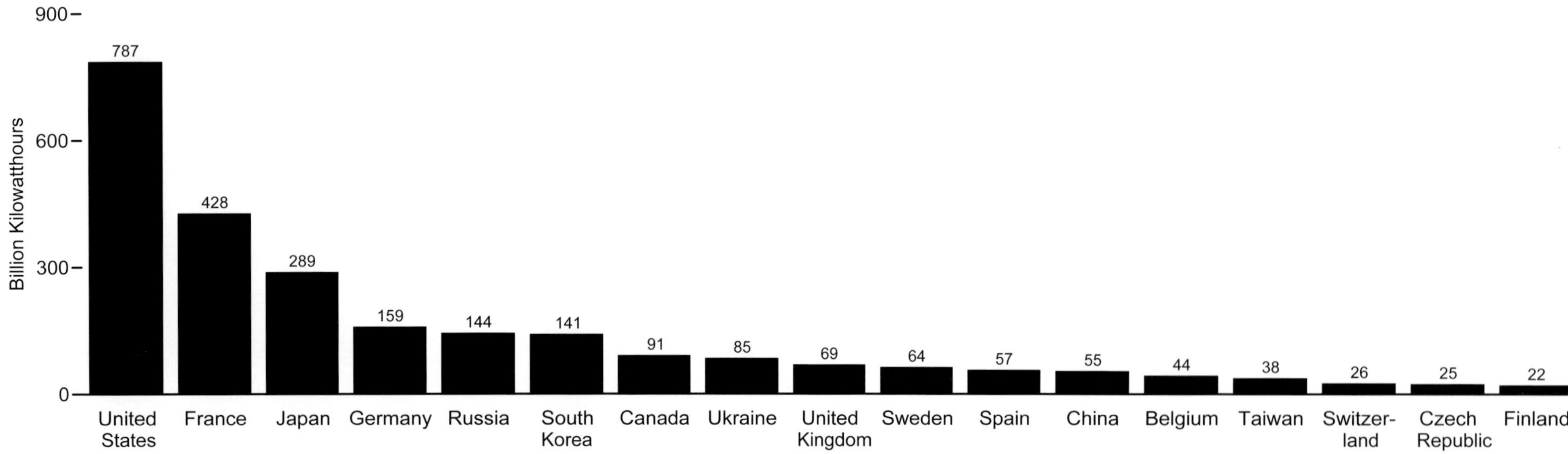

[1] Excludes countries that were part of the former U.S.S.R. See "U.S.S.R." in Glossary.
[2] Includes only countries that were part of the former U.S.S.R. See "U.S.S.R." in Glossary.

Note: Because vertical scales differ, graphs should not be compared.
Source: Table 11.18.

Table 11.18 World Nuclear Electricity Net Generation, 1997-2006

(Billion Kilowatthours)

Region and Country	1997	1998	1999	2000	2001	2002	2003	2004	2005	2006
North America	**716.4**	**750.2**	**807.6**	**830.9**	**850.0**	**861.1**	**844.9**	**883.1**	**879.7**	**888.9**
Canada	77.9	67.7	69.8	69.2	72.9	71.8	71.1	85.9	87.4	91.4
Mexico	9.9	8.8	9.5	7.8	8.3	9.3	10.0	8.7	10.3	10.3
United States	628.6	673.7	728.3	753.9	768.8	780.1	763.7	788.5	782.0	787.2
Central and South America	**10.5**	**10.3**	**10.5**	**10.9**	**20.8**	**19.2**	**20.4**	**18.9**	**16.3**	**20.9**
Argentina	7.5	7.1	6.7	6.0	6.5	5.4	7.0	7.3	6.4	7.2
Brazil	3.0	3.1	3.8	4.9	14.3	13.8	13.4	11.6	9.9	13.8
Europe [1]	**902.8**	**897.7**	**911.3**	**914.9**	**944.1**	**953.7**	**957.0**	**966.6**	**957.3**	**955.6**
Belgium	45.0	43.9	46.6	45.7	44.0	45.0	45.0	44.9	45.2	44.2
Bulgaria	16.4	16.1	15.0	17.3	18.2	20.2	16.0	15.6	17.3	18.1
Czech Republic	12.5	12.5	12.7	12.9	14.0	17.8	24.6	25.0	23.5	24.7
Finland	19.0	20.8	21.8	21.4	21.6	21.2	21.6	21.6	22.1	21.7
France	375.7	368.6	374.5	394.4	400.0	414.9	419.0	425.8	429.0	427.7
Germany	161.8	153.6	161.5	161.1	162.7	156.6	156.8	158.7	154.9	158.9
Hungary	13.3	13.3	13.4	13.5	13.4	13.3	10.5	11.3	13.1	12.9
Netherlands	2.3	3.6	3.6	3.7	3.8	3.7	3.8	3.6	3.8	3.3
Romania	5.1	4.9	4.8	5.2	5.0	5.1	4.5	5.3	5.3	5.2
Slovakia	10.5	10.8	12.5	15.7	16.2	17.1	17.0	16.2	16.8	17.1
Slovenia	4.8	4.8	4.5	4.5	5.0	5.3	5.0	5.2	5.6	5.3
Spain	52.5	56.0	55.9	59.1	60.5	59.9	58.8	60.4	54.7	57.1
Sweden	66.4	69.9	69.5	54.5	68.5	64.2	64.0	73.6	68.6	63.6
Switzerland	24.1	24.5	24.5	25.1	25.5	25.9	26.1	25.6	22.2	26.4
United Kingdom	93.2	94.5	90.4	80.8	85.4	83.6	84.3	73.7	75.2	69.2
Eurasia [2]	**192.5**	**183.4**	**189.7**	**203.4**	**209.8**	**223.0**	**234.4**	**236.7**	**235.8**	**240.2**
Armenia	1.4	1.4	2.1	1.8	2.0	2.1	1.8	2.2	2.5	2.4
Kazakhstan	.3	.1	(s)	.0	.0	.0	.0	.0	.0	.0
Lithuania	10.9	12.9	9.4	8.0	10.8	13.4	14.7	14.3	9.8	8.7
Russia	104.5	98.3	110.9	122.5	125.4	134.1	141.2	137.5	140.2	144.3
Ukraine	75.4	70.6	67.4	71.1	71.7	73.4	76.7	82.7	83.3	84.8
Africa	**12.6**	**13.6**	**12.8**	**13.0**	**10.7**	**12.0**	**12.7**	**14.3**	**12.2**	**10.1**
South Africa	12.6	13.6	12.8	13.0	10.7	12.0	12.7	14.3	12.2	10.1
Asia and Oceania [1]	**436.5**	**460.8**	**461.2**	**476.8**	**481.3**	**476.2**	**448.4**	**495.4**	**524.3**	**541.5**
China	11.4	13.5	14.1	15.9	16.6	25.2	41.7	47.9	50.3	54.8
India	10.5	10.6	11.5	14.1	18.2	17.8	16.4	15.0	15.7	15.6
Japan	306.2	315.7	300.8	305.9	303.9	280.3	228.0	268.3	278.4	288.9
Pakistan	.4	.4	.1	.4	2.0	1.8	1.8	1.9	2.4	2.5
South Korea	73.2	85.2	97.9	103.5	106.5	113.1	123.2	124.2	139.4	141.3
Taiwan	34.8	35.4	36.9	37.0	34.1	38.0	37.4	37.9	38.0	38.3
World	**2,271.3**	**2,316.0**	**2,393.1**	**2,449.9**	**2,516.7**	**2,545.3**	**2,517.8**	**2,615.0**	**2,625.6**	**2,657.3**

[1] Excludes countries that were part of the former U.S.S.R. See "U.S.S.R." in Glossary.
[2] Includes only countries that were part of the former U.S.S.R. See "U.S.S.R." in Glossary.
(s) = Less than 0.05 billion kilowatthours.
Note: Totals may not equal sum of components due to independent rounding.

Web Page: For related information, see http://www.eia.doe.gov/international.
Sources: **United States:** Table 8.2a. **All Other Data:** Energy Information Administration, "International Energy Annual 2006" (June-July 2008), Table 2.7.

Figure 11.19 World Carbon Dioxide Emissions From Energy Consumption

World, 1996-2005

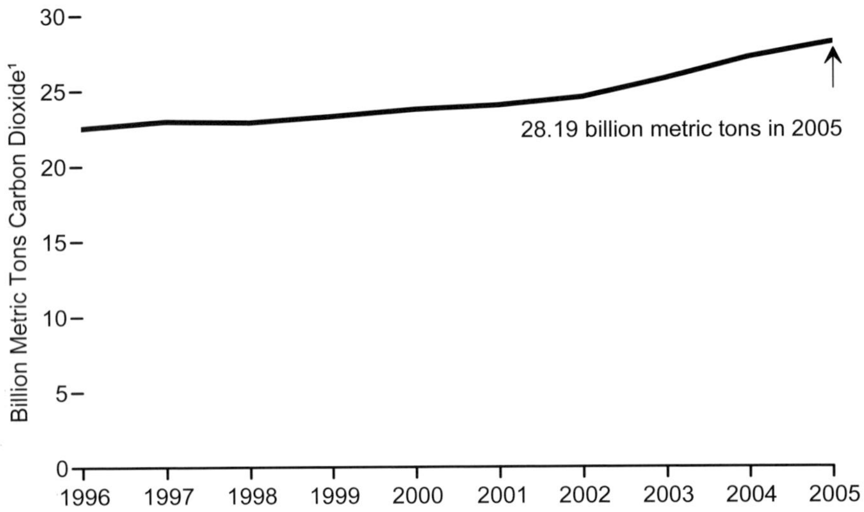

World by Region, 2005

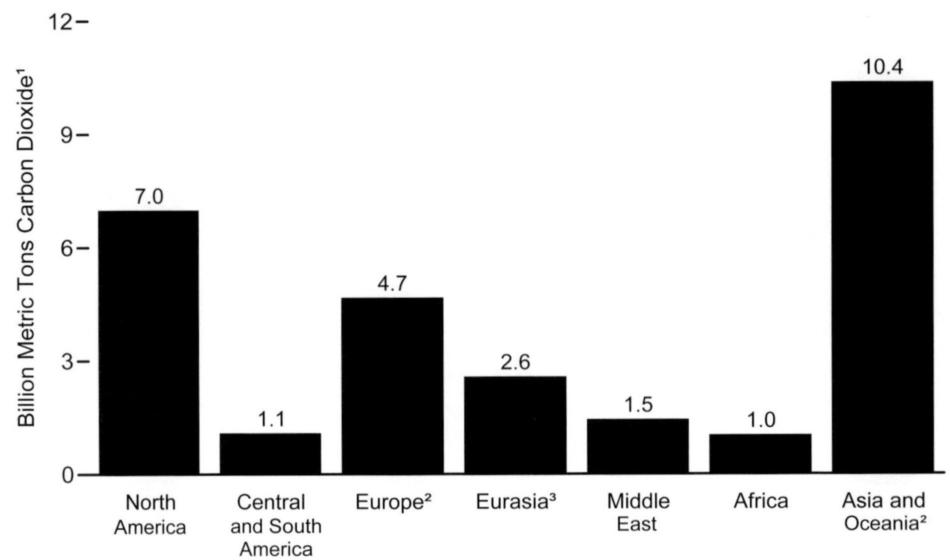

Top Countries, 2005

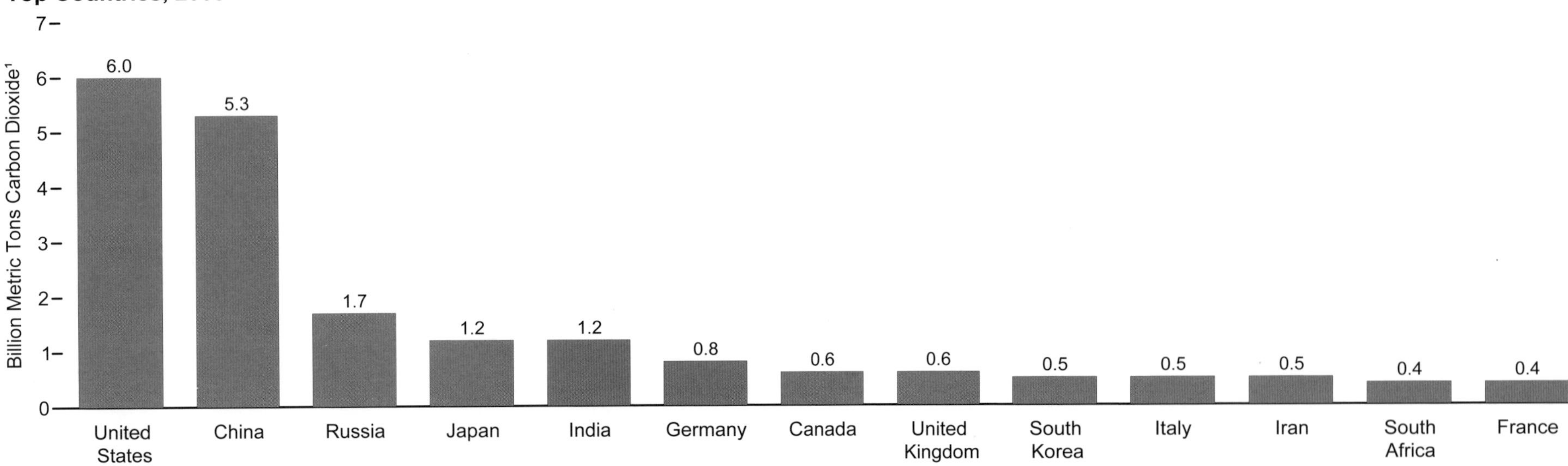

[1] Metric tons of carbon dioxide can be converted to metric tons of carbon equivalent by multiplying by 12/44.
[2] Excludes countries that were part of the former U.S.S.R. See "U.S.S.R." in Glossary.
[3] Includes countries that were part of the former U.S.S.R. See "U.S.S.R." in Glossary.

Notes: • Data include carbon dioxide emissions from fossil-fuel energy consumption and natural gas venting and flaring. • Because vertical scales differ, graphs should not be compared.
Source: Table 11.19.

Table 11.19 World Carbon Dioxide Emissions From Energy Consumption, 1996-2005

(Million Metric Tons of Carbon Dioxide [1])

Region and Country	1996	1997	1998	1999	2000	2001	2002	2003	2004	2005P
North America	**R6,325**	**6,436**	**6,499**	**R6,576**	**R6,764**	**R6,667**	**R6,735**	**R6,817**	**R6,945**	**6,988**
Canada	R517	R543	R548	R561	R558	R566	R586	R613	R626	631
Mexico	R330	347	R369	R361	R381	377	R384	R389	R382	398
United States [2]	R5,476	R5,544	R5,581	R5,653	R5,823	R5,723	R5,763	R5,813	R5,935	5,957
Other	1	1	1	1	1	1	1	1	1	1
Central and South America	**R896**	**R941**	**R967**	**R978**	**R988**	**R1,011**	**R997**	**R1,014**	**R1,057**	**1,096**
Argentina	129	R130	135	139	137	127	119	133	R140	147
Brazil	R303	R321	R320	334	R342	R346	R344	R343	R353	361
Venezuela	R133	134	R142	R133	R134	148	146	133	R142	151
Other	R331	R356	370	R372	375	R389	R388	R405	R422	438
Europe [3]	**R4,429**	**R4,446**	**R4,434**	**R4,380**	**R4,445**	**R4,503**	**R4,468**	**R4,622**	**R4,666**	**4,675**
Belgium	137	141	R147	138	144	142	138	145	R144	136
France	388	R381	R406	R400	R399	403	R399	R405	R412	415
Germany	R883	R881	R863	832	R848	869	R841	R865	R867	844
Italy	420	420	R438	437	444	442	R448	470	R466	467
Netherlands	R227	R238	R240	R237	249	R276	R257	R259	R268	270
Poland	R341	R333	311	324	290	274	272	R285	R289	285
Romania	R126	119	100	91	R93	101	99	99	R99	99
Spain	R242	R269	R279	R305	R323	R328	R345	R353	R374	387
Turkey	R168	181	R183	180	200	183	193	204	R209	230
United Kingdom	R585	R562	R558	R551	R555	R565	R556	R567	R576	577
Other	R913	R920	R910	R883	R900	R922	R920	R969	R962	965
Eurasia [4]	**R2,404**	**R2,236**	**R2,227**	**R2,310**	**R2,340**	**R2,323**	**R2,343**	**R2,458**	**R2,511**	**2,578**
Kazakhstan	140	118	R115	131	R135	R146	R150	R159	R169	198
Russia	R1,602	R1,482	R1,483	R1,558	R1,580	R1,570	R1,569	R1,627	R1,669	1,696
Ukraine	R362	340	R329	R323	323	R314	R322	R353	R342	343
Uzbekistan	103	103	101	R103	106	111	113	114	121	118
Other	R197	R193	200	194	196	182	188	R204	210	223
Middle East	**R928**	**R982**	**R1,013**	**R1,055**	**R1,081**	**R1,114**	**R1,168**	**R1,231**	**1,320**	**1,451**
Iran	261	R289	R293	315	R319	332	R362	R384	R404	451
Saudi Arabia	R249	R254	R257	R263	R289	R300	R310	345	R386	412
Other	R418	R439	R464	R477	R473	R482	R496	R502	R531	588
Africa	**R836**	**R861**	**R850**	**R864**	**R881**	**R914**	**R912**	**R960**	**R1,010**	**1,043**
Egypt	R108	111	114	R115	119	R132	R133	R142	R152	162
South Africa	349	380	362	R373	R383	R391	R377	R409	R438	424
Other	R379	R369	R374	R376	R379	391	402	R408	R420	457
Asia and Oceania [3]	**R6,713**	**R7,068**	**R6,910**	**R7,128**	**R7,252**	**R7,480**	**R7,924**	**R8,678**	**R9,677**	**10,362**
Australia	298	327	R334	351	353	367	R375	R375	R381	407
China	R2,878	R3,069	R2,968	R2,932	R2,913	R3,051	R3,376	R3,983	R4,753	5,323
India	R820	R862	R898	R954	R994	R1,017	R1,015	R1,029	R1,129	1,166
Indonesia	235	245	239	263	R271	297	R311	315	R342	359
Japan	R1,105	R1,138	R1,098	R1,147	1,190	R1,178	1,186	R1,234	R1,242	1,230
Malaysia	101	R102	102	106	111	124	138	R149	R164	156
South Korea	R399	R429	R370	R427	R440	R446	465	R479	R488	500
Taiwan	R196	R208	R220	R221	R249	245	269	286	R282	284
Thailand	169	R176	161	170	R161	171	186	R204	R224	234
Other	R511	R513	R519	R557	R571	R585	R603	R623	R672	704
World	**R22,531**	**R22,969**	**R22,901**	**R23,290**	**R23,751**	**R24,012**	**R24,546**	**R25,780**	**R27,186**	**28,193**

[1] Metric tons of carbon dioxide can be converted to metric tons of carbon equivalent by multiplying by 12/44. For data in million metric tons carbon equivalent, see the "International Energy Annual 2005" (June-October 2007), Table H.1.

[2] Data for the United States differ from those in Table 12.1 because U.S. data in this table include emissions from bunker fuels consumption; exclude emissions from geothermal power generation, cement production and other industrial processes, and municipal solid waste combustion; and exclude data for the U.S. Territories.

[3] Excludes countries that were part of the former U.S.S.R. See "U.S.S.R." in Glossary.

[4] Includes only countries that were part of the former U.S.S.R. See "U.S.S.R." in Glossary.

R=Revised. P=Preliminary.

Notes: • Data in this table do not include recent updates (see the Energy Information Administration's "International Energy Annual 2006"). • Data include carbon dioxide emissions from fossil-fuel energy consumption and natural gas venting and flaring. • See Note 2, "World Carbon Dioxide Emissions," at end of section. • Totals may not equal sum of components due to independent rounding.

Web Page: For related information, see http://www.eia.doe.gov/international.

Source: Energy Information Administration, "International Energy Annual 2005" (June-October 2007), Table H.1co2.

International Energy

Note 1. World Primary Energy Production. World primary energy production includes production of crude oil (including lease condensate), natural gas plant liquids, dry natural gas, and coal; and net electricity generation from nuclear electric power, hydroelectric power, wood, waste, geothermal, solar, and wind. Data for the United States also include other renewable energy. Crude oil production is measured at the wellhead and includes lease condensate. Natural gas plant liquids are products obtained from processing natural gas at natural gas processing plants, including natural gas plants, cycling plants, and fractionators. Dry natural gas production is that amount of natural gas produced that is available to be marketed and consumed as a gas. Coal (anthracite, bituminous coal, subbituminous coal, and lignite) production is the sum of sales, mine consumption, issues to miners, and issues to coking, briquetting, and other ancillary plants at mines. Coal production data include quantities extracted from surface and underground mines and normally exclude wastes removed at mines or associated preparation plants. The data on generation of electricity from nuclear electric power, hydroelectric power, wood, waste, geothermal, solar, and wind include data reported on a net basis, thus excluding electricity that is generally used by the electric power plant for its own operating purposes or electricity losses in the transformers that are considered integral parts of the station.

Note 2. World Carbon Dioxide Emissions. In Table 11.19, data for carbon dioxide emissions include anthropogenic (human-caused) emissions from the consumption of petroleum, natural gas, and coal, and also from natural gas venting and flaring. They do not include carbon dioxide emissions from geothermal power generation, cement production and other industrial processes, and municipal solid waste combustion. Fossil-fuel consumption and natural gas flaring statistics for each country have been reduced to account for the fraction of fuels not combusted and, in the case of petroleum, for the fraction of sequestration of non-fuel uses. Carbon dioxide emissions have been determined by applying carbon emission coefficients to the adjusted consumption and flaring data. Carbon emission coefficients for petroleum and natural gas consumption and natural gas flaring are from Energy Information Administration (EIA), *Documentation for Emissions of Greenhouse Gases in the United States 2005* (October 2007), Table 6.1. Carbon emission coefficients for coal consumption are from EIA, *Emissions of Greenhouse Gases in the United States 1985-1990* (October 1993), Table 11.

12 Environmental Indicators

“Harpers Ferry, Junction of the Rivers Shenandoah and Potomac.” Engraving by W. Goodacre and James Archer, published in *The History and Topography of the United States of North America*, by John Howard Hinton, 1852. From the collection of the National Park Service, Harpers Ferry National Historical Park, Accession #1297.

Figure 12.1 Emissions of Greenhouse Gases

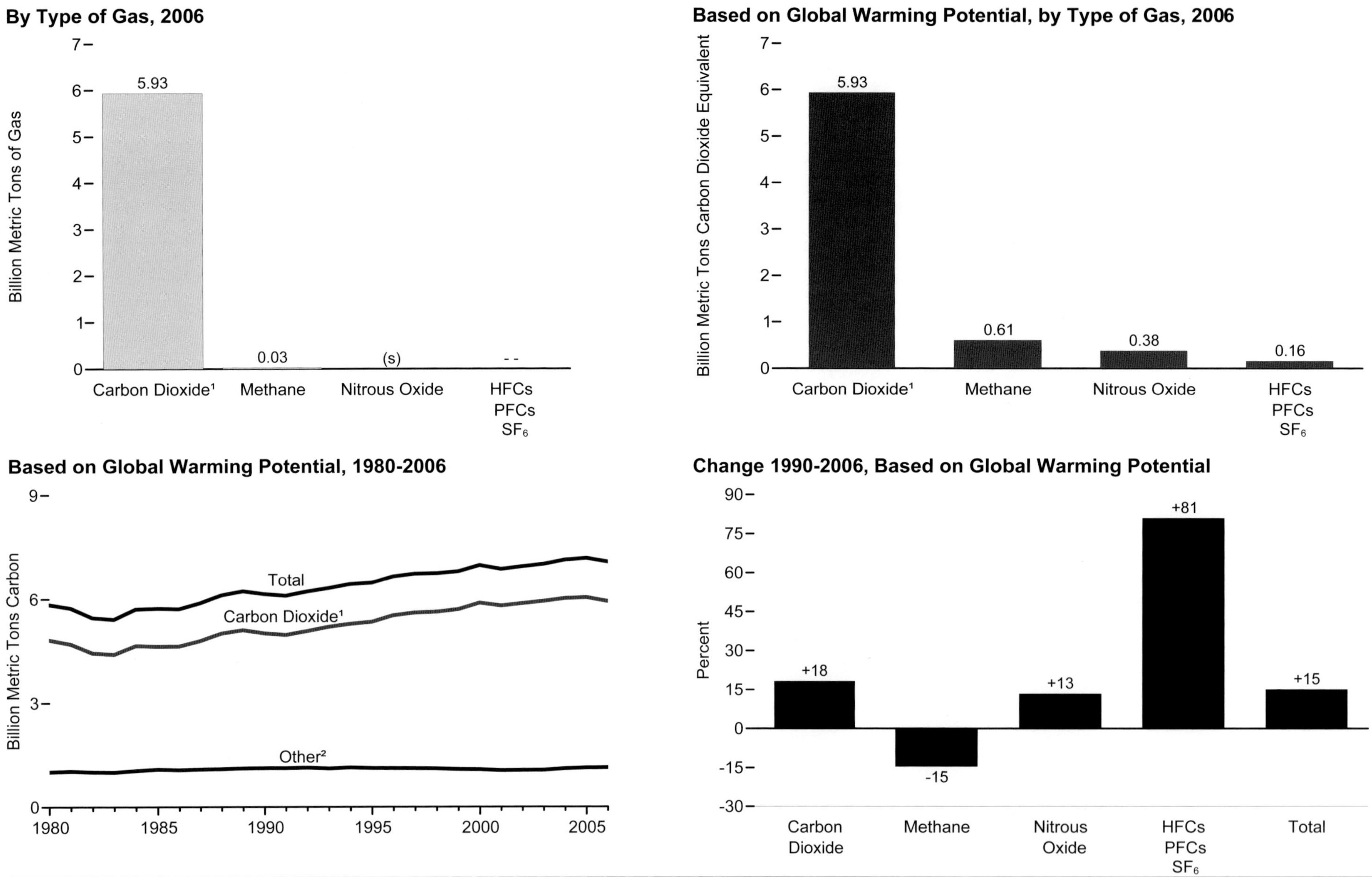

[1] Metric tons of carbon dioxide can be converted to metric tons of carbon equivalent by multiplying by 12/44.

[2] Methane, nitrous oxide, hydrofluorocarbons (HFCs), perfluorocarbons (PFCs), and sulfur hexafluoride (SF_6).

(s)=Less than 0.005 billion metric tons of gas.

- - = Not applicable because these gases cannot be summed in native units.

Notes: • HFCs=hydrofluorocarbons; PFCs=perfluorocarbons; and SF_6=sulfur hexafluoride. • Emissions by type of gas should not be compared; for comparison, see emissions based on global warming potential by type of gas. • Because vertical scales differ, graphs should not be compared.

Source: Table 12.1.

Table 12.1 Emissions of Greenhouse Gases, 1980-2006

Year	Greenhouse Gases				Greenhouse Gases, Based on Global Warming Potential [1]				
	Carbon Dioxide [2,3]	Methane	Nitrous Oxide	HFCs PFCs SF_6	Carbon Dioxide [2]	Methane	Nitrous Oxide	HFCs PFCs SF_6	Total
	Million Metric Tons of Gas				Million Metric Tons Carbon Dioxide Equivalent [2]				
1980	4,824.7	28.6	1.0	– –	4,824.7	658.0	287.0	70.4	5,840.0
1981	4,704.3	29.2	1.0	– –	4,704.3	671.1	292.0	74.0	5,741.3
1982	4,448.8	29.4	1.0	– –	4,448.8	676.8	282.6	55.4	5,463.7
1983	4,408.0	29.1	.9	– –	4,408.0	669.9	270.2	67.1	5,415.3
1984	4,655.8	29.8	1.0	– –	4,655.8	684.5	294.0	75.5	5,709.9
1985	4,638.3	30.0	1.1	– –	4,638.3	689.7	330.7	70.5	5,729.3
1986	4,642.5	29.4	1.1	– –	4,642.5	676.5	323.8	75.0	5,717.8
1987	4,800.2	29.9	1.1	– –	4,800.2	688.3	323.4	77.8	5,889.8
1988	5,012.6	30.1	1.1	– –	5,012.6	692.0	316.9	91.3	6,112.8
1989	5,105.8	30.2	1.1	– –	5,105.8	693.8	332.8	94.5	6,226.9
1990	[R]5,017.5	[R]30.8	1.1	– –	[R]5,017.5	[R]708.4	[R]333.7	87.1	[R]6,146.7
1991	[R]4,969.4	[R]30.8	1.2	– –	[R]4,969.4	[R]707.7	[R]342.9	79.0	[R]6,098.9
1992	[R]5,078.7	[R]30.9	1.2	– –	[R]5,078.7	[R]709.7	[R]350.0	83.7	[R]6,222.1
1993	[R]5,203.0	[R]29.8	1.2	– –	[R]5,203.0	[R]684.8	[R]349.5	82.9	[R]6,320.2
1994	[R]5,288.3	[R]29.8	1.3	– –	[R]5,288.3	[R]685.6	[R]374.9	[R]85.3	[R]6,434.0
1995	[R]5,343.4	[R]29.4	1.2	– –	[R]5,343.4	[R]675.9	[R]357.1	94.9	[R]6,471.2
1996	[R]5,531.0	[R]28.5	1.2	– –	[R]5,531.0	[R]656.0	[R]357.6	110.6	[R]6,655.2
1997	[R]5,606.7	[R]28.5	1.2	– –	[R]5,606.7	[R]654.6	[R]350.5	[R]118.0	[R]6,729.8
1998	[R]5,632.5	27.4	1.2	– –	[R]5,632.5	[R]631.3	[R]347.9	[R]134.4	[R]6,746.1
1999	[R]5,703.1	26.8	1.2	– –	[R]5,703.1	[R]615.8	[R]346.3	133.9	[R]6,799.1
2000	[R]5,890.5	[R]26.4	1.2	– –	[R]5,890.5	[R]608.0	[R]341.9	138.0	[R]6,978.4
2001	[R]5,806.3	[R]25.8	1.1	– –	[R]5,806.3	[R]593.9	[R]336.6	[R]128.6	[R]6,865.4
2002	[R]5,875.9	[R]26.0	1.1	– –	[R]5,875.9	[R]598.6	[R]332.5	137.8	[R]6,944.9
2003	[R]5,940.4	26.2	1.1	– –	[R]5,940.4	[R]603.7	[R]331.7	136.6	[R]7,012.4
2004	[R]6,019.9	[R]26.3	1.2	– –	[R]6,019.9	[R]605.9	[R]358.3	[R]149.4	[R]7,133.5
2005	[R]6,045.0	[R]26.4	1.2	– –	[R]6,045.0	[R]607.3	[R]368.0	[R]161.2	[R]7,181.4
2006[P]	5,934.4	26.3	1.3	– –	5,934.4	605.1	378.6	157.6	7,075.6

[1] Emissions of greenhouse gases are weighted based upon their relative global warming potential (GWP), with carbon dioxide equal to a weight of one. The use of updated estimates of GWP resulted in a number of revisions to previously published data. It is also important to note that revisions in estimated emissions result from revisions in energy consumption as well.

[2] Metric tons of carbon dioxide can be converted to metric tons of carbon equivalent by multiplying by 12/44.

[3] Carbon dioxide data in this table differ from those for the United States in Table 11.19 because data in this table exclude emissions from international bunker fuels consumption; include emissions from geothermal power generation, cement production and other industrial processes, and municipal solid waste combustion; and include data for the U.S. Territories.

R=Revised. P=Preliminary. – – = Not applicable because these gases cannot be summed in native units.

Notes: • HFCs = hydrofluorocarbons; PFCs = perfluorocarbons; and SF_6 = sulfur hexafluoride. • Emissions are from anthropogenic sources. "Anthropogenic" means produced as the result of human activities, including emissions from agricultural activity and domestic livestock. Emissions from natural sources, such as wetlands and wild animals, are not included. • Because of the continuing goal to improve estimation methods for greenhouse gases, data are frequently revised on an annual basis in keeping with the latest findings of the international scientific community. • For information on units for measuring greenhouse gases, see http://www.eia.doe.gov/oiaf/1605/archive/gg06rpt/pdf/executive_summary.pdf, page 2, box titled "Units for Measuring Greenhouse Gases. • Totals may not equal sum of components due to independent rounding.

Web Page: For related information, see http://www.eia.doe.gov/environment.html.

Sources: **1990, 1995, and 1999-2006:** Energy Information Administration (EIA), *Emissions of Greenhouse Gases in the United States 2006* (November 2007), Table 1. **All Other Data:** EIA, *Emissions of Greenhouse Gases in the United States,* annual reports and unpublished revisions.

Figure 12.2 Carbon Dioxide Emissions From Energy Consumption by Sector

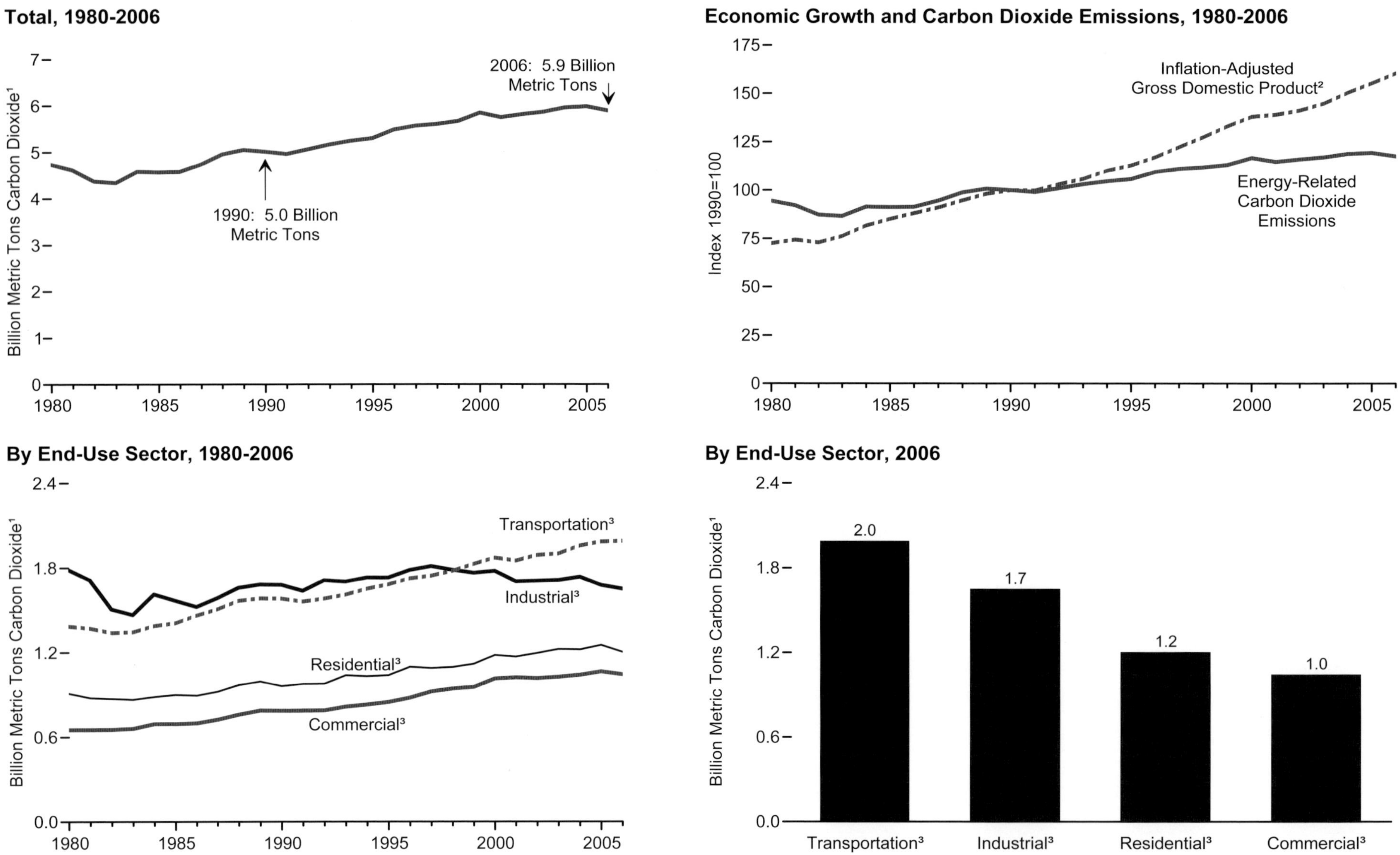

[1] Metric tons of carbon dioxide can be converted to metric tons of carbon equivalent by multiplying by 12/44.
[2] Based on chained (2000) dollars.
[3] Electric power sector emissions are allocated to end-use sectors in proportion to each sector's share of total electricity retail sales (see Table 8.9).
Note: Because vertical scales differ, graphs should not be compared.
Sources: Tables 1.5 and 12.2.

Table 12.2 Carbon Dioxide Emissions From Energy Consumption by Sector, 1980-2006

(Million Metric Tons of Carbon Dioxide [1])

Year	End-Use Sectors								Electric Power Sector [4]	Total [7]
	Residential		Commercial [2]		Industrial [3]		Transportation			
	Primary [5]	Total [6]	Primary [5]	Total [6]	Primary [5]	Total [6]	Primary [5]	Total [6]	Primary [5]	
1980	385.2	909.0	244.5	652.5	1,192.8	1,787.7	1,383.9	1,386.2	1,529.0	4,735.4
1981	360.8	877.8	225.8	652.2	1,123.3	1,714.2	1,369.4	1,371.7	1,536.7	4,616.0
1982	359.1	872.2	226.1	654.1	983.2	1,506.9	1,338.3	1,340.5	1,467.1	4,373.8
1983	340.4	866.4	225.7	660.5	923.2	1,466.7	1,343.0	1,345.3	1,506.5	4,338.8
1984	348.8	885.8	236.2	693.7	1,036.0	1,612.6	1,387.2	1,389.6	1,573.5	4,581.7
1985	351.4	899.7	217.9	694.0	990.0	1,567.6	1,406.3	1,408.9	1,604.6	4,570.3
1986	342.5	895.2	216.2	698.8	963.2	1,523.4	1,460.2	1,462.9	1,598.2	4,580.3
1987	345.8	921.9	220.0	724.6	1,004.3	1,585.6	1,504.4	1,506.9	1,664.5	4,738.9
1988	366.7	969.6	230.1	760.0	1,054.1	1,659.3	1,564.1	1,566.8	1,740.7	4,955.7
1989	371.6	994.8	229.9	788.5	1,045.4	1,682.3	1,581.5	1,584.3	1,821.4	5,049.8
1990	R341.6	R961.6	R225.0	R787.5	R1,045.4	R1,679.9	R1,579.4	R1,582.6	R1,820.2	R5,011.6
1991	R348.5	R977.1	R225.7	R788.4	R1,014.9	R1,637.7	R1,558.1	R1,561.3	R1,817.3	R4,964.5
1992	R358.4	R978.6	R226.8	R790.0	R1,067.6	R1,712.2	R1,579.0	R1,582.1	R1,831.2	R5,063.0
1993	R374.4	R1,039.2	R224.5	R815.8	R1,049.5	R1,702.6	R1,607.4	R1,610.6	R1,912.3	R5,168.2
1994	R365.8	R1,032.2	R226.9	R830.3	R1,065.6	R1,731.7	R1,648.5	R1,651.8	R1,939.2	R5,246.0
1995	R362.8	R1,039.2	R230.0	R848.4	R1,073.8	R1,730.9	R1,679.0	R1,682.2	R1,955.0	R5,300.6
1996	R391.2	R1,098.4	R238.6	R879.0	R1,109.5	R1,784.8	R1,722.2	R1,725.4	R2,026.1	R5,487.6
1997	R372.8	R1,089.7	R238.8	R922.9	R1,120.6	R1,812.4	R1,740.9	R1,744.2	R2,096.0	R5,569.1
1998	R340.3	R1,096.9	R221.8	R943.5	R1,082.5	R1,786.2	R1,776.2	R1,779.5	R2,185.3	R5,606.1
1999	R360.9	R1,120.0	R223.6	R955.5	R1,063.1	R1,764.8	R1,824.9	R1,828.3	R2,196.3	R5,668.6
2000	R379.7	R1,181.5	R235.5	R1,015.1	R1,062.4	R1,778.1	R1,868.9	R1,872.6	R2,300.7	R5,847.2
2001	R368.1	R1,171.1	R227.0	R1,023.3	R1,045.6	R1,703.8	R1,847.3	R1,851.0	R2,261.1	R5,749.1
2002	R367.2	R1,196.2	R228.5	R1,018.1	R1,059.1	R1,707.8	R1,887.2	R1,890.9	R2,270.9	R5,813.0
2003	R385.1	R1,224.1	R238.4	R1,027.1	R1,046.4	R1,712.8	R1,896.8	R1,901.4	R2,298.8	R5,865.5
2004	R371.7	R1,221.5	234.2	R1,041.6	R1,066.6	R1,735.7	R1,953.9	R1,958.6	R2,331.0	R5,957.4
2005	R362.9	R1,253.0	R230.5	R1,065.4	R1,009.8	R1,677.1	R1,981.2	R1,986.2	R2,397.1	R5,981.6
2006P	338.2	1,204.2	213.3	1,045.2	1,010.1	1,650.8	1,984.9	1,990.1	2,343.9	5,890.3

[1] Metric tons of carbon dioxide can be converted to metric tons of carbon equivalent by multiplying by 12/44.

[2] Commercial sector, including commercial combined-heat-and-power (CHP) and commercial electricity-only plants.

[3] Industrial sector, including industrial combined-heat-and-power (CHP) and industrial electricity-only plants.

[4] Electricity-only and combined-heat-and-power (CHP) plants within the NAICS 22 category whose primary business is to sell electricity, or electricity and heat, to the public.

[5] Carbon dioxide emissions from the combustion of fossil fuels. The electric power sector also has a small amount of emissions from geothermal power generation and the combustion of the plastics component of municipal solid waste.

[6] In addition to "Primary" emissions, also includes emissions from energy consumption (for electricity and a small amount of useful thermal output) in the electric power sector, which are allocated to the end-use sectors in proportion to each sector's share of total electricity retail sales (see Table 8.9).

[7] The sum of "Primary" emissions in the five energy-use sectors equals the sum of "Total" emissions in the four end-use sectors.

R=Revised. P=Preliminary.

Notes: • See Note 2, "Classification of Power Plants Into Energy-Use Sectors," at end of Section 8.
• Because of the continuing goal to improve estimation methods for greenhouse gases, data are frequently revised on an annual basis in keeping with the latest findings of the international scientific community.
• Totals may not equal sum of components due to independent rounding.

Web Page: For related information, see http://www.eia.doe.gov/environment.html.

Sources: **1990, 1995, and 1999-2006:** Energy Information Administration (EIA), *Emissions of Greenhouse Gases in the United States 2006* (November 2007), Tables 5-9. **All Other Data:** EIA, *Emissions of Greenhouse Gases in the United States,* annual reports and unpublished revisions.

Figure 12.3 Carbon Dioxide Emissions From Energy Consumption by Sector by Energy Source, 2006

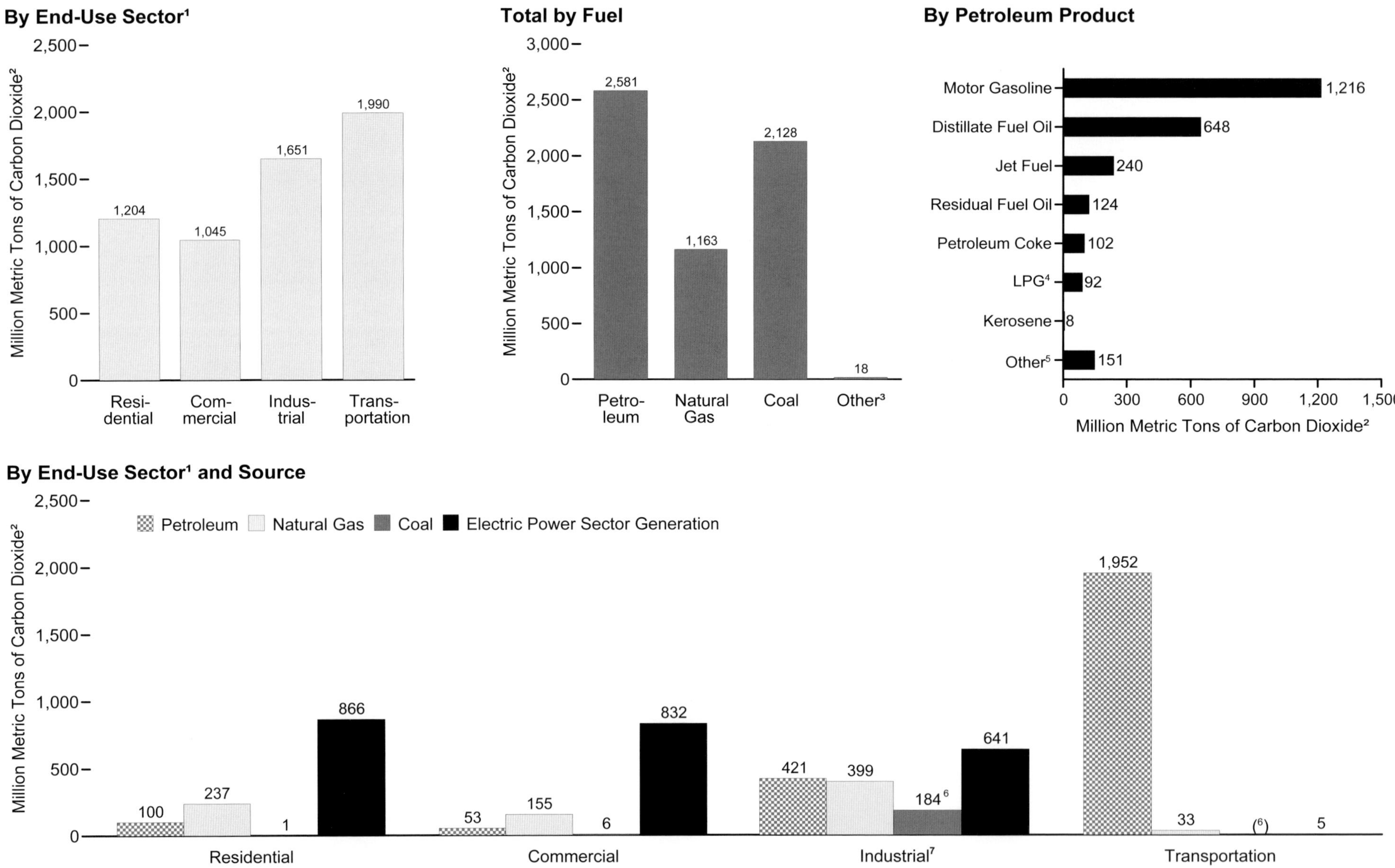

[1] Emissions from energy consumption in the electric power sector are allocated to the end-use sectors in proportion to each sector's share of total electricity retail sales (see Table 8.9).
[2] Metric tons of carbon dioxide can be converted to metric tons of carbon equivalent by multiplying by 12/44.
[3] Coal coke net imports, the plastics component of municipal solid waste, and geothermal.
[4] Liquefied petroleum gases.
[5] Aviation gasoline, lubricants, and other products.
[6] Small amounts of coal consumed for transportation are reported as industrial consumption.
[7] The industrial sector also includes 4 million metric tons of coal coke net imports.

Note: Because scales differ, graphs should not be compared

Source: Table 12.3.

Table 12.3 Carbon Dioxide Emissions From Energy Consumption by Sector by Energy Source, 2006

(Million Metric Tons of Carbon Dioxide [1])

Energy Source	End-Use Sectors					Electric Power Sector [4]	Total
	Residential	Commercial [2]	Industrial [3]	Transportation	Total		
Petroleum	100.3	52.6	421.4	1,952.4	2,526.6	54.5	2,581.2
Aviation Gasoline	– –	– –	– –	2.3	2.3	– –	2.3
Distillate Fuel Oil	63.6	33.3	93.4	452.2	642.5	5.4	647.9
Jet Fuel	– –	– –	– –	239.5	239.5	– –	239.5
Kerosene	5.1	1.2	1.7	– –	8.0	– –	8.0
Liquefied Petroleum Gases	31.6	5.6	54.1	1.1	92.4	– –	92.4
Lubricants	– –	– –	– –	[5] 5.5	5.5	– –	5.5
Motor Gasoline	– –	3.5	26.7	1,186.2	1,216.4	– –	1,216.4
Petroleum Coke	– –	– –	81.9	– –	81.9	20.5	102.4
Residual Fuel Oil	– –	8.9	20.3	65.6	94.8	28.7	123.5
Other	– –	– –	143.3	– –	143.3	– –	143.3
Natural Gas	237.3	154.6	399.2	32.5	823.6	339.5	1,163.1
Coal	0.6	6.2	[6] 183.8	([6])	190.5	1,937.9	2,128.4
Coal Coke Net Imports	– –	– –	5.7	– –	5.7	– –	5.7
Municipal Solid Waste [7]	– –	– –	– –	– –	– –	11.5	11.5
Geothermal	– –	– –	– –	– –	– –	0.4	0.4
Primary	**338.2**	**213.3**	**1,010.1**	**1,984.9**	**3,546.4**	**2,343.9**	**5,890.3**
Electric Power Sector Generation [8]	866.0	831.9	640.7	5.2	2,343.9	– –	– –
Total	**1,204.2**	**1,045.2**	**1,650.8**	**1,990.1**	**5,890.3**	**– –**	**5,890.3**

[1] Metric tons of carbon dioxide can be converted to metric tons of carbon equivalent by multiplying by 12/44.
[2] Commercial sector, including commercial combined-heat-and-power (CHP) and commercial electricity-only plants.
[3] Industrial sector, including industrial combined-heat-and-power (CHP) and industrial electricity-only plants.
[4] Electricity-only and combined-heat-and-power (CHP) plants within the NAICS 22 category whose primary business is to sell electricity, or electricity and heat, to the public.
[5] Includes emissions from nonfuel use of lubricants.
[6] Small amounts of coal consumed for transportation are reported as industrial sector consumption.
[7] The plastics component of municipal solid waste.
[8] Emissions from energy consumption (for electricity and a small amount of useful thermal output) in the electric power sector are allocated to the end-use sectors in proportion to each sector's share of total electricity retail sales (see Table 8.9).

– – = Not applicable.

Notes: • Data are preliminary estimates. • Emissions from blast furnace gas, propane gas, and other manufactured and waste gases derived from fossil fuels are counted under their primary energy source—i.e., petroleum, natural gas, or coal. • See Note 2, "Classification of Power Plants Into Energy-Use Sectors," at end of Section 8. • Totals may not equal sum of components due to independent rounding.

Web Page: For related information, see http://www.eia.doe.gov/environment.html.

Source: Energy Information Administration, *Emissions of Greenhouse Gases in the United States 2006* (November 2007), Tables 5-9.

Figure 12.4 Carbon Dioxide Emissions From Consumption of Energy for All Purposes in the Manufacturing Sector, 2002

Carbon Dioxide Emissions by Top Industry Groups

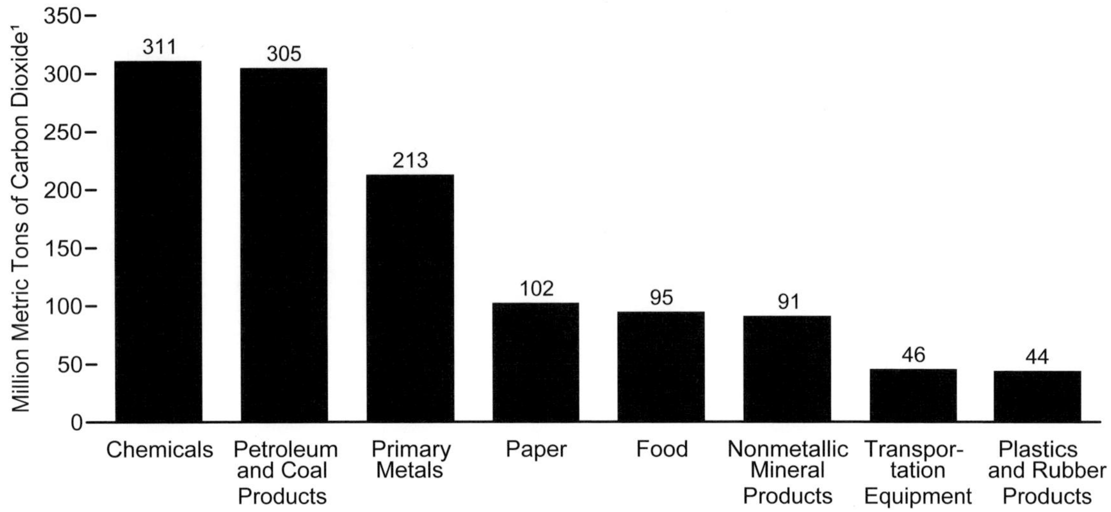

Carbon Dioxide Emissions by Energy Source

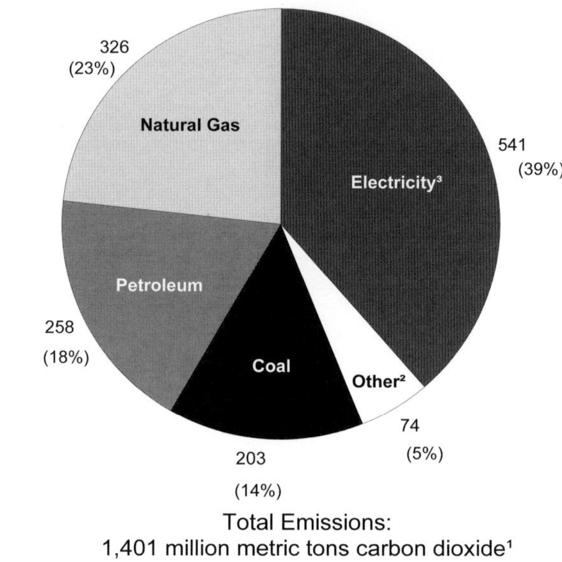

Carbon Dioxide Emissions per Unit of Primary Consumption, Top Industry Groups

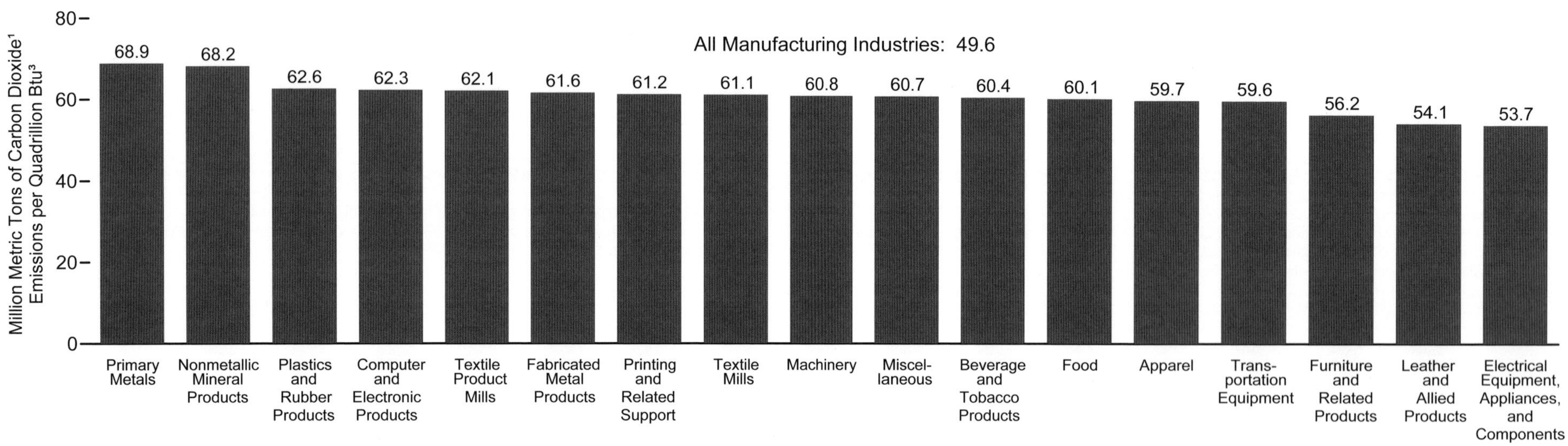

[1] Metric tons of carbon dioxide can be converted to metric tons of carbon equivalent by multiplying by 12/44.

[2] All other types of energy that respondents indicated were consumed or allocated.

[3] From energy inputs used to produce electricity, including associated losses.

Source: Table 12.4.

Table 12.4 Carbon Dioxide Emissions From Consumption of Energy for All Purposes in the Manufacturing Sector, 2002

(Million Metric Tons of Carbon Dioxide,[1] Except as Noted)

NAICS[2] Code	Major Group	Carbon Dioxide Emissions						Carbon Dioxide Emissions per Unit of Primary Consumption[5]	Carbon Dioxide Emissions per Chained Dollar of Shipments[6]
		Coal	Natural Gas	Petroleum	Electricity[3]	Other[4]	Total		
311	Food	17.3	30.7	2.9	43.8	0.1	94.7	60.1	215.2
312	Beverage and Tobacco Products	1.6	2.4	0.4	4.9	(s)	9.4	60.4	93.1
313	Textile Mills	2.1	4.0	0.6	16.4	0.0	23.0	61.1	518.3
314	Textile Product Mills	0.7	1.5	0.3	3.2	0.0	5.8	62.1	170.7
315	Apparel	0.0	0.8	0.1	2.3	0.0	3.2	59.7	59.3
316	Leather and Allied Products	0.0	0.2	0.0	0.4	0.0	0.6	54.1	59.1
321	Wood Products	0.1	3.0	1.2	13.7	0.4	18.4	35.6	205.7
322	Paper	22.5	26.6	10.0	42.4	0.8	102.4	36.6	661.3
323	Printing and Related Support	0.0	2.4	0.1	9.5	0.0	12.0	61.2	125.9
324	Petroleum and Coal Products	19.3	46.4	153.9	24.6	60.8	304.8	43.2	1,301.1
325	Chemicals	32.8	106.2	70.2	99.4	2.4	311.0	41.5	738.1
326	Plastics and Rubber Products	2.1	6.8	0.9	34.5	(s)	44.2	62.6	249.4
327	Nonmetallic Mineral Products	30.1	22.3	11.4	26.8	0.4	91.1	68.2	1,046.0
331	Primary Metals	72.4	37.2	2.4	93.8	7.0	212.8	68.9	1,511.1
332	Fabricated Metal Products	0.8	11.1	0.9	30.6	0.0	43.4	61.6	173.4
333	Machinery	0.1	4.3	0.4	16.0	(s)	20.8	60.8	82.3
334	Computer and Electronic Products	0.0	3.4	0.2	24.9	(s)	28.5	62.3	59.9
335	Electrical Equipment, Appliances, and Components	0.0	2.8	0.1	8.9	2.3	14.2	53.7	135.3
336	Transportation Equipment	1.0	10.7	1.2	32.7	0.1	45.7	59.6	74.1
337	Furniture and Related Products	0.1	1.3	0.1	4.6	0.1	6.3	56.2	91.5
339	Miscellaneous	0.0	1.7	0.1	6.7	0.0	8.5	60.7	71.7
—	Total Manufacturing	202.8	325.9	257.6	540.7	74.2	1,401.2	49.6	352.7

[1] Metric tons of carbon dioxide can be converted to metric tons of carbon equivalent by multiplying by 12/44.

[2] North American Industry Classification System (NAICS).

[3] Carbon dioxide emitted from energy inputs used to produce electricity (including associated losses), derived by calculating the manufacturing subsector share of the electric power sector's total carbon dioxide emissions based upon the weighted share of electricity retail sales to (receipts by) the manufacturing subsector.

[4] Includes all other types of energy that respondents indicated were consumed or allocated, such as asphalt and road oil, lubricants, naphtha < 401° F, other oils >= 401° F, special naphthas, waxes, and miscellaneous nonfuel products, which are nonfuel products assigned to the petroleum refining industry group (NAICS 324110).

[5] Data are in million metric tons of carbon dioxide per quadrillion Btu of energy (including allocated electricity losses).

[6] Data are in metric tons of carbon dioxide per million chained (2000) dollars.

(s)=Less than 0.05 million metric tons.

Notes: • Data are estimates for the first use of energy for heat and power and as feedstocks or raw material inputs. "First use" is the consumption of energy that was originally produced offsite or was produced onsite from input materials not classified as energy. Minor revisions to the 2002 Manufacturing Energy Consumption Survey (MECS) consumption data have been made since the estimates in this table have been computed. The revisions would likely not have a discernible effect on the estimates shown. • Electricity was converted from point-of-use to primary electricity using Table A6 of this report. • See Table 2.2 for manufacturing energy use. • Totals may not equal sum of components due to independent rounding.

Web Page: For related information, see http://www.eia.doe.gov/emeu/mecs.

Sources: Energy Information Administration, Form EIA-846, "2002 Manufacturing Energy Consumption Survey," Form EIA-810, "Monthly Refinery Report" (for 2002), and *Documentation for Emissions of Greenhouse Gases in the United States 2003* (May 2005).

Figure 12.5 Methane Emissions

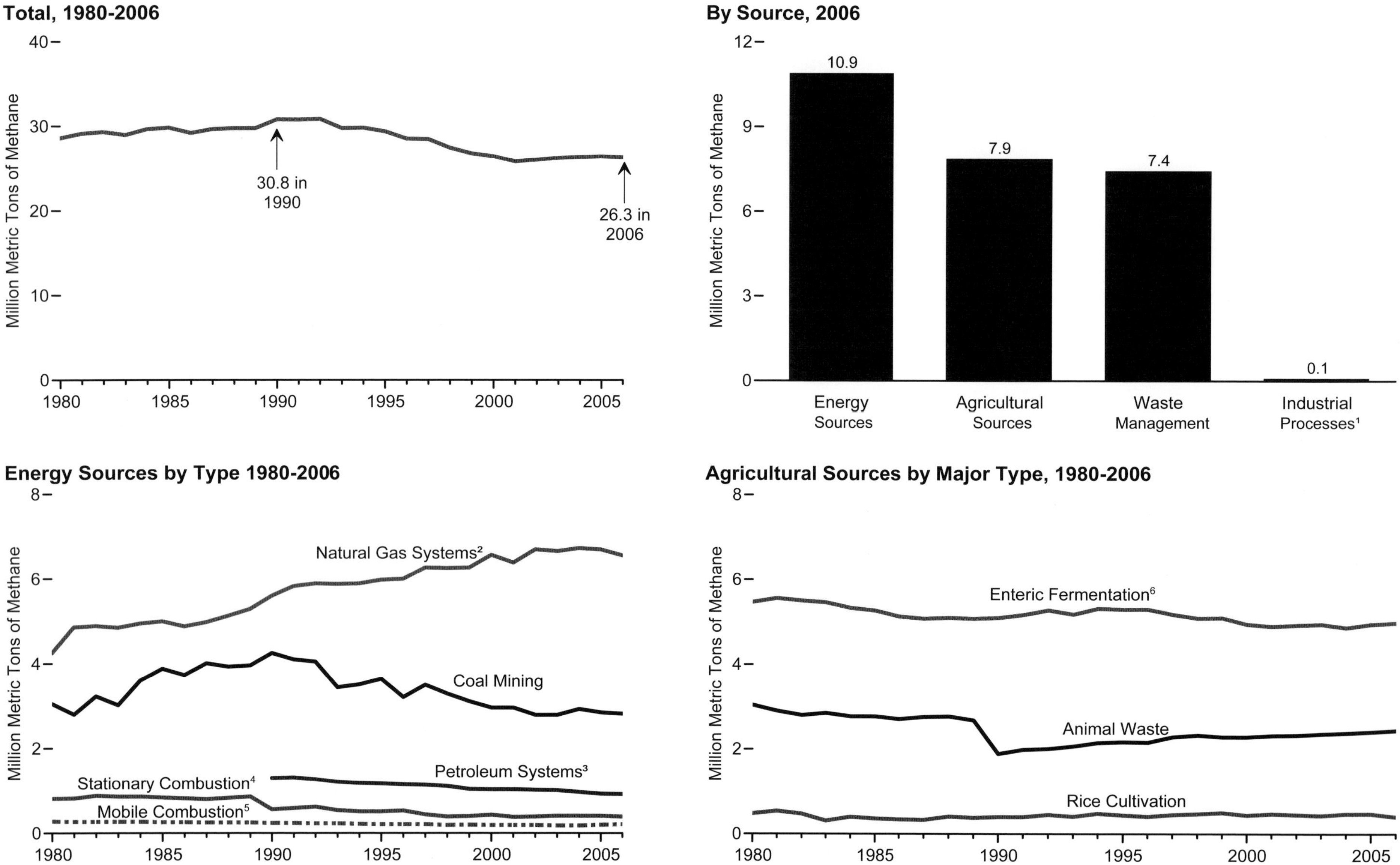

[1] Chemical production, and iron and steel production.
[2] Natural gas production, processing, and distribution.
[3] Petroleum production, refining, and distribution.
[4] Consumption of coal, petroleum, natural gas, and wood for heat or electricity.
[5] Emissions from passenger cars, trucks, buses, motorcycles, and other transport.
[6] Methane emitted as a product of digestion in animals such as cattle, buffalo, sheep, goats, and camels.

Note: Because vertical scales differ, graphs should not be compared.
Source: Table 12.5.

Table 12.5 Methane Emissions, 1980-2006

(Million Metric Tons of Methane)

Year	Energy Sources						Waste Management			Agricultural Sources					Industrial Processes [9]	Total [5]
	Coal Mining	Natural Gas Systems [1]	Petroleum Systems [2]	Mobile Combustion [3]	Stationary Combustion [4]	Total [5]	Landfills	Wastewater Treatment [6]	Total [5]	Enteric Fermentation [7]	Animal Waste [8]	Rice Cultivation	Crop Residue Burning	Total [5]		
1980	3.05	4.25	NA	0.28	0.81	8.39	R10.50	R0.52	R11.02	5.47	3.04	0.48	0.04	9.04	0.13	R28.59
1981	2.80	4.86	NA	.27	.82	8.75	R10.67	.53	R11.20	5.56	2.90	.54	.05	9.05	.14	R29.13
1982	3.23	4.89	NA	.27	.88	9.26	R10.61	.54	R11.15	5.50	2.80	.47	.05	8.81	.10	R29.32
1983	3.02	4.85	NA	.27	.86	9.00	R10.65	.54	R11.19	5.46	2.85	.31	.03	8.66	.11	R28.96
1984	3.60	4.95	NA	.27	.86	9.68	R10.66	R.66	R11.32	5.33	2.77	.40	.04	8.55	.11	R29.66
1985	3.88	5.00	NA	.26	.84	9.98	R10.63	R.67	R11.30	5.27	2.77	.36	.05	8.45	.11	R29.83
1986	3.73	4.88	NA	.26	.82	9.69	R10.51	R.68	R11.18	5.13	2.71	.34	.04	8.22	.10	R29.20
1987	4.01	4.98	NA	.25	.80	10.04	R10.61	R.68	R11.29	5.08	2.76	.33	.04	8.21	.11	R29.65
1988	3.93	5.13	NA	.25	.83	10.14	R10.49	R.69	R11.18	5.10	2.77	.41	.03	8.31	.12	R29.76
1989	3.96	5.29	NA	.25	.86	10.36	R10.41	R.70	R11.11	5.08	2.68	.38	.04	8.18	.12	R29.77
1990	4.25	5.60	1.30	.24	.56	11.96	R10.40	R.89	R11.29	R5.10	1.89	.40	.04	R7.44	.12	R30.80
1991	4.10	5.83	1.31	.24	.59	12.06	R10.11	R.90	R11.01	5.17	1.99	.40	.04	7.59	.11	R30.77
1992	4.05	5.89	1.27	.23	.62	12.05	R9.97	R.92	R10.90	5.28	2.01	.45	.05	7.79	.12	R30.86
1993	3.44	5.88	1.21	.23	.54	11.29	R9.73	R.93	R10.66	5.18	2.07	.41	.04	7.70	.12	R29.77
1994	3.51	5.89	1.18	.22	.51	11.32	R9.41	R.95	R10.36	5.32	2.15	.48	.05	8.00	.13	R29.81
1995	3.64	5.98	1.17	.21	.51	R11.51	R8.82	R.97	R9.79	R5.30	2.17	.44	.04	R7.95	.13	R29.39
1996	3.21	6.00	1.15	.21	.53	11.10	R8.39	R.98	R9.37	5.30	2.16	.41	.05	7.92	.13	R28.52
1997	3.50	6.27	1.14	.21	.44	R11.56	R7.80	R.99	R8.80	5.18	2.29	.45	.05	7.97	.13	R28.46
1998	3.29	6.26	1.11	R.19	.39	11.24	R7.14	R1.00	R8.14	5.09	2.33	.47	.05	7.94	.13	R27.45
1999	3.11	6.27	1.04	.20	R.40	R11.02	R6.67	R1.02	R7.69	5.10	2.29	.50	.05	7.94	.13	R26.77
2000	2.96	6.57	1.03	R.19	.43	R11.18	R6.36	R1.02	R7.38	R4.95	2.29	.44	.05	R7.74	.13	R26.43
2001	2.96	6.39	1.03	R.19	.38	10.95	R6.01	R1.02	R7.03	R4.90	2.32	.47	.05	R7.74	.11	R25.82
2002	2.79	6.70	1.02	.19	.39	R11.08	R6.05	R1.03	R7.08	R4.93	2.33	.45	.05	R7.76	R.12	R26.03
2003	2.79	6.66	1.01	.18	.41	R11.04	R6.28	R1.03	R7.31	R4.95	2.36	.43	.05	R7.78	R.12	R26.25
2004	2.93	6.73	.97	R.18	.41	11.23	R6.18	R1.04	R7.22	R4.87	2.38	.47	.06	R7.78	.12	R26.34
2005	2.85	6.70	R.93	.20	R.41	R11.09	R6.27	R1.04	R7.32	R4.95	2.41	.47	.05	R7.89	.11	R26.41
2006P	2.82	6.56	.92	.21	.39	10.89	6.38	1.06	7.44	4.98	2.44	.40	.05	7.87	.10	26.31

[1] Natural gas production, processing, and distribution; processing is not included in 1980 and is incompletely covered in 1981-1989.
[2] Petroleum production, refining, and distribution.
[3] Emissions from passenger cars, trucks, buses, motorcycles, and other transport.
[4] Consumption of coal, petroleum, natural gas, and wood for heat or electricity.
[5] See notes on components for specific coverage, which is inconsistent prior to 1990 in some cases.
[6] 1980-1984, domestic wastewater only; 1985 forward, industrial and domestic wastewater.
[7] Methane emitted as a product of digestion in animals such as cattle, buffalo, sheep, goats, and camels.
[8] Estimation methods for 1990 forward reflect a shift in waste management away from liquid systems to dry-lot systems, thus lowering emissions.
[9] Chemical production, and iron and steel production.
R=Revised. P=Preliminary. NA=Not available.

Notes: • Emissions are from anthropogenic sources. "Anthropogenic" means produced as the result of human activities, including emissions from agricultural activity and domestic livestock. Emissions from natural sources, such as wetlands and wild animals, are not included. • Under certain conditions, methane may be produced via anaerobic decomposition of organic materials in landfills, animal wastes, and rice paddies. • Because of the continuing goal to improve estimation methods for greenhouse gases, data are frequently revised on an annual basis in keeping with the latest findings of the international scientific community. • For information on units for measuring greenhouse gases, see http://www.eia.doe.gov/oiaf/1605/archive/gg06rpt/pdf/executive_summary.pdf, page 2, box titled "Units for Measuring Greenhouse Gases. • Totals may not equal sum of components due to independent rounding.

Web Page: For related information, see http://www.eia.doe.gov/environment.html.

Sources: Energy Information Administration, *Emissions of Greenhouse Gases in the United States*, annual reports and unpublished revisions.

Figure 12.6 Nitrous Oxide Emissions

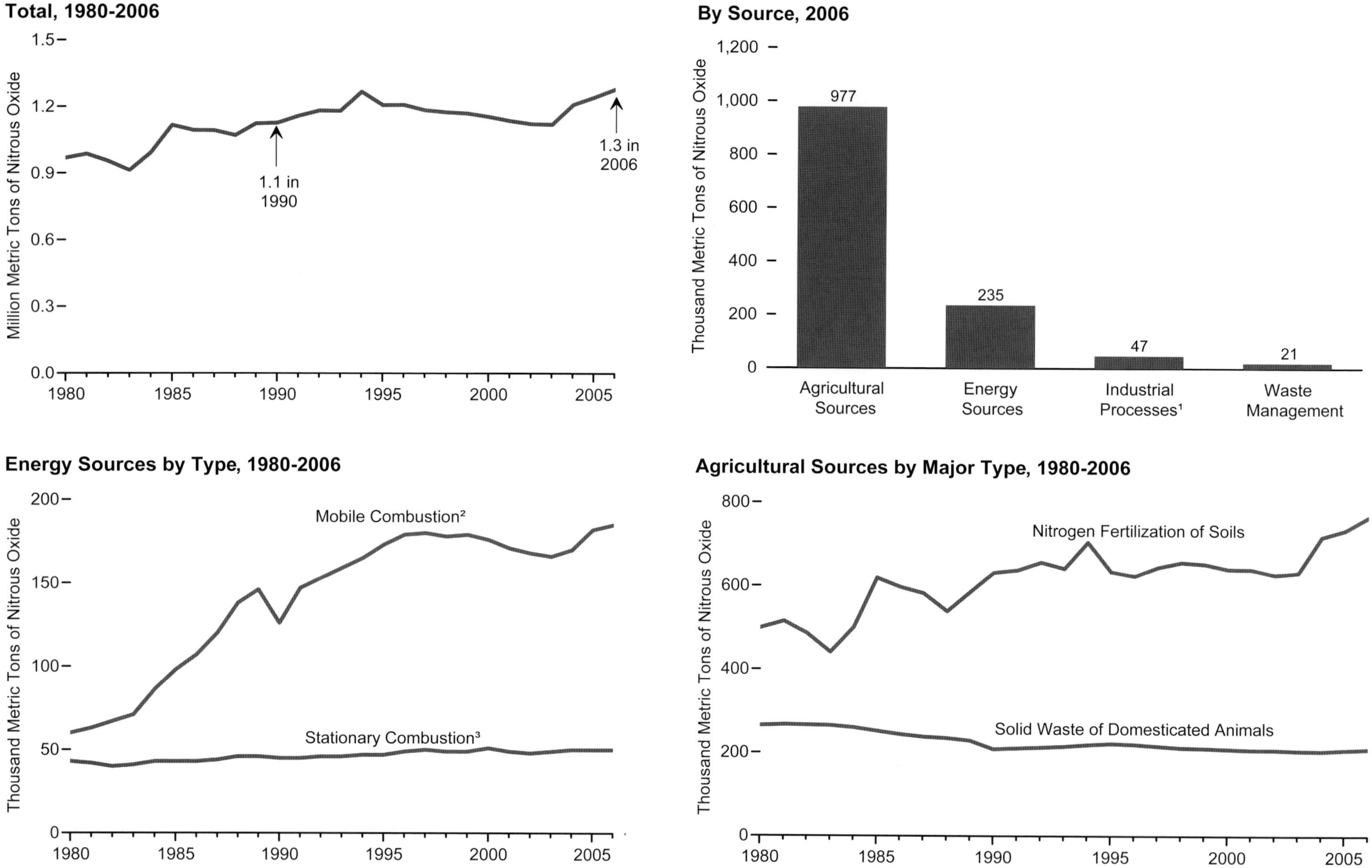

[1] Adipic acid production (primarily for the manufacture of nylon fibers and plastics) and nitric acid production (primarily for fertilizers).

[2] Emissions from passenger cars and trucks; air, rail, and marine transportation; and farm and construction equipment.

[3] Consumption of coal, petroleum, natural gas, and wood for heat or electricity.

Note: Because vertical scales differ, graphs should not be compared.

Source: Table 12.6.

Table 12.6 Nitrous Oxide Emissions, 1980-2006

(Thousand Metric Tons of Nitrous Oxide)

Year	Energy Sources			Waste Management			Agricultural Sources				Industrial Processes [3]	Total
	Mobile Combustion [1]	Stationary Combustion [2]	Total	Waste Combustion	Human Sewage in Wastewater	Total	Nitrogen Fertilization of Soils	Crop Residue Burning	Solid Waste of Domesticated Animals	Total		
1980	60	43	102	(s)	13	13	499	1	265	766	88	969
1981	63	42	105	(s)	13	14	515	2	267	783	85	987
1982	67	40	107	(s)	13	14	486	2	266	754	81	955
1983	71	41	112	(s)	14	14	441	1	265	707	80	913
1984	86	43	130	(s)	14	14	500	2	260	762	88	993
1985	98	43	141	(s)	15	15	619	2	252	872	89	1,117
1986	107	43	150	(s)	15	15	597	2	244	842	87	1,094
1987	120	44	164	1	15	16	582	1	238	822	91	1,093
1988	138	46	183	1	15	16	539	1	235	775	96	1,071
1989	146	46	192	(s)	15	16	586	2	229	817	99	1,124
1990	126	45	172	1	16	17	R632	2	209	R843	96	1,127
1991	147	45	192	1	16	17	R638	2	211	R850	99	1,158
1992	153	46	198	1	16	17	R657	2	213	R872	95	R1,182
1993	159	46	205	1	17	17	R642	1	215	R858	100	R1,181
1994	165	47	212	1	17	18	R705	2	219	R926	110	R1,267
1995	R173	47	R220	1	17	18	R634	2	222	R857	111	R1,207
1996	R179	49	R228	1	17	18	R624	2	220	R845	116	R1,208
1997	180	50	230	1	18	19	R645	2	216	R862	74	R1,184
1998	R178	49	R227	1	18	19	R657	2	212	R871	58	R1,175
1999	R179	49	R229	1	19	20	R653	2	211	R865	57	R1,170
2000	R176	51	R227	1	19	20	R641	2	209	R852	56	R1,155
2001	R171	49	R220	1	19	20	R640	2	207	R850	47	R1,137
2002	R168	48	R216	1	19	20	R627	2	207	R835	51	R1,123
2003	R166	49	R215	1	19	20	R632	2	205	R839	R46	R1,121
2004	R170	50	R220	1	20	21	R718	2	204	R924	R46	R1,210
2005	R182	50	R232	1	20	21	R734	2	207	R943	R47	R1,243
2006P	185	50	235	1	20	21	766	2	209	977	47	1,279

[1] Emissions from passenger cars and trucks; air, rail, and marine transportation; and farm and construction equipment.

[2] Consumption of coal, petroleum, natural gas, and wood for heat or electricity.

[3] Adipic acid production (primarily for the manufacture of nylon fibers and plastics), and nitric acid production (primarily for fertilizers).

R=Revised. P=Preliminary. (s)=Less than 0.5 thousand metric tons.

Notes: • Emissions are from anthropogenic sources. "Anthropogenic" means produced as the result of human activities, including emissions from agricultural activity and domestic livestock. Emissions from natural sources, such as wetlands and wild animals, are not included. • Under certain conditions, methane may be produced via anaerobic decomposition of organic materials in landfills, animal wastes, and rice paddies. • Because of the continuing goal to improve estimation methods for greenhouse gases, data are frequently revised on an annual basis in keeping with the latest findings of the international scientific community. • For information on units for measuring greenhouse gases, see http://www.eia.doe.gov/oiaf/1605/archive/gg06rpt/pdf/executive_summary.pdf, page 2, box titled "Units for Measuring Greenhouse Gases. • Totals may not equal sum of components due to independent rounding.

Web Page: For related information, see http://www.eia.doe.gov/environment.html.

Sources: Energy Information Administration, *Emissions of Greenhouse Gases in the United States,* annual reports and unpublished revisions.

Figure 12.7 Emissions From Energy Consumption for Electricity Generation and Useful Thermal Output

Emissions by Type of Generating Unit, 2006

Carbon Dioxide

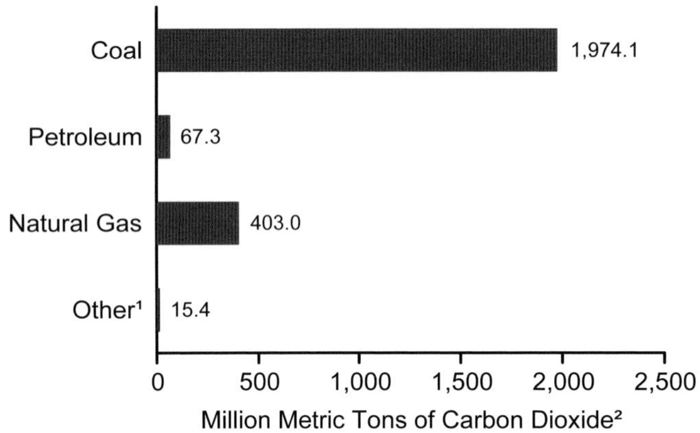

Sulfur Dioxide

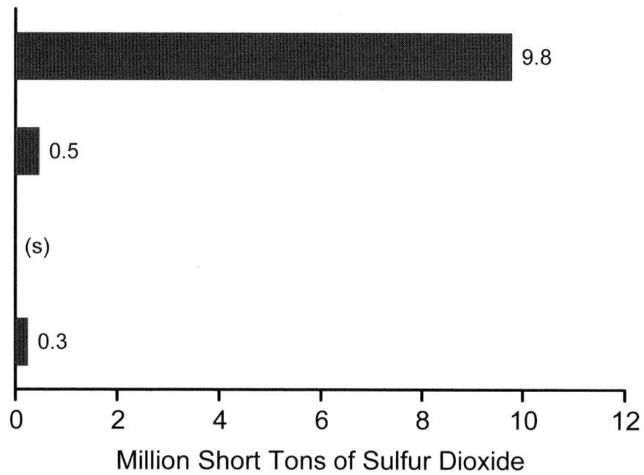

Nitrogen Oxides

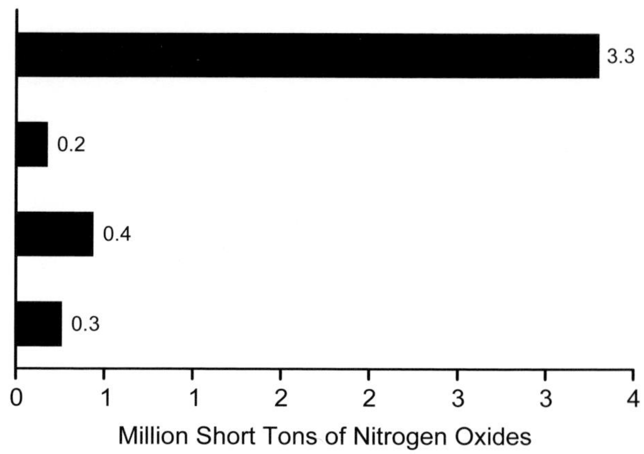

Emissions by Sector, 1989-2006

Carbon Dioxide

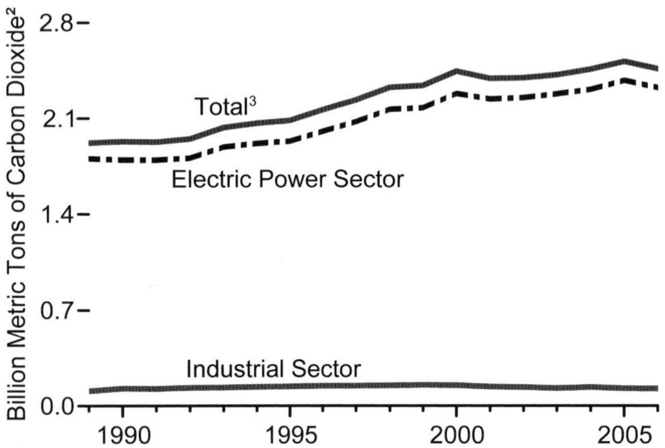

Sulfur Dioxide

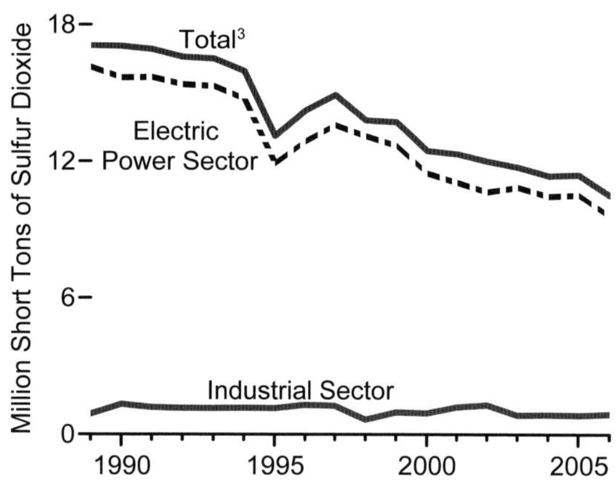

Nitrogen Oxides

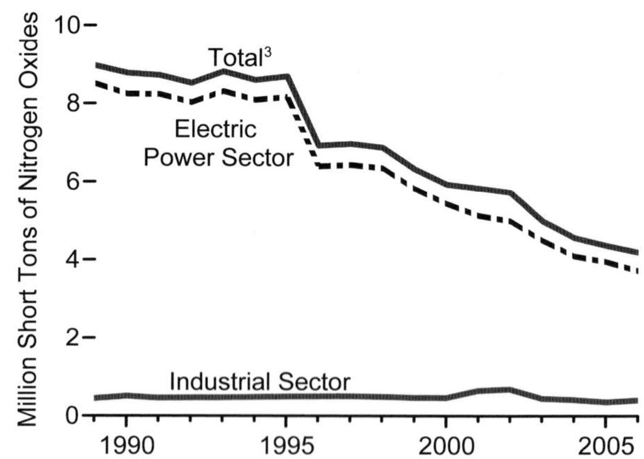

[1] For carbon dioxide: municipal solid waste (only the estimated plastics portion of municipal solid waste is included); tire-derived fuel, and geothermal. For sulfur dioxide and nitrogen oxides: blast furnace gas, propane gas, and other manufactured and waste gases derived from fossil fuels; wood, black liquor, and other wood waste; municipal solid waste, landfill gas, sludge waste, tires, agricultural byproducts, and other biomass; and chemicals, hydrogen, pitch, sulfur, and tar coal.

[2] Metric tons of carbon dioxide can be converted to metric tons of carbon equivalent by multiplying by 12/44.

[3] Includes Commercial Sector.

(s)=Less than 0.05 million short tons.

Sources: Tables 12.7a-12.7c.

Table 12.7a Emissions From Energy Consumption for Electricity Generation and Useful Thermal Output: Total (All Sectors), 1989-2006 (Sum of Tables 12.7b and 12.7c)

Year	Carbon Dioxide						Sulfur Dioxide					Nitrogen Oxides				
	Coal [1]	Petroleum [2]	Natural Gas [3]	MSW [4]	Geo-thermal [5]	Total	Coal [1]	Petroleum [2]	Natural Gas [3]	Other [6]	Total	Coal [1]	Petroleum [2]	Natural Gas [3]	Other [6]	Total
	Million Metric Tons of Carbon Dioxide [7]						Thousand Short Tons of Sulfur Dioxide					Thousand Short Tons of Nitrogen Oxides				
1989	1,553.9	143.9	217.3	5.5	0.4	1,921.0	15,949	1,085	1	43	17,079	8,026	296	545	102	8,969
1990	1,572.4	118.4	232.7	7.4	.4	1,931.2	15,742	1,033	1	268	17,043	7,847	229	565	134	8,776
1991	1,572.1	110.2	236.9	8.4	.4	1,928.0	15,696	943	1	271	16,912	7,836	213	549	125	8,723
1992	1,596.7	95.7	246.9	9.9	.4	1,949.6	15,499	776	1	291	16,568	7,688	175	525	131	8,519
1993	1,666.2	107.1	249.2	10.3	.4	2,033.1	15,259	938	1	299	16,497	7,964	191	523	137	8,815
1994	1,675.7	101.8	274.9	11.1	.4	2,063.9	14,769	875	1	307	15,953	7,722	176	565	136	8,599
1995	1,698.0	76.3	297.1	11.8	.3	2,083.5	12,133	676	2	303	13,113	7,755	135	660	142	8,692
1996	1,788.7	83.2	276.5	12.6	.4	2,161.3	13,034	878	3	312	14,226	6,036	164	578	146	6,924
1997	1,834.9	92.6	291.7	13.2	.4	2,232.7	13,689	949	2	264	14,904	6,050	177	591	153	6,971
1998	1,862.9	122.4	325.8	12.6	.4	2,324.1	12,569	1,065	2	153	13,789	5,800	253	670	152	6,875
1999	1,869.6	114.7	341.4	12.6	.4	2,338.7	12,408	1,054	7	248	13,717	5,294	232	653	140	6,319
2000	1,960.0	107.5	361.7	12.2	.4	2,441.7	11,338	866	3	246	12,453	4,944	193	657	136	5,930
2001	1,895.2	116.1	365.3	12.8	.3	2,389.7	10,918	1,105	2	293	12,318	4,515	324	696	295	5,831
2002	1,912.7	90.2	377.1	14.8	.4	2,395.0	10,787	852	2	353	11,994	4,472	248	689	317	5,725
2003	1,947.2	111.0	343.4	13.8	.4	2,415.7	10,679	791	2	263	11,735	3,976	265	499	256	4,995
2004	1,962.7	114.7	365.2	14.0	.4	2,456.9	10,402	698	2	261	11,364	3,622	248	458	239	4,567
2005	2,001.2	115.9	381.9	14.1	.4	2,513.6	10,471	648	2	277	11,397	3,456	244	422	244	4,366
2006	1,974.1	67.3	403.0	15.0	.4	2,459.8	9,774	471	2	250	10,498	3,303	180	440	264	4,188

[1] Anthracite, bituminous coal, subbituminous coal, lignite, waste coal, and synthetic coal.
[2] Distillate fuel oil, residual fuel oil, petroleum coke, jet fuel, kerosene, other petroleum, and waste oil.
[3] Natural gas, plus a small amount of supplemental gaseous fuels.
[4] Municipal solid waste (only the estimated plastics portion of municipal solid waste is included) and tire-derived fuel.
[5] Carbon dioxide in geothermal steam.
[6] Blast furnace gas, propane gas, and other manufactured and waste gases derived from fossil fuels; wood and wood-derived fuels; municipal solid waste, landfill gas, sludge waste, tires, agricultural byproducts, and other biomass; and chemicals, hydrogen, pitch, sulfur, and tar coal.
[7] Metric tons of carbon dioxide can be converted to metric tons of carbon equivalent by multiplying by 12/44.

Notes: • Data are for emissions from energy consumption for electricity generation and useful thermal output. • See "Useful Thermal Output" in Glossary. • Totals may not equal sum of components due to independent rounding.

Web Page: For related information, see http://www.eia.doe.gov/fuelelectric.html.

Sources: Tables 12.7b and 12.7c.

Table 12.7b Emissions From Energy Consumption for Electricity Generation and Useful Thermal Output: Electric Power Sector, 1989-2006 (Subset of Table 12.7a)

Year	Carbon Dioxide						Sulfur Dioxide					Nitrogen Oxides				
	Coal [1]	Petroleum [2]	Natural Gas [3]	MSW [4]	Geo-thermal [5]	Total	Coal [1]	Petroleum [2]	Natural Gas [3]	Other [6]	Total	Coal [1]	Petroleum [2]	Natural Gas [3]	Other [6]	Total
	Million Metric Tons of Carbon Dioxide [7]						Thousand Short Tons of Sulfur Dioxide					Thousand Short Tons of Nitrogen Oxides				
1989	1,501.1	132.2	168.8	4.3	0.4	1,806.8	15,229	893	1	8	16,130	7,777	271	430	28	8,506
1990	1,514.8	100.8	176.3	5.7	.4	1,798.0	14,965	692	1	14	15,672	7,582	193	430	40	8,245
1991	1,515.2	94.2	179.6	7.1	.4	1,796.6	14,981	684	1	16	15,682	7,590	182	423	46	8,241
1992	1,537.1	78.4	186.8	8.4	.4	1,811.0	14,743	616	1	13	15,373	7,440	141	395	51	8,027
1993	1,605.4	89.5	187.3	8.5	.4	1,891.2	14,476	810	1	14	15,302	7,712	157	393	54	8,316
1994	1,613.3	84.2	210.1	9.2	.4	1,917.2	13,994	733	1	12	14,741	7,470	141	430	52	8,094
1995	1,635.4	60.5	227.5	9.9	.3	1,933.6	11,377	527	1	11	11,917	7,501	104	508	54	8,167
1996	1,725.3	65.5	204.2	9.8	.4	2,005.2	12,266	601	2	11	12,881	5,784	131	423	56	6,394
1997	1,771.3	74.4	219.1	10.2	.4	2,075.3	12,874	692	2	14	13,582	5,796	141	429	66	6,432
1998	1,801.3	104.6	248.6	10.1	.4	2,164.9	12,161	926	2	11	13,099	5,561	220	503	68	6,352
1999	1,807.8	97.0	261.1	10.2	.4	2,176.4	11,844	832	7	10	12,692	5,067	198	491	59	5,817
2000	1,897.7	91.3	281.6	10.0	.4	2,281.1	10,770	696	3	7	11,476	4,724	167	483	60	5,434
2001	1,837.8	101.9	289.6	10.8	.3	2,240.4	10,230	831	2	5	11,069	4,274	279	469	106	5,128
2002	1,853.7	78.0	305.9	12.6	.4	2,250.6	10,037	606	2	17	10,661	4,203	206	468	115	4,992
2003	1,891.5	97.2	277.9	11.3	.4	2,278.4	10,202	639	2	15	10,857	3,853	228	311	108	4,500
2004	1,903.7	99.2	296.2	11.1	.4	2,310.6	9,911	544	2	10	10,466	3,508	212	266	111	4,098
2005	1,944.2	101.5	318.9	11.1	.4	2,376.2	9,999	508	2	11	10,520	3,363	208	268	113	3,952
2006	1,918.4	54.8	337.9	11.4	.4	2,322.9	9,277	291	2	9	9,579	3,199	149	253	118	3,719

[1] Anthracite, bituminous coal, subbituminous coal, lignite, waste coal, and synthetic coal.
[2] Distillate fuel oil, residual fuel oil, petroleum coke, jet fuel, kerosene, other petroleum, and waste oil.
[3] Natural gas, plus a small amount of supplemental gaseous fuels.
[4] Municipal solid waste (only the estimated plastics portion of municipal solid waste is included) and tire-derived fuel.
[5] Carbon dioxide in geothermal steam.
[6] Blast furnace gas, propane gas, and other manufactured and waste gases derived from fossil fuels; wood and wood-derived fuels; municipal solid waste, landfill gas, sludge waste, tires, agricultural byproducts, and other biomass; and chemicals, hydrogen, pitch, sulfur, and tar coal.
[7] Metric tons of carbon dioxide can be converted to metric tons of carbon equivalent by multiplying by 12/44.

Notes: • Data are for emissions from energy consumption for electricity generation and useful thermal output. • The electric power sector comprises electricity-only and combined-heat-and-power (CHP) plants within the NAICS 22 category whose primary business is to sell electricity, or electricity and heat, to the public. • See Table 12.7c for commercial and industrial CHP and electricity-only data. • See Note 2, "Classification of Power Plants Into Energy-Use Sectors," at end of Section 8. • See "Useful Thermal Output" in Glossary. • Totals may not equal sum of components due to independent rounding.

Web Page: For related information, see http://www.eia.doe.gov/fuelelectric.html.

Sources: **Carbon Dioxide:** • 1989-1997—Energy Information Administration (EIA), Form EIA-759, "Monthly Power Plant Report," and Form EIA-867, "Annual Nonutility Power Producer Report." • 1998-2000—EIA, Form EIA-759, "Monthly Power Plant Report," and Form EIA-860B, "Annual Electric Generator Report—Nonutility." • 2001-2003—EIA, Form EIA-906, "Power Plant Report." • 2004 forward—EIA, Form EIA-906, "Power Plant Report," and Form EIA-920,"Combined Heat and Power Plant Report." **Sulfur Dioxide** and **Nitrogen Oxides:** EIA, Form EIA-767, "Steam-Electric Plant Operation and Design Report." Data were adjusted by the Environmental Protection Agency's Continuous Emission Monitoring System.

Table 12.7c Emissions From Energy Consumption for Electricity Generation and Useful Thermal Output: Commercial and Industrial Sectors, 1989-2006 (Subset of Table 12.7a)

Year	Carbon Dioxide						Sulfur Dioxide					Nitrogen Oxides				
	Coal [1]	Petroleum [2]	Natural Gas [3]	MSW [4]	Geo-thermal [5]	Total	Coal [1]	Petroleum [2]	Natural Gas [3]	Other [6]	Total	Coal [1]	Petroleum [2]	Natural Gas [3]	Other [6]	Total
	Million Metric Tons of Carbon Dioxide [7]						Thousand Short Tons of Sulfur Dioxide					Thousand Short Tons of Nitrogen Oxides				
	Commercial Sector [8]															
1989	2.3	0.6	1.5	0.8	–	5.3	41	6	(s)	1	48	10	2	4	3	19
1990	2.4	.7	2.3	.9	–	6.3	43	5	(s)	1	49	11	1	6	4	23
1991	2.6	.5	2.3	1.0	–	6.5	35	3	(s)	1	39	11	1	7	4	23
1992	2.5	.5	2.8	1.2	–	7.0	35	3	(s)	1	39	11	1	8	4	24
1993	3.0	.6	3.3	1.3	–	8.1	44	4	(s)	1	48	13	1	8	4	26
1994	2.9	.6	3.7	1.3	–	8.5	43	3	(s)	(s)	47	13	1	8	4	26
1995	3.1	.5	4.0	1.4	–	9.1	43	2	(s)	(s)	46	13	1	10	5	29
1996	3.6	.5	4.3	2.0	–	10.5	48	2	(s)	1	51	16	1	10	7	34
1997	3.8	.7	4.6	2.3	–	11.4	62	4	(s)	3	69	17	1	11	9	37
1998	3.3	.8	4.7	2.1	–	10.9	36	4	(s)	2	42	14	1	11	8	34
1999	3.4	.7	4.5	2.0	–	10.7	51	3	(s)	(s)	54	15	1	10	7	32
2000	3.6	.7	4.6	1.7	–	10.6	47	3	(s)	1	51	14	1	10	6	31
2001	3.3	.8	4.3	1.4	–	9.8	48	4	(s)	2	53	14	3	21	17	55
2002	3.0	.6	4.0	1.5	–	9.1	46	3	(s)	2	50	14	3	22	14	52
2003	3.9	.7	3.2	1.7	–	9.4	35	3	(s)	2	40	10	6	18	16	50
2004	4.0	.9	3.9	1.9	–	10.7	33	4	(s)	2	39	9	8	20	18	54
2005	4.0	.8	4.1	1.9	–	10.7	36	3	(s)	1	40	10	7	26	17	59
2006	3.9	.4	4.4	1.9	–	10.7	36	3	(s)	1	40	10	3	39	18	70
	Industrial Sector [9]															
1989	50.4	11.1	47.0	0.4	–	108.9	679	186	(s)	35	901	241	24	110	69	444
1990	55.2	16.9	54.1	.7	–	126.9	734	335	(s)	252	1,322	257	34	128	88	508
1991	54.3	15.5	55.0	.2	–	125.0	681	256	(s)	254	1,191	237	30	119	73	459
1992	57.1	16.8	57.3	.3	–	131.6	722	157	(s)	277	1,156	240	32	122	74	468
1993	57.8	17.0	58.5	.5	–	133.8	739	124	(s)	283	1,147	241	32	122	77	472
1994	59.5	17.0	61.1	.6	–	138.2	732	139	(s)	294	1,165	242	33	126	78	479
1995	59.5	15.3	65.5	.5	–	140.8	713	146	(s)	291	1,150	243	30	142	81	496
1996	59.7	17.2	67.9	.7	–	145.6	720	273	(s)	300	1,294	238	33	144	82	497
1997	59.7	17.5	68.0	.7	–	145.9	753	253	(s)	247	1,253	238	35	150	79	502
1998	58.3	17.0	72.5	.5	–	148.4	372	135	(s)	141	648	225	32	157	75	490
1999	58.4	17.0	75.7	.5	–	151.6	514	219	(s)	237	971	214	32	151	73	470
2000	58.7	15.4	75.5	.5	–	150.1	520	167	(s)	239	927	208	24	164	69	465
2001	54.1	13.4	71.4	.6	–	139.5	640	270	(s)	285	1,196	227	42	206	172	648
2002	56.0	11.6	67.1	.6	–	135.3	704	243	(s)	334	1,282	255	39	199	188	681
2003	51.8	13.0	62.3	.8	–	127.9	443	149	(s)	247	839	112	31	171	131	446
2004	55.1	14.5	65.1	1.0	–	135.6	458	150	(s)	250	859	105	27	173	110	415
2005	53.0	13.7	58.9	1.1	–	126.7	435	137	(s)	265	837	83	29	129	114	355
2006	51.8	12.1	60.7	1.7	–	126.3	461	177	(s)	241	880	94	28	148	129	399

[1] Anthracite, bituminous coal, subbituminous coal, lignite, waste coal, and synthetic coal.
[2] Distillate fuel oil, residual fuel oil, petroleum coke, jet fuel, kerosene, other petroleum, and waste oil.
[3] Natural gas, plus a small amount of supplemental gaseous fuels.
[4] Municipal solid waste (only the estimated plastics portion of municipal solid waste is included) and tire-derived fuel.
[5] Carbon dioxide in geothermal steam.
[6] Blast furnace gas, propane gas, and other manufactured and waste gases derived from fossil fuels; wood and wood-derived fuels; municipal solid waste, landfill gas, sludge waste, tires, agricultural byproducts, and other biomass; and chemicals, hydrogen, pitch, sulfur, and tar coal.
[7] Metric tons of carbon dioxide can be converted to metric tons of carbon equivalent by multiplying by 12/44.
[8] Commercial combined-heat-and-power (CHP) and commercial electricity-only plants.
[9] Industrial combined-heat-and-power (CHP) and industrial electricity-only plants.

– = No data reported. (s)=Less than 0.05 million metric tons or less than 500 short tons.

Notes: • Data are for emissions from energy consumption for electricity generation and useful thermal output. • See Table 12.7b for electric power sector data. • See Note 2, "Classification of Power Plants Into Energy-Use Sectors," at end of Section 8. • See "Useful Thermal Output" in Glossary. • Totals may not equal sums of components due to independent rounding.

Web Page: For related information, see http://www.eia.doe.gov/fuelelectric.html.

Sources: **Carbon Dioxide:** • 1989-1997—Energy Information Administration (EIA), Form EIA-867, "Annual Nonutility Power Producer Report." • 1998-2000—EIA, Form EIA-860B, "Annual Electric Generator Report—Nonutility." • 2001-2003—EIA, Form EIA-906, "Power Plant Report." • 2004 forward—EIA, Form EIA-906, "Power Plant Report," and Form EIA-920, "Combined Heat and Power Plant Report." **Sulfur Dioxide** and **Nitrogen Oxides:** EIA, Form EIA-767, "Steam-Electric Plant Operation and Design Report." Data were adjusted by the Environmental Protection Agency's Continuous Emission Monitoring System.

Figure 12.8 Installed Nameplate Capacity of Steam-Electric Generators With Environmental Equipment

By Fuel and Equipment Type, 2005

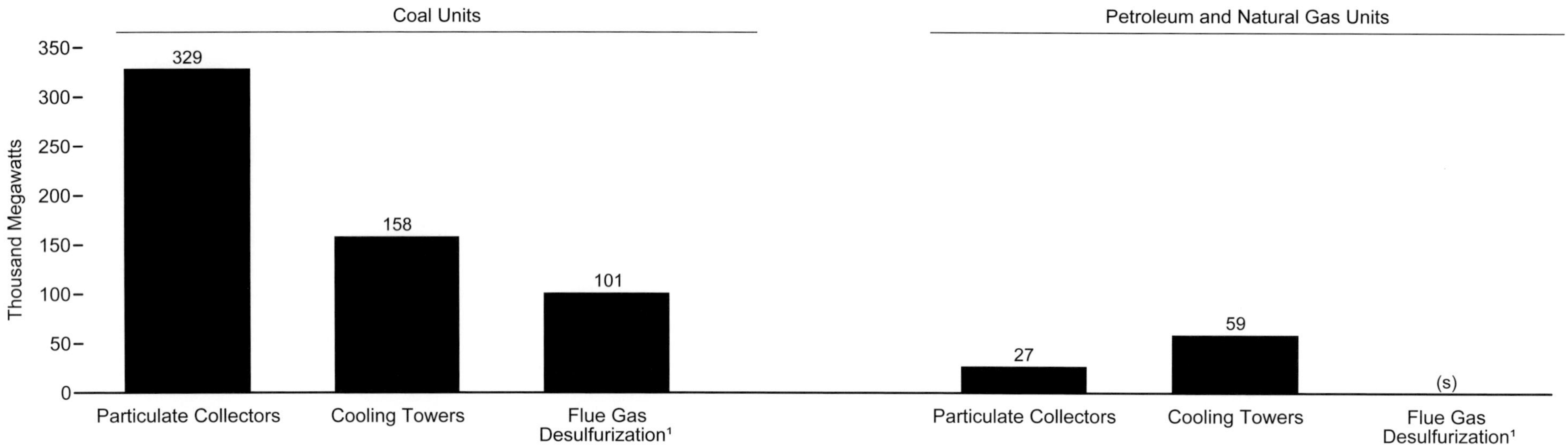

Total Units by Equipment Type, 1985-2005[2]

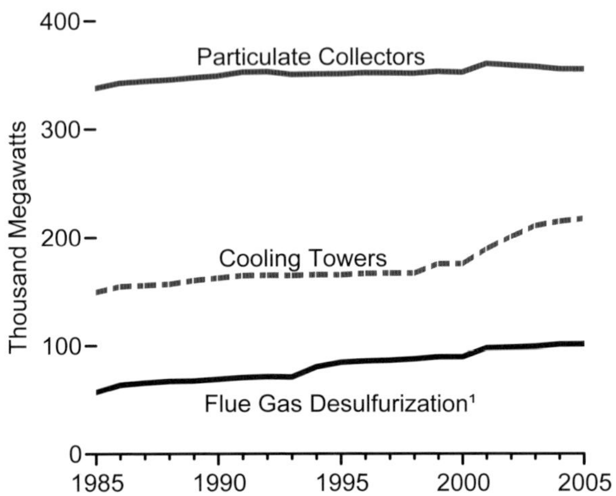

Coal Units by Equipment Type, 1985-2005[2]

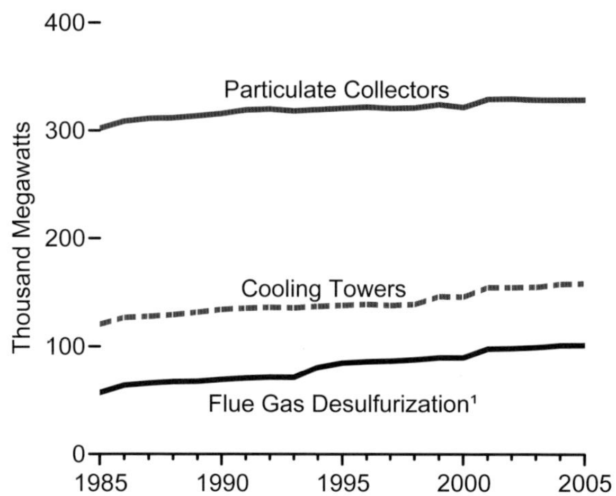

Petroleum and Natural Gas Units by Equipment Type, 1985-2005[2]

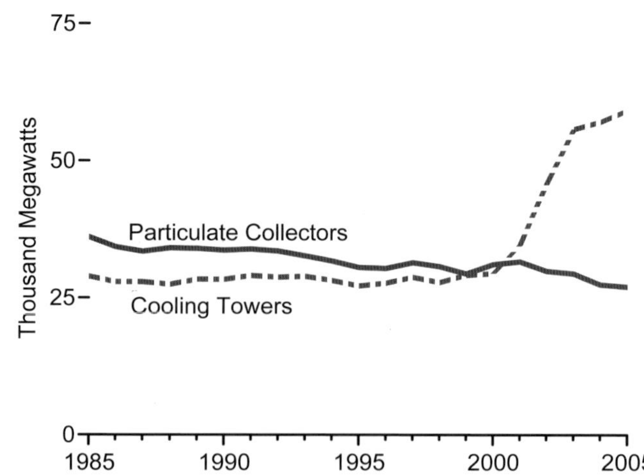

(s)=Less than 0.5 thousand megawatts.

[1] Also called "scrubbers."

[2] Through 2000, data are for electric utility plants with fossil-fueled steam-electric capacity of 100 megawatts or greater. Beginning in 2001, data are for electric utility and unregulated generating plants (independent power producers, commercial plants, and industrial plants) in operating or standby status, with fossil-fueled steam-electric capacity of 100 megawatts or greater, or combustible-renewable steam electric capacity of 10 megawatts or greater.

Notes: • Components are not additive because some generators are included in more than one category. • Because vertical scales differ, graphs should not be compared.

Source: Table 12.8.

Table 12.8 Installed Nameplate Capacity of Steam-Electric Generators With Environmental Equipment, 1985-2005

(Megawatts)

Year	Coal				Petroleum and Natural Gas				Total			
	Particulate Collectors	Cooling Towers	Flue Gas Desulfurization (Scrubbers)	Total [1]	Particulate Collectors	Cooling Towers	Flue Gas Desulfurization (Scrubbers)	Total [1]	Particulate Collectors	Cooling Towers	Flue Gas Desulfurization (Scrubbers)	Total [1]
1985	302,056	120,591	56,955	304,706	36,054	28,895	65	62,371	338,110	149,486	57,020	367,078
1986	308,566	126,731	63,735	311,217	34,258	27,919	65	59,618	342,825	154,650	63,800	370,835
1987	311,043	127,875	65,688	312,885	33,431	27,912	65	58,783	344,474	155,786	65,753	371,668
1988	311,776	129,366	67,156	313,618	34,063	27,434	65	58,937	345,839	156,800	67,221	372,555
1989	313,680	131,701	67,469	315,521	33,975	28,386	65	59,736	347,655	160,087	67,534	375,257
1990	315,681	134,199	69,057	317,522	33,639	28,359	65	59,372	349,319	162,557	69,122	376,894
1991	319,046	135,565	70,474	319,110	33,864	29,067	260	59,773	352,910	164,632	70,734	378,883
1992	319,856	136,266	71,336	319,918	33,509	28,764	195	59,116	353,365	165,030	71,531	379,034
1993	318,188	135,885	71,106	318,251	32,620	28,922	–	58,580	350,808	164,807	71,106	376,831
1994	319,485	137,266	80,617	319,776	31,695	28,186	–	57,123	351,180	165,452	80,617	376,899
1995	320,685	138,108	84,677	320,749	30,513	27,187	–	54,942	351,198	165,295	84,677	375,691
1996	321,805	139,065	85,842	321,869	30,349	27,685	–	55,275	352,154	166,749	85,842	377,144
1997	320,646	138,120	86,605	320,710	31,422	28,766	–	56,485	352,068	166,886	86,605	377,195
1998	321,082	139,082	87,783	321,353	30,708	27,814	–	55,764	351,790	166,896	87,783	377,117
1999	324,109	146,377	89,666	331,379	29,371	29,142	–	55,812	353,480	175,520	89,666	387,192
2000	321,636	146,093	89,675	328,741	31,090	29,427	–	57,697	352,727	175,520	89,675	386,438
2001[2]	329,187	154,747	97,804	329,187	31,575	34,649	184	61,634	360,762	189,396	97,988	390,821
2002	329,459	154,750	98,363	329,459	29,879	45,920	310	72,008	359,338	200,670	98,673	401,341
2003	328,587	155,158	99,257	328,587	29,422	55,770	310	81,493	358,009	210,928	99,567	409,954
2004	328,506	157,968	101,182	328,506	27,402	57,082	310	81,450	355,782	214,989	101,492	409,769
2005	328,720	158,493	101,338	328,720	27,005	59,214	310	83,307	355,599	217,646	101,648	411,840

[1] Components are not additive because some generators are included in more than one category.

[2] Through 2000, data are for electric utility plants with fossil-fueled steam-electric capacity of 100 megawatts or greater. Beginning in 2001, data are for electric utility and unregulated generating plants (independent power producers, commercial plants, and industrial plants) in operating or standby status, with fossil-fueled steam-electric capacity of 100 megawatts or greater, or combustible-renewable steam-electric capacity of 10 megawatts or greater.

– = No data reported.

Note: Beginning in 2006, data are not available.

Web Page: For related information, see http://www.eia.doe.gov/fuelelectric.html.

Sources: • 1985-1993—Energy Information Administration (EIA), Form EIA-767, "Steam-Electric Plant Operation and Design Report." • 1994 forward—EIA, *Electric Power Annual 2005* (November 2006), Table 5.2, and EIA, Form EIA-767, "Steam-Electric Plant Operation and Design Report."

Appendix A

Thermal Conversion Factors

Using Thermal Conversion Factors

The thermal conversion factors presented in the following tables can be used to estimate the heat content in British thermal units (Btu) of a given amount of energy measured in physical units, such as barrels or cubic feet. For example, 10 barrels of asphalt has a heat content of approximately 66.36 million Btu (10 barrels x 6.636 million Btu per barrel = 66.36 million Btu).

The heat content rates (i.e., thermal conversion factors) provided in this section represent the gross (or higher or upper) energy content of the fuels. Gross heat content rates are applied in all Btu calculations for the *Annual Energy Review* and are commonly used in energy calculations in the United States; net (or lower) heat content rates are typically used in European energy calculations. The difference between the two rates is the amount of energy that is consumed to vaporize water that is created during the combustion process. Generally, the difference ranges from 2 percent to 10 percent, depending on the specific fuel and its hydrogen content. Some fuels, such as unseasoned wood, can be more than 40 percent different in their gross and net heat content rates. See "Heat Content" and "British thermal unit (Btu)" in the Glossary for more information.

Thermal conversion factors for hydrocarbon mixes (Table A1) are weighted averages of the thermal conversion factors for each hydrocarbon included in the mix. For example, in calculating the thermal conversion factor for a 60-40 butane-propane mixture, the thermal conversion factor for butane is weighted 1.5 times the thermal conversion factor for propane.

In general, the annual thermal conversion factors presented in Tables A2 through A6 are computed from final annual data or from the best available data and are labeled "preliminary." Often, the previous year's factor is used as the preliminary value until data become available to calculate the factor appropriate to the year. The source of each factor is described in the section entitled "Thermal Conversion Factor Source Documentation," which follows Table A6 in this appendix.

Table A1. Approximate Heat Content of Petroleum Products
(Million Btu per Barrel)

Product	Million Btu per Barrel
Asphalt	6.636
Aviation Gasoline	5.048
Butane	4.326
Butane-Propane Mixture (60 percent-40 percent)	4.130
Distillate Fuel Oil	5.825
Ethane	3.082
Ethane-Propane Mixture (70 percent-30 percent)	3.308
Isobutane	3.974
Jet Fuel, Kerosene-Type	5.670
Jet Fuel, Naphtha-Type	5.355
Kerosene	5.670
Lubricants	6.065
Motor Gasoline	
Conventional[1]	5.253
Oxygenated[1]	5.150
Reformulated[1]	5.150
Fuel Ethanol[2]	3.539
Natural Gasoline	4.620
Pentanes Plus	4.620
Petrochemical Feedstocks	
Naphtha less than 401° F	5.248
Other Oils equal to or greater than 401° F	5.825
Still Gas	6.000
Petroleum Coke	6.024
Plant Condensate	5.418
Propane	3.836
Residual Fuel Oil	6.287
Road Oil	6.636
Special Naphthas	5.248
Still Gas	6.000
Unfinished Oils	5.825
Unfractionated Stream	5.418
Waxes	5.537
Miscellaneous	5.796

[1]See Table A3 for motor gasoline annual weighted averages beginning in 1994.
[2]Fuel ethanol, which is derived from agricultural feedstocks (primarily corn), is not a petroleum product but is blended into motor gasoline.
Web Page: For related information, see http://www.eia.doe.gov/emeu/aer/append_a.html.
Sources: See "Thermal Conversion Factor Source Documentation," which follows Table A6.

Table A2. Approximate Heat Content of Petroleum Production, Imports, and Exports, Selected Years, 1949-2007
(Million Btu per Barrel)

Year	Production		Imports			Exports		
	Crude Oil [1]	Natural Gas Plant Liquids	Crude Oil [1]	Petroleum Products	Total	Crude Oil [1]	Petroleum Products	Total
1949	5.800	4.544	5.952	6.261	6.059	5.800	5.651	5.692
1950	5.800	4.522	5.943	6.263	6.080	5.800	5.751	5.766
1955	5.800	4.406	5.924	6.234	6.040	5.800	5.765	5.768
1960	5.800	4.295	5.911	6.161	6.021	5.800	5.835	5.834
1965	5.800	4.264	5.872	6.123	5.997	5.800	5.742	5.743
1970	5.800	4.146	5.822	6.088	5.985	5.800	5.811	5.810
1971	5.800	4.117	5.824	6.062	5.961	5.800	5.775	5.775
1972	5.800	4.070	5.809	6.045	5.935	5.800	5.741	5.741
1973	5.800	4.049	5.817	5.983	5.897	5.800	5.752	5.752
1974	5.800	4.011	5.827	5.959	5.884	5.800	5.773	5.774
1975	5.800	3.984	5.821	5.935	5.858	5.800	5.747	5.748
1976	5.800	3.964	5.808	5.980	5.856	5.800	5.743	5.745
1977	5.800	3.941	5.810	5.908	5.834	5.800	5.796	5.797
1978	5.800	3.925	5.802	5.955	5.839	5.800	5.814	5.808
1979	5.800	3.955	5.810	5.811	5.810	5.800	5.864	5.832
1980	5.800	3.914	5.812	5.748	5.796	5.800	5.841	5.820
1981	5.800	3.930	5.818	5.659	5.775	5.800	5.837	5.821
1982	5.800	3.872	5.826	5.664	5.775	5.800	5.829	5.820
1983	5.800	3.839	5.825	5.677	5.774	5.800	5.800	5.800
1984	5.800	3.812	5.823	5.613	5.745	5.800	5.867	5.850
1985	5.800	3.815	5.832	5.572	5.736	5.800	5.819	5.814
1986	5.800	3.797	5.903	5.624	5.808	5.800	5.839	5.832
1987	5.800	3.804	5.901	5.599	5.820	5.800	5.860	5.858
1988	5.800	3.800	5.900	5.618	5.820	5.800	5.842	5.840
1989	5.800	3.826	5.906	5.641	5.833	5.800	5.869	5.857
1990	5.800	3.822	5.934	5.614	5.849	5.800	5.838	5.833
1991	5.800	3.807	5.948	5.636	5.873	5.800	5.827	5.823
1992	5.800	3.804	5.953	5.623	5.877	5.800	5.774	5.777
1993	5.800	3.801	5.954	5.620	5.883	5.800	5.777	5.779
1994	5.800	3.794	5.950	5.534	5.861	5.800	5.777	5.779
1995	5.800	3.796	5.938	5.483	5.855	5.800	5.740	5.746
1996	5.800	3.777	5.947	5.468	5.847	5.800	5.728	5.736
1997	5.800	3.762	5.954	5.469	5.862	5.800	5.726	5.734
1998	5.800	3.769	5.953	5.462	5.861	5.800	5.710	5.720
1999	5.800	3.744	5.942	5.421	5.840	5.800	5.684	5.699
2000	5.800	3.733	5.959	5.432	5.849	5.800	5.651	5.658
2001	5.800	3.735	5.976	5.443	5.862	5.800	5.751	5.752
2002	5.800	3.729	5.971	5.451	5.863	5.800	5.687	5.688
2003	5.800	3.739	5.970	5.438	5.857	5.800	5.739	5.740
2004	5.800	3.724	5.981	5.475	5.863	5.800	5.753	5.754
2005	5.800	3.724	5.977	5.474	5.845	5.800	5.741	5.743
2006	5.800	3.712	5.980	R5.454	R5.842	5.800	R5.723	R5.724
2007P	5.800	3.701	5.981	5.500	5.858	5.800	5.745	5.746

[1] Includes lease condensate.
R=Revised. P=Preliminary.
Note: The values in this table are for gross heat contents. See "Heat Content" in Glossary.

Web Page: For all data beginning in 1949, see http://www.eia.doe.gov/emeu/aer/append_a.html.
Sources: See "Thermal Conversion Factor Source Documentation," which follows Table A6.

Table A3. Approximate Heat Content of Petroleum Consumption and Biofuels Production, Selected Years, 1949-2007

(Million Btu per Barrel)

Year	Total Petroleum [1] Consumption by Sector						Liquefied Petroleum Gases Consumption [5]	Motor Gasoline Consumption [6]	Fuel Ethanol	Fuel Ethanol Feed-stock [7]	Biodiesel	Biodiesel Feed-stock [8]
	Residential	Commercial [2]	Industrial [2]	Transportation [2]	Electric Power [3,4]	Total [2]						
1949	5.493	5.858	5.946	5.465	6.254	5.649	4.011	5.253	NA	NA	NA	NA
1950	5.482	5.865	5.940	5.461	6.254	5.649	4.011	5.253	NA	NA	NA	NA
1955	5.480	5.832	5.867	5.408	6.254	5.591	4.011	5.253	NA	NA	NA	NA
1960	5.430	5.849	5.800	5.388	6.267	5.555	4.011	5.253	NA	NA	NA	NA
1965	5.380	5.837	5.728	5.387	6.267	5.532	4.011	5.253	NA	NA	NA	NA
1970	5.216	5.773	5.603	5.393	6.252	5.503	[5]3.779	5.253	NA	NA	NA	NA
1971	5.212	5.758	5.598	5.389	6.245	5.504	3.772	5.253	NA	NA	NA	NA
1972	5.193	5.733	5.563	5.388	6.233	5.500	3.760	5.253	NA	NA	NA	NA
1973	5.205	5.749	5.569	5.395	6.245	5.515	3.746	5.253	NA	NA	NA	NA
1974	5.196	5.740	5.538	5.394	6.238	5.504	3.730	5.253	NA	NA	NA	NA
1975	5.192	5.704	5.527	5.392	6.250	5.494	3.715	5.253	NA	NA	NA	NA
1976	5.215	5.726	5.536	5.395	6.251	5.504	3.711	5.253	NA	NA	NA	NA
1977	5.213	5.733	5.554	5.400	6.249	5.518	3.677	5.253	NA	NA	NA	NA
1978	5.213	5.716	5.554	5.404	6.251	5.519	3.669	5.253	NA	NA	NA	NA
1979	5.298	5.769	5.419	5.428	6.258	5.494	3.680	5.253	NA	NA	NA	NA
1980	5.245	5.803	5.374	5.440	6.254	5.479	3.674	5.253	NA	NA	NA	NA
1981	5.191	5.751	5.312	5.432	6.258	5.448	3.643	5.253	3.539	6.486	NA	NA
1982	5.167	5.751	5.263	5.422	6.258	5.415	3.615	5.253	3.539	6.428	NA	NA
1983	5.022	5.642	5.275	5.415	6.255	5.406	3.614	5.253	3.539	6.388	NA	NA
1984	5.184	5.705	5.223	5.418	6.251	5.395	3.599	5.253	3.539	6.356	NA	NA
1985	5.153	5.661	5.215	5.422	6.247	5.387	3.603	5.253	3.539	6.331	NA	NA
1986	5.169	5.694	5.283	5.425	6.257	5.418	3.640	5.253	3.539	6.310	NA	NA
1987	5.144	5.661	5.248	5.429	6.249	5.403	3.659	5.253	3.539	6.291	NA	NA
1988	5.165	5.661	5.241	5.433	6.250	5.410	3.652	5.253	3.539	6.275	NA	NA
1989	5.105	5.621	5.234	5.437	[3]6.240	5.410	3.683	5.253	3.539	6.260	NA	NA
1990	5.027	5.621	5.270	5.442	6.244	5.411	3.625	5.253	3.539	6.247	NA	NA
1991	4.968	5.599	5.186	5.440	6.246	5.384	3.614	5.253	3.539	6.235	NA	NA
1992	5.004	5.589	5.185	5.442	6.238	5.378	3.624	5.253	3.539	6.224	NA	NA
1993	4.975	[2]5.580	[2]5.196	[2]5.436	6.230	[2]5.379	3.606	5.253	3.539	6.214	NA	NA
1994	4.983	5.592	5.166	5.424	6.213	5.361	3.635	[6]5.230	3.539	6.204	NA	NA
1995	4.940	5.554	5.137	5.417	6.188	5.341	3.623	5.215	3.539	6.196	NA	NA
1996	4.869	5.498	5.133	5.420	6.195	5.336	3.613	5.216	3.539	6.187	NA	NA
1997	4.859	5.459	5.138	5.416	6.199	5.336	3.616	5.213	3.539	6.180	NA	NA
1998	4.837	5.446	5.155	5.413	6.210	5.349	3.614	5.212	3.539	6.172	NA	NA
1999	4.761	5.369	5.113	5.413	6.205	5.328	3.616	5.211	3.539	6.165	NA	NA
2000	4.761	5.394	5.082	5.421	6.189	5.326	3.607	5.210	3.539	6.159	NA	NA
2001	4.796	5.403	5.164	5.412	6.199	5.345	3.614	5.210	3.539	6.152	[9,R]5.359	[9,R]5.433
2002	4.742	5.364	5.116	5.410	6.173	5.324	3.613	5.208	3.539	6.146	[9,R]5.359	[9,R]5.433
2003	4.763	5.407	5.161	5.408	6.182	5.340	3.629	5.207	3.539	6.141	[9,R]5.359	[9,R]5.433
2004	4.807	5.434	5.164	5.420	6.192	5.350	3.618	5.215	3.539	6.135	[9,R]5.359	[9,R]5.433
2005	[R]4.783	[R]5.427	[R]5.200	[R]5.426	6.188	5.365	3.620	5.218	3.539	6.130	[9,R]5.359	[9,R]5.433
2006	[RE]4.667	[RE]5.343	[RE]5.197	[RE]5.430	[R]6.143	[R]5.353	[R]3.605	5.218	3.539	6.125	[9,R]5.359	[9,R]5.433
2007	[E]4.640	[E]5.340	[E]5.167	[E]5.432	[P]6.150	[P]5.347	[P]3.592	[P]5.219	3.539	5.987	[9,R]5.359	[9,R]5.433

[1] Petroleum products supplied, including natural gas plant liquids and crude oil burned directly as fuel. Quantity-weighted averages of the petroleum products included in each category are calculated by using heat content values shown in Table A1.

[2] Beginning in 1993, includes ethanol blended into motor gasoline.

[3] Electricity-only and combined-heat-and-power (CHP) plants within the NAICS 22 category whose primary business is to sell electricity, or electricity and heat, to the public. Through 1988, data are for electric utilities only; beginning in 1989, data are for electric utilities and independent power producers.

[4] Electric power sector factors are weighted average heat contents for distillate fuel oil, petroleum coke, and residual fuel oil; they exclude other liquids.

[5] There is a discontinuity in this time series between 1966 and 1967; beginning in 1967, the single constant factor is replaced by a quantity-weighted factor—quantity-weighted averages of the major components of liquefied petroleum gases are calculated by using heat content values shown in Table A1.

[6] There is a discontinuity in this time series between 1993 and 1994; beginning in 1994, the single constant factor is replaced by a quantity-weighted factor—quantity-weighted averages of the major components of motor gasoline, including fuel ethanol, are calculated by using heat content values shown in Table A1.

[7] Corn input to the production of fuel ethanol (million Btu corn per barrel denatured ethanol), used as the approximate heat content for total biomass inputs to the production of fuel ethanol.

[8] Soybean oil input to the production of biodiesel (million Btu soybean oil per barrel biodiesel), used as the approximate heat content for total biomass inputs to the production of biodiesel.

[9] Gross heat content (higher heating value).

R=Revised. P=Preliminary. E=Estimate. NA=Not available.

Note: The values in this table are for gross heat contents. See "Heat Content" in Glossary.

Web Page: For all data beginning in 1949, see http://www.eia.doe.gov/emeu/aer/append_a.html.

Sources: See "Thermal Conversion Factor Source Documentation," which follows Table A6.

Table A4. Approximate Heat Content of Natural Gas, Selected Years, 1949-2007

(Btu per Cubic Foot)

Year	Production		Consumption [1]			Imports	Exports
	Marketed	Dry	End-Use Sectors [2]	Electric Power Sector [3]	Total		
1949	1,120	1,035	1,035	1,035	1,035	– –	1,035
1950	1,119	1,035	1,035	1,035	1,035	– –	1,035
1955	1,120	1,035	1,035	1,035	1,035	1,035	1,035
1960	1,107	1,035	1,035	1,035	1,035	1,035	1,035
1965	1,101	1,032	1,032	1,032	1,032	1,032	1,032
1970	1,102	1,031	1,031	1,031	1,031	1,031	1,031
1971	1,103	1,031	1,031	1,031	1,031	1,031	1,031
1972	1,100	1,027	1,027	1,027	1,027	1,027	1,027
1973	1,093	1,021	1,020	1,024	1,021	1,026	1,023
1974	1,097	1,024	1,024	1,022	1,024	1,027	1,016
1975	1,095	1,021	1,020	1,026	1,021	1,026	1,014
1976	1,093	1,020	1,019	1,023	1,020	1,025	1,013
1977	1,093	1,021	1,019	1,029	1,021	1,026	1,013
1978	1,088	1,019	1,016	1,034	1,019	1,030	1,013
1979	1,092	1,021	1,018	1,035	1,021	1,037	1,013
1980	1,098	1,026	1,024	1,035	1,026	1,022	1,013
1981	1,103	1,027	1,025	1,035	1,027	1,014	1,011
1982	1,107	1,028	1,026	1,036	1,028	1,018	1,011
1983	1,115	1,031	1,031	1,030	1,031	1,024	1,010
1984	1,109	1,031	1,030	1,035	1,031	1,005	1,010
1985	1,112	1,032	1,031	1,038	1,032	1,002	1,011
1986	1,110	1,030	1,029	1,034	1,030	997	1,008
1987	1,112	1,031	1,031	1,032	1,031	999	1,011
1988	1,109	1,029	1,029	1,028	1,029	1,002	1,018
1989	1,107	1,031	1,031	[1]1,028	1,031	1,004	1,019
1990	1,105	1,029	1,030	1,027	1,029	1,012	1,018
1991	1,108	1,030	1,031	1,025	1,030	1,014	1,022
1992	1,110	1,030	1,031	1,025	1,030	1,011	1,018
1993	1,106	1,027	1,028	1,025	1,027	1,020	1,016
1994	1,105	1,028	1,029	1,025	1,028	1,022	1,011
1995	1,106	1,026	1,027	1,021	1,026	1,021	1,011
1996	1,109	1,026	1,027	1,020	1,026	1,022	1,011
1997	1,107	1,026	1,027	1,020	1,026	1,023	1,011
1998	1,109	1,031	1,033	1,024	1,031	1,023	1,011
1999	1,107	1,027	1,028	1,022	1,027	1,022	1,006
2000	1,107	1,025	1,026	1,021	1,025	1,023	1,006
2001	1,105	1,028	1,029	1,026	1,028	1,023	1,010
2002	1,106	1,027	1,029	1,020	1,027	1,022	1,008
2003	1,106	1,031	1,033	1,025	1,031	1,025	1,009
2004	1,105	1,027	1,027	1,027	1,027	1,025	1,009
2005	R1,105	1,029	1,029	1,028	1,029	1,025	1,009
2006	R1,103	R1,028	R1,028	1,028	R1,028	1,025	1,009
2007	E1,103	E1,028	E1,028	P1,028	E1,028	E1,025	E1,009

[1] Consumption factors are for natural gas, plus a small amount of supplemental gaseous fuels.
[2] Residential, commercial, industrial, and transportation sectors.
[3] Electricity-only and combined-heat-and-power (CHP) plants within the NAICS 22 category whose primary business is to sell electricity, or electricity and heat, to the public. Through 1988, data are for electric utilities only; beginning in 1989, data are for electric utilities and independent power producers.

R=Revised. P=Preliminary. E=Estimate. – – – = Not applicable.
Note: The values in this table are for gross heat contents. See "Heat Content" in Glossary.
Web Page: For all data beginning in 1949, see http://www.eia.doe.gov/emeu/aer/append_a.html.
Sources: See "Thermal Conversion Factor Source Documentation," which follows Table A6.

Table A5. Approximate Heat Content of Coal and Coal Coke, Selected Years, 1949-2007

(Million Btu per Short Ton)

	Coal									Coal Coke
			Consumption							
			Residential and Commercial Sectors	Industrial Sector		Electric Power Sector [4,5]	Total			
Year	Production [1]	Waste Coal Supplied [2]		Coke Plants	Other [3]			Imports	Exports	Imports and Exports
1949	24.916	NA	24.263	26.797	24.612	23.761	24.793	25.000	26.759	24.800
1950	25.090	NA	24.461	26.798	24.820	23.937	24.989	25.020	26.788	24.800
1955	25.201	NA	24.373	26.794	24.821	24.056	24.982	25.000	26.907	24.800
1960	24.906	NA	24.226	26.791	24.609	23.927	24.713	25.003	26.939	24.800
1965	24.775	NA	24.028	26.787	24.385	23.780	24.537	25.000	26.973	24.800
1970	23.842	NA	23.203	26.784	22.983	22.573	23.440	25.000	26.982	24.800
1971	23.507	NA	23.090	26.784	22.670	22.301	23.124	25.000	26.981	24.800
1972	23.389	NA	22.998	26.782	22.550	22.204	23.036	25.000	26.979	24.800
1973	23.376	NA	22.831	26.780	22.586	22.246	23.057	25.000	26.596	24.800
1974	23.072	NA	22.479	26.778	22.419	21.781	22.677	25.000	26.700	24.800
1975	22.897	NA	22.261	26.782	22.436	21.642	22.506	25.000	26.562	24.800
1976	22.855	NA	22.774	26.781	22.530	21.679	22.498	25.000	26.601	24.800
1977	22.597	NA	22.919	26.787	22.322	21.508	22.265	25.000	26.548	24.800
1978	22.248	NA	22.466	26.789	22.207	21.275	22.017	25.000	26.478	24.800
1979	22.454	NA	22.242	26.788	22.452	21.364	22.100	25.000	26.548	24.800
1980	22.415	NA	22.543	26.790	22.690	21.295	21.947	25.000	26.384	24.800
1981	22.308	NA	22.474	26.794	22.585	21.085	21.713	25.000	26.160	24.800
1982	22.239	NA	22.695	26.797	22.712	21.194	21.674	25.000	26.223	24.800
1983	22.052	NA	22.775	26.798	22.691	21.133	21.576	25.000	26.291	24.800
1984	22.010	NA	22.844	26.799	22.543	21.101	21.573	25.000	26.402	24.800
1985	21.870	NA	22.646	26.798	22.020	20.959	21.366	25.000	26.307	24.800
1986	21.913	NA	22.947	26.798	22.198	21.084	21.462	25.000	26.292	24.800
1987	21.922	NA	23.404	26.799	22.381	21.136	21.517	25.000	26.291	24.800
1988	21.823	NA	23.571	26.799	22.360	20.900	21.328	25.000	26.299	24.800
1989	21.765	[2]10.391	23.650	26.800	22.347	[4]20.898	21.307	25.000	26.160	24.800
1990	21.822	9.303	23.137	26.799	22.457	20.779	21.197	25.000	26.202	24.800
1991	21.681	10.758	23.114	26.799	22.460	20.730	21.120	25.000	26.188	24.800
1992	21.682	10.396	23.105	26.799	22.250	20.709	21.068	25.000	26.161	24.800
1993	21.418	10.638	22.994	26.800	22.123	20.677	21.010	25.000	26.335	24.800
1994	21.394	11.097	23.112	26.800	22.068	20.589	20.929	25.000	26.329	24.800
1995	21.326	11.722	23.118	26.800	21.950	20.543	20.880	25.000	26.180	24.800
1996	21.322	12.147	23.011	26.800	22.105	20.547	20.870	25.000	26.174	24.800
1997	21.296	12.158	22.494	26.800	22.172	20.518	20.830	25.000	26.251	24.800
1998	21.418	12.639	21.620	27.426	23.164	20.516	20.881	25.000	26.800	24.800
1999	21.070	12.552	23.880	27.426	22.489	20.490	20.818	25.000	26.081	24.800
2000	21.072	12.360	25.020	27.426	22.433	20.511	20.828	25.000	26.117	24.800
2001	[1]20.772	12.169	24.909	27.426	22.622	20.337	20.671	25.000	25.998	24.800
2002	20.673	12.165	22.962	27.426	22.562	20.238	20.541	25.000	26.062	24.800
2003	20.499	R12.360	22.242	27.425	22.468	20.082	20.387	25.000	25.972	24.800
2004	20.424	R12.266	22.324	27.426	22.473	19.980	20.290	25.000	26.108	24.800
2005	R20.348	R12.093	22.342	26.279	22.178	19.988	R20.246	25.000	25.494	24.800
2006	R20.310	R12.080	R22.066	26.271	22.050	R19.931	R20.181	25.000	25.453	24.800
2007P	20.341	12.616	22.034	26.329	22.371	19.911	20.169	25.000	25.466	24.800

[1] Beginning in 2001, includes a small amount of refuse recovery (coal recaptured from a refuse mine, and cleaned to reduce the concentration of noncombustible materials).

[2] Waste coal (including fine coal, coal obtained from a refuse bank or slurry dam, anthracite culm, bituminous gob, and lignite waste) consumed by the electric power and industrial sectors. Beginning in 1989, waste coal supplied is counted as a supply-side item to balance the same amount of waste coal included in "Consumption."

[3] Includes transportation. Excludes coal synfuel plants.

[4] Electricity-only and combined-heat-and-power (CHP) plants within the NAICS 22 category whose primary business is to sell electricity, or electricity and heat, to the public. Through 1988, data are for electric utilities only; beginning in 1989, data are for electric utilities and independent power producers.

[5] Electric power sector factors are for anthracite, bituminous coal, subbituminous coal, lignite, waste coal, and, beginning in 1998, coal synfuel.

R=Revised. P=Preliminary. NA=Not available.

Note: The values in this table are for gross heat contents. See "Heat Content" in Glossary.

Web Page: For all data beginning in 1949, see http://www.eia.doe.gov/emeu/aer/append_a.html.

Sources: See "Thermal Conversion Factor Source Documentation," which follows Table A6.

Table A6. Approximate Heat Rates for Electricity, and Heat Content of Electricity, Selected Years, 1949-2007

(Btu per Kilowatthour)

Year	Approximate Heat Rates [1] for Electricity Net Generation			Heat Content [6] of Electricity [7]
	Fossil-Fueled Plants [2,3]	Nuclear Plants [4]	Geothermal Energy Plants [5]	
1949	15,033	– –	– –	3,412
1950	14,030	– –	– –	3,412
1955	11,699	– –	– –	3,412
1960	10,760	11,629	23,200	3,412
1965	10,453	11,804	22,182	3,412
1970	10,494	10,977	21,606	3,412
1971	10,478	10,837	21,655	3,412
1972	10,379	10,792	21,668	3,412
1973	10,389	10,903	21,674	3,412
1974	10,442	11,161	21,674	3,412
1975	10,406	11,013	21,611	3,412
1976	10,373	11,047	21,611	3,412
1977	10,435	10,769	21,611	3,412
1978	10,361	10,941	21,611	3,412
1979	10,353	10,879	21,545	3,412
1980	10,388	10,908	21,639	3,412
1981	10,453	11,030	21,639	3,412
1982	10,454	11,073	21,629	3,412
1983	10,520	10,905	21,290	3,412
1984	10,440	10,843	21,303	3,412
1985	10,447	10,622	21,263	3,412
1986	10,446	10,579	21,263	3,412
1987	10,419	10,442	21,263	3,412
1988	10,324	10,602	21,096	3,412
1989	10,432	10,583	21,096	3,412
1990	10,402	10,582	21,096	3,412
1991	10,436	10,484	20,997	3,412
1992	10,342	10,471	20,914	3,412
1993	10,309	10,504	20,914	3,412
1994	10,316	10,452	20,914	3,412
1995	10,312	10,507	20,914	3,412
1996	10,340	10,503	20,960	3,412
1997	10,213	10,494	20,960	3,412
1998	10,197	10,491	21,017	3,412
1999	10,226	10,450	21,017	3,412
2000	10,201	10,429	21,017	3,412
2001	[2]10,333	10,448	21,017	3,412
2002	10,173	10,439	21,017	3,412
2003	10,241	10,421	21,017	3,412
2004	10,022	10,427	21,017	3,412
2005	9,999	10,435	21,017	3,412
2006	R9,919	R10,434	21,017	3,412
2007	E9,919	E10,434	E21,017	3,412

[1] The values in columns 1-3 of this table are for net heat rates. See "Heat Rate" in Glossary.

[2] Used as the thermal conversion factor for hydro, solar/photovoltaic, and wind electricity net generation to approximate the quantity of fossil fuels replaced by these sources. Through 2000, also used as the thermal conversion factor for wood and waste electricity net generation at electric utilities; beginning in 2001, Btu data for wood and waste at electric utilities are available from surveys.

[3] Through 2000, heat rates are for fossil-fueled steam-electric plants at electric utilities. Beginning in 2001, heat rates are for all fossil-fueled plants at electric utilities and independent power producers.

[4] Used as the thermal conversion factor for nuclear electricity net generation.

[5] Used as the thermal conversion factor for geothermal electricity net generation.

[6] See "Heat Content" in Glossary.

[7] The value of 3,412 Btu per kilowatthour is a constant. It is used as the thermal conversion factor for electricity retail sales, and electricity imports and exports.

R=Revised. E=Estimate. – – = Not applicable.

Web Page: For all data beginning in 1949, see http://www.eia.doe.gov/emeu/aer/append_a.html.

Sources: See "Thermal Conversion Factor Source Documentation," which follows this table.

Thermal Conversion Factor Source Documentation

Approximate Heat Content of Petroleum and Natural Gas Plant Liquids

Asphalt. The Energy Information Administration (EIA) adopted the thermal conversion factor of 6.636 million British thermal units (Btu) per barrel as estimated by the Bureau of Mines and first published in the *Petroleum Statement, Annual, 1956*.

Aviation Gasoline. EIA adopted the thermal conversion factor of 5.048 million Btu per barrel as adopted by the Bureau of Mines from the Texas Eastern Transmission Corporation publication *Competition and Growth in American Energy Markets 1947-1985*, a 1968 release of historical and projected statistics.

Butane. EIA adopted the Bureau of Mines thermal conversion factor of 4.326 million Btu per barrel as published in the *California Oil World and Petroleum Industry*, First Issue, April 1942.

Butane-Propane Mixture. EIA adopted the Bureau of Mines calculation of 4.130 million Btu per barrel based on an assumed mixture of 60 percent butane and 40 percent propane. See **Butane** and **Propane**.

Crude Oil Exports. Assumed by EIA to be 5.800 million Btu per barrel or equal to the thermal conversion factor for crude oil produced in the United States. See **Crude Oil Production**.

Crude Oil Imports. Calculated annually by EIA as the average of the thermal conversion factors for each type of crude oil imported weighted by the quantities imported. Thermal conversion factors for each type were calculated on a foreign country basis, by determining the average American Petroleum Institute (API) gravity of crude oil imported from each foreign country from Form ERA-60 in 1977 and converting average API gravity to average Btu content by using National Bureau of Standards, Miscellaneous Publication No. 97, *Thermal Properties of Petroleum Products*, 1933.

Crude Oil Production. EIA adopted the thermal conversion factor of 5.800 million Btu per barrel as reported in a Bureau of Mines internal memorandum, "Bureau of Mines Standard Average Heating Values of Various Fuels, Adopted January 3, 1950."

Distillate Fuel Oil. EIA adopted the Bureau of Mines thermal conversion factor of 5.825 million Btu per barrel as reported in a Bureau of Mines internal memorandum, "Bureau of Mines Standard Average Heating Values of Various Fuels, Adopted January 3, 1950."

Ethane. EIA adopted the Bureau of Mines thermal conversion factor of 3.082 million Btu per barrel as published in the *California Oil World and Petroleum Industry*, First Issue, April 1942.

Ethane-Propane Mixture. EIA calculation of 3.308 million Btu per barrel based on an assumed mixture of 70 percent ethane and 30 percent propane. See **Ethane** and **Propane**.

Isobutane. EIA adopted the Bureau of Mines thermal conversion factor of 3.974 million Btu per barrel as published in the *California Oil World and Petroleum Industry*, First Issue, April 1942.

Jet Fuel, Kerosene-Type. EIA adopted the Bureau of Mines thermal conversion factor of 5.670 million Btu per barrel for "Jet Fuel, Commercial" as published by the Texas Eastern Transmission Corporation in the report *Competition and Growth in American Energy Markets 1947-1985*, a 1968 release of historical and projected statistics.

Jet Fuel, Naphtha-Type. EIA adopted the Bureau of Mines thermal conversion factor of 5.355 million Btu per barrel for "Jet Fuel, Military" as published by the Texas Eastern Transmission Corporation in the report *Competition and Growth in American Energy Markets 1947-1985*, a 1968 release of historical and projected statistics.

Kerosene. EIA adopted the Bureau of Mines thermal conversion factor of 5.670 million Btu per barrel as reported in a Bureau of Mines internal memorandum, "Bureau of Mines Standard Average Heating Values of Various Fuels, Adopted January 3, 1950."

Liquefied Petroleum Gases Consumption. • 1949-1966: U.S. Department of the Interior, Bureau of Mines, Mineral Industry Surveys, "Crude Petroleum and Petroleum Products, 1956," Table 4 footnote, constant value of 4.011 million Btu per barrel. • 1967 forward: Calculated annually by EIA as the average of the thermal conversion factors for all liquefied petroleum gases consumed (see Table A1) weighted by the quantities consumed. The component products of liquefied petroleum gases are ethane (including ethylene), propane (including propylene), normal-butane (including butylene), butane-propane mixtures, ethane-propane mixtures, and isobutane. For 1967-1980, quantities consumed are from EIA, Energy Data Reports, "Petroleum Statement, Annual," Table 1. For 1981 forward, quantities consumed are from EIA, *Petroleum Supply Annual*, Table 2.

Lubricants. EIA adopted the thermal conversion factor of 6.065 million Btu per barrel as estimated by the Bureau of Mines and first published in the *Petroleum Statement, Annual, 1956*.

Miscellaneous Products. EIA adopted the thermal conversion factor of 5.796 million Btu per barrel as estimated by the Bureau of Mines and first published in the *Petroleum Statement, Annual, 1956*.

Motor Gasoline Consumption. • 1949-1993: EIA adopted the Bureau of Mines thermal conversion factor of 5.253 million Btu per barrel for "Gasoline, Motor Fuel" as published by the Texas Eastern Transmission Corporation in Appendix V of *Competition and Growth in American Energy Markets 1947-1985*, a 1968 release of historical and projected statistics. • 1994 forward: EIA calculated national annual quantity-weighted average conversion factors for conventional, reformulated, and oxygenated motor gasolines (see Table A3). The factor for conventional motor gasoline is 5.253 million Btu per barrel, as used for previous years. The factors for reformulated and oxygenated gasolines, both currently 5.150 million Btu per barrel, are based on data published in Environmental Protection Agency, Office of Mobile Sources, National Vehicle and Fuel Emissions Laboratory report EPA 420-F-95-003, "Fuel Economy Impact Analysis of Reformulated Gasoline." See **Fuel Ethanol (Blended Into Motor Gasoline).**

Natural Gas Plant Liquids Production. Calculated annually by EIA as the average of the thermal conversion factors for each natural gas plant liquid produced weighted by the quantities produced.

Natural Gasoline. EIA adopted the thermal conversion factor of 4.620 million Btu per barrel as estimated by the Bureau of Mines and first published in the *Petroleum Statement, Annual, 1956*.

Pentanes Plus. EIA assumed the thermal conversion factor to be 4.620 million Btu or equal to that for natural gasoline. See **Natural Gasoline**.

Petrochemical Feedstocks, Naphtha less than 401° F. Assumed by EIA to be 5.248 million Btu per barrel, equal to the thermal conversion factor for special naphthas. See **Special Naphthas**.

Petrochemical Feedstocks, Other Oils equal to or greater than 401° F. Assumed by EIA to be 5.825 million Btu per barrel, equal to the thermal conversion factor for distillate fuel oil. See **Distillate Fuel Oil**.

Petrochemical Feedstocks, Still Gas. Assumed by EIA to be 6.000 million Btu per barrel, equal to the thermal conversion factor for still gas. See **Still Gas**.

Petroleum Coke. EIA adopted the thermal conversion factor of 6.024 million Btu per barrel as reported in Btu per short ton in the Bureau of Mines internal memorandum, "Bureau of Mines Standard Average Heating Values of Various Fuels, Adopted January 3, 1950." The Bureau of Mines calculated this factor by dividing 30.120 million Btu per short ton, as given in the referenced Bureau of Mines internal memorandum, by 5.0 barrels per short ton, as given in the Bureau of Mines Form 6-1300-M and successor EIA forms.

Petroleum Consumption, Commercial Sector. Calculated annually by EIA as the average of the thermal conversion factors for all petroleum products consumed by the commercial sector weighted by the estimated quantities consumed by the commercial sector. The quantities of petroleum products consumed by the commercial sector are estimated in the State Energy Data System—see documentation at http://www.eia.doe.gov/emeu/states/sep_use/notes/use_petrol.pdf.

Petroleum Consumption, Electric Power Sector. Calculated annually by EIA as the average of the thermal conversion factors for all petroleum products consumed by the electric power sector weighted by the quantities consumed by the electric power sector. Data are from Form EIA-906, "Power Plant Report"; Form EIA-920, "Combined Heat and Power Plant Report"; and predecessor forms.

Petroleum Consumption, Industrial Sector. Calculated annually by EIA as the average of the thermal conversion factors for all petroleum products consumed by the industrial sector weighted by the estimated quantities consumed by the industrial sector. The quantities of petroleum products consumed by the industrial sector are estimated in the State Energy Data System—see documentation at http://www.eia.doe.gov/emeu/states/sep_use/notes/use_petrol.pdf.

Petroleum Consumption, Residential Sector. Calculated annually by EIA as the average of the thermal conversion factors for all petroleum products consumed by the residential sector weighted by the estimated quantities consumed by the residential sector. The quantities of petroleum products consumed by the residential sector are estimated in the State Energy Data System—see documentation at http://www.eia.doe.gov/emeu/states/sep_use/notes/use_petrol.pdf.

Petroleum Consumption, Total. Calculated annually by EIA as the average of the thermal conversion factors for all petroleum products consumed weighted by the quantities consumed.

Petroleum Consumption, Transportation Sector. Calculated annually by EIA as the average of the thermal conversion factors for all petroleum products consumed by the transportation sector weighted by the estimated quantities consumed by the transportation sector. The quantities of petroleum products consumed by the transportation sector are estimated in the State Energy Data System—see documentation at http://www.eia.doe.gov/emeu/states/sep_use/notes/use_petrol.pdf.

Petroleum Products Exports. Calculated annually by EIA as the average of the thermal conversion factors for each petroleum product exported weighted by the quantities exported.

Petroleum Products Imports. Calculated annually by EIA as the average of the thermal conversion factors for each petroleum product imported weighted by the quantities imported.

Plant Condensate. Estimated to be 5.418 million Btu per barrel by EIA from data provided by McClanahan Consultants, Inc., Houston, Texas.

Propane. EIA adopted the Bureau of Mines thermal conversion factor of 3.836 million Btu per barrel as published in the *California Oil World and Petroleum Industry, First Issue*, April 1942.

Residual Fuel Oil. EIA adopted the thermal conversion factor of 6.287 million Btu per barrel as reported in the Bureau of Mines internal memorandum, "Bureau of Mines Standard Average Heating Values of Various Fuels, Adopted January 3, 1950."

Road Oil. EIA adopted the Bureau of Mines thermal conversion factor of 6.636 million Btu per barrel, which was assumed to be equal to that of asphalt (see **Asphalt**) and was first published by the Bureau of Mines in the *Petroleum Statement, Annual, 1970*.

Special Naphthas. EIA adopted the Bureau of Mines thermal conversion factor of 5.248 million Btu per barrel, which was assumed to be equal to that of the total gasoline (aviation and motor) factor and was first published in the *Petroleum Statement, Annual, 1970*.

Still Gas. EIA adopted the Bureau of Mines estimated thermal conversion factor of 6.000 million Btu per barrel, first published in the *Petroleum Statement, Annual, 1970*.

Total Petroleum Exports. Calculated annually by EIA as the average of the thermal conversion factors for crude oil and each petroleum product exported weighted by the quantities exported. See **Crude Oil Exports** and **Petroleum Products Exports**.

Total Petroleum Imports. Calculated annually by EIA as the average of the thermal conversion factors for each type of crude oil and petroleum product imported weighted by the quantities imported. See **Crude Oil Imports** and **Petroleum Products Imports**.

Unfinished Oils. EIA assumed the thermal conversion factor to be 5.825 million Btu per barrel or equal to that for distillate fuel oil (see **Distillate Fuel Oil**) and first published it in EIA's *Annual Report to Congress, Volume 3, 1977*.

Unfractionated Stream. EIA assumed the thermal conversion factor to be 5.418 million Btu per barrel or equal to that for plant condensate (see **Plant Condensate**) and first published it in EIA's *Annual Report to Congress, Volume 2, 1981*.

Waxes. EIA adopted the thermal conversion factor of 5.537 million Btu per barrel as estimated by the Bureau of Mines and first published in the *Petroleum Statement, Annual, 1956*.

Approximate Heat Content of Biofuels

Biodiesel. EIA estimated the gross heat content (higher heating value) for biodiesel to be 5.359 million Btu per barrel.

Biodiesel Feedstock. EIA estimated the soybean oil input to the production of biodiesel to be 5.433 million Btu soybean oil per barrel biodiesel, which is used as the approximate gross heat content (higher heating value) for total biomass inputs to the production of biodiesel.

Fuel Ethanol. EIA adopted the thermal conversion factor of 3.539 million Btu per barrel published in "Oxygenate Flexibility for Future Fuels," a paper presented by William J. Piel of the ARCO Chemical Company at the National Conference on Reformulated Gasolines and Clean Air Act Implementation, Washington, D.C., October 1991.

Fuel Ethanol Feedstock. EIA estimated the corn input to the production of fuel ethanol (million Btu corn per barrel denatured ethanol), which is used as the approximate heat content for total biomass inputs to the production of ethanol.

Approximate Heat Content of Natural Gas

Natural Gas Consumption, Electric Power Sector. Calculated annually by EIA by dividing the heat content of natural gas consumed by the electric power sector by the quantity consumed. Data are from Form EIA-906, "Power Plant Report"; Form EIA-920, "Combined Heat and Power Plant Report"; and predecessor forms.

Natural Gas Consumption, End-Use Sectors. Calculated annually by EIA by dividing the heat content of natural gas consumed by the end-use sectors (residential, commercial, industrial, and transportation) by the quantity consumed. Data are from Form EIA-176, "Annual Report of Natural and Supplemental Gas Supply and Disposition."

Natural Gas Consumption, Total. • 1949-1962: EIA adopted the thermal conversion factor of 1,035 Btu per cubic foot as estimated by the Bureau of Mines and first published in the *Petroleum Statement, Annual, 1956*. • 1963-1979: EIA adopted the thermal conversion factor calculated annually by the American Gas Association (AGA) and published in *Gas Facts*, an AGA annual publication. • 1980 forward: Calculated annually by EIA by dividing the total heat content of natural gas consumed by the total quantity consumed.

Natural Gas Exports. • 1949-1972: Assumed by EIA to be equal to the thermal conversion factor for dry natural gas consumed (see **Natural Gas Consumption, Total**). • 1973 forward: Calculated annually by EIA by dividing the heat content of natural gas exported by the quantity exported. For 1973-1995, data are from Form FPC-14, "Annual Report for Importers and Exporters of Natural Gas." Beginning in 1996, data are from U.S. Department of Energy, Office of Fossil Energy, *Natural Gas Imports and Exports*.

Natural Gas Imports. • 1949-1972: Assumed by EIA to be equal to the thermal conversion factor for dry natural gas consumed (see **Natural Gas Consumption, Total**). • 1973 forward: Calculated annually by EIA by dividing the heat content of natural gas imported by the quantity imported. For 1973-1995, data are from Form FPC-14, "Annual Report for Importers and Exporters of Natural Gas." Beginning in 1996, data are from U.S. Department of Energy, Office of Fossil Energy, *Natural Gas Imports and Exports*.

Natural Gas Production, Dry. Assumed by EIA to be equal to the thermal conversion factor for dry natural gas consumed. See **Natural Gas Consumption, Total**.

Natural Gas Production, Marketed. Calculated annually by EIA by dividing the heat content of dry natural gas produced (see **Natural Gas Production, Dry**) and liquids produced (see **Natural Gas Plant Liquids Production**) by the total quantity of marketed natural gas produced.

Approximate Heat Content of Coal and Coal Coke

Coal Coke Imports and Exports. EIA adopted the Bureau of Mines estimate of 24.800 million Btu per short ton.

Coal Consumption, Electric Power Sector. Calculated annually by EIA by dividing the heat content of coal consumed by the electric power sector by the quantity consumed. Data are from Form EIA-906, "Power Plant Report"; Form EIA-920, "Combined Heat and Power Plant Report"; and predecessor forms.

Coal Consumption, Industrial Sector, Coke Plants. Calculated annually by EIA by dividing the heat content of coal consumed by coke plants by the quantity consumed. Data are from Form EIA-5, "Quarterly Coal Consumption and Quality Report—Coke Plants."

Coal Consumption, Industrial Sector, Other. Calculated annually by EIA by dividing the heat content of coal consumed by manufacturing plants by the quantity consumed. Data are from Form EIA-3, "Quarterly Coal Consumption and Quality Report—Manufacturing Plants."

Coal Consumption, Residential and Commercial Sectors. Calculated annually by EIA by dividing the heat content of coal consumed by the residential and commercial sectors by the quantity consumed. Through 1999, data are from Form EIA-6, "Coal Distribution Report." Beginning in 2000, data are for commercial combined-heat-and-power (CHP) plants from Form EIA-920, "Combined Heat and Power Plant Report," and predecessor forms.

Coal Consumption, Total. Calculated annually by EIA by dividing the total heat content of coal consumed by all sectors by the total quantity consumed.

Coal Exports. Calculated annually by EIA by dividing the heat content of steam coal and metallurgical coal exported by the quantity exported. Data are from U.S. Department of Commerce, Bureau of the Census, "Monthly Report EM 545."

Coal Imports. • 1949-1963: Calculated annually by EIA by dividing the heat content of coal imported by the quantity imported. • 1963 forward: Assumed by EIA to be 25.000 million Btu per short ton.

Coal Production. Calculated annually by EIA to balance the heat content of coal supply (production and imports) and the heat content of coal disposition (exports, stock change, and consumption).

Approximate Heat Rates for Electricity

Electricity Net Generation, Fossil-Fueled Plants. There is no generally accepted practice for measuring the thermal conversion rates for power plants that generate electricity from hydro, wind, photovoltaic, or solar thermal energy sources. Therefore, EIA calculates a rate factor that is equal to the annual average heat rate factor for fossil-fueled power plants in the United States. By using that factor, it is possible to evaluate fossil fuel requirements for replacing those sources during periods of interruption, such as droughts. The heat content of a kilowatthour of electricity produced, regardless of the generation process, is 3,412 Btu. • 1949-1955: The weighted annual average heat rate for fossil-fueled steam-electric power plants in the United States, as published by EIA in *Thermal-Electric Plant Construction Cost and Annual Production Expenses—1981* and *Steam-Electric Plant Construction Cost and Annual Production Expenses—1978*. • 1956-1988: The weighted annual average heat rate for fossil-fueled steam-electric power plants in the United States, as published in EIA, *Electric Plant Cost and Power Production Expenses 1991*, Table 9. • 1989-2000: Calculated annually by EIA by using heat rate data reported on Form EIA-860, "Annual Electric Generator Report" (and predecessor forms); and net generation data reported on Form EIA-759, "Monthly Power Plant Report." The computation includes data for all electric utility steam-electric plants using fossil fuels. 2001 forward: Calculated annually by EIA by using fuel consumption and net generation data reported on Form EIA-906, "Power Plant Report," and Form EIA-920, "Combined Heat and Power Plant Report." The computation includes data for all electric utilities and electricity-only independent power producers using fossil fuels.

Electricity Net Generation, Geothermal Energy Plants. • 1960-1981: Calculated annually by EIA by weighting the annual average heat rates of operating geothermal units by the installed nameplate capacities as reported on Form FPC-12, "Power System Statement." • 1982 forward: Estimated annually by EIA on the basis of an informal survey of relevant plants.

Electricity Net Generation, Nuclear Plants. • 1957-1984: Calculated annually by dividing the total heat content consumed in nuclear generating units by the total (net) electricity generated by nuclear generating units. The heat content and electricity generation were reported on Form FERC-1, "Annual Report of Major Electric Utilities, Licensees, and Others"; Form EIA-412, "Annual Report of Public Electric Utilities"; and predecessor forms. For 1982, the factors were published in EIA, *Historical Plant Cost and Annual Production Expenses for Selected Electric Plants 1982*, page 215. For 1983 and 1984, the factors were published in EIA, *Electric Plant Cost and Power Production Expenses 1991*, Table 13. • 1985 forward: Calculated annually by EIA by using the heat rate reported on Form EIA-860, "Annual Electric Generator Report" (and predecessor forms); and the generation reported on Form EIA-906, "Power Plant Report," and Form EIA-920, "Combined Heat and Power Plant Report" (and predecessor forms).

Appendix B. Metric and Other Physical Conversion Factors

Data presented in the *Annual Energy Review* and in other Energy Information Administration publications are expressed predominately in units that historically have been used in the United States, such as British thermal units, barrels, cubic feet, and short tons. However, because U.S. commerce involves other nations, most of which use metric units of measure, the U.S. Government is committed to the transition to the metric system, as stated in the Metric Conversion Act of 1975 (Public Law 94–168), amended by the Omnibus Trade and Competitiveness Act of 1988 (Public Law 100–418), and Executive Order 12770 of July 25, 1991.

The metric conversion factors presented in Table B1 can be used to calculate the metric-unit equivalents of values expressed in U.S. customary units. For example, 500 short tons are the equivalent of 453.6 metric tons (500 short tons x 0.9071847 metric tons/short ton = 453.6 metric tons).

In the metric system of weights and measures, the names of multiples and subdivisions of any unit may be derived by combining the name of the unit with prefixes, such as deka, hecto, and kilo, meaning, respectively, 10, 100, 1,000, and deci, centi, and milli, meaning, respectively, one-tenth, one-hundredth, and one-thousandth. Common metric prefixes can be found in Table B2.

The conversion factors presented in Table B3 can be used to calculate equivalents in various physical units commonly used in energy analyses. For example, 10 barrels are the equivalent of 420 U.S. gallons (10 barrels x 42 gallons/barrel = 420 gallons).

Table B1. Metric Conversion Factors

Type of Unit	U.S. Unit		Equivalent in	Metric Units
Mass	1 short ton (2,000 lb)	=	0.907 184 7	metric tons (t)
	1 long ton	=	1.016 047	metric tons (t)
	1 pound (lb)	=	0.453 592 37[a]	kilograms (kg)
	1 pound uranium oxide (lb U_3O_8)	=	0.384 647[b]	kilograms uranium (kgU)
	1 ounce, avoirdupois (avdp oz)	=	28.349 52	grams (g)
Volume	1 barrel of oil (bbl)	=	0.158 987 3	cubic meters (m^3)
	1 cubic yard (yd^3)	=	0.764 555	cubic meters (m^3)
	1 cubic foot (ft^3)	=	0.028 316 85	cubic meters (m^3)
	1 U.S. gallon (gal)	=	3.785 412	liters (L)
	1 ounce, fluid (fl oz)	=	29.573 53	milliliters (mL)
	1 cubic inch (in^3)	=	16.387 06	milliliters (mL)
Length	1 mile (mi)	=	1.609 344[a]	kilometers (km)
	1 yard (yd)	=	0.914 4[a]	meters (m)
	1 foot (ft)	=	0.304 8[a]	meters (m)
	1 inch (in)	=	2.54[a]	centimeters (cm)
Area	1 acre	=	0.404 69	hectares (ha)
	1 square mile (mi^2)	=	2.589 988	square kilometers (km^2)
	1 square yard (yd^2)	=	0.836 127 4	square meters (m^2)
	1 square foot (ft^2)	=	0.092 903 04[a]	square meters (m^2)
	1 square inch (in^2)	=	6.451 6[a]	square centimeters (cm^2)
Energy	1 British thermal unit (Btu)[c]	=	1,055.055 852 62[a]	joules (J)
	1 calorie (cal)	=	4.186 8[a]	joules (J)
	1 kilowatthour (kWh)	=	3.6[a]	megajoules (MJ)
Temperature[d]	32 degrees Fahrenheit (°F)	=	0[a]	degrees Celsius (°C)
	212 degrees Fahrenheit (°F)	=	100[a]	degrees Celsius (°C)

[a]Exact conversion.
[b]Calculated by the Energy Information Administration.
[c]The Btu used in this table is the International Table Btu adopted by the Fifth International Conference on Properties of Steam, London, 1956.
[d]To convert degrees Fahrenheit (°F) to degrees Celsius (°C) exactly, subtract 32, then multiply by 5/9.

Notes: • Spaces have been inserted after every third digit to the right of the decimal for ease of reading. • Most metric units belong to the International System of Units (SI), and the liter, hectare, and metric ton are accepted for use with the SI units. For more information about the SI units, see http://physics.nist.gov/cuu/Units/index.html.

Web Page: For related information, see http://www.eia.doe.gov/emeu/aer/append_b.html.

Sources: • General Services Administration, Federal Standard 376B, *Preferred Metric Units for General Use by the Federal Government* (Washington, DC, January 1993), pp. 9-11, 13, and 16. • U.S. Department of Commerce, National Institute of Standards and Technology, Special Publications 330, 811, and 814. • American National Standards Institute/Institute of Electrical and Electronic Engineers, ANSI/IEEE Std 268-1992, pp. 28 and 29.

Table B2. Metric Prefixes

Unit Multiple	Prefix	Symbol	Unit Subdivision	Prefix	Symbol
10^{1}	deka	da	10^{-1}	deci	d
10^{2}	hecto	h	10^{-2}	centi	c
10^{3}	kilo	k	10^{-3}	milli	m
10^{6}	mega	M	10^{-6}	micro	µ
10^{9}	giga	G	10^{-9}	nano	n
10^{12}	tera	T	10^{-12}	pico	p
10^{15}	peta	P	10^{-15}	femto	f
10^{18}	exa	E	10^{-18}	atto	a
10^{21}	zetta	Z	10^{-21}	zepto	z
10^{24}	yotta	Y	10^{-24}	yocto	y

Web Page: For related information, see http://www.eia.doe.gov/emeu/aer/append_b.html.
Source: U.S. Department of Commerce, National Institute of Standards and Technology, *The International System of Units (SI)*, NIST Special Publication 330, 1991 Edition (Washington, DC, August 1991), p.10.

Table B3. Other Physical Conversion Factors

Energy Source	Original Unit		Equivalent in Final Units	
Petroleum	1 barrel (bbl)	=	42[a]	U.S. gallons (gal)
Coal	1 short ton	=	2,000[a]	pounds (lb)
	1 long ton	=	2,240[a]	pounds (lb)
	1 metric ton (t)	=	1,000[a]	kilograms (kg)
Wood	1 cord (cd)	=	1.25[b]	shorts tons
	1 cord (cd)	=	128[a]	cubic feet (ft^3)

[a]Exact conversion.
[b]Calculated by the Energy Information Administration.
Web Page: For related information, see http://www.eia.doe.gov/emeu/aer/append_b.html.
Source: U.S. Department of Commerce, National Institute of Standards and Technology, *Specifications, Tolerances, and Other Technical Requirements for Weighing and Measuring Devices*, NIST Handbook 44, 1994 Edition (Washington, DC, October 1993), pp. B-10, C-17 and C-21.

Appendix C

Figure C1. U.S. Census Regions and Divisions

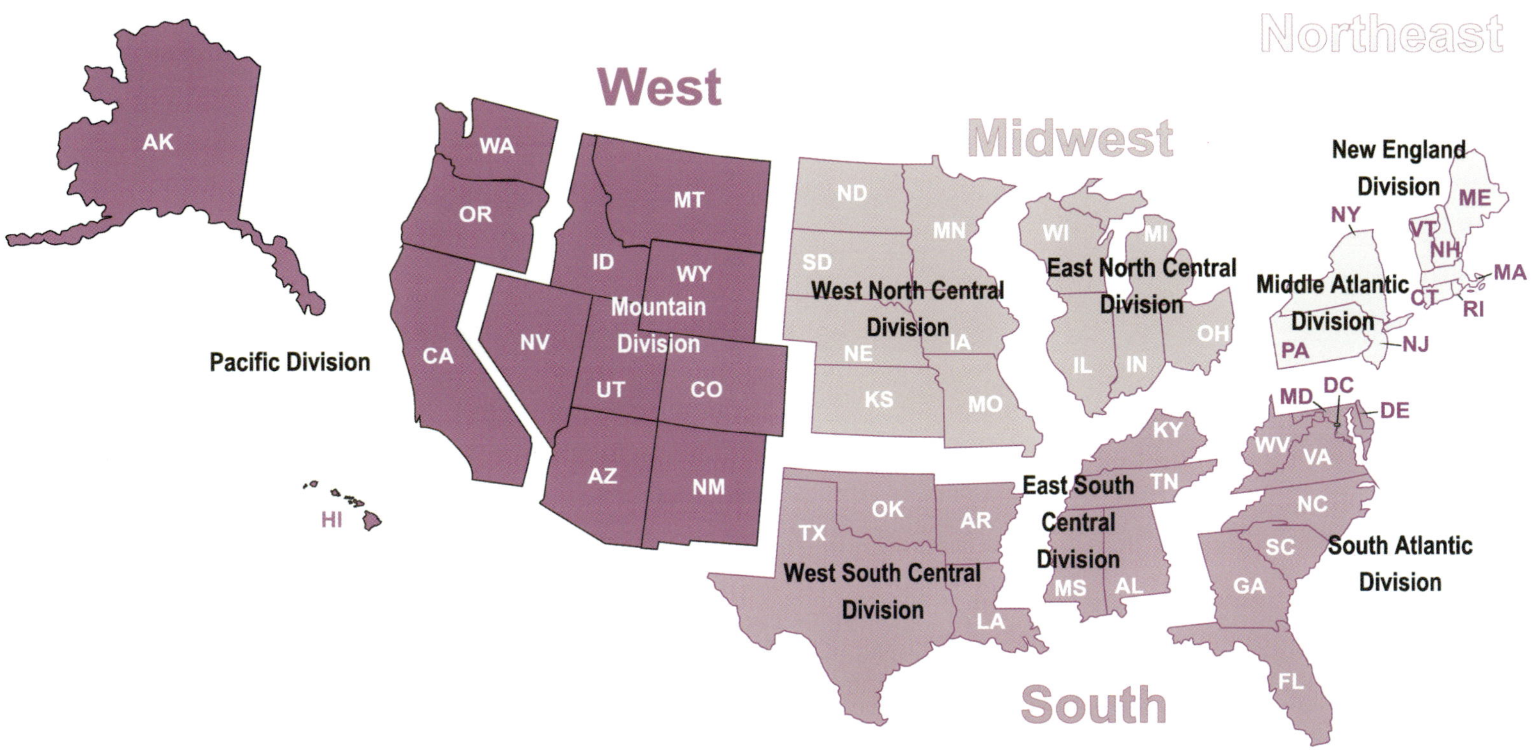

Note: Map not to scale.
Web Page: See www.census.gov/geo/www/us_regdiv.pdf.
Source: U.S. Department of Commerce, Bureau of the Census.

Appendix D

Table D1. Population, U.S. Gross Domestic Product, and Implicit Price Deflator, Selected Years, 1949-2007

Year	Population: United States [1] (Million People)	Population: World (Million People)	Population: United States as Share of World (Percent)	U.S. Gross Domestic Product: Billion Nominal Dollars [2]	U.S. Gross Domestic Product: Billion Chained (2000) Dollars [3]	U.S. Gross Domestic Product: Implicit Price Deflator [4] (2000 = 1.00000)
1949	R149.2	NA	NA	267.3	1,634.6	0.16352
1950	R152.3	2,556.5	R6.0	293.8	1,777.3	.16531
1955	R165.9	2,781.0	R6.0	414.8	2,212.8	.18743
1960	R180.7	3,040.6	5.9	526.4	2,501.8	.21041
1965	R194.3	3,346.8	5.8	719.1	3,191.1	.22535
1970	R205.1	3,707.9	5.5	1,038.5	3,771.9	.27534
1971	R207.7	3,785.3	5.5	1,127.1	3,898.6	.28911
1972	R209.9	3,861.3	5.4	1,238.3	4,105.0	.30166
1973	R211.9	3,936.8	5.4	1,382.7	4,341.5	.31849
1974	R213.9	4,011.3	5.3	1,500.0	4,319.6	.34725
1975	R216.0	R4,083.7	5.3	1,638.3	4,311.2	.38002
1976	R218.0	4,154.7	5.2	1,825.3	4,540.9	.40196
1977	R220.2	4,226.3	5.2	2,030.9	4,750.5	.42752
1978	R222.6	4,297.7	5.2	2,294.7	5,015.0	.45757
1979	R225.1	4,372.0	5.1	2,563.3	5,173.4	.49548
1980	R227.2	4,447.1	5.1	2,789.5	5,161.7	.54043
1981	229.5	R4,522.5	5.1	3,128.4	5,291.7	.59119
1982	231.7	R4,601.6	5.0	3,255.0	5,189.3	.62726
1983	233.8	R4,682.7	5.0	3,536.7	5,423.8	.65207
1984	235.8	R4,762.5	5.0	3,933.2	5,813.6	.67655
1985	237.9	R4,843.7	4.9	4,220.3	6,053.7	.69713
1986	240.1	R4,926.8	4.9	4,462.8	6,263.6	.71250
1987	242.3	R5,012.7	4.8	4,739.5	6,475.1	.73196
1988	244.5	R5,099.3	4.8	5,103.8	6,742.7	.75694
1989	246.8	R5,185.7	4.8	5,484.4	6,981.4	.78556
1990	R249.6	R5,273.4	4.7	5,803.1	7,112.5	.81590
1991	253.0	R5,357.2	4.7	5,995.9	7,100.5	.84444
1992	256.5	R5,440.5	4.7	6,337.7	7,336.6	.86385
1993	259.9	R5,521.3	4.7	6,657.4	7,532.7	.88381
1994	263.1	R5,601.0	4.7	7,072.2	7,835.5	.90259
1995	266.3	R5,681.7	4.7	7,397.7	8,031.7	.92106
1996	269.4	R5,761.6	4.7	7,816.9	8,328.9	.93852
1997	272.6	R5,840.6	4.7	8,304.3	8,703.5	.95414
1998	275.9	R5,918.7	4.7	8,747.0	9,066.9	.96472
1999	279.0	R5,995.6	4.7	9,268.4	9,470.3	.97868
2000	R282.2	R6,071.7	4.6	9,817.0	9,817.0	1.00000
2001	R285.1	R6,147.5	4.6	10,128.0	9,890.7	1.02399
2002	R287.9	R6,222.6	4.6	10,469.6	10,048.8	1.04187
2003	R290.4	R6,297.4	4.6	10,960.8	10,301.0	1.06404
2004	R293.2	R6,372.9	4.6	R11,685.9	R10,675.8	R1.09462
2005	R295.9	R6,449.1	4.6	R12,433.9	R11,003.4	R1.13000
2006	R298.8	R6,525.6	4.6	R13,194.7	R11,319.4	R1.16567
2007	301.6	6,602.3	4.6	13,841.3	11,566.8	1.19664

[1] Resident population of the 50 States and the District of Columbia estimated for July 1 of each year.
[2] See "Nominal Dollars" in Glossary.
[3] See "Chained Dollars" in Glossary.
[4] The gross domestic product implicit price deflator is used to convert nominal dollars to chained (2000) dollars.
R=Revised. NA=Not available.
Web Pages: • For all data beginning in 1949, see http://www.eia.doe.gov/emeu/aer/append_d.html. • For related information, see http://www.census.gov/ and http://www.bea.doc.gov/.

Sources: **U.S. Population:** • 1949-1989—Department of Commerce (DOC), U.S. Bureau of the Census, Current Population Reports Series P-25, June 2000. • 1990-1999—DOC, U.S. Bureau of the Census, State Population Estimates, April 11, 2002. • 2000 forward—DOC, U.S. Bureau of the Census, State Population Estimates, December 27, 2007. **World Population:** • 1950 forward—DOC, U.S. Bureau of the Census, International Database (July 16, 2007). **U.S. Gross Domestic Product:** • 1949 forward—DOC, Bureau of Economic Analysis, National Income and Product Accounts (March 27, 2008), Tables 1.1.5, 1.1.6, and 1.1.9.

Appendix E

Table E1. Estimated Primary Energy Consumption in the United States, Selected Years, 1635-1945

(Quadrillion Btu)

Year	Fossil Fuels				Renewable Energy			Electricity Net Imports	Total
	Coal	Natural Gas	Petroleum	Total	Conventional Hydroelectric Power	Biomass: Wood [1]	Total		
1635	NA	– –	– –	NA	– –	(s)	(s)	– –	(s)
1645	NA	– –	– –	NA	– –	0.001	0.001	– –	0.001
1655	NA	– –	– –	NA	– –	.002	.002	– –	.002
1665	NA	– –	– –	NA	– –	.005	.005	– –	.005
1675	NA	– –	– –	NA	– –	.007	.007	– –	.007
1685	NA	– –	– –	NA	– –	.009	.009	– –	.009
1695	NA	– –	– –	NA	– –	.014	.014	– –	.014
1705	NA	– –	– –	NA	– –	.022	.022	– –	.022
1715	NA	– –	– –	NA	– –	.037	.037	– –	.037
1725	NA	– –	– –	NA	– –	.056	.056	– –	.056
1735	NA	– –	– –	NA	– –	.080	.080	– –	.080
1745	NA	– –	– –	NA	– –	.112	.112	– –	.112
1755	NA	– –	– –	NA	– –	.155	.155	– –	.155
1765	NA	– –	– –	NA	– –	.200	.200	– –	.200
1775	NA	– –	– –	NA	– –	.249	.249	– –	.249
1785	NA	– –	– –	NA	– –	.310	.310	– –	.310
1795	NA	– –	– –	NA	– –	.402	.402	– –	.402
1805	NA	– –	– –	NA	– –	.537	.537	– –	.537
1815	NA	– –	– –	NA	– –	.714	.714	– –	.714
1825	NA	– –	– –	NA	– –	.960	.960	– –	.960
1835	NA	– –	– –	NA	– –	1.305	1.305	– –	1.305
1845	NA	– –	– –	NA	– –	1.757	1.757	– –	1.757
1850	0.219	– –	– –	0.219	– –	2.138	2.138	– –	2.357
1855	.421	– –	– –	.421	– –	2.389	2.389	– –	2.810
1860	.518	– –	0.003	.521	– –	2.641	2.641	– –	3.162
1865	.632	– –	.010	.642	– –	2.767	2.767	– –	3.409
1870	1.048	– –	.011	1.059	– –	2.893	2.893	– –	3.952
1875	1.440	– –	.011	1.451	– –	2.872	2.872	– –	4.323
1880	2.054	– –	.096	2.150	– –	2.851	2.851	– –	5.001
1885	2.840	0.082	.040	2.962	– –	2.683	2.683	– –	5.645
1890	4.062	.257	.156	4.475	0.022	2.515	2.537	– –	7.012
1895	4.950	.147	.168	5.265	.090	2.306	2.396	– –	7.661
1900	6.841	.252	.229	7.322	.250	2.015	2.265	– –	9.587
1905	10.001	.372	.610	10.983	.386	1.843	2.229	– –	13.212
1910	12.714	.540	1.007	14.261	.539	1.765	2.304	– –	16.565
1915	13.294	.673	1.418	15.385	.659	1.688	2.347	0.002	17.734
1920	15.504	.813	2.676	18.993	.738	1.610	2.348	.003	21.344
1925	14.706	1.191	4.280	20.177	.668	1.533	2.201	.004	22.382
1930	13.639	1.932	5.897	21.468	.752	1.455	2.207	.005	23.680
1935	10.634	1.919	5.675	18.228	.806	1.397	2.203	.005	20.436
1940	12.535	2.665	7.760	22.960	.880	1.358	2.238	.007	25.205
1945	15.972	3.871	10.110	29.953	1.442	[1]1.261	2.703	.009	32.665

[1] There is a discontinuity in the "Wood" time series between 1945 and 1949. Through 1945, data are for fuelwood only; beginning in 1949, data are for wood and wood-derived fuels (see Table 10.1).

NA=Not available. – – = Not applicable. (s)=Less than 0.0005 quadrillion Btu.

Notes: • For years not shown, there are no data available. • See Tables 1.3 and 10.1 for continuation of these data series from 1949 forward. • See Note, "Geographic Coverage of Statistics for 1635-1945," at end of section.

Sources: **Coal, Natural Gas, and Petroleum:** *Energy in the American Economy, 1850-1975,* Table VII. **Conventional Hydroelectric Power:** *Energy in the American Economy, 1850-1975,* Table II. **Wood:** • 1635-1845: U.S. Department of Agriculture Circular No. 641, *Fuel Wood Used in the United States 1630-1930,* February 1942. This source estimates fuelwood consumption in cords per decade, which were converted to Btu using the conversion factor of 20 million Btu per cord. The annual average value for each decade was assigned to the fifth year of the decade on the assumption that annual use was likely to increase during any given decade and the average annual value was more likely to reflect mid-decade yearly consumption than use at either the beginning or end of the decade. Values thus begin in 1635 and are plotted at 10-year intervals. • 1850-1945: *Energy in the American Economy, 1850-1975,* Table VII. **Electricity Net Imports:** *Energy in the American Economy, 1850-1975,* Tables I and VI. Calculated as the difference between hydroelectric consumption and hydroelectric production times 3,412 Btu per kilowatthour.

Appendix E

Note: Geographic Coverage of Statistics for 1635-1945. Table E1 presents estimates of U.S. energy consumption by energy source for a period that begins a century and a half before the original 13 colonies formed a political union and continues through the decades during which the United States was still expanding territorially. The question thus arises, what exactly is meant by "U.S. consumption" of an energy source for those years when the United States did not formally exist or consisted of less territory than is now encompassed by the 50 States and the District of Columbia?

The documents used to assemble the estimates, and (as far as possible) the sources of those documents, were reviewed carefully for clues to geographic coverage. For most energy sources, the extent of coverage expanded more rapidly than the Nation, defined as all the official States and the District of Columbia. Estimates or measurements of consumption of each energy source generally appear to follow settlement patterns. That is, they were made for areas of the continent that were settled enough to have economically significant consumption even though those areas were not to become States for years. The wood data series, for example, begins in 1635 and includes 12 of the original colonies (excepting Georgia), as well as Maine, Vermont, and the area that would become the District of Columbia. By the time the series reaches 1810, the rest of the continental States are all included, though the last of the 48 States to achieve statehood did not do so until 1912. Likewise, the coal data series begins in 1850 but includes consumption in areas, such as Utah and Washington (State), which were significant coal-producing regions but had not yet attained statehood. (Note: No data were available on State-level historical coal consumption. The coal data shown in Table E1 through 1945 describe *apparent* consumption, i.e., production plus imports minus exports. The geographic coverage for coal was therefore based on a tally of coal-*producing* States listed in various historical issues of *Minerals Yearbook*. It is likely that coal was consumed in States where it was not mined in significant quantities.)

By energy source, the extent of coverage can be summarized as follows: • **Coal**—35 coal-producing States by 1885. • **Natural Gas**—All 48 contiguous States, the District of Columbia, and Alaska by 1885. • **Petroleum**—All 48 contiguous States, the District of Columbia, and Alaska by 1885. • **Conventional Hydroelectric Power**—Coverage for 1890 and 1895 is uncertain, but probably the 48 contiguous States and the District of Columbia. Coverage for 1900 through 1945 is the 48 contiguous States, and the District of Columbia. • **Wood**—All 48 contiguous States and the District of Columbia by 1810.

Glossary

Alcohol: The family name of a group of organic chemical compounds composed of carbon, **hydrogen**, and oxygen. The series of molecules vary in chain length and are composed of a **hydrocarbon** plus a hydroxyl group: CH_3-(CH_2)n-OH (e.g., **methanol**, **ethanol**, and tertiary butyl alcohol). See **Fuel Ethanol**.

Alternative Fuel: As defined pursuant to the Energy Policy Act of 1992 (EPACT), **methanol**, denatured **ethanol**, and other **alcohols**, separately or in mixtures of 85 percent by volume or more (or other percentage not less than 70 as determined by DOE rule) with **motor gasoline** or other fuels, **compressed natural gas** (CNG), **liquefied natural gas** (LNG), **liquefied petroleum gases** (LPG), **hydrogen**, coal-derived liquid fuels, fuels other than alcohols derived from biological materials, **electricity**, or any other fuel determined to be substantially not **petroleum** and yielding substantial energy security benefits and substantial environmental benefits.

Alternative-Fueled Vehicle (AFV): A vehicle either designed and manufactured by an original equipment manufacturer or a converted vehicle designed to operate in either dual-fuel, flexible-fuel, or dedicated modes on fuels other than **motor gasoline** or **diesel fuel**. This does not include a conventional vehicle that is limited to operation on blended or **reformulated motor gasoline** fuels.

Anthracite: The highest rank of **coal**; used primarily for residential and commercial **space heating**. It is a hard, brittle, and black lustrous coal, often referred to as hard coal, containing a high percentage of fixed carbon and a low percentage of volatile matter. The moisture content of fresh-mined anthracite generally is less than 15 percent. The heat content of anthracite ranges from 22 to 28 million **Btu** per **short ton** on a moist, mineral-matter-free basis. The heat content of anthracite consumed in the United States averages 25 million Btu per short ton, on the as-received basis (i.e., containing both inherent moisture and mineral matter). *Note*: Since the 1980's, anthracite refuse or mine waste has been used for steam-electric power generation. This fuel typically has a heat content of 15 million Btu per short ton or less. See **Coal Rank**.

Anthracite Culm: Waste from Pennsylvania **anthracite** preparation plants, consisting of coarse rock fragments containing as much as 30 percent small-sized **coal**; sometimes defined as including very fine coal particles called silt. Its heat value ranges from 8 to 17 million **Btu** per **short ton**.

Anthropogenic: Made or generated by a human or caused by human activity. The term is used in the context of global **climate change** to refer to gaseous emissions that are the result of human activities, as well as other potentially climate-altering activities, such as deforestation.

API: The American Petroleum Institute, a trade association.

API Gravity: American Petroleum Institute measure of specific gravity of **crude oil** or condensate in degrees. An arbitrary scale expressing the gravity or density of liquid **petroleum products**. The measuring scale is calibrated in terms of degrees API; it is calculated as follows:
Degrees API = (141.5 / sp.gr.60 deg.F/60 deg.F) - 131.5.

Asphalt: A dark-brown to black cement-like material obtained by **petroleum** processing and containing bitumens as the predominant component; used primarily for road construction. It includes crude asphalt as well as the following finished products: cements, fluxes, the asphalt content of emulsions (exclusive of water), and petroleum distillates blended with asphalt to make cutback asphalts. *Note*: The conversion factor for asphalt is 5.5 **barrels** per **short ton**.

ASTM: The American Society for Testing and Materials.

Aviation Gasoline Blending Components: **Naphthas** that will be used for blending or compounding into finished **aviation gasoline** (e.g., straight run gasoline, alkylate, reformate, benzene, toluene, and xylene). Excludes **oxygenates** (**alcohols**, **ethers**), **butane**, and **pentanes plus**. Oxygenates are reported as other hydrocarbons, **hydrogen**, and **oxygenates**.

Aviation Gasoline, Finished: A complex mixture of relatively volatile **hydrocarbons** with or without small quantities of additives, blended to form a fuel suitable for use in aviation reciprocating engines. Fuel specifications are provided in ASTM Specification D910 and Military Specification MIL-G-5572. *Note*: Data on blending components are not counted in data on finished aviation gasoline. See **Jet Fuel**; **Jet Fuel, Kerosene-Type**; and **Jet Fuel, Naphtha-Type**.

Barrel (Petroleum): A unit of volume equal to 42 U.S. Gallons.

Barrels per Calendar Day: The amount of input that a distillation facility can process under usual operating conditions. The amount is expressed in terms of capacity during a 24-hour period and reduces the maximum processing capability of all units at the facility under continuous operation to account for the following limitations that may delay, interrupt, or slow down production: 1) the capability of downstream processing units to absorb the output of **crude oil** processing facilities of a given refinery (no reduction is necessary for intermediate streams that are distributed to other than downstream facilities as part of a refinery's normal operation); 2) the types and grades of inputs to be processed; 3) the types and grades of products expected to be manufactured; 4) the environmental constraints associated with refinery operations; 5) the reduction of capacity for scheduled downtime due to such conditions as routine inspection, maintenance, repairs, and turnaround; and 6) the reduction of capacity for unscheduled downtime due to such conditions as mechanical problems, repairs, and slowdowns.

Base Gas: The volume of gas needed as a permanent inventory to maintain adequate underground storage reservoir pressures and deliverability rates throughout the withdrawal season. All native gas is included in the base gas volume.

Biodiesel: Any liquid **biofuel** suitable as a **diesel fuel** substitute or diesel fuel additive or extender. Biodiesel fuels are typically made from oils such as soybean, rapeseed, or sunflower, or from animal tallow. Biodiesel can also be made from **hydrocarbons** derived from agricultural products such as rice hulls.

Biofuels: Liquid fuels and blending components produced from **biomass** (plant) feedstocks, used primarily for transportation. See **Biodiesel** and **Fuel Ethanol**.

Biogenic: Produced by biological processes of living organisms. *Note:* EIA uses the term "biogenic" to refer only to organic nonfossil material of biological origin.

Biomass: Organic nonfossil material of biological origin constituting a **renewable energy** source. See **Biodiesel, Biofuels, Biomass Waste, Fuel Ethanol,** and **Wood and Wood-Derived Fuels**.

Biomass Waste: Organic non-fossil material of biological origin that is a byproduct or a discarded product. "Biomass waste" includes municipal solid waste from **biogenic** sources, landfill gas, sludge waste, agricultural crop byproducts, straw, and other **biomass** solids, liquids, and gases; but excludes **wood and wood-derived fuels** (including **black liquor**), **biofuels** feedstock, **biodiesel**, and **fuel ethanol**. *Note:* EIA "biomass waste" data also include energy crops grown specifically for energy production, which would not normally constitute waste.

Bituminous Coal: A dense **coal**, usually black, sometimes dark brown, often with well-defined bands of bright and dull material, used primarily as fuel in steam-electric power generation, with substantial quantities also used for heat and power applications in manufacturing and making **coke**. Bituminous coal is the most abundant coal in active U.S. mining regions. Its moisture content usually is less than 20 percent. The heat content of bituminous coal ranges from 21 to 30 million **Btu** per **short ton** on a moist, mineral-matter-free basis. The heat content of bituminous coal consumed in the United States averages 24 million Btu per short ton, on the as-received basis (i.e., containing both inherent moisture and mineral matter). See **Coal Rank**.

Black Liquor: A byproduct of the paper production process, alkaline spent liquor, that can be used as a source of energy. Alkaline spent liquor is removed from the digesters in the process of chemically pulping wood. After evaporation, the residual "black" liquor is burned as a fuel in a recovery furnace that permits the recovery of certain basic chemicals.

Breeze: The fine screenings from crushed coke. Usually breeze will pass through a 1/2-inch or 3/4-inch screen opening. It is most often used as a fuel source in the process of agglomerating iron ore.

British Thermal Unit (Btu): The quantity of heat required to raise the temperature of 1 pound of liquid water by 1 degree Fahrenheit at the temperature at which water has its greatest density (approximately 39 degrees Fahrenheit). See **Heat Content**.

Btu: See **British Thermal Unit.**

Bunker Fuels: Fuel supplied to ships and aircraft, both domestic and foreign, consisting primarily of **residual fuel oil** and **distillate fuel oil** for ships and **kerosene-type jet fuel** for aircraft. The term "international bunker fuels" is used to denote the consumption of fuel for international transport activities. *Note*: For the purposes of **greenhouse gas** emissions inventories, data on emissions from combustion of international bunker fuels are subtracted from national emissions totals. Historically, bunker fuels have meant only ship fuel.

Butane: A normally gaseous straight-chain or branched-chain **hydrocarbon** (C_4H_{10}) extracted from **natural gas** or **refinery gas** streams. It includes isobutane and normal butane and is designated in ASTM Specification D1835 and Gas Processors Association Specifications for commercial butane.

> *Isobutane*: A normally gaseous branched-chain hydrocarbon. It is a colorless paraffinic gas that boils at a temperature of 10.9 degrees Fahrenheit. It is extracted from natural gas or refinery gas streams.
>
> *Normal Butane*: A normally gaseous straight-chain hydrocarbon. It is a colorless paraffinic gas that boils at a temperature of 31.1 degrees Fahrenheit. It is extracted from natural gas or refinery gas streams.

Butylene: An olefinic **hydrocarbon** (C_4H_8) recovered from refinery processes.

Capacity: See **Generator Capacity.**

Capacity Factor: See **Generator Capacity Factor.**

Carbon Dioxide: A colorless, odorless, non-poisonous gas (CO_2) that is a normal part of Earth's atmosphere. Carbon dioxide is a product of **fossil-fuel** combustion as well as other processes. It is considered a **greenhouse gas** as it traps heat (infrared energy) radiated by the Earth into the atmosphere and thereby contributes to the potential for **global warming**. The **global warming potential** (GWP) of other greenhouse gases is measured in relation to that of carbon dioxide, which by international scientific convention is assigned a value of one (1).

Carbon Dioxide Equivalent: The amount of **carbon dioxide** by weight emitted into the atmosphere that would produce the same estimated radiative forcing as a given weight of another radiatively active gas. Carbon dioxide equivalents are computed by multiplying the weight of the gas being measured (for example, **methane**) by its estimated **global warming potential** (which is 21 for methane). "Carbon equivalent units" are defined as carbon dioxide equivalents multiplied by the carbon content of carbon dioxide (i.e., 12/44).

Chained Dollars: A measure used to express **real prices**. Real prices are those that have been adjusted to remove the effect of changes in the purchasing power of the dollar; they usually reflect buying power relative to a reference year. Prior to 1996, real prices were expressed in constant dollars, a measure based on the weights of goods and services in a single year, usually a recent year. In 1996, the U.S. Department of Commerce introduced the chained-dollar measure. The new measure is based on the average weights of goods and services in successive pairs of years. It is "chained" because the second year in each pair, with its weights, becomes the first year of the next pair. The advantage of using the chained-dollar measure is that it is more closely related to any given period covered and is therefore subject to less distortion over time.

Chlorofluorocarbon (CFC): Any of various compounds consisting of carbon, **hydrogen**, chlorine, and flourine used as refrigerants. CFCs are now thought to be harmful to the Earth's atmosphere.

City Gate: A point or measuring station at which a distribution gas utility receives gas from a **natural gas pipeline** company or transmission system.

Climate Change: A term used to refer to all forms of climatic inconsistency, but especially to significant change from one prevailing climatic condition to another. In some cases, "climate change" has been used synonymously with the term "**global warming**"; scientists, however, tend to use the term in a wider sense to include natural changes in climate as well as climatic cooling.

Coal: A readily combustible black or brownish-black rock whose composition, including inherent moisture, consists of more than 50 percent by weight and more than 70 percent by volume of carbonaceous material. It is formed from plant remains that have been compacted, hardened, chemically altered, and metamorphosed by heat and pressure over geologic time. See **Coal Rank**.

Coal Coke: See **Coke, Coal**.

Coal Rank: The classification of **coals** according to their degree of progressive alteration from lignite to anthracite. In the United States, the standard ranks of coal include **lignite**, **subbituminous coal**, **bituminous coal**, and **anthracite** and are based on fixed carbon, volatile matter, heating value, and agglomerating (or caking) properties.

Coal Stocks: **Coal** quantities that are held in storage for future use and disposition. *Note*: When coal data are collected for a particular reporting period (month, quarter, or year), coal stocks are commonly measured as of the last day of this period.

Coal Synfuel: Coal-based solid fuel that has been processed by a **coal synfuel plant**; and coal-based fuels such as briquettes, pellets, or extrusions, which are formed from fresh or recycled coal and binding materials.

Coal Synfuel Plant: A plant engaged in the chemical transformation of **coal** into **coal synfuel**.

Coke, Coal: A solid carbonaceous residue derived from low-ash, low-sulfur **bituminous coal** from which the volatile constituents are driven off by baking in an oven at temperatures as high as 2,000 degrees Fahrenheit so that the fixed carbon and residual ash are fused together. Coke is used as a fuel and as a reducing agent in smelting iron ore in a blast furnace. Coke from coal is gray, hard, and porous and has a heating value of 24.8 million **Btu** per **short ton**.

Coke, Petroleum: A residue high in carbon content and low in **hydrogen** that is the final product of thermal decomposition in the condensation process in cracking. This product is reported as marketable coke or catalyst coke. The conversion is 5 **barrels** (of 42 U.S. gallons each) per **short ton**. Coke from **petroleum** has a heating value of 6.024 million **Btu** per barrel.

Combined-Heat-and-Power (CHP) Plant: A plant designed to produce both heat and **electricity** from a single heat source. *Note*: This term is being used in place of the term "cogenerator" that was used by EIA in the past. CHP better describes the facilities because some of the plants included do not produce heat and power in a sequential fashion and, as a result, do not meet the legal definition of cogeneration specified in the Public Utility Regulatory Polices Act (PURPA). See **Electricity-Only Plant**.

Commercial Building: A building with more than 50 percent of its floorspace used for commercial activities. Commercial buildings include, but are not limited to, stores, offices, schools, churches, gymnasiums, libraries, museums, hospitals, clinics, warehouses, and jails. Government buildings are included, except buildings on military bases or reservations.

Commercial Sector: An **energy**-consuming sector that consists of service-providing facilities and equipment of: businesses; Federal, State, and local governments; and other private and public organizations, such as religious, social, or fraternal groups. The commercial sector includes institutional living quarters. It also includes sewage treatment facilities. Common uses of energy associated with this sector include **space heating**, water heating, air conditioning,lighting, refrigeration, cooking, and running a wide variety of other equipment. *Note*: This sector includes **generators** that produce **electricity** and/or **useful thermal output** primarily to support the activities of the above-mentioned commercial establishments. Various EIA programs differ in sectoral coverage—for more information see http://www.eia.doe.gov/neic/datadefinitions/Guideforwebcom.htm. See **End-Use Sectors** and **Energy-Use Sectors.**

Completion (Crude Oil/Natural Gas Production): The term refers to the installation of permanent equipment for the production of **crude oil** or **natural gas**. If a **well** is equipped to produce only crude oil or natural gas from one zone or reservoir, the definition of a "well" (classified as a **crude oil well** or **natural gas well**) and the definition of a "completion" are identical. However, if a well is equipped to produce crude oil and/or natural gas separately from more than one reservoir, a "well" is not synonymous with a "completion."

Compressed Natural Gas (CNG): Natural gas compressed to a volume and density that is practical as a portable fuel supply (even when compressed, natural gas is not a liquid).

Conventional Hydroelectric Power: See **Hydroelectric Power, Conventional**.

Conventional Motor Gasoline: See **Motor Gasoline, Conventional**.

Conversion Factor: A number that translates units of one measurement system into corresponding values of another measurement system. *Note:* For specific conversion factors, see EIA data products.

Cooling Tower: A common type of environmental equipment installed at **electric power plants** used to transfer heat, produced by burning fuel, to the atmosphere. Cooling towers are installed where there is insufficient cooling water available or where waste heat discharged into cooling water would affect marine life.

Criteria Pollutant: A pollutant determined to be hazardous to human health and regulated under the Environmental Protection Agency's (EPA) National Ambient Air Quality Standards. The 1970 amendments to the Clean Air Act require EPA to describe the health and welfare impacts of a pollutant as the "criteria" for inclusion in the regulatory regime.

Crude Oil: A mixture of **hydrocarbons** that exists in liquid phase in natural underground reservoirs and remains liquid at atmospheric pressure after passing through surface separating facilities. Depending upon the characteristics of the crude stream, it may also include: 1) small amounts of hydrocarbons that exist in gaseous phase in natural underground reservoirs but are liquid at atmospheric pressure after being recovered from oil well (casinghead) gas in lease separators and are subsequently commingled with the crude stream without being separately measured. Lease condensate recovered as a liquid from **natural gas wells** in lease or field separation facilities and later mixed into the crude stream is also included; 2) small amounts of nonhydrocarbons produced with the oil, such as sulfur and various metals; and 3) drip gases, and liquid hydrocarbons produced from tar sands, oil sands, gilsonite, and oil shale. Liquids produced at natural gas processing plants are excluded. Crude oil is refined to produce a wide array of **petroleum products**, including heating oils; gasoline, **diesel** and **jet fuels**; **lubricants**; **asphalt**; **ethane**, **propane**, and **butane**; and many other products used for their **energy** or chemical content.

Crude Oil Domestic First Purchase Price: The price for domestic **crude oil** reported by the company that owns the crude oil the first time it is removed from the lease boundary.

Crude Oil Landed Cost: The price of **crude oil** at the port of discharge, including charges associated with purchasing, transporting, and insuring a cargo from the purchase point to the port of discharge. The cost does not include charges incurred at the discharge port (e.g., import tariffs or fees, wharfage charges, and demurrage).

Crude Oil Refiner Acquisition Cost: The cost of **crude oil** to the refiner, including transportation and other fees. The composite cost is the weighted average of domestic and imported crude oil costs. The refiner acquisition cost does not include the cost of crude oil purchased for the **Strategic Petroleum Reserve**.

Crude Oil Refinery Input: The total **crude oil** put into processing units at refineries.

Crude Oil Stocks: Stocks of **crude oil** and **lease condensate** held at refineries, in **petroleum pipelines**, at pipeline terminals, and on leases.

Crude Oil Used Directly: **Crude oil** consumed as fuel by **petroleum pipelines** and on crude oil leases.

Crude Oil Well: A **well** completed for the production of **crude oil** from one or more crude oil zones or reservoirs. Wells producing both crude oil and **natural gas** are classified as crude oil wells.

Cubic Foot (Natural Gas) The amount of **natural gas** contained at standard temperature and pressure (60 degrees Fahrenheit and 14.73 pounds standard per square inch) in a cube whose edges are one foot long.

Degree-Day Normals: Simple arithmetic averages of monthly or annual **degree-days** over a long period of time (usually the 30-year period 1971–2000). The averages may be simple degree-day normals or population-weighted degree-day normals.

Degree-Days, Cooling (CDD): A measure of how warm a location is over a period of time relative to a base temperature, most commonly specified as 65 degrees Fahrenheit. The measure is computed for each day by subtracting the base temperature (65 degrees) from the average of the day's high and low temperatures, with negative values set equal to zero. Each day's cooling degree-days are summed to create a cooling degree-day measure for a specified reference period. Cooling degree-days are used in energy analysis as an indicator of air conditioning energy requirements or use.

Degree-Days, Heating (HDD): A measure of how cold a location is over a period of time relative to a base temperature, most commonly specified as 65 degrees Fahrenheit. The measure is computed for each day by subtracting the average of the day's high and low temperatures from the base temperature (65 degrees), with negative values set equal to zero. Each day's heating degree-days are summed to create a heating degree-day measure for a specified reference period. Heating degree-days are used in energy analysis as an indicator of space heating energy requirements or use.

Degree-Days, Population-Weighted: Heating or cooling **degree-days** weighted by the population of the area in which the degree-days are recorded. To compute State population-weighted degree-days, each State is divided into from one to nine climatically homogeneous divisions, which are assigned weights based on the ratio of the population of the division to the total population of the State. Degree-day readings for each division are multiplied by the corresponding population weight for each division and those products are then summed to arrive at the State population-weighted degree-day figure. To compute national population-weighted degree-days, the Nation is divided into nine Census regions, each comprising from three to eight States, which are assigned weights based on the ratio of the population of the region to the total population of the Nation. Degree-day readings for each region are multiplied by the corresponding population weight for each region and those products are then summed to arrive at the national population-weighted degree-day figure.

Demand-Side Management: The planning, implementation, and monitoring of **electric utility** activities designed to encourage consumers to modify patterns of **electricity** usage, including the timing and level of electricity demand.

Demonstrated Reserve Base (Coal): A collective term for the sum of **coal** in both measured and indicated resource categories of reliability, representing 100 percent of the in-place coal in those categories as of a certain date. Includes beds of **bituminous coal** and **anthracite** 28 or more inches thick and beds of **subbituminous coal** 60 or more inches thick that can occur at depths of as much as 1,000 feet. Includes beds of **lignite** 60 or more inches thick that can be surface mined. Includes also thinner and/or deeper beds that currently are being mined or for which there is evidence that they could be mined commercially at a given time. Represents that portion of the identified coal resource from which reserves are calculated.

Development Well: A **well** drilled within the proved area of a **crude oil** or **natural gas** reservoir to the depth of a stratigraphic horizon known to be productive.

Diesel Fuel: A fuel composed of **distillate fuel oils** obtained in **petroleum** refining operation or blends of such distillate fuel oils with **residual fuel oil** used in motor vehicles. The boiling point and specific gravity are higher for diesel fuels than for gasoline.

Direct Use: Use of **electricity** that 1) is self-generated, 2) is produced by either the same entity that consumes the power or an affiliate, and 3) is used in direct support of a service or industrial process located within the same facility or group of facilities that house the generating equipment. Direct use is exclusive of **station use**.

Distillate Fuel Oil: A general classification for one of the **petroleum** fractions produced in conventional distillation operations. It includes **diesel fuels** and fuel oils. Products known as No. 1, No. 2, and No. 4 diesel fuel are used in on-highway diesel engines, such as those found in cars and trucks, as well as off-highway engines, such as those in railroad locomotives and agricultural machinery. Products known as No. 1, No. 2, and No. 4 fuel oils are used primarily for **space heating** and **electricity generation**.

Distillation Unit (Atmospheric): The primary distillation unit that processes **crude oil** (including mixtures of other hydrocarbons) at approximately atmospheric conditions. It includes a pipe still for vaporizing the crude oil and a **fractionation** tower for separating the vaporized hydrocarbon components in the crude oil into fractions with different boiling ranges. This is done by continuously vaporizing and condensing the components to separate higher boiling point material. The selected boiling ranges are set by the processing scheme, the properties of the crude oil, and the product specifications.

District Heat: Steam or hot water from an outside source used as an **energy source** in a building. The steam or hot water is produced in a central plant and is piped into the building. District heat may be purchased from a utility or provided by a physical plant in a separate building that is part of the same facility (for example, a hospital complex or university).

Dry Hole: An **exploratory well** or **development well** found to be incapable of producing either **crude oil** or **natural gas** in sufficient quantities to justify completion as a **crude oil well** or **natural gas well**.

Dry Natural Gas: See **Natural Gas, Dry**.

Dry Natural Gas Production: See **Natural Gas (Dry) Production**.

Electric Energy: The ability of an electric current to produce work, heat, light, or other forms of **energy**. It is measured in **kilowatthours**.

Electric Non-Utility: Any entity that generates, transmits, or sells **electricity**, or sells or trades electricity services and products, where costs are not established and recovered by regulatory authority. Examples of these entities include, but are not limited to, **independent power producers**, power marketers and aggregators (both wholesale and retail), merchant transmission service providers, self-generation entities, and cogeneration firms with Qualifying Facility Status. See **Electric Utility.**

Electric Power Plant: A station containing **prime movers**, electric **generators**, and auxiliary equipment for converting mechanical, chemical, and/or fission **energy** into **electric energy**.

Electric Power Sector: An **energy**-consuming sector that consists of **electricity-only** and **combined-heat-and-power (CHP)** plants within the **NAICS** (North American Industry Classification System) 22 category whose primary business is to sell **electricity**, or electricity and heat, to the public. *Note*: This sector includes **electric utilities** and **independent power producers**. See **Energy-Use Sectors.**

Electric Utility: Any entity that generates, transmits, or distributes **electricity** and recovers the cost of its generation, transmission or distribution assets and operations, either directly or indirectly, through cost-based rates set by a separate regulatory authority (e.g., State Public Service Commission), or is owned by a governmental unit or the consumers that the entity serves. Examples of these entities include: investor-owned entities, public power districts, public utility districts, municipalities, rural electric cooperatives, and State and Federal agencies. Electric utilities may have Federal Energy Regulatory Commission approval for interconnection agreements and wholesale trade tariffs covering either cost-of-service and/or market-based rates under the authority of the Federal Power Act. See **Electric Non-Utility**.

Electrical System Energy Losses: The amount of **energy** lost during generation, transmission, and distribution of **electricity**, including plant and unaccounted-for uses.

Electricity: A form of **energy** characterized by the presence and motion of elementary charged particles generated by friction, induction, or chemical change.

Electricity Generation: The process of producing **electric energy**, or the amount of electric energy produced by transforming other forms of **energy**; commonly expressed in **kilowatthours** (kWh) or megawatthours (MWh). See **Electricity Generation, Gross** and **Electricity Generation, Net**.

Electricity Generation, Gross: The total amount of **electric energy** produced by **generating units** and measured at the generating terminal.

Electricity Generation, Net: The amount of **gross electricity generation** less **station use** (the **electric energy** consumed at the generating station(s) for station service or auxiliaries). *Note*: Electricity required for pumping at **hydroelectric pumped-storage** plants is regarded as electricity for station service and is deducted from gross generation.

Electricity Retail Sales: The amount of **electricity** sold by **electric utilities** and other **energy service providers** to customers purchasing electricity for their own use and not for resale.

Electricity-Only Plant: A plant designed to produce **electricity** only. See **Combined-Heat-and-Power (CHP) Plant**.

Emissions: **Anthropogenic** releases of gases to the atmosphere. In the context of global **climate change**, they consist of radiatively important **greenhouse gases** (e.g., the release of **carbon dioxide** during fuel combustion).

End-Use Sectors: The **residential**, **commercial**, **industrial**, and **transportation** sectors of the economy. See **Energy-Use Sectors.**

Energy: The capacity for doing work as measured by the capability of doing work (potential energy) or the conversion of this capability to motion (kinetic energy). Energy has several forms, some of which are easily convertible and can be changed to another form useful for work. Most of the world's convertible energy comes from **fossil fuels**

that are burned to produce heat that is then used as a transfer medium to mechanical or other means in order to accomplish tasks. **Electric energy** is usually measured in **kilowatthours**, while heat energy is usually measured in **British thermal units**.

Energy Consumption: The use of **energy** as a source of heat or power or as an input in the manufacturing process.

Energy Expenditures: The money spent directly by consumers to purchase **energy**. Expenditures equal the amount of energy used by the consumer times the price per unit paid by the consumer.

Energy Service Provider: An **energy** entity that provides service to a retail or end-use customer.

Energy Source: Any substance or natural phenomenon that can be consumed or transformed to supply heat or power. Examples include **petroleum**, **coal**, **natural gas**, **nuclear**, **wood**, **waste**, **electricity**, **wind**, **geothermal**, sunlight (**solar energy**), water movement, and **hydrogen** in fuel cells.

Energy-Use Sectors: A group of major **energy**-consuming components of U.S. society developed to measure and analyze energy use. The sectors most commonly referred to in EIA are: **residential**, **commercial**, **industrial**, **transportation**, and **electric power**.

Ethane: A normally gaseous straight-chain **hydrocarbon** (C_2H_6). It is a colorless, paraffinic gas that boils at a temperature of -127.48 degrees Fahrenheit. It is extracted from **natural gas** and **refinery gas** streams.

Ether: The family name applied to a group of organic chemical compounds composed of carbon, **hydrogen**, and oxygen, and which are characterized by an oxygen atom attached to two carbon atoms (for example, **methyl tertiary butyl ether**).

Ethanol (CH_3-CH_2OH): A clear, colorless, flammable oxygenated **hydrocarbon**. Ethanol is typically produced chemically from **ethylene**, or biologically from fermentation of various sugars from carbohydrates found in agricultural crops and cellulosic residues from crops or wood. It is used in the United States as a gasoline octane enhancer and **oxygenate** (blended up to 10 percent concentration). Ethanol can also be used in high concentrations (E85) in vehicles designed for its use. See **Alcohol** and **Fuel Ethanol.**

Ethyl Tertiary Butyl Ether (ETBE): A colorless, flammable, oxygenated hydrocarbon blend stock, **$(CH_3)_3COC_2H_5$**, formed by the catalytic etherification of **isobutylene** with **ethanol**. See **Oxygenates**.

Ethylene: An olefinic **hydrocarbon** recovered from refinery processes or petrochemical processes. Ethylene is used as a **petrochemical feedstock** for numerous chemical applications and the production of consumer goods.

Exploratory Well: A **well** drilled to find and produce **crude oil** or **natural gas** in an area previously considered unproductive, to find a new reservoir in a known field (i.e., one previously producing crude oil or natural gas in another reservoir), or to extend the limit of a known crude oil or natural gas reservoir.

Exports: Shipments of goods from within the 50 States and the District of Columbia to U.S. possessions and territories or to foreign countries.

Extraction Loss: The reduction in volume of **natural gas** due to the removal of **natural gas liquid** constituents such as **ethane**, **propane**, and **butane** at natural gas processing plants.

Federal Energy Administration (FEA): A predecessor of the Energy Information Administration.

Federal Energy Regulatory Commission (FERC): The Federal agency with jurisdiction over interstate **electricity** sales, wholesale electric rates, hydroelectric licensing, **natural gas** pricing, **petroleum pipeline** rates, and **natural gas pipeline** certification. FERC is an independent regulatory agency within the Department of Energy and is the successor to the Federal Power Commission.

Federal Power Commission (FPC): The predecessor agency of the **Federal Energy Regulatory Commission**. The Federal Power Commission was created by an Act of Congress under the Federal Water Power Act on June 10, 1920. It was charged originally with regulating the electric power and **natural gas** industries. It was abolished on September 30, 1977, when the Department of Energy was created. Its functions were divided between the Department of Energy and the Federal Energy Regulatory Commission, an independent regulatory agency.

Financial Reporting System (FRS): The Energy Information Administration's statutory requirement to identify major **energy**-producing companies and develop and implement a data-reporting program for energy financial and operating information from these companies. Companies are selected if they are within the top 50 publicly-owned U.S. **crude oil** producers that have at least 1 percent of either production or reserves of crude oil, **natural gas**, **coal**, or **uranium** in the United States, or 1 percent of either refining capacity or **petroleum product** sales in the United States.

Finished Motor Gasoline: See **Motor Gasoline, Finished**.

First Purchase Price: See **Crude Oil Domestic First Purchase Price**.

First Use: Manufacturing establishments' consumption of the **energy** that was originally produced offsite or was produced onsite from input materials not classified as energy.

Fiscal Year: The U.S. Government's fiscal year runs from October 1 through September 30. The fiscal year is designated by the calendar year in which it ends; e.g., fiscal year 2002 began on October 1, 2001, and ended on September 30, 2002.

Flared Natural Gas: See **Natural Gas, Flared.**

F.O.B.: See **Free on Board**.

Footage Drilled: Total footage for **wells** in various categories, as reported for any specified period, includes (1) the deepest total depth (length of well bores) of all wells drilled from the surface, (2) the total of all bypassed footage drilled in connection with reported wells, and (3) all new footage drilled for directional sidetrack wells. Footage reported for directional sidetrack wells does not include footage in the common bore, which is reported as footage for the original well. In the case of old wells drilled deeper, the reported footage is that which was drilled below the total depth of the old well.

Former U.S.S.R.: See **U.S.S.R.**

Forward Costs (Uranium): The operating and capital costs that will be incurred in any future production of **uranium** from in-place reserves. Included are costs for labor, materials, power and fuel, royalties, payroll taxes, insurance, and general and administrative costs that are dependent upon the quantity of production and, thus, applicable as variable costs of production. Excluded from forward costs are prior expenditures, if any, incurred for property acquisition, exploration, mine development, and mill construction, as well as income taxes, profit, and the cost of money. *Note*: By use of forward costing, estimates of reserves for **uranium ore** deposits in differing geological settings can be aggregated and reported as the maximum amount that can theoretically be extracted to recover the specified costs of **uranium oxide** production under the listed forward cost categories.

Fossil Fuel: An **energy source** formed in the Earth's crust from decayed organic material, such as **petroleum**, **coal**, and **natural gas**.

Fossil-Fueled Steam-Electric Power Plant: An **electric power plant** in which the **prime mover** is a turbine rotated by high-pressure steam produced in a boiler by heat from burning **fossil fuels**.

Fractionation: The process by which saturated **hydrocarbons** are removed from **natural gas** and separated into distinct parts, or "fractions" such as **propane**, **butane**, and **ethane**.

Free Alongside Ship (F.A.S.): The value of a commodity at the port of exportation, generally including the purchase price plus all charges incurred in placing the commodity alongside the carrier at the port of exportation.

Free on Board (F.O.B.): A sales transaction in which the seller makes the product available for pick up at a specified port or terminal at a specified price and the buyer pays for the subsequent transportation and insurance.

Free on Board (F.O.B.) Rail/Barge Price: The **free on board** price of coal at the point of first sale. It excludes freight or shipping and insurance costs.

Fuel Ethanol (C_2H_5OH): An anhydrous **alcohol** (**ethanol** with less than 1% water) intended for gasoline blending. See **Oxygenates**.

Full-Power Operation: Operation of a nuclear **generating unit** at 100 percent of its design capacity. Full-power operation precedes commercial operation.

Gasohol: A blend of **finished motor gasoline** containing **alcohol** (generally **ethanol** but sometimes **methanol**) at a concentration between 5.7 percent and 10 percent by volume. See **Oxygenates**.

Generating Unit: Any combination of physically connected **generators**, reactors, boilers, combustion turbines, or other **prime movers** operated together to produce electric power.

Generator: A machine that converts mechanical **energy** into **electric energy**.

Generator Capacity: The maximum output, commonly expressed in megawatts (MW), that generating equipment can supply to system load, adjusted for ambient conditions. See **Generator Nameplate (Installed) Capacity** and **Generator Net Summer Capacity.**

Generator Capacity Factor: The ratio of the **electric energy** produced by a **generating unit** for a given period of time to the electric energy that could have been produced at continuous full-power operation during the same period.

Generator Nameplate (Installed) Capacity: The maximum rated output of a **generator**, **prime mover**, or other electric power production equipment under specific conditions designated by the manufacturer. Installed generator nameplate capacity is commonly expressed in megawatts (MW) and is usually indicated on a nameplate physically attached to the generator.

Generator Net Summer Capacity: The maximum output, commonly expressed in megawatts (MW), that generating equipment can supply to system load, as demonstrated by a multi-hour test, at the time of summer peak demand (period of June 1 through September 30). This output reflects a reduction in capacity due to electricity use for station service or auxiliaries.

Geothermal Energy: Hot water or steam extracted from geothermal reservoirs in the Earth's crust and used for geothermal heat pumps, water heating, or **electricity generation**.

Global Warming: An increase in the near-surface temperature of the Earth. Global warming has occurred in the distant past as the result of natural influences, but the term is today most often used to refer to the warming some scientists predict will occur as a result of increased **anthropogenic** emissions of **greenhouse gases**. See **Climate Change**.

Global Warming Potential (GWP): An index used to compare the relative radiative forcing of different gases without directly calculating the changes in atmospheric concentrations. GWPs are calculated as the ratio of the radiative forcing that would result from the emission of one kilogram of a **greenhouse gas** to that from the emission of one kilogram of **carbon dioxide** over a period of time, such as 100 years.

Greenhouse Gases: Those gases, such as water vapor, **carbon dioxide**, nitrous oxide, **methane**, **hydrofluorocarbons** (HFCs), **perfluorocarbons** (PFCs), and **sulfur hexafluoride**, that are transparent to solar (short-wave) radiation but opaque to long-wave radiation, thus preventing long-wave radiant energy from leaving the Earth's atmosphere. The net effect is a trapping of absorbed radiation and a tendency to warm the planet's surface.

Gross Domestic Product (GDP): The total value of goods and services produced by labor and property located in the United States. As long as the labor and property are located in the United States, the supplier (that is, the workers and, for property, the owners) may be either U.S. residents or residents of foreign countries.

Gross Domestic Product (GDP) Implicit Price Deflator: A measure used to convert **nominal prices** to **real prices**. See **Chained Dollars**.

Gross Electricity Generation: See **Electricity Generation, Gross.**

Gross Withdrawals: See **Natural Gas Gross Withdrawals**.

Gross Input to Atmospheric Crude Oil Distillation Units: Total input to atmospheric crude oil distillation units. Includes all **crude oil**, **lease condensate**, **natural gas plant liquids**, **unfinished oils**, **liquefied refinery gases**, slop oils, and other liquid **hydrocarbons** produced from tar sands, gilsonite, and oil shale.

Heat Content: The amount of heat **energy** available to be released by the transformation or use of a specified physical unit of an energy form (e.g., a **short ton** of **coal**, a **barrel** of **crude oil**, a **kilowatthour** of **electricity**, a **cubic foot** of **natural gas**, or a pound of steam). The amount of heat energy is commonly expressed in **British thermal units (Btu)**. *Note*: Heat content of combustible energy forms can be expressed in terms of either gross heat content (higher or upper heating value) or net heat content (lower heating value), depending upon whether or not the available heat energy includes or excludes the energy used to vaporize water (contained in the original energy form or created during the combustion process). The Energy Information Administration typically uses gross heat content values.

Heat Rate: A measure of generating station thermal efficiency commonly stated as **Btu** per **kilowatthour**. *Note:* Heat rates can be expressed as either gross or net heat rates, depending whether the electricity output is gross or net generation. Heat rates are typically expressed as net heat rates.

Household: A family, an individual, or a group of up to nine unrelated persons occupying the same housing unit. "Occupy" means the housing unit was the person's usual or permanent place of residence.

Housing Unit: A house, an apartment, a group of rooms, or a single room if it is either occupied or intended for occupancy as separate living quarters by a family, an individual, or a group of one to nine unrelated persons. Separate living quarters means the occupants (1) live and eat separately from other persons in the house or apartment and (2) have direct access from the outside of the buildings or through a common hall—that is, they can get to it without going through someone else's living quarters. Housing units do not include group quarters such as prisons or nursing homes where ten or more unrelated persons live. A common dining area used by residents is an indication of group quarters. Hotel and motel rooms are considered housing units if occupied as the usual or permanent place of residence.

Hydrocarbon: An organic chemical compound of **hydrogen** and carbon in the gaseous, liquid, or solid phase. The molecular structure of hydrocarbon compounds varies from the simplest (**methane**, a constituent of **natural gas**) to the very heavy and very complex.

Hydroelectric Power: The production of **electricity** from the kinetic **energy** of falling water. See **Hydroelectric Power, Conventional** and **Hydroelectric Pumped Storage**.

Hydroelectric Power, Conventional: **Hydroelectric power** generated from flowing water that is not created by **hydroelectric pumped storage**.

Hydroelectric Pumped Storage: **Hydroelectric power** that is generated during peak load periods by using water previously pumped into an elevated storage reservoir during off-peak periods when excess generating capacity is available to do so. When additional generating capacity is needed, the water can be released from the reservoir through a conduit to turbine **generators** located in an **electric power plant** at a lower level.

Hydrofluorocarbons (HFCs): A group of man-made chemicals composed of one or two carbon atoms and varying numbers of **hydrogen** and fluorine atoms. Most HFCs have 100-year **global warming potentials** in the thousands.

Hydrogen (H): The lightest of all gases, hydrogen occurs chiefly in combination with oxygen in water. It also exists in acids, bases, **alcohols**, **petroleum**, and other **hydrocarbons**.

Implicit Price Deflator: The implicit price deflator, published by the U.S. Department of Commerce, Bureau of Economic Analysis, is used to convert **nominal prices** to **real prices**.

Imports: Receipts of goods into the 50 States and the District of Columbia from U.S. possessions and territories or from foreign countries.

Independent Power Producer: A corporation, person, agency, authority, or other legal entity or instrumentality that owns or operates facilities for the generation of electricity for use primarily by the public, and that is not an **electric utility**. Independent power producers are included in the **electric power sector.**

Indicated Resources, Coal: **Coal** for which estimates of the **coal rank**, quality, and quantity are based partly on sample analyses and measurements and partly on reasonable geologic projections. Indicated resources are computed partly from specified measurements and partly from projection of visible data for a reasonable distance on the basis of geologic evidence. The points of observation are ½ to 1½ miles apart. Indicated coal is projected to extend as a ½-mile-wide belt that lies more than ¼ mile from the outcrop or points of observation or measurement.

Industrial Sector: An **energy**-consuming sector that consists of all facilities and equipment used for producing, processing, or assembling goods. The industrial sector encompasses the following types of activity: manufacturing (**NAICS** codes 31-33); agriculture, forestry, fishing and hunting (NAICS code 11); mining, including oil and gas extraction (NAICS code 21); and construction (NAICS code 23). Overall energy use in this sector is largely for process heat and cooling and powering machinery, with lesser amounts used for facility heating, air conditioning, and lighting. Fossil fuels are also used as raw material inputs to manufactured products. *Note:* This sector includes **generators** that produce **electricity** and/or **useful thermal output** primarily to support the above-mentioned industrial activities. Various EIA programs differ in sectoral coverage—for more information see http://www.eia.doe.gov/neic/datadefinitions/Guideforwebind.htm. See **End-Use Sectors** and **Energy-Use Sectors**.

Isobutane: See **Butane.**

Isobutylene: An olefinic **hydrocarbon** recovered from refinery processes or petrochemical processes.

Isopentane: A saturated branched-chain **hydrocarbon** obtained by **fractionation** of **natural gasoline** or isomerization of normal pentane.

Jet Fuel: A refined petroleum product used in jet aircraft engines. See **Jet Fuel, Kerosene-Type** and **Jet Fuel, Naphtha-Type**.

Jet Fuel, Kerosene-Type: A **kerosene**-based product with a maximum distillation temperature of 400 degrees Fahrenheit at the 10-percent recovery point and a final maximum boiling point of 572 degrees Fahrenheit and meeting ASTM Specification 1655 and Military Specifications MIL-T-5624P and MIL-T-83133D (Grades JP-5 and JP-8). It is used for commercial and military turbojet and turboprop aircraft engines.

Jet Fuel, Naphtha-Type: A fuel in the heavy **naphtha** boiling range, with an average gravity of 52.8 degrees API, 20 to 90 percent distillation temperature of 290 to 470 degrees Fahrenheit, and meeting Military Specification MIL-T-5624L (Grade JP-4). It is used primarily for military turbojet and turboprop aircraft engines because it has a lower freeze point than other aviation fuels and meets engine requirements at high altitudes and speeds.

Kerosene: A light **petroleum** distillate that is used in space heaters, cook stoves, and water heaters and is suitable for use as a light source when burned in wick-fed lamps. Kerosene has a maximum distillation temperature of 400 degrees Fahrenheit at the 10-percent recovery point, a final boiling point of 572 degrees Fahrenheit, and a minimum flash point of 100 degrees Fahrenheit. Included are No. 1-K and No. 2-K, the two grades recognized by ASTM Specification D3699 as well as all other grades of kerosene called range or stove oil, which have properties similar to those of No. 1 fuel oil. See **Jet Fuel, Kerosene-Type**.

Kerosene-Type Jet Fuel: See **Jet Fuel, Kerosene-Type**.

Kilowatt: A unit of electrical power equal to 1,000 **watts**.

Kilowatthour (kWh): A measure of **electricity** defined as a unit of work or **energy**, measured as 1 **kilowatt** (1,000 **watts**) of power expended for 1 hour. One kilowatthour is equivalent to 3,412 **Btu**. See **Watthour**.

Landed Cost: See **Crude Oil Landed Cost**.

Lease and Plant Fuel: **Natural gas** used in **well**, field, and lease operations (such as natural gas used in drilling operations, heaters, dehydrators, and field compressors) and used as fuel in natural gas processing plants.

Lease Condensate: A mixture consisting primarily of pentanes and heavier **hydrocarbons** which is recovered as a liquid from **natural gas** in lease separation facilities. This category excludes **natural gas plant liquids**, such as **butane** and **propane**, which are recovered at downstream natural gas processing plants or facilities.

Lignite: The lowest rank of **coal**, often referred to as brown coal, used almost exclusively as fuel for steam-electric power generation. It is brownish-black and has a high inherent moisture content, sometimes as high as 45 percent. The heat content of lignite ranges from 9 to 17 million **Btu** per **short ton** on a moist, mineral-matter-free basis. The heat content of lignite consumed in the United States averages 13 million Btu per short ton, on the as-received basis (i.e., containing both inherent moisture and mineral matter). See **Coal Rank**.

Liquefied Natural Gas (LNG): **Natural gas** (primarily **methane**) that has been liquefied by reducing its temperature to -260 degrees Fahrenheit at atmospheric pressure.

Liquefied Petroleum Gases (LPG): A group of **hydrocarbon**-based gases derived from **crude oil** refining or **natural gas fractionation**. They include **ethane**, **ethylene**, **propane**, **propylene**, **normal butane**, **butylene**, **isobutane**, and **isobutylene**. For convenience of transportation, these gases are liquefied through pressurization.

Liquefied Refinery Gases (LRG): **Liquefied petroleum gases** fractionated from refinery or **still gases**. Through compression and/or refrigeration, they are retained in the liquid state. The reported categories are **ethane**/**ethylene**, **propane**/**propylene**, **normal butane**/**butylene**, and **isobutane**. Excludes still gas.

Losses: See **Electrical System Energy Losses**.

Low-Power Testing: The period of time between a nuclear **generating unit**'s initial fuel loading date and the issuance of its operating (full-power) license. The maximum level of operation during that period is 5 percent of the unit's design thermal rating.

Lubricants: Substances used to reduce friction between bearing surfaces or incorporated into other materials used as processing aids in the manufacture of other products, or used as carriers of other materials. **Petroleum** lubricants may be produced either from distillates or residues. Lubricants include all grades of lubricating oils, from spindle oil to cylinder oil, and those used in greases.

Manufacturing: An energy-consuming subsector of the **industrial sector** that consists of all facilities and equipment engaged in the mechanical, physical, chemical, or electronic transformation of materials, substances, or components into new products. Assembly of component parts of products is included, except for that which is included in construction.

Marketed Production (Natural Gas): See **Natural Gas Marketed Production**.

Measured Resources, Coal: **Coal** resources for which estimates of the **coal rank**, quality, and quantity have been computed, within a margin of error of less than 20 percent, from sample analyses and measurements from closely spaced and geologically well known sample sites. Measured resources are computed from dimensions revealed in outcrops, trenches, mine workings, and drill holes. The points of observation and measurement are so closely spaced and the thickness and extent of coals are so well defined that the tonnage is judged to be accurate within 20 percent. Although the spacing of the point of observation necessary to demonstrate continuity of the coal differs from region to region, according to the character of the coal-beds, the points of observation are no greater than ½ mile apart. Measured coal is projected to extend as a belt ¼ mile wide from the outcrop or points of observation or measurement.

Methane: A colorless, flammable, odorless **hydrocarbon** gas (CH_4), which is the major component of **natural gas**. It is also an important source of **hydrogen** in various industrial processes.

Methanol: A light, volatile **alcohol** (CH_3OH) eligible for **motor gasoline blending**. See **Oxygenates**.

Methyl Tertiary Butyl Ether (MTBE): An ether, $(CH_3)_3COCH_3$, intended for **motor gasoline blending**. See **Oxygenates.**

Miscellaneous Petroleum Products: All finished **petroleum products** not classified elsewhere—for example, petrolatum, lube refining byproducts (aromatic extracts and tars), absorption oils, ram-jet fuel, petroleum rocket fuels, synthetic natural gas feedstocks, and specialty oils.

Motor Gasoline Blending: Mechanical mixing of **motor gasoline blending components** and **oxygenates** as required, to produce **finished motor gasoline**.

Finished motor gasoline may be further mixed with other motor gasoline blending components or oxygenates, resulting in increased volumes of finished motor gasoline and/or changes in the formulation of finished motor gasoline (e.g., **conventional motor gasoline** mixed with **MTBE** to produce **oxygenated motor gasoline**).

Motor Gasoline Blending Components: **Naphthas** (e.g., straight-run gasoline, alkylate, reformate, benzene, toluene, xylene) used for blending or compounding into **finished motor gasoline**. These components include reformulated gasoline blendstock for oxygenate blending (RBOB) but exclude **oxygenates** (**alcohols**, **ethers**), **butane**, and **pentanes plus**. *Note:* Oxygenates are reported as individual components and are included in the total for other hydrocarbons, **hydrogen**, and oxygenates.

Motor Gasoline, Conventional: **Finished motor gasoline** not included in the **oxygenated** or **reformulated** motor gasoline categories. *Note*: This category excludes reformulated gasoline blendstock for oxygenate blending (RBOB) as well as other blendstock. Conventional motor gasoline can be leaded or unleaded; regular, midgrade, or premium. See **Motor Gasoline Grades**.

Motor Gasoline, Finished: A complex mixture of relatively volatile **hydrocarbons** with or without small quantities of additives, blended to form a fuel suitable for use in spark-ignition. Motor gasoline, as defined in ASTM Specification D-4814 or Federal Specification VV-G-1690C, is characterized as having a boiling range of 122 to 158 degrees Fahrenheit at the 10-percent recovery point to 365 to 374 degrees Fahrenheit at the 90-percent recovery point. "Motor gasoline" includes **conventional motor gasoline**, all types of **oxygenated motor gasoline** including **gasohol**, and **reformulated motor gasoline**, but excludes **aviation gasoline**. *Note:* Volumetric data on **motor gasoline blending components**, as well as **oxygenates**, are not counted in data on finished motor gasoline until the blending components are blended into the gasoline.

Motor Gasoline Grades: The classification of gasoline by octane ratings. Each type of gasoline (**conventional**, **oxygenated**, and **reformulated**; leaded or unleaded) is classified by three grades: regular, midgrade, and premium. *Note*: Motor gasoline sales are reported by grade in accordance with their classification at the time of sale. In general, automotive octane requirements are lower at high altitudes. Therefore, in some areas of the United States, such as the Rocky Mountain States, the octane ratings for the gasoline grades may be 2 or more octane points lower.

Regular Gasoline: Gasoline having an antiknock index, i.e., octane rating, greater than or equal to 85 and less than 88.

Midgrade Gasoline: Gasoline having an antiknock index, i.e., octane rating, greater than or equal to 88 and less than or equal to 90.

Premium Gasoline: Gasoline having an antiknock index, i.e., octane rating, greater than 90.

Motor Gasoline, Oxygenated: **Finished motor gasoline** other than **reformulated motor gasoline**, having an oxygen content of 2.7 percent or higher by weight and required by the U.S. Environmental Protection Agency (EPA) to be sold in areas designated by EPA as carbon monoxide (CO) nonattainment areas. *Note*: Oxygenated gasoline excludes oxygenated fuels program reformulated gasoline (OPRG) and reformulated gasoline blendstock for oxygenate blending (RBOB). Data on gasohol that has at least 2.7 percent oxygen, by weight, and is intended for sale inside CO nonattainment areas are included in data on oxygenated gasoline. Other data on gasohol (for use outside of nonattainment areas) are included in data on conventional gasoline.

Motor Gasoline, Reformulated: **Finished motor gasoline** formulated for use in motor vehicles, the composition and properties of which meet the requirements of the reformulated gasoline regulations promulgated by the U.S. Environmental Protection Agency under Section 211(k) of the Clean Air Act. It includes gasoline produced to meet or exceed emissions performance and benzene content standards of federal-program reformulated gasoline even though the gasoline may not meet all of the composition requirements (e.g. oxygen content) of federal-program reformulated gasoline. *Note*: This category includes oxygenated fuels program reformulated gasoline (OPRG). Reformulated gasoline excludes reformulated blendstock for oxygenate blending (RBOB) and gasoline treated as blendstock (GTAB).

MTBE: See **Methyl Tertiary Butyl Ether**.

NAICS: See **North American Industry Classification System**.

Naphtha: A generic term applied to a **petroleum** fraction with an approximate boiling range between 122 and 400 degrees Fahrenheit.

Naphtha-Type Jet Fuel: See **Jet Fuel, Naphtha-Type**.

Natural Gas: A gaseous mixture of **hydrocarbon** compounds, primarily **methane**, used as a fuel for **electricity generation** and in a variety of ways in buildings, and as raw material input and fuel for industrial processes.

Natural Gas, Dry: **Natural gas** which remains after: 1) the liquefiable **hydrocarbon** portion has been removed from the gas stream (i.e., gas after lease, field, and/or plant separation); and 2) any volumes of **nonhydrocarbon gases** have been removed where they occur in sufficient quantity to render the gas unmarketable. *Note:* Dry natural gas is also known as consumer-grade natural gas. The parameters for measurement are cubic feet at 60 degrees Fahrenheit and 14.73 pounds per square inch absolute.

Natural Gas (Dry) Production: The process of producing consumer-grade **natural gas**. Natural gas withdrawn from reservoirs is reduced by volumes used at the production (lease) site and by processing losses. Volumes used at the production site include 1) the volume returned to reservoirs in cycling, repressuring of oil reservoirs, and conservation operations; and 2) **vented natural gas** and **flared natural gas**. Processing losses include 1) **nonhydrocarbon gases** (e.g., water vapor, **carbon dioxide**, helium, hydrogen sulfide, and nitrogen) removed from the gas stream; and 2) gas converted to liquid form, such as **lease condensate** and **natural gas plant liquids**. Volumes of dry gas withdrawn from gas storage reservoirs are not considered part of production. Dry natural gas production equals **natural gas marketed production** less **extraction loss**.

Natural Gas, Flared: **Natural gas** burned in flares on the base site or at gas processing plants.

Natural Gas Gross Withdrawals: Full well stream volume of produced **natural gas**, excluding **lease condensate** separated at the lease.

Natural Gas Liquids (NGL): Those **hydrocarbons** in **natural gas** that are separated from the gas as liquids through the process of absorption, condensation, adsorption, or other methods in gas processing or cycling plants. Generally such liquids consist of **propane** and heavier hydrocarbons and are commonly referred to as **lease condensate**, **natural gasoline**, and **liquefied petroleum gases**. Natural gas liquids include **natural gas plant liquids** (primarily **ethane**, **propane**, **butane**, and **isobutane)** and lease condensate (primarily pentanes produced from natural gas at lease separators and field facilities).

Natural Gas Marketed Production: **Natural gas gross withdrawals** from production reservoirs, less gas used for reservoir repressuring; **nonhydrocarbon gases** removed in treating or processing operations; and quantities of **vented natural gas** and **flared natural gas**. Includes all quantities of natural gas used in field and processing operations.

Natural Gas Pipeline: A continuous pipe conduit, complete with such equipment as valves, compressor stations, communications systems, and meters, for transporting **natural gas** and/or **supplemental gaseous fuels** from one point to another, usually from a point in or beyond the producing field or processing plant to another pipeline or to points of utilization. Also refers to a company operating such facilities.

Natural Gas Plant Liquids (NGPL): Those **hydrocarbons** in **natural gas** that are separated as liquids at natural gas processing plants, fractionating and cycling plants, and, in some instances, field facilities. **Lease condensate** is excluded. Products obtained include **ethane**; **liquefied petroleum gases** (**propane**, **butanes**, propane-butane mixtures, ethane-propane mixtures); **isopentane**; and other small quantities of finished products, such as **motor gasoline**, **special naphthas**, **jet fuel**, **kerosene**, and **distillate fuel oil**. See **Natural Gas Liquids**.

Natural Gas Processing Plant: A surface installation designed to separate and recover **natural gas liquids** from a stream of produced **natural gas** through the processes of condensation, absorption, refrigeration, or other methods, and to control the quality of natural gas marketed or returned to oil or gas reservoirs for pressure maintenance, repressuring, or cycling.

Natural Gas, Vented: **Natural gas** released into the air on the production site or at processing plants.

Natural Gas Well: A **well** completed for the production of **natural gas** from one or more natural gas zones or reservoirs. (Wells producing both **crude oil** and natural gas are classified as **crude oil wells**.)

Natural Gas Wellhead Price: Price of **natural gas** calculated by dividing the total reported value at the wellhead by the total quantity produced as reported by the appropriate agencies of individual producing States and the U.S. Mineral Management Service. The price includes all costs prior to shipment from the lease, including gathering and compression costs, in addition to State production, severance, and similar charges.

Natural Gasoline: A mixture of **hydrocarbons** (mostly pentanes and heavier) extracted from **natural gas** that meets vapor pressure, end-point, and other specifications for natural gasoline set by the Gas Processors Association. Includes **isopentane**, which is a saturated branch-chain hydrocarbon obtained by **fractionation** of natural gasoline or isomerization of normal pentane.

NERC: See **North American Electric Reliability Council**.

Net Electricity Generation: See **Electricity Generation, Net**.

Net Summer Capacity: See **Generator Net Summer Capacity.**

Neutral Zone: A 6,200 square-mile area shared equally between Kuwait and Saudi Arabia under a 1992 agreement.

Nitrogen Oxides (NO_x): Compounds of nitrogen and oxygen produced by the burning of **fossil fuels**.

Nominal Dollars: A measure used to express **nominal price**.

Nominal Price: The price paid for a product or service at the time of the transaction. Nominal prices are those that have not been adjusted to remove the effect of changes in the purchasing power of the dollar; they reflect buying power in the year in which the transaction occurred.

Non-Biomass Waste: Material of non-biological origin that is a byproduct or a discarded product. "Non-biomass waste" includes municipal solid waste from non-biogenic sources, such as plastics, and tire-derived fuels.

Noncoincident Peak Load: The sum of two or more peak loads on individual systems that do not occur in the same time interval. Meaningful only in the context of loads within a limited period of time, such as day, week, month, a heating or cooling season, and usually for not more than 1 year.

Nonhydrocarbon Gases: Typical nonhydrocarbon gases that may be present in reservoir **natural gas**, such as **carbon dioxide**, helium, hydrogen sulfide, and nitrogen.

Normal Butane: See **Butane.**

North American Electric Reliability Council (NERC): A council formed in 1968 by the **electric utility** industry to promote the reliability and adequacy of bulk power supply in the electric utility systems of North America. NERC consists of regional reliability councils and encompasses essentially all the power regions of the contiguous United States, Canada, and Mexico. See the various NERC Regional Reliability Councils at:
http://www.eia.doe.gov/cneaf/electricity/chg_str_fuel/html/fig02.html.

North American Industry Classification System (NAICS): A classification scheme, developed by the Office of Management and Budget to replace the Standard Industrial Classification (SIC) System, that categorizes establishments according to the types of production processes they primarily use.

Nuclear Electric Power (Nuclear Power): Electricity generated by the use of the thermal energy released from the fission of nuclear fuel in a reactor.

Nuclear Electric Power Plant: A single-unit or multi-unit facility in which heat produced in one or more reactors by the fissioning of nuclear fuel is used to drive one or more steam turbines.

Nuclear Reactor: An apparatus in which a nuclear fission chain reaction can be initiated, controlled, and sustained at a specific rate. A reactor includes fuel (fissionable material), moderating material to control the rate of fission, a heavy-walled pressure vessel to house reactor components, shielding to protect personnel, a system to conduct heat away from the reactor, and instrumentation for monitoring and controlling the reactor's systems.

OECD: See **Organization for Economic Cooperation and Development**.

Offshore: That geographic area that lies seaward of the coastline. In general, the coastline is the line of ordinary low water along with that portion of the coast that is in direct contact with the open sea or the line marking the seaward limit of inland water. If a State agency uses a different basis for classifying onshore and offshore areas, the State classification is used (e.g., Cook Inlet in Alaska is classified as offshore; for Louisiana, the coastline is defined as the Chapman Line, as modified by subsequent adjudication).

Oil: See **Crude Oil**.

OPEC: See **Organization of the Petroleum Exporting Countries**.

Operable Nuclear Unit: In the United States, a nuclear **generating unit** that has completed low-power testing and is in possession of a full-power operating license issued by the Nuclear Regulatory Commission.

Operable Refineries: Refineries that were in one of the following three categories at the beginning of a given year: in operation; not in operation and not under active repair, but capable of being placed into operation within 30 days; or not in operation, but under active repair that could be completed within 90 days.

Operating Income: Operating revenues less operating expenses. Excludes items of other revenue and expense, such as equity in earnings of unconsolidated affiliates, dividends, interest income and expense, income taxes, extraordinary items, and cumulative effect of accounting changes.

Organization for Economic Cooperation and Development (OECD): An international organization helping governments tackle the economic, social and governance challenges of a globalized economy. Its membership comprises about 30 member countries. With active relationships with some 70 other countries, non-governmental organizations (NGOs) and civil society, it has a global reach. For details about the organization, see http://www.oecd.org.

Organization of the Petroleum Exporting Countries (OPEC): An intergovernmental organization whose stated objective is to coordinate and unify petroleum policies among member countries. It was created at the Baghdad Conference on September 10–14, 1960, by Iran, Iraq, Kuwait, Saudi Arabia and Venezuela. The five founding members were later joined by nine other members: Qatar (1961); Indonesia (1962); Libya (1962); United Arab Emirates (1967); Algeria (1969); Nigeria (1971); Ecuador (1973–1992, 2007); Gabon (1975–1994); and Angola (2007).

Oxygenated Motor Gasoline: See **Motor Gasoline, Oxygenated**.

Oxygenates: Substances which, when added to **motor gasoline**, increase the amount of oxygen in that gasoline blend. **Ethanol**, **methyl tertiary butyl ether (MTBE), ethyl tertiary butyl ether (ETBE)**, and **methanol** are common oxygenates. See **Motor Gasoline, Oxygenated**.

Ozone: A molecule made up of three atoms of oxygen. Occurs naturally in the stratosphere and provides a protective layer shielding the Earth from harmful ultraviolet radiation. In the troposphere, it is a chemical oxidant, a greenhouse gas, and a major component of photochemical smog.

PAD Districts: Petroleum Administration for Defense Districts. Geographic aggregations of the 50 States and the District of Columbia into five districts for the Petroleum Administration for Defense in 1950. The districts were originally instituted for economic and geographic reasons as Petroleum Administration for War (PAW) Districts, which were established in 1942.

Particulate Collectors: Equipment used to remove fly ash from the combustion gases of a boiler plant before discharge to the atmosphere. Particulate collectors include electrostatic precipitators, mechanical collectors (cyclones, fabric filters [baghouses]), and wet scrubbers.

Peak Kilowatt: thousand **peak watts**.

Peak Watt: A manufacturer's unit indicating the amount of power a photovoltaic cell or module will produce at standard test conditions (normally 1,000 **watts** per square meter and 25 degrees Celsius).

Pentanes Plus: A mixture of **hydrocarbons**, mostly pentanes and heavier, extracted from **natural gas**. Includes **isopentane**, **natural gasoline**, and **plant condensate**.

Perfluorocarbons (PFCs): A group of man-made chemicals composed of one or two carbon atoms and four to six flourine atoms, containing no chlorine. PFCs have no commercial uses and are emitted as a byproduct of aluminum smelting and semiconductor manufacturing. PFCs have very high 100-year **global warming potentials** and are very long-lived in the atmosphere.

Petrochemical Feedstocks: Chemical feedstocks derived from **petroleum** principally for the manufacture of chemicals, synthetic rubber, and a variety of plastics.

Petroleum: A broadly defined class of liquid **hydrocarbon** mixtures. Included are **crude oil**, **lease condensate**, **unfinished oils**, refined products obtained from the processing of crude oil, and **natural gas plant liquids**. *Note*: Volumes of finished **petroleum products** include nonhydrocarbon compounds, such as additives and detergents, after they have been blended into the products.

Petroleum Coke: See **Coke, Petroleum**.

Petroleum Consumption: See **Products Supplied (Petroleum)**.

Petroleum Imports: Imports of **petroleum** into the 50 States and the District of Columbia from foreign countries and from Puerto Rico, the Virgin Islands, and other U.S. territories and possessions. Included are imports for the **Strategic Petroleum Reserve** and withdrawals from bonded warehouses for onshore consumption, offshore bunker use, and military use. Excluded are receipts of foreign petroleum into bonded warehouses and into U.S. territories and U.S. Foreign Trade Zones.

Petroleum Pipeline: Crude oil and product pipelines used to transport **crude oil** and **petroleum products**, respectively (including interstate, intrastate, and intracompany pipelines), within the 50 States and the District of Columbia.

Petroleum Products: Petroleum products are obtained from the processing of **crude oil** (including **lease condensate**), **natural gas**, and other **hydrocarbon** compounds. Petroleum products include **unfinished oils**, **liquefied petroleum gases**, **pentanes plus**, **aviation gasoline**, **motor gasoline**, **naphtha-type jet fuel**, **kerosene-type jet fuel**, **kerosene**, **distillate fuel oil**, **residual fuel oil**, **petrochemical feedstocks**, **special naphthas**, **lubricants**, **waxes**, **petroleum coke**, **asphalt**, **road oil**, **still gas**, and **miscellaneous petroleum products**.

Petroleum Stocks, Primary: For individual **petroleum products**, quantities that are held at refineries, in **petroleum pipelines**, and at bulk terminals that have a capacity of 50,000 barrels or more, or that are in transit thereto. Stocks held by product retailers and resellers, as well as tertiary stocks held at the point of consumption, are excluded. Stocks of individual products held at gas processing plants are excluded from individual product estimates but are included in other oil estimates and total.

Photovoltaic Energy: Direct-current **electricity** generated from sunlight through solid-state semiconductor devices that have no moving parts.

Photovoltaic Module: An integrated assembly of interconnected photovoltaic cells designed to deliver a selected level of working voltage and current at its output terminals, packaged for protection against environmental degradation, and suited for incorporation in photovoltaic power systems.

Pipeline Fuel: **Natural gas** consumed in the operation of pipelines, primarily in compressors.

Plant Condensate: One of the **natural gas liquids**, mostly pentanes and heavier **hydrocarbons**, recovered and separated as liquids at gas inlet separators or scrubbers in processing plants.

Prime Mover: The engine, turbine, water wheel, or similar machine that drives an electric **generator**; or, for reporting purposes, a device that converts **energy** to **electricity** directly.

Process Fuel: All **energy** consumed in the acquisition, processing, and transportation of energy. Quantifiable process fuel includes three categories: natural gas lease and plant operations, **natural gas pipeline** operations, and oil refinery operations.

Processing Gain: The volumetric amount by which total output is greater than input for a given period of time. This difference is due to the processing of **crude oil** into **petroleum products** which, in total, have a lower specific gravity than the crude oil processed.

Processing Loss: The volumetric amount by which total refinery output is less than input for a given period of time. This difference is due to the processing of **crude oil** into **petroleum products** which, in total, have a higher specific gravity than the crude oil processed.

Products Supplied (Petroleum): Approximately represents consumption of **petroleum products** because it measures the disappearance of these products from primary sources, i.e., **refineries**, **natural gas processing plants**, blending plants, pipelines, and bulk terminals. In general, product supplied of each product in any given period is computed as follows: field production, plus refinery production, plus imports, plus unaccounted-for crude oil (plus net receipts when calculated on a PAD District basis) minus stock change, minus crude oil losses, minus refinery inputs, and minus exports.

Propane: A normally gaseous straight-chain **hydrocarbon** (C_3H_8). It is a colorless paraffinic gas that boils at a temperature of -43.67 degrees Fahrenheit. It is extracted from **natural gas** or **refinery gas** streams. It includes all products designated in ASTM Specification D1835 and Gas Processors Association Specifications for commercial propane and HD-5 propane.

Propylene: An olefinic **hydrocarbon** (C_3H_6) recovered from refinery processes or petrochemical processes.

Proved Reserves, Crude Oil: The estimated quantities of all liquids defined as **crude oil** that geological and engineering data demonstrate with reasonable certainty to be recoverable in future years from known reservoirs under existing economic and operating conditions.

Proved Reserves, Lease Condensate: The volumes of **lease condensate** expected to be recovered in future years in conjunction with the production of proved reserves of **natural gas** based on the recovery efficiency of lease and/or field separation facilities installed.

Proved Reserves, Natural Gas: The estimated quantities of **natural gas** that analysis of geological and engineering data demonstrates with reasonable certainty to be recoverable in future years from known reservoirs under existing economic and operating conditions.

Proved Reserves, Natural Gas Liquids: Those volumes of **natural gas liquids** (including **lease condensate**) demonstrated with reasonable certainty to be separable in the future from proved **natural gas** reserves, under existing economic and operating conditions.

Pumped Storage: See **Hydroelectric Pumped Storage**.

Real Price: A price that has been adjusted to remove the effect of changes in the purchasing power of the dollar. Real prices, which are expressed in constant dollars, usually reflect buying power relative to a base year. See **Chained Dollars**.

Refiner Acquisition Cost of Crude Oil: See **Crude Oil Refiner Acquisition Cost**.

Refinery Gas: See **Still Gas**.

Refinery and Blender Net Inputs: Raw materials, **unfinished oils**, and blending components processed at **refineries**, or blended at refineries or petroleum storage terminals to produce finished **petroleum products**. Included are gross inputs of **crude oil**, **natural gas plant liquids**, other hydrocarbon raw materials, **hydrogen**, and **oxygenates**. Also included are net inputs of unfinished oils, **motor gasoline blending components**, and **aviation gasoline blending components**. Net inputs are calculated as gross inputs minus gross production. Negative net inputs indicate gross inputs are less than gross production. Examples of negative net inputs include reformulated gasoline blendstock for oxygenate blending (RBOB) produced at refineries for shipment to blending terminals, and unfinished oils produced and added to inventory in advance of scheduled maintenance of a refinery crude oil distillation unit.

Refinery and Blender Net Production: Liquefied refinery gases, and finished **petroleum products** produced at a **refinery** or petroleum storage terminal blending facility. Net production equals gross production minus gross inputs. Negative net production indicates gross production is less than gross inputs for a finished petroleum product. Examples of negative net production include reclassification of one finished product to another finished product, or reclassification of a finished product to **unfinished oils** or blending components.

Refinery (Petroleum): An installation that manufactures finished **petroleum products** from **crude oil**, **unfinished oils**, **natural gas liquids**, other hydrocarbons, and **alcohol.**

Reformulated Motor Gasoline: See **Motor Gasoline, Reformulated.**

Refuse Mine: A surface mine where **coal** is recovered from previously mined coal. It may also be known as a silt bank, culm bank, refuse bank, slurry dam, or dredge operation.

Refuse Recovery: The recapture of **coal** from a **refuse mine** or the coal recaptured by that process. The resulting product has been cleaned to reduce the concentration of noncombustible materials.

Renewable Energy: Energy obtained from sources that are essentially inexhaustible (unlike, for example, **fossil fuels**, which are in finite supply). Renewable sources of energy include **conventional hydroelectric power**, **geothermal**, **solar, wind,** and **biomass.**

Replacement Fuel: The portion of any motor fuel that is **methanol**, **ethanol**, or other **alcohols**, **natural gas**, **liquefied petroleum gases**, **hydrogen**, coal-derived liquid fuels, **electricity** (including electricity from **solar energy**), **ethers**, **biodiesel**, or any other fuel the Secretary of Energy determines, by rule, is substantially not **petroleum** and would yield substantial energy security benefits and substantial environmental benefits.

Repressuring: The injection of gas into **crude oil** or **natural gas** formations to effect greater ultimate recovery.

Residential Sector: An **energy**-consuming sector that consists of living quarters for private households. Common uses of energy associated with this sector include **space heating**, water heating, air conditioning, lighting, refrigeration, cooking, and running a variety of other appliances. The residential sector excludes institutional living quarters. *Note*: Various EIA programs differ in sectoral coverage—for further explanation see http://www.eia.doe.gov/neic/datadefinitions/Guideforwebres.htm. See **End-Use Sectors** and **Energy-Use Sectors.**

Residual Fuel Oil: The heavier oils, known as No. 5 and No. 6 fuel oils, that remain after the **distillate fuel oils** and lighter **hydrocarbons** are distilled away in refinery operations. It conforms to ASTM Specifications D396 and D975 and Federal Specification VV-F-815C. No. 5, a residual fuel oil of medium viscosity, is also known as Navy Special and is defined in Military Specification MIL-F-859E, including Amendment 2 (NATO Symbol F-770). It is used in steam-powered vessels in government service and inshore **electric power plants**. No. 6 fuel oil includes Bunker C fuel oil and is used for **electricity generation**, **space heating**, **vessel bunkering**, and various industrial purposes.

Road Oil: Any heavy **petroleum** oil, including residual asphaltic oil, used as a dust palliative and surface treatment on roads and highways. It is generally produced in six grades, from 0, the most liquid, to 5, the most viscous.

Rotary Rig: A machine used for drilling wells that employs a rotating tube attached to a bit for boring holes through rock.

Royalty Interest: An interest in a mineral property provided through a royalty contract.

Short Ton (Coal): A unit of weight equal to 2,000 pounds.

Solar Energy: See **Solar Thermal Energy** and **Photovoltaic Energy**.

Solar Thermal Collector: A device designed to receive solar radiation and convert it to thermal **energy**. Normally, a solar thermal collector includes a frame, glazing, and an absorber, together with appropriate insulation. The heat collected by the solar thermal collector may be used immediately or stored for later use. Solar collectors are used for **space heating**, domestic hot water heating, and heating swimming pools, hot tubs, or spas.

Solar Thermal Energy: The radiant **energy** of the sun that can be converted into other forms of energy, such as heat or **electricity**.

Space Heating: The use of **energy** to generate heat for warmth in housing units using space-heating equipment. The equipment could be the main space-heating equipment or secondary space-heating equipment. It does not include the use of energy to operate appliances (such as lights, televisions, and refrigerators) that give off heat as a byproduct.

Special Naphthas: All finished **petroleum products** within the **naphtha** boiling range that are used as paint thinners, cleaners, or solvents. Those products are refined to a specified flash point. Special naphthas include all commercial hexane and cleaning solvents conforming to ASTM Specifications D1836 and D484, respectively. Naphthas to be blended or marketed as **motor gasoline** or **aviation gasoline** or that are to be used as **petrochemical feedstocks** or synthetic natural gas (SNG) feedstocks are excluded.

Spent Liquor: The liquid residue left after an industrial process; can be a component of waste materials used as fuel.

Spot Market Price: See **Spot Price**.

Spot Price: The price for a one-time open market transaction for immediate delivery of the specific quantity of product at a specific location where the commodity is purchased "on the spot" at current market rates.

Station Use: Energy that is used to operate an **electric power plant**. It includes energy consumed for plant lighting, power, and auxiliary facilities, regardless of whether the energy is produced at the plant or comes from another source.

Steam-Electric Power Plant: An **electric power plant** in which the **prime mover** is a steam turbine. The steam used to drive the turbine is produced in a boiler where **fossil fuels** are burned.

Still Gas (Refinery Gas): Any form or mixture of gases produced in refineries by distillation, cracking, reforming, and other processes. The principal constituents are **methane**, **ethane**, **ethylene**, **normal butane**, **butylene**, **propane**, **propylene**, etc. Still gas is used as a refinery fuel and a **petrochemical feedstock**. The conversion factor is 6 million **Btu** per fuel oil equivalent **barrel**.

Stocks: Inventories of fuel stored for future use. See **Crude Oil Stocks, Coal Stocks,** and **Petroleum Stocks, Primary.**

Strategic Petroleum Reserve (SPR): **Petroleum** stocks maintained by the Federal Government for use during periods of major supply interruption.

Subbituminous Coal: A **coal** whose properties range from those of **lignite** to those of **bituminous coal** and used primarily as fuel for steam-electric power generation. It may be dull, dark brown to black, soft and crumbly, at the lower end of the range, to bright, jet black, hard, and relatively strong, at the upper end. Subbituminous coal contains 20 to 30 percent inherent moisture by weight. The heat content of subbituminous coal ranges from 17 to 24 million **Btu** per **short ton** on a moist, mineral-matter-free basis. The heat content of subbituminous coal consumed in the United States averages 17 to 18 million Btu per short ton, on the as-received basis (i.e., containing both inherent moisture and mineral matter). See **Coal Rank**.

Sulfur Dioxide (SO_2): A toxic, irritating, colorless gas soluble in water, **alcohol**, and **ether**. Used as a chemical intermediate, in paper pulping and ore refining, and as a solvent.

Sulfur Hexafluoride (SF_6): A colorless gas soluble in **alcohol** and **ether**, and slightly less soluble in water. It is used as a dielectric in electronics. It possesses the highest 100-year **global warming potential** of any gas (23,900).

Supplemental Gaseous Fuels: Any gaseous substance introduced into or commingled with **natural gas** that increases the volume available for disposition. Such substances include, but are not limited to, propane-air, **refinery gas**, coke-oven gas, manufactured gas, biomass gas, or air or inerts added for Btu stabilization.

Synthetic Natural Gas (SNG): (Also referred to as substitute natural gas.) A manufactured product, chemically similar in most respects to **natural gas**, resulting from the conversion or reforming of **hydrocarbons** that may easily be substituted for or interchanged with pipeline-quality natural gas.

Thermal Conversion Factor: See **Conversion Factor**.

Transportation Sector: An **energy**-consuming sector that consists of all vehicles whose primary purpose is transporting people and/or goods from one physical location to another. Included are automobiles; trucks; buses; motorcycles; trains, subways, and other rail vehicles; aircraft; and ships, barges, and other waterborne vehicles. Vehicles whose primary purpose is not transportation (e.g., construction cranes and bulldozers, farming vehicles, and warehouse tractors and forklifts) are classified in the sector of their primary use. *Note*: Various EIA programs differ in sectoral coverage—for more information see http://www.eia.doe.gov/neic/datadefinitions/Guideforwebtrans.htm. See **End-Use Sectors** and **Energy-Use Sectors.**

Unaccounted-for Crude Oil: Represents the arithmetic difference between the calculated supply and the calculated disposition of **crude oil**. The calculated supply is the sum of crude oil production plus imports minus changes in crude oil stocks. The calculated disposition of crude oil is the sum of crude oil input to refineries, crude oil exports, crude oil burned as fuel, and crude oil losses.

Unaccounted-for Natural Gas: Represents differences between the sum of the components of **natural gas** supply and the sum of components of natural gas disposition. These differences may be due to quantities lost or to the effects of data reporting problems. Reporting problems include differences due to the net result of conversions of flow data metered at varying temperatures and pressure bases and converted to a standard temperature and pressure base; the effect of variations in company accounting and billing practices; differences between billing cycle and calendar-period time frames; and imbalances resulting from the merger of data reporting systems that vary in scope, format, definitions, and type of respondents.

Underground Natural Gas Storage: The use of sub-surface facilities for storing **natural gas** that has been transferred from its original location. The facilities are usually hollowed-out salt domes, geological reservoirs (depleted **crude oil** or natural gas fields) or water-bearing sands topped by an impermeable cap rock (aquifer).

Undiscovered Recoverable Reserves (Crude Oil and Natural Gas): Those economic resources of **crude oil** and **natural gas**, yet undiscovered, that are estimated to exist in favorable geologic settings.

Unfinished Oils: All oils requiring further processing, except those requiring only mechanical blending. Unfinished oils are produced by partial refining of **crude oil** and include **naphthas** and lighter oils, **kerosene** and light gas oils, heavy gas oils, and residuum.

Unfractionated Stream: Mixtures of unsegregated **natural gas liquid** components, excluding those in **plant condensate**. This product is extracted from **natural gas**.

United States: The 50 States and the District of Columbia. *Note*: The United States has varying degrees of jurisdiction over a number of territories and other political entities outside the 50 States and the District of Columbia, including Puerto Rico, the U.S. Virgin Islands, Guam, American Samoa, Johnston Atoll, Midway Islands, Wake Island, and the Northern Mariana Islands. EIA data programs may include data from some or all of these areas in U.S. totals. For these programs, data products will contain notes explaining the extent of geographic coverage included under the term "United States."

Uranium: A heavy, naturally radioactive, metallic element (atomic number 92). Its two principally occurring isotopes are uranium-235 and uranium-238. Uranium-235 is indispensable to the nuclear industry because it is the only isotope existing in nature, to any appreciable extent, that is fissionable by thermal neutrons. Uranium-238 is also important because it absorbs neutrons to produce a radioactive isotope that subsequently decays to the isotope plutonium-239, which also is fissionable by thermal neutrons.

Uranium Concentrate: A yellow or brown powder obtained by the milling of **uranium ore**, processing of in situ leach mining solutions, or as a byproduct of phosphoric acid production. See **Uranium Oxide**.

Uranium Ore: Rock containing **uranium** mineralization in concentrations that can be mined economically, typically one to four pounds of U_3O_8 (**uranium oxide**) per ton or 0.05 percent to 0.2 percent U_3O_8.

Uranium Oxide: **Uranium concentrate** or **yellowcake**. Abbreviated as U_3O_8.

Uranium Resource Categories: Three categories of **uranium** resources defined by the international community to reflect differing levels of confidence in the existence of the resources. Reasonably assured resources (RAR), estimated additional resources (EAR), and speculative resources (SR) are described below.

> *Reasonably assured resources (RAR)*: **Uranium** that occurs in known mineral deposits of such size, grade, and configuration that it could be recovered within the given production cost ranges, with currently proven mining and processing technology. Estimates of tonnage and grade are based on specific sample data and measurements of the deposits and on knowledge of deposit characteristics. *Note*: RAR corresponds to DOE's uranium reserves category.
>
> *Estimated additional resources (EAR)*: **Uranium** in addition to RAR that is expected to occur, mostly on the basis of geological evidence, in extensions of well-explored deposits, in little-explored deposits, and in undiscovered deposits believed to exist along well-defined geological trends with known deposits. This uranium can subsequently be recovered within the given cost ranges. Estimates of tonnage and grade are based on available sampling data and on knowledge of the deposit characteristics, as determined in the best-known parts of the deposit or in similar deposits. *Note*: EAR corresponds to DOE's probable potential resources category.
>
> *Speculative resources (SR)*: **Uranium** in addition to EAR that is thought to exist, mostly on the basis of indirect evidence and geological extrapolations, in deposits discoverable with existing exploration techniques. The location of deposits in this category can generally be specified only as being somewhere within given regions or geological trends. The estimates in this category are less reliable than estimates of RAR and EAR. *Note*: SR corresponds to the combination of DOE's possible potential resources and speculative potential resources categories.

Useful Thermal Output: The thermal **energy** made available in a **combined-heat-and-power** system for use in any industrial or commercial process, heating or cooling application, or delivered to other end users, i.e., total thermal energy made available for processes and applications other than **electricity generation**.

U.S.S.R.: The Union of Soviet Socialist Republics consisted of 15 constituent republics: Armenia, Azerbaijan, Belarus, Estonia, Georgia, Kazakhstan, Kyrgyzstan, Latvia, Lithuania, Moldova, Russia, Tajikistan, Turkmenistan, Ukraine, and Uzbekistan. As a political entity, the U.S.S.R. ceased to exist as of December 31, 1991.

Vented Natural Gas: See **Natural Gas, Vented.**

Vessel Bunkering: Includes sales for the fueling of commercial or private boats, such as pleasure craft, fishing boats, tugboats, and ocean-going vessels, including vessels operated by oil companies. Excluded are volumes sold to the U.S. Armed Forces.

Waste: See **Biomass Waste** and **Non-Biomass Waste.**

Waste Coal: Usable material that is a byproduct of previous **coal** processing operations. Waste coal is usually composed of mixed coal, soil, and rock (mine waste). Most waste coal is burned as-is in unconventional fluidized-bed combustors. For some uses, waste coal may be partially cleaned by removing some extraneous noncombustible constituents. Examples of waste coal include fine coal, coal obtained from a refuse bank or slurry dam, **anthracite culm**, bituminous gob, and lignite waste.

Watt (W): The unit of electrical power equal to one ampere under a pressure of one volt. A watt is equal to 1/746 horsepower.

Watthour (Wh): The **electric energy** unit of measure equal to one **watt** of power supplied to, or taken from, an electric circuit steadily for one hour.

Wax: A solid or semi-solid material at 77 degrees Fahrenheit consisting of a mixture of **hydrocarbons** obtained or derived from **petroleum** fractions, or through a Fischer-Tropsch type process, in which the straight chained paraffin series predominates. This includes all marketable wax, whether crude or refined, with a congealing point (ASTM D 938) between 80 (or 85) and 240 degrees Fahrenheit and a maximum oil content (ASTM D 3235) of 50 weight percent.

Well: A hole drilled in the Earth for the purpose of (1) finding or producing **crude oil** or **natural gas**; or (2) producing services related to the production of crude oil or natural gas. See **Completion (Crude Oil/Natural Gas Production), Crude Oil Well, Development Well, Dry Hole, Exploratory Well,** and **Natural Gas Well**.

Wellhead: The point at which the **crude oil** (and/or **natural gas**) exits the ground. Following historical precedent, the volume and price for crude oil production are labeled as "wellhead," even though the cost and volume are now generally measured at the lease boundary. In the context of domestic crude price data, the term "wellhead" is the generic term used to reference the production site or lease property.

Wellhead Price: The value of **crude oil** or **natural gas** at the mouth of the well. See **Natural Gas Wellhead Price**.

Well Servicing Unit: Truck-mounted equipment generally used for downhole services after a **well** is drilled. Services include well completions and recompletions, maintenance, repairs, workovers, and well plugging and abandonments. Jobs range from minor operations, such as pulling the rods and rod pumps out of a **crude oil well**, to major workovers, such as milling out and repairing collapsed casing. Well depth and characteristics determine the type of equipment used.

Western Europe: Includes Austria, Belgium, Bosnia and Herzegovina, Croatia, Denmark, Finland, France, Germany, Greece, Iceland, Ireland, Italy, Luxembourg, Macedonia (The Former Yugoslav Republic of), Malta, Netherlands, Norway, Portugal, Serbia and Montenegro, Slovenia, Spain, Sweden, Switzerland, Turkey, and the United Kingdom.

Wind Energy: Kinetic **energy** present in wind motion that can be converted to mechanical energy for driving pumps, mills, and electric power **generators**

Wood and Wood-Derived Fuels: Wood and products derived from wood that are used as fuel, including round wood (cord wood), limb wood, wood chips, bark, sawdust, forest residues, charcoal, paper pellets, railroad ties, utility poles, **black liquor**, red liquor, sludge wood, spent sulfite liquor, and other wood-based solids and liquids.

Working Gas: The volume of gas in the reservoir that is in addition to the cushion or **base gas**. It may or may not be completely withdrawn during any particular withdrawal season. Conditions permitting, the total working capacity could be used more than once during any season.

Yellowcake: A natural **uranium concentrate** that takes its name from its color and texture. Yellowcake typically contains 70 to 90 percent U_3O_8 (**uranium oxide**) by weight. It is used as feedstock for **uranium** fuel enrichment and fuel pellet fabrication